Lecture Notes in Artificial Intelligence 9810

Subseries of Lecture Notes in Computer Science

More information about this series at http://www.springer.com/series/1244

Richard Booth · Min-Ling Zhang (Eds.)

PRICAI 2016: Trends in Artificial Intelligence

14th Pacific Rim
International Conference on Artificial Intelligence
Phuket, Thailand, August 22–26, 2016
Proceedings

 Springer

Editors
Richard Booth
Cardiff University
Cardiff
UK

Min-Ling Zhang
Southeast University
Nanjing
China

ISSN 0302-9743   ISSN 1611-3349   (electronic)
Lecture Notes in Artificial Intelligence
ISBN 978-3-319-42910-6   ISBN 978-3-319-42911-3   (eBook)
DOI 10.1007/978-3-319-42911-3

Library of Congress Control Number: 2016945115

LNCS Sublibrary: SL7 – Artificial Intelligence

Preface

This volume contains the papers presented at the 14th Pacific Rim International Conference on Artificial Intelligence (PRICAI 2016) held during August 22–26, 2016, in Phuket, Thailand.

PRICAI is a biennial conference inaugurated in Tokyo in 1990. It provides a common forum for researchers and practitioners in various branches of artificial intelligence (AI) to exchange new ideas and share experience and expertise. Over the past 26 years the conference has grown, both in participation and scope, to be a premier international AI event for all major Pacific Rim nations as well as countries from further afield. This year marked the first time that the conference was held in Thailand. In addition to the main track, PRICAI 2016 featured a special track on "Smart Modelling and Simulation."

This year, we received 161 high-quality submissions from 25 countries to both the main and special tracks. From these, 53 papers (33 %) were accepted as regular papers, with a further 15 accepted as short papers. Each submitted paper was considered by the Program Committee members and external reviewers, and evaluated against criteria such as relevance, significance, technical soundness, novelty, and clarity. Every paper received at least three reviews, and in some cases up to four, supplemented by rigorous discussion among the reviewers. Finally, the program co-chairs read the reviews and discussion among reviewers, and made the final decision to ensure fairness and consistency in the paper selection.

The technical program began with two days of workshops and tutorials, followed by the main conference program. The workshops included the Pacific Rim Knowledge Acquisition Workshop (PKAW), co-chaired by Hayato Ohwada (Tokyo University of Science, Japan) and Kenichi Yoshida (University of Tsukuba, Japan), which has long enjoyed a successful co-location with PRICAI. Authors of short papers presented their results during poster sessions, but were also given the opportunity to present shortened talks to introduce their work. As in previous years, participants at PRICAI were also able to attend the co-located 19th International Conference on Principles and Practice of Multi-Agent Systems (PRIMA 2016). We were honored to have three outstanding keynote speakers, whose contributions have crossed discipline boundaries: Phan Minh Dung (Asian Institute of Technology, Thailand), Sheng-Chuan Wu (Franz Inc., USA), and Zhi-Hua Zhou (Nanjing University, China). We are grateful to them for sharing their insights on their latest research with us.

It would not have been possible to organize the technical program without the considerable help of various people who committed their time and effort toward making PRICAI 2016 a success. We would like to thank the Program Committee members and external reviewers for their engagements in providing rigorous and timely reviews. It is because of them that the quality of the papers in this volume is maintained at a high level.

We wish to express our gratitude to the General Co-chairs, Dickson Lukose (MIMOS Berhad, Malaysia) and Thanaruk Theeramunkong (Thammasat University, Thailand), for their continued support and guidance. We are also thankful to the Tutorial Co-chairs, Sankalp Khanna (CSIRO, Australia), Manabu Okumura (Tokyo Institute of Technology, Japan) and Kritsada Sriphaew (Rangsit University, Thailand), for selecting the fruitful tutorials, and the Workshop Co-chairs, Masayuki Numao (Osaka University, Japan), Boonserm Kijsirikul (Chulalongkorn University, Thailand), and Sanparith Marukatat (NECTEC, Thailand), for coordinating the attractive workshops.

We gratefully acknowledge the support of the organizing institutions Artificial Intelligence Association of Thailand, Thammasat University, Prince of Songkla University, and NECTEC, as well as the financial support from Artificial Intelligence Journal, Air Force Office of Scientific Research, Asian Office of Aerospace Research and Development, Thammasat University, Thailand Convention and Exhibition Bureau, SERTIS Co., Ltd., Defence Technology Institute, Provincial Electricity Authority of Thailand, Electronic Government Agency, Franz Inc., MIMOS Berhad, and Springer Publishing. Special thanks to Easychair, whose paper submission platform we used to organize reviews and collate the files for this proceedings. We are also grateful to Springer, and in particular Alfred Hofmann and Anna Kramer, for their assistance in publishing the PRICAI 2016 proceedings as a volume in its *Lecture Notes in Artificial Intelligence* series.

Last but not least, we also want to thank all authors and all conference participants for their contribution and support. We hope all participants took this opportunity to share and exchange ideas and thoughts with one another and enjoyed PRICAI 2016.

August 2016   Richard Booth
                                                                   Min-Ling Zhang

Organization

Steering Committee

Tru Hoang Cao	Ho Chi Minh City University of Technology, Vietnam
Tu Bao Ho	Japan Advanced Institute of Science and Technology, Japan
Mitsuru Ishizuka	Waseda University, Japan
Byeong-Ho Kang	University of Tasmania, Australia (PKAW representative)
Dickson Lukose	MIMOS Berhad, Malaysia
Hideyuki Nakashima	Future University Hakodate, Japan
Duc Nghia Pham	MIMOS Berhad, Malaysia
Abdul Sattar	Griffith University, Australia (Chair)
Makoto Yokoo	Kyushu University, Japan (PRIMA representative)
Byoung-Tak Zhang	Seoul National University, Korea
Chengqi Zhang	University of Technology Sydney, Australia
Zhi-Hua Zhou	Nanjing University, China (Secretary)

Organizing Committee

Honorary Co-chairs

Wai Kiang (Albert) Yeap	Auckland University of Technology, New Zealand
Abdul Sattar	Griffith University, Australia
Hiroshi Motoda	Osaka University, Japan
Vilas Wuwongse	Mahidol University, Thailand
Somnuk Tangtermsirikul	Thammasat University, Thailand
Sarun Sumriddetchkajorn	NECTEC, Thailand
Pun Thongchunum	Prince of Songkla University, Thailand

General Co-chairs

Dickson Lukose	MIMOS Berhad, Malaysia
Thanaruk Theeramunkong	Thammasat University, Thailand

Program Committee Co-chairs

Richard Booth	Cardiff University, UK
Min-Ling Zhang	Southeast University, China

Tutorial Co-chairs

Sankalp Khanna	CSIRO, Australia
Manabu Okumura	Tokyo Institute of Technology, Japan
Kritsada Sriphaew	Rangsit University, Thailand

Workshop Co-chairs

Masayuki Numao	Osaka University, Japan
Boonserm Kijsirikul	Chulalongkorn University, Thailand
Sanparith Marukatat	NECTEC, Thailand

Special Session and Research Student Symposium Co-chairs

Abdul Sattar	Griffith University, Australia
Mahasak Ketcham	King Mongkut's University of Technology North Bangkok, Thailand
Chuleerat Jaruskulchai	Kasetsart University, Thailand
Rachada Kongkachandra	Thammasat University, Thailand
Pokpong Songmuang	Thammasat University, Thailand
Patiyuth Pramkeaw	King Mongkut's University of Technology Thonburi, Thailand
Narit Hnoohom	Mahidol University, Thailand

Publicity Co-chairs

Waralak Vongdoiwang Siricharoen	University of the Thai Chamber of Commerce, Thailand
Vincent C.S. Lee	Monash University, Australia

Sponsorship Co-chairs

Chai Wutiwiwatchai	NECTEC, Thailand
Mahasak Ketcham	King Mongkut's University of Technology North Bangkok, Thailand

Financial Co-chairs

Chutima Beokhaimook	Rangsit University, Thailand
Nongnuch Ketui	Rajamangala University of Technology Lanna, Thailand
Choermath Hongakkaraphan	Thammasat University, Thailand

Local Organizing Co-chairs

Rattana Wetprasit	Prince of Songkla University, Thailand
Virach Sortlertlamvanich	Thammasat University, Thailand
Thepchai Supnithi	NECTEC, Thailand
Nattapong Tongtep	Prince of Songkla University, Thailand

Secretaries General

Thatsanee Chareonporn	Burapa University, Thailand
Choermath Hongakkaraphan	Thammasat University, Thailand
Kiyota Hashimoto	Prince of Songkla University, Thailand

Webmasters

Thanasan Tanhermhong	Thammasat University, Thailand
Wirat Chinnan	Thammasat University, Thailand

Program Committee

Mohd Sharifuddin Ahmad	Universiti Tenaga Nasional, Malaysia
Eriko Aiba	The University of Electro-Communications, Japan
Pakinee Aimmanee	Thammasat University, Thailand
Akiko Aizawa	National Institute of Informatics, Japan
David Albrecht	Monash University, Australia
Arun Anand Sadanandan	MIMOS Berhad, Malaysia
Patricia Anthony	Lincoln University, New Zealand
Judith Azcarraga	De La Salle University, Philippines
Quan Bai	Auckland University of Technology, New Zealand
Ghassan Beydoun	University of Wollongong, Australia
Ateet Bhalla	Independent consultant, India
Mehul Bhatt	The University of Bremen, Germany
Patrice Boursier	University of La Rochelle, France
Khalil Bouzekri	MIMOS Berhad, Malaysia
The Duy Bui	Vietnam National University, Vietnam
Marut Buranarach	NECTEC, Thailand
Rafael Cabredo	De La Salle University, Philippines
Tru Cao	Ho Chi Minh City University of Technology, Vietnam
Songcan Chen	Nanjing University of Aeronautics and Astronautics, China
Wu Chen	Southwest University, China
Yi-Ping Phoebe Chen	La Trobe University, Australia
Wai Khuen Cheng	Universiti Tunku Abdul Rahman, Malaysia
William K. Cheung	Hong Kong Baptist University, China
Krisana Chinnasarn	Burapha University, Thailand
Seungjin Choi	Pohang University of Science and Technology, Korea
Phatthanaphong Chomphuwiset	Mahasarakham University, Thailand
Jirapun Daengdej	Assumption University, Thailand
Matthew Dailey	Asian Institute of Technology, Thailand
Enrique de La Hoz	University of Alcalá, Spain
Andreas Dengel	German Research Center for Artificial Intelligence, Germany

Xiangjun Dong	Qilu University of Technology, China
Shyamala Doraisamy	Universiti Putra Malaysia, Malaysia
Duc Duong	Ho Chi Minh City University of Information Technology, Vietnam
Vlad Estivill-Castro	Griffith University, Australia
Christian Freksa	University of Bremen, Germany
Katsuhide Fujita	Tokyo University of Agriculture and Technology, Japan
Naoki Fukuta	Shizuoka University, Japan
Dragan Gamberger	Ruđer Bošković Institute, Croatia
Wei Gao	Nanjing University, China
Xiaoying Gao	Victoria University of Wellington, New Zealand
Guido Governatori	NICTA, Australia
Michael Granitzer	University of Passau, Germany
Fikret Gürgen	Bogazici University, Turkey
Peter Haddawy	Mahidol University, Thailand
Bing Han	Xidian University, China
Choochart Haruechaiyasak	NECTEC, Thailand
Tomomichi Hayakawa	Nagoya Institute of Technology, Japan
Tessai Hayama	Nagaoka University of Technology, Japan
Juhua Hu	Simon Fraser University, Canada
Sheng-Jun Huang	Nanjing University of Aeronautics and Astronautics, China
Van Nam Huynh	Japan Advanced Institute of Science and Technology, Japan
Masashi Inoue	Yamagata University, Japan
Sanjay Jain	National University of Singapore, Singapore
Yuan Jiang	Nanjing University, China
Geun Sik Jo	Inha University, Korea
Hideaki Kanai	Japan Advanced Institute of Science and Technology, Japan
Ryo Kanamori	Nagoya University, Japan
Kee-Eung Kim	Korea Advanced Institute of Science and Technology, Korea
Canasai Kruengkrai	National Institute of Information and Communications Technology, Japan
Alfred Krzywicki	University of New South Wales, Australia
Satoshi Kurihara	The University of Electro-Communications, Japan
Young-Bin Kwon	Chung-Ang University, Korea
Weng Kin Lai	Tunku Abdul Rahman University College, Malaysia
Ho-Pun Lam	Data61, CSIRO, Australia
Roberto Legaspi	The Institute of Statistical Mathematics, Japan
Chun-Hung Li	Hong Kong Baptist University, China
Gang Li	Deakin University, Australia
Li Li	Southwest University, China
Ming Li	Nanjing University, China

Tony Smith	University of Waikato, New Zealand
Chattrakul Sombattheera	Mahasarakham Unversity, Thailand
Safeeullah Soomro	Indus University, Pakistan
Kritsada Sriphaew	Rangsit University, Thailand
Biplav Srivastava	IBM Research, USA
Markus Stumptner	University of South Australia, Australia
Xing Su	Beijing University of Technology, China
Merlin Suarez	Center for Empathic Human-Computer Interactions, Philippines
Wing-Kin Sung	National University of Singapore, Singapore
Boontawee Suntisrivaraporn	Sirindhorn International Institute of Technology, Thailand
Thepchai Supnithi	NECTEC, Thailand
David Taniar	Monash University, Australia
Satoshi Tojo	Japan Advanced Institute of Science and Technology, Japan
Kuniaki Uehara	Kobe University, Japan
Ventzeslav Valev	Institute of Mathematics and Informatics, Bulgarian Academy of Sciences, Bulgaria
Miroslav Velev	Aries Design Automation, USA
Waralak Vongdoiwang Siricharoen	University of the Thai Chamber of Commerce, Thailand
Toby Walsh	NICTA and UNSW, Australia
Kewen Wang	Griffith University, Australia
Qi Wang	Northwestern Polytechnical University, China
Wei Wang	Nanjing University, China
Wayne Wobcke	University of New South Wales, Australia
Guandong Xu	University of Technology Sydney, Australia
Ming Xu	Xi'an Jiaotong-Liverpool University, China
Roland Yap	National University of Singapore, Singapore
Dayong Ye	Swinburne University of Technology, Australia
Chao Yu	Dalian University of Technology, China
Yang Yu	Nanjing University, China
Zhiwen Yu	South China University of Technology, China
Takaya Yuizono	Japan Advanced Institute of Science and Technology, Japan
Yi Zeng	Institute of Automation, Chinese Academy of Sciences, China
De-Chuan Zhan	Nanjing University, China
Zhi-Hui Zhan	South China University of Technology, China
Chengqi Zhang	University of Technology Sydney, Australia
Daoqiang Zhang	Nanjing University of Aeronautics and Astronautics, China
Du Zhang	Macau University of Science and Technology, Macau, China
Junping Zhang	Fudan University, China

Shichao Zhang	Guangxi Normal University, China
Wen Zhang	Institute of Software, Chinese Academy of Sciences, China
Yu Zhang	Hong Kong Baptist University, China
Yanchang Zhao	RDataMining.com
Xiaofeng Zhu	Guangxi Normal University, China
Xingquan Zhu	Florida Atlantic University, USA
Fuzhen Zhuang	Institute of Computing Technology, Chinese Academy of Sciences, China
Quan Zou	Tianjin University, China
Dominik Ślęzak	University of Warsaw and Infobright Inc., Poland

External Reviewers

Afzal, Muhammad Zeshan	Bukhari, Syed Saqib	Chen, Weiyang
Bizid, Imen	Demirović, Emir	Gao, Ping
Chen, Xiaohong	Haryanto, Anasthasia	Kim, Saehoon
Gao, Qian	Agnes	Lee, Kwanyong
Kong, Jie	Lee, Juho	Liu, Mingxia
Li, Weihua	Li, Xin	Lye, Guang Xing
Liu, Xiaofang	Lv, Guohua	Nguyen, Van Doan
Matsubara, Takashi	Munir, Mohsin	Polash, Md. Masbaul
Ou, Wei	Pan, Shirui	Alam
Riveret, Regis	Suh, Suwon	Tajvidi, Masoumeh
Tian, Qing	Tischer, Peter	van de Ven, Jasper
Vu, Huy Quan	Wang, Guixiang	Wang, Hanmo
Wang, Liping	Wang, Yuwei	Wang, Zhe
Wang, Zi-Jia	Wu, Jia	Xu, Feng
Zhang, Heng	Zhang, Lefeng	Zhu, Qi
Zhuang, Zhiqiang	Zu, Chen	
Ahmad, Riaz	Aziz, Tarique	

Sponsoring Organizations

 Artificial Intelligence Journal

 Air Force Office of Scientific Research

 Asian Office of Aerospace Research and Development

 Thammasat University

Thailand Convention and Exhibition Bureau

SERTIS Co., Ltd.

Defence Technology Institute

Provincial Electricity Authority of Thailand

Electronic Government Agency

 Franz Inc.

 MIMOS Berhad

 Springer Publishing

Contents

Special Track: Smart Modelling and Simulation

PRICAI 2016 Main Track

A Study of Players' Experiences During Brain Games Play

Faizan Ahmad[1,2,3], Yiqiang Chen[1(✉)], Shuangquan Wang[1], Zhenyu Chen[1,2],
Jianfei Shen[1,2], Lisha Hu[1,2], and Jindong Wang[1,2]

[1] Beijing Key Laboratory of Mobile Computing and Pervasive Device,
Institute of Computing Technology, Chinese Academy of Sciences, Beijing, China
{yqchen,wangshuangquan,chenzhenyu,shenjianfei,hulisha,
wangjindong}@ict.ac.cn
[2] University of Chinese Academy of Sciences, Beijing, China
[3] COMSATS Institute of Information Technology, Lahore, Pakistan
faizan.ahmad.1988@gmail.com

Abstract. Much of the experience of videogame players remains hidden. This paper presents an empirical study that assesses the experience of 50 participants (i.e. 25 children and 25 adults) during brain games play. Results from the empirical study show a number of significant correlations among diverse kinds of players' experiences (i.e. engagement, enjoyment, anxiety, usability, adaptability and noninvasiveness). It is further identified by the study that the similarities and differences exist among the experiences of children and adults. Consequently, the observations of presenting study provide an insight against the experience of players during brain games play, which was previously unknown. Besides, we exploit these insights to successfully narrow down the complexity of user feedback process for brain games playing activity.

Keywords: Gamification · Experiences · Smart assessment · Children · Adults

1 Introduction

Various questions arise when we talk about the experience of videogame players. Why do players of one generation like some videogames while the others don't? Which are the elements required by the players of a specific generation to accept the videogame? What are the similarities and differences among the players' psyche of different generations that reflect upon their perception towards videogames? In this paper, we present an empirical study to seek answers to some of these critical questions. However, the focus of presenting study is limited to the experience of two generations (i.e. children and adults) in brain games play. The term "brain games" refers to the category of videogames that are specifically designed to enhance the mental fitness of players. Besides, these games (e.g. [12, 24, 36]) also contain specific content, dynamics, and mechanics that determines their effects on the brain [4].

Nonetheless, in order to investigate both generations' (i.e. children's and adults') experience an empirical study has been carried out in two-fold, a brain games play and questionnaire based feedback (see Fig. 1). In the first fold of an empirical study, participants have been asked to play with the *"BrainStorm"* game suite (see Fig. 2), which

© Springer International Publishing Switzerland 2016
R. Booth and M.-L. Zhang (Eds.): PRICAI 2016, LNAI 9810, pp. 3–15, 2016.
DOI: 10.1007/978-3-319-42911-3_1

contains three brain games. Subsequently in the second fold, they have been requested to provide their feedback on the gameplay activity, by filling out a questionnaire. A questionnaire was compiled to cover four aspects of players' engagement (i.e. immersion, presence, flow and absorption), two aspects of players' emotion (i.e. enjoyment and anxiety) as well as usability, adaptability and noninvasiveness of the players. A questionnaire based feedback data has been statistically analyzed to study the experience of the players during *"BrainStorm"* gameplay. An understanding of players' experiences has been further exploited to successfully position the videogame assessment measures (i.e. engagement, enjoyment, anxiety, usability, adaptability and noninvasiveness) in relation to each other. Besides, a hypothesis has been tested that whether the positioning of videogame assessment measures in relation to each other support in narrowing down user feedback process, which will lead to broad scale assessment with less measurement efforts. The results of an empirical study support the argument.

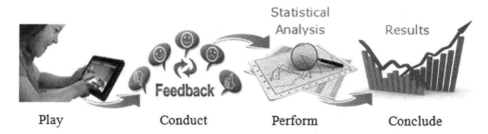

Fig. 1. Flow diagram of an empirical study

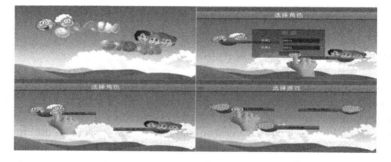

Fig. 2. Main user interfaces of *"BrainStorm"*

The presented study makes a number of contributions. First of all, to the best of our knowledge, this is a first attempt to analyze the experience of players (i.e. children and adults) during brain games play activity. Secondly, a current study reports significant correlations among the different experiences (i.e. engagement, enjoyment, anxiety, usability, adaptability and noninvasiveness) of the players. Thirdly, it identifies the similarities and differences among the experience of both generations. Finally, it exploits the above insights to narrow down a user feedback process.

2 Definitional Issues

There exists a theoretical framework [31] that serves to define the "experience" of videogame players; however, in the presence of previously existing 250 state-of-the-art publications elaborating the multidimensional perspective of players' experiences, the framework concludes itself as groundwork. It is complex to cover all the attributes of players' experience (i.e. associated in major or minor extent), as the term has been casually used several times to measure the diverse aspects of players that include emotions [17, 25], usability [32, 33], pleasure [21], fun [15], motivation [22] and play-ability [13, 37]. Thus, no definitional agreement certainly exists regarding how to comprehensively model the experience of players during the gameplay activity [31].

In this paper, the term "experience" refers to those personal states of players during the game playing situations that have been observed under the ongoing study. These observed personal states include players' engagement, enjoyment, anxiety, usability, adaptability and noninvasiveness. We explain these terms as follows.

2.1 Engagement

Engagement implies a general involvement of the players in videogames; however, it further technically includes "immersion", "presence", "flow" and "absorption", which can be understood as representing a continuation of ever-deeper engagement while playing videogames [19]. Their explanation is stated as follows.

Immersion. Immersion is a term used to define the capability of a videogame to induce a feeling in the player of actually being a part of it [44]. Immersion has also been considered to measure the experience of getting engaged in a gameplay activity while keeping some consciousness of one's surroundings [1, 39].

Presence. Presence is a term employed to describe the awareness of being inside a virtual environment [26, 28, 34, 42]. Another term "spatial presence" has been proposed [44] to describe the awareness of being integrated into a mediated environment. Unlike previous formulations, this definition includes both, a new media (e.g. videogames) as well as conventional media (e.g. books).

Flow. Flow is a term utilized to express the feelings that occur when a balance between skill and challenge is achieved to perform an activity [8, 29, 30]. Therefore, flow also includes a feeling of being in control, being one with the activity, and experiencing time distortion.

Absorption. Absorption is a term that describes the total engagement in a present experience [18]. Unlike immersion and presence, and like flow, being in a state of absorption induces a modified state of consciousness. In this modified state there is a separation of feelings, thoughts, and experiences and effect is less accessible to consciousness [16].

2.2 Enjoyment

Enjoyment is a term that describes the positive emotions of an individual in general. The definition of enjoyment during a gameplay activity was previously fuzzy [41] and not well differentiated from other potentially related perspectives [31]; however, recently it is defined as a multi-dimensional construct, made up of entertainment, challenge, competence, (minimum) frustration and one's interest [9].

2.3 Anxiety

Anxiety is an emotion that describes one's worriedness, nervousness, or uneasiness. It is also characterized by an unpleasant state of inner turmoil, often accompanied by the nervous behavior [38]. Similarly, the anxiety of a player during the gameplay activity refers to its unpleasant mood often take place due to the unanticipated gameplay experience.

2.4 Usability

Usability is not a characteristic that exists in any absolute sense; however, it can be best summarized as one's appropriateness towards the purpose [3]. ISO 9241-11 suggests that a usability measurement should cover effectiveness, efficiency and satisfaction. Likewise, the usability of a videogame refers to the effectiveness, efficiency and satisfaction w. r. t. its context.

2.5 Adaptability

Adaptability is a broad term that describes one's ability of being flexible to fit in changed circumstances. In just the same way, the adaptability of a videogame refers to the characteristics of being acceptable by its diverse target users [14].

2.6 Noninvasiveness

Noninvasiveness is a term commonly used in medical sciences in order to refer a certain treatment that is performed without cutting a body or putting something into the body [43]. Likewise, in the field of videogame interaction, noninvasiveness refers to a technique that achieves its goal without having any visible or tactile interaction with its target user [23].

3 Experiments and Data Collection

The empirical study has been carried out with children and adults, respectively, which includes an activity of brain games play and questionnaire based feedback. In total 50 participants, equally distributed as 25 children and adults have been recruited to voluntarily take part in the designed study. The recruited children (i.e. 15 male and 10 female) have been reported as 8 to 9 years old with the mean age of 8.7 years, whereas the adults (i.e. 13 male and 12 female) were 30 to 45 years of age with the mean age of 33.4 years. The recruitment process has been carefully made based on the adequate gameplay experience of the participants (i.e. habitual to gameplay, at least once a week).

To perform the gameplay activity, we employed *"BrainStorm"* that includes *"Picture Puzzle"*, *"Letter and Number"* and *"Find the Difference"* brain games. *"BrainStorm"* is the game suite that was previously developed for noninvasive cognitive capabilities assessment. However, a functionality of the brain games of *"BrainStorm"* is as follows. In *"Picture Puzzle"* brain game, an image of famous and/or historical personalities or places show on the screen, and the player has to choose its correct name among the different options (see Fig. 3(a)). A core mechanism of *"Picture Puzzle"* requires player attention to receive the data from a visual source and passes it to the short-term memory, short-term memory then processes it and retrieves its correct information by communicating with the long-term memory.

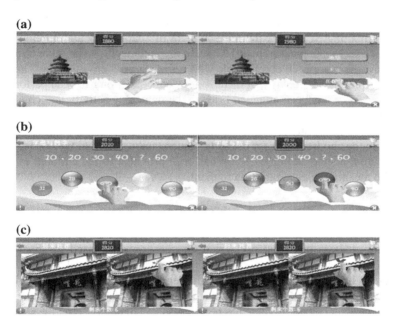

Fig. 3. User interface of (a). *"Picture Puzzle"* (b). *"Letter and Number"* (c). *"Find the Difference"*

Whereas in *"Letter and Number"* brain game, an incomplete sequence of letters or numbers show on the screen, and the player has to analyze its pattern and complete the sequence by selecting a correct option among the given options (see Fig. 3(b)). A core mechanism of *"Letter and Number"* requires player to perform information visualization, articulation, analysis and decision making based on their personal understanding.

Moreover, in *"Find the Difference"* brain game, two similar images show on the screen, and the player has to find six differences between both images (see Fig. 3(c)).

A core mechanism of *"Find the Difference"* requires player to select the relevant information and filtering out the irrelevant information from a visual space by using spotlight [10] and zoom-lens [11] models.

It took participants 20 min (on average) to complete the gameplay activity on 19.5 in. touch screen, subsequently they were requested to provide their feedback on the event, by filling out the questionnaire (i.e. 5-level likert scale measurement). To collect feedback regarding the four aspects of players' engagement (i.e. immersion, presence, flow and absorption), Game Engagement Questionnaire (GEQ) [19] has been employed. To collect feedback regarding the two aspects of players' emotion (i.e. enjoyment and anxiety), 11 the most frequently used terms have been utilized [9]. Feedback on the usability factor of brain games has been collected by exploiting System Usability Scale (SUS) [3]. Besides, to collect feedback about the adaptability of the players and noninvasiveness of data collection, Adaptability, Social interaction, Children education and Noninvasiveness Questionnaire (ASCNQ) has been partly utilized. ASCNQ is a multidimensional construct that has been equally distributed for the measurement of its four aspects; therefore it doesn't affect the results if the questionnaire gets partially used. The designed terms of ASCNQ for the measurement of adaptability and noninvasiveness (i.e. 5-level likert scale measurement) respectively include T1: "The more time I spent in gameplay, the more I felt comfortable with *"BrainStorm"* environment." and T2: "I didn't feel while the gameplay that there is any data collection has been performed.".

Nonetheless, it has been assumed prior to the gameplay activity that children will face a certain level of difficulty in understanding the terms of employed questionnaires. As no appropriate questionnaire is publically available to measure the above stated aspects from the children; thus for the better understanding of employed questionnaires to the children, a short training session has been provided to the children in order to explain the meaning of each question to them.

4 Statistical Analysis

A statistical analysis has been performed on questionnaire based feedback data to analyze the experience of the players. In the initial phase of statistical analysis, Pearson correlation [6] has been applied to calculate the degree of correlation among the different experiences, where r ($-1 \leq r \leq 1$) indicates the direction and strength of the correlation. Whilst in the second phase, p-value has been calculated to demonstrate the significance ($p \leq 0.05$) of the findings [6]. It is well-known that the correlation doesn't imply causation, yet an approach has been used by the vast range of literature, which also includes a recently done research on *"StudentLife"* [35]. It is nearly impossible in a real world scenario to find the element(s) that has a causal relationship with the other element, as there always exist un(known/addressed) factor(s) that affects causality between the associated elements. Therefore, a reason behind the use of correlation technique was not to find the causal relationship but to understand the significance of one element in relation to the other, while acknowledging they are not causal. Apart from the correlation analysis technique, we further performed a regression analysis [7] on each significantly correlated dependent (i.e. "Y") and independent (i.e. "X") experiences, in order to develop their regression function (see Eq. 1). The purpose of function development was to predict each dependent experience based on their significantly correlated independent experience(s) [34].

$$Y = constant + (a_1 * X_1) + (a_2 * X_2) + ... + (a_n * X_n) \qquad (1)$$

In what follows, we draw the main observations of the presented study. (1) A number of significant correlations among the different experiences (i.e. engagement, enjoyment, anxiety, usability, adaptability and noninvasiveness) of the players (i.e. children and adults) have been found. (2) Similarities and differences among the experience of both generations have been identified. (3). A possibility to successfully predict players' experience, based on the other significantly correlated experience(s), has been validated (see Fig. 4). Their details are as follows.

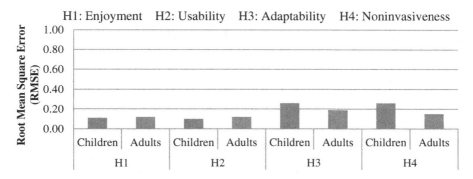

Fig. 4. Root Mean Square Error (RMSE) results of the predictions

4.1 Correlation Between Enjoyment and Other Experiences

Table 1 shows a number of significant correlations between the players' enjoyment and other gameplay experiences. The stated results highlight a fact that an existence of immersion is not significantly relevant for children's enjoyment, as it does for adults; however, the importance of presence, flow and absorption over enjoyment are common among both generations. Besides, an impact of anxiety over enjoyment is also common and almost to the same degree among both generations. Thus, we hypothesized (i.e. H1) that an enjoyment of both generations depends upon their respectively correlated gameplay experiences. To analyze the hypothesis, we developed a regression function (see Eqs. 2 & 3) for each generation (i.e. children and adults) in order to estimate their predictive enjoyment. Subsequently, we calculated Root Mean Square Error (RMSE) (i.e. by using leave-one-out cross validation technique) between their actual enjoyment and predicted enjoyment. Results concluded that the correlated experiences are significant to predict an enjoyment of both generations (i.e. children (RMSE = 0.11) and adults (RMSE = 0.12)) (see Fig. 4).

$$Enjoyment_{children} = 2.44 + (.24 * X_1) + (.19 * X_2) + (.33 * X_3) + (-.35 * X_4) \qquad (2)$$

$$Enjoyment_{adults} = 2.32 + (.05 * X_1) + (.28 * X_2) + (.43 * X_3) + (-.19 * X_4) + (-.02 * X_5) \tag{3}$$

Table 1. Correlation of enjoyment with other gameplay experiences

Children				Adults			
ID	Experiences	r	p-value	ID	Experiences	r	p-value
–	–	–	–	X_1	Immersion	0.26	0.019
X_1	Presence	0.42	<0.001	X_2	Presence	0.53	0.023
X_2	Flow	0.47	<0.001	X_3	Flow	0.57	<0.001
X_3	Absorption	0.46	<0.001	X_4	Absorption	0.37	<0.001
X_4	Anxiety	−0.30	<0.001	X_5	Anxiety	−0.29	<0.001

Note: "ID" w. r. t. the corresponding "experiences" of children and adults are referred in Eqs. 2 & 3, respectively.

4.2 Correlation Between Usability and Other Experiences

Table 2 shows a significant correlation between the players' usability and anxiety, which is almost of same degree among both generations. Thus, we hypothesized (i.e. H2) that a usability of both generations depends upon their feeling of anxiety. To analyze the hypothesis, we developed a regression function (see Eqs. 4 & 5) for each generation (i.e. children and adults) in order to estimate their predictive usability. Subsequently, we calculated RMSE (i.e. by using leave-one-out cross validation technique) between their actual usability and predicted usability. Results concluded that the feeling of anxiety is significant to predict a usability of both generations (i.e. children (RMSE = 0.10) and adults (RMSE = 0.12)) (see Fig. 4).

$$Usability_{children} = 3.65 + (-.21 * X_1) \tag{4}$$

$$Usability_{adults} = 3.83 + (-.26 * X_1) \tag{5}$$

Table 2. Correlation of usability with other gameplay experiences

Children				Adults			
ID	Experiences	r	p-value	ID	Experiences	r	p-value
X_1	Anxiety	−0.32	<0.001	X_1	Anxiety	−0.29	<0.001

Note: "ID" w. r. t. the corresponding "experiences" of children and adults are referred in Eqs. 4 & 5, respectively.

4.3 Correlation Between Adaptability and Other Experiences

Table 3 shows a number of significant correlations between the players' adaptability and other gameplay experiences. However, unlike Tables 1 and 2 where majority of the correlations were common among both generations, Table 3 demonstrates the diversity of perception among both generations regarding an adaptability of brain games.

The stated results highlight a fact that the positive experience (i.e. enjoyment and (minimal) anxiety) of gameplay is significantly relevant for adults to be adaptive towards brain games; however, it is trivial for children. Nevertheless, the deeper aspect of engagement (i.e. absorption) is important for children's adaptability. Thus, we hypothesized (i.e. H3) that an adaptability of both generations depends upon their respectively correlated gameplay experience(s). To analyze the hypothesis, we developed a regression function (see Eqs. 6 & 7) for each generation (i.e. children and adults) in order to estimate their predictive adaptability. Subsequently, we calculated RMSE (i.e. by using leave-one-out cross validation technique) between their actual adaptability and predicted adaptability. Results concluded that the correlated experience(s) is significant to predict an adaptability of both generations (i.e. children (RMSE = 0.26) and adults (RMSE = 0.19)) (see Fig. 4).

$$Adaptability_{children} = 1.73 + (.70 * X_1) \tag{6}$$

$$Adaptability_{adults} = 2.24 + (.42 * X_1) + (-.40 * X_2) \tag{7}$$

Table 3. Correlation of adaptability with other gameplay experiences

Children				Adults			
ID	Experiences	r	p-value	ID	Experiences	r	p-value
X_1	Absorption	0.38	0.010	X_1	Enjoyment	0.36	0.017
–	–	–	–	X_2	Anxiety	−0.35	<0.001

Note: "ID" w. r. t. the corresponding "experiences" of children and adults are referred in Eqs. 6 & 7, respectively.

4.4 Correlation Between Noninvasiveness and Other Experiences

Table 4 shows a number of significant correlations between the players' noninvasiveness and other gameplay experiences. Similar with Table 3, Table 4 also demonstrates the diversity of perception (i.e. to a certain level) among both generations regarding the noninvasiveness during brain games play. The stated results highlight a fact that an enjoyment is significantly relevant for children's noninvasiveness during brain games play; however, it is trivial for adults. Nevertheless, the deeper aspect(s) of engagement for children (i.e. immersion) and adults (i.e. flow and absorption) are also important in achieving noninvasiveness. Thus, we hypothesized (i.e. H4) that a noninvasiveness of both generations depends upon their respectively correlated gameplay experiences. To analyze the hypothesis, we developed a regression function (see Eqs. 8 & 9) for each generation (i.e. children and adults) in order to estimate their predictive noninvasiveness. Subsequently, we calculated RMSE (i.e. by using leave-one-out cross validation technique) between their actual noninvasiveness and predicted noninvasiveness. Results concluded that the correlated experiences are significant to predict a noninvasiveness of both generations (i.e. children (RMSE = 0.26) and adults (RMSE = 0.15)) (see Fig. 4).

$$Noninvasiveness_{children} = 1.54 + \left(.45 * X_1\right) + \left(-.05 * X_2\right) \tag{8}$$

$$Noninvasiveness_{adults} = 1.87 + \left(-.13 * X_1\right) + \left(.74 * X_2\right) \tag{9}$$

Table 4. Correlation of noninvasiveness with other gameplay experiences

Children				Adults			
ID	Experiences	r	p-value	ID	Experiences	r	p-value
X_1	Immersion	0.38	0.048	X_1	Flow	0.23	0.020
X_2	Enjoyment	0.21	<0.001	X_2	Absorption	0.41	0.016

Note: "ID" w. r. t. the corresponding "experiences" of children and adults are referred in Eqs. 8 & 9, respectively.

5 Discussion

In view of theoretically defined players' experiences (i.e. in Sect. 2), we elucidate the results of statistical analysis (i.e. stated in Sect. 4) as follows. The results of statistical analysis indicate that in order to ensure an appropriateness of brain games in terms of providing an effective, efficient and satisfactory gameplay experience to the players (i.e. children and adults), it is significant that the games shouldn't stimulate the feeling of worriedness, nervousness, or uneasiness within the players (i.e. derived from Table 2). However, a balance between the challenged environment of brain games and players' skills make the gameplay experience more enjoyable, which eventually assist in reducing the feeling of worriedness, nervousness, or uneasiness. Besides, this balanced gameplay experience along with the associated positive emotions (i.e. enjoyment and (minimal) anxiety), successfully integrate players into the mediate environment of brain games as well as totally engage them in the gameplay activity (i.e. derived from Table 1). The experience of total engagement is also significant for children to accept the brain games, whereas it contradicts with the adults' psyche as positive emotions (i.e. enjoyment and (minimal) anxiety) are more likely required by the adults in order to accept the brain games (i.e. derived from Table 3). Nonetheless, in order to keep children unaware from the hidden goals of brain games play activity, it is necessary that they experience certain level of engagement along with the feeling of enjoyment; however, for adults this unawareness is more likely dependent on the balanced experience of gameplay as well as total engagement (i.e. derived from Table 4).

6 Conclusion and Future Work

In this paper, we presented an empirical study on the experiences of children and adults during brain games play activity. We discussed a number of insights into behavioral trends, and importantly, correlations between players' engagement, enjoyment, anxiety, usability, adaptability and noninvasiveness. Consequently, the presented study attempts to provide an insight against the pressing questions, which were highlighted at the beginning of "Introduction" section of this paper. Besides, we exploited the insights to successfully predict players' enjoyment, usability, adaptability and noninvasiveness, which lead to broad scale assessment with less measurement efforts.

There exist several research studies that investigated the experience of players in terms of immersion [5], anxiety [27], usability [40] and flow [2, 20]; however, no scientific study employed brain games for the investigation of players' experiences. Subsequently, it is also evidenced that the experience of players differs w. r. t. the game genre [34]. The absence of literature on brain games' experiences as well as an individuality of genre-specific game experiences make the current empirical study first-of-its-kind as well as incomparable with the results of existing literature.

A future intent is to exploit the presented understanding of players' experiences in order to develop brain games design guideline, which will assist to achieve the exact goal (i.e. engagement, enjoyment, (minimal) anxiety, usability, adaptability and noninvasiveness) by targeting the corresponding aspects.

Acknowledgments. This work is supported in part by Natural Science Foundation of China under Grant No. 61572471, 61572466, 61572004, 61472399, 61502456, Chinese Academy of Sciences Research Equipment Development Project under Grant No. YZ201527, Science and Technology Planning Project of Guangdong Province under Grant No. 2015B010105001, Innovation Project of Institute of Computing Technology under Grant No. 20156010, Beijing Natural Science Foundation under Grant No. 4162059 and International Science & Technology Cooperation Program of China under Grant No. 2014DFG12750, CAS-TWAS President's PhD Fellowship Programme.

References

1. Banos, R.M., Botella, C., Alcaniz, M., Liano, V., Guerrero, B., Rey, B.: Immersion and emotion: their impact of sense of presence. CyberPsychol. Behav. 7(6), 734–741 (2004)
2. Ben, C., Darryl, C., Michaela, B., Ray, H.: Toward an understanding of flow in video games. ACM Comput. Entertainment – Theor. Pract. Comput. Appl. Entertainment 6(2) (2008)
3. Brooke, J.: SUS: a quick and dirty usability scale. In: Usability Evaluation in Industry. Taylor and Francis (1996)
4. Green, C.S., Seitz, A.R.: The impacts of video games on cognition. Policy Insights Behav. Brain Sci. (2015)
5. Charlene, J., Anna, L.C., Paul, C., Samira, D., Andrew, E., Tim, T., Alison, W.: Measuring and defining the experience of immersion in games. Int. J. Hum. Comput. Stud. 66, 641–661 (2008)
6. Cohen, J.: Statistical Power Analysis for the Behavioral Sciencies. Routledge, New York (1988)
7. David, A.F.: Statistical Models: Theory and Practice. Cambridge University Press, New York (2005)
8. Csikszentmihalyi, M., Csikszentmihalyi, I.S.: Optimal Experience. Psychological Studies of Flow in Consciousness. Cambridge University Press, Cambridge (1988)
9. Elisa, D.M., et al.: A systematic review of quantitative studies on the enjoyment of digital entertainment games. In: Proceedings of the SIGCHI Conference on Human Factors in Computing Systems (2014)

10. Eriksen, C., Hoffman, J.: Temporal and spatial characteristics of selective encoding from visual displays. Percept. Psychophys. **2**, 201–204 (1972)
11. Eriksen, C., St. James, J.: Visual attention within and around the field of focal attention: a zoom lens model. Percept. Psychophys. **40**(4), 225–240 (1986)
12. Matsushima, F., Vilar, R.G., Mitani, K., Hoshino, Y.: Touch screen rehabilitation system prototype based on cognitive exercise therapy. In: Stephanidis, C. (ed.) HCI 2014, Part II. CCIS, vol. 435, pp. 361–365. Springer, Heidelberg (2014)
13. Fernandez, A.: Fun experience with digital games. In: Extending Experiences: Structure, Analysis and Design of Computer Game Player *Experience*. Lapland University Press (2008)
14. Gallagher, P., Prestwich, S.: Supporting cognitive adaptability through game design. In: 6th European Conference on Games Based Learning. Academic Publishing International Limited (2012)
15. Garrett, J.J.: The Elements of User Experience: User-Centered Design for the Web. New Riders, CA (2003)
16. Glicksohn, J., Avnon, M.: Explorations in virtual reality: absorption, cognition and altered state of consciousness. Imagin. Cogn. Pers. **17**(2), 141–151 (1997)
17. Hassenzahl, M., Tractinsky, N.: User experience - a research agenda. Behav. IT **25**(2), 91–97 (2006)
18. Irwin, H.J.: Pathological and nonpathological dissociation: the relevance of childhood trauma. J. Psychol. **133**(2), 157–164 (1999)
19. Jeanne, H.B., Christine, M.F., Kathleen, A.C., Evan, M., Kimberly, M.B., Jacquelyn, N.P.: The development of the game engagement questionnaire: a measure of engagement in video game-playing. J. Exp. Soc. Psychol. **45**(4), 624–634 (2009)
20. John, L.S.: Flow and media enjoyment. Commun. Theor. **14**(4), 328–347 (2004)
21. Jordan, P.W.: Pleasure with products: human factors for body, mind and soul. In: Human Factors in Product Design: Current Practise and Future Trends. Taylor & Francis (1999)
22. Kankainen, A.: UCPCD: user-centered product concept design. In: Proceedings of Conference on Designing for UX (2003)
23. Kickmeier-Rust, M.D., Hockemeyer, C., Albert, D., Augustin, T.: Micro adaptive, non-invasive knowledge assessment in educational games. In: IEEE International Conference on Digital Game and Intelligent Toys Based Education (2008)
24. Lopez-Samaniego, L., et al.: Cognitive rehabilitation based on working brain reflexes using computer games over iPad. In: Computer Games: AI, Animation, Mobile, Multimedia, Educational and Serious Games (2014)
25. Law, E., Vermeeren, A.P.O.S., Hassenzahl, M., Blythe, M.: Towards a UX manifesto. In: Proceedings of British HCI Group Annual Conference. British Computer Society (2007)
26. Mania, K., Chalmers, A.: The effects of levels of immersion on memory and presence in virtual environments: a reality-centered approach. Cyberpsychol. Behav. **4**(2), 247–264 (2001)
27. Mehwash, M., Mark, D.G.: Online gaming addiction: the role of sensation seeking, self-control, neuroticism, aggression, state anxiety, and trait anxiety. Cyberpsychol. Behav. Soc. Netw. **13**(3) (2009)
28. Mikropoulos, T., Strouboulis, V.: Factors that influence presence in educational virtual environments. Cyberpsychol. Behav. **7**(5), 582–591 (2004)
29. Moneta, G.B., Csikszentmihalyi, M.: The effect of perceived challenges and skills on the quality of subjective experience. J. Pers. **64**(2), 275–310 (1996)
30. Moneta, G.B., Csikszentmihalyi, M.: Models of concentration in natural environments: a comparative approach based on streams of experiential data. Soc. Behav. Pers. **27**(6), 603–637 (1999)

31. Nacke, L., Drachen, A.: Towards a framework of player experience research. In: EPEX 2011 (2011)
32. Pagulayan, R., Keeker, K., Wixon, D., Romero, R.L., Fuller, T.: User-centered design in games. In: The Human-Computer Interaction Handbook. L. Erlbaum Associates Inc. (2003)
33. Pagulayan, R., Steury, K.: Beyond usability in games. Interactions **11**(5), 70–71 (2004)
34. Richard, M.R., Rigby, C.S., Andrew, P.: The motivational pull of video games: a self-determination theory approach. Motiv. Emot. **30**(4), 344–360 (2006)
35. Rui, W., et al.: StudentLife: assessing mental health, academic performance and behavioral trends of college students using smartphones. In: Proceedings of the ACM Conference on Ubiquitous Computing (2014)
36. Byun, S., Park, C.: Serious game for cognitive testing of elderly. In: Stephanidis, C. (ed.) Posters, Part I, HCII 2011. CCIS, vol. 173, pp. 354–357. Springer, Heidelberg (2011)
37. Sánchez, J.L.G., Zea, N.P., Gutiérrez, F.L.: From usability to playability: introduction to player-centred video game development process. In: Kurosu, M. (ed.) HCD 2009. LNCS, vol. 5619, pp. 65–74. Springer, Heidelberg (2009)
38. Seligman, M.P., Walker, E.F., Rosenhan, D.L.: Abnormal psychology, 4th edn. W.W. Norton & Company, New York
39. Singer, M.J., Witmer, B.G.: On selecting the right yardstick. Presence **8**(5), 566–573 (1999)
40. Steve, C.: The usability of massively multiplayer online roleplaying games: designing for new users. In: Proceedings of the SIGCHI (2004)
41. Sweetser, P., Wyeth, P.: GameFlow: a model for evaluating player enjoyment in games. Comput. Entertainment **3**(3)
42. Tamborini, R., Skalski, P.: The role of presence in the experience of electronic games. In: Playing Video Games: Motives, Responses and Consequences (2006)
43. Topalo, V., Chele, N.: Minimally invasive method of early dental implant placement in two surgical steps. Revista de chirurgie oro-maxilo-facială și implantologie **3**(1) (2012)
44. Wirth, W., et al.: A process model of the formation of spatial presence experiences. Media Psychol. **9**(3), 493–525 (2007)

Faster Convergence to Cooperative Policy by Autonomous Detection of Interference States in Multiagent Reinforcement Learning

Sachiyo Arai[✉] and Haichi Xu

Faculty of Engineering, Chiba University,
P1-33 Yayoi-cho, Inage-ku, Chiba 263-8522, Japan
arai@tu.chiba-u.ac.jp

Abstract. In this paper, we propose a method for ameliorating the state-space explosion that can occur in the context of multiagent reinforcement learning. In our method, an agent considers other agents' states only when they interfere with each other in attaining their goals. Our idea is that the initial state-space of each agent does not include information about other spaces. Agents then automatically expand their state-space if they detect interference states. We adopt the information theory measure of entropy to detect the interference states for which agents should consider the state information of other agents. We demonstrate the advantage of our method with respect to the efficiency of global convergence.

Keywords: Multiagent system · Reinforcement learning · Conflict resolution

1 Introduction

In general, multiagent systems are applied to large and complex problems. It is often difficult to design the behavior rules of each agent beforehand, so they are expected to learn adaptively and autonomously. Multiagent Reinforcement Learning (MARL) is an effective approach for this design problem, and has attracted the attention of many researchers. In many existing MARL studies, it is assumed that the state of the environment is observable, including the selected actions and states of all other agents [3–7]. This method does not scale well, because the state-space that agents must learn is usually exponential with respect to the number of agents. In addition, the solution will often be sub-optimal or instable because of the agents' simultaneous and independent learning [16].

To solve these problems, recent research has proposed the observation of only a limited part of the other agents' state-spaces [9,11]. However, the detection of which part of the state-space should be considered relies on a heuristic approach, which requires different knowledge for different problems. In this paper, we propose a novel approach to detect the states in which extra information from

© Springer International Publishing Switzerland 2016
R. Booth and M.-L. Zhang (Eds.): PRICAI 2016, LNAI 9810, pp. 16–29, 2016.
DOI: 10.1007/978-3-319-42911-3_2

other agents is necessary. Our method adopts the information theory measure of entropy to reduce the state-space properly, thus achieving good performance in both learning speed and solution quality.

The remainder of this paper is organized as follows. In Sect. 2, we give the basic background information needed to understand our approach. In Sect. 3, we explain the class of problems, introduce some related work in this domain, and explain the main differences with our approach. Section 4 presents the proposed method. Experimental results are presented in Sect. 5, before our conclusions and ideas for future work are given in Sect. 6.

2   Glossary

In this section, we review the Markov Decision Process framework, Q-learning, Markov Game framework, and multiagent Q-learning, which are the basic concepts behind our proposed method.

2.1   Markov Decision Processes and Q-learning

Markov Decision Processes (MDPs) provide a theoretical framework for single-agent decision making, and are the basis on which reinforcement learning is built. An MDP can be described as a tuple $< S, A, P, R >$, where S is a finite set of states, A is a set of actions available to the agent, $P : S \times A \times S \to [0, 1]$ is the transition function that describes the probability $P(s'|s, a)$ of ending up in state s', and $R : S \times A \to \mathbb{R}$ is a reward function that returns the reward $R(s, a)$ for taking action a in state s.

An agent's policy is defined as a mapping $\pi : S \to A$. The objective is to find the optimal policy π^* that maximizes the expected discounted future reward $U^*(s) = \max_\pi E[\sum_{t=0}^{\infty} \gamma^t R(s_t)|\pi, s_0 = s]$ for each state s. The expectation operator $E[\cdot]$ is averaged over the reward and stochastic transitions, and $\gamma \in [0, 1)$ is the discount factor. This objective can also be expressed using Q-values, which store the expected discounted future reward for each state s and action a:

$$Q^*(s, a) = R(s, a) + \gamma \sum_{s'} P(s'|s, a) \max_{a'} Q^*(s', a') \qquad (1)$$

The optimal policy for a state s is the argument of $\max_a Q^{*(s', a')}$ that maximizes the expected future discounted reward. Watkins [2] described a Q-learning algorithm to iteratively approximate Q^*. Q-learning starts from some initial estimate $Q(s, a)$ for each state-action pair. When an exploration action a is taken in state s, the reward $R(s, a)$ is received and the next state s' is observed. The Q-values are updated according to the following update rule:

$$Q(s, a) \leftarrow (1 - \alpha)Q(s, a) + \alpha(R(s, a) + \gamma V(s')) \qquad (2)$$

$$V(s') \leftarrow \max_{a'} Q(s', a') \qquad (3)$$

where $\alpha \in (0, 1)$ is an appropriate learning rate. Under certain conditions, Q-learning is known to converge to the optimal $Q^*(s, a)$ [2].

2.2   Markov Games and Multiagent Q-learning

Markov Games (MGs) [3] multiagent decision making. An MG can be described by a tuple $< \mathcal{N}, \mathcal{S}, \mathcal{A}_1, \ldots, \mathcal{A}_n, \mathcal{P}, \mathcal{R}_1, \ldots, \mathcal{R}_n >$, where $\mathcal{N}$ is the set of ($n = |\mathcal{N}|$) agents, $\mathcal{S}$ is the finite set of states, and $\mathcal{A}_i (i \in 1, \ldots, n)$ is the set of actions available to agent i. The transition function $\mathcal{P} : \mathcal{S} \times \mathcal{A}_1 \times \cdots \times \mathcal{A}_n \times \mathcal{S} \to [0, 1]$ represents the probability $P(s'|s, a)$ that the state will transit from state s to s' after performing the joint action $a \in \mathcal{A}_1 \times \cdots \times \mathcal{A}_n$, and $\mathcal{R}_i : \mathcal{S} \times \mathcal{A}_1 \times \cdots \times \mathcal{A}_n \to \mathbb{R}$ is the reward function that returns the reward $R_i(s, a)$ for agent i after joint action a is taken in state s. Note that, when $n = |\mathcal{N}| = 1$, the MG is equivalent to an MDP.

Table 1 classifies MGs by the relationship between $\mathcal{R}_i$, and summarizes their existing multiagent Q-learning algorithms. In these algorithms, it is necessary for an agent to observe the state-space consisting of its own state and those of other agents at all times.

Table 1. Markov Games and existing algorithms

Reward	Classification	Algorithms
$R_1 + R_2 = 0$	Zero-sum MGs	Minimax Q-learning [3]
$\Sigma_{k=1}^n R_k = Const.$	General-sum MGs	Nash Q-learning [6], Correlated-Q [7]
$R_1 = \cdots = R_n = R$	Team MGs	Team-Q [4], OAL [5]

3   Problem Domain

In the existing MARL research discussed in Sect. 2.2, the states of other agents must be observed at all times. However, as mentioned in Sect. 1, such approaches do not scale well, as the state-space often becomes exponential in the number of agents. In addition, the experimental results of Busoniu et al. [1] showed that agents tend to learn slowly, and the solution will often be sub-optimal due to this explosion in the state-space.

In reality, it is difficult to observe the states of other agents at all times [8]. Further, in general, as the multiagent system relies upon the sparse interaction between agents, there should be no need to observe the complete state of the other agents. If complete observation is necessary, we can achieve a control system via the central management of a super-agent. Therefore, in a multiagent system with sparse interaction between agents, recognizing which state-spaces should be considered becomes an important issue.

3.1   Definition of Terminology

In this paper, we divide the state-space of the agent into **collision states**, **interference states**, and **non-interference states**. In collision states and interference states, agents interact with each other. There is no interaction among agents in the non-interference state. The set of **collision states** is defined as the set in which agents may be present in the same state transition simultaneously. The set of **interference states** is defined as the set of states prior to the transition to collision states. In addition, we call the actions such that agents transit from interference states to collision states. A conceptual diagram of the collision states and interference states is shown in Fig. 1.

Consider the state transition shown in Fig. 1. Here, the observation of agent i is s^i, and that of other agents'is s^{-i}. Because the destination state of the transition from interference states is dependent on the action of other agents, different rewards will be given depending on the transition to a collision state or non-interference state. These are marked by r_{ic}, r_{in}, respectively, in Fig. 1. An agent recognizes a transition to the same state, and the action value in the interference state will be updated by the different rewards. Therefore, incorrect Q-values will propagate through the interference states and their previous states, and lead to uncertain policies in these states. Thus, we should observe the state of other agents in interference states, but not necessarily in the other states. Recently, research has shown that it is possible to detect these interference states. In this paper, we propose a novel method for detecting the interference states, and develop a learning algorithm for different types of states.

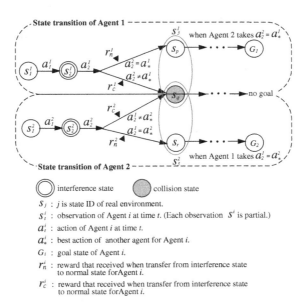

Fig. 1. Image of interference states and collision states

3.2   Related Work

In this section, we present a brief overview of the most relevant existing work in this problem domain, and describe the differences, as well as some similarities, between our proposed method and these approaches.

Kok et al. [9] described an approach in which agents know which states are interference states beforehand, allowing the successful reduction of state-spaces. In [10], Kok et al. use Coordination Graphs (CGs) to represent the relationship between agents in interference states and interference actions. They propose an approach for learning the CGs during the learning process, which enables the detection of interference states and reduction of state-spaces. However, this approach is limited to Team MGs.

Melo et al. introduced interaction-driven MGs, which contain a set of states in which interaction occurs between agents. The list of interference states was then applied to the agents beforehand [11]. In their later work, an algorithm was proposed in which agents learn to detect the interference states autonomously, rather than being given them [12]. In other words, agents detect the states in which interaction is necessary via a learning process. To achieve this, the action space of each agent is augmented with a pseudo-coordination action that performs an active perception step. In this perception step, agents observe the states of other agents, and determine whether the other agents' states should be considered. Because the penalty for collision is bigger than the cost of this active perception, the agents learn to take this action in interference states. However, determining the cost of using active perception is a challenge, because it is necessary to consider the penalty of collision comprehensively.

De Hauwere et al. succeeded in detecting the interference states using a Generalized Learning Automaton (GLA) [13]. The GLA receives the Manhattan distance between two agents as input, and the agents can then learn which are the interference states. In addition, De Hauwere et al. introduced an effective statistical method that focuses on the reward sequence given in each state for detecting the interference states [14,15]. Although similar to the approach of [10], this method further reduces the state-space. However, although this approach is effective when considering immediate rewards, it is not applicable under a delayed-reward environment.

Busoniu et al. [8] proposed a method for detecting interference states by focusing on the differences in Q-value convergence between the interference states and non-interference states. Their approach does not consider rewards, unlike [14,15]. However, to analyze the convergence of the Q-values, many parameters must be set appropriately, which is a challenging task.

In this paper, we propose an approach in which agents learn to detect interference states autonomously during the learning process, but have no knowledge of the interference states beforehand. Our approach focuses on the fluctuation of the Q-values, which is similar to the approach of [8] and different from those of [10,14,15]. Our approach formulates the variation of the Q-values as the entropy of information theory, and detects the interference states by observing fluctuations in the entropy.

4   Proposed Method

4.1   Entropy Based Approach

In information theory, entropy is a measure of the uncertainty in a random variable. The information entropy is defined as $H(S) = -\sum_{i=1}^{m} p_i \log p_i$, where $\{p_1, p_2, \ldots, p_m\}$ is the probability distribution of random variable S on a discrete set with m elements. We developed a measure based on information theory [16] for evaluating the degree of interaction in a multiagent system during the learning process. In this paper, we detect the interference states by adopting the information theory measure of entropy, in accordance with this quantification. The quantification of the uncertainty of a specific policy is summarized as follows.

Consider the state set $\mathcal{S} = \{s_1, \ldots, s_h, \ldots, s_m\}$, let the action set that can be selected by an agent be $\mathcal{A} = \{a_1, \ldots, a_i, \ldots, a_n\}$, and assume that the policy is $\pi(\mathcal{S}, \mathcal{A})$. In this context, the policy in state s_h is then $\sum_{i=1}^{n} \pi(s_h, a_i) = 1$. According to the definition of information entropy, the entropy of policy π in state s_h can be calculated by Eq. (4). In the following, $H(\pi(s_h, \mathcal{A}))$ is abbreviated to $H(s_h)$ for simplification.

$$H(\pi(s_h, \mathcal{A})) = -\sum_{i=1}^{n} \pi(s_h, a_i) \log \pi(s_h, a_i) \tag{4}$$

4.2   Detection of Interference States by Entropy

In terms of the probability distribution of policy $\pi(s_h, \mathcal{A})$, the entropy in state s_h has the following properties.

– if $\pi(s_h, a_1) = \cdots = \pi(s_h, a_i) = \cdots = \pi(s_h, a_n)$, $H(s_h)$ becomes the maximum value
– if $\exists a_i \in \mathcal{A}$, $\pi(s_h, a_i) = 1$, $H(s_h) = 0$.

Therefore, during the learning process, if the policy convergences to a "deterministic policy," then $H(s_h) = 0$. However, $H(s_h)$ does not decrease monotonically. In the early learning stages, to avoid convergence to a local solution, action a in which $Q(s, a)$ is maximized may not be selected because of the exploration process. In addition, in the multiagent environment, the influence of simultaneous learning between agents means that the magnitude relation between the value of $Q(s, a)$ varies frequently, and the optimal action also changes.

Further, as mentioned above, with the progress of reinforcement learning under single agent MDPs, the policy converges to a "deterministic policy," and $H(s_h)$ should become 0. In MGs, however, more than one action may be effective, and the Q-values of these actions will be similar. Hence, the policy might convergence to a "deterministic policy" even when the learning is at an advanced stage. This is because the agent cannot observe the internal states of other agents in the multiagent environment if they learn independently. In this case, although the entropy is not 0, the fluctuation of entropy is eliminated. On the other hand, the entropy of a policy in the interference states does not tend to 0, and the

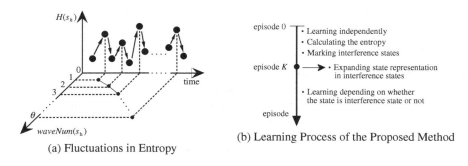

(a) Fluctuations in Entropy

(b) Learning Process of the Proposed Method

Fig. 2. Basic idea to find interferences

value fluctuates frequently. In the non-interference states, the entropy of a policy may fluctuate in the early learning process due to the randomness of the action selection, but it should become 0, or at least remain stable.

Based on the description above, the entropy of a policy in the interference states fluctuates frequently for long periods of time during the learning process, but this phenomenon does not occur in other states. As shown in Fig. 2(a), the transition to the collision state from the interference state varies because of the influence of other agents' actions. In particular, the higher the state-value in the posterior state of the collision state, the more the entropy of the policy in the interference state prior to the collision state fluctuates. This is because of the increase during the transition to the collision state. Therefore, to detect interference states, we examine the frequency of the entropy fluctuation. That is, as shown in Fig. 2(a), we aggregate the increase or decrease in entropy over a certain learning time. If this exceeds a predetermined threshold, the state is determined to be an interference state. Here, we denote the increase or decrease in the entropy of a policy in state $s_h \in \mathcal{S}$ as $waveNum(s_h)$, and the predetermined threshold value as θ. Equation (4) calculates the entropy of agent i's in state $s_h \in \mathcal{S}$ using Q values. Here, τ is a parameter for the exploration.

$$\pi(s_h, a_i) = \frac{e^{Q(s_h, a_i)/\tau}}{\sum_{b \in \mathcal{A}} e^{Q(s_h, b)/\tau}} \qquad (5)$$

4.3   Algorithm

An outline of the proposed method is shown in Fig. 2(b). The proposed learning algorithm is shown in Figs. 3 and 5. Note that the superscripts i and $-i$ in Figs. 3 and 5 refer to agent i and the other agents, respectively.

First, the set of interference states is initialized to the empty set $\mathcal{S}_o^i = \emptyset$. Before the K-th episode, agents learn independently by single-agent Q-learning without observing the other agents' states, and we calculate $H(s^i)$, which is the entropy of a policy in each state s^i. When the entropy $H(s^i)$ increases or decreases, $waveNum(s^i)$ is updated as $waveNum(s^i) \leftarrow waveNum(s^i) + 1$.

Initialize Q^i and Q^i_o
Set $episode = 0$, $S^i_o = \emptyset$, and θ
Set waveNum(s^i)=0 for $\forall s^i \in S^i$
while $2 \le episode < K$ **do**
    Repeat for steps:
        Observe s^i
        Choose a^i from Q^i
        Observe $r^i, s^{i\,'}$
        Update $Q^i(s^i, a^i) \leftarrow (1 - \alpha)Q^i(s^i, a^i)+$
            $\alpha(r^i + \gamma \max_{a^i \in \mathcal{A}^i} Q(s^{i\,'}, a^i))$
    Calculate $H(s^i)_{episode}$
    if $H(s^i)_{episode-1} > H(s^i)_{episode-2}$ and
    $H(s^i)_{episode-1} < H(s^i)_{episode}$ **then**
        $waveNum(s^i) \leftarrow waveNum(s^i) + 1$
    end if
    if $waveNum(s^i) \ge \theta$
        Mark s^i as an interference state and
        add s^i to S^i_o
    end if
    $episode \leftarrow episode + 1$
end while
if $episode = K$ **then**
    Expand Q-table in interference state
        $Q^i_o(s^i, s^{-i}, a^i) \leftarrow Q^i(s^i, a^i)$
        for $\forall s^i \in S^i_o, s^{-i} \in S^{-i}, a^i \in \mathcal{A}$
end if

Fig. 3. Agent Algorithm before K-th Episode in the Detection Phase of Interference States

When $waveNum(s_h)$ exceeds the threshold value θ, s_h is marked as an interference state and added to the set of interference states S^i_o.

At the end of episode K, to take advantage of the learning results up to that point, we use Eq. (6) to initialize the Q-value in the interference states.

$$Q^i_o(s^i, s^{-i}, a^i) \leftarrow Q^i(s^i, a^i)$$
$$\forall s^i \in S^i_o, s^{-i} \in S^{-i}, a^i \in \mathcal{A}^i \tag{6}$$

where $Q^i(s^i, a^i)$ denotes the Q-table learned by episode K, and $Q^i_o(s^i, s^{-i}, a^i)$ represents the combination of interference state s^i and the state of other agents s^{-i}. In Fig. 4, we show an example of this extension of the Q-table. The states of agent 2 are added to the state representation of agent 1 in states 4 and 6, because states 4 and 6 are detected as interference states. By this extension of the Q-table, we are able to reduce the state-space.

After episode K, the process proceeds as follows. When an agent selects an action and updates the Q-values, it checks whether its current state is an interference state. If not, it will use $Q^i(s^i, a^i)$ to select an action, and the rule

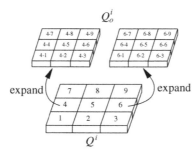

Fig. 4. Expanding the Q-table in the Interference States

Repeat for episodes:
    Observe s^i
    if $s^i \notin S^i_o$ **then** //s^i is not an interference state
        Choose a^i from Q^i
        Observe $r^i, s^{i'}$
        Update $Q^i(s^i, a^i) \leftarrow (1 - \alpha)Q^i(s^i, a^i) +$
                $\alpha(r^i + \gamma \max_{a^i \in \mathcal{A}^i} Q(s^{i'}, a^i))$
    else //s^i is an interference state
        Observe s^{-i}
        Choose a^i from Q^i_o
        Observe $r^i, s^{i'}$
        Update $Q^i_o(s^i, s^{-i}, a^i) \leftarrow (1 - \alpha)Q^i_o(s^i, s^{-i}, a^i) +$
                $\alpha(r^i + \gamma \max_{a^i \in \mathcal{A}^i} Q(s^{i'}, a^i))$
    end if

Fig. 5. Agent Algorithm after $(K + 1)$-th Episode for Updating Q-values

represented as Eq. (7) will be used to update the Q-values.

$$Q^i(s^i, a^i) = (1 - \alpha)Q^i(s^i, a^i) +$$
$$\alpha(r^i + \gamma \max_{a^i \in \mathcal{A}^i} Q(s^{i'}, a^i)) \tag{7}$$

On the other hand, if the agent is currently in an interference state, it will select an action by $Q^i_o(s^i, s^{-i}, a^i)$, and the rule represented as Eq. (8) will be used to update the Q-values.

$$Q^i_o(s^i, s^{-i}, a^i) = (1 - \alpha)Q^i_o(s^i, s^{-i}, a^i) +$$
$$\alpha(r^i + \gamma \max_{a^i \in \mathcal{A}^i} Q(s^{i'}, a^i)) \tag{8}$$

Also note that when the value of $Q(s^{i'}, a^i)$ in Eqs. (7) and (8), is calculated, $s^{i'}$ is considered as a non-interference states. This is because each agent iterates a cycle of observation, action, then transit a next state, the agent cannot identify whether new (transited) state is a non-interference state or not, when agent updates $Q^i(s^i, a^i)$. In other words, the Q value of the transited state, is a previous value of expansion.

5   Experiments

5.1   Experimental Environment and Settings

Our experimental environment is shown in the left of Fig. 6. It is a maze environment containing two agents, and their start states and goal states are denoted by S_1, S_2, and G_1, G_2, respectively. The black blocks represent walls that agents cannot pass through. The agents' task is to find the shortest path to their goals. However, the task is not completed when only one agent reaches its goal.

The action set available to the agents is < UP, DOWN, LEFT, RIGHT, STOP >. These cause the agent to move simultaneously one cell up, down, left, or right, or to stop. The transitions are deterministic. Initially, each agent is placed in their start state. An episode is defined as the period from the start time of the initial state to the time at which both agents have reached their goal states. Once an agent reaches its goal state, it remains there. When the length of an episode exceeds some upper limit, the episode is terminated.

Each agent receives a reward of $R_i = 500$ when they complete the task, and a reward of -50 when they collide with each other. When an agent collides with a wall, it receives a reward of -20. In all other cases, the reward is 0. Note that, when an agent collides with another agent or a wall, it returns to the state in the previous time step. This settings of reward correspond to the Markov Game situation as mentioned in Sect. 2.2.

Agents learn for 400 episodes, and the upper limit of time steps in an episode is set to 500. We use an ϵ-greedy action selection strategy, where ϵ is set to 0.2 in the first 300 episodes and $\epsilon = 0$ from then on. The learning rate is set to $\alpha = 0.3$, the discount factor is set to $\gamma = 0.9$, and the initial Q-values are set to 0.1. We set the parameters related to the detection of interference states as $K = 50$ and the threshold $\theta = 2$.

5.2   Experimental Results

Acquired Behavior: First, in the right of Fig. 6, we show which states were detected as the interference states by the agents. We can see that agents have correctly learned to detect the states in which collisions are frequent or most likely. Two typical examples of acquired behaviors are shown in Fig. 7(a), and Fig. 8(a) where the red and blue colored lines indicate the behaviors of agent1's and agent2's, respectively. It is found that the both behaviors are globally optimal.

Fluctuation of Entropy in Interference States. Figure 7(a) shows that agent2 wait around $(4, 4)$ to avoid collision while Fig. 8(a) shows that agent1 wait around $(4, 2)$ to avoid collision. The fluctuation of entropy of both situations are

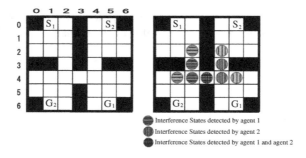

Fig. 6. Detected Interference States. Sparse lines with circles indicate the interference states with a low frequency of detection. (Color figure online)

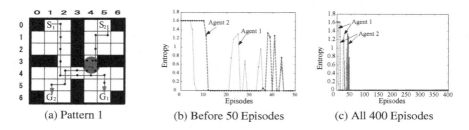

(a) Pattern 1   (b) Before 50 Episodes   (c) All 400 Episodes

Fig. 7. Pattern1: Finally Obtained Path. (The red and blue lines indicate the path of agent 1 and agent 2 respectively.), Fluctuations of Entropy in State (4, 2) of Agent 1 (Color figure online)

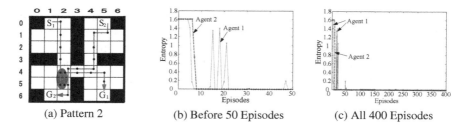

(a) Pattern 2   (b) Before 50 Episodes   (c) All 400 Episodes

Fig. 8. Pattern2: Finally Obtained Path. (The red and blue lines indicate the path of agent 1 and agent 2 respectively.), Fluctuations of Entropy in State (4, 4) of Agent 2 (Color figure online)

shown in Figs. 7(b) and 8(b). The vertical axis shows the value of entropy of each state, and the horizontal axis records the number of episodes. Figures 7(c) and 8 show the convergence of learning after 400 episodes.

5.3   Evaluation of Learning Efficiency

Next, to evaluate the performance of the proposed method, we compared it with two other multiagent Q-learning methods. One is **Independent learners**, which learn without any information about the state of other agents, and the second is **Joint-state learners**, which receive the joint location of the agents as state information, but choose their actions independently.

Figure 9 shows the number of steps taken to complete each episode, averaged over 10 runs. The vertical axis shows the number of time steps needed to complete one episode, and the horizontal axis records the number of episodes. We also show the size of the state-space, the converged steps, and the required steps to acquire the optimal/best solution in each method in Table 2.

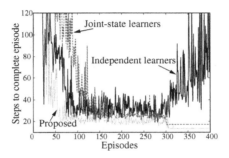

Fig. 9. Experiment on the Efficiency of the Proposed Method, Compared with the Independent Learners and Joint-state Learners, Averaged over 10 Runs.

Table 2. Comparison of learning performance. Average and s.d 10 series of experiments

Algorithm	State space ± s.d	Steps ± s.d	Required Episodes to 17 steps ± s.d
Independent learners	35	not converge	not converge
Joint-state learners	1225	17.1±1.22	164.90±45.66
Proposed method	100.8±20.4	13.5±0.92	79.4±33.26

Both the Independent learners and our proposed method learn quickly in early episodes (<= 50) compared with the Joint-state learners, because learning is based on a smaller state-space. However, the Independent learners do not converge to a stable policy, but instead oscillate because the policies in the interference states are uncertain. Our proposed method converges to a stable policy, because the agents observe the states of the other agents after the 50th episode. The Joint-state learners were consistently able to converge to a stable policy, but learned slowly due to the large size of the state-space. Our proposed method learns quickly and converges to a better stable policy than the Joint-state learners, because of the reduction in the state-space brought about by detecting the interference states.

6   Conclusions

In this paper, we presented an approach for reducing the state-space in the MARL domain. An overly large state-space usually causes agents to learn slowly and reach sub-optimal solutions. We defined the states in which agents should consider the influence of other agents' actions as interference states. Our proposed method is made up of two phases: one for detecting the interference states, and another for learning different types of states.

We noted that if the states of other agents are ignored when they should be considered, agents' policies became uncertain. The interference states were detected by calculating the degree of this uncertainty via the entropy of information theory. We reduced the state-space by limiting the observations of other agents' states, thus improving the learning speed. In addition, it was possible to avoid the incomplete perception caused by the actions of agents, which led to an improvement in the optimality of the solution.

However, as shown in Eq. (6) and Fig. 4 in Sect. 4.3, by expanding the Q-table in the interference states, all possible states of other agents are considered comprehensively. In fact, we can speculate that the influence of other agents' states on a given agent's own actions only arise in a part therein. In future work, we will focus on finding a method to achieve a more compact state-space representation, whereby an agent expands its own state information using that of other agents. Furthermore, we need to investigate methods to set a suitable threshold value for the number of entropy fluctuations. This was set by preliminary experiments in our current implementation.

References

1. Busoniu, L., De Schutter, B., Babuška, R.: A comprehensive survey of multiagent reinforcement learning. IEEE Trans. Syst. Man Cybern. Part C Appl. Rev. **38**(2), 156–172 (2008)
2. Watkins, C., Dayan, P.: Q-learning. Mach. Learn. **8**(3), 279–292 (1992)
3. Littman, M.: Markov games as a framework for multi-agent reinforcement learning. In: Proceedings of the Eleventh International Conference on Machine Learning, pp. 242–250 (1994)
4. Littman, M.: Value-function reinforcement learning in Markov games. Cogn. Syst. Res. **2**(1), 55–66 (2001)
5. Wang, X., Sandholm, T.: Reinforcement learning to play an optimal Nash equilibrium in team Markov games. In: Advances in Neural Information Processing Systems vol. 15, pp. 1571–1578 (2002)
6. Hu, J., Wellman, M.: Nash Q-learning for general-sum stochastic games. J. Mach. Learn. Res. **4**, 1039–1069 (2003)

7. Greenwald, A., Zinkevich, M., Kaelbling, P.: Correlated Q-learning. In: Proceedings of the Twentieth International Conference on Machine Learning, pp. 242–249 (2003)
8. Busoniu, L., De Schutter, B., Babuška, R.: Multiagent reinforcement learning with adaptive state focus. In: Proceedings of the Seventeenth Belgian-Dutch Conference on Artificial Intelligence, pp. 35–42 (2005)
9. Kok, J.R., Vlassis, N.: Sparse cooperative Q-learning. In: Proceedings of the Twenty-First International Conference on Machine Learning, pp. 61–68 (2004)
10. Kok, J.R., Hoen, P., Bakker, B., Vlassis, N.: Utile coordination: learning interdependencies among cooperative agents. In: Proceedings of the IEEE Symposium on Computational Intelligence and Games (CIG), pp. 29–36 (2005)
11. Spaan, M.T.J., Melo, F.S.: Interaction-driven Markov games for decentralized multiagent planning under uncertainty. In: Proceedings of the 7th International Conference on Autonomous Agents and Multiagent Systems, pp. 525–532 (2008)
12. Melo, F.S., Veloso, M.: Learning of coordination: exploiting sparse interactions in multiagent systems. In: Proceedings of the 8th International Conference on Autonomous Agents and Multiagent Systems, pp. 773–780 (2009)
13. De Hauwere, Y., Vrancx, P., Nowé, A.: Learning what to observe in multi-agent systems. In: Proceedings of the Twentieth Belgian-Dutch Conference on Artificial Intelligence, pp. 83–90 (2009)
14. De Hauwere, Y., Vrancx, P., Nowé, A.: Learning multi-agent state space representations. In: Proceedings of the 9th International Conference on Autonomous Agents and Multiagent Systems, pp. 715–722 (2010)
15. De Hauwere, Y., Vrancx, P., Nowé, A.: Adaptive state representations for multiagent reinforcement learning. In: Proceedings of the 3rd International Conference on Agents and Artificial Intelligence, pp. 181–189 (2011)
16. Arai, S., Ishigaki, Y.: Information theoretic approach for measuring interaction in multiagent domain. J. Adv. Comput. Intell. Intell. Inform. **13**(6), 649–657 (2009)

SWARM: An Approach for Mining Semantic Association Rules from Semantic Web Data

Molood Barati[1](✉), Quan Bai[1](✉), and Qing Liu[2]

[1] Auckland University of Technology, Auckland, New Zealand
{mbarati,qbai}@aut.ac.nz
[2] Data61, CSIRO, 15 College Road, Sandy Bay, Tasmania 7005, Australia
q.liu@csiro.au

Abstract. The ever growing amount of Semantic Web data has made it increasingly difficult to analyse the information required by the users. Association rule mining is one of the most useful techniques for discovering frequent patterns among RDF triples. In this context, some statistical methods strongly rely on the user intervention that is time-consuming and error-prone due to a large amount of data. In these studies, the rule quality factors (e.g. Support and Confidence measures) consider only knowledge in the instance-level data. However, Semantic Web data contains knowledge in both instance-level and schema-level. In this paper, we introduce an approach called SWARM (Semantic Web Association Rule Mining) to automatically mine Semantic Association Rules from RDF data. We discuss how to utilize knowledge encode in the schema-level to enrich the semantics of rules. We also show that our approach is able to reveal common behavioral patterns associated with knowledge in the instance-level and schema-level. The proposed rule quality factors (Support and Confidence) consider knowledge not only in the instance-level but also schema-level. Experiments performed on the DBpedia Dataset (3.8) demonstrate the usefulness of the proposed approach.

Keywords: Semantic Web data · Association rule mining · Ontology · Knowledge discovery

1 Introduction

The Semantic Web is an effort to make knowledge on the Web both human-understandable and machine-readable [1]. Semantic Web data is normally structured in triple formats called Resource Description Framework (RDF). By emerging RDF/S, OWL and SPARQL standardization, the number of large KBs such as YAGO, DBpedia and Freebase[1] is growing so fast. Although these KBs suffer many issues such as incompleteness and inconsistencies, they already contain millions of facts which raise new opportunities for data mining community. In recent years, researchers have been working on developing methods and tools

---

[1] http://freebase.com.

© Springer International Publishing Switzerland 2016
R. Booth and M.-L. Zhang (Eds.): PRICAI 2016, LNAI 9810, pp. 30–43, 2016.
DOI: 10.1007/978-3-319-42911-3_3

for mining hidden patterns from Semantic Web data that promise more potential for Semantic Web applications [2]. In this regard, association rule mining is one of the most common Data Mining (DM) techniques for extracting frequent patterns.

There are several methods in mining associations from large RDF-style KBs. Most existing methods focus on Inductive Logic Programming (ILP) to mine association rules. ILP usually requires counterexamples. AMIE [3,4] is a multi-threaded approach where the KB is kept and indexed in the memory. High memory usage is one of the drawbacks of this approach. This method is restricted to a complete ontology structure. To compute support and confidence values, the method only considers knowledge in the instance-level and removes *rdf:type* relations from datasets. A recent statistical approach for mining association rules in RDF data is [5]. It automatically generates three forms of $s_i \Rightarrow s_j$, $p_i \Rightarrow p_j$, and $o_i \Rightarrow o_j$ rules. This approach does not require counterexamples. However, the method discovers the sequence of subjects, predicates, or objects which are correlated independently. Additionally, the rule quality factors (Support and Confidence) only assess instance-level data.

In comparison with [5], we propose a statistical approach to automatically mine rules from RDF data. Our approach is based on the methodology that adapts association rule mining to RDF data. The rules reveal common behavioural patterns associated with knowledge in the instance-level and schema-level. Consider the RDF triples shown in Table 1. From the triples, our approach generates the following rule:

$$\{Person\}: (instrument, \ Guitar) \Rightarrow (occupation, \ Songwriter)$$

The above rule shows that most of the time persons who play a musical instrument such as Guitar, they are probably Songwriters. Mining such regularities help us to gain a better understanding of Semantic Web data. In order to elaborate the semantics of the rules, our approach considers *rdf:type* and *rdf:subClassOf* relations in the ontology. As seen in Table 1, both *John Lennon* and *George Harrison* are guitarists and songwriters. Consider Fig. 2 as a small fragment of DBpedia ontology. *George Harrison*[2] is an instance of Musical Artist while *John Lennon*[3] belongs to the Person class. Regarding the concept of hierarchy in the ontology, if Musical Artist class is a subclass of Artist class and the Artist class is a subclass of Person class, then *George Harrison* belongs to the Person class as well. But *John Lennon* is not an instance of Musical Artist class. In the context of Semantic Web data, it is not reasonable to interpret the discovered rules without considering such relationships between instance-level and schema-level. As far as we know, the proposed approach in [3–5] do not cover such issues on mining Semantic Web data.

Under this motivation, in this paper, we proposed a novel approach called SWARM (Semantic Web Association Rule Mining) to automatically mine and

---

[2] http://dbpedia.org/page/George_Harrison.
[3] http://dbpedia.org/page/John_Lennon.

Table 1. RDF triples from DBpedia

Subject	Predicate	Object
John Lennon	instrument	Guitar
John Lennon	spouse	Yoko Ono
John Lennon	occupation	Songwriter
George Harrison	instrument	Guitar
George Harrison	occupation	Songwriter
Jimmy Carter	office	President of the USA
Jimmy Carter	party	Democratic
Bill Clinton	office	President of the USA
Bill Clinton	party	Democratic
George W. Bush	office	President of the USA
George W. Bush	party	Republic
John Lennon	*rdf:type*	dbo:Person
George Harrison	*rdf:type*	dbo:MusicalArtist
George Harrison	*rdf:type*	dbo:Person
Jimmy Carter	*rdf:type*	dbo:Person
Bill Clinton	*rdf:type*	dbo:Person
George W. Bush	*rdf:type*	dbo:Person

generate semantically-enriched rules from RDF data. The main contribution of this paper is threefold:

1. The SWARM is an approach that automatically mines association rules from RDF data without the need of domain experts.
2. The SWARM measures the quality of rules (Support and confidence) by utilizing knowledge not only in the instance-level but also schema-level.
3. The SWARM reveals common behavioural patterns associated with knowledge in the instance-level and schema-level.

The remainder of this paper is organized as follows. Section 2 gives a general overview of the related works in this context. In Sect. 3, the SWARM approach is introduced in detail. Both framework architecture and algorithms are presented in this section. Section 4 shows the experimental results. Finally, the conclusion and future work are presented in Sect. 5.

2   Related Works

In the following, we discuss state-of-the-art approaches in the context of Semantic Web data mining.

Logical Rule Mining. Most related research on mining Semantic Web data relies on ILP techniques. ALEPH [6] is an ILP system implemented in the Prolog. WARMeR [7] used a declarative language to mine association rules correspondent to conjunctive queries from a relational database. Galárraga, et al. [3] proposed a multi-threaded approach called AMIE for mining association rules from RDF-style KBs. Galárraga, et al. [4] extends AMIE to AMIE$^+$ using pruning and query rewriting techniques. Similar to AMIE, [8] proposed another approach for extracting horn rules. The proposed methods in [3,4,8] measures the quality of the discovered rules using instance-level data. However, instances in a rule might belong to different classes of the ontologies. To express a broader meaning of the rules, the SWARM considers instance-level data along with *rdf:type* and *rdf:subClassOf* relations in the schema-level.

Association Rule Mining. Association rule mining was originally proposed for shopping basket problems [9]. It reflects high correlation between multiple objects and extracts interesting relationships between data [10]. Nebot and Berlanga [11] proposed a rule mining approach over RDF-based medical data. Transactions have been generated using mining patterns developed by SPARQL queries. This approach heavily relies on the domain experts. Namely, the user should have background knowledge of vocabularies used in the ontology. In comparison to [11], SWARM approach does not require the domain experts.

Abedjan and Naumann [5,12] developed an approach to identify schema and value dependencies between RDF data using six different Configurations. Any part of Subject-Predicate-Object (SPO) statement can be considered as a *context*, which is used for grouping one of the two remaining parts of the statement as the *target* of mining. This approach mines three forms of $s_i \Rightarrow s_j$, $p_i \Rightarrow p_j$, and $o_i \Rightarrow o_j$ rules. The discovered rules shows the correlation among subjects, predicates or objects independently. In comparison with this approach, SWARM generates common behavioural patterns associated with knowledge in instance-level and schema-level.

3   The SWARM Approach

In this section, we describe a detailed view of SWARM approach along with the definitions. The overall framework is shown in Fig. 1. The main goal of SWARM approach is to tie instance-level to schema-level to attach more semantics to the rules. The SWARM generates Semantic Association Rules from RDF data.

The RDF triples are automatically processed via Pre-processing Module consisting two sub-modules: Semantic Item Generation and Common Behaviour Set Generation. The Mining Module receives Common Behaviour Sets to generate Semantic Association Rules. The SWARM approach evaluates the importance of rules by using *rdf:type* and *rdf:subClassOf* relations in the ontology. The proposed rule quality factors (Support and confidence) consider knowledge not only in the instance-level bust also schema-level.

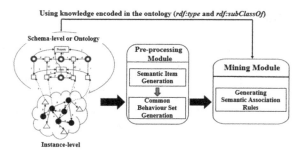

Fig. 1. The SWARM framework

3.1 Pre-processing Module

The concept of association rule mining was first introduced in [9]. Let $I = \{i_1, i_2, ..., i_n\}$ be a set of items and $D = \{t_1, t_2, ..., t_m\}$ be a set of transactions. Each transaction contains a subset of items in I. An association rule represents a frequent pattern of the occurrence of some items in transactions. In addition, association rules reveal behavioural patterns of some particular entities. For example, in the traditional shopping basket problem, the rule $\{butter, bread\} \Rightarrow \{milk\}$ shows a behavioural pattern of customers. Namely, if a customer buys butter and bread together, she is likely to buy milk as well.

Traditional association rule mining algorithms are suited for homogeneous repositories, where items and transactions play significant roles in the mining process [13]. However, most Semantic Web data are not transactional data, and there exists no items or transactions. To generate association rules in the context of Semantic Web data, we need to model such notions.

As mentioned earlier, Semantic Web data are normally structured in triple format. The assertion of a triple (i.e., subject, predicate, object) indicates a meaningful relationship between entities (subject and object) provided by the predicate. A triple can also be considered as the description of one particular behaviour of entities.

For example, suppose we have a triple $t1$ in Table 1: *(John Lennon, instrument, Guitar)*. If we consider the subject in $t1$ (i.e., John Lennon) as the entity, the other two elements in the triple (i.e., instrument Guitar) can be considered the description of a particular behaviour of John Lennon. Based on this concept, in the SWARM approach, we target at exploring behavioural patterns among entities. Under this motivation, we define Semantic Item and Common Behaviour Set to summarize common behaviours of entities.

3.2 Semantic Item Generation

Consider the example presented in the previous paragraph with triple $t4$ in Table 1: *(George Harrison, instrument, Guitar)*. These two subjects in $t1$ and $t4$ {*John Lennon, George Harrison*} have a common activity, i.e., *(instrument*

Guitar). Namely, playing guitar is a common behaviour taken by a group of entities, i.e., {*John Lennon, George Harrison*}. In the SWARM approach, such combinations are represented as Semantic Items.

Definition 1 *(Semantic Item).* A Semantic Item si is a 2-tuple, i.e., $si = (es, pa)$. es is an Element Set of si. It contains a list of subjects, i.e., $\{s_1, s_2, ..., s_n\}$. pa is a Pair of si. Corresponding with the content in es, pa contains a combination of predicate-object, i.e., (p, o).

According to Definition 1, triples in a triple store can be converted to a set of Semantic Items *i.e.* $SI = \{si_1, si_2, ..., si_n\}$. Each Semantic Item contains a Pair, which can be considered as a common behaviour taken by entities in the Element Set.

Example 1. Consider the triples shown in Table 1. Table 2 shows some of the Semantic Items generated by Definition 1. For example, the Element Set of si_2 including {*JohnLennon, GeorgeHarrison*} represents all subjects that contain (*occupation, Songwriter*) as a Pair.

Table 2. Semantic Items

Semantic Items	
si_1	{John Lennon, George Harrison}(instrument, Guitar)
si_2	{John Lennon, George Harrison}(occupation, Songwriter)
si_3	{Jimmy Carter, Bill Clinton, George W. Bush}(office, President of the USA)
si_4	{Jimmy Carter, Bill Clinton}(party, Democratic)

3.3 Common Behaviour Set Generation

As introduced in the previous subsection, a Semantic Item indicates a common behaviour (i.e., the Pair) taken by a group of entities in the Element Set. We define a Common Behaviour Set that represents all common activities taken by similar groups of entities in the Element Sets.

Definition 2 *(Common Behaviour Set).* A Common Behaviour Set cbs contains a set of Semantic Items with similar Element Sets, i.e., $\{(es, pa)_1, (es, pa)_2, ..., (es, pa)_n\}$. Items can be aggregated into the same cbs, if the similarity degree of their *Element Sets* are greater than or equal to Similarity Threshold *SimTh*. The Similarity Degree of *Element Sets* can be calculated by using Eq. 1.

$$sim(es_a, es_b, ..., es_m) = \frac{|es_a \cap es_b \cap ... \cap es_m|}{|es_a \cup es_b \cup ... \cup es_m|} \tag{1}$$

According to Definition 2, the cbs is a set of Semantic Items aggregated through the similarity of entities in their *Element Sets*. Namely, a cbs shows a collection of common occurrence of some activities taken by entities in the Element Sets.

Example 2. Table 3 shows Common Behaviour Sets generated by Semantic Items in Table 2. Semantic Items si_1, si_2, and Semantic Items si_3, si_4 generate Common Behaviour Sets cbs_1 and cbs_2, when the $SimTh$ among Element Sets is greater than or equal to 50 %.

Table 3. Common Behaviour Sets

Common Behaviour Sets	
$cbs1$	{John Lennon, George Harrison}(instrument, Guitar)
	{John Lennon, George Harrison}(occupation, Songwriter)
$cbs2$	{Jimmy Carter, Bill Clinton, George W. Bush}(office, President of the USA)
	{Jimmy Carter, Bill Clinton}(party, Democratic)

3.4 Mining Module

To generate Semantic Association Rules, we need to have a notion of frequency. As we discussed in the previous subsection, each particular Common Behaviour Set cbs is a unique set and reveals the common occurrence of some activities taken by entities (subjects) in its Element Sets. In fact, it is a particular form of transaction in the context of Semantic Web data. Under this motivation, we first generate Semantic Association Rules from Common Behaviour Sets and then we evaluate the quality of rules (Support and confidence measures) by extracting knowledge encoded in the ontology.

Definition 3 *(Semantic Association Rule).* A Semantic Association Rule r is composed by two different sets of Pairs pa_{ant} and pa_{con}, where pa_{ant} is called Pairs of Antecedent and pa_{con} is called Pairs of Consequent. pa_{ant} is a set including the number of Pairs in cbs_j, i.e., $\{pa_1, ..., pa_n\}$. pa_{con} is a set including the remaining number of Pairs in the cbs_j, i.e., $\{pa_{n+1}, ..., pa_m\}$. Rule r contains a common Rule's Element Set res where res is a set including union of the Element Sets in cbs_j, i.e., $\{es_1 \cup ... \cup es_m\}$. Each Element Set es_i is a set of instances, i.e., $es_i = \{ins_1, ins_2, ..., ins_k\}$. We indicate a rule r with the antecedent and consequent by an implication

$$res: pa_{ant} \implies pa_{con}$$

where res is a common Rule's Element Set containing $\bigcup\limits_{si_i \in cbs_j} si_i.es$ and pa_{ant}, $pa_{con} \in cbs_j$ and $pa_{ant} \cap pa_{con} = \emptyset$.

Table 4 shows two examples of rules generated from Common Behaviour Sets in Table 3. For example, rule r_1 contains a common Rule's Element Set res generated by union of Element Sets in cbs_1, i.e., *{John Lennon, George Harrison}*. The antecedent and the consequent of r_1 holds the Pair *(instrument, Guitar)* and *(occupation, Songwriter)*, respectively.

Our goal is to measure the quality of rules by using knowledge in the instance-level and schema-level. *rdf:type* is basically an RDF property that ties an instance to a class in the ontology. In the traditional association rule mining, all instances often have one type (class) of actors, i.e., shopping customers. Namely, particular activities have always done by customers. However, here instances in the Rule's Element Set may belong to different types/classes in the ontology. Consider all instances in the Rule's Element Set of rule r_1, i.e., *John Lennon* and *George Harrison*. Figure 2 shows a small fragment of DBpedia ontology. In this ontology, *George Harrison* belongs to the Musical Artist, while *John Lennon* is an instance of Person class. As we discussed in the introduction section, in the context of Semantic Web data, it does not make sense to measure the quality of rules by only considering knowledge in the instance-level. This observation leads us to assess *rdf:type* and *rdfs:subClassOf* relations in the schema-level. In this paper, we focus on interpreting rules through having a single ontological structure, e.g. DBpedia ontology. Furthermore, we assume that each instance belongs to a single class in the ontology.

Table 4. Semantic Association Rules

Semantic Association Rules
r_1 {*John Lennon, George Harrison*}: (*instrument, Guitar*) $\Rightarrow$ (*occupation, Songwriter*)
r_2 {*Jimmy Carter, Bill Clinton, George W. Bush*}: (*office, President of the USA*) $\Rightarrow$ (*party, Democratic*)

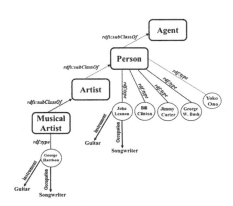

Fig. 2. A fragment of the DBpedia ontology

Figure 3 shows three different hierarchical structures of an ontology. As previously mentioned, in Fig. 3(a), if Class c_1 is subclass of Class c_3 through middle

Class c_2 ($c_1 \subseteq c_3$), then the Instance I_a belongs to c_3 as well. However, in Fig. 3(b), Class c_1 and Class c_5 are not in the same hierarchy ($c_1 \nsubseteq c_5$). Even if we consider Class c_3 as a lowest common class for c_1 and c_5, we reduce their semantics. Because classes on the upper levels illustrate more general descriptions to compare with Lower level classes which provide more special descriptions. Therefore, in case that classes are not in the same hierarchy, we just consider the Lowest Level Class (LLC) for each instance in a Rule's Element Set. For example, in Fig. 3(b), I_a and I_b belong to c_1 and c_5, respectively. In Fig. 3(c), in the LLC, I_a and I_b belong to c_2, while I_c is an instance of c_9. Consider again instances in the Rule's Element Set r_1. Based on our assumption that each instance belongs to a single class, the LLC for both *George Harrison* and *George Harrison* is the Person class.

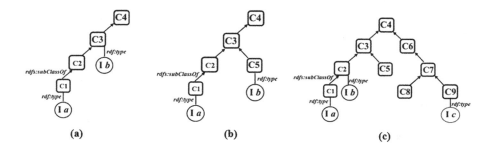

Fig. 3. Examples of different hierarchical structures of an ontology

Support. Consider the Semantic Association Rule r in the form of $res : pa_{ant} \Longrightarrow pa_{con}$. The support $Sup(r)$ is defined as:

$$Sup(r) = \frac{\left| \bigcup\limits_{ins_j \in c_i \wedge ins_j \in res_k \wedge ins_j.pa_{ant_k}} c_i \right|}{\left| \bigcup\limits_{ins_j \in c_i \wedge ins_j \in res_k} c_i \right|} \tag{2}$$

The numerator of support fraction is the total number of instances of Class c_i that contains pa_{ant} as the Pairs. The denominator is the total number of instances of c_i.

Example 3. Regarding three different schemas shown in Fig. 3, rules generated from Schema a, b, and c are $r_a = \{I_a, I_b\}: pa_{ant} \Rightarrow pa_{con}$, $r_b = \{I_a, I_b\}: pa_{ant} \Rightarrow pa_{con}$, and $r_c = \{I_a, I_b, I_c\}: pa_{ant} \Rightarrow pa_{con}$, respectively. The supports of rules can be calculated by the following fractions:

$$Sup(r_a) = \frac{|c_3 \cap pa_{ant}|}{|c_3|}$$

$$Sup(r_b) = \frac{|(c_1 \cup c_3) \cap pa_{ant}|}{|c_1 \cup c_3|}$$

$$Sup(r_c) = \frac{|(c_2 \cup c_9) \cap pa_{ant}|}{|c_2 \cup c_9|}$$

Example 4. The support of rule r_1 in Table 4 can be calculated by the following fraction. The numerator of support fraction shows the total number of instances belong to *Person* class that contain *instrumentGuitar* as a Pair. Based on the existing ontology shown in Fig. 2, there is only two instances that contain *instrumentGuitar* as a Pair. The denominator of the fraction also is total number of instances belong to the *Person* class which is six in this example ($Sup. = 0.33$).

$$Sup(r_1) = \frac{|Person \cap instrument\ Guitar|}{|Person|}$$

Confidence. Consider the Semantic Association Rule r in the form of $res : pa_{ant} \Longrightarrow pa_{con}$. The confidence $Conf\ (r)$ is defined as:

$$Conf(r) = \frac{\left| \bigcup\limits_{ins_j \in c_i \wedge ins_j \in res_k \wedge ins_j.pa_{ant_k} \wedge ins_j.pa_{con_k}} c_i \right|}{\left| \bigcup\limits_{ins_j \in c_i \wedge ins_j \in res_k \wedge ins_j.pa_{ant_k}} c_i \right|} \tag{3}$$

The numerator of confidence fraction is the total number of instances of Class c_i that contains pa_{ant} and pa_{con} as the Pairs. The denominator of the fraction is the total number of instances of c_i that contains pa_{ant} as the Pairs.

Example 5. The numerator of confidence fraction of rule r_1 shows the total number of instances belong to the *Person* class that contain *instrumentGuitar* and *occupationSongwriter* as the Pairs. The denominator of the fraction also is the total number of instances belong to the *Person* class along with *instrumentGuitar* as a Pair ($Conf. = 1.0$). The rule shows that most of the time persons who play Guitar, they probably work as Songwriters. The rule shows that at least 50 % of instances in the Rule's Element Set satisfy the rule.

$$Conf(r_1) = \frac{|Person \cap instrument\ Guitar \cap occupation\ Songwriter|}{|Person \cap instrument\ Guitar|}$$

Example 6. Rule r_2 in Table 4 shows that most of the time people who are President of the USA, they are probably members of Democratic party ($Sup. = 0.5, Conf. = 0.66$).

4 Experiments

4.1 Overview

Dataset. As a proof of concept, we ran the SWARM on DBpedia (3.8)[4]. The DBpedia datasets usually provide the A-Box and T-Box in two separate files:

---

[4] http://wiki.dbpedia.org/services-resources/datasets/data-set-38/downloads-38/.

Ontology Infobox Properties and *Ontology Infobox Types*. The *Ontology Infobox Properties* provides instance-level data and the *Ontology Infobox Types* contains triples in the form of (*subject, rdf : type, ClassName*). The *ClassName* declares the name of classes for each subject in the DBpedia ontology. For example, *Anton Drexler* belongs to the Politician, Person, and Agent classes. In this paper, we filtered out the *Ontology Infobox Types* based on the person class and its subclasses. In the DBpedia ontology, the Person class contains 26 subclasses. By using triples filtered from *Ontology Infobox Types*, we extracted about 50,000 triples of *Ontology Infobox Properties*. We also removed some triples with literals (numbers and strings) from the subset dataset. Literal values such *Birthdate* information are less interesting for rule mining.

Goal. The main goal of this research is to automatically tie instance-level to schema-level to attach more semantics to the rules. To the best of our knowledge, this issue has not yet been considered by the existing methods. In comparison to [5,12] that mentioned their approach is more granular in considering predicate correlations and object correlations independently, our approach is able to automatically mine common behavioural patterns associated with knowledge in the instance-level and schema-level.

Evaluations. Table 5 represents some Semantic Association Rules of Person class generated by $SimTh = 60\%$. For example, rule r_1 shows that the Scientists who are known for Natural selection theory, they were probably awarded with the Copley and Royal Medals. Note that in this table, at least 60% of instances

Table 5. Semantic Association Rules from Person class ($SimTh = 60\%$)

Rule	Rule's Element Set	Semantic Association Rule	Sup.	Conf.
r_1	{*Alfred Russel Wallace, Charles Darwin, Andrew Wiles, Robert Bunsen*}	{*Scientist*}: (*knownFor, Natural selection*) ⇒ (*award,Copley Medal*), (*award,Royal Medal*)	0.01	1.0
r_2	{*Lodewijk Asscher, Eberhard van der Laan*}	{*Politician*}: (*residence, Amsterdam*), (*party, Labour Party (Netherlands)*) ⇒ (*residence, Netherlands*)	0.04	1.0
r_3	{*Augustine of Hippo, Saint Titus, Bernard of Clairvaux, Athanasius of Alexandria*}	{*Saint*}: (*veneratedIn, Lutheranism*) ⇒ (*veneratedIn, Anglican Communion*)	0.04	0.87
r_4	{*Amyntas I of Macedon, Alcetas I of Macedon, Alexander I of Macedon, Alcetas II of Macedon, Perdiccas II of Macedon*}	{*Person*}: (*title, King of Macedon*) ⇒ (*religion, Religion in ancient Greece*)	0.02	1.0

Table 6. Semantic Association Rules from Person class $(SimTh = 80\%)$

Rule	Rule's Element Set	Semantic Association Rule	Sup.	Conf.
r_1	{*Afonso VI of Portugal, Peter II of Portugal*}	{**BritishRoyalty**}: (*birthPlace, Ribeira Palace*), (*parent, Luisa of Guzman*), (*parent, John IV of Portugal*) ⇒ (*restingPlace, Royal Pantheon of the House of Braganza*)	0.04	1.0
r_2	{*Alfonso V of Aragon, John II of Aragon*}	{**BritishRoyalty**}: (*parent, Ferdinand I of Aragon*), (*birthPlace, Medina del Campo*) ⇒ (*parent, Eleanor of Alburquerque*)	0.04	1.0

of Element Sets satisfy rules. Rule r_2 illustrates that Politicians who are residents of Amsterdam and works in the Labour Party of Netherlands, they are probably residents of Netherlands. Rule r_3 shows that the Saints who venerate in Lutheranism which is a major branch of Protestant Christianity, they are more likely to be venerated in Anglican Communion as well. Rule r_4 represents that people who were Kings of Macedon, they probably had Ancient Greek religion.

Table 6 also shows some rules generated by $SimTh = 80\%$. Based on the DBpedia ontology, we identify some inconsistent patterns. Rule r_1 represents that some members of British Royal family who were born in Ribeira Palace and whose parents are Luisa of Guzman and John IV of Portugal, they more likely to be buried in the Royal Pantheon of the House of Braganza. Although the instances of Rule's Element Set r_1 satisfy the rule, none of them belongs to the British Royal family. In fact, they are members of Portugal Royal family. Rule r_2 also suffers from the same issue as r_1 does. In the DBpedia ontology, all royalties belong to the BritishRoyalty and PolishKing classes. The ontology does not define any other classes for these instances. Such inconsistencies between ontology definitions and underlying data lead to an ambiguous interpretation. Sometimes the ontology has been created independently before actual data usage. In the case of DBpedia project, revising existing class definitions might be helpful to obtain a better understanding of data.

We observe that the generated rules tend to have low support rates. The intuition behind this is that the denominator of support fraction usually contains the total number of instances of particular classes. In the real world KBs, the number of instances is too large and it leads to low support rates. Figure 4 shows the number of strong Semantic Association Rules with different minimum Similarity Thresholds. The $SimTh$ has a direct effect on the number of generated rules. As seen in Fig. 4, the number of high confidence rules from 0.6 to 0.8 has

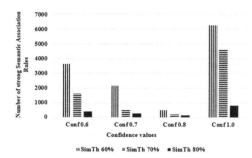

Fig. 4. Number of strong Semantic Association Rules with different minimum Similarity Thresholds

been decreased by increasing the *SimTh*. The reason that SWARM discovers a large number of rules with confidence 1.0 is because of filtering mechanism for generating Common Behaviour Sets. Note that this approach is implemented in the Eclipse Java with 3.20 GHz Intel Core i5 processors and 16 GB memory. The time complexity of SWARM algorithm belongs to the $O(n^2)$ class (including the time for generating the Semantic Items, Common Behaviour Sets, and mining Semantic Association Rules by utilizing the instance-level and schema-level knowledge).

5   Conclusion

In this paper, we propose an approach to automatically mine Semantic Association Rules from Semantic Web data by utilizing knowledge in the instance-level and schema-level. We believe that this type of learning will become important in the future of Semantic Web data mining especially for re-engineering ontology definitions. In comparison with the existing methods [5,12] that evaluate the quality of rules by only using instance-level data, the SWARM approach takes advantage of *rdf:type* and *rdfs:subClassOf* relations to interpret the association rules. Future work will aim to test the approach on different classes of DBpedia Ontology. In order to produce more precise rules, we target at developing this approach wherein each instance of a Rule's Element Set belongs to the multiple classes in the ontologies.

References

1. Bizer, C., Heath, T., Berners-Lee, T.: Linked data-the story so far. Int. J. Semant. Web Inf. Syst. **5**(3), 1–22 (2009)
2. Kabir, S., Ripon, S., Rahman, M., Rahman, T.: Knowledge-based data mining using semantic web. IERI Procedia **7**, 113–119 (2014)

3. Galárraga, L.A., Teflioudi, C., Hose, K., Suchanek, F.: AMIE: association rule mining under incomplete evidence in ontological knowledge bases. In: Proceedings of the 22nd International Conference on World Wide Web, pp. 413–422. International World WideWeb Conferences Steering Committee (2013)
4. Galárraga, L., Teflioudi, C., Hose, K., Suchanek, F.M.: Fast rule mining in ontological knowledge bases with AMIE+. VLDB J. **24**(6), 707–730 (2015)
5. Abedjan, Z., Naumann, F.: Improving rdf data through association rule mining. Datenbank-Spektrum **13**(2), 111–120 (2013)
6. Muggleton, S.: Inverse entailment and progol. New Gener. Comput. **13**(3–4), 245–286 (1995)
7. Goethals, B., Van den Bussche, J.: Relational association rules: getting WARM ER. In: Hand, D.J., Adams, N.M., Bolton, R.J. (eds.) Pattern Detection and Discovery. LNCS (LNAI), vol. 2447, pp. 125–139. Springer, Heidelberg (2002)
8. Yang, B., Yih, W., He, X., Gao, J., Deng, L.: Embedding entities and relations for learning and inference in knowledge bases. arXiv preprint arXiv:1412.6575 (2014)
9. Agrawal, R., Imieliński, T., Swami, A.: Mining association rules between sets of items in large databases. In ACM SIGMOD Record, vol. 22, no. 2. ACM, pp. 207–216 (1993)
10. Han, J., Cheng, H., Xin, D., Yan, X.: Frequent pattern mining: current status and future directions. Data Min. Knowl. Disc. **15**(1), 55–86 (2007)
11. Nebot, V., Berlanga, R.: Finding association rules in semantic web data. Knowl. Based Syst. **25**(1), 51–62 (2012)
12. Abedjan, Z., Naumann, F.: Amending RDF entities with new facts. In: Presutti, V., Blomqvist, E., Troncy, R., Sack, H., Papadakis, I., Tordai, A. (eds.) ESWC Satellite Events 2014. LNCS, vol. 8798, pp. 131–143. Springer, Heidelberg (2014)
13. Han, J., Pei, J., Yin, Y.: Mining frequent patterns without candidate generation. In: ACM SIGMOD Record, vol. 29, no. 2, pp. 1–12. ACM (2000)

Information Retrieval from Unstructured Arabic Legal Data

Imen Bouaziz Mezghanni[✉] and Faiez Gargouri

MIRACL Laboratory, ISIM Sfax, Sfax, Tunisia
imen_bouaziz_miracl@yahoo.com, faiez.gargouri@isimsf.rnu.tn
http://www.miracl.rnu.tn/

Abstract. Given the steady increase of published and stored information in the form of Arabic unstructured texts, current Information Retrieval (IR) systems must be able to suit the nature and requirements of this language for an accurate and efficient search. This paper sheds light on the challenges in Arabic IR (AIR) and proposes an approach for enhancing the process of AIR based on transforming these texts into structured documents in XML format through a document ontology as well as a set of linguistic grammars. The IR system hence is done on the XML documents. The aim of such system is to incorporate the knowledge on the document structure and on specific content elements in computing the relevance of an information element. A query expansion module mainly based on domain ontology as well as user profile is proposed for the enhancement of the search results.

Keywords: Information retrieval · Arabic information retrieval · Unstructured data · Structured data

1 Introduction

Over the past few years, the amount of electronic information is increasing tremendously by the rapid and continual flow of information through independent internet media. Dealing with this perceived explosion information requires Information Retrieval systems (IRS). The main goal of IRS is to retrieve the relevant and only the relevant documents in response to user's queries from mainly unstructured textual data.

Specialized techniques such as Text Mining specifically operating on textual data are becoming inevitable to extract information from such kind of texts. Text Mining techniques are dedicated to discover the implicit structure in the documents. The discovered structure is called Index which contains the most significant terms known as descriptors. Natural Language Processing (NLP) can then be seen as a powerful technology for the vital tasks of IR and knowledge discovery as it allows both facilitating descriptions of document content as well as presenting the user's query usually formulated as a set of natural language keywords.

© Springer International Publishing Switzerland 2016
R. Booth and M.-L. Zhang (Eds.): PRICAI 2016, LNAI 9810, pp. 44–54, 2016.
DOI: 10.1007/978-3-319-42911-3_4

In IRS, index terms play the connecting role between documents and user queries. Usually, the queries fail to match the index terms contained in the relevant documents. Dealing with Arabic legal data doubles the challenge of satisfying the user's need. The challenges arise from the language itself since Arabic is known by its complexity and likewise from the process of IR. Current IRS that access legal databases offer the possibility of a full text search, in which every term can act as a search term, and returns a ranked list of information answers. This answer list can be filtered through a deterministic fulfillment of extra conditions set by the structured information. Thus, on one hand we can search the free texts of the documents and on the other hand the structured information. However, few attempts have yet been made to exploit the structured information in retrieval models. For obtaining more relevant results, our idea is to search on structured documents instead of unstructured ones, we consider in fact that the underlying information should be in a structured form. Furthermore, the integration of a query expansion mechanism can bring up promising answer list and thus improve the retrieval performance.

The rest of the paper is organized as follows, where Sect. 2 addresses the challenges of Arabic language in IR and reviews related works dealing with Arabic IRS, Sect. 3 presents our system architecture for AIR, while Sect. 4 provides conclusions and presents plans for further work.

2   Challenges of Arabic Language in IR

Arabic is the official language of over 420 million people, across 22 countries throughout the Middle East, Europe, and Asia making it the sixth-most spoken language in the world. It is classified into three variants [Ibrahim et al. 2015]:

- Classical Arabic which is the form of the Arabic language used in literary texts and Holy Quran.
- Modern Standard Arabic (MSA) which is the standard and the literary variety of Arabic used in writing and in most formal speech.
- Colloquial Arabic which refers to the many national or regional varieties of spoken Arabic. These differ significantly from MSA and Classical Arabic, as well as from each other and are usually unwritten.

The retrieval of Arabic content, primarily written in MSA, is affected greatly by the properties of Arabic language. We expound these properties in the following section. In the remainder of this article, the given examples are extracted from our corpora. They are given in Arabic along with their English translation and their transliteration using Xerox Morphology System[1].

---

[1] https://open.xerox.com/Services/arabic-morphology/Consume/
Morphological%20Analysis-218.

2.1   Arabic Specificities

Unlike Indo-European languages, Arabic includes specific pecularities effecting IR. First of all, Arabic is written in horizontal lines from right to left, and there are no capital letters in the Arabic alphabet making the task of text splitting into sentences difficult since capital letters are considered as cues for text segmentation.

Secondly, it is characterized by orthographic variations (The different typographical forms for one letter) such as ALEF (like إ and آ and أ and ا), YAA with dots or without dots (like ﻲ and ى) and also HAA (like ه and ة). Indeed, the substitution of one of these forms with another alter the meaning of the words. For example (علي/Ely) which indicates a proper noun and (علی/ElY/on) which is a preposition.

Besides this, Arabic is a highly agglutinative language with the a rich set of clitics aglutinated to words. Its inflectional and derivational productions introduce a big increase in the number of possible word forms. For instance, prepositions (like ل/li/for), conjunctions (like و/w/and), articles (like ال/Al/the) and pronouns (like ه/h/he), can be agglutinated to nouns, adjectives, particles and verbs which causes several lexical ambiguities.

Moreover, long and complex sentences are frequently used in its discourses. Punctuation marks are narrowly used in such a way that we can simply find an entire long paragraph without any punctuation.

Furthermore, the lack of vowels in current texts and the multiplicity of the vowel forms lead to that a word can have different meanings which make the analysis and the comprehension of Arabic texts more difficult. For example, the word (وهن/whn/weakness) can correspond in English to the noun "Illusion" or to a conjunction (و/wa/and) followed by the pronoun (هن/hun a/they). In Arabic there are principal four categories of words which are noun, proper noun, verbs and prepositions. The absence of short vowels can likewise cause ambiguities within the same category or across different categories. For example, the word (بعد/bEd) corresponds to a preposition with the meaning "After", a Noun with the meaning "Remoteness", a Verb with the meaning "go away".

2.2   Arabic IR

Nowadays, the rising number of Arabic users as well as the amount of digital information available in document repositories has motivated researchers to develop many different Arabic IRS in order to enhance the Arabic documents retrieval process. Indeed, availability of information doesn't mean it is helpful as the user may not always find the needed information.

The classical IRS comprises three main processes, namely Indexing, Query processing and Matching. In indexing phase, documents are indexed using keywords that represent each document in the collection. Indexing mechanism is based on the "inverted index" in which information symbols and all documents containing that symbol are listed. Thus its structure is composed of two elements: the vocabulary defined by the set of all words in the text, and the term occurrences defining the set of lists comprising the positions of appearance of each word in the text.

Then, the query that is entered by the user is generally pre-processed by the same algorithms used to select the index terms. Additional query processing as query expansion can be performed by using external resources like thesaurus, taxonomies or ontology. A query is reformulated to comply with the information retrieval model as well as to add other keywords or modify the weights of the existent words to achieve better search accuracy.

Finally, in the matching step, the query entered by user will be matched with index. The results of matching between existing index and query will be sorted based on the ranking algorithms which depend on their similarity.

Several information retrieval models exist, of which the most common models include the Boolean model, fuzzy model, and vector space model. Various studies have been elaborated intending the Arabic IR improvement. Among the recent studies, [Maitah et al. 2013] address improving the effectiveness of information retrieval system using adaptive genetic algorithm (AGA) under the vector space model, Extended Boolean model, and Language model in IR.

The main goal of the [Yousef and Khafajeh 2013] research is to design and build an automatic Arabic thesaurus using Local Context Analysis technique to improve the expansion process and so to get more relevance documents for the user's query. The results of this study showed that it improved the retrieval in a remarkable way better than the classical retrieval method. In 2014, the authors developed an algorithm for Arabic word root extraction based on N-gram [Yousef et al. 2014]. Morphological rules aren't used in order to avoid the complexity arising from the morphological richness of the language on one hand and the multiplicity of morphological rules in the Arabic language on the other hand.

[Mahgoub et al. 2014] introduced a query expansion approach using an ontology built from Wikipedia pages in addition to other thesaurus to improve search accuracy. The proposed approach outperformed the traditional keyword based approach in terms of both F-score and NDCG measures.

[Hanandeh and Mabreh 2015] is based on the Genetic Algorithm (GA) to improve the effectiveness such systems. This work uses the Vector Space Model (VSM) and the Extended Boolean Model (EBM) to compute the similarities between queries and documents. Two fitness functions are proposed in this paper: One as fitness function and the other as adaptive mutation. Then comparing each of these functions with a number of ratio mutations that have been introduced to get better results. The experimental results reveal that the proposed cosine function outperformed other fitness models.

A pre-retrieval (offline) method is proposed by [Mohamed 2015] to build a statistical based dictionary for query expansion which is based on a statistical methods (co-occurrence technique and Latent Semantic Analysis (LSA) model) to improve the effectiveness of the search result by retrieving the most relevant documents regardless of their dialect. The evaluation was done using the average recall (Avg-R), average precision (Avg-P) and average F-measure (Avg-F) and the proposed method proved to be efficient for improving retrieval via expands the query by regional variation's synonyms, with accuracy 83 % in form of Avg-F.

A more recent research of [Atwan et al. 2016] aims to enhance an AIR by improving the processes in a conventional IR framework. An enhanced stop-word list is introduced in the pre-processing level and several Arabic stemmers are investigated. In addition, an Arabic WordNet was utilized in the corpus and query expansion levels. A semantic information for the Pseudo Relevance Feedback was adopted. The enhanced Arabic IR framework was built and evaluated and demonstrated an improvement by 49 % in terms of mean average precision, with an increase of 7.3 % in recall compared with the baseline framework.

In summary, the majority of the above reviewed works has tried to handle Arabic challenges for improving the process of AIR. However they reached significant results, we note that they completely ignore document structure. The information is usually searched by means of a full text search, every term in the texts of the documents can function as a search key, missing a great opportunity for a more effective search which motivate us to perform our research work.

3   Proposed Approach

Searching on structured documents can be more fruitful than on unstructured ones. Indeed, the indexing process can be not only by content, but also by their structure which likewise enables users to formulate versatile queries mixing keywords and structural information. Thus, the matching between texts in the corpus with a structured query generates a more precise answer to an information query by returning a structural element (or several elements) instead of the complete document. This way can meet greatly user's need who looks forward to precise answers.

In the legal domain, integrating document structure into a retrieval system for accessing legal documents has a lot of potential especially that legal documents are characterized by diverse structural pecularities:

- **Document's Content:** which describes the content of the document. This property is relevant because it contains and refers to the document's element level (as concepts, facts, people, etc.) and relations that draw on them.
- **Document's logical structure:** which refers to the document's hierarchical outline structure (book, chapter, section, and so on).
- **Document's Metadata:** Metadata are particular information that are not in the original content of the document and that are added to improve comprehension and classification of the document.
- **Document's collections structure:** which refers to references and links existing in and between legal documents.

The whole approach that we follow to take into account these pecularities is articulated around two modules as illustrated in Fig. 1. The former focus on detecting the structure of the documents while the later concerns the process of IR in the output of the first module.

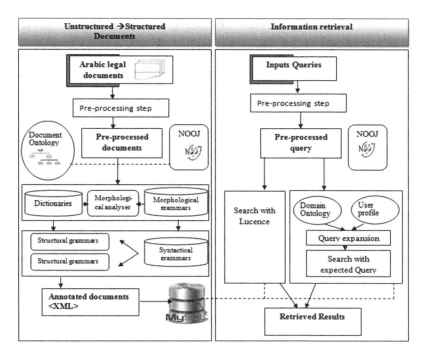

Fig. 1. Approach architecture

3.1 Annotation Module

These particular structures, however are interested in retrieving information, aren't currently explicit in the available legal documents, for this reason before indexation process, an annotation step must be elaborated regrouping five main steps:

– Corpus Selection and Preparation. Faced to the obvious lack of Arabic resources and the unavailability of the existent free corpora, it is primordial to build our own corpus "PenalAr". Dealing with Arabic texts starts in this step, which introduces the beginning of handling Arabic challenges. Firstly, the documents are cleaned from orthographic errors, then segmented to handle the absence of capital letters. Splitting is done using a Clauses splitter, a cascade of finite-state transducers elaborated by [Keskes et al. 2012], which proceed to split extended sentences into a set of clauses. Normalization of documents in a standard format is then applied for easy manipulation as well as orthographic variation resolution. The normalization includes the suppression of special characters if exist, the replacement rules of some Arabic letters:

  • Replacement of إ, آ and أ with ا
  • Replacement of ة with ه
  • Replacement of ي with ى

- Identification of documents Types. We have developed an Arabic Legal Document ontology (ALDO) defining a general legal document class, specifying all types of elements that can be used in the logical structure of a document of this class. Relations between these sub-classes are likewise modeled. The cross-reference between document are also taken into account in the ontology. Identification process is simply carried out by comparing the document title with the ontology concepts. The title is often comprised by the first words in the document.
- Derivation of corresponding structural grammars. Once the document type is identified, the corresponding path in the ontology will be translated by a syntactic grammar in the linguistic NooJ plateform[2] that will be applied to the document for extracting its title and contents table including the books, chapters and sections titles, enumeration lists, and the references... In order to recognize them, we have elaborate grammars represented as directed graphs and apply them to documents of the corpus.
- Application of morphological and syntactic grammars. To handle agglutination issue, a morphological grammar of agglutination is constructed since these forms are not present in the NooJ's dictionary of inflected forms. An excerpt from the used grammars is showed in Fig. 2.
  An Arabic dictionary for legal compound terms is created to tread both simple and compound word to recognize them later. Noting that Arabic NooJ's dictionary generally associates each lexical entry with an inflectional and/or derivational paradigm to describe its syntactic and semantic and inflectional (gender, number, conjugation...), and derivational features. Compound terms are added manually to the dictionary with its semantic attribute referring to the lemma of the term, its category of the compound noun in order to treat each category separately in the syntactical grammar according to NooJ entry format. Each of these features is translated by Inflectional/Derivational grammars based on generic commands to generate the different voweled forms of the dictionary entries and syntactical grammars to extract all related derived and agglutinated forms.
- Projection of grammars on Documents After having applied the grammars to the documents NooJ to produce an annotated text by applying grammar to the documents, the resulting text can be exported as an XML documents, in which XML tags have been inserted.

All the annotated documents are stored in MySQL database to begin the process of indexing and retrieval.

3.2 Indexing and Retrieval

The aim of the XML retrieval system is to incorporate the knowledge on the document structure and on specific content elements in computing the relevance of an information element. This field is fairly fresh. Like any new field, a number

---

[2] http://www.nooj4nlp.net/.

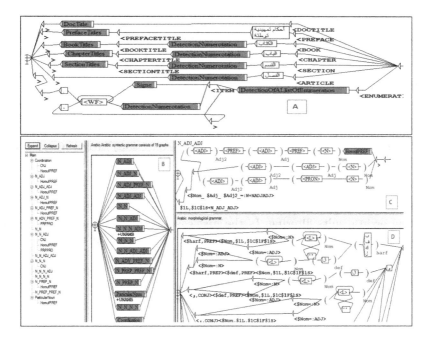

Fig. 2. A. Example of a Structural grammar for legal Code B. C. Graph of Syntactic grammar, D. Graph of morphological grammar

of questions remained unanswered is posed. The most important of these questions, concerns the element to retrieve in response to a query. This is different from traditional IR which considered whole documents to be the only retrievable entity and did not take into account the document's internal structure. The difference here is that the retrieval system must sweep the document at various levels of granularity and compare elements at different levels and by this way it can identify the most specific document element that answers the requested information in the query. Our proposition in this point is to index only elements at the leaf nodes of the XML document. This designates considering each leaf node as a distinct document and indexing it separately.

After the identification step and processing of the query, we propose to the user two alternatives for information retrieval:

– Simple Retrieval without reformulation.
  This alternative is applied when the user wants to send his query directly without enhancement.
  For that, we implement an initial prototype of a search process as illustrated in Fig. 3. We depend on LUCENE, which is free open source information retrieval library released under the Apache Software License with full text indexing and searching capabilities. LUCENE is a high-performance, full-featured text search engine library written entirely in Java. It depends on the Vector Space

Model (VSM) of information retrieval, and the Boolean model to determine how relevant a retrieval element is to a user's query. The search process starts over the annotated documents and the results that match the query will be displayed.

– Retrieval with reformulation.

The quality of responses by IRS depends not only on the quality of the matching step but also of the query made by the user, hence the interest of the reformulation. This alternative is applied when the user wants to enrich his query. The use of ontologies for enrichment (expansion) of the query constitutes a solution among others to solve the problem of semantic variations as they provide resources in the form of semantic relations extending the search field of a query, which improved the search results.

The use of ontologies in IRS may be used either before sending the query, in indexing or in results filtering. In the first case, the query can be enriched by the near concepts of the ontology through the use of different relations such as the generalization/specialization, synonyms. The indexing of documents can be done using the concepts of ontology and not using keywords and Filtering according to a particular domain for users profile.

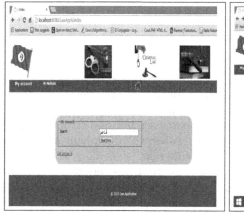

Fig. 3. Initial interface prototype

For very specific queries, a single element may contain the right answer, whereas more general queries could be answered by returning a general element. The models also allow that a group of related elements such as the articles in a section or chapter, contribute to the overall relevance of a single element for example an article.

The retrieval results obtained by the two types alternatives are then saved in various files and various values of recall and precision of the system are calculated for each type of retrieval and query in order to be compared.

3.3   Evaluation

The experiments of the annotation module is done on small corpus consisting of the new-tunisian-constitution's draft released on June 2013 after Tunisia's "Jasmine Revolution" published on the Tunisia OpenGov site[3] and 20 decisions 20 Criminal Law Decisions of the cassation Court available on the web portal site of Justice and Human Rights[4].

We achieved best results through recall and precision which confirm its performance since as regards testing and results on this data, satisfactory ratio are reported for syntactic annotation with 85 % as precision as well as for structural annotation with 73.7 %. The incorrect matches correspond mostly to the syntactic annotation due to sequences containing anaphora. The provided results by the approach have been assessed and validated by two legal experts, a lawyer and a Professor of Law. The former assessed the structural annotation while the latter evaluated the second kind of annotation: the syntactic annotation. Their relevance judgments have achieved an acceptable level of agreement.

4   Conclusion

The present paper has given an overview of how to improve the Arabic legal information retrieval. In our view, it holds great promise in making legal information more transparent and available to more legal professionals. The exploitation of structure requires transforming the data from unstructured to structured in a first stage then applying a retrieval model incorporating the structural information in his ranking algorithm in second stage. The answer list would consist of elements that are highly specific in addition to being relevant to the query and may be found at any level in the document structure. We will be charged of achieving the second stage as perspective.

References

Atwan, J., Mohd, M., Rashaideh, H., Kanaan, G.: Semantically enhanced pseudo relevance feedback for arabic information retrieval. J. Inf. Sci. **42**(2), 246–260 (2016)

Hanandeh, E., Mabreh, K.: Effective information retrieval method based on matching adaptative genetic algorithm. J. Theoret. Appl. Inf. Technol. **81**(3), 446–452 (2015)

Ibrahim, H., Abdou, S., Gheith, M.: Idioms-proverbs lexicon for modern standard arabic and colloquial sentiment analysis. Int. J. Comput. Appl. **118**(11), 26–31 (2015)

Keskes, I., Benamara, F., Belguith, L.H.: Clause-based discourse segmentation of arabic texts. In: Proceedings of the Eight International Conference on Language Resources and Evaluation, LREC 2012, Istanbul, Turkey, pp. 2826–2832 (2012)

Mahgoub, A., Rashwan, M., Raafat, H., Zahran, M., Fayek, M.: Semantic query expansion for arabic information retrieval. In: Proceedings of the EMNLP 2014 Workshop on Arabic Natural Language Processing (ANLP), Doha, Qatar, pp. 87–92 (2014)

---

[3] http://www.nooj4nlp.net/.

[4] http://ejustice.tn.

Maitah, W., Al-Rababaa, M., Kannan, G.: Improving the effectiveness of information retrieval system using adaptive genetic algorithm. Int. J. Comput. Sci. Inf. Technol. **5**(5), 91–105 (2013)

Mohamed, A.: Design of arabic dialects information retrieval model for solving regional variation problem. Thesis, Sudan University of Science and Technology, Sudan (2015)

Yousef, N., Abu-Errub, A., Odeh, A., Khafajeh, H.: An improved arabic words roots extraction method using n-gram technique. J. Comput. Sci. **10**(4), 716–719 (2014)

Yousef, N., Khafajeh, H.: Evaluation of different query expansion techniques by using different similarity measures in arabic documents. Int. J. Comput. Sci. Inf. Technol. **10**(4), 160–166 (2013)

Generalized Extreme Value Filter to Remove Mixed Gaussian-Impulse Noise

Sakon Chankhachon and Sathit Intajag[(✉)]

Artificial Intelligence Research Laboratory,
Department of Computer Science, Faculty of Science,
Prince of Songkla University, Hat Yai, Thailand
{5810230038,sathit.i}@psu.ac.th

Abstract. Noise removal in image restoration is an important technique of image processing. In this paper, a new efficient approach is proposed for removing the mixed Gaussian-impulse noise in a color image. The proposed method utilizes the concept of local rank ordered absolute distances to measure similarity between a processing pixel in the small window and their neighborhood pixels in the processing block. The generalized extreme value distribution was employed to estimate weighted averages of the pixels in the processing block for filtering the mixed Gaussian-impulse noise. From the experimental results, our filter has yielded the better results in suppressing high density levels of the mixed noise in the color images than the state-of-the-art denoising methods.

Keywords: Generalized extreme value · Mixed Gaussian-impulse noise · Local rank ordered absolute distances

1 Introduction

Noise reduction in digital color images is the most importance pre-processing operation of the image restoration. In real world applications, noise usually results from many sources such as introduced by sensor malfunction, transmission errors, and electronic interference [1]. These noise sources can produce a mixed noise.

In this paper, we address the problem of noise reduction in digital color images those are corrupted by the mixed Gaussian-impulse noise. Suppression this type of noise is quite difficult, as designed filters to deal with Gaussian noise are mostly ineffective to remove impulse noise; conversely, the technique designs for impulsive noise are unable to remove Gaussian noise.

The method to address the mixed noise is designed the filter for removing impulses first and then designed another filter to remove Gaussian noise. However, this method produces image distortions and considerable increases computational complexity [2].

The well-known method to remove the mixed noise in multi-dimensional data such as a color image is Vector Median Filter (VMF) [3]. The filter uses the vector ordering concept of a set of pixels in processing windows. VMF is an efficient approach to remove impulse noise, but inadequate for removing Gaussian noise.

© Springer International Publishing Switzerland 2016
R. Booth and M.-L. Zhang (Eds.): PRICAI 2016, LNAI 9810, pp. 55–67, 2016.
DOI: 10.1007/978-3-319-42911-3_5

Another efficient filter family is based on switching filter [5–7, 14] which identifies the impulsive pixels and replaces them with output from the appropriated filter. However, these filters fail to detect the impulses which are masked by the Gaussian process.

An interesting type of the mixed noise reduction techniques is based on a peer group concept [8]. This concept denotes as a set of similarity pixels when compares with the central pixel under the processing window. The peer group methods replace contaminated pixels with the averaging value of its peer group. Morilas et al. [4] extended to a fuzzy peer group.

An efficient method for suppression the mixed noise called Robust Local Similarity Filter (RLSF) was proposed in [9]. RLSF is based on Bilateral Filter (BF) [10] with incorporated rank-order absolute difference (ROAD) statistics for impulse detection [11]. RLSF measures the similarity of a pixel from the region and the centered pixel from the small window, then used kernel functions to assign weights for each pixel in the region. However, this filter used various kernel functions which depend on the image structure.

In this paper, we propose a new robust filter to remove the mixed noise, which combines Generalize Extreme Value (GEV) function [15] and local ROAD which introduced in RLSF. Because of RLSF [9] have many kernel functions to design the average weight filter, which is not easy to select the appropriated kernel for each image. Thus, the proposed method uses GEV distribution in the small sample to design the optimal weight average filter from ROAD statistics [11].

The rest of paper is structured as follows. Section 2 describes related works. GEV distribution describes in Sect. 3. Section 4 explains the proposed filter designs. The experimental results show in Sect. 5. Finally, conclusions are described.

2 Related Works

Methods to remove the mixed noise consist of several steps. An important step is impulse noise detection; however, the popular technique is the ROAD statistics [11]. The other steps are used to formulate weighted matrix to filter the mixed noise.

2.1 Rank Order Absolute Distance Calculation

ROAD statistics [11] could measure how close a pixel value is to its most similar neighbors is that the impulse pixels will differ from their neighborhood pixels; because, most pixels composing the actual image should have at least half of their neighboring pixels of similar intensity, even pixels on an edge. ROAD can formulate as the follows.

Let F be an image. F_x represents intensity value at location $x = x_1, x_2$, which locates at a center of $(2N + 1) \times (2N + 1)$ window, for some positive integer N. $d_{x,y}$ in Eq. (1) is an absolute difference between the intensity value, F_x, and it's neighborhood pixels, F_y, which are memberships in the window.

$$d_{x,y} = |F_x - F_y| \tag{1}$$

$r_i(x)$ represents sorted $d_{x,y}$ in the ascending order; thus, the ROAD is given by

$$\text{ROAD}_m(x) = \sum_{i=1}^{m} r_i(x) \tag{2}$$

where m is positive value in interval [2, 7], generally $m = 4$ [11]. Figure 1 shows how the ROAD was calculated with $m = 4$ and $N = 1$.

$$\begin{bmatrix} 143 & 106 & 122 \\ 117 & 132 & 145 \\ 112 & 117 & 153 \end{bmatrix} \longrightarrow \begin{bmatrix} 11 & 26 & 10 \\ 15 & 0 & 13 \\ 20 & 15 & 21 \end{bmatrix}$$
$$\text{(1) Window} \qquad\qquad \text{(2) ROAD}$$

$$r_1 = 10 , r_2 = 11$$
$$r_3 = 13 , r_4 = 15$$
$$\text{(3) Four Smallest Values}$$

$$ROAD_{m=4}(F_x = 132) = \sum_{i=1}^{4} r_i = 10 + 11 + 13 + 15 = 49$$
$$\text{(4) ROAD of four smallest values}$$

Fig. 1. ROAD Calculation.

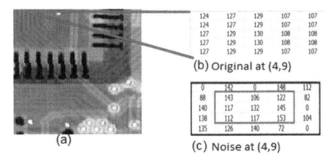

(b) Original at (4,9)

(c) Noise at (4,9)

(a)

Fig. 2. Processing block β and small window W at coordinate $(x_1, x_2) = (4, 9)$ of (a) Circuit board image [16], (b) Original data, and (c) The mixed noise with the noise levels [9] of the Gaussian noise ($\rho = 10$) and the impulse noise ($\rho = 10\ \%$). (Color figure online)

2.2 Local Rank Order Distance

RLSF method [9] was developed to remove the mixed noise in color images by using weighted averaging. RLSF operates pixels, $F_j, j = 1, 2, \dots, M$, in the processing block

β of the size $(2N+1) \times (2N+1)$ pixels, in this case, $M = (2N+1) \times (2N+1)$. F_i, $i = 1, 2, \ldots, 9$, represents neighbor pixels of the central pixel $F_{i=1}$ in the small window W with the size 3×3. As seen in Fig. 2(c), W (small red rectangle) contains in the center block of the processing block β (big black rectangle) with $N = 2$, and the center pixel, $F_{i=1} = 132$.

The local ROAD evaluates the similarity between each pixel, F_j, and its neighboring pixels of the pixel $F_{i=1}$ by RLSF. The similarity was measured by Euclidean distance of the color images.

RLSF identifies the uncorrupted pixels between F_j and all of F_i in W by the local ROAD with the smallest distance value, α. The value α has the same meaning as m in Eq. (2). The local ROAD calculations of the data from Fig. 2 can be demonstrated as the follows.

The absolute distances between $F_{j=0}$ and $F_{i=1,\ldots,9}$ are given by

$$L_2\left(F_{i=1}, F_j = 0\right) = \sqrt{(143-0)^2} = 143$$

$$L_2\left(F_{i=2}, F_j = 0\right) = \sqrt{(117-0)^2} = 117$$

$$\vdots$$

$$L_2\left(F_{i=9}, F_j = 0\right) = \sqrt{(153-0)^2} = 153$$

The ascending ordered vector, $L_2\left(F_i, F_j\right) = \{106, 112, \ldots, 153\}$. Local ROAD of $ROAD_{\alpha=4}\left(F_j = 0\right)$ is determined as

$$ROAD_{\alpha=4}(F_j = 0) = \sum_{p=1}^{4} L_2(F_i, F_j) = 106 + 112 + 117 + 117 = 452.$$

The other pixels in Fig. 2(c) can be formulated as $F_j = 0$. Figure 3. illustrates the local ROAD of Fig. 2(c).

$$\begin{bmatrix} 452 & 25 & 452 & 29 & 16 \\ 100 & 23 & 28 & 20 & 124 \\ 29 & 10 & 34 & 23 & 452 \\ 33 & 16 & 10 & 23 & 36 \\ 34 & 28 & 29 & 164 & 452 \end{bmatrix}$$

Fig. 3. ROAD of the data pixels in Fig. 2(c).

2.3   Calculate Weight by Kernel Function

RLSF method uses various kernel functions to estimate weight, w_j, of the pixels, F_j, in the processing block β as given by

$$w_j = K(\alpha^{-1} ROAD_\alpha(F_j)) \qquad (3)$$

where K denotes the kernel functions [9], which consist of Triangular, Epanechnikov, Biweight, Triweight, Tricube, Gaussian and Cosine as seen in the Fig. 4(a). These kernels were defined threshold at 100 and provided the weight matrix as shown in Fig. 5(a)-(g).

2.4 Weight Average Filter

The output pixel, $y(x_1, x_2)$, is provided by the weighted average filter as given by

$$y(x_1, x_2) = \sum_{j=1}^{N} w_j F_j \bigg/ \sum_{j=1}^{N} w_j. \qquad (4)$$

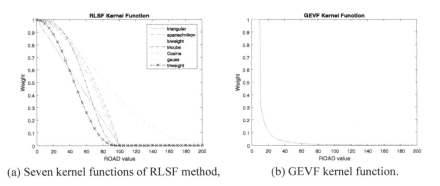

(a) Seven kernel functions of RLSF method,   (b) GEVF kernel function.

Fig. 4. Weighted kernel functions of ROAD.

Triweight function provides good results $y(x_1 = 4, x_2 = 9) = 124.37$, which approaches to original data at coordinate $(4,9) = 130$ as shown in Fig. 2(b). Figure 5 shows weigh matrices and the output pixel, $y(4, 9)$, of the kernel functions from RLSF method. However, the problem is how to select the appropriated kernel functions for designing the weight, w_j, to produce the good output pixel. From this problem, we could formulate only one kernel by using GEV distribution to replace the seven kernel functions of RLSF.

3 Generalized Extreme Value Distribution

GEV distribution was introduced by [12]. It was applied in many applications for describing floods, rainfall, snow depths, and other maxima etc. Probability density function (PDF) of GEV is given by:

$$\begin{bmatrix} 0.00 & 0.94 & 0.00 & 0.93 & 0.96 \\ 0.75 & 0.94 & 0.93 & 0.96 & 0.69 \\ 0.92 & 0.98 & 0.92 & 0.94 & 0.00 \\ 0.92 & 0.96 & 0.98 & 0.90 & 0.91 \\ 0.91 & 0.93 & 0.93 & 0.59 & 0.00 \end{bmatrix} \begin{bmatrix} 0.00 & 0.99 & 0.00 & 0.99 & 0.99 \\ 0.93 & 0.99 & 0.99 & 0.99 & 0.90 \\ 0.99 & 0.99 & 0.99 & 0.99 & 0.00 \\ 0.99 & 0.99 & 0.99 & 0.99 & 0.99 \\ 0.99 & 0.99 & 0.99 & 0.83 & 0.00 \end{bmatrix} \begin{bmatrix} 0.00 & 0.99 & 0.00 & 0.99 & 0.99 \\ 0.87 & 0.99 & 0.99 & 0.99 & 0.81 \\ 0.99 & 0.99 & 0.98 & 0.99 & 0.00 \\ 0.98 & 0.99 & 0.99 & 0.98 & 0.98 \\ 0.98 & 0.99 & 0.99 & 0.69 & 0.00 \end{bmatrix}$$

(a) Triangular, $y(4,9)$=124.24,(b) Epanechnikov, $y(4,9)$=123.24, (c) Biweight, $y(4,9)$=123.83,

$$\begin{bmatrix} 0.00 & 0.98 & 0.00 & 0.98 & 0.99 \\ 0.82 & 0.99 & 0.98 & 0.99 & 0.73 \\ 0.98 & 0.99 & 0.97 & 0.99 & 0.00 \\ 0.98 & 0.99 & 0.99 & 0.97 & 0.97 \\ 0.97 & 0.98 & 0.98 & 0.57 & 0.00 \end{bmatrix} \begin{bmatrix} 0.00 & 0.99 & 0.00 & 0.99 & 1.00 \\ 0.95 & 0.99 & 0.99 & 1.00 & 0.91 \\ 0.99 & 1.00 & 0.99 & 0.99 & 0.00 \\ 0.99 & 1.00 & 1.00 & 0.99 & 0.99 \\ 0.99 & 0.99 & 0.99 & 0.80 & 0.00 \end{bmatrix} \begin{bmatrix} 0.27 & 0.99 & 0.27 & 0.99 & 0.99 \\ 0.93 & 0.99 & 0.99 & 0.99 & 0.90 \\ 0.99 & 0.99 & 0.99 & 0.99 & 0.27 \\ 0.99 & 0.99 & 0.99 & 0.98 & 0.99 \\ 0.99 & 0.99 & 0.99 & 0.86 & 0.27 \end{bmatrix}$$

(d) Triweight, $y(4,9)$=124.37, (e) Tricube, $y(4,9) = 123.28$, (f) Gaussian, $y(4,9) = 116.87$,

$$\begin{bmatrix} 0.00 & 0.99 & 0.00 & 0.99 & 0.99 \\ 0.92 & 0.99 & 0.99 & 0.99 & 0.88 \\ 0.99 & 0.99 & 0.99 & 0.99 & 0.00 \\ 0.99 & 0.99 & 0.99 & 0.98 & 0.99 \\ 0.99 & 0.99 & 0.99 & 0.80 & 0.00 \end{bmatrix} \begin{bmatrix} 0.00 & 1.00 & 0.04 & 1.00 & 1.00 \\ 0.07 & 1.00 & 1.00 & 1.00 & 0.04 \\ 1.00 & 1.00 & 0.41 & 1.00 & 0.00 \\ 0.49 & 1.00 & 1.00 & 0.35 & 0.41 \\ 0.41 & 1.00 & 1.00 & 0.02 & 0.00 \end{bmatrix}$$

(g) Cosine, $y(4,9)$= 123.37,   (h) GEVF, $y(4,9) = 128.35$.

Fig. 5. Weighted average filters and the output pixel, $y(x_1 = 4, x_2 = 9)$ of RLSF (a)-(g), and the proposed method, (h).

$$p(X) = \begin{cases} \exp\{-(1+\zeta(\dfrac{X-\mu}{\sigma}))^{-\frac{1}{\zeta}}\}\sigma^{-1}[1+\zeta(\dfrac{X-\mu}{\sigma})]^{-(\frac{1}{\zeta})-1} & if\ \zeta \neq 0 \\[2ex] \exp\{-\exp(-\dfrac{X-\mu}{\sigma})\}\sigma^{-1}\exp(-\dfrac{X-\mu}{\sigma}) & if\ \zeta = 0 \end{cases} \quad (5)$$

where X is a random variable; in our method, X was given by the local $ROAD_\alpha(F_j)$ as described in the Sect. 2.2. Three GEV parameters, μ, σ, and ζ, denote the location, scale, and shape parameters, respectively.

GEV function consists of three type distributions. Gumbel's type I for $\zeta = 0$ and $-\infty < X < \infty$. Frechet's type II for $\zeta < 0$, and $\mu + \sigma/\zeta \leq X \leq \infty$. Weibull's type III for $\zeta > 0$, and $-\infty < X < \mu + \sigma/\zeta$. The GEV parameters were estimated in small sample by maximum likelihood estimation.

3.1   Maximum Likelihood Estimation for Small Sample

GEV parameters in (5) were estimated by Generalized Maximum Likelihood Estimators (GMLE) [15]. GEV log-likelihood function of n observations, $X = \{x_1, x_2, x_3, \ldots, x_N\}$ is given by

$$\ln[L(\mu, \sigma, \zeta|X)] = -n\ln(\sigma) + \sum_{i=1}^{n}\left[\left(\frac{1}{\zeta} - 1\right)\ln(y_i) - (y_i)^{1/\zeta}\right] \qquad (6)$$

where $y_i = [1 - (\zeta - \sigma)(X - \mu)]$. Maximum likelihood of the parameters, μ, σ, and ζ could be estimated by solving the first-order partial derivatives of (6) with respect to each parameter, which was given by:

$$\frac{\partial \ln L}{\partial \mu} = \frac{1}{\sigma}\sum_{i=1}^{n}\left[\frac{1 - \zeta - (y_i)^{\frac{1}{\zeta}}}{y_i}\right] = 0,$$

$$\frac{\partial \ln L}{\partial \sigma} = -S\sigma^{-1} + \sigma^{-1}\sum_{i=1}^{n}\left[\frac{1 - \zeta - (y_i)^{\frac{1}{\zeta}}}{y_i}\left(\frac{X - \zeta}{\sigma}\right)\right] = 0, \qquad (7)$$

$$\frac{\partial \ln L}{\partial \zeta} = -\frac{1}{\sigma}\zeta^2\sum_{i=1}^{S}\ln(y_i)\left[1 - \zeta - (y_i)^{\frac{u}{\zeta}}\right] + \frac{1 - \zeta - (y_i)^{\frac{1}{\zeta}}}{y_i}\zeta\left(\frac{X - \zeta}{\sigma}\right) = 0.$$

GMLE [15] for small sample will modify (6), which is given by

$$\ln[GL(\mu, \sigma, \zeta|x)] = \ln[L(\mu, \sigma, \zeta|x)] + \ln[\pi(\zeta)] \qquad (8)$$

where $\pi(\zeta)$ denotes prior distribution of the shape parameter. $\pi(\zeta)$ is beta distribution which is given by:

$$\pi(\zeta) = (0.5 + \zeta)^{p-1}(0.5 - \zeta)^{q-1}\Big/B(p, q) \qquad (9)$$

where $B(p, q) = (\Gamma(p)\Gamma(q))/\Gamma(p + q)$. Γ denotes gamma function. In small data, the parameters p and q were assigned to 6 and 9 [15], respectively.

4   Generalized Extreme Value Filter

GEV filter (GEVF) to remove the mixed Gaussian-impulse noise consists of two steps. The first step provides to estimate the three parameters of GEV in a small sample. The second step uses the parameters to design the weighting functions.

4.1   Estimate GEV Parameters

In our scheme, ROAD statistics [11] was used to measure how close a pixel value is to its four most similar neighbors. The local ROAD values in the operation block were the sample data to estimate the GEV parameters. The parameters were estimated by Newton-Raphson method to optimize the Eq. (8).

The Newton-Raphson method was used to solve the likelihood Eq. (8) [15]. In our scheme, convergence criteria for the three parameters sets to 10^{-5}, maximum steps for the location, scale, and shape parameters are 0.5, 0.5, and 0.1, respectively. Maximum number of iteration is 30.

4.2   Weighting Function

The optimal GEV parameters were provided to calculate probability curve by Eq. (5) with $X = 0, 1, 2, \ldots, 255$. The weighting function, GEVF(X), was formulated by the normalized probability curve as defined by

$$\text{GEVF}(X) = p(X)/p(X = \hat{\mu}) \tag{10}$$

where $\hat{\mu}$ denotes the estimated location parameter. GEVF(X) was regulated at the location parameter that is if $X \leq \hat{\mu}$ the GEVF(X) was assigned to 1. Weighted matrix was mapped by the normalized probability curve. The matrix was given by

$$M_{GEV} = \text{GEVF}\big(X = ROAD_{\alpha=4}(F_j)/4\big). \tag{11}$$

Kernel function of GEVF from the local ROAD in Fig. 3 was shown in Fig. 4(b) with $\hat{\mu} = 10$.

Finally, the weighted matrix, M_{GEV}, was employed for calculating the output pixel $y(x_1, x_2)$ by Eq. (4). The weighted matrix and the output pixel, $y(x_1 = 4, x_2 = 9) = 128.35$, from the image data in Fig. 2(c) were illustrated in Fig. 5(h). Considered responses of the weighted matrices from Fig. 5, GEVF could produce the minimum error when compared with the other kernels.

5   Experimental Results

Testing images depicted in Fig. 6 were used to evaluate the proposed method. The images were contaminated by the mixed Gaussian-impulse noise. The noise density levels, ρ, denote standard deviation of Gaussian in the range 10-50. Percentage of the density levels, ρ, represents the corrupted probability of random impulsive noise [9] with the impulsive noise values in the range [0, 255].

The image results were evaluated by peak signal to noise ratio (*PSNR*) as defined by

$$RMSE = \sqrt{\frac{1}{3MN} \sum_{i=1}^{M} \sum_{j=1}^{N} \sum_{l=1}^{3} \big(x_{i,j,l} - o_{i,j,l}\big)^2} \tag{12}$$

$$PSNR = 20log_{10}\{255 \,/\, RMSE\} \tag{13}$$

where M and N are size of images. $x_{i,j,l}$ denotes the output pixel at spatial coordinate (i,j). in the color channel l of RGB color space and $o_{i,j,l}$ stand for the original pixel.

χ^2 goodness-of-fit (GOF) [18] was used to evaluate performance of the kernel functions. GOF compares original image pixel, $o_{i,j,l}$, with estimated pixel, $x_{i,j,l}$, that

(a) LENNA   (b) GIRL

(c) HOUSE   (d) CANDY

Fig. 6. Test images in visual evaluations. (Color figure online)

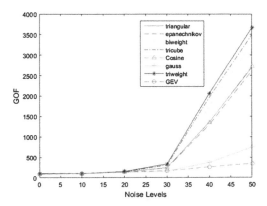

Fig. 7. Average GOF values of Kodak test images to compare filter functions of GEV and seven kernels of RLSF method.

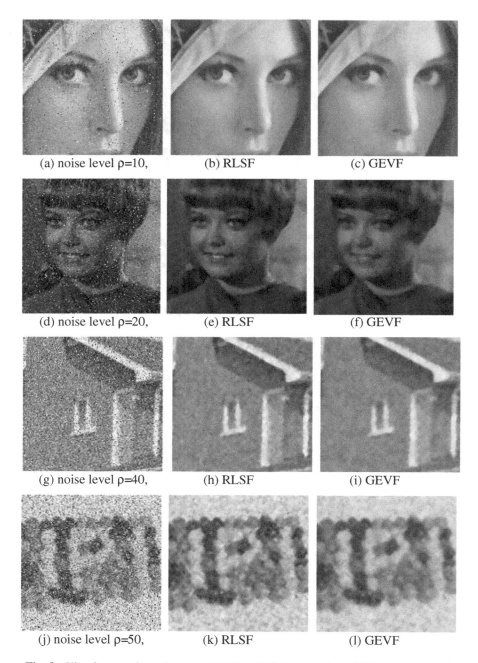

(a) noise level ρ=10,   (b) RLSF   (c) GEVF

(d) noise level ρ=20,   (e) RLSF   (f) GEVF

(g) noise level ρ=40,   (h) RLSF   (i) GEVF

(j) noise level ρ=50,   (k) RLSF   (l) GEVF

Fig. 8. Visual comparisons between GEVF and RLSF methods at different noise levels.

Table 1. Quantitative comparison results between GEVF and RLSF by PSNR and QSSIM.

ρ	RLSF kernels														GEVF	
	Triangle		Epanech.		Biweight		Tricube		Triweight		Gaussian		Cosine			
	PSNR	QSSIM	PSNR	QSSIM	PSNR	QSSIM	PSNR	QSSIM	PSNR	QSSIM	PSNR	QSSIM	PSNR	QSSIM	PSNR	QSSIM
GIRL																
10	29.00	0.76	28.63	0.74	29.22	0.76	29.16	0.76	29.58	0.78	27.80	0.71	28.74	0.74	29.59	0.81
20	28.57	0.75	28.40	0.74	28.73	0.77	28.73	0.77	28.82	0.77	27.46	0.71	28.47	0.75	28.43	0.75
30	26.90	0.69	26.93	0.70	26.64	0.68	26.52	0.67	26.27	0.66	26.54	0.69	26.91	0.69	26.97	0.70
40	24.12	0.55	24.24	0.56	23.55	0.52	23.29	0.51	23.00	0.49	25.24	0.63	24.18	0.55	25.45	0.63
50	20.93	0.39	21.03	0.39	20.38	0.36	20.12	0.35	19.92	0.34	23.59	0.56	21.01	0.39	23.78	0.56
CANDY																
10	28.81	0.88	28.58	0.87	08.94	0.88	28.91	0.88	29.12	0.89	27.61	0.85	28.66	0.87	28.48	0.88
20	28.52	0.87	28.46	0.87	28.56	0.87	28.56	0.87	28.47	0.86	27.31	0.84	28.49	0.87	27.89	0.84
30	26.26	0.77	26.39	0.78	25.79	0.74	25.59	0.73	25.26	0.71	26.38	0.81	26.32	0.77	26.47	0.78
40	22.68	0.59	22.82	0.60	22.04	0.56	21.75	0.54	21.49	0.60	24.92	0.73	22.75	0.60	24.86	0.70
50	18.94	0.42	19.02	0.43	18.47	0.40	18.23	0.39	18.09	0.38	23.18	0.62	19.11	0.43	23.31	0.63
LENNA																
10	30.57	0.84	30.22	0.83	30.80	0.84	30.74	0.84	31.12	0.85	29.23	0.81	30.33	0.83	30.32	0.84
20	29.06	0.81	28.73	0.80	29.35	0.81	29.35	0.81	29.66	0.82	27.58	0.78	28.86	0.80	29.45	0.80
30	27.66	0.72	27.83	0.73	27.13	0.69	26.93	0.68	26.51	0.66	27.89	0.77	27.75	0.73	27.76	0.75
40	23.42	0.51	23.58	0.52	22.74	0.47	22.43	0.45	22.13	0.43	25.88	0.68	23.56	0.51	25.74	0.68
50	30.57	0.84	30.22	0.83	30.80	0.84	30.74	0.84	31.12	0.85	29.23	0.81	30.33	0.83	30.32	0.84
HOUSE																
10	29.49	0.74	29.24	0.72	29.57	0.73	29.50	0.74	29.70	0.74	28.40	0.71	29.32	0.73	29.35	0.75
20	28.86	0.72	28.83	0.72	28.79	0.73	28.75	0.73	28.60	0.72	27.96	0.70	28.85	0.72	28.37	0.72
30	26.51	0.63	26.66	0.64	25.99	0.61	25.78	0.60	25.42	0.58	26.97	0.68	26.59	0.64	26.70	0.65
40	22.82	0.46	22.97	0.46	22.16	0.42	21.85	0.41	21.57	0.40	25.45	0.60	22.92	0.46	25.03	0.57
50	18.86	0.31	18.94	0.31	18.38	0.29	18.13	0.28	17.99	0.27	23.47	0.51	19.01	0.31	23.28	0.51

employs to measures the fitting function as given by $\sum_{i=1}^{n} \sum_{j=1}^{n} \sum_{l=1}^{3} \frac{(o_{i,j,l} - x_{i,j,l})^2}{x_{i,j,l}}$. The smaller value of GOF is the better fit.

Results of GEVF were compared with RLSF with the same size of the processing block β and the small window W. The kernel functions of RLSF were used to estimate the weight values consisting of Triangular, Epanechnikov, Biweight, Tricube, Gaussian, and Cosine.

To evaluate the kernel fitting functions, Fig. 7 shows the average χ^2 values of the image database from Kodak test images [19], which consists of 24 pictures. The proposed method provides the best GOF values of the higher density of noise levels.

Quantitative comparisons were presented in Table 1 consisted of *PSNR* and Quaternion Structural SIMilarity (QSSIM) [14, 17] to assess the preservation of color image details. QSSIM [17] is a good image measurement for color distortion, contrast and blur that based on the principle of human visual system. It operates by engaging the structure dissimilarity of image differences in each color band. If QSSIM value approach to 1 then the output image step toward to good quality.

As seen from Table 1, our method could provide good results in LENNA and GIRL images. They had small error as seen from *PSNR* values. From QSSIM index, GEVF had shown good; especially, the images with high detail such as LENNA and GIRL images. These comparisons could verify from Fig. 8, which was compared at different noise density levels. RLSF method with Triweight kernel provides some good results for high noise levels with the low detailed images or homogeneous background as seen in HOUSE and CANDY images.

6 Conclusion

Our paper was proposed the efficient method for removing the mixed Gaussian-impulse noise in color images by using generalized extreme value filter named as GEVF. The proposed method utilizes statistics of local rank order absolute distance for impulse detection to estimate appropriated weights with the only one kernel, which designs by PDF of GEV distribution. From the experimental results, GEVF could provide good kernel matrix to estimate for each output pixel. The proposed method could remove the mixed noise better than RLSF in some high detailed images. In future work, we will investigate the influence of ROAD statistics to improve the technique of GEV parameter estimation for small sample.

References

1. Platoniotis, N., Venetsanopoulos, A.: Color Image Processing and Applications. Springer, Heidelberg (2000)
2. Camarena, J., Gregori, S., Morillas, V., Sapena, A.: A simple fuzzy method to remove mixed noise Gaussian-impulsive noise from color images. Fuzzy Syst. **21**(5), 971–978 (2013). IEEE Press

3. Astola, J., Haavisto, P., Neuvo, Y.: Vector median filters. Proc. IEEE **78**(4), 678–689 (1990)
4. Morillas, S., Gregori, V., Hervas, A.: Fuzzy peer groups for reducing mixed Gaussian-impulse noise from color images. Image Process. **18**(7), 1452–1466 (2009). IEEE Press
5. Chin-Hsing, L., Jia-Shiuan, T., Ching-Te, C.: Switch bilateral filter with a texture/noise detector for universal noise removal. Image Process. **19**(9), 2307–2320 (2010). IEEE Press
6. Lukac, R., Plataniotis, K.N., Venetsanopoulos, A.N., Smolka, B.: A statistically-switched adaptive vector median filter. J. Intell. Robot. Syst. Theor. Appl. **42**, 361–391 (2005)
7. Smolka, B.: Soft switching technique for impulsive noise removal in color images. In: 2013 Fifth International Conference on Computational Intelligence, Communication systems and Networks, pp. 222–227 (2013)
8. Kenney, C., Deng, Y., Manjunath, B.S., Hewer, G.: Peer group image enhancement. Image Process. **10**, 326–334 (2001). IEEE Press
9. Smolka, B., Kusnik, D.: Robust local similarity filter for reduction of mixed Gaussian and impulsive noise in color images. SIViP **9**, 49–56 (2015)
10. Tomasi, C., Manduchi, R.: Bilateral filtering for gray and color images. In: Sixth Internaional Conference on Computer Vision, pp. 839–846 (1998)
11. Garnett, R., Huegerich, T., Chui, C., He, W.: A universal noise removal algorithm with an impulse detector. Image Process. **14**, 1747–1754 (2005). IEEE Press
12. Jenkinson, A.F.: The frequency distribution of the annualo maximum (or minimum) values of meteorological elements. Q. J. R. Meteorol. Soc. **81**, 158–171 (1955)
13. Bednar, J., Watt, T.: Alpha-trimmed means and their relationship to median filters. Acoust. Speech Signal Process. **32**(1), 145–153 (1984). IEEE Press
14. Chankhachon, S., Intajag, S.: Resourceful method to remove mixed gaussian-impulse noise in color images. In: JCSSE, 2015 12th International Joint Conference, pp. 18–23 (2015)
15. Martins, E.S., Stedinger, J.R.: Generalized maximum-likelihood extreme-value quantile estimators for hydrologic data. Water Resour. Res. **36**(3), 737–744 (2000)
16. Image Database. http://www.imageprocessingplace.com/DIP-3E/dip3e_book_images_downloads.htm
17. Kolaman, A., Yadid-Pecht, O.: Quaternion structural similarity: a new quality index for color images. Image Process. **21**(4), 1526–1536 (2012). IEEE Press
18. Papoulis, A.: Probability, Random Variables, and Stochastic Processes, vol. 3. McGrawHill, New York (1991)
19. Kodak test images. http://r0k.us/graphics/kodak/

Restricted Four-Valued Semantics for Answer Set Programming

Chen Chen and Zuoquan Lin$^{(\boxtimes)}$

Department of Information Science, School of Mathematical Sciences,
Peking University, Beijing 100871, China
skydark2@gmail.com, linzuoquan@pku.edu.cn

Abstract. In answer set programming, an extended logic program may
have no answer set, or only one trivial answer set. In this paper, we
propose a new stable model semantics based on the restricted four-valued
logic to overcome both inconsistences and incoherences in answer set
programming. Our stable models coincide with classical answer sets when
reasoning on consistent and coherent logic programs, and can be solved
by transformation in existing ASP solvers. We also show the connection
between our stable models and the extensions of default logic.

1 Introduction

Answer set programming (ASP) is a nonmonotonic logic programming paradigm.
Its stable model semantics provides solutions of logic programs by answer sets.
One limitation of the stable model semantics is that some logic programs lack
answer sets, which is called incoherent programs. In extended logic programs
which support both negation as failure and classical negation [12], an inconsistent program have only one trivial answer set that can infer every formula.
Although it seems intuitive to detect and revise contradictions in logic programs,
sometimes it is not practical and may lose information during revision. Since it
is not always feasible to keep every knowledge base consistent and coherent, a
paraconsistent and coherent reasoning method will be helpful since it can extract
meaningful conclusions from the fragile information.

Some researchers introduced paraconsistency to answer set programming
[6,14,18]. Among those, Belnap's four-valued logic [3–5], which is a paraconsistent logic that does not imply everything from inconsistent information, has
been used [11,17,18] due to its intuitive semantics based on a bilattice structure.

Reiter's default logic [16] is a well-known nonmonotonic logic, which has a
close relationship to answer set semantics of extended logic programs [12]. The
study of handling both inconsistent and incoherent information in default logic is
under investigation. The restricted four-valued default logic [8], which is based
on four-valued logic, has been proposed in the presence of inconsistency and
incoherence. By restricting atoms that allowed to be inconsistent, this approach
can reason on both inconsistent and incoherent default theories, and also hold the
same result as classical default logic while reasoning on consistent and coherent

© Springer International Publishing Switzerland 2016
R. Booth and M.-L. Zhang (Eds.): PRICAI 2016, LNAI 9810, pp. 68–79, 2016.
DOI: 10.1007/978-3-319-42911-3_6

theories. In this paper, we will introduce the restricted four-valued stable models inspired by the similar intuition. Our main contributions are in the following:

(1) to present a new semantics of extended logic programs which is based on the restricted four-valued logic.
(2) to show that our models can handle both inconsistencies and incoherences.
(3) to prove that our models have an 1-1 correspondence with answer set semantics on consistent and coherent programs.
(4) to give a transformation for solving our models in modern ASP solvers which supports weak constraints like DLV [13].
(5) to show that our semantics can be embedded into restricted four-valued default logic.

This paper is structured as follows. In Sect. 2, we review preliminaries. In Sect. 3, we describe our restricted four-valued semantics for extended logic programs. We show how to solve our stable models by transformation in Sect. 4. In Sect. 5, we present the connection between our models and the restricted four-valued default logic. In Sect. 6, we compare our semantics with related works. Finally, we summarize in the concluding section.

2   Preliminaries

Throughout this paper, let $\mathcal{L}$ be a propositional language, and $\mathcal{A}$ the finite set of all atoms in $\mathcal{L}$. For a literal l, we denote $atom(l)$ as the atom of l, and denote $\neg l$ as $\neg a$ if $l = a$ is an atom or a if $l = \neg a$ is a negation of an atom.

2.1   Extended Logic Programs

In [12], Gelfond and Lifschitz extend logic programming by introducing classical negation. An *extended logic program* is constructed by rules. Each rule r has the form of

$$l_0 \leftarrow l_1, \cdots, l_m, \text{not } l_{m+1}, \cdots, \text{not } l_n,$$

where all l_i are literals and $n \geq m \geq 0$. The negations in literals are classical negations and the nots in rules are default negations. We denote $H(r) = \{l_0\}$ as the *head* of r, $B^+(r) = \{l_1, \cdots, l_m\}$ as the *positive body* of r, and $B^-(r) = \{l_{m+1}, \cdots, l_n\}$ as the *negative body* of r. A logic program is positive if all its rules have empty negative bodies.

An *interpretation* I, which is a literal set, satisfies a rule r, denoted as $I \models r$, iff $I \cap H(r) \neq \emptyset$ if $B^+(r) \subseteq I$ and $B^-(r) \cap I = \emptyset$. We say that an interpretation I is a *model* of a logic program P if I satisfies every rule of P. We say that I is *minimal* if there is no other model J of P that $J \subsetneq I$.

For each extended logic program P, the *Gelfond-Lifschitz reduct* of P w.r.t. an interpretation I, denoted as P^I, is an extended logic program obtained from P by deleting:

(1) each rule r that $B^-(r) \cap I$ is not an empty set;
(2) all other not l in remaining rules.

An interpretation I is an *answer set* of P, if I is the minimal model of P^I.
   A logic program may have no answer set or only one trivial answer set.

Example 1. Consider the following logic programs:

(1) $P_1 = \{a \leftarrow, \neg a \leftarrow\}$;
(2) $P_2 = \{a \leftarrow \text{not } b, \neg a \leftarrow \text{not } b\}$;
(3) $P_3 = \{c \leftarrow b, \text{not } c; b \leftarrow \text{not } a\}$.

P_1 has only one trivial answer set, P_2 and P_3 have no answer set.

   The extended logic programs can be embedded into Reiter's default logic [16].
A *default* d is an inference rule of form $d = \frac{\alpha:\beta_1,\ldots,\beta_n}{\gamma}$, where $\alpha, \beta_1, \ldots, \beta_n, \gamma$ are
all propositional formulas. We define $Pre(d) = \alpha$ as *prerequisite* of d, $Just(d) = \{\beta_1, \ldots, \beta_n\}$ as *justification* of d, and $Con(d) = \gamma$ as *consequence* of d. A *default theory* is a pair $T = (D, W)$, where D is a set of defaults and W is a set of formulas.
   An *extension* of a default theory is defined as follows.

Definition 1 ([16]). *Let* $T = (D, W)$ *be a default theory. For any set of formulas* E, *let* $\Gamma(E)$ *be the smallest set of formulas such that*

(1) $W \subseteq \Gamma(E)$;
(2) $Th(\Gamma(E)) = \Gamma(E)$;
(3) *For any* $d \in D$, *if* $\Gamma(E) \models_2 Pre(d)$ *and* $\neg\beta \notin E$ *for all* $\beta \in Just(d)$, *then* $\Gamma(E) \models_2 Con(d)$, *where* $\models_2$ *is the classical 2-valued propositional consequence relation.*

   A set of formulas E *is an (default) extension of* T *iff* $\Gamma(E) = E$, *i.e.* E *is a fixed point of the operator* Γ.

Proposition 1 ([12]). *We identify a rule* $r = l_0 \leftarrow l_1 \cdots, l_m, \text{not } l_m, \cdots, \text{not } l_n$
with the default $d(r) = \frac{l_1,\cdots,l_m:\neg l_{m+1},\cdots,\neg l_n}{l_0}$. *For any extended program* P, *I is an answer set of* P *iff the deductive closure of* I *is an extension of* $T(P)$, *where* $T(P) = (\{d(r)|r \in P\}, \emptyset)$ *is a default theory generated by* P.

2.2   Four-Valued Logic

Belnap's four-valued logic [3–5] is based on the bilattice structure $FOUR = \{t, f, \top, \bot\}$, in which the truth values represent true, false, inconsistent and unknown respectively. The bilattice structure has two partial orderings: $\leq_t$ measures the degree of truth and $\leq_k$ measures the degree of knowledge. In this paper, we use the knowledge ordering $\leq_k$ if the using ordering is not explicitly indicated. According to the knowledge ordering, $\bot$ is the minimal element, $\top$ is the maximal element, while t and f are not comparable. Since both classical negations and default negations appear in extended logic programming, it is necessary to distinguish explicit negative information and implicit ones. In four-valued logic, the truth value f has the explicit negative information, while $\perp$ means unknown, which has no information about whether it is true or false.

The *set of designated elements* is chosen as $\mathcal{D} = \{t, \top\}$. A *four-valued valuation* is a function that assigns a truth value from $FOUR$ to each atomic formula. A valuation v *satisfies* a formula ϕ if $v(\phi) \in \mathcal{D}$. We say that v is a *model* of a set of formula S if v satisfies every formula in S. The ordering between valuations can be defined by the ordering of value on each atom, i.e. let u and v be valuations, $u \leq v$ if for each atom a we have $u(a) \leq v(a)$. A four-valued interpretation I of a logic program P can be defined by a four-valued valuation.

Definition 2 ([18]). *Let P be a positive extended program and I a four-valued interpretation, then for any rule r, $I \models r$ iff $I \models H(r)$ or there is a literal $l \in B^+(r)$ that $I \not\models l$. An interpretation I is a four-valued model of P if $I \models r$ for every rule r of P.*

A four-valued interpretation can be characterized by the literals that it satisfies, denoted as I_L. For example, we can use $I_L = \{a, \neg a, b, \neg c\}$ to denote the interpretation I which maps a to $\top$, b to t, c to f and other atoms to $\perp$. Also, let I and J be two interpretations, $I \leq J$ iff $I_L \subseteq J_L$.

3   Restricted Four-valued Semantics for Extended Logic Programs

The restricted four-valued logic is presented in [8] as the underlying logic for default reasoning. The intuition of restricting inconsistent atoms in a restricting set is a trade-off between the reasoning ability of classical logic and paraconsistency of four-valued logic.

Definition 3 ([8]). *Let S be a set of atoms. A four-valued valuation v is restricted by S, if $\{a \in \mathcal{A}|v(a) \notin \{t, f\}\} \subseteq S$. A four-valued valuation v is a four-valued model of Γ restricted by S if v is a four-valued model of Γ and restricted by S.*

Let Γ, Σ be sets of formulas. $\Gamma \models_S \Sigma$ if every four-valued model of Γ restricted by S is a four-valued model of Σ.

Based on the restricted four-valued logic, we define the *restricted reduct* as follows.

Definition 4. *Let P be an extended logic program, I an interpretation, and S a restricting set of atoms. We construct the extended logic program P_S^I obtained from P by deleting:*

(1) all not l for each literal l that $atom(l) \in S$, and
(2) each rule r that $l \in B^-(r)$ with any $l \in I$, and
(3) all other not l in remaining rules.

By the restricted reduct, we define the restricted four-valued stable models of extended logic programs in a way similar to the stable models of answer sets.

Definition 5. *Let P be an extended logic program, I an interpretation, and S a restricting set of atoms. I is a restricted four-valued stable model of P restricted by S, if I is the minimal model of P_S^I, and is restricted by S.*

I is a preferred restricted four-valued stable model of P, if I is a restricted four-valued stable model of P restricted by some restricting set S, and there is no restricted four-valued stable model of P restricted by R and R has less atoms than S by cardinality.

Notice that an interpretation I is restricted by S means that $\{a, \neg a\} \in I$ iff $a \in S$ for each atom a. The reason to compare restricting sets by cardinality is easier to calculate by existing ASP solvers as we shall see in later.

It is not surprising that the existence of restricted four-valued stable models can not be guaranteed if we set a fixed restricting set S. However, it is different if we consider all possible restricting sets.

Example 2. (Continuation of Example 1) All the logic programs in Example 1 have nontrivial and intuitive (preferred) restricted four-valued answer sets.

(1) The only preferred restricted four-valued stable model of P_1 is $\{a, \neg a\}$ restricted by $\{a\}$. This model is not trivial due to the paraconsistent underlying logic.
(2) The only preferred restricted four-valued stable model of P_2 is $\{a, \neg a\}$ restricted by $\{a\}$. It is confused that P_1 has only trivial answer set but P_2 has no answer set though they are very similar. But, P_1 and P_2 share the same preferred restricted four-valued stable model.
(3) The only preferred restricted four-valued stable model of P_3 is $\{b, c\}$ restricted by $\{c\}$. In this model, we accept b as the result of applying the second rule of P_3. The head of the first rule c is also accepted but annotated as problematic since it is included in the restricting set.

In fact, we have the following theorem to guarantee the existence of restricted four-valued stable models.

Theorem 1. *Every extended logic program P has a restricted four-valued stable model. As a result, P also has a preferred restricted four-valued stable model.*

Proof. By choosing the full atom set $\mathcal{A}$ as the restricting set, we get a positive logic program $P_I^{\mathcal{A}}$ which is independent of I. The minimal model I' of $P_I^{\mathcal{A}}$ is a restricted four-valued stable model of P restricted by $\mathcal{A}$.   □

The above theorem reveals that our semantics is paraconsistent and coherent, which always has nontrivial models for every program. It is also a good news that we can keep classical stable models if they are meaningful.

Theorem 2. *Let P be an extended logic program. P has an answer set I iff I is a restricted four-valued stable model of P restricted by $\emptyset$.*

Proof. Notice that when the restricting set is an empty set, the definitions of restricted reduct and restricted four-valued stable models are all reduced to their counterparts of general answer set semantics.   □

Corollary 1. *Let P be an extended logic program which has answer sets. An interpretation I is an answer set of P iff I is a preferred restricted four-valued stable model of P.*

Proof. It is easy to see by Theorem 2 and the fact that every stable model of P restricted by an empty set is preferred and only preferred.   □

4   Reduction to Classical Programs

In this section we discuss how to solve (preferred) restricted four-valued stable models by transformation.

In [12], the positive form of literals has been introduced to transform extended logic programs to general logic programs with no classical negations.

Proposition 2 ([12]). *Let l be a literal. We denote the new atom l^+ as the positive form of l, The positive form of a literal set I is defined as $I^+ = \{l^+ | l \in I\}$. Let P be an extended logic program. For each rule r of P,*

$$r = l_0 \leftarrow l_1, \cdots, l_m, \text{not } l_{m+1}, \cdots, \text{not } l_n,$$

we define

$$trans(r) = l_0^+ \leftarrow l_1^+, \cdots, l_m^+, \text{not } l_{m+1}^+, \cdots, \text{not } l_n^+.$$

We define P^+ as the program by replacing each rule r by $trans(r)$. A consistent set of literals I is an answer set of P iff I^+ is an answer set of P^+.

We define our transformation in a similar way.

Definition 6. *Let P be an extended logic program. For each rule r of P,*

$$r = l_0 \leftarrow l_1, \cdots, l_m, \text{not } l_{m+1}, \cdots, \text{not } l_n,$$

we define

$$trans_r(r) = l_0^+ \leftarrow l_1^+, \cdots, l_m^+, \text{not } \neg(\neg l_{m+1})^+, \cdots, \text{not } \neg(\neg l_n)^+.$$

For each literal l, we define $neg(l) = \neg l^+ \leftarrow (\neg l)^+$. We denote $P_S^+ = \{trans_r(r) | r \in P\} \cup \{neg(l) | atom(l) \notin S\}$ as the restricting transformation *of P restricted by S.*

By the restricting transformation, we want to transform a solving problem of restricted four-valued stable models to a solving problem of answer set, which can be solved by mature ASP solvers.

Proposition 3. *Let P be an extended logic program. An interpretation I is a restricted four-valued stable model of P restricted by S iff $comp_S(I^+)$ is a consistent answer set of P_S^+, where $comp_S(I^+) = I^+ \cup \{\neg l^+ | (\neg l)^+ \in I^+ \text{ and } atom(l) \notin S\}$.*

Proof. Let I be a consistent answer set of P_S^+. We know $\neg l^+ \notin I$ for each $atom(l) \in S$, since $\neg l^+$ does not occur in any heads of rules and I is consistent. We also know that $\neg l^+ \in I$ iff $(\neg l)^+ \in I$ for each $atom(l) \notin S$, since $\neg l^+$ can only be inferred by applying rule $neg(l)$. As a result, the function $comp_S$ is an 1-1 correspondence.

By Proposition 2 and the definition of the restricted reduct, I is an answer set of P iff I^+ is a consistent answer set of P_0^+, where P_0^+ is constructed by deleting every $not\ l$ in P^+ for each literal l that $atom(l) \in S$ when the restricting set S is given. So we only need to prove that $comp_S(I^+)$ is a consistent answer set of P_S^+ iff I^+ is a consistent answer set of P_0^+.

For each rule r of P,

(1) if $trans(r)$ is applied in the reduct $(P_0^+)^{I^+}$, then $B^+(trans(r)) \subseteq I^+$, and $B^-(trans(r)) \cap I^+ = \emptyset$, also $H(trans(r)) \subseteq I^+$. Since $\neg(\neg l)^+ \in comp_S(I^+)$ iff $l^+ \in I^+$, together with $B^-(trans(r)) \cap I^+ = \emptyset$, we have $B^-(trans_r(r)) \cap comp_S(I^+) = \emptyset$. Combined with that $B^+(trans_r(r)) = B^+(trans(r)) \subseteq I^+ \subseteq comp_S(I^+)$, we know that the rule $trans_r(r)$ is also appliable in the reduct $(P_S^+)^{comp_S(I^+)}$. Also $H(trans_r(r)) = H(trans(r)) \subseteq I^+ \subseteq comp_S(I^+)$;

(2) if $trans(r)$ is not applied in the reduct $(P_0^+)^{I^+}$, then
   (a) $B^+(trans(r)) \not\subseteq I^+$. We have $B^+(trans_r(r)) \not\subseteq comp_S(I^+)$ since $B^+(trans(r)) = B^+(trans_r(r))$ and $B^+(trans_r(r))$ only includes positive forms. So $trans_r(r)$ is not appliable in the reduct $(P_S^+)^{comp_S(I^+)}$ either.
   (b) or $B^-(trans(r)) \cap I^+ \neq \emptyset$. Then there is a literal l that $l^+ \in B^-(trans(r))$ and $l^+ \in I^+ \subseteq comp_S(I^+)$. We ensure that $atom(l) \notin S$ because of the construction of P_0^+. So $B^-(trans_r(r)) \cap comp_S(I^+) \supseteq \{\neg(\neg l)^+\}$.
   As a result, $trans_r(r)$ is not appliable in the reduct $(P_S^+)^{comp_S(I^+)}$ either.

Altogether, we have proved that $trans(r)$ is appliable in the reduct $(P_0^+)^{I^+}$ iff $trans_r(r)$ is appliable in the reduct $(P_S^+)^{comp_S(I^+)}$, which share the same head of rules. Therefore, I^+ is a consistent answer set of P_0^+ implies that I^+ is exactly the positive forms included in the minimal model of $(P_S^+)^{comp_S(I^+)}$, which implies $comp_S(I^+)$ is an answer set of P_S^+.

The opposite direction can be proved by the same approach. □

To solve preferred restricted four-valued stable models, we need another technique called weak constraint [7], which is a construct that extends logic programs and has been implied by some modern answer set solvers like DLV [13].

In a logic program, a *constraint* is a rule with an empty head. The only syntactic difference between constraints and weak constraints is that the symbol ←

is replaced by $:\sim$ in the weak constraints. The semantics of weak constraints is defined as follows.

Definition 7 ([13]). *Let P be an extended logic program, An interpretation I is an (optimal) answer set of P iff*

(1) I is an answer set of P without weak constraints;
(2) $H^P(I)$ is minimal over all the answer sets of P.

The objective function $H^P(I)$ maps an answer set I to a weight number by checking weak constraints. Here, we only need the simplest function that counts the number of weak constraints r that $I \supseteq B^+(r)$ and $I \cap B^-(r) = \emptyset$.

Definition 8. *Let P be an extended logic program. We define P^p as the pre-ferred restricting transformation of P, where P^p contains the following rules:*

(1) $trans_r(r)$ for each rule r of P;
(2) $neg_p(l) = \neg l^+ \leftarrow (\neg l)^+, \neg(atom(l))^p$ for each literal l;
(3) $\neg a^p \leftarrow not\ a^p$ and $a^p \leftarrow not\ \neg a^p$ for each atom $a \in \mathcal{A}$;

We define P^{pp} as the program P^p with weak constraints $:\sim a^p$ for each atom $a \in \mathcal{A}$.

Theorem 3. *Let P be an extended logic program. An interpretation I is a pre-ferred restricted four-valued stable model of P restricted by S iff $comp_S^p(I^+)$ is a consistent answer set of P^{pp}, where $comp_S^p(I^+) = comp_S(I^+) \cup \{a^p|a \in S\} \cup \{\neg a^p|a \notin S\}$.*

Proof. Notice that $a^p \in comp_S^p(I^+)$ iff $a \in S$, so $comp_S^p(I^+)$ and $comp_S(I^+)$ have an 1-1 corresponding relation. It is very direct to verify that $comp_S(I^+)$ is a consistent answer set of P_S^+ iff $comp_S^p(I^+)$ is a consistent answer set of P^p.

According to Proposition 3, we show that I is a restricted four-valued stable model of P restricted by S iff $comp_S^p(I^+)$ is a consistent answer set of P^p. The weak constraints of a^p in P^{pp} ensure that we only accept those answer sets of P^p with the minimal cardinality of S, which coincides with the definition of preferred restricted four-valued stable models.   ☐

5   Connection with Default Logic

In this section, we show that our models have very close relation with restricted four-valued default extensions [8], which are also based on the restricted four-valued logic. The restricted four-valued extensions can be calculated by the for-mula transformation approach, which also introduces new atoms like positive forms. The following theorem is a limited version that only considering literals, and using symbol $(\neg a)^+$ instead of a^- in [8].

Definition 9. *For any literal set* E*, let* $\overline{E}_S^+ = \{l^+|l \in E\} \cup \{l^+ \leftrightarrow \neg(\neg l)^+|atom(l) \notin S\})$. *For any default rule* $d = \frac{\alpha_1,...,\alpha_m:\beta_1,...,\beta_n}{\gamma}$*, let* $\overline{d}^+ = \frac{\alpha_1^+,...,\alpha_m^+:\beta_1^+,...,\beta_n^+}{\gamma^+}$. *Let* $T = (D,W)$ *be a default theory. The transformed default theory* $\overline{T}_S^+$ *of* T *restricted by* S*, is defined as* $\overline{T}_S^+ = (\overline{D}^+, \overline{W}_S^+)$*, where* $\overline{D}^+ = \{\overline{d}^+|d \in D\}$.

Theorem 4 ([8]). *Let* $T = (D,W)$ *be a default theory.* E *is a restricted four-valued extension of* T *restricted by* S*, iff* $\overline{E}_S^+$ *is a consistent extension of* $\overline{T}^+$.

We can prove that our stable model semantics can be embedded into restricted four-valued default logic, which is a paraconsistent and coherent expansion of Reiter's default logic.

Theorem 5. *Let* $Th_S(I)$ *be the restricted four-valued deductive closure of* I *restricted by* S*. For any extended program* P*,* I *is a restricted four-valued stable model of* P *restricted by* S *iff* $Th_S(I)$ *is a restricted four-valued extension of* $T(P)$ *restricted by* S*, where* $T(P)$ *is the default theory generated by* P *defined in Proposition 1.*

Proof. It can be verified by definitions that the deductive closure of $comp_S(I^+)$ is $Th(\overline{I}_S^+)$. By Proposition 3, I is a restricted four-valued stable model of P restricted by S iff $comp_S(I^+)$ is a consistent answer set of P_S^+, iff $Th(\overline{I}_S^+)$ is a consistent extension of $T(P_S^+)$ by Proposition 1. The default theory $T(P_S^+)$ can be expanded as $(\{\frac{l_1^+,\cdots,l_m^+:(\neg l_m)^+,\cdots,(\neg l_n)^+}{l_0^+}|r = l_0 \leftarrow l_1 \cdots, l_m, \text{not } l_m, \cdots, \text{not } l_n \in P\} \cup \{\frac{(\neg l)^+}{\neg l^+}|l \text{ is a literal and } atom(l) \notin S\}, \emptyset)$.

On the other hand, by formula transformation in Theorem 4, $Th_S(I)$ is a restricted four-valued extension of $T(P)$ restricted by S, iff $Th(\overline{I}_S^+)$ is a consistent extension of $\overline{T(P)}_S^+$. The default theory $\overline{T(P)}_S^+$ can be expanded as $(\{\frac{l_1^+,\cdots,l_m^+:(\neg l_m)^+,\cdots,(\neg l_n)^+}{l_0^+}|r = l_0 \leftarrow l_1 \cdots, l_m, \text{not } l_m, \cdots, \text{not } l_n \in P\}, \{\neg l^+ \leftrightarrow (\neg l)^+|l \text{ is a literal and } atom(l) \notin S\})$, which has the same extensions of $T(P_S^+)$. $\square$

As a corollary, our preferred restricted four-valued stable models can be embedded into a subset of preferred restricted four-valued extensions. The different is only caused by choosing preferred models by cardinality.

Proposition 4. *For any extended program* P*, If* I *is a preferred restricted four-valued stable model of* P *restricted by* S*, then* $Th_S(I)$ *is a preferred restricted four-valued extension of* $T(P)$ *restricted by* S*, where* $T(P)$ *is a default theory generated by* P *defined in Proposition 1.*

Proof. By Theorem 5, if I is a preferred restricted four-valued stable model of P restricted by S, $Th_S(I)$ is a restricted four-valued extension of $T(P)$ restricted by S. If $Th_S(I)$ is not preferred, there is a restricted four-valued extension

$Th_R(J)$ of $T(P)$ restricted by R and $R \subsetneq S$. By Theorem 5, J is a restricted four-valued stable model of P restricted by R which has a smaller cardinality than I restricted by S. This fact contradicts with that I is preferred.   □

6   Related Works

The research on paraconsistent logic programming is even earlier than answer set semantics [6,10]. We focus on the answer set semantics of extended logic programs, which contains both nonmonotonic negation and classical negation that cause incoherences and incoherences together.

Routley semantics for answer set [14] is a paraconsistent semantics with Nelson's logic N_9 as its underlying logic. Different from their semantics, our works are based on Belnap's four-valued logic. Moreover, we do not only focus on handling inconsistent information, and also on handling incoherent information. This also distinguishes our works with other paraconsistent answer set semantics like [1].

In [18], the authors present their paraconsistent stable model semantics for extended disjunctive programs. They show that their semantics can reason with inconsistent information. For handling incoherent programs, they also provide a semi-stable (SST) model semantics by transforming original programs to disjunctive positive programs and choosing the maximally canonical models. Although this approach has many benefits, it allows redundant models even on consistent and coherent programs. For example, the program $P = \{a, b, c, d \leftarrow \text{not } a, \text{not } b; d \leftarrow \text{not } b, \text{not } c\}$ has two semi-stable models $\{a, b, c, Kb\}$ and $\{a, b, c, Ka, Kc\}$. Despite that, we have only one preferred restricted four-valued stable model $\{a, b, c\}$ which coincides with the only answer set of P.

The semi-equilibrium (SEQ) models [9,15] aim at providing an alternative coherent semantics for logic programming, which is characterized using here-and-there models. In [2], the authors figure out that SEQ-models may not respect modular structure in the rules and solve the issue by refining SEQ-models to use splitting sets. For example, the program $P = \{c \leftarrow b, \text{not } c; b \leftarrow \text{not } a\}$ has two SEQ-models (b, bc) and $(\emptyset, a)$, obviously the second model which rejects b should not be accepted since a does not occur in any heads of rules. Despite that, the only preferred restricted four-valued stable model is $\{b, c\}$ restricted by $\{c\}$, since it is not possible to reject b without proving a first. The other difference is that we treat inconsistent and incoherent as two sides of the coin. As a result, we use the same approach to solve the two problems simultaneously, but do not use paraconsistent logic to solve inconsistences only and use preferences to solve incoherences separatingly. Also, our stable models are strongly relevant to restricted four-valued default logic.

7   Conclusion

In this paper, we present the restricted four-valued stable model semantics on extended logic programs. Our semantics is paraconsistent and coherent which benefit from the paraconsistent underlying logic. The preferred stable models correspond to the stable models if they exist and coherent. To calculate our stable models, we show a transformation which can be solved by existing ASP solvers. We reveal the relation between our models and the restricted four-valued default extensions.

We consider that the idea of restricted semantics could be applied to more applications in logic programming.

Acknowledgments. This work is partially supported by the Advance Programs Fund of Ministry of Education of China and Natural Science Foundation of China.

References

1. Alcântara, J., Damásio, C.V., Pereira, L.M.: A declarative characterisation of disjunctive paraconsistent answer sets. In: ECAI, vol. 16, p. 951. Citeseer (2004)
2. Amendola, G., Eiter, T., Leone, N.: Modular paracoherent answersets. In: Proceedings of Logics in Artificial Intelligence - 14th European Conference, JELIA 2014, Funchal, Madeira, Portugal, 24–26 September 2014, pp. 457–471 (2014). http://dx.doi.org/10.1007/978-3-319-11558-0_32
3. Arieli, O., Avron, A.: The value of the four values. Artif. Intell. **102**(1), 97–141 (1998)
4. Belnap, N.: How a computer should think. In: Ryle, G. (ed.) Contemporary Aspects of Philosophy. Oriel Press Ltd., London (1977)
5. Belnap Jr., N.D.: A useful four-valued logic. In: Michael Dunn, J., Epstein, G. (eds.) Modern Uses of Multiple-valued Logic, pp. 5–37. Springer, Netherlands (1977)
6. Blair, H.A., Subrahmanian, V.: Paraconsistent logic programming. Theor. Comput. Sci. **68**(2), 135–154 (1989)
7. Buccafurri, F., Leone, N., Rullo, P.: Enhancing disjunctive datalog by constraints. IEEE Trans. Knowl. Data Eng. **12**(5), 845–860 (2000)
8. Chen, C., Lin, Z.: Restricted four-valued logic for default reasoning. Knowledge Science, Engineering and Management. LNCS, vol. 9403, pp. 40–52. Springer International Publishing, Cham (2015). http://dx.doi.org/10.1007/978-3-319-25159-2_4
9. Eiter, T., Fink, M., Moura, J.: Paracoherent answer set programming. In: KR, pp. 486–496 (2010)
10. Fitting, M.: Bilattices and the semantics of logic programming. J. Logic Program. **11**(2), 91–116 (1991)
11. Fitting, M.: Fixpoint semantics for logic programming a survey. Theor. Comput. Sci. **278**(1), 25–51 (2002)
12. Gelfond, M., Lifschitz, V.: Classical negation in logic programs and disjunctive databases. New Gener. Comput. **9**(3–4), 365–385 (1991)
13. Leone, N., Pfeifer, G., Faber, W., Eiter, T., Gottlob, G., Perri, S., Scarcello, F.: The DLV system for knowledge representation and reasoning. ACM Trans. Comput. Logic (TOCL) **7**(3), 499–562 (2006)
14. Odintsov, S., Pearce, D.J.: Routley semantics for answer sets. In: Baral, C., Greco, G., Leone, N., Terracina, G. (eds.) LPNMR 2005. LNCS (LNAI), vol. 3662, pp. 343–355. Springer, Heidelberg (2005)

15. Pearce, D.J., Valverde, A.: Quantified equilibrium logic and foundations for answer set programs. In: Garcia de la Banda, M., Pontelli, E. (eds.) ICLP 2008. LNCS, vol. 5366, pp. 546–560. Springer, Heidelberg (2008)
16. Reiter, R.: A logic for default reasoning. Artif. Intell. **13**(1), 81–132 (1980)
17. Ruet, P., Fages, F.: Combining explicit negation and negation by failure via belnap's logic. Theor. Comput. Sci. **171**(1), 61–75 (1997)
18. Sakama, C., Inoue, K.: Paraconsistent stable semantics for extended disjunctive programs. J. Logic Comput. **5**(3), 265–285 (1995)

Combining RDR-Based Machine Learning Approach and Human Expert Knowledge for Phishing Prediction

Hyunsuk Chung, Renjie Chen, Soyeon Caren Han, and Byeong Ho Kang[(✉)]

School of Engineering and ICT, Hobart, TAS 7005, Australia
{David.Chung,renjiec,Soyeon.Han,Byeong.Kang}@utas.edu.au

Abstract. Detecting phishing websites has been noted as complex and dynamic problem area because of the subjective considerations and ambiguities of detection mechanism. We propose a novel approach that uses Ripple-down Rule (RDR) to acquire knowledge from human experts with the *modified RDR* model-generating algorithm (Induct RDR), which applies machine-learning approach. The modified algorithm considers two different data types (numeric and nominal) and also applies information theory from decision tree learning algorithms. Our experimental results showed the proposing approach can help to deduct the cost of solving over-generalization and over-fitting problems of machine learning approach. Three models were included in comparison: RDR with machine learning and human knowledge, RDR machine learning only and J48 machine learning only. The result shows the improvements in prediction accuracy of the knowledge acquired by machine learning.

Keywords: Phishing prediction, RDR · Knowledge-based system · Machine learning · Decision tree

1 Introduction

An accelerative growth of Internet-based financing increases online fraudulent activity in which malicious people tries to reveal sensitive information of Internet users, also called as phishing. Phishing detection has received great attention but there has been limited research on a way of overall success due to the nature of problems. The problems of detecting phishing websites are very complex and hard to analyze as technical and social problems are joining each other [1]. Either machine learning technique and human expert system has been applied to acquire and maintain the knowledge for phishing website detection and prediction while the results do not show significance. A large number of knowledge-based systems are built for acquiring and maintaining the knowledge for detecting and predicting the phishing website. Phishing website detection knowledge was originally acquired from domain experts. However, acquiring knowledge from an expert in a slow pace cannot meet the demand of the expanding systems since a sophisticated expert system may require an extremely large number of rules. This leads to machine learning based approach as a solution to manage knowledge-based systems. Although machine learning technique can acquire knowledge from data without the help of a domain expert and an abundance of classifier models exist and

© Springer International Publishing Switzerland 2016
R. Booth and M.-L. Zhang (Eds.): PRICAI 2016, LNAI 9810, pp. 80–92, 2016.
DOI: 10.1007/978-3-319-42911-3_7

decision tree based algorithms provide the best performance, over-generalization and over-fitting are still significant problems when sufficient training data are not available so there are not enough patterns which can be found by machine learning. Therefore, large effort usually has to be undertaken to cover those abnormal cases arising from this problem and the cost usually results in repeating reconstruction of the knowledge base [4, 5]. The aim of the research is to find a way to optimize the knowledge acquired by machine learning to deduct the large cost of solving over-generalization and over-fitting problems. Machine learning is used because acquiring human expert knowledge becomes insufficient when a system expands swiftly. In the opposite direction, if machine learning is not perfect, human knowledge might help to improve the knowledge acquired by machine learning. We assume combining two different mechanisms of having machine learning and expert system-style knowledge acquisition will optimize knowledge engineering process. Hence, we focused on developing phishing website detection model by applying Induct RDR approach. The proposed induct RDR (Ripple Down Rules) approach allows to acquire the phishing detection knowledge by machine learning, and maintained by human domain expert.

2   Related Work

2.1   Human Knowledge Acquisition

Human knowledge is the knowledge acquired from a domain expert to manage a related knowledge-based system. In typical knowledge-based systems or expert systems, transferring knowledge of the experts is bottleneck of building knowledge base for the systems. This is because the process of transferring expert knowledge into knowledge base of the system requires rich resources and the knowledge engineer requires to fully-understand the domain expert knowledge to construct the knowledge base [6]. These knowledge-based systems are built with large structures of concepts and rules, and the difficulty of interacting new arising circumstances with existing rules. In Ripple-down Rules (RDR), its unique knowledge acquisition process solves the problems lie on knowledge engineering process. RDR is built with rules of hierarchical exceptions [16]. It is a knowledge acquisition and representation technique that allows knowledge of a certain domain to be interpreted as rules. The RDR structure is a finite binary tree where each node can have two distinct branches, which are called *except* and *if-not*. Cases are evaluated from the root node of the RDR tree. Each node in the tree is a rule with the form of if α then β (α is the condition and β is the conclusion). When the system encounters an incorrect classification, a new exception rule is added based on experts' judgment [7] with the given case. Therefore, RDR can incrementally develop a relatively accurate knowledge base, provided the domain is fixed and the experts provides the correct judgments [8]. Since RDR based knowledge base depends on experts' judgment, the correctness of the used language expressed by the expert is the key of developing a good knowledge base [17]. According to Pham and Hoffmann [8], it may cost a long time to classify most of the relevant cases correctly, if the target is linear threshold in the numerical input space then an expert is only allowed to use axis-parallel cuts, since it is unsuitable for him to express accurately.

2.2 Knowledge Acquisition by Machine Learning

Knowledge is traditionally collected as rules through sustained interaction between domain specialists and knowledge engineers. However, acquiring knowledge from an expert in a slow pace cannot meet the demand of the expanding systems since a sophisticated expert system may require an extremely large number of rules. This leads to machine learning based approach as a solution to developing knowledge.

Decision Tree Learning. According to Quinlan [2], as one of the technology for building knowledge-based systems by learning from cases, decision tree has been demonstrated successfully. Decision tree based classifiers are used in many areas such as radar signal classification, character recognition, remote sensing, medical diagnosis, expert systems, and speech recognition etc. [3]. ID3 (Iterative Dichotomiser 3) is a typical algorithm to synthesize decision trees in knowledge-based systems. Since there are usually many attributes and the training dataset contains many cases, different decision trees can be created from the same dataset. The fundamental idea is to disperse a complicated decision into a collection of several simpler decision, with the final solution resembling the intended desired solution [3]. A subset of the training dataset is chosen randomly to form a decision tree in which all objects in the subset is correctly classified. Then all other objects are classified using the tree. A subset of those objects is chosen randomly in the same way and the process continues. The random selection is based on two assumptions: (1) The probability of an arbitrary object being determined to belong to class P equals to the proportion of class P in the dataset, and (2) The decision tree is generated by the expected information of the chosen objects. The mentioned expected information is measured by entropy, information gain or gain ratio of the features (attributes) of the chosen objects [9] Compared with categorical attributes, numeric attributes seem to be more difficult to evaluate since they are continuous and the threshold can be arbitrary. In C4.5, the training cases are first sorted by values of the attribute. If the number of the attribute values is m, and the values are sorted as $v_1, v_2, ..., v_m$, there are only $m - 1$ ways to split the dataset into two subsets. Each possible split is examined by their information gain or gain ratio to find the best split. The midpoint $(vi + vi + 1)/2$, in which vi is the largest value of the first subset and vi + 1 is the smallest value of the second, is not chosen for the threshold in C4.5. It chooses the largest value of the attribute in the entire training dataset that does not exceed the midpoint [10].

Induct RDR. Induct RDR was introduced by Gaines when illustrating a fundamental relation between techniques that transfer existing knowledge from human experts and those that create new expertise through machine learning [11]. He mentioned a sequence of dispersing knowledge partially from the view of a human expert, which consists of the following seven stages: (1) Minimal Rules, (2) Adequate Rules, (3) Critical Cases, (4) Source of Cases, (5) Irrelevant Attributes, (6) Incorrect Decisions, and (7) Irrelevant Attributes & Incorrect Decisions. The first stage is a complete, minimal set of correct decision rules so no data is required for knowledge acquisition since the correct answer is available from the expert. On the contrary, the last stage is a source of data from which the correct answer might be derived with the greatest probability of correct decisions so the expert has provided little. The stages in the middle from top to bottom show a decrease in existing knowledge though human intervention but an increase in new expertise through machine learning [11]. The main use of existing RDR is close to the top stage. Therefore, Induct RDR which derives rules directly from an extension of Cendrowska's Prism algorithm [12] was made to be close to the bottom. This Induct RDR sums standard binomial distribution as the possibility of selecting correct data at random to measure the correctness of a rule. In supervised learning, there is a risk of over-fitting the noise by memorizing the peculiarities of the training data [4]. Pruning methods are commonly applied to solve the problem. Although Induct RDR recognizes the importance of pruning, it only removes redundant clauses and compresses the structure to some extent. Reducing over-fitting and improving generalization prediction capability has not been considered [13]. Ripple-Down Rules classifier (Ridor) is an implementation of Induct RDR in Weka (Weka is a tool which provides a collection of machine learning algorithms). It first creates the default rule. The exceptions are created for the default rule with the lowest (weighted) error rate [14]. Different from the original Induct RDR, Ridor applies information gain to evaluate each rule and it prunes a rule by reduced error pruning.

3 Methodology

3.1 Overview

Although machine learning can acquire knowledge from data without the help of a domain expert, over-generalization and over-fitting is still a significant problem when sufficient training data are not available. Because insufficient data may end up with small amount of patterns that cannot be used for rule generation. Therefore, it takes a huge effort to cover abnormal cases arising from the problem [4, 5]. If machine learning is not perfect, human knowledge might help to improve the knowledge base constructed by machine learning approach. In order to see the results following experiments were conducted:

- Although human knowledge acquired can be incorrect or different among different experts, but the knowledge base built can still be further refined. Therefore, for the experiment we allow that the human knowledge we use is not perfect.
- Ideally speaking, when given all the patterns, we can obtain the correct knowledge base by machine learning. However, it is practically not very possible. For the experiment, we consider the dataset we use as whole is sufficient in patterns and the knowledge based on the entire dataset is correct.
- Knowledge based on part of the dataset tends to be incorrect because insufficient data results in difficulties in finding all necessary patterns.

The flow of how to combine human knowledge and machine learning is illustrated in the following way:

- Generate rules by machine learning via training dataset. (Modified Induct RDR)
- Given testing dataset, find out incorrectly classified data.
- Acquire rules from the expert.

- Use those rules (human knowledge) to add exception rules for the machine learning rules where data are incorrectly classified.

If we need to use human knowledge to improve machine learning, we need to find an intermediate between them. That is the reason that we propose RDR as the method (Fig. 1).

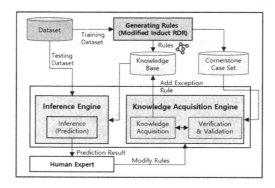

Fig. 1. The flow of combining human knowledge and machine learning

3.2   Induct RDR Modification (Modified Induct RDR)

The modified Induct RDR is based on Induct RDR algorithm that is the third generation of a family of rule induction algorithms [12]. The big picture of the algorithm is a rule generating function where a RDR structure is constructed through the process. A rule at one single RDR node is called a clause and a clause is a collection of one or more terms with the form of attribute-relation-value. Rule generating function follows the following three steps:

- The class that occurs most frequently in the training data is selected as the default class value for the top-level empty rule.
- It finds the clause at each node by searching any class other than the default class that has the smallest m-value (based on standard binomial distribution) to split the training set into two subsets: all true cases for the rule, and all false cases.
- If either of these subsets contains more than one class, rule-generating function is called recursively on the subset. The selected class is used on the first subset as the default class and the current default class is used on the second subset.

Best Clause Selection.   The best clause function is the core of Induct RDR algorithm. Given a specified class, the original function searches all possible combinations of terms to find a particular set of terms which fit the best for covering the class. m function is called to assess the quality of a term. The result of m function is called m-value. The number of cases selected by the combination of terms and z is the number of those examples that are actually needed (true positive). Terms are assessed and qualified terms are added to the clause (combination) until it only selects true positive examples (when

$z = s$). However this procedure might be very computationally intensive because when the number of attributes is large, the number of combinations increases exponentially. In order to solve the problem, terms are first ordered in the modified Induct RDR. Since m-values are the criteria for whether or not adding a term to the clause and the m-value should be minimized, terms can be sorted by m-value in ascending order. Therefore, the possible best terms will be always combined and assessed first, which contributes to finding the best clause within shorter time. This way is much more efficient but it does not prefer combinations such as a good term with a bad term. It might miss the best clause formed by a good term combined with a bad term, but if this combination does not perform better than others, the impact is small.

Best Clause Evaluation. At the end of best clause function, terms are removed from the clause until the m-value is minimized (Induct RDR pruning). Due to the above change, there is no need to search every combination so this part has been moved into the loop to stop the searching at an early age if a new term makes the clause worse than before adding it. The final part of the algorithm is m function, which assesses the correctness of the combination of terms (the clause). m function uses the following formula which sums the standard binomial distribution to calculate m-value.

$$\cdot \quad m'(S) = \sum_{i=z}^{\min_{(s,k)}} {}^sC_i \left(\frac{k}{n}\right)^i \left(1 - \frac{k}{n}\right)^{s-i} \tag{1}$$

n is the number of the whole training set.
k is the number of data which need to be selected.
s is the number of data which are actually selected.
z is the number of data which are correctly selected.

According to Gaines [11], the advantage of using m-value as a measure of the correctness of a rule (terms or clause) is because the probability that the rule could be this good at random, and that it involves no assumptions about the problem such as sampling distributions. He also points out when $s = k$ in the above formula which means all data selected are correct, then $\log(m) = s \log(k/n)$ which seems to be the basis of 'information-theoretic' measures. However, when the dataset is too large, m-values of all rules become to 0, which means that there is no difference in choosing different rules. In this case, the best clauses are just chosen randomly. Verified by decision tree learning algorithms, in fact choosing attributes having larger information gain can still help improving prediction accuracy. In the modified Induct RDR, when the training dataset is too large so that m-value equals to 0, information gain is used to evaluate the best clause instead.

Numeric Number Handling. When the dataset is too large, m-values of all rules become to 0, which means that there is no difference in choosing different rules. In this case, the best clauses are just chosen randomly. Verified by decision tree learning algorithms, in fact choosing attributes having larger information gain can still help improving prediction accuracy. In the modified Induct RDR, when the training dataset is too large so that m-value equals to 0, information gain is used to evaluate the best clause instead.

In the original ID3 algorithm, information gain is used not only to divide the numeric data but also to generate classification rules to make the decision tree through both nominal and numeric data. Due to the different nature of Induct RDR algorithm, the method can only be used to handle the numeric data; hence the following two points were proposed.

- Decision tree focuses on each attribute so it only has one rule (term) at each node while Induct RDR focuses on the combination of terms so it has a clause, which contains one or more terms.
- Induct RDR already uses m-value to measure the quality of the clause so it is not suitable to use another method.

3.3   Knowledge Acquisition

For the dataset used in the research, the actual human experts are less available so we propose to use a simulated expert. Apart from the knowledge acquired by RDR, the knowledge acquired from the simulated expert is treated as human knowledge and the incorrectly or insufficient conclusions will be replaced or supplemented. The whole dataset is divided into a training dataset and a test dataset. The training dataset is used to generate the original knowledge base. The test dataset is used to examine the prediction accuracy of the knowledge. Figure 2 shows the RDR rule tree, where red-circled nodes indicate those nodes with poor prediction of accuracy.

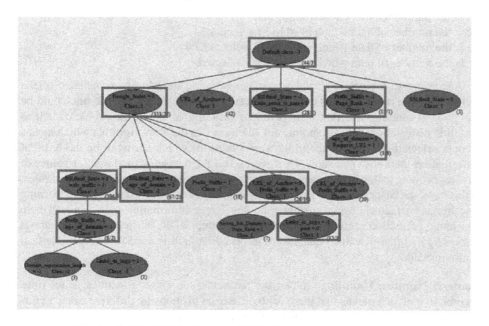

Fig. 2. Original RDR rule tree with highlighted nodes to be modified

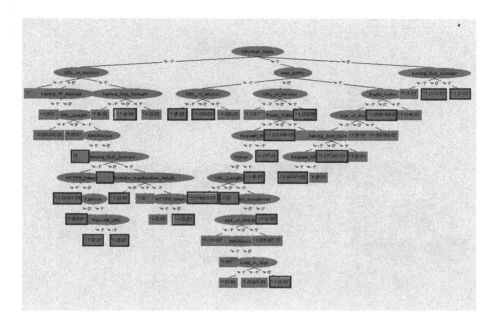

Fig. 3. Knowledge tree built from a simulated expert

These nodes should be modified by adding a new branch to the node, deleting the node, or deleting one of the branches of the node. It is supposed that when using the whole dataset, the knowledge acquired is correct so the human knowledge is acquired from the simulated expert based on the whole dataset. In Fig. 3, the highlighted nodes are the rules, which have same or similar conditions as the ones above.

Not all of the human knowledge can be applied. There are two reasons summarized in the following list.

- There are data, which have the same vector of attributes but belong to different classes. This is because the existing attributes are not enough to tell the difference. Therefore, the class which the majority belong to will be decided as the conclusion and it is less possible to correct the minority.
- Some rules applied might affect other correctly classified data. The knowledge created by the simulated expert gives a hint about how these rules affect the whole dataset. If a rule has more incorrectly classified data than correctly classified data, it should not be applied (Fig. 4).

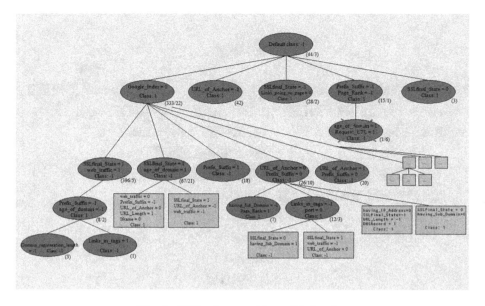

Fig. 4. RDR rule tree with modified nodes

4 Evaluation

4.1 Dataset Selection

We chose 75 datasets from UCI machine learning repository for evaluating the modified Induct RDR with other machine learning models, including Ridor, C4.5 Decision Tree, and NB Tree. The performance of the RDR algorithm combined with human knowledge was verified for the purpose of our experiment. UCI has been published the training dataset that includes important 31 features in detecting and predicting phishing websites. The training dataset contains 11063 websites [15].

– Features/Attributes: having_IP_Address, URL_Length, Shortining_Service, having_At_Symbol, double_slash_redirecting, Prefix_Suffix, having_Sub_Domain, SSLfinal_State, Domain_registeration_length, Favicon, port, HTTPS_token, Request_URL, URL_of_Anchor, Links_in_tags, SFH, Submitting_to_email, Abnormal_URL, Redirect, on_mouseover, RightClick, popUpWidnow, Iframe, age_of_domain, DNSRecord, web_traffic, Page_Rank, Google_Index, Links_pointing_to_page, and Statistical_report.
– Class: Phishing/Non-Phishing.

4.2   The Modified RDR

For the evaluation of the *modified RDR*, we chose the following machine learning models to be included in the comparison. For the *modified RDR*, the minimum number of a subset was set to 2 and the result was collected from its output (Table 1).

Table 1. Models for comparison

Model name	Based algorithm
Modified RDR	Induct RDR
Ridor	Induct RDR
J48	C4.5 (decision tree)
NBTree	Naive Bayes & decision tree

The result of comparing the *modified RDR*, Ridor, J48 and NBTree is listed in Table 2. It can be concluded that the *modified RDR* works better on 66 % of the datasets than Ridor, an existing algorithm based on Induct RDR. Although it is 39 % for J48 and 43 % for NBTree, some datasets with the same accuracy are not counted. Therefore, at least the *modified RDR* has a comparable performance to J48 and NBTree. Besides, it is an inspiring result that the *modified RDR* performs best on 30 % (more than one out of four) of the entire datasets.

Table 2. Comparison result

When	Number of datasets	Proportion
Modified RDR is the best	22	30 % (out of 75 datasets)
Modified RDR is better than Ridor	45	66 % (out of 68 datasets)
Modified RDR is better than J48	29	39 % (out of 75 datasets)
Modified RDR is better than NBTree	27	43 % (out of 63 datasets)

4.3   Combining Human Knowledge with Machine Learning

In order to solve over-generalization and over-fitting problems, which usually affect the prediction accuracy of the knowledge base when unrecognized patterns occur, new knowledge needs to be added to the knowledge base.

The first task of this evaluation was to compare the prediction accuracy between the hybrid way (combing human knowledge and machine learning) and the pure machine learning way. It was because it would become meaningless if adding human knowledge did not help improving prediction. Prediction accuracy of three different models (algorithms) were compared each other: RDR machine learning modified by human knowledge, RDR machine learning only, and J48.

The second task was adding new knowledge to the existing knowledge base. In a knowledge-based system, knowledge is stored in a tree-like structure which consists of nodes, conditions, conclusions and branches. The amount of knowledge can be quantified as the numbers of nodes and conditions which are the main components of a knowledge base, so how much knowledge are reconstructed or added can be quantified as how many nodes and conditions are reconstructed or added. Therefore, the numbers of nodes and conditions can be the objects for the comparison purpose since the larger the numbers are, the more cost has to be spent on constructing knowledge bases. For RDR (machine learning and human rules), the comparison object was the increased number of nodes and conditions by adding new human rules. Human rules were acquired from a simulated expert. For RDR (machine learning only) and J48, the comparison object was the reconstructed number of nodes and conditions by adding new data. Practically speaking, new data are usually found gradually, so we added data cases to the knowledge base one by one and it was reconstructed several times. The total reconstructed number of nodes and conditions was our comparison object. We found that RDR with machine learning only achieved 93.18 % of prediction accuracy, while after adding human rules, the result was improved up to 95.09 %. Although J48 had the best prediction accuracy (94.45 %), RDR with machine learning and human rules outperformed it eventually. Therefore, it can be concluded that adding human knowledge to the knowledge base created by machine learning does improve the quality of the knowledge base (Table 3).

Table 3. Result of prediction accuracy

Models	Prediction accuracy
RDR (ML and human rules)	95.09 %
RDR (ML only)	93.18 %
J48	94.45 %

Table 4. Evaluation result of knowledge increased

Models	RDR (ML and human rules)	RDR (ML only)	J48
Number of nodes original	16	16	28
Number of conditions original	26	26	73
Number of nodes after solving the stated problems	27	77	80
Number of conditions after solving the stated problems	63	119	210
Improved ratio of predication accuracy	2.05 %	3.41 %	0.96 %
Increased ratio of nodes	68.75 %	381.25 %	185.71 %
Increased ratio of conditions	142.30 %	340.74 %	187.67 %
Increased ratio of nodes per 1 % of accuracy improvement	33.54 %	111.80 %	193.45 %
Increased ratio of conditions per 1 % of accuracy improvement	69.41 %	99.92 %	195.49 %

Table 4 summarizes the result of reconstructed or increased nodes and conditions by comparing the above mentioned three models. By applying human knowledge, the increased ratio of nodes for improving 1 % of accuracy was 33.54 %, much smaller than those of RDR (machine learning only) and J48, 111.80 % and 99.92 % respectively.

Similarly the increased ratio of conditions for improving 1 % of accuracy was 69.41 %, much smaller than those of RDR (machine learning only) and J48, 193.45 % and 195.49 % respectively. As mentioned above, the reason that pure machine learning models cost much, is because they abandon the existing knowledge base and create a new one every single time that it encounters a new data case which cannot be explained by the existing knowledge base.

5 Conclusion

We aimed at finding how to optimize the knowledge acquired by machine learning to deduct the large cost of solving over-generalization and over-fitting problems for having the better knowledge base of phishing prediction. The experiment investigated how an approach based on RDR can be the intermediate between human knowledge and machine learning. First comparing with existing machine learning models, our experiments show some interesting facts. These are:

- The *modified RDR* performs better than Ridor which is also based on Induct RDR, especially when a dataset has both numeric and nominal attributes.
- For some datasets, the *modified RDR* performs much better than decision tree learning algorithms.
- The *modified RDR* tends to perform slightly better when a dataset has only numeric or only nominal attributes.
- The *modified RDR* tends to perform better when the ratio of training data to test data is small.

Therefore, as whole, the *modified RDR* performs better than the existing RDR model Ridor. It is used to improve prediction accuracy which might be worsened by over-generalization and over-fitting problems.

Three models were compared: RDR with ML and human knowledge, RDR ML only and J48 ML only. The result shows that applying human knowledge do improve prediction accuracy of the knowledge acquired by machine learning. Our example shows the increased ratio of nodes for improving 1 % of accuracy is 33.54 %, much smaller than using RDR alone and J48 (111.80 % and 99.92 % respectively).

Acknowledgement. This paper was supported by the grant FA2386-15-1-6061, funded by Asian Office of Aerospace Research and Development (AOARD), Japan. This work was supported by the Industrial Strategic Technology Development Program, 10052955, Experiential Knowledge Platform Development Research for the Acquisition and Utilization of Field Expert Knowledge, funded by the Ministry of Trade, Industry & Energy (MI, Korea).

References

1. Aburrous, M., Khelifi, A.: Phishing detection plug-in toolbar using intelligent Fuzzy-classification mining techniques. In: The International Conference on Soft Computing and Software Engineering [SCSE 2013]. San Francisco State University, San Francisco (2013)

2. Quinlan, J.R.: Induction of decision trees. Mach. Learn. **1**(1), 81–106 (1986)
3. Safavian, S.R., Landgrebe, D.: A survey of decision tree classifier methodology (1990)
4. Dietterich, T.: Overfitting and undercomputing in machine learning. ACM Comput. Surv. (CSUR) **27**(3), 326–327 (1995)
5. Pham, H.N.A., Triantaphyllou, E.: The impact of overfitting and overgeneralization on the classification accuracy in data mining. In: Soft Computing for Knowledge Discovery and Data Mining, pp. 391–431. Springer (2008)
6. Compton, P., Jansen, R.: Knowledge in context: a strategy for expert system maintenance. In: Barter, C.J., Brooks, M.J. (eds.) AI 1988. LNCS, vol. 406, pp. 292–306. Springer, Heidelberg (1998)
7. Nguyen, D.Q., Nguyen, D.Q., Pham, S.B., Pham, D.D.: Ripple down rules for part-of-speech tagging. In: Gelbukh, A.F. (ed.) CICLing 2011, Part I. LNCS, vol. 6608, pp. 190–201. Springer, Heidelberg (2011)
8. Pham, S.B., Hoffmann, A.: A new approach for scientific citation classification using cue phrases. In: Gedeon, T(.D., Fung, L.C.C. (eds.) AI 2003. LNCS (LNAI), vol. 2903, pp. 759–771. Springer, Heidelberg (2003)
9. Mazid, M.M., Ali, S., Tickle, K.S.: Improved C4. 5 algorithm for rule based classification. In: Proceedings of the 9th WSEAS International Conference on Artificial intelligence, knowledge Engineering and Data Bases. World Scientific and Engineering Academy and Society (WSEAS) (2010)
10. Ruggieri, S.: Efficient C4. 5 [classification algorithm]. IEEE Trans. Knowl. Data Eng. **14**(2), 438–444 (2002)
11. Gaines, B.R.: An ounce of knowledge is worth a ton of data: quantitative studies of the trade-off between expertise and data based on statistically well-founded empirical induction. In: ML (1989)
12. Cendrowska, J.: PRISM: an algorithm for inducing modular rules. Int. J. Man Mach. Stud. **27**(4), 349–370 (1987)
13. Joshi, M.V., Kumar, V.: CREDOS: classification using ripple down structure (a case for rare classes). In: SDM. SIAM (2004)
14. Devasena, C.L., et al.: Effectiveness evaluation of rule based classifiers for the classification of iris data set. Bonfring Int. J. Man Mach. Interface **1**, 5 (2011)
15. Mohammad, R.M., Thabtah, F., McCluskey, L.: Predicting phishing websites based on self-structuring neural network. Neural Comput. Appl. **25**(2), 443–458 (2014). Aug 1
16. Han, S.C., Yoon, H.G., Kang, B.H., Park, S.B.: Using MCRDR based agile approach for expert system development. Computing **96**(9), 897–908 (2014). Sep 1
17. Han, S.C., Mirowski, L., Kang, B.H.: Exploring a role for MCRDR in enhancing telehealth diagnostics. Multimedia Tools Appl. **74**(19), 8467–8481 (2015). Oct 1

Early Detection of Osteoarthritis Using Local Binary Patterns: A Study Directed at Human Joint Imagery

Kwankamon Dittakan[1(✉)] and Frans Coenen[2]

[1] Faculty of Technology and Environment, Prince of Songkla University, Phuket Campus, 80 Moo 1, Vichit-Songkram Road, Kathu, Phuket, Thailand
`kwankamon.d@phuket.psu.ac.th`
[2] Department of Computer Science, University of Liverpool, Liverpool L69 3BX, UK
`coenen@liverpool.ac.uk`

Abstract. Osteoarthritis (OA) is a chronic health condition that causes severe joint pain and stiffness; it is a major cause of disability in older people. The risk of OA increases from age 45 and older. Early diagnosis is typically made using X-ray imagery. In this paper an automated mechanism for OA screening is proposed. The fundamental idea is to generate a classifier that is able to distinguish between OA or non-OA images. The challenge is how bast to translate an X-ray image into a form that serves to both captures key information while remaining compatible with the classification process. It is suggested that image texture is the most desirable feature to be considered. The process is filly described and evaluated. The data used for the evaluation was obtained from the right Tibia of 50 female subjects. Excellent results were obtained, recorded AUC values of 1.0.

Keywords: Data mining · Image classification · Medical image analysis and mining · Osteoarthritis screening

1 Introduction

Osteoarthritis (OA) is a degenerative joint disease causing joint inflammation and consequently joint pain and stiffness. OA is a major cause of disability amongst older people (over the age of 45). In (Arthritic Research UK 2013) it was reported that 8.75 millions people in the UK have sought treatment for OA (5 millions in women and 3.5 million in men). World wide figures are not available as not all countries specifically record instance of OA. However, in (Tanna 2004), it was estimated that (worldwide) 9.6 % of men and 18 % of women aged over 60 have "symptomatic OA". Given the ageing global population, the prevalence of OA is anticipated to increase. The OA condition is typically diagnosed by a doctor or clinician by clinical examination supported by X-ray or Magnetic resonance imaging (MRI). Early diagnosis, before external symptoms present themselves, can be conducted using X-ray and MRI imagery. The physical signs

© Springer International Publishing Switzerland 2016
R. Booth and M.-L. Zhang (Eds.): PRICAI 2016, LNAI 9810, pp. 93–105, 2016.
DOI: 10.1007/978-3-319-42911-3_8

of OA include: joint tenderness, creaking or grating (crepitus) sounds, bony swelling, excess fluid, reduced movement, joint instability and muscle thinning (Zahurul et al. 2010).

In this paper a mechanism for automating the process is proposed founded on the concept of medical image analysis. More specifically a process for the early detection of OA is proposed founded on the use of the idea classification applied to human bone imagery. To act as a focus for the work X-ray images of the right Tibia were considered. The fundamental idea is to generate a classifier, using labelled X-ray image data, which can then be applied to detect OA. The dataset used to evaluate the proposed framework was composed of a set of X-ray images obtained from the right Tibia of 50 women, of which 25 were from control (non-OA) individuals and the remaining 25 were from clinically diagnosed OA patients. The main objective of the work presented in this paper is thus to classify X-ray image as being either OA or non-OA with respect to the nature of the presented X-ray image data. The proposed approach offers a number of advantages: (i) speed of processing, (ii) automation (unlikely to to be subject to human error) and (iii) low cost.

The main challenge for X-ray image classification (and medical image mining in general) is identifying the most appropriate representation for the input X-ray images; the representation needs to capture key information while remaining compatible with the classification process. Note that without modern Multicore Computing the medical image of interests are typically too large to be used in their entirety. The idea presented in this paper is thus to use the a texture representation to capture the image data. More specifically the Local Binary Pattern (LBP) concept (Pietikäinen 2005). The proposed process is filly described and evaluated.

The reminder of the paper is organised as follows. In Sect. 2 some related work is briefly presented. An overview of the proposed framework for the early detection of OA is presented in Sect. 3. Section 4 describes the adopted image texture-based representation. The nature of the classification mechanism used is presented in Sect. 5. Section 6 reports on the evaluation of the proposed framework. Finally, a summary, some conclusions and suggestions for future work are discussed in Sect. 7.

2   Related Work

Digital imagery has been increasingly use for medical diagnosis; consequently medical digital images have become of increasing significance to health care. There are a number of image capture mechanisms that can are commonly used, including: (i) X-ray, (ii) Computerised Tomography (CT), (iii) Magnetic Resonance Imaging (MRI), (iv) Ultrasound, (v) Positron Emission Testing (PET) and (vi) Single Photon Emission Computed Tomography (SPECT) (Alattas and Barkana 2015). Example medical application domains, where medical image data has been utilised, include: (i) the identification conditions such a epilepsy using

MRI brain data (Udomchaiporn et al. 2014) and detection and (ii) the classification of brain tumours according to whether they are malignant or benign (Nasir et al. 2014).

Computer-based screening for OA, using medical image data, has not been widely available. In (Soh et al. 2014) a mechanism was reported, with respect to knee OA (the most common form of OA), whereby images were segmented using shape analysis algorithms. A similar approach was used in (Ababneh and Gurcan 2010) but in this case a graph-cut strategy was adopted. In both cases MRI images were used, work has also been reported where X-Ray image data was used (Shamir et al. 2009), as in the case of this paper.

It is worth noting that image analysis is also an important and fundamental component with respect to other analysis domains such as computer vision and pattern recognition where the main goal is to understand the characteristics of an image and interpret its semantic meaning. Image classification is an emerging image analysis technique whereby image classifiers (predictors) are built using pre-labelled training data. The performance of such classifiers depends on: (i) the quality of the training data and (ii) the nature of the preprocessing applied to capture the embedded image information. With respect to the latter the idea is to identify features within the training set in order to define a feature vector space representation. Image features of interest can be categorised into two groups: (i) general features and (ii) domain-specific features (Han and Ma 2002). General features are application independent features such as: (i) colour, (ii) texture and (iii) shape. Domain-specific features are application dependent features, such as elements of the human face when considering face recognition.

With respect to the work presented in this paper a representation used in the context of texture analysis was used. Texture is an important feature with respect to both human and computer vision. There are three principle mechanisms that may be adopted to describe texture in digital images: (i) statistical, (ii) structural and (iii) spectral. The statistical mechanism is concerned with the capture of texture using quantitative measures such as "smoothness", "coarseness" and "graininess". Using structural mechanisms image texture is described using a set of texture primitives or elements (texels) that occur as regularly spaced or repeating patterns. In the case of the spectral mechanism the image texture features are extracted by using the properties of the Fourier spectrum domain so that "high-energy narrow peaks" in the spectrum can be identified (Gonzalez and Woods 2007). The usage of Local Binary Patterns (LBPs), the mechanism adopted with respect to the work presented in this paper, is a texture representation strategy which is both statistical and structural in nature (Pietikäinen 2005). Using the LBP approach a binary number is produced, for each pixel, by thresholding its value with its neighbouring pixels. LBPs offer advantages of robustness with respect to illumination changes and computational simplicity. The LBP method has been used with respect to many image analysis application domains such as face recognition (Hadid 2008), (Zhao and Pietikainen 2007). Because of their significance in the context of the work presented in this paper, LBPs are discussed in further detail in Sect. 4.

3   Proposed Framework

The proposed framework for the generation of the desired OA classifier is presented in this section. A schematic of the framework is given in Fig. 1. The framework comprised three phases (as represented by the rectangular boxes): (i) image representation, (ii) feature discretisation and selection and (iii) classifier generation. The input to the first phase is a collection of X-ray images that have been pre-labeled using domain experts. Recall that each X-ray image has to be represented in an appropriate manner that ensures that the salient features are maintained while at the same time supporting classifier generation. The left rectangular box in Fig. 1 is where the image translation is conducted; the input data is translated into a LBP-based representation, which can then be combined to define a feature space from which feature vectors can be generated. Once the feature vectors have been generated feature discretisation and selection are applied (the centre rectangular box in Fig. 1) in order to reduce the overall size of the feature space by reducing the number of attribute-values to be considered and by pruning those features that do not serve to discriminate significantly between classes. Feature discretisation and selection is considered in further detail in Sect. 5 below. With respect to the work presented in this paper, the Chi-squared feature selection method was used. Classifier model generation is the third phase in the proposed framework (the right rectangular box in Fig. 1) during which the desired classifier is generated (also discussed further in Sect. 5 below).

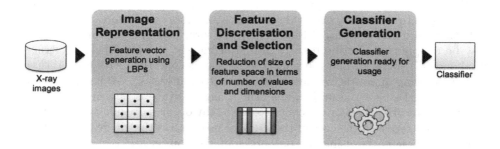

Fig. 1. Schematic illustration the proposed OA classifier generation framework

4   Local Binary Pattern Based Representation

Local Binary Patterns (LBPs) have been widely applied in the context of image processing and computer vision applications where it is important to capture texture information. Examples include: (i) texture segmentation and classification (Chen and Chen 2002), (ii) object detection (Heikkila and Pietikainen 2006) (in images) and (iii) image segmentation (Kachouie and Fieguth 2007),

(Wu et al. 2014). The LBP concept, in the context of image texture analysis, was first introduced by (Ojala et al. 2002). The major advantages pf LBPs are there simplicity, ease of generation and robustness to rotation and monotonic transformation of the adopted colour scale. The fundamental idea is to define each pixel in an image according to its eight cardinal and sub-cardinal neighbours. To generate a set of LBPs from an X-ray image the image is first converted into a linear colour representation, the greyscale representation in our case, if not already in this form. A 3×3 pixel window is then used and moved over the image, with the current pixel at the centre, the centre pixel's greyscale value is then compared with its eight neighbouring greyscale values. If the neighbouring greyscale value is greater than the centre pixel greyscale value a 1 is recorded, otherwise a 0 is recorded (Heusch et al. 2006). In this manner an eight digit binary number is defined describing each pixel according to its 3×3 pixel neighbourhood. In other words 256 (2^8) different patterns can be described. The process is demonstrated in Fig. 2.

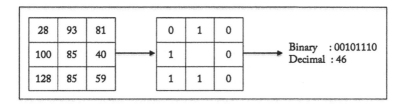

Fig. 2. LBP generation

More formally we can describe an LBP in terms of an ordered set of binary comparisons of greyscale values between the centre pixel of the window with its eight surrounding neighbourhoods. The process can be expressed as shown in Eq. 1:

$$LPB(x_c, y_c) = \sum_{n=0}^{7} f(i_n - i_c)2^n \tag{1}$$

where i_c is the greyscale value of the centre pixel (x_c, y_c), i_n is the greyscale value of a neighbourhood pixel and $f(x)$ is defined according to Eq. 2 below:

$$f(x) = \begin{cases} 1, & if \ x \geq 0 \\ 0, & if \ x < 0 \end{cases} \tag{2}$$

In the example given in Fig. 2, a 3×3 immediate neighbourhood is considered. However, variations of the basic LBP concept can be produced by: (i) using different *radii of neighbourhoods* (R) and/or (ii) different numbers of sampling points (P). The variations can be described using the notation $LBP_{P,R}$. With respect to the work presented in this paper $P = 8$ and a range of values for R was used. Some examples variations are shown in Fig. 3. With reference to the figure: (i) LBP$_{8,1}$ equates to 8 sampling points within a radius of 1 (Fig. 3(a)), (ii) LBP$_{8,2}$, equates to 8 sampling points within a radius of 2 (Fig. 3(b)), and (iii) LBP$_{8,3}$, equates to 8 sampling points within a radius of 3 (Fig. 3(c)). Note that we kept the value of P constant at 8 because this conveniently resulted in a 8 bit integer. Alternatives might have been $P = 4$ or $P = 16$.

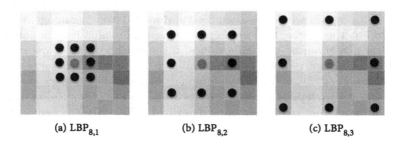

(a) LBP$_{8,1}$ (b) LBP$_{8,2}$ (c) LBP$_{8,3}$

Fig. 3. LBP variations

A set of LBPs can be visualised as a histogram with 256 "bins" on the X-axis, and occurrence count on the Y-axis. Alternatively we can think of a set of LBPs as defining a 256 dimensional feature space, where each element represents a potential LBP value which has an occurrence count associated with it. The latter is the conceptualisation used in this paper.

5   Feature Discretisation, Selection and Classification

Once a collection of LBPs have been identified (as described above) the next step in the framework (Fig. 1) is feature discretisation and selection. However, before this could be commenced data discretisation was applied so as to reduce the number of values in each dimension in the feature space; in other words the continuously valued attributes were converted into a set of ranged attributes. The number of ranges was set to 10 as this was felt to provide a sufficient level of distinction while still considerably reducing the overall number of values to be considered. Feature selection was then applied so as to reduce the number of dimensions in the feature space, but in such a way that the reduced set still provided for a good discrimination between classes. The Chi-squared feature selection mechanism was used with respect to the work presented later in this paper. Note that, in common with many other feature selection techniques, Chi-squared feature selection requires a parameter k, the maximum number of "best" features to be selected.

When the feature selection was completed the images were represented in terms of a set of feature vectors drawn from the reduced feature space to which any number of different classifier generators could be applied. With respect to the evaluation presented later in this paper five classifier generation machine learning methods were considered: (i) Naive Bayes, (ii) Decision Tree (C4.5), (iii) Sequential Minimal Optimisation (SMO), (iv) Back Propagation Neural Networks and (v) Logistic Regression. The implementations used were those available in the Waikato Environment for Knowledge Analysis (WEKA) machine learning workbench (Witten et al. 2011).

6 Evaluation

The evaluation of the proposed approach to OA screening is presented in this section. Extensive evaluation was conducted with respect to the proposed approach. This section reports on only the most significant results obtained (there is insufficient space to allow for the presentation of all the results obtained). The evaluation was conducted by considering a specific case study directed at digital X-ray images taken from the right Tibia of 50 women. In each case the entire X-ray image was not used for the analysis, as it was known that clinicians consider only two particular areas within these X-rays. These two different areas were identified and extracted from the X-rays in the form of sub-images which could be used in isolation or combination. Further detail concerning this image pre-processing is provided in Subsect. 6.1. The overall aim of the evaluation was to provide evidence that the OA condition can be easily detected using the proposed approach. To this end four sets of experiments were conducted with the following objectives:

1. To determine whether if it was better to use the derived sub-images in isolation or in combination.
2. To determined the most appropriate LBP representation with respect to the three variations described in Subsect. 6.3.
3. To identify the most appropriate value for the Chi-squared feature selection k parameter, the maximum number of features to be retained.
4. To provide a comparison of the effectiveness of a number of classifier generation methods. To this end a selection of different classifier generators, taken from the Waikato Environment for Knowledge Analysis (WEKA) machine learning workbench (Witten et al. 2011), was used.

Each of the above objectives is discussed in further detail in Subsects. 6.2 to 6.5 below in the context of the results obtained. Ten fold Cross-Validation (TCV) was applied throughout and performances recorded in terms of: (i) accuracy, (ii) area under the ROC curve (AUC), (iii) sensitivity, (iv) specificity and (v) precision. The results presented in the tables included in the following four subsections are thus all averages over ten runs of cross validation.

6.1 Tibia X-ray Image Data Sets

For the evaluation, as already noted, a training set comprising 50 X-ray images was used represented in greyscale using the DICOM (Digital Image and Communication on medicine) format for X-ray images, a commonly used format in Computed Radiography (CR). An example is given in Fig. 4. The X-ray images were obtained from the right Tibia of 50 women. The data set comprised: (i) 25 control images and (ii) 25 OA images. For each image two sub-images were extracted referred to as the medial and lateral sub images (highlighted in Fig. 4. Detail of these two sub images are given in Fig. 5(a) and (b). From Fig. 5(a) and (b) it is noticeable that it is not obvious that we can detect anything very much from these images. However, it was clear that the sub-images could be used in isolation or in combination. Therefore three individual data sets were created: (i) medial, (ii) lateral and (iii) medial and lateral. The later was generated simply by concatenating the feature vectors from the first two.

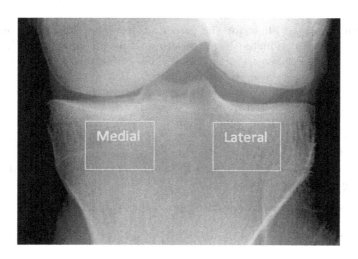

Fig. 4. Examples of right Tibia X-ray image.

6.2   Sub-Image Usage (Individual v. Combination)

This sub-section considers the first of the evaluation objectives, whether it is better to consider the derived sub-images in isolation or in combination. For the experiments the $LBP_{8,3}$ representation was used together with $k = 50$ for the "Chi-squared" feature selection and Logistic regression classification (because experiments reported on later in this paper had indicated that these produced good results). The results are presented in Table 1 (highest values indicated in bold font). From the table it can be observed that best result was obtained by using the sub-images in combination, a best recorded AUC value of 1.000 and sensitivity value of 1.000. Although the sub-images used in isolation also produced reasonable results. Thus, in conclusion, pairing the sub-images makes sense (the more data the better). It should also be noted here that clinicians who analyse Tibia X-ray images typically consider both areas (medial and lateral), thus the results confirm this practice.

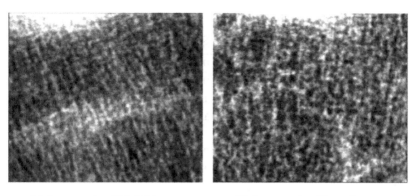

(a) An example of Medial image   (b) An example of Lateral image

Fig. 5. Example sub-images, (a) medial and (b) lateral (b).

Table 1. Results using sub-images in isolation and in combination.

Image set	Accuracy	AUC	Sensitivity	Specificity	Precision
Medial dataset	0.960	0.984	0.960	0.960	0.960
Lateral dataset	0.940	0.994	0.940	0.940	0.941
Medial and Lateral	**1.000**	**1.000**	**1.000**	**1.000**	**1.000**

6.3 LBP Representation

In Subsect. 4 it was noted that variations of the LBP representation can be defined in terms of the values P and R, where P is the number of sampling points and R is the neighbourhood radius; the notation $LBP_{P,R}$ was used to indicate particular variations. This section reports on the evaluation conducted to compare the operation of a range of LBP variations with increasing values of R (P was kept constant because this conveniently resulted in an eight bit integer). For the experiments the combined sub-image dataset was used because previous experiments, reported above in Subsect. 6.2, had indicated that this produced the best performance. Again $k = 50$ was used for the Chi-squared feature selection, together with Logistic regression classification. The results are presented in Table 2 (best results shown in bold font). From the table it can be seen that best results started to be encountered using LBP$_{8,3}$ ($R = 3$); a recorded AUC value of 1.000 and a sensitivity value of 1.000. This continued until $LBP_{8,8}$ was reached when effectiveness started to diminish. There is no clear reason for the anomalous results when using $LPBP_{8,6}$ other than some undefined vagary of the data set. The results indicating that lower values of R did not serve to encapsulate the most effective level of detail. Similarly when R gets too large detail is again lost. It was thus concluded that $LBP_{8,3}$ was the most appropriate LBP variation to adopt.

Table 2. Results using alternative LBP representations

Representation	Accuracy	AUC	Sensitivity	Specificity	Precision
$LBP_{8,1}$	0.980	0.989	0.980	0.980	0.981
$LBP_{8,2}$	0.980	**1.000**	0.980	0.980	0.981
$LBP_{8,3}$	**1.000**	**1.000**	**1.000**	**1.000**	**1.000**
$LBP_{8,4}$	**1.000**	**1.000**	**1.000**	**1.000**	**1.000**
$LBP_{8,5}$	**1.000**	**1.000**	**1.000**	**1.000**	**1.000**
$LBP_{8,6}$	0.980	**1.000**	0.980	0.980	0.981
$LBP_{8,7}$	**1.000**	**1.000**	**1.000**	**1.000**	**1.000**
$LBP_{8,8}$	0.980	0.988	0.980	0.980	0.981

6.4 Best Value for k

This section reports on the evaluation conducted to identify a best value for K with respect to the adopted Chi-squared feature selection method. For the evaluation a sequence of experiments was conducted using a range of k values from 30 to 70 incrementing in steps of 10. For the evaluation the combined sub-image dataset was again adopted together with the the $LBP_{8,3}$ variation (because experiments reported in Subsects. 6.2 and 6.3 had shown that these produced the best results). Logistic regression classification was also again used. The obtained results are presented in Table 3 (best results highlighted in bold font). From the table it can be observed that $k = 50$ produced the best outcomes with respect to all the evaluation metrics considered (a best recorded AUC value of 1.000 and sensitivity of 1.000). Note that accuracy, sensitivity, specificity and precision "drop off" either side of $k = 50$.

Table 3. Results using a range of different values for k with respect to Chi-squared feature selection

k	Accuracy	AUC	Sensitivity	Specificity	Precision
$k = 30$	0.960	0.997	0.960	0.960	0.963
$k = 40$	0.960	0.980	0.960	0.960	0.960
$k = 50$	**1.000**	**1.000**	**1.000**	**1.000**	**1.000**
$k = 60$	0.980	**1.000**	0.980	0.980	0.981
$k = 70$	0.980	**1.000**	0.980	0.980	0.981

6.5 Classification Learning Methods

There are many different kinds of classification paradigms to select from, with no obvious best paradigm for texture-based image classification. This sub-section presents a comparative analysis of a number of these in the context texture-based Tibia X-ray image analysis: (i) Naive Bayes, (ii) Decision Tree, (iii) Sequential Minimal Optimisation (SMO), (iv) Back Propagation Neural Networks and (v) Logistic Regression. For the analysis the combined sub-image dataset was used together with the the $LBP_{8,3}$ variation and $k = 50$ (because earlier reported experiments had shown that these produced the best results). The results of the comparative analysis are presented in Table 4. From the Table it can be observed that, with the exception of decision tree classification, the remaining four selected classifiers all produced good performances. However, best results were obtained using Logistic Regression (AUC value of 1.000 and sensitivity value of 1.000).

Table 4. Results using a range of Classification Learning methods

Classifier generation methods	Accuracy	AUC	Sensitivity	Specificity	Precision
Naive Bayes	0.980	**1.000**	0.980	0.980	0.981
Decision Tree (C4.5)	0.680	0.725	0.680	0.680	0.685
SMO	0.980	0.980	0.980	0.980	0.981
Neural Networks	0.960	0.998	0.960	0.960	0.960
Logistic Regression	**1.000**	**1.000**	**1.000**	**1.000**	**1.000**

7   Conclusions

A framework for the early detection of Osteoarthritis (OA) from X-ray imagery has been presented. The significance is that this can be used for national screening programmes. The main idea presented in this paper is to construct a classification model based on the texture features of X-ray sub-images, which can then be used to predict the OA condition. The key element with respect to this process is the way in which individual X-ray sub-images are represented so that an effective classifier can be generated. To this end the Local Binary Pattern (LBP) texture representation mechanism was adopted with which to encode the sub-image content. To evaluate the proposed approach experiments were conducted using a collection of 50 test datasets obtained from the right Tibia of 50 women (these had been hand-labelled by trained clinicians). The main findings evidenced by the reported evaluation were:

- The proposed approach was extremely effective, AUC and sensitivity scores of 1.000 were regularly recorded.
- Practitioners, in the context of Tibia X-ray images, concentrate on two areas within such images (referred to as the medial and lateral areas). Sub-images were extracted with respect to both these areas and experiments conducted using these sub-images in isolation and in combination; as anticipated using these sub-images in combination produced the best results (a recorded average AUC value of 1.000).

- The most appropriate LBP variation, out of the variations considered, was $LBP_{8,3}$ (recorded AUC value of 1.000).
- The most appropriate k value to be used with respect to Chi-squared feature selection was found to be $k = 50$, in other words the best fifty LBP dimensions should be selected.
- In the context of the most appropriate classification methods to be adopted experiments were conducted using five different classification paradigms. The best performing was to be Logistic regression (again evidenced by AUC results of 1.000).

Thus, in conclusion, the proposed approach produced excellent results indicating a good "way forward" with respect to OA screening. A criticism that might be directed at the presented work is that the evaluation data set was relatively small comprising 50 X-ray images from which 100 sub-images were extracted (the evaluation data set featured a 50 : 50 class distribution). For the future work the research team thus intends to conduct further experiments using larger datasets.

Acknowledgments. We would like to thank Christian Schön from Braincon Handles GmbH, Vienna, for providing us with the image data used for evaluation purposes with respect to the work presented in this paper, and for his valuable comments regrading our results.

References

Ababneh, S., Gurcan, M.: An efficient graph-cut segmentation for knee bone osteoarthritis medical images. In: 2010 IEEE International Conference on Electro/Information Technology (EIT), pp. 1–4 (2011)

Alattas, R., Barkana, B.: A comparative study of brain volume changes in alzheimer's disease using MRI scans. In: 2015 IEEE Long Island, Systems, Applications and Technology Conference (LISAT), pp. 1–6 (2011)

Arthritic Research UK.: Osteoarthritis in general practice – data and perspectives. Technical report, Arthritic Research UK, Copeman House, St Marys Gate, Chesterfield, S41 7TD (2013)

Chen, K.-M., Chen, S.-Y.: Color texture segmentation using feature distributions. Pattern Recogn. Lett. **23**(7), 755–771 (2002)

Gonzalez, R.C., Woods, R.E.: Digital Image Processing, 3rd edn. Pearson Prentice Hall, Upper Saddle River (2007)

Hadid, A.: The local binary pattern approach and its applications to face analysis. In: Proceedings of the First Workshop on Image Processing Theory, Tools and Applications (IPTA), pp. 1–9. IEEE Computer Society (2008)

Han, J., Ma, K.: Fuzzy color histogram and its use in color image retrieval. IEEE Trans. Image Process. **11**(8), 944–952 (2002)

Heikkila, M., Pietikainen, M.: A texture-based method for modeling the background and detecting moving objects. IEEE Trans. Pattern Anal. Mach. Intell. **28**(4), 657–662 (2006)

Heusch, G., Rodriguez, Y., Marcel, S.: Local binary patterns as an image preprocessing for face authentication. In: Proceedings of the Seventh IEEE International Conference on Automatic Face and Gesture Recognition (FGR), pp. 9–14. IEEE Computer Society (2006)

Kachouie, N., Fieguth, P.A.: Medical texture local binary pattern for trus prostate segmentation. In: Engineering in Medicine and Biology Society, EMBS 2007, pp. 5605–5608 (2007)

Nasir, M., Khanum, A., Baig, A.: Classification of brain tumor types in MRI scans using normalized cross-correlation in polynomial domain. In: 2014 12th International Conference on Frontiers of Information Technology (FIT), pp. 280–285 (2014)

Ojala, T., Pietikainen, M., Maenpaa, T.: Multiresolution gray-scale and rotation invariant texture classification with local binary patterns. IEEE Trans. Pattern Anal. Mach. Intell. **24**(7), 971–987 (2002)

Pietikäinen, M.: Image analysis with local binary patterns. In: Kalviainen, H., Parkkinen, J., Kaarna, A. (eds.) SCIA 2005. LNCS, vol. 3540, pp. 115–118. Springer, Heidelberg (2005)

Shamir, L., Ling, S., Scott, W., Bos, A., Orlov, N., Macura, T., Eckley, D., Ferrucci, L., Goldberg, I.: Knee x-ray image analysis method for automated detection of osteoarthritis. IEEE Trans. Biomed. Eng. **56**(2), 407–415 (2009)

Soh, S.S., Swee, T.T., Ying, S.S., En, C.Z., Bin Mazenan, M., Meng, L.K.: Magnetic resonance image segmentation for knee osteoarthritis using active shape models. In: 2014 7th Biomedical Engineering International Conference (BMEiCON), pp. 1–5 (2014)

Tanna, A.: Osteoarthritis "opportunities to address pharmaceutical gap". Technical report, World Health Organisation Archive (2004)

Udomchaiporn, A., Coenen, F., García-Fiñana, M., Sluming, V.: 3-D MRI brain scan classification using a point series based representation. In: Bellatreche, L., Mohania, M.K. (eds.) DaWaK 2014. LNCS, vol. 8646, pp. 300–307. Springer, Heidelberg (2014)

Witten, I.H., Frank, E., Hall, M.A.: Data Mining: Practical Machine Learning Tools and Techniques, 3rd edn. Morgan Kaufmann, Amsterdam (2011)

Wu, C., Lu, L., Li, Y.: A study on pattern encoding of local binary patterns for texture-based image segmentation. In: 2014 International Conference on Machine Learning and Cybernetics (ICMLC), vol. 2, pp. 592–596 (2014)

Zahurul, S., Zahidul, S., Jidin, R.: An adept edge detection algorithm for human knee osteoarthritis images. In: International Conference on Signal Acquisition and Processing, ICSAP 2010, pp. 375–379 (2010)

Zhao, G., Pietikainen, M.: Dynamic texture recognition using local binary patterns with an application to facial expressions. Proc. IEEE Int. Conf. Pattern Anal. Mach. Intell. (TPAMI) **29**(6), 915–928 (2007)

Thai Printed Character Recognition Using Long Short-Term Memory and Vertical Component Shifting

Taweesak Emsawas[✉] and Boonserm Kijsirikul

Department of Computer Engineering, Chulalongkorn University,
Phayathai Rd., Phathumwan, Bangkok 10330, Thailand
taweesak.e@student.chula.ac.th, boonserm.k@chula.ac.th

Abstract. The segmentation-based approach for Optical Character Recognition (OCR) works by first segmenting a text line image into individual character images and then recognizing the characters. The approach relies heavily on the performance of the segmentation process and thus suffers from the problem of touching and broken characters. On the other hand, the unsegmented approach for OCR processes the text line image without segmenting the image into individual characters, and the approach is more suitable for languages such as Thai that contains a lot of touching characters in nature. This paper proposes an application of Long Short-Term Memory (LSTM), which is an unsegmented method, to Thai OCR. The paper also introduces a method called *vertical component shifting* to solve the problem of a large number of vertically occurring character combinations that occur in four-level writing system of Thai, and pose difficulty for standard LSTM networks. The experimental results demonstrate the better accuracy of our proposed method over standard LSTM networks and other commercial software for Thai OCR.

Keywords: Thai printed character recognition · Recurrent neural network · Long Short-Term Memory · Vertical component shifting

1 Introduction

Optical Character Recognition (OCR) is the process that converts text images into machine-encoded text. Approaches for character recognition can be divided into segmented and unsegmented approaches. A segmented approach segments each character image and classifies the character image using a classifier such as support vector machines, feed-forward neural networks, k-nearest neighbor, etc. Then, the classified characters are combined into text lines. This approach needs pre-processing to segment character images and post-processing to combine classified characters into the encoded text. The segmented approach has the problem of connected and broken characters that often appear in the image. To improve the accuracy of results, post-processing with some techniques such as language models are used. In an unsegmented approach, the main idea is to use the contextual information. The connected and broken characters or unclear characters can be solved with the contextual characters. This approach has an advantage

© Springer International Publishing Switzerland 2016
R. Booth and M.-L. Zhang (Eds.): PRICAI 2016, LNAI 9810, pp. 106–115, 2016.
DOI: 10.1007/978-3-319-42911-3_9

that it does not require pre-processing and post-processing. Examples of unsegmented techniques are Hidden Markov Models (HMMs) and Recurrent Neural Network (RNN).

A recurrent neural network (Jaeger 2002) is an artificial neural network that allows the network to contain cyclical connections. An RNN can use its internal connections to memory and process an arbitrary sequence of inputs. The main idea of the RNN is to memorize previous information in the network's internal states by using self-connections. The dependencies of the inputs can remain in the network, However, when processing a long sequence, the information from previous states vanishes in time. This problem is called "Vanishing Gradient Problem" (Hochreiter et al. 2001). Long Short-Term Memory (LSTM) (Hochreiter and Schmidhuber 1997) is designed for solving the problem. LSTM networks can memorize long range dependency by using memory cells with three types of gates (input, output and forget gates). LSTM works well in recognition tasks such as speech recognition, handwriting recognition and character recognition. Several implementations of LSTM networks for character recognition tasks have been proposed such as online and offline handwriting recognition using LSTM in comparison to HMM (Schmidhuber 2008), multidimensional LSTM in Arabic handwriting recognition (Graves 2012) and bidirectional LSTM networks in English and Fraktur character recognition (Breuel et al. 2013).

This paper proposes an application of LSTM to printed Thai character recognition. In Thai language, each line of text is written in four levels, and this causes a lot of vertically-occurring character combinations which pose difficulty for an LSTM network to learn and produce good generalization models for unseen combinations. Here, we propose a method called vertical component shifting to rearrange the positions of vertically occurring characters so that the number of vertically-occurring character combinations is reduced and thus this will help an LSTM network generalize better. We evaluate our proposed method by comparing LSTM networks with and without vertical component shifting. We also conduct experiments to compare our method with commercial software for Thai OCR, i.e., ABBYY (2012) and ArnThai (NECTEC 2008). The experimental results show that the proposed method outperforms the others.

This paper is organized as follows. Section 2 introduces Thai character recognition and the problem of vertically-occurring character combinations. Section 3 describes LSTM networks, bidirectional LSTM networks and vertical component shifting. Section 4 provides three experiments to evaluate the performance of LSTM networks with/without vertical component shifting and to compare the results to commercial software for Thai OCR. Finally, Sect. 5 provides the conclusion and direction for future works.

2 Thai Character Recognition

Thai language has 44 consonants, 22 vowels, 4 tones, 10 numbers and symbols. Each line of text is written in (e.g. อรุณสวัสดิ์) four-levels including top, upper, middle and lower levels as shown in Fig. 1. Consonants are placed the in the middle level from left to right. Vowel or tone characters can appear above, below, to the left, to the right of a consonant.

Top level
Upper level
Middle level
Lower level

Fig. 1. An example of Thai words written in four levels: "อรุณสวัสดิ์" (Good Morning)

This four-level language structure causes the problem for Thai character recognition; there are a lot of vertically-occurring character combinations which pose difficulty for recognition. Another problem of Thai character recognition is that Thai text often contains a lot of touching characters. For example, in word "ไม้ไผ่" (Bamboo), vowel "◌้" is connected to vowel "ไ", which is difficult to be classified. Word segmentation (Bheganan et al. 2009) is another problem for the recognition of Thai text.

Most printed Thai character recognition research worked on the segmented approach. For example, "Benchmark for Enhancing the Standard of Thai language processing (BEST)" 2013 and 2014 (Methasate and Marukatut 2013) is a Thai-printed character-recognition competition. In this competition, several techniques are used such as histogram of gradient (HOG) with SVM (Siriteerakul 2013) and the baseline method which uses k-NN, KD-tree (Methasate and Marukatut 2013) and approximate Euclidean distance (Marukatat and Methasate 2013). In addition, HMM was used in many recognition tasks such as word segmentation and text line recognition.

3   Long Short-Term Memory and Vertical Component Shifting

In this section, we propose the use of bidirectional LSTM networks for Thai character recognition. First, the text image is split into lines using y-derivative of a Gaussian kernel (see (Rashid et al. 2012) for details). A text line image is then binarized into foreground and background pixels and the input vector sequence is constructed for feeding into an LSTM network. To construct the input sequence, the text line image is divided into vertical windows, or frames. The width of a window is set to be one pixel, and the height is the height of the text line image. The sliding window is shifted along the text line image to construct an input vector of each frame. After having been trained, the network can be used to predict the output sequence. Finally, the Connectionist Temporal Classification (CTC) (Grave et al. 2006) is employed to map the output sequence into the encoded text.

We first briefly describe Long Short-Term Memory (LSTM), and then propose our method, i.e. vertical component shifting, that rearranges a text line image to help LSTM efficiently learn for Thai four-level writing system.

3.1   Long Short-Term Memory

The benefit of an RNN is the ability to use contextual information from the input sequence to label the output sequence. The vanishing gradient problem occurs in RNN when processing a long sequence input. Long Short-Term Memory (LSTM) is an architecture that preserves information overtime by memory cells and gate units. The illustration of memory blocks is shown in Fig. 2. A memory block has three gates: input gate, output gate and forget gate. An input gate controls information from the current state input to the memory cell. An output gate controls information from the memory cell to other nodes. A forget gate controls information between the memory cell states. The memory cell keeps the constant value or information with internal self-connections. The network training uses a forward-backward algorithm (Grave 2008). In the forward pass, the input vector is applied to nodes and its information is propagated through the LSTM network layer by layer until the network produces the output. In the backward pass, the network weights are tuned by using Back-Propagation Through Time (BPTT).

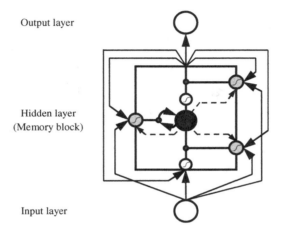

Output layer

Hidden layer
(Memory block)

Input layer

Fig. 2. A simple LSTM network containing one node in each layer. The center rectangle represents a memory block. The memory block contains a memory cell (black circle) and three gates: input gate (bottom right gray circle), output gate (top right gray circle) and forget gate (left gray circle). This network has 4 weights between the input layer and the memory block, 5 weights for self-loop connections and 1 weight between the memory block and the output node. The dash line represents an internal weighted connection. The information from the input node is inputted into the memory block and multiplied by multiplication (small black circle). Then, the information is multiplied by the previous state information from the forget gate and sent to external nodes as output information.

3.2 Bidirectional LSTM Network

To classify a particular character, sometimes it is useful to know not only the information of the characters coming after it, but also the ones before it. Standard LSTM networks can use only the previous information in recognizing a character in consideration. To use two-ways information (both the past and the future contexts), we choose bidirectional LSTM networks for our task. A bidirectional LSTM network is separated into two unidirectional LSTM networks which process the sequence inputs from left to right and right to left respectively. Both of them are connected to the same output layer. A bidirectional LSTM network has been shown to be more efficient than a unidirectional LSTM network in many tasks (Grave et al. 2005).

In our task, the sliding windows for constructing input vectors of the text line images process from left to right and right to left. The illustration of an unfold bidirectional LSTM network overtime with the sliding window of one pixel width is shown in Fig. 3. When a sliding window processes on a frame containing more than one character, the network has to learn a combination of vertically-occurring characters. Since the number of possible patterns of vertically-occurring characters in Thai language is very large, a huge set of training data is necessary that is very difficult to be collected and labeled in practice. Below, we propose a method of vertical component shifting to remedy the problem.

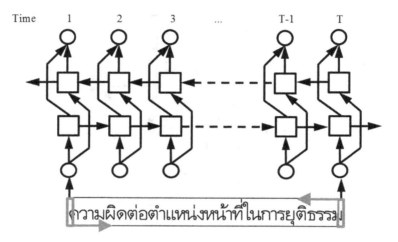

Fig. 3. An unfold bidirectional LSTM network with three layers containing one node in each input and output layers and two memory blocks (square) in the hidden layer. Two sliding windows (grey rectangles) of one pixel width are used to construct input vectors from time 1 to T (the width of the text line image), and T to 1.

3.3 Vertical Component Shifting

To solve the problem of a large number of vertically-occurring character combinations in Thai four-level writing system, we propose a method called vertical component shifting. The main idea is to reduce the number of the character combinations. Though an LSTM network is able to learn touching characters, too many combinations of vertically-occurring characters decrease the generalization ability of the network; the network should to be trained by all possible combinations which is very difficult to be achieved in practice. Vertical component shifting attempts to reduce the patterns of vertically-occurring characters by separating the vertically-occurring characters into individual components. The algorithm is shown in Fig. 4.

```
Algorithm: Vertical Component Shifting

1. Find the connected components of the input image with
   4-connectivity.
2. For Each connected component DO
      center_line = 1/2 x the height of the line image

      IF the component is above center_line THEN
            Label it as upper_and_top_component
      ELSE IF the component is below center_line THEN
            Label it as lower_component
      ELSE Label is as base_component.
3. For Each lower_component and upper_and_top_component DO
      3.1 Find the corresponding base_component
      3.2 Reorder the components to be a sequence of
          (base_component, lower_component,
           upper_and_top_component)
      3.3 Expand the image of vertically occurring
          components with the sequence of non-overlapping
          reordered components.
```

Fig. 4. The algorithm of vertical component shifting

Using a standard connected-component detection algorithm, a text line image is separated into connected components. Each connected component is then labeled as *base_component* (a component in the middle level), *lower_component, upper_and_top_component* according to its position in the text line. Vertical component shifting then scans for vertically occurring components composed of (1) *base_component* and (2) components, which vertically overlap with the base component, with labeled as *lower_component* and/or *upper_and_top_component*. These components are then horizontally shifted such that the components do not vertically overlap with each other. The order of the shifted components is *base_component, lower_component, upper_and_top_component*. An example of vertical component shifting is shown in Fig. 5.

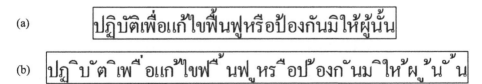

(a)

(b)

Fig. 5. Thai text line images (a) before and (b) after vertical component shifting

Note that the objective of vertical component shifting is not to segment the touching characters, but instead it aims to reduce the combination patterns of vertically occurring characters for the increase of generalization ability of the trained LSTM network. The advantage of LSTM as an unsegmented approach still remains the same as before.

4   Experiments

In our experiments, we used OCRopus (Breuel 2008) to implement LSTM networks. A dataset was prepared using 11 Thai standard fonts. Each image in the dataset was scanned with 300 dpi and then binarized into black text and white background. The error rate is computed by the Levenshtein distance for measuring the difference between the obtained output text and the ground-truth text at a character level.

4.1   Experiment 1: LSTM Network and Vertical Component Shifting

To evaluate the performance of LSTM networks and vertical component shifting, we trained two bidirectional LSTM networks with and without vertical component shifting for comparison. In our dataset, the training set contained 187 text line images or 3,436 characters and the test set was composed of 4,298 text line images or 260,250 characters. The text line images were size-normalized to 48 pixels height. In the experiment, each bidirectional LSTM network had 48 input nodes corresponding to the input vector and 131 output nodes corresponding to the output character.

To determine an appropriate number of hidden nodes in the hidden layer, we run preliminary experiments to tune the nodes by varying the number of memory blocks (50, 100, 150, 200, 400 or 800 per unidirectional network). The results showed that 50 memory blocks achieved the worst error rate while the other numbers of memory blocks achieved almost the same error rate with different numbers of iterations for convergence. We decided to use a network with 150 memory blocks as it was able to achieve the best error rate and did not consume a long training-time before it converged. The network setup is shown in Fig. 6.

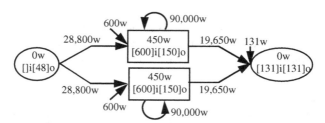

Fig. 6. The structure of the bidirectional LSTM network in our experiments. The network has 48 nodes in the input layer (represented by one big circle on the left) for the input vector of size 48 and two sets of hidden nodes (two big rectangles in the center), each of which represents 150 memory blocks. 131 nodes in the output layer (represented by one big circle on the right) are for the output vector of size 131. The number of weights between the input layer to each memory block is 28,800. The self-loop connections in each memory block have 90,000 weights and 600 bias weights. The number of weights between each memory block to the output layer is 19,650 and there are 131 bias weights at the output layer.

Using the obtained structure of the network, we then run experiments for training bidirectional LSTM networks 5 rounds. In each round, we selected 80 percent of the training set to train the network and used the remaining training-data as the validation set to determine when to stop training. The trained network was then evaluated by 20 percent of the test set as shown in Fig. 7. Table 1 summarizes the error rates on the test set.

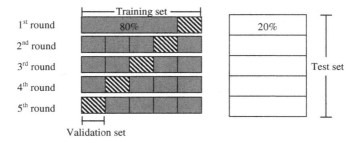

Fig. 7. The five rounds of the experiment. In each round, data in the black blocks and in the lined-pattern block are used as real-training set and validation set, respectively. Twenty percent of the test set is used for each round.

Table 1. Error rate and the number of error characters of three experiments in the same dataset

Experiment	Recognition techniques	Test set (260,250 characters)	
		Error rate	No. of error characters
1	LSTM network	7.012 %	18,248
	LSTM network with VCS	5.024 %	13,075
2	LSTM network	0.607 %	1,578
	LSTM network with VCS	0.564 %	1,467
3	ABBYY	6.739 %	17,538
	ArnThai	18.840 %	49,032

The error rate of the LSTM network without vertical component shifting was 7.012 %. Most of errors were caused by unseen vertically-occurring character-combinations. On the other hand, the LSTM networks with vertical component shifting generalized well by reducing the number of vertically-occurring combinations and its error rate was able to achieve 5.024 %.

4.2 Experiment 2: Large Training Dataset

This experiment also applied the bidirectional LSTM network and vertical component shifting on Thai character recognition. In this experiment, we want to see if vertical component shifting is still useful when we supply the network a large dataset which cover a lot of patterns of vertically-occurring combinations. Therefore, in this experiment, we used 4,298 text line images (the test set from the first experiment), to be the dataset and trained and evaluated the network using 5-fold cross validation. The configuration of LSTM networks was the same as ones in the first experiment. The result of this experiment was shown in Table 1. The experimental result shows that both LSTM networks were able to achieve 0.607 % and 0.564 % error rates on the test set. Most of error characters of LSTM without vertical component shifting were confused characters in vertical direction such as the wrong word "ผู้" from the correct one "ผู้", "ใหญ่" from "ใหญ่", "กิ่ง" from "กิ่ง" and "เรื่อง" from "เรื่อง". The error characters of LSTM network with vertical component shifting were the touching characters.

4.3   Experiment 3: Comparison to Commercial Software

To compare the results with other commercial software, we used the same test set of the text line images as that of the first experiment. For the commercial software, ABBYY (2012) is the best one of optical character recognition software that provides the collection tools in many languages. In this experiment, we used ABBYY version 12 with the default setting of English and Thai languages. ArnThai 2.5 (2008) is also an application that allows a user to convert Thai and English text images into encoded-text. The result was shown in Table 1. The ABBYY system achieved 6.739 % error rate while ArnThai provided 18.840 % error rate with high error rates on KodchiangUPC and LilyUPC fonts. The error rate without these two fonts was 7.344 %. Examples of error characters are shown in Fig. 8.

Fig. 8. Examples of error characters. Input, LSTM w/o VCS, LSTM with VCS, ABBYY, ArnThai are the text line image, the output text of LSTM without vertical component shifting, the output of LSTM with vertical component shifting, the output of ABBYY, and the output of ArnThai, respectively. Words with errors are shown by boldface letters.

5   Conclusion

This paper has shown the performance of an unsegment approach with LSTM networks in printed Thai character recognition. The LSTM networks, which were trained on unsegmented sequence data, achieved the good result without connected and broken characters problem and needed no language model or post-processing. Vertical component shifting has also been proposed to solve the problem of the four-level structure in Thai. The LSTM networks with vertical component shifting achieved the better results than the networks without vertical component shifting. The proposed method also provided significantly higher accuracy than the commercial software packages, i.e., ABBYY and ArnThai. In the future work, we will investigate the use the other techniques for the input of text line images in multi-level structure languages and apply the LSTM network to Thai handwriting recognition.

References

Jaeger, H.: Tutorial on training recurrent neural networks, covering BPPT, RTRL, EKF and the echo state network approach. GMD Report 159, Fraunhofer Institute AIS (2002)

Hochreiter S., Bengio Y., Frasconi P., Schmidhuber J.: Gradient flow in recurrent nets: the difficulty of learning long-term dependencies. In: Kremer, S.C., Kolen, J.F. (eds.) A Field Guide to Dynamical Recurrent Neural Networks. IEEE Press (2001)

Hochreiter, S., Schmidhuber, J.: Long Short-Term Memory. Nueral Comput. **9**(8), 1735–1780 (1997)

Schmidhuber, J.: A novel connectionist system for unconstrained handwriting recognition. IEEE Trans. Pattern Anal. Mach. Intell. **31**(5), 855–868 (2008)

Graves, A.: Offline Arabic handwriting recognition with multidimensional recurrent neural networks. In: Guide to OCR for Arabic Scripts, pp. 297–313 (2012)

Breuel, T.M., Ul-Hasan, A., Azawi, M.A., Shafait, F.: High-performance OCR for printed English and Fraktur using LSTM networks. In: 12th International Conference on Document Analysis and Recognition, ICDAR, pp. 683–687 (2013)

ABBYY (2012). http://www.abbyy.com/ocr-sdk/

ArnThai (2008). http://arnthai-lite.software.informer.com/2.5/

Bheganan, P., Nayak, R., Xu, Y.: Thai word segmentation with hidden markov model and decision tree. In: Theeramunkong, T., Kijsirikul, B., Cercone, N., Ho, T.-B. (eds.) PAKDD 2009. LNCS, vol. 5476, pp. 74–85. Springer, Heidelberg (2009)

Methasate, I., Marukatut, S.: BEST 2013: Thai Printed Character Recognition Competition. National Electronics and Computer Technology Center, Image Technology Laboratory, Thailand (2013)

Siriteerakul T.: Mixed Thai-English character classification based on histogram of oriented gradient feature. In: 12th International Conference on Document Analysis and Recognition, ICDAR, pp. 847–851 (2013)

Marukatat, S., Methasate, I.: Fast nearest neighbor retrieval using randomized binary codes and approximate euclidean distance. Pattern Recogn. Lett. **34**, 1101–1107 (2013)

Rashid, S.F., Shafait, F., Breuel, T.M.: Scanning Nerual Network for Text Line Recognition. In DAS, Gold Coast (2012)

Graves, A., Fernandez, S., Gomes, F., Schmidhuber, J.: Connectionist Temporal Classification: Labeling Unsegemented Sequence Data with Recurrent Nerual Networks, pp. 369–376. In ICML, Pennsylvania (2006)

Graves, A.: Supervised Sequence Labelling with Recurrent Neural Networks. TU Munchen (2008)

Graves, A., Fernandez, S., Schmidhuber, J.: Bidirectional LSTM networks for improved phoneme classification and recognition. In: International Conference on Artificial Neural Networks, Warsaw, Poland, pp. 799–804 (2005)

Breuel, T.M.: The OCRopus open source OCR system. In: DRR XV, vol. 6815, p. 68150F (2008)

Learning Sentimental Weights of Mixed-gram Terms for Classification and Visualization

Tszhang Guo[1,2], Bowen Li[1], Zihao Fu[3], Tao Wan[1,4(✉)], and Zengchang Qin[1(✉)]

[1] Intelligent Computing and Machine Learning Lab, School of ASEE,
Beihang University, Beijing 100191, China
guozikeng@foxmail.com, libowen.ne@gmail.com, tao.wan.wan@gmail.com,
zengchang.qin@gmail.com
[2] Department of Automation, Tsinghua University, Beijing, China
[3] Alibaba Group, Beijing 100022, China
fuzihaofzh@163.com
[4] School of Biological Science and Medical Engineering,
Beihang University, Beijing, China

Abstract. Sentimental analysis is an important topic in natural language processing and opinion mining. Many previous studies have reported to judge whether a term is with emotion or not. However, little work has been done in measuring degrees of sentiment for these terms. For example, the word *excellent* has stronger positive sentiment than the word *good* and *okay*. In this paper, we investigate how to model this intricate sentimental difference by assigning sentimental weights. A simple and effective model is proposed based on logistic regression to extract emotional terms associated with sentiment weights. Weighted terms can be used in sentiment classification and visualization by drawing emotional clouds of texts. The new model is tested using uni-gram, bi-gram and mixed-gram language models on two benchmark datasets. The empirical results show that the new model is highly efficient with comparable accuracy to other sentiment classifiers.

1 Introduction

Sentiment analysis is an important task in natural language processing (NLP) for analyzing people's opinions, attitudes and emotions towards certain services and products [1]. When we think of emotive reviews or comments, we are inclined to think of predicates of personal taste (*boring, fun*), exclamatives (*awesome, damn*) and other *emotional* words or terms that are more or less contributed to convey sentimental information of our opinions. Once we get the knowledge of these dominant emotional terms, we could judge the sentimental polarity (positive or negative) of a given sentence or document. Unfortunately, for particular utterance in a given context, such significant emotional terms are not always so apparent. We hope to find a way to automatically extract these emotional terms from corpus.

Sentiment classification can be simply considered as a text classification task. Most previous studies focused on using supervised machine learning methods

© Springer International Publishing Switzerland 2016
R. Booth and M.-L. Zhang (Eds.): PRICAI 2016, LNAI 9810, pp. 116–124, 2016.
DOI: 10.1007/978-3-319-42911-3_10

based on *N*-gram language model to do sentiment classification [2,3]. In movie reviews, we can often see terms like *worth watching*, that expresses strong positive emotion. Though the word *watching* is emotionally neutral, it will be assigned with positive weight alone with *worth*. It may become troublesome in sentences containing *watching* only. In this paper, we develop a mixed-gram model by embedding both uni-gram and bi-gram together to tackle this problem. Therefore, the bi-gram term *worth watching* will be assigned with large positive weight without separating them. We propose a simple but effective approach to extract sentiment terms by assuming that every distinct term in the corpus has an unique sentiment weight and the sentiment of a sentence or document is determined by a nonlinear combination of sentiment weights. We employ logistic regression to estimate sentiment weights of terms and use these weights for predictions.

For sentiment weight extraction, some studies have been done include the following: [4] did the work on tackling the problem of determining the semantic orientation (or polarity) of words; [5] proposed an approach to find subjective adjectives using the results of word clustering according to their distributional similarity; [6] proposed a double propagation algorithm to expand opinion words and extract target words. There are also some related work in sentiment lexicon extraction: [7,8] showed that supervised learning methods can achieve state-of-the-art results for lexicon extraction. In the domain-specific lexicon extraction, [9] got significant improvement by using active learning method. [10] used logistic regression (LR) to get the terms' weights corresponding to different ratings. [11] focused on the relevant weights of sentences in a given document for aspect rating prediction.

In terms of representation of documents, BoW could nicely reduce a piece of text with arbitrary length to a fixed length vector. In recent years, by exploiting the co-occurrence pattern of words, embedding model was employed to gain lower-dimensional, compact and meaningful vectors for words or documents [12]. Deep neural network approaches like convolutional neural network (CNN) can also bring significant improvement to sentiment classification task [13]. However, deep learning (DL) methods are always computationally expensive. In this paper, we hope to develop a simple and efficient learning model for intuitive sentiment visualization of a text with comparable results to the DL methods in sentiment classification.

2   Polarity Model of Sentiment

Given a sentimental text of being positive or negative, it is easy to see that the sentiment contributions of its consisting terms are different. Some terms like *excellent, good* occur more often in positive documents, and terms like *bad, horrible* occur more often in negative documents. This implies that such terms have high sentimental contributions and should be assigned with large sentiment weights. For some objective nouns and action verbs like *take, walk*, very likely they appear equally in both positive or negative documents, therefore, they contribute less to the sentiment of a text. Such terms are neutral and should be assigned with small sentiment weights. In this polarity model, we assume that the sentiment of a sentence is a function of the sentimental weights of its consisting terms. A term can be either one word in uni-gram model or two words in bi-gram model, or even mixed-gram of both.

Given a sentence of N different terms, the associated sentiment weight of term t_i is denoted by w_i for $1 \le i \le N$. The sentiment score h of the given sentence is:

$$h = f \left(\sum_{i=1}^{N} w_i x_i \right) = f(\mathbf{w}^T \mathbf{x}) \tag{1}$$

where x_i is the feature value of a given term t_i. It could be term frequency, binary value (appears in the particular document or not) or the TF-IDF value of the term.

Moreover, the sentiment polarity of a term x_i can be defined according to its sentiment weight w_i. It is not easy to set thresholds among positive, neutral and negative, as it is quite data dependent. We will test thresholding based on sentiment weights in Sect. 3. Function $f(\cdot)$ is a nonlinear function to smooth the linear combination of the sentimental weights.

2.1   Sentiment Weight Learning

In this section, we are going to use logistic regression (LR) and Gradient Descent algorithm (GD) to learn the sentiment weight based on given training data. We use $\mathbf{w} = \{w_1, w_2, \ldots, w_N\}$ to represent the weight vector where $w_i : i \in \{1, 2, \ldots, N\}$ is the sentimental weight of term t_i, where $T = \{t_1, t_2, \ldots, t_N\}$ is a set of terms in a corpus based on uni-gram or bi-gram. We use $y \in \{0, 1\}$ to represent the document's sentiment label (0 for negative and 1 for positive), and h to represent the sentiment score of the given document. $\mathbf{x}_j = \{x_{j1}, x_{j2}, \ldots, x_{ji}, \ldots, x_{jN}\}$ denotes the jth document's feature vector where x_{ji} represents the ith term's feature value in the jth document. We can initialize $\mathbf{w}$ randomly and calculate sentiment score by using logistic function:

$$h_j = \frac{1}{1 + \exp\left(-\mathbf{w}^T \mathbf{x}_j\right)} \quad ; \quad 1 \le j \le M \tag{2}$$

where M is the total number of documents in the training corpus. In order to minimize the squared error: $E = \frac{1}{2} \sum_{j=1}^{M} (h_j - y_j)^2$. we can update $\mathbf{w}$ given h_i, y_i and x_i using the Gradient Descent algorithm: $\mathbf{w} := \mathbf{w} - \alpha(h_j - y_j)\mathbf{x}_j$ where α is the learning rate. We iterate the process until convergence and use the final $\mathbf{w}$ to predict the new unseen document's score $\hat{h}$ by Eq. (2). We then can predict the sentiment label by:

$$y = \begin{cases} 0 & if \quad \hat{h} < 0.5 \\ 1 & otherwise \end{cases} \tag{3}$$

In addition, the computational complexity of LR is $O(|V|)$, where $|V|$ is the size of vocabulary.

2.2 Sparsity Constraints

As we have discussed, given a text, there are few words with strong sentiment. Most words including nouns and verbs are neutral. Therefore, we can put a constraint of sparsity in learning sparse sentimental weights using logistic regression. We can use L-BFGS method to minimize the following cost function in ℓ_1 (or ℓ_2) norms:

$$\min_{\mathbf{w}} \left[C\|\mathbf{w}\|_1 - \frac{1}{M} \sum_{j=1}^{M} (y_j \log(h_j) + (1 - y_j) \log(1 - h_j)) \right] \qquad (4)$$

By such constraints, we hope to learn a sparse vector $\mathbf{w}$ in which zero values represent neutral terms and terms with non-zero values carry the sentiment.

2.3 Mixed-gram Model

Previous work shows that bi-gram generally performs better than uni-gram, since bi-gram has more semantic information of word order or word position [3]. As we have seen before, terms like *good movie* or *bad script* often appear in pairs in reviews or comments. In uni-gram model, since the word *movie* appears together with *good* in positive examples, it is likely to be assigned sentimentally positive value. The neutral word *movie* may become problematic in classification. However, if we consider the same example in bi-gram model, the positive sentiment value will be assigned to *good movie*. The word comes after *moive* could be very random and it won't be biased on certain special bi-gram terms starting with *movie*. Ideally, weights should be assigned to sentimentally segmented terms. For example, *This is a good movie, actors are excellent!* should be segmented to: *This|is a|good movie|actors|are|excellent|*! This may bring a new problem of sentimental segmentation and it is beyond the scope of this paper. In this paper, we simply use mixed-gram model that is a mixture of both uni-gram and bi-gram.

2.4 Visualization of Emotional Word Cloud

Word cloud is a visualization form for text that is recognized for its aesthetic, social, and analytical values. Here, we are concerned with deepening its analytical value for visual comparison of documents. A word is expected to have the same color and position across word clouds. This aims to reduce the cognitive effort needed for comparing word clouds. However, it has two shortcomings. First, it only seeks to synchronize the appearance of each distinct word. This is problematic, as text frequently uses different words to refer to the same concept. Second, its synchronization of all word clouds imposes sizeable runtime requirement that prevents real-time generation of word clouds. These issues arise because word clouds are still high-dimensional representations, with dimensionality the size of the vocabulary. In this paper, we use the color (from green to red) feature to represent sentimental weights and the size of a word is determined by its TF-IDF value.

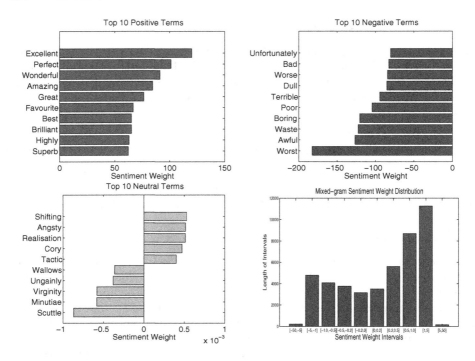

Fig. 1. Upper: Top 10 positive and negative terms in the IMDB corpus. *Excellent* is the strongest positive term and *Worst* is the strongest negative term. Below: Top 10 neutral terms and weight distribution of mixed-gram on the IMDB dataset (Color figure online)

3   Experimental Studies

We choose two benchmark datasets on sentiment classification in our experiments. The first is the IMDB dataset of online movie reviews [14], it contains 25000 reviews (12500 positives and 12500 negatives) for training and 25000 (12500 positives and 12500 negatives) for test. The second dataset is the Product Review including DVD, electronics, books and kitchens, and each of them contains 1000 positive and 1000 negative reviews. Stop words are not removed in our text preprocessing.

3.1   Sentiment Term Extraction

We first test the model on the IMDB dataset. Sentiment weight for each term is estimated based on uni-gram, bi-gram and mixed-gram, respectively. The top 10 results of uni-gram are shown in Fig. 1 and top 20 sentiment terms of mixed-gram are listed in Table 1. It is easy to see that our model successfully extracts terms with strong sentiment and the results are intuitively agree with human perceptions. Weights for positives terms are positive values and weights for negative terms are negative values. Weights for neutral terms are close to zero. The learned term weights have a big variance, which indicates that a small set of terms carrying strong sentiment than other terms.

Table 1. Top 20 sentiment terms based on the mixed-gram model.

Polarity	Sentiment terms (Mixed-gram)
Positive	great, the best, excellent, perfect, wonderful, amazing, a bit, a great well worth, is a, a must, fun, my favorite, today, very good brilliant, definitely worth, is great, very well, superb
Negative	the worst, bad, worst, awful, boring, poor, no, terrible waste, nothing, waste of, at all, worse, not even, dull, horrible poorly, stupid, annoying, lame

Fig. 2. Emotional word clouds of positive (first 2 figures) and negative reviews (the last figure) generated from the IMDB dataset. (Color figure online)

Particularly, in the mixed-gram model, *well worth*, *a must*, *not even* and *at all(not ... at all)* are valuable patterns to determine the sentiment orientation. It should be noted that stop words cannot be filtered to recognize such patterns. Though some neutral uni-grams or bi-grams are wrongly classified as positive (e.g. *today*, *is a*) or negative (e.g. *script*). One possible reason is that these neutral uni-grams or bi-grams occur frequently in some sentimental contexts. For example, *job* is more likely to occur in the forms of *good job* or *great job*; *script* is more likely to be discussed when people want to criticize a movie. Some misclassified neural words (e.g. *today*) may simply come from the bias of the dataset. And top 20 terms also demonstrate that mixed-gram behave better than uni-gram intuitively.

Furthermore, sentiment weights can help to draw *emotional word clouds* of positives reviews and negative reviews. The color of term shows its sentiment; the deeper the color is, the stronger sentiment the term has. We use cold colors (blue, cyan) to represent negativity and warm colors (red, yellow and orange) to

Table 2. Classification results under sparse constraints using logistic regressions with different N-gram language models.

N-gram	No constraints	ℓ_1 Norm	ℓ_2 Norm
Uni-gram	11.22 %	11.14 %	10.94 %
Bi-gram	11.03 %	11.00 %	10.84 %
Mixed-gram	9.77 %	9.61 %	9.58 %

Table 3. Performance comparison of LR+mixed-gram-ℓ_2 to other approaches on the IMDB dataset.

Sentiment classification model	Error	Sentiment classification model	Error
BoW(bnc) [14]	12.20 %	LDA [14]	32.58 %
Full+Unlabeled+BoW [14]	11.11 %	WRRBM [15]	12.58 %
WRRBM+BoW(bnc) [15]	10.77 %	MNB-uni [3]	16.45 %
MNB-bi [3]	13.41 %	SVM-uni [3]	13.05 %
SVM-bi [3]	10.84 %	NBSVM-uni [3]	11.71 %
NBSVM-bi [3]	**8.78 %**	PV+Unlabled [12]	**7.42 %**
LR-mixed-gram-ℓ_2 (Our model)	**9.58 %**		

represent positivity. The fontsize of a term is proportional to its TF-IDF value. Two sample emotional word clouds of positive and negative reviews in IMDB are shown in Fig. 2.

Table 2 shows the classification results under sparse constraints of ℓ_1 and ℓ_2 norms, respectively. Classification results of ℓ_2 norm is generally better than ℓ_1 and classical logistic regress without sparse constraints. Many researchers have reported their results on the IMDB dataset. In particular, one of the most significant improvement recently was the work of [3] in which they found that bi-gram feature works the best and yields a considerable improvement of 2 % in error rate. Another important contribution is [15] in which they combine a Restricted Boltzmann Machines model with BoW. The best result so far was reported by [12] in which deep learning was used and it involves a big computing resources. The method we proposed (LR+mixed-gram+ℓ_2) is the simplest one and with least computational time (Table 3).

We also conduct experiments on the Product Review dataset. In our experiments, like what we have done to the IMDB dataset, we do not remove any stopwords or apply any stemming in preprocessing for fair comparison to approaches proposed in [2]. We don't handle the problem of orthographic mistakes, abbreviations, idiomatic expressions or ironic sentences either. From Table 4, we can find that our model also has comparable performance to the baseline approaches. Though it performs slightly worse than SVM in classifying Books and DVDs, it performs well in Electronics and Kitchen.

Table 4. Comparisons to baseline approaches on the Product Review dataset.

Category	ANN	SVM	LR-mixed-gram
Books	18.3 %	17.2 %	19.8 %
DVD	18.4 %	16.3 %	19.9 %
Electronics	16.3 %	15.1 %	15.6 %
Kitchen	14.8 %	13.6 %	13.8 %

4 Conclusions

In this paper, we propose a model using Logistic Regression with Gradient Descent algorithm for extracting sentiment terms and learning sentiment weights. We assume the sentiment of a sentence or documents is a function of sentiment weights of consisting terms. The extracted sentiment terms can be drawn as emotional clouds. In sentiment classification, we have tested our model based on different N-gram models and find that the mixed-gram model outperforms both uni-gram and bi-gram models. Extensive experimental results show our proposed method can extract precise sentiment terms and achieve a high level accuracy in classification on given benchmark datasets.

References

1. Pang, B., Lee, L., Vaithyanathan, S.: Thumbs up? sentiment classification using machine learning techniques. In: Proceedings of EMNLP, pp. 79–86 (2002)
2. Fattah, M.A.: New term weighting schemes with combination of multiple classifiers for sentiment analysis. Neurocomputing **167**, 434–442 (2015)
3. Wang, S., Manning, C.D.: Baselines and bigrams: simple, good sentiment and text classification. In: Proceedings of ACL, pp. 90–94 (2012)
4. Kaji, N., Kitsuregawa, M.: Building lexicon for sentiment analysis from massive collection of HTML documents. In: Proceedings of EMNLP, pp. 1075–1083 (2007)
5. Wiebe, J.M.: Learning subjective adjective from corpora. In: Proceedings of AAAI, pp. 735–740 (2000)
6. Qiu, G., Liu, B., Bu, J., Chen, C.: Opinion word expansion and target extraction through double propagation. In: Proceedings of ACL, pp. 9–27 (2011)
7. Jin, W., Ho, H.H.: A novel lexicalized hmm-based learning framework for web opinion mining. In: Proceedings of ICML, pp. 465–472 (2009)
8. Li, F., Huang, M., Zhu, X.: Sentiment analysis with global topics and local dependency. In: Proceedings of AAAI, pp. 1371–1376 (2010)
9. Park, S., Lee, W., Moon, I.C.: Efficient extraction of domain specific sentiment lexicon with active learning. Pattern Recogn. Lett. **56**, 38–44 (2015)
10. Potts, C., Schwarz, F.: Affective 'this'. Linguist. Issues Lang. Technol. **3**(1), 1–30 (2010)
11. Pappas, N., Popescu-Belis, A.: Efficient extraction of domain specific sentiment lexicon with active learning. In: Proceedings of EMNLP, pp. 455–466 (2014)
12. Le, Q., Mikolov, T.: Distributed representations of sentences and documents. In: Proceedings of ACL (2014). arXiv:1405.4053

13. Kim, Y.: Convolutional neural networks for sentence classification. In: Proceedings of EMNLP, pp. 1746–1751 (2014)
14. Maas, A.L., Daly, R.E., Pham, P.T.: Learning word vectors for sentiment analysis. In: Proceedings of ACL, pp. 142–150 (2011)
15. Dahl, G.E., Adams, R.P., Larochelle, H.: Training restricted boltzmann machines on word observations (2012). arXiv:1202.5695

Prediction with Confidence in Item Based Collaborative Filtering

Tadiparthi V.R. Himabindu[1], Vineet Padmanabhan[1(✉)], Arun K. Pujari[1], and Abdul Sattar[2]

[1] School of Computer and Information Sciences,
University of Hyderabad, Hyderabad, India
himaworld_06@yahoo.com, {vineetcs,akpcs}@uohyd.ernet.in
[2] Griffith University, Brisbane, Australia
a.sattar@griffith.edu.au

Abstract. Recommender systems can be viewed as prediction systems where we can predict the ratings which represent users' interest in the corresponding item. Typically, items having the highest predicted ratings will be recommended to the users. But users do not know how *certain* these predictions are. Therefore, it is important to associate a confidence measure to the predictions which tells users how certain the system is in making the predictions. Many different approaches have been proposed to estimate confidence of predictions made by recommender systems. But none of them provide guarantee on the error rate of these predictions. Conformal Prediction is a framework that produces predictions with a guaranteed error rate. In this paper, we propose a conformal prediction algorithm with item-based collaborative filtering as the underlying algorithm which is a simple and widely used algorithm in commercial applications. We propose different nonconformity measures and empirically determine the best nonconformity measure. We empirically prove validity and efficiency of proposed algorithm. Experimental results demonstrate that the predictive performance of conformal prediction algorithm is very close to its underlying algorithm with little uncertainty along with the measures of confidence and credibility.

Keywords: Recommender systems · Conformal prediction · Confidence · Nonconformity measure

1 Introduction

Collaborative filtering (CF) is a very promising approach in recommender systems and is the most widely adopted technique both in academic research and commercial applications. CF algorithms can be classified in two ways: in *neighborhood based approaches* prediction and recommendation can be done either by computing the similarities between users (user-based collaborative filtering (UBCF) [16]) or similarities between items (item-based collaborative filtering

© Springer International Publishing Switzerland 2016
R. Booth and M.-L. Zhang (Eds.): PRICAI 2016, LNAI 9810, pp. 125–138, 2016.
DOI: 10.1007/978-3-319-42911-3_11

(IBCF) [2]) and *model-based approaches* [17] use mathematical models for making predictions. Many model-based algorithms are very complex which involves estimation of large number of parameters. Moreover if the assumptions of the model do not hold, it may lead to wrong predictions. On the other hand, neighborhood approaches are very simple both in terms of underlying principles and implementation while achieving reasonably accurate results. But UBCF does not perform well when the active user is having too few neighbors and neighbors with very low correlation to the active user [8]. In our paper, for the proposed conformal prediction algorithm we have chosen IBCF as the underlying algorithm because of its large potential in research and commercial applications [12].

Most of the CF algorithms are limited to making only single point predictions. Metrics such as MAE and RMSE [21] were proposed in the literature to measure the prediction accuracy. But accuracy of individual predictions can not be estimated using these measures, as these measures are used to predict the overall accuracy of the recommendation algorithm. Some confidence estimation algorithms have been proposed in the literature to estimate the confidence of each prediction. But none of these algorithms provide an upper bound on the error rate. In contrast, conformal predictors are able to produce confidence measures specific to each individual prediction with guaranteed error rate.

Conformal Prediction (CP) [4,5] is the framework used in machine learning (ML) to make reliable predictions with known level of significance or error probability. Moreover, CP is increasingly becoming popular due to the fact that it can be built on top of any conventional point prediction algorithms like K-NN [6], SVM [18], decision trees [19], neural networks [20] etc. The confidence measures produced by CPs are not only useful in practice, but also their accuracy is comparable to, and sometimes even better than that of their underlying algorithms. So we use CP to associate a confidence measure for each individual prediction made by our chosen underlying algorithm.

The regions produced by any CP algorithm are automatically valid. But efficiency in terms of tightness and usefulness of prediction regions depends on the nonconformity measure (NCM) used by the CP algorithm. Moreover, we can define many different NCMs for a given underlying algorithm and each of these measures defines a different CP. So determining an efficient NCM based on the underlying algorithm is one important step in CP. In this work, we define different NCMs based on the underlying algorithm and empirically demonstrates that the CP with simple NCM which is a variant of NCM used in K-NN [6] performs well from both accuracy and efficiency perspectives.

Major contributions of this paper are: 1. Adaptation of conformal prediction to Item-based collaborative filtering. 2. Define NCMs based on similarity measure used in IBCF and empirically determine the best NCM. 3. Empirically demonstrate validity and efficiency of our conformal prediction algorithm.

The rest of the paper is organized as follows: In the next section we discuss related work. In Sect. 3, we discuss general idea of conformal prediction. In Sect. 4, we describe Item-based collaborative filtering. Section 5 describes our proposed algorithm which apply conformal prediction on top of Item-based collaborative filtering and defines different NCMs based on IBCF. Section 6 details our experimental results and show the validity and efficiency of our proposed conformal prediction algorithm. Finally, Sect. 7 gives our conclusions.

2   Related Work

In this section we review existing methods proposed in the literature to estimate confidence of CF algorithms. McNee et al. [7] estimate the confidence of an item as support for the item. Mclaughlin and Herlocker [8] proposed an algorithm for UBCF, which generates belief distributions for each prediction. Although their algorithm is good at achieving good precision by making sure that more popular items are recommended, they did not demonstrate the accuracy of their algorithm. Adomavicius et al. [1] estimate the confidence based on rating variance of each item. Shani and Gunawardana [9] defines confidence as the system's trust in its recommendations. They proposed a method to estimate the confidence of recommendation algorithms, but no experiments are provided due to the broader scope of the chapter. Koren and Sill [10] formulated the problem of confidence estimation as a binary classification problem and find whether the predicted rating is within one rating level of the true rating. Although confidence is associated with each item in the recommendation list, the proposed confidence estimation algorithm is applicable only for their proposed CF algorithm. Mazurowski [3] introduced three confidence estimation algorithms based on resampling and standard deviation of predictions to predict confidence of individual predictions.

But all of the above algorithms failed to provide an upper bound on the error rate i.e., the probability of excluding the correct class label is guaranteed to be smaller than a predetermined significance level. On the other hand, our CP algorithms produce prediction regions with a bound on the probability of error. When forced to make point predictions, the confidence of a prediction is $1-$ the second largest p-value and this *second largest p-value* becomes the upper bound on the probability of error. Moreover, with CP we can control the number of erroneous predictions by varying the significance level, thus making it suitable to different kinds of applications.

3   Conformal Prediction

In this section, we introduce the general idea behind conformal prediction [4,5]. We have a training set of examples $Z = \{z_1, z_2, ..., z_l\}$. Each $z_i \in Z$ is a pair $(x_i, y_i); x_i \in \mathbb{R}^d$ is the set of attributes for i^{th} example and y_i is the class label for that example. Our only assumption in CP is that all $Z_i's$ are independently and identically distributed (i.i.d.). Given a new object our task is to predict the class label y_{l+1}. We try out all possible class labels y_j for the label y_{l+1} and append z_{l+1} $(=(x_{l+1}, y_{l+1}))$ to Z. Then estimate the typicalness of the sequence $Z \cup z_{l+1}$ with respect to i.i.d by using p-value function. Our prediction for y_{l+1} is the set of y_j for which p-value $> \epsilon$, where ϵ is the significance level.

One way of obtaining p-value function is by considering how strange each example in our sequence is from all other examples. To measure these strangeness values we use NCM. NCM $\mathcal{A}$ is a family of functions which assigns a numerical score to each example z_i indicating how different it is from the examples in the set $\{z_1, ..., z_{i-1}, z_{i+1}, ..., z_n\}$.

NCM has to satisfy the following properties [4]:

1. Nonconformity score of an example is invariant w.r.t. permutations. i.e., for any permutation π of $1, 2..., n$

$$\mathcal{A}(z_1, z_2, ..., z_n) = (\alpha_1, \alpha_2, ..., \alpha_n) \implies$$
$$\mathcal{A}(z_{\pi(1)}, z_{\pi(2)}, ..., z_{\pi(n)}) = (\alpha_{\pi(1)}, \alpha_{\pi(2)}, ..., \alpha_{\pi(n)}). \tag{1}$$

2. $\mathcal{A}$ is chosen such that larger the value of α_i stranger is z_i to other examples. p-value is computed by comparing α_{l+1} with all other nonconformity scores.

$$p(y_j) = \frac{\#\{i = 1, 2, ..., l+1 : \alpha_i \geq \alpha_{l+1}\}}{l+1}. \tag{2}$$

An important property of p-value is that $\forall \epsilon \in [0, 1]$ and for all probability distributions P on Z,

$$P\{\{z_1, z_2, ..., z_{l+1}\} : p(y_{l+1}) \leq \epsilon\} \leq \epsilon. \tag{3}$$

This original approach to CP is called Transductive Conformal Prediction (TCP). The p-values obtained from CP for each possible classification can be used in two different modes: *Point prediction:* for each test example, predict the classification with the highest p-value. The confidence of this prediction is $1-$ the second largest p-value and credibility is the highest p-value(credibility tells how well the new item with the assumed label conforms to the training set of items). *Region prediction:* given the $\epsilon > 0$, output the prediction as the set of all classifications whose p-value $> \epsilon$ with $1 - \epsilon$ confidence that the true label will be in this set. A method for finding $(1 - \epsilon)$ prediction set is said to be *valid* if it has atleast $1 - \epsilon$ probability of containing the true label. *Efficiency* of CP is the tightness of prediction regions it produces. The narrower (small number of labels) the prediction region the more efficient the conformal predictor is.

4   Item Based Collaborative Filtering Algorithm

In IBCF, prediction and recommendation are based on item to item similarity. The key motivation behind this scheme is that a user will more likely purchase items that are similar to items he already purchased. This can be done as follows: Assume that I is the set of all available items. For every target user u_t, first the algorithm looks into the set of items that he has rated (training set C_t) and computes how similar they are to the target item $i_t \in S_t$ (S_t set of test items for the target user u_t) using a similarity measure, and then selects k most similar items $\{i_1, i_2, ..., i_k\}$. Once the most similar items are found, the prediction is then computed by taking weighted average of the target user's ratings on these similar items. So two main tasks in IBCF are: computing item similarities and rating prediction.

In order to apply CP to IBCF, it is appropriate to convert all ratings to binary i.e., the user likes or dislikes the item. The reason for this is twofold: first, uneven distribution of ratings in the data sets: For instance, more than 80 % of all ratings in MovieLens 100 K are greater than 2 and nearly 70 % of all ratings in Eachmovie are greater than 3. As a result, it becomes very difficult to identify k most similar items consumed by the target user which are rated as 1 when the target item rating is assumed as 1. Second, in [14] authors have shown that user's rating as the noisy evidence of user's true rating. Therefore identifying k most similar items which are rated as 1 when the new item rating is assumed as 1 does not make sense.

The simple way to convert all ratings to binary is as follows: take the middle of the rating scale as the threshold (for instance, 3 in the rating scale of 1–6) and assume all ratings greater than the threshold as liked and all other ratings as disliked. But this approach works fine when the distribution of all ratings is even which is not the case in most of the data sets. Other approach to convert the ratings into binary is, compute the average rating for every user and consider all ratings whose rating is greater than the average as liked and the all ratings below this average as disliked. This is the best approach to deal with all types of users including pessimistic, optimistic and strict users. For example, pessimistic (optimistic) users who usually give low (high) rating to every item they consume, we assume that they like items rated above their average rating and dislike items rated below the average. Similarly, in case of strict users who rate every item correctly according to whether they like that item or not (gives high rating when they like and low rating when they do not like) in which case we assume all ratings above the average (approximately equal to the middle of the rating scale) as like and other ratings as dislike [13]. This ensures that our CP algorithm have a reasonable number of liked and disliked items in the data set which makes it easier to find k similar items when the target item is assumed as like or dislike. Item similarity computation and prediction are done as follows:

– Item Similarity Computation: In our IBCF we have chosen cosine based similarity measure to compute similarities among the items. Since, we do not have rating values, our algorithm uses binary cosine similarity measure [15] that finds the number of common users between the two items i and j and is defined as follows:

$$similarity(i,j) = \frac{\#common\_users(i,j)}{\sqrt{\#users(i)}.\sqrt{\#users(j)}}. \tag{4}$$

The above equation will give the value in between [0,1]. $Similarity(i,j) = 1$ when these two items are rated by exactly same set of users. For simplicity, we use this simple cosine similarity measure as our aim is not to improve the accuracy of the algorithm, but to provide confidence to the predictions generated by the algorithm. We can use efficient similarity measures instead of this simple measure to obtain more accurate results.

– Predicting the label (like or dislike): Once the similarities are computed, find the k most similar items (k nearest neighbors) of the target item among the consumed items of the target user. Then predict the label for the target item as the most common label among its k nearest neighbors.

5   Application of TCP to IBCF

In this section we discuss how to build TCP on top of IBCF algorithm (a variant of the algorithm in [6]). We first discuss the algorithm setup and then define different NCMs based on IBCF algorithm. Finally, we present our TCP algorithm with IBCF as an underlying algorithm.

5.1   Algorithm Setup

In order to apply CP, we need a training set of examples $\{z_1, z_2, ..., z_l\}$ and a test example z_{l+1} for which we want to make the prediction. Here we discuss how this is formulated in the context of IBCF. For every target user u_t, there is a set of items W_t rated by this user and we consider a part of W_t as the training set C_t. For every item $i \in C_t$, user has assigned a label which tells whether the user liked($+1$) or disliked(-1) the item i. Therefore, $Y = \{+1, -1\}$. We also have a test set of items S_t ($W_t - C_t$) for which we hide the actual labels assigned by the user. Now, our task is to assign a label (which makes the current test item conforms to the training set) to each of the test set items with an associated confidence measure which is valid according to Eq. (3).

5.2   Nonconformity Measures (NCMs)

We propose different NCMs based on IBCF. First we introduce the terminology to define NCMs. Since we are having only two labels, $Y = \{+1, -1\}$, we are assuming that if $y = 1$ $\bar{y} = -1$ and vice-versa. Assume that $similarity_i^y$ is vector which is a sorted sequence (in descending order) of similarity of an item i with items $\in C_t$ with the same label y and $similarity_i^{\bar{y}}$ is a sorted sequence (in descending order) of similarity of an item i with items $\in C_t$ with the label $\bar{y}$. The weight for the item i with label y, w_i^y is defined as the sum of similarity values of k most similar items with the label y among the set of rated items C_t of the target user u_t. Similarly $w_i^{\bar{y}}$ is defined as the sum of similarity values of k most similar items with the label $\bar{y}$ among the items in C_t of the target user u_t.

$$w_i^y = \sum_{j=1}^{k} similarity(i, j)^y. \quad (5) \qquad w_i^{\bar{y}} = \sum_{j=1}^{k} similarity(i, j)^{\bar{y}}. \quad (6)$$

where k is the number of most similar items, $similarity(i, j)^y$ is jth most similar item in $similarity_i^y$, y is the label of the item i, $similarity(i, j)^{\bar{y}}$ is jth most similar item in $similarity_i^{\bar{y}}$, $\bar{y}$ is the label other than the label of item i.

In what follows we define NCMs based on IBCF:

$$NCM1 = k - w_i^y. \quad (7) \qquad NCM2 = \frac{1}{w_i^y}. \quad (8) \qquad NCM3 = \frac{w_i^{\bar{y}}}{w_i^y}. \quad (9)$$

1. NCM1 & NCM2: The simple NCMs for an item i are as follows: The maximum value of the similarity function defined in Eq. (4) is 1. As a result, the maximum value of w_i^y becomes k. The higher the value of w_i^y, the more conforming the item i is with respect to the other items. NCM1 will be high when w_i^y is small and smaller value of w_i^y indicates that the item with label y is nonconforming with other items of the same label. As a consequence, the higher the value of NCM1, the stranger the item i is with respect to the other examples with the same label according to the second property of NCM. Similarly, smaller the value of w_i^y the higher the value of NCM2. The higher the value of NCM2, the more nonconforming the item i is with respect to the other examples.

2. NCM3: A more efficient and a variant of NCM proposed in [6] can be defined by taking into consideration $w_i^{\bar{y}}$ along with w_i^y in computing the nonconformity score. According to NCM3, example i with label y is nonconforming when it is very similar to the items with label $\bar{y}$ (high value of $w_i^{\bar{y}}$) and dissimilar to the items with label y (low value of w_i^y).

5.3 Item-Based Collaborative Filtering with TCP (IBCFTCP)

Algorithm 1 describes the application of TCP to IBCF in detail. For every item i in S_t of the target user u_t, try all possible labels in Y and compute the typicalness of the sequence E resulting from appending i with the assumed label to C_t using p-value function which in turn uses nonconformity values (calculated using any of the NCMs discussed above) of all items in E. For region predictions, output the prediction as the set of all labels whose p-values are $> \epsilon$ with confidence $1 - \epsilon$ or in case of point predictions, output the label with the highest p-value with confidence $1-$ the second highest p-value and credibility as the highest p-value.

6   Experimental Results

We tested our algorithm on four data sets: MovieLens 100K, MovieLens 1M, MovieLens-latest-small and EachMovie. We randomly selected 50 users and for each user first 60 % of the data is considered as training set and remaining 40 % is taken as the test set. Details of data sets is given in Table 1. As TCP is a time consuming approach and it increases with the number of items when applied to IBCF, we do not conduct our experiments on data sets in other domains such as books and music where there are large number of items. Moreover, we do not compare our results with other state-of-the-art algorithms, as our aim is not to improve the performance of the algorithm but to associate confidence to the predictions made by the algorithm without compromising the performance.

Algorithm 1. Item-based Collaborative Filtering with TCP

Input: $k, u_t, C_t, S_t, I, Y, \epsilon$
for each $i \in I$ **do**
  for each $j \in I$ **do**
    Compute the similarity between i and j using Eq. (4)
  end for
end for
for each $i \in C_t$ **do**
  Compute w_i^y and $w_i^{\bar{y}}$
  Compute nonconformity scores using any of the NCMs discussed above.
end for
for each $i \in S_t$ **do**
  for each $y \in Y$ **do**
    Update the similarity values of item i with other items
    Compute w_i^y and $w_i^{\bar{y}}$
    Compute nonconformity score of i using any of the NCMs discussed above.
    for each $j \in C_t$ **do**
      if label(j) = y **then**
        if $similarity(i,j)^y > similarity(j,k)^y$ **then**
          recompute the nonconformity score of item j
        end if
      else if $label(j) = \bar{y}$ **then**
        if $similarity(i,j)^{\bar{y}} > similarity(j,k)^{\bar{y}}$ **then**
          recompute the nonconformity score of item j
        end if
      end if
    end for
    Compute the p-value(p(y)) of item i with label y
  end for
  prediction region = $\{y|p(y) > \epsilon\}$ with confidence 1-ϵ
  OR
  Assign the label y to item i such that max{p(+1),p(-1)} = p(y)
  confidence = 1 - second highest p-value
  credibility = highest p-value
end for

Here we compare performance of CP with the underlying algorithm (IBCF) in terms of percentage of correct classifications (%CC) for different data sets and for different k values. In our experiments, we considered 4 different k values: 5,10,15 and 20. In order to do this comparison, we have to make single point predictions, since the underlying algorithm make only single point predictions. In single point predictions we output a label with the highest p-value. In case, if both labels share this p-value then we can take any one of these labels randomly. In our experiments we take both labels: one that is same as the true label (conforming one) and other one which is other than the true label (nonconforming one) and compute the performance of CP algorithm separately for both cases. In this way we can measure the certainty in predictions. In Table 2 we compare

Table 1. Summary of data sets

Dataset	#Users	#Items	#Ratings
MovieLens 100K	943	1682	100000
MovieLens 1M	6040	3952	1000209
MovieLens-latest-small	718	8915	100234
EachMovie	36656	1621	279983

the performance in terms of %CC of IBCFTCP with that of IBCF with both conforming labels (CL) and nonconforming labels (NL). We got same results for both NCMs1&2 in terms of %CC, validity and efficiency. So we do not show their results separately. We also calculate the uncertainty in the predictions as the percentage of items having more than one label (in our case both labels) shares the highest p-value. As the percentage of uncertainty is increased the deviation between performance of CP with conforming label and nonconforming label will also be increased as shown in Table 2.

Table 2. Performance comparison of IBCF with IBCFTCP with CLs and NLs

Dataset	Number of nearest neighbors	Algorithm								
		IBCF			IBCFTCP					
					NCMs1&2			NCM3		
		%CC		Uncertainty	%CC		Uncertainty	%CC		Uncertainty
		CL	NL		CL	NL		CL	NL	
MovieLens 100 K	5	0.668	0.668	0	0.6964	0.6492	0.0472	**0.6764**	**0.6708**	0.0056
	10	0.7178	0.6086	0.1092	0.7002	0.652	0.0481	0.6793	**0.6715**	**0.0077**
	15	0.6654	0.6654	0	0.6917	0.6494	0.0423	**0.6741**	**0.668**	0.0061
	20	0.6861	0.6232	0.0629	0.6868	0.6492	0.0376	0.6715	**0.6668**	**0.0047**
MovieLens 1M	5	0.6899	0.6899	0	0.7226	0.675	0.0476	**0.6993**	**0.6944**	0.0049
	10	0.7469	0.6317	0.1152	0.7181	0.6739	0.0442	0.7002	**0.6948**	**0.0055**
	15	0.6903	0.6903	0	0.7189	0.6746	0.0442	**0.7014**	**0.6967**	0.0047
	20	0.7269	0.6592	0.0677	0.7179	0.6739	0.044	0.6989	**0.694**	**0.0049**
MovieLens-latest-small	5	0.612	0.612	0	0.6421	0.5929	0.0492	**0.6246**	**0.6113**	0.0133
	10	0.6948	0.5388	0.1559	0.6494	0.6026	0.0468	0.6353	**0.6244**	**0.0109**
	15	0.6148	0.6148	0	0.6462	0.6122	0.034	**0.6338**	**0.6265**	0.0073
	20	0.6678	0.5636	0.1042	0.6441	0.6092	0.0349	0.6314	**0.6235**	**0.0079**
EachMovie	5	0.6798	0.6798	0	0.732	0.6344	0.0976	**0.6941**	**0.6842**	0.0099
	10	0.7454	0.6114	0.134	0.723	0.6376	0.0855	0.6904	**0.6839**	**0.0065**
	15	0.6758	0.6758	0	0.7146	0.6419	0.0727	**0.686**	**0.6776**	0.0084
	20	0.7053	0.6242	0.0811	0.701	0.645	0.056	0.6792	**0.6742**	**0.005**

From Table 2 when CLs are considered, the CP algorithm (for all NCMs) is outperforming IBCF for odd k values (5 & 15) due to its slightly higher uncertainty values compared to IBCF, whereas IBCF is performing better for even k values (10 & 20) because of its significantly higher uncertainty values.

In case of NLs, CP with NCMs1&2 is showing performance improvement for even k values and lower performance for odd k values (5 &15) compared to IBCF. On the other hand, CP with NCM3 is outperforming IBCF for all k values when NLs are taken into account. From Table 2 we can say that NCM3 is the best NCM (shown in bold numbers) as it is showing good performance with less uncertainty compared to NCMs1&2. Although NCMs1&2 are outperforming NCM3 in terms of %CC when CLs are taken into consideration, these are not good NCMs, as this improvement is due to uncertain predictions produced by these NCMs and though IBCF with CLs is performing well for even values of k compared to CP with NCM3, this is also because of high uncertainty involved in predictions made by IBCF. Uncertainty of IBCF for odd k values is 0 (because in this case there is no possibility of obtaining equal number of +1 s and −1 s) and for even k values its uncertainty is even greater than IBCFTCP with NCMs1&2.

All CPs are automatically valid. Notice that in IBCFTCP the training set and test set of items differ from user to user in contrast to CP in ML where the training set and test sets are fixed for the whole algorithm. So, it is necessary to show that the validity is satisfied for each user. The validity of CP with NCM3 for each user for different data sets and for $k = 5$ is shown in Fig. 1 (We got similar results for validity and efficiency for other k values. Because of space limitations we are not showing their results). Although validity for each user is also satisfied for NCMs1&2 we did not show the results here because NCM3 is the best NCM in terms of prediction accuracy as shown in Table 2 and efficiency (which will be discussed in the following paragraphs). From Fig. 2 we can see that the error values (validity) of most of the users are within the bounds of ϵ.

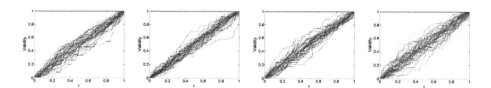

Fig. 1. Validity of IBCFTCP for different data sets (left to right:Movielens 100 K, Movielens 1M, Movielens-small-latest and Eachmovie)

Figure 2 shows average validity of IBCFTCP for different data sets and for different NCMs for $k = 5$. From Fig. 2, it is clear that prediction regions produced by CP with all NCMs are valid and follows a straight line as required.

Usefulness of CPs depend on its efficiency. Vovk et al. [11] proposed ten different ways of measuring the efficiency of CPs. In our experiments we use Observed Excess criterion which gives the the average number of false labels (ANFL) included in the prediction set at significance level ϵ. The formula to calculate ANFL is:

$$ANFL = \frac{\sum_{i=1}^{|S_t|} |\Gamma_i^\epsilon \setminus T_i|}{|S_t|}. \tag{10}$$

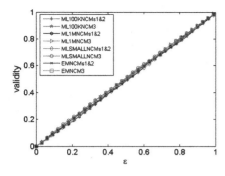

Fig. 2. Validity

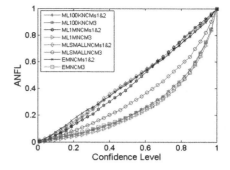

Fig. 3. ANFL

where Γ_i^ϵ is the prediction region for the ith test item at significance level ϵ, T_i is the true label of the ith test element, $|S_t|$ is the total number of test items.

The efficiency of CP algorithm in terms of ANFL for different data sets, different NCMs and for $k = 5$ is shown in Fig. 3 shows that CP with NCM3 is producing less number of false labels compared to NCMs1&2. From Fig. 3, we can observe that the number of false labels are decreasing with the decrease of confidence level (if we go from higher to lower confidence levels).

In addition to ANFL we use three other criteria to compute the efficiency of CP: 1. % of test elements having prediction regions with single label. 2. % of test elements having prediction regions with more than one label. 3. % of test elements having empty prediction region. A CP is said to be efficient when the % of second and third criteria are relatively small whereas the % of first should be high especially at higher confidence levels (50–99%). Our algorithm is optimal in the sense that it produces 0 % empty prediction regions from 70–99% confidence levels and < 20 % from 50–70 % confidence levels, while showing moderate performance in minimizing the second one and maximizing the first one at these confidence levels. The % of test elements having prediction regions producing single labels at different confidence levels for different data sets, different NCMs and for $k = 5$ is shown in Fig. 4. From Fig. 4 we can observe that NCM3 is outperforming NCMs1&2, in producing the higher number of prediction regions with single labels. The % of single labels produced by NCMs1&2 never exceed 20 % at any confidence level, whereas NCM3 is producing sufficiently large % of single labels especially from 30 %–90 % confidence levels. Figure 5 shows the % of correct predictions among the % of single labels produced by NCM3 at each confidence level. From Fig. 5 we can see that the % of correct predictions at any confidence level is > 60 %. The % of test elements having prediction regions with more than one rating at different confidence levels is shown in Fig. 6. In this case also NCM3 is performing better than NCM1&2 as NCM3 is producing less number of prediction regions with multiple labels compared to NCMs1&2 at all confidence levels and in case of NCM3 this number is zero from 0–60 % confidence levels. The % of test elements having empty prediction regions at different confidence levels is shown in Fig. 7. In this case also NCM3 is giving good results compared to NCMs1&2 by producing less number of empty prediction regions compared to NCMs1&2 and this number is zero from 70–99% confidence levels.

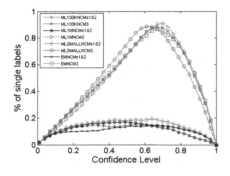

Fig. 4. % of single labels

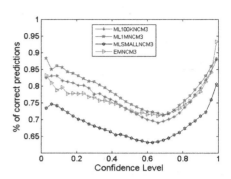

Fig. 5. % of correct predictions

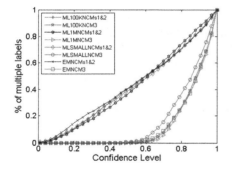

Fig. 6. % of multiple labels

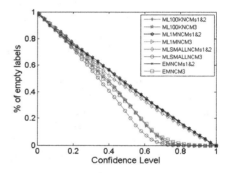

Fig. 7. % of empty labels

The mean confidence and mean credibility of single point predictions produced by IBCFTCP is shown in Table 3. We also calculated mean difference between highest p-value and lowest p-value for all test ratings. We will get confidence predictions when this difference is high. From Table 3, we can observe that this difference for NCM3 is around 50 %, whereas it is only around 15 % with NCMs1&2. Also, the mean confidence and mean credibility values produced by CP with NCM3(shown in bold numbers) are better than that of NCMs1&2.

In summary, NCM3 is the best NCM compared to NCMs1&2 in terms of prediction accuracy and efficiency. Moreover, when restricted to make single point predictions, mean confidence and credibility values produced by CP with NCM3 are higher than that of NCMs1&2.

Table 3. Mean confidence and mean credibility of IBCFTCP

Number of nearest neighbors	Performance measure	Dataset							
		MovieLens 100K		MovieLens 1M		MovieLens-latest-small		EachMovie	
		NCMs 1&2	NCM3	NCMs 1&2	NCM3	NCMs 1&2	NCM3	NCMs 1&2	NCM3
5	mean confidence	0.5841	**0.8643**	0.5864	**0.8718**	0.5881	**0.8346**	0.5738	**0.865**
	mean p1 - p2	0.1183	**0.4949**	0.1144	**0.4933**	0.1379	**0.4965**	0.1064	**0.4942**
	mean credibility	0.5347	**0.6306**	0.5284	**0.6216**	0.5503	**0.662**	0.5335	**0.6293**
10	mean confidence	0.5902	**0.8663**	0.5903	**0.8761**	0.5907	**0.8397**	0.585	**0.8681**
	mean p1 - p2	0.125	**0.4905**	0.1191	**0.4933**	0.1392	**0.4982**	0.1218	**0.496**
	mean credibility	0.5353	**0.6242**	0.5292	**0.6172**	0.5489	**0.6586**	0.5377	**0.6279**
15	mean confidence	0.598	**0.8666**	0.5947	**0.8782**	0.595	**0.8446**	0.5939	**0.8666**
	mean p1 - p2	0.1352	**0.4896**	0.1256	**0.4932**	0.1443	**0.5012**	0.1357	**0.4977**
	mean credibility	0.5376	**0.623**	0.5313	**0.615**	0.5496	**0.6567**	0.5425	**0.6311**
20	mean confidence	0.6063	**0.8675**	0.5994	**0.8783**	0.6	**0.8468**	0.6021	**0.8662**
	mean p1 - p2	0.147	**0.4927**	0.1327	**0.4931**	0.1505	**0.5017**	0.1499	**0.5003**
	mean credibility	0.541	**0.6253**	0.5337	**0.6148**	0.5509	**0.655**	0.5484	**0.6341**

7 Conclusions

In this work, we show the adaptation of CP to IBCF and proposed different NCMs for CP based on the underlying algorithm. Using our CP algorithm we are associating confidence values to each prediction along with the guaranteed error rate unlike IBCF which produces only bare predictions. Our algorithm is tested on different data sets and we experimentally proved that NCM3 is the best NCM compared to NCMs1&2 in achieving the prediction accuracy as good as the underlying algorithm with little uncertainty and also in producing efficient prediction regions. When making single point predictions, the mean confidence and credibility values produced by the proposed algorithm are reasonably high. Although our algorithm failed in producing large percentage of single labels at 90–99 % confidence levels (desired confidence levels in medical applications) we can use this algorithm to make predictions in certain recommendation domains such as movies, books, news articles, restaurants, music and in tourism where the confidence level of 50 %–90 % is acceptable.

References

1. Adomavicius, G., Kamireddy, S., Kwon, Y.: Towards more confident recommendations:improving recommender systems using filtering approach based onrating variance. In: Proceedings of (WITS 2007) (2007)
2. Sarwar, B.M., Karypis, G., Konstan, J.A., Riedl, J.: Item-based collaborative filtering recommendation algorithms. In: Proceedings of WWW, pp. 285–295 (2001)

3. Mazurowski, M.A.: Estimating confidence of individual rating predictions in collaborative filtering recommender systems. Expert Syst. Appl. **40**(10), 3847–3857 (2013)
4. Shafer, G., Vovk, V.: A tutorial on conformal prediction. J. Mach. Learn. Res. **9**, 371–421 (2008)
5. Vovk, V., Gammerman, A., Shafer, G.: Algorithmic Learning in a Random World. Springer, New York (2005)
6. Proedrou, K., Nouretdinov, I., Vovk, V., Gammerman, A.J.: Transductive confidence machines for pattern recognition. In: Elomaa, T., Mannila, H., Toivonen, H. (eds.) ECML 2002. LNCS (LNAI), vol. 2430, p. 381. Springer, Heidelberg (2002)
7. McNee, S.M., Lam, S.K., Guetzlaff, C., Konstan, J.A., Riedl, J.: Confidence displays and training in recommender systems. In: International Conference on Human-Computer Interaction (2003)
8. McLaughlin, M.R., Herlocker, J.L.: A collaborative filtering algorithm and evaluation metric that accurately model the user experience. SIGIR **2004**, 329–336 (2004)
9. Shani, G., Gunawardana, A.: Evaluating recommendation systems. In: Ricci, F., Rokach, L., Shapira, B., Kantor, P.B. (eds.) Recommender Systems Handbook, pp. 257–297. Springer, New York (2011)
10. Koren, Y., Sill, J.: Ordrec: an ordinal model for predicting personalized item rating distributions. In: RecSys, pp. 117–124 (2011)
11. Vovk, V., Fedorova, V., Nouretdinov, I., Gammerman, A.: Criteria of Efficiency for Conformal Prediction. Technical report (2014)
12. Linden, G., Smith, B., York, J.: Amazon.com recommendations: item-to-item collaborative filtering. IEEE Internet Comput. **7**(1), 76–80 (2003)
13. Lemire, D., Maclachlan, A.: Slope One Predictors for Online Rating-Based Collaborative Filtering. CoRR abs/cs/0702144 (2007)
14. Hill, W., Stead, L., Rosenstein, M., Furnas, G.W.: Recommending and evaluating choices in a virtual community of use. In: Proceedings of ACM CHI95 Conference on Human Factors in Computing Systems, pp. 194–201 (1995)
15. Gunawardana, A., Shani, G.: A survey of accuracy evaluation metrics of recommendation tasks. J. Mach. Learn. Res. **10**, 2935–2962 (2009)
16. Johansson, U., Bostrom, H., Lofstrom, T.: Conformal prediction using decision-trees. In: IEEE 13th International Conference on Data Mining, pp. 330–339 (2013)
17. Koren, Y., Bell, R.M., Volinsky, C.: Matrix factorization techniques for recommender systems. IEEE Comput. **42**(8), 30–37 (2009)
18. Saunders, C., Gammerman, A., Vovk, V.: Transduction with confidence and credibility. In: Proceedings of IJCAI 1999, pp. 722–726 (1999)
19. Johansson, U., Bostrom, H., Lofstrom, T.: Conformal prediction using decision trees. In: IEEE 13th International Conference on Data Mining, pp. 330–339 (2013)
20. Papadopoulos, H., Vovk, V., Gammerman, A.: Conformal prediction with neural networks. In: 19th IEEE ICTAI 2007, pp. 388–395 (2007)
21. Breese, J. S., Heckerman, D., and Kadie, C. Empirical analysis of predictive algorithms for collaborative filtering. In: Proceedings of the 14th Conference on UAI 1998, pp. 43–52 (1998)

Multi-view Representative and Informative Induced Active Learning

Huaxi Huang, Changqing Zhang$^{(\boxtimes)}$, Qinghua Hu, and Pengfei Zhu

School of Computer Science and Technology, Tianjin University, Tianjin, China
{hhx10,zhangchangqing,huqinghua,zhupengfei}@tju.edu.cn

Abstract. Most existing active learning methods often manually label samples and train models with labeled data in an iterative way. Unfortunately, at the early stage of the experiment, few labeled data are available, hence, selecting the most valuable data points to label is necessary and important. To this end, we propose a novel method, called *Multi-view Representative and Informative-induced Active Learning (**MRI-AL**)*, which selects samples of both representativeness and informativeness with the help of complementarity of multiple views. Specifically, subspace reconstruction with structure sparsity technique is employed to ensure the selected samples to be representative, while the global similarity constraint guarantees the informativeness of the selected samples. The proposed method is solved efficiently by alternating direction method of multipliers (ADMM). We empirically show that our method outperforms existing early experimental design approaches.

Keywords: Active learning · Multi-view · Representative · Informative · Subspace learning · Structure sparsity

1 Introduction

Labeling training data (e.g., gene tagging) is a tedious and time consuming task for supervised learning. To alleviate this problem, many researchers have taken significant interests in active learning. Generally, active learning aims to pick up the most valuable examples from massive unlabeled data to label and then add the labeled data to the training set, which reduces the cost for users' involvement in the data labeling. Depending on the stage of experiment, there are two main types of sampling algorithms, i.e., mid-stage experimental design and early-stage experimental design methods. The first one usually selects the informative data points that are easily misclassified. The typical methods include uncertainty sampling [22,24], SG-net [4] and its extensions [17]. The second type of methods tends to select the representative data points which can capture the intrinsic information of the whole data collection (e.g., the cluster centroid in clustering structures). Holding this statistical point, Yu et al. [25] proposed the transductive experimental design (TED). Based on TED Cai et al. [3] proposed manifold adaptive experimental design (MAED) to select the data points in the data manifold adaptive kernel space, Nie et al. [20] proposed robust representation and

© Springer International Publishing Switzerland 2016
R. Booth and M.-L. Zhang (Eds.): PRICAI 2016, LNAI 9810, pp. 139–151, 2016.
DOI: 10.1007/978-3-319-42911-3_12

structured sparsity (RRSS), Hu et al. [11] proposed active learning via neighborhood reconstruction (ALNR). In ALNR, each point is reconstructed by the linear combination of only its neighbors to capture the local geometrical information of data, it is obviously a method to find the representative data points.

All the above methods are designed for the single view setting. The representative algorithm for active learning in multi-view setting is co-testing [18,19]. It employs a two-step iterative strategies. Firstly, it uses the labeled examples to learn a hypothesis in each view, then it applies the learned hypotheses to all unlabeled examples and detects the set of contention points (i.e., unlabeled examples on which the views predict a different label). It is reported that the co-testing algorithm outperforms existing active learners on a variety of real-world domains such as wrapper induction, Web page classification, advertisement removal and discourse tree parsing. The co-testing algorithm is the first multi-view active learning algorithm [19] and it is obviously a mid-stage experimental design algorithm which need a number of labeled data in advance. Actually, most of current multi-views active learning methods belong to the mid-stage experimental design like [5,23] but few methods consider the early stage experimental design. However, in practice, at the very beginning, there are often no labeled samples to use.

To address the above problems, we propose a novel early-stage active learning algorithm for multi-view setting, which aims to make full use of the complementary information across different views at the start of the experiment. As stated in some other mid-stage active learning methods like Huang et al. [12], only considering the representativeness or informativeness are insufficient for selecting the most valuable data points. Hence, our method enforces the selected samples to be both representative and informative in the early stage as shows in Fig. 1. Specifically, the structured sparsity constraint enforces the representativeness of the selected samples, while the global similarity term gives the guarantee for the informativeness. An efficient optimization algorithm is given to solve the problem and we also give the convergence guarantee of the algorithm. We evaluate our method using several benchmark data sets and the experimental results show that our method outperforms other state-of-the-art methods.

Notations. In order to make the reading and understanding more clearly for the proposed method, we write matrices as bold uppercase letters and write

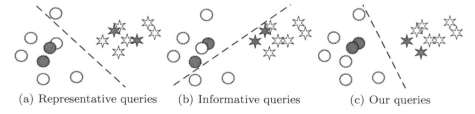

(a) Representative queries   (b) Informative queries   (c) Our queries

Fig. 1. An example of comparison between representative, informative and our proposed methods for selecting queries. Notice that the circles and stars represent two different classes, colored points are selected samples using different methods.

vector as bold lowercase letters. Given a matrix $\mathbf{D} = \{d_{ij}\}$, we denote its i-th row, j-th column as $\mathbf{d}^i, \mathbf{d}_j$, respectively. The ℓ_2-norm of a vector $\mathbf{v}$ is defined as $\|\mathbf{v}\|_2 = \sqrt{\mathbf{v}^T \mathbf{v}}$. The $\ell_{2,1}$-norm of a matrix is defined as $\|\mathbf{D}\|_{2,1} = \sum_{j=1}^m \sqrt{\sum_{i=1}^n d_{ij}^2} = \sum_{j=1}^m \|\mathbf{d}_j\|_2$. For consistency, the $\ell_{2,0}$-norm of a matrix $\mathbf{D}$ is defined as the number of the nonzero rows of $\mathbf{D}$.

2   The Proposed Method

2.1   Background

For a given unlabeled data collection $\mathbf{X} = [\mathbf{x}_1, ..., \mathbf{x}_n] \in \mathbb{R}^{d \times n}$, the goal of active learning is selecting m ($m < n$) most valuable samples for the "Oracle" to label. With these labeled samples, a maximized latent performance could be expected. To this end, Yu et al. [25] propose the Transductive Experimental Design (TED) method as follows

$$\min_{\mathbf{A}, \mathbf{T}} \sum_{s=1}^n \left(\|\mathbf{x}_s - \mathbf{T}\boldsymbol{a}_s\|_2^2 + \lambda \|\boldsymbol{a}_s\|_2^2 \right)$$
$$s.t. \quad \mathbf{A} = [\boldsymbol{a}_1, ..., \boldsymbol{a}_n] \in \mathbb{R}^{m \times n}, \mathbf{T} \subset \mathbf{X}, |\mathbf{T}| = m. \tag{1}$$

where $\mathbf{A}$ is the reconstruction coefficient matrix. The TED solves the problem in a linear way in which each data point can be represented by m samples in a linear combination. However, TED has two shortcomings. First, Eq. (1) leads a combinatorial optimization problem which is NP-hard. Secondly, the least square error could always result in sensitiveness of the outliers.

To solve the above deficiencies, Nie et al. [20] proposed Robust Representation and Structured Sparsity (RRSS) method using $\ell_{2,1}$-norm in both the loss and the regularization item:

$$\min_{\mathbf{A}} \|\mathbf{X} - \mathbf{X}\mathbf{A}\|_{2,1} + \lambda \|\mathbf{A}^T\|_{2,1}$$
$$s.t. \quad \mathbf{A} = [\boldsymbol{a}_1, ..., \boldsymbol{a}_n] \in \mathbb{R}^{n \times n}. \tag{2}$$

In the loss item of RRSS, ℓ_2-norm is applied to all features and ℓ_1-norm is used to all data samples. In this way, the sensitiveness of the data outliers is alleviated with a robust norm. As a regulation item, $\ell_{2,1}$-norm is used to get a column-sparse performance which is aimed to select the most representative samples.

2.2   MRI-AL: Multi-view Representative and Informative-Induced Active Learning

However, RRSS also has the following shortages. First, it only applied to single view data, as a result. It cannot deal with multi-view learning problems which are ubiquitous in real world applications. Second, it only considers the most representative samples, while neglects the informativeness of samples. To extend it to the multi-view setting, we first propose a new multi-view extension. In our method, we aim to select the most representatives from different views. Specifically, selected samples should be the most representative ones across these multiple views. Based on the above, our method has the following formulation:

$$\min_{\mathbf{A}} \|\mathbf{E}\|_{2,1} + \lambda \left\|\mathbf{A}^T\right\|_{2,1}$$
$$s.t. \quad \mathbf{X}_i = \mathbf{X}_i \mathbf{A} + \mathbf{E}_i, i = 1, ..., k$$
$$\mathbf{E} = [\mathbf{E}_1; \mathbf{E}_2; ...; \mathbf{E}_k], \tag{3}$$

where $\mathbf{X}_i$ is the i-th view of k views, while $\mathbf{A}$ is the reconstruction coefficient matrix and $\mathbf{E}_i$ is the loss of reconstruction under the i-th view. $\mathbf{E}$ is the vertical arrangement of all loss matrix $\mathbf{E}_i$.

To integrate these multiple views seamlessly, we impose restrictions in two directions. On the one hand, we enforce the model to learn a shared reconstruction coefficient matrix $\mathbf{A}$ for different views, and on the other hand, by minimizing the $\ell_{2,1}$-norm of $\mathbf{E}$ we enforce the columns of $\mathbf{E}_1, ..., \mathbf{E}_k$ to have the similar sparsity patterns. Hence, we take full advantage of these multiple views to select the most representative samples for active learning. After obtaining the optimal reconstruction coefficient matrix $\mathbf{A}$ in Eq. (3), the most m samples are often obtained by selecting the samples with the top largest row-sum values of $\mathbf{A}$.

In our Multi-View Representative-induced Sparsity Pursuit method (**MVRSP**), the selected points by our method are the most representative ability for the data collection. However, for the classification task, the informativeness is usually critical for classification performance. Hence, we enforce the model to select the samples with both representativeness and informativeness. Holding this consideration, we propose a new regularization term and propose the *Multi-view Representative and Informative-induced Active Learning* (**MRI-AL**) method. As Fig. 1(a) and (b) shows, it is difficult to classify the data well with only considering representativeness or informativeness alone, while our proposed method can deal with the classification task well with the samples of both representativeness or informativeness, as shows in Fig. 1(c).

For the given dataset $\mathbf{X} = [\mathbf{x}_1, ..., \mathbf{x}_n] \in \mathbb{R}^{d \times n}$, we define a set of nonnegative dissimilarities $d_{ij}(i, j = 1, ..., n)$ between every pair of data samples i and j. Then we get a dissimilarity matrix as follows:

$$\mathbf{D} = \begin{bmatrix} \mathbf{d}^1 \\ \vdots \\ \mathbf{d}^n \end{bmatrix} = \begin{bmatrix} d_{11} & d_{12} & ... & d_{1n} \\ \vdots & \vdots & & \vdots \\ d_{n1} & d_{n2} & ... & d_{nn} \end{bmatrix} \in \mathbb{R}^{n \times n}. \tag{4}$$

The dissimilarity matrix $\mathbf{D}$ can indicate the spatial distance of sample pairs. More specifically, we aim to choose the samples which are close to different classes, since these samples are usually difficult to classify. Then, we have the following objective function:

$$\min_{\mathbf{A}} \|\mathbf{E}\|_{2,1} + \lambda \left\|\mathbf{A}^T\right\|_{2,1} + \beta tr(\mathbf{D}^T \mathbf{A})$$

$$s.t. \quad \mathbf{X}_i = \mathbf{X}_i \mathbf{A} + \mathbf{E}_i, i = 1, ..., k \tag{5}$$
$$\mathbf{E} = [\mathbf{E}_1; \mathbf{E}_2; ...; \mathbf{E}_k], \mathbf{A} \geq \mathbf{0}.$$

where $tr(\cdot)$ denotes the trace operator. Notice that, in Eq. (5). We let the coefficient matrix $\mathbf{A}$ is no-negative as Elhamifar et al. [7] did to choose the samples that have a close distance to all other samples. Thus, we use our method aims to select the samples of both representativeness and informativeness from multiple views.

3   Optimization and Analysis

In this section, we will give an algorithm to solve the problem (3) and (5) and a theoretical analysis of the algorithms to show they can be convergence.

3.1   Optimization Procedure of MVRSP

Problem Eq. (3) is convex and can be solved by the ALM method [15]. We convert it into the following equivalent problem firstly:

$$\min_{\mathbf{J}, \mathbf{E}, \mathbf{A}} \left\| \mathbf{J}^T \right\|_{2,1} + \lambda \left\| \mathbf{E} \right\|_{2,1}$$
$$s.t. \quad \mathbf{X}_i = \mathbf{X}_i \mathbf{A} + \mathbf{E}_i, i = 1, ..., k \tag{6}$$
$$\mathbf{A} = \mathbf{J}, \mathbf{E} = [\mathbf{E}_1; \mathbf{E}_2; ...; \mathbf{E}_k].$$

By minimizing the following Augmented Lagrange function, this problem can be solved efficiently:

$$\mathcal{L} = \left\| \mathbf{J}^T \right\|_{2,1} + \lambda \left\| \mathbf{E} \right\|_{2,1} + \sum_{i=1}^{k} \left(\langle \mathbf{Y}_i, \mathbf{X}_i - \mathbf{X}_i \mathbf{A} - \mathbf{E}_i \rangle + \langle \mathbf{W}_i, \mathbf{A} - \mathbf{J} \rangle + \right.$$
$$\left. \frac{\mu}{2} \left\| \mathbf{X}_i - \mathbf{X}_i \mathbf{A} - \mathbf{E}_i \right\|_F^2 + \frac{\mu}{2} \left\| \mathbf{A} - \mathbf{J} \right\|_F^2 \right), \tag{7}$$

where $\mathbf{Y}_i, \mathbf{W}_i$ $(i = 1, ..., k)$ are Lagrange multipliers and μ is a penalty parameter which is a positive value, to get the optimal solution, we first introduce a lemma [16]:

Lemma 1. *Let $\mathbf{Q}$ be a given matrix. If the optimal solution to*

$$\min_{W} \alpha \left\| \mathbf{W} \right\|_{2,1} + \frac{1}{2} \left\| \mathbf{W} - \mathbf{Q} \right\|_F^2$$

is $\mathbf{W}^*$, then the ith column of $\mathbf{W}^*$ is

$$[\mathbf{W}^*]_{:,i} = \begin{cases} \frac{\left\| [\mathbf{Q}]_{:,i} \right\|_2 - \alpha}{\left\| [\mathbf{Q}]_{:,i} \right\|_2} \mathbf{Q}_{:,i}, & if \left\| [\mathbf{Q}]_{:,i} \right\|_2 > \alpha; \\ 0, & otherwise. \end{cases}$$

For Augmented Lagrange function Eq. (7), we fixed other variables except $\mathbf{J}$, then we can get the updating rule of $\mathbf{J}$ for each $\mathbf{W}_i(i = 1, ..., k)$:

$$\mathbf{J} = arg \min_{\mathbf{J}} \frac{1}{k\mu} \left\|\mathbf{J}^T\right\|_{2,1} + \frac{1}{2} \left\|\mathbf{J} - \left(\mathbf{A} + \frac{\mathbf{W}_i}{\mu}\right)\right\|_F^2. \tag{8}$$

Notice that to get a similar result with lemma [16], we just have:

$$[\mathbf{J}^*]_{i,:} = \begin{cases} \frac{\|[\mathbf{J}]_{i,:}\|_2 - \frac{1}{k\mu}}{\|[\mathbf{J}]_{i,:}\|_2} \mathbf{J}_{i,:}, & if \left\|[\mathbf{J}]_{i,:}\right\|_2 > \frac{1}{k\mu}; \\ 0, & otherwise, \end{cases} \tag{9}$$

Like the updating rule of $\mathbf{J}$, we get the updating rule of $\mathbf{A}$ for each $\mathbf{X}_i$, $\mathbf{Y}_i$, $\mathbf{E}_i$, $\mathbf{W}_i (i = 1, ..., k)$ and $\mathbf{I}$ is a $n \times n$ identity matrix:

$$\mathbf{A} = \frac{1}{k} \left\{ \sum_{i=0}^{k} \left(\mathbf{I} + \mathbf{X}_i^T \mathbf{X}_i\right)^{-1} \left(\mathbf{X}_i^T \mathbf{X}_i - \mathbf{X}_i^T \mathbf{E}_i + \mathbf{J} + \frac{\mathbf{X}_i^T \mathbf{Y}_i - \mathbf{W}_i}{\mu}\right) \right\}. \tag{10}$$

As the similar way with $\mathbf{J}$ and $\mathbf{A}$, the updating rule of $\mathbf{E}$ is:

$$\mathbf{E} = arg \min_{\mathbf{E}} \frac{\lambda}{\mu} \|\mathbf{E}\|_{2,1} + \frac{1}{2} \|\mathbf{E} - \mathbf{G}\|_F^2$$
$$\mathbf{G} = [\mathbf{G}_1; \mathbf{G}_2; ...; \mathbf{G}_k] \tag{11}$$
$$\mathbf{G}_i = \mathbf{X}_i - \mathbf{X}_i \mathbf{A} + \frac{\mathbf{Y}_i}{\mu}, i = 1, ..., k.$$

The inexact ALM method can be also called the Alternating Direction Mehtod (ADM), is outlined in Algorithm 1. The solution of $\mathbf{J}$ and $\mathbf{E}$ is solved via Lemma 1. Because of the subproblems of the algorithm are convex. They have close-form solution. Notice that ρ is a parameter to control the convergence speed.

Algorithm 1. Solving Problem (6) by ADM

Input: Data matrics $\{\mathbf{X}_i\}$, parameter λ.
Output: Matric $\mathbf{A}$
 1: $\mathbf{A} \leftarrow 0, \mathbf{J} \leftarrow 0, k \leftarrow NumberOfModals$
 2: $\mathbf{Y}_i \leftarrow 0, \mathbf{W}_i \leftarrow 0, \mathbf{E}_i \leftarrow 0 \quad for \ i = 1, ..., k$
 3: **while** not converged **do**
 4:     **for** $i = 1 \rightarrow k$ **do**
 5:         Solve $\mathbf{J}$ according to the solution of $Eq.$ (8)
 6:         Solve $\mathbf{A}$ according to the solution of $Eq.$ (10)
 7:         Solve $\mathbf{E}$ according to the solution of $Eq.$ (11)
 8:         $\mathbf{Y}_j \leftarrow \mathbf{Y}_j + \mu (\mathbf{X}_j - \mathbf{X}_j \mathbf{A} - \mathbf{E}_j)$
 9:         $\mathbf{W}_j \leftarrow \mathbf{W}_j + \mu (\mathbf{A} - \mathbf{J})$
 10:     **end for**
 11:     $\mu = \min (\rho\mu, \mu_{max})$
 12: **end while**
 13: **return** $\mathbf{A}$

3.2   Optimization Procedure of MRI-AL

Problem Eq. (5) has similar properties with Eq. (3) so we also use ADM to solve it. As before we have:

$$\min_{\mathbf{J},\mathbf{E},\mathbf{A}} \left\|\mathbf{J}^T\right\|_{2,1} + \lambda\left\|\mathbf{E}\right\|_{2,1} + \beta tr(\mathbf{D}^T\mathbf{J})$$

$$s.t. \quad \mathbf{X}_i = \mathbf{X}_i\mathbf{A} + \mathbf{E}_i, i = 1, ..., k \tag{12}$$

$$\mathbf{A} = \mathbf{J}, \mathbf{E} = [\mathbf{E}_1; \mathbf{E}_2; ...; \mathbf{E}_k], \mathbf{J} \geq \mathbf{0}.$$

Then we get the Augmented Lagrange function:

$$\mathcal{L}' = \left\|\mathbf{J}^T\right\|_{2,1} + \lambda\left\|\mathbf{E}\right\|_{2,1} + \sum_{i=1}^{k}(\langle\mathbf{Y}_i, \mathbf{X}_i - \mathbf{X}_i\mathbf{A} - \mathbf{E}_i\rangle + \langle\mathbf{W}_i, \mathbf{A} - \mathbf{J}\rangle +$$

$$\frac{\mu}{2}\left\|\mathbf{X}_i - \mathbf{X}_i\mathbf{A} - \mathbf{E}_i\right\|_F^2 + \frac{\mu}{2}\left\|\mathbf{A} - \mathbf{J}\right\|_F^2) + \beta tr(\mathbf{D}^T\mathbf{J}). \tag{13}$$

The updating rule of $\mathbf{A}$ and $\mathbf{E}$ is same as MVRSP's optimization. For $\mathbf{J}$ we have:

$$\mathbf{J} = arg \min_{\mathbf{J}} \frac{1}{k\mu}\left\|\mathbf{J}^T\right\|_{2,1} + \frac{1}{2}\left\|\mathbf{J} - \left(\mathbf{A} + \frac{\mathbf{W}_i}{\mu}\right)\right\|_F^2 + \frac{\beta tr\left(\mathbf{D}^T\mathbf{J}\right)}{k\mu}. \tag{14}$$

To get an optimal solution of each iteration in ADM in Eq. (14) we use the similar rule of Lemma 1:

$$[\mathbf{J}^*]_{i,:} = \begin{cases} \frac{\left\|[\mathbf{J}]_{i,:}\right\|_2 - \gamma}{\left\|[\mathbf{J}]_{i,:}\right\|_2}\mathbf{J}_{i,:}, & if \left\|[\mathbf{J}]_{i,:}\right\|_2 > \gamma; \\ 0, & otherwise, \end{cases} \quad where\,\gamma = \frac{1 + \beta\left\|[\mathbf{D}]_{i,:}\right\|_2}{k\mu}. \tag{15}$$

Solving Problem (12) is similarly with Algorithm 1, just need to replace the solution of $\mathbf{J}$ with Eq. (14).

3.3   On the Convergence Properties

The exact ALM algorithm will converge when the objective function is smooth, which is proved by Bertsekas [2]. On the inexact ALM also called as the ADM which is a variation of the exact ALM. Its convergence has also been proved when the blocks number is less than or equal two by Afonso et al. [1] and Lin et al. [15]. For more blocks, it is difficult to ensure the convergence of the ADM [26]. In our Algorithm, there are seven blocks and the objective function Eq. (3) is not smooth. So proving it's convergence strictly in theory will be difficult. Nevertheless, according to Eckstein and Bertsekas [6], they give some theoretical results to ensure the ADM to converge. In particular, there exist two sufficient but unnecessary conditions for our Algorithm to converge, one is the matrixes $\mathbf{X}_i(i = 1, ..., k)$ which represent the different views of the data are full column rank, the other is that during each optimization iteration, the optimality gap is monotone decreasing like Lang et al. [14] discussed. What's more, the inexact ALM is well known to perform generally well in reality just as illustrated in Zhang [26].

4   Experiments

In this section, we empirically evaluate the proposed methods on both synthetic data and real-world data.

4.1   Experiment on Synthetic Data

In our experiments, we firstly give an experimental result on a multi-view synthetic dataset to validate the effectiveness of MVRSP. The synthetic $\mathbf{A}$ has 10 out of 100 nonzero rows. The groundtruth of $\mathbf{A}^T$ can be seen from Fig. 2(c). Data observations have two views. The first view has 100 dimensions and the second view has 150 dimensions which can be seen from Fig. 2(a) and Fig. 2(b). Figure 2(d) is the learned $\mathbf{A}^T$ after using our multi-view active learning algorithm. The result clearly indicates that our method could select the underlying most representative samples that the learned matrix $\mathbf{A}^T$ gets the same column-sparse with groundtruth which means our method selects the candidate samples.

4.2   Experiment on Real-World Data

We conducted our method on three public image data sets as follows:

  □ Yale [8] contains 165 images of 15 individuals. For each person, there are 11 images with different facial expression and shooting conditions, we exact Gray, Lbp and Gabor features of each image as their views.

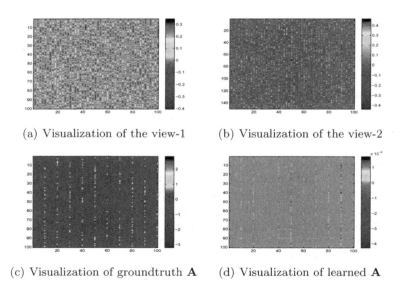

(a) Visualization of the view-1   (b) Visualization of the view-2

(c) Visualization of groundtruth $\mathbf{A}$ (d) Visualization of learned $\mathbf{A}$

Fig. 2. The experimental result on the noised multi-view synthetic data. (a) and (b) are the first and second views, respectively. (c) shows groundtruth matrix $\mathbf{A}$. (d) is the learned matrix $\mathbf{A}$. (Color figure online)

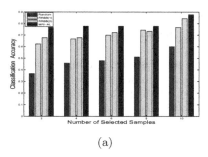
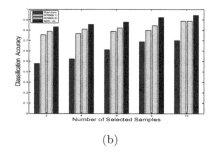

(a)   (b)

Fig. 3. Accuracy of binary classification. The left figure is the result of the first two classes and the right figure is the result of the last two classes. (Color figure online)

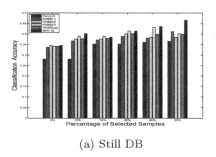
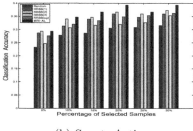

(a) Still DB   (b) Sports Action

Fig. 4. Accuracy of multiple classification on the Still DB (left) and Sports Action (right). Notice that RRSS(1) refer to the method that we use RRSS on the first modality of the dataset so as to RRSS(2) and RRSS(3), RRSS(ag) refer to the method that we average the three classification accuracies above. (Color figure online)

☐ Still DB [13] is a still images dataset which is for recognizing actions. It contains 467 images and six classes. We extract Sift Bow, Color Sift Bow and Shape context Bow as their features.

☐ Sports [10] is also an images dataset for recognizing actions which contain 300 images and six classes. We also extract Sift Bow, Color Sift Bow and Shape context Bow as their features.

☐ Corel [9] is a widely used image dataset in the computer version, we randomly selected 4 categories, each contains 90 photographs with two views: DenseSift and Gist as our Binary Multi-view Classification dataset.

We first evaluate the proposed method MRI-AL in a multi-view binary classification task with Corel dataset. Then we evaluate the proposed method MRI-AL in multi-view multiple classification tasks on the Still DB and Sports Action data sets. In the multi-view experiment, we use the MKL method [21] as the classification model, which is an extension of SVM for multi-view setting. We implemented the MKL learning algorithm for multi-class classification based on simpleMKL [21]. For MKL method, an independent kernel was set for each individual feature group. We have tried some different kernels including RBF kernel,

Polynomial kernel and Linear kernel. We selected the best kernel for every view (feature groups) depending on the classification performance on a small validation set.

Since most existing multi-view approaches are designed for mid-stage task, it is difficult to compare them with ours. So we employ the randomly selection method as baseline. We also compare MRI-AL with RRSS [20]. RRSS is a single-view active learning method, so use RRSS to select samples in each single view. Moreover we also report the average performance of the three views. For MRI-AL, we first use the Euclidean distance to compute the dissimilarity matrix on each single view, then we use the average number of the computed matrixes as the final dissimilarity matrix $\mathbf{D}$.

We conduct our experiments as follows. For each database, we first randomly pick up half of the data samples as a candidate set for training, in which we use the MRI-AL to select some samples as training set. Then we train a classification model. Finally, with the classifier we classify the rest of the samples which are considered as the test set.

4.3   Experiments Results Analysis

According to Fig. 3. We can observe that MRI-AL have the obvious advantage over the compared methods. The classification accuracy with different proportions of selected samples is shown as Fig. 4. As a whole, like the results shows, the classification performance of active learning methods includes our MRI-AL on multi-view data and RRSS on every view data is generally superior to the Random method that is also called as passive learning method. So it reflects the value of active learning.

Specifically, it is observed from Fig. 4 that different views have different representative ability for the samples. For example, on the Still DB, view-3 shows a better classification performance than other views, while on Sports Action, view-2 has a good classification performance. Therefore, we can draw a conclusion that different views have different values for different data, which impose the necessity of the fusion of different views. From Still DB, our MRI-AL also obtains a significant improvement over the single view ones. For instance, our method achieves 46.58 % compared to Random (36.54 %), RRSS(1) (41.18 %), RRSS(2) (38.46 %), RRSS(3) (40.17 %) and RRSS(ag) (39.94 %) in terms of accuracy when selected 30 % samples to be labeled and train a classifier.

4.4   Discussion

When we use Yale to evaluate the proposed method MRI-AL, we find it is hard to select all the categories using both passive learning and active learning to select the samples in this dataset. So we use the class coverage rate (number of selected samples' category / number of all category) to evaluate the performance of the selecting methods. We compare four methods: Random selects samples, RRSS, MVRSP and MRI-AL. For RRSS, we use each view of the data to get the sample coverage rate. For each experiment, we select 15,25,35,45,50 samples respectively to estimate the compared methods. Table 1 shows the class coverage rates with different methods for different amount of samples. Although our method performs better than other methods it still cannot cover all categories in this dataset when only selected a few samples.

Table 1. Sample coverage rate which is the number of selected category divided by the number of the all category, we show the results when selected 15,25,35,45,50 samples on different learning methods.

	15	25	35	45	50
Random	67 %	84 %	93 %	97 %	99 %
RRSS(1)	47 %	67 %	93 %	100 %	100 %
RRSS(2)	53 %	73 %	87 %	97 %	100 %
RRSS(3)	47 %	73 %	97 %	100 %	100 %
MVRSP	53 %	67 %	97 %	100 %	100 %
MRI-AL	60 %	87 %	100 %	100 %	100 %

5   Conclusion

In this paper, we introduce an early-stage multi-view active learning method to deal with the multi-view data. Firstly, we proposed a multi-view method named MVRSP using the robust sparse representation. Secondly, we improved our method by adding a global similarity term to ensure the selected samples of both good representativeness and informativeness. Our method achieves promising performance on binary and multiple classification tasks.

In future work, we want to deal with the problem like Yale dataset that the current methods cannot select all the categories. In addition, we will conduct our method on large dataset to further evaluate the efficiency of our method.

Acknowledgments. This work was supported by the National Program on Key Basic Research Project under Grant 2013CB329304, the National Natural Science Foundation of China under Grants 61502332, 61432011, 61222210.

References

1. Afonso, M.V., Bioucas-Dias, J.M., Figueiredo, M.A.T.: An augmented lagrangian approach to the constrained optimization formulation of imaging inverse problems. IEEE Trans. Image Process. **20**(3), 681–695 (2011)
2. Bertsekas, D.P.: Constrained Optimization and Lagrange Multiplier Methods. Academic Press, New York (2014)

3. Cai, D., He, X.: Manifold adaptive experimental design for text categorization. IEEE Trans. Knowl. Data Eng. **24**(4), 707–719 (2012)
4. Cohn, D., Atlas, L., Ladner, R.: Improving generalization with active learning. Mach. Learn. **15**(2), 201–221 (1994)
5. Di, W., Crawford, M.M.: Active learning via multi-view and local proximity co-regularization for hyperspectral image classification. IEEE J. Sel. Top. Sign. Process. **5**(3), 618–628 (2011)
6. Eckstein, J., Bertsekas, D.P.: On the DouglasRachford splitting method and the proximal point algorithm for maximal monotone operators. Math. Program. **55**(1–3), 293–318 (1992)
7. Elhamifar, E., Sapiro, G., Vidal, R.: Finding exemplars from pairwise dissimilarities via simultaneous sparse recovery. In: Advances in Neural Information Processing Systems, pp. 19–27 (2012)
8. Georghiades, A.S., Belhumeur, P.N., Kriegman, D.J.: From few to many: Illumination cone models for face recognition under variable lighting and pose. IEEE Trans. Pattern Anal. Mach. Intell. **23**(6), 643–660 (2001)
9. Guillaumin, M., Mensink, T., Verbeek, J., Schmid, C.: Tagprop: Discriminative metric learning in nearest neighbor models for image auto-annotation. In: IEEE 12th International Conference on Computer Vision, pp. 309–316 (2009)
10. Gupta, A., Kembhavi, A., Davis, L.S.: Observing human-object interactions: Using spatial and functional compatibility for recognition. IEEE Trans. Pattern Anal. Mach. Intell. **31**(10), 1775–1789 (2009)
11. Hu, Y., Zhang, D., Jin, Z., Cai, D., He, X.: Active learning via neighborhood reconstruction. In: Proceedings of the Twenty-Third International Joint Conference on Artificial Intelligence, vol. 31(10), pp. 1775–1789 (2009)
12. Huang, S.-J., Jin, R., Zhou, Z.-H.: Active learning by querying informative and representative examples. In: Advances in Neural Information Processing Systems, pp. 1415–1421 (2013)
13. Ikizler, N., Cinbis, R.G., Pehlivan, S., Duygulu, P.: Recognizing actions from still images. In: 19th International Conference on Pattern Recognition, pp. 1–4. IEEE (2008)
14. Lang, C., Liu, G., Yu, J., Yan, S.: Saliency detection by multitask sparsity pursuit. IEEE Trans. Image Process. **21**(3), 1327–1338 (2012)
15. Lin, Z., Chen, M., Ma, Y.: The augmented lagrange multiplier method for exact recovery of corrupted low-rank matrices (2010). arXiv preprint arXiv:1009.5055
16. Liu, G., Lin, Z., Yu, Y.: Robust subspace segmentation by low-rank representation. In: ICML, pp. 663–670 (2010)
17. Melville, P., Mooney, R.J.: Diverse ensembles for active learning. In: ICML, p. 74 (2004)
18. Muslea, I., Minton, S., Knoblock, C.A.: Selective sampling with redundant views. In: AAAI/IAAI, pp. 621–626 (2000)
19. Muslea, I., Minton, S., Knoblock, C.A.: Active learning with multiple views. J. Artif. Intell. Res. **27**, 203–233 (2006)
20. Nie, F., Wang, H., Huang, H., Ding, C.: Early active learning via robust representation and structured sparsity. In: Proceedings of the Twenty-Third International Joint Conference on Artificial Intelligence, IJCAI 13, pp. 1572–1578. AAAI Press (2013)
21. Rakotomamonjy, A., Bach, F., Canu, S., Grandvalet, Y.: SimpleMKL. J. Mach. Learn. Res. **9**, 2491–2521 (2008)
22. Schohn, G., Cohn, D.: Less is more: Active learning with support vector machines. In: ICML, pp. 839–846 (2000)

23. Sun, S.: Semantic features for multi-view semi-supervised and active learning of text classification. In: IEEE International Conference on Data Mining Workshops, pp. 731–735. IEEE (2008)

24. Thompson, C.A., Califf, M.E., Mooney, R.J.: Active learning for natural language parsing and information extraction. In: ICML, pp. 406–414 (1999)

25. Yu, K., Bi, J., Tresp, V.: Active learning via transductive experimental design. In: Proceedings of the 23rd International Conference on Machine Learning, pp. 1081–1088 (2006)

26. Zhang, Y.: Recent advances in alternating direction methods: Practice and theory. In: IPAM Workshop: Numerical Methods for Continuous Optimization. UCLA, Los Angeles (2010)

Computing Probabilistic Assumption-Based Argumentation

Nguyen Duy Hung[(✉)]

Sirindhorn International Institute of Technology, Bangkok, Thailand
hung.nd.siit@gmail.com

Abstract. We develop inference procedures for a recently proposed model of probabilistic argumentation called PABA, taking advantages of well-established dialectical proof procedures for Assumption-based Argumentation and Bayesian Network algorithms. We establish the soundness and termination of our inference procedures for a general class of PABA frameworks. We also discuss how to translate other models of probabilistic argumentation into this class of PABA frameworks so that our inference procedures can be used for these models as well.

Keywords: Probabilistic argumentation · Inference procedures · Bayesian networks

1 Introduction

Standard Abstract Argumentation (AA [3]) is inadequate in capturing argumentation processes involved probabilities such as the following.

Example 1 (Borrowed from [5]). John sued Henry for the damage caused to him when he drove off the road to avoid hitting Henry's cow.

- John: Henry should pay damage because Henry is the owner of the cow and the cow caused the accident (J_1).
- Henry: John was negligent as evidences at the accident location show that John was driving fast. Hence the cow was not the cause of the accident (H_1).

Let's try to construct an AA framework $\mathcal{F} = (AR, Att)$ to represent the judge's beliefs. The judge may consider J_1 as an argument proper, but not H_1 because according to him, the evidences at the accident location gives only some probability (p_0) that John was driving fast; and even if John was driving fast, the accident is caused by his fast-driving with some other probability p_1. Hence while the representation of J_1 is quite simple: $J_1 \in AR$, there is no perfect representation for H_1. $H_1 \notin AR$ (resp. $H_1 \in AR$) would mean that the judge would undoubtedly find for John (resp. Henry). However, in fact the chance that a party wins depends on the values that the judge assigns to p_0 and p_1.

© Springer International Publishing Switzerland 2016
R. Booth and M.-L. Zhang (Eds.): PRICAI 2016, LNAI 9810, pp. 152–166, 2016.
DOI: 10.1007/978-3-319-42911-3_13

To remedy the above situation, several authors extend AA with probability theory, resulting in different models of Probabilistic Argumentation. Of our interest is the Probabilistic Assumption-based Argumentation framework of [5] ($PABA$) extending an instance of AA called Assumption-based Argumentation (ABA [2,4]). To anchor our contributions, let's loosely recall some technicalities. An ABA framework comprises inference rules in the form $c \leftarrow b_1, \ldots b_n$, representing that proposition c holds whenever propositions $b_1, \ldots b_n$ hold (b_i can be an assumption but not c). An $PABA$ framework is a triple $(\mathcal{A}_p, \mathcal{R}_p, \mathcal{F})$ where $\mathcal{A}_p$ is a set of (positive) probabilistic assumptions, $\mathcal{R}_p$ is a set of probabilistic rules and $\mathcal{F}$ is an ABA framework. A probabilistic rule in $PABA$ also has the same form as an inference rule, except that its head is a proposition of the form $[\alpha : x]$ representing that the probability of probabilistic assumption α is x.

Example 2 (Cont. Example 1). The judge's beliefs is representable [5] by $PABA \, \mathcal{P} = (\mathcal{A}_p, \mathcal{R}_p, \mathcal{F})$ where $\mathcal{A}_p = \{p_0, p_1\}$; $\mathcal{F}$ consists of assumptions $\sim forceMajeure$, $\sim johnNegligent$ (with contraries $forceMajeure$, $johnNegligent$) and inference rules $r_1, \ldots, r_5$; while $\mathcal{R}_p$ consists of probabilistic rules $r_6, \ldots, r_9$ where[1]

r_1 : $henryPay \leftarrow henryOwnerOfCow, cowCauseAccident, \sim forceMa$
$jeure$

r_2 : $cowCauseAccident \leftarrow \sim johnNegligent$ r_3 : $henryOwnerOfCow \leftarrow$

r_4 : $johnNegligent \leftarrow drivingFast, p_1$ r_5 : $drivingFast \leftarrow p_0$

r_6 : $[p_0 : 0.8] \leftarrow$ r_7 : $[\neg p_0 : 0.2] \leftarrow$ r_8 : $[p_1 : 0.75] \leftarrow$ r_9 : $[\neg p_1 : 0.25] \leftarrow$

A possible world of $PABA \, \mathcal{P} = (\mathcal{A}_p, \mathcal{R}_p, \mathcal{F})$ is a complete truth assignment over $\mathcal{A}_p$. For a possible world ω, $P(\omega)$ refers to the probability of ω generated by $\mathcal{P}$ (see Definition 6); and $\mathcal{F}_\omega$ denotes the revised version of $\mathcal{F}$ assuming that ω is the actual world. Each ABA semantics sem induces a $PABA$ semantics $Prob_{sem}$ stating the probability of the acceptability of a given proposition as follows.

$$Prob_{sem}(\pi) = \sum_{\omega \in \mathcal{W}: ABA \, \mathcal{F}_\omega \vdash_{sem} \pi} P(\omega)$$

where $\mathcal{W}$ is the set of all possible worlds; $ABA \, \mathcal{F}_\omega \vdash_{sem} \pi$ states that π is acceptable in $ABA \, \mathcal{F}_\omega$ under semantics sem.

Example 3 (Cont. Example 2). There are four possible worlds, in which the acceptability of proposition $henryPay$ under any ABA semantics sem is shown below. Clearly $Prob_{sem}(henryPay) = 1 - P(\{p_0, p_1\}) = 1 - 0.8 \times 0.75 = 0.4$.

Possible world	$\{p_0, p_1\}$	$\{\neg p_0, p_1\}$	$\{p_0, \neg p_1\}$	$\{\neg p_0, \neg p_1\}$
$henryPay$ is acceptable?	no	yes	yes	yes

---

[1] Probabilistic values are made up for demonstration.

Inference procedures for $PABA$ are procedures computing $Prob_{sem}(.)$ which have been unexplored. Note that since an $ABA \mathcal{F}$ can be represented by an $PABA$ framework with empty sets $\mathcal{A}_p$ and $\mathcal{R}_p{}^2$, inference procedures for $PABA$ subsume proof procedures for ABA, which have been developed in [2,4]. On the other hand, given a proof procedure for ABA semantics sem, one can computing $Prob_{sem}(\pi)$ by checking if $ABA \mathcal{F}_\omega \vdash_{sem} \pi$ for each possible world ω. Unfortunately this naive approach *always* results in an exponential blowup since there are as many as $2^{|\mathcal{A}_p|}$ possible worlds. It turns out that in the worst case we can not avoid this exponential blowup since $PABA$ subsumes Bayesian networks known to be exponentially complex in the worst case. However, as there are many inference algorithms Bayesian networks working efficiently in the average case, they may exist inference procedures for $PABA$ working efficiently in the average case. In this paper, we aim at developing such inference procedures. We establish the soundness and termination of our procedures for a general class of PABA frameworks. We also implement them to obtain an PABA inference engine capable of computing the credulous semantics and the ideal semantics of PABA[3]. Empirical evaluations of the engine, however, remains a future work.

The paper is organized as follows: Sect. 2 is a review of abstract argumentation and probabilistic assumption-based argumentation; Sect. 3 presents the theoretical basis of our inference procedures; Sect. 4 presents our inference procedures (due to space limitation, we present only the computation of PABA's credulous semantics and skip proofs of lemmas and theorems); Sect. 5 discusses translations of other models of probabilistic argumentation [7,8] into PABA in order to widen the applicability of our contributions and concludes.

2   Background on Argumentation

2.1   Abstract Argumentation

An AA framework [3] is a pair (AR, Att) where AR is a set of arguments, $Att \subseteq AR \times AR$ and $(A, B) \in Att$ means that A attacks B. $S \subseteq AR$ attacks $A \in AR$ iff $(B, A) \in Att$ for some $B \in S$. $A \in AR$ is acceptable wrt to S iff S attacks every argument attacking A. S is *conflict-free* iff S does not attack itself; *admissible* iff S is conflict-free and each argument in S is acceptable wrt S; *complete* iff S is admissible and contains every arguments acceptable wrt S; a *preferred* extension iff S is a maximal (wrt set inclusion) complete set; the *grounded* extension iff S is the least complete set; the *ideal* extension iff it is the maximal admissible set contained in every preferred extensions. An argument A is accepted under semantics $sem \in \{cr, gr, id\}^4$, denoted $AA \mathcal{F} \vdash_{sem} A$, iff A is in a sem extension.

---

[2] $ABA \mathcal{F} \vdash_{sem} \pi$ iff wrt this $PABA$ framework, $Prob_{sem}(\pi) = 1$.

[3] See https://pengine.herokuapp.com.

[4] Preferred/grounded/ideal semantics.

2.2   Assumption-Based Argumentation

As AA ignores the internal structure of argument, an instance of AA called Assumption-Based Argumentation (ABA [2,4]) defines arguments by deductive proofs based on assumptions and inference rules. Assuming a language $\mathcal{L}$ consisting of countably many sentences, an ABA framework is a triple $\mathcal{F} = (\mathcal{R}, \mathcal{A}, ^{-})$ where $\mathcal{R}$ is a set of inference rules of the form $r : l_0 \leftarrow l_1, \ldots, l_n$ $(n \geq 0)^5$, $\mathcal{A} \subseteq \mathcal{L}$ is a set of assumptions, and $^{-}$ is a (total) one-to-one mapping from $\mathcal{A}$ into $\mathcal{L}$, where $\overline{x}$ is referred to as the *contrary* of x. Assumptions do not appear in the heads of inference rules and contraries of assumptions are not assumptions.

A *(backward) deduction* of a conclusion π supported by a set of premises Q is a sequence of sets $S_1, S_2, \ldots, S_n$ where $S_i \subseteq \mathcal{L}$, $S_1 = \{\pi\}$, $S_n = Q$, and for every i, where σ is the selected proposition in S_i: $\sigma \notin Q$ and $S_{i+1} = S_i \setminus \{\sigma\} \cup body(r)$ for some inference rule $r \in \mathcal{R}$ with $head(r) = \sigma$.

An argument for $\pi \in \mathcal{L}$ supported by a set of assumptions Q is a deduction d from π to Q and denoted by (Q, d, π). An argument (Q, d, π) attacks an argument (Q', d', π') if π is the contrary of some assumption in Q'. For simplicity, we often refer to an argument (Q, d, π) by (Q, π) if there is no possibility for mistake.

A proposition π is said to be credulously/groundedly/ideally accepted in ABA $\mathcal{F}$, denoted ABA $\mathcal{F} \vdash_{cr} \pi$ (resp. ABA $\mathcal{F} \vdash_{gr} \pi$ and ABA $\mathcal{F} \vdash_{id} \pi$) if in the AA framework consisting of above defined arguments and attacks, there is an argument for π accepted under the credulous/grounded/ideal semantics.

2.3   Probabilistic Assumption-Based Argumentation

For clarity and modification, we break down the original definition of PABA (Definition 2.1 of [5] into two Definitions 1 and 2 below, where Definition 2 in fact slightly relaxes Definition 2.1 of [5], and as a result, our class of PABA frameworks subsumes the class of PABA frameworks in [5][6]. We also extend the definition of PABA's grounded semantics of [5] to define other semantics of PABA.

Definition 1 [5]. *A probabilistic assumption-based argumentation (PABA) framework $\mathcal{P}$ is a triple $(\mathcal{A}_p, \mathcal{R}_p, \mathcal{F})$ satisfying the following properties*

1. $\mathcal{F} = (\mathcal{R}, \mathcal{A}, ^{-})$ *is an ABA framework.*
2. $\mathcal{A}_p$ *is a finite set of* **positive probabilistic assumptions**. *Elements of* $\neg \mathcal{A}_p = \{\neg p \mid p \in \mathcal{A}_p\}$ *are called* **negative probabilistic assumptions**[7].
3. $\mathcal{R}_p$ *is a set of probabilistic rules of the form*

$$[\alpha : x] \leftarrow \beta_1, \ldots, \beta_n \; n \geq 0, x \in [0,1], \alpha \in \mathcal{A}_p \cup \neg \mathcal{A}_p.$$

where $[\alpha : x]$, called a **probabilistic proposition**, *represents that the probability of probabilistic assumption α is x.*

---

[5] For convenience, define $head(r) = l_0$ and $body(r) = \{l_1, \ldots l_n\}$.
[6] Any PABA framework in [5] is also an PABA framework in our extended definition, but the reverse may not hold.
[7] $\neg$ is the classical negation operator.

Definition 2 [5]. *PABA* $\mathcal{P} = (\mathcal{A}_p, \mathcal{R}_p, \mathcal{F})$ *is said to be **well-formed** if the following syntactic constraints are satisfied.*

1. *For each probabilistic assumption* $\alpha \in \mathcal{A}_p \cup \neg\mathcal{A}_p$
   (a) α *does not occur in* $\mathcal{A}$ *as well as in the head of any rule in* $\mathcal{R}$, *and*
   (b) $[\alpha : x]$ *does not occurs in the body of any rule in* $\mathcal{R}$ *or* $\mathcal{R}_p$.
2. *If a rule of the form* $[\alpha : x] \leftarrow \beta_1, \ldots, \beta_n$ *appears in* $\mathcal{R}_p$, *then* $\mathcal{R}_p$ *also contains a complementary rule* $[\neg\alpha : 1 - x] \leftarrow \beta_1, \ldots, \beta_n$[8].
3. *For each probabilistic assumption* α, *there exists a set of probabilistic assumptions* $Pa_\alpha \subseteq \mathcal{A}_p$ *such that for each maximally consistent subset* $\{\beta_1, \ldots, \beta_m\}$ *of* $Pa_\alpha \cup \neg Pa_\alpha$, $\mathcal{R}_p$ *contains a rule* $[\alpha : x] \leftarrow \beta_1, \ldots, \beta_m$ *(and complementary rule* $[\neg\alpha : 1 - x] \leftarrow \beta_1, \ldots, \beta_m$).
4. *If two rules of the form* $r_1 : [\alpha : x] \leftarrow \ldots$ *and* $r_2 : [\alpha : y] \leftarrow \ldots$ *appear in* $\mathcal{R}_p$ *and* $x \neq y$, *then either conditions below holds*
   (a) $body(r_1) \subset body(r_2)$ *or* $body(r_2) \subset body(r_1)$.
   (b) *There is a probabilistic assumption* $\alpha \in body(r_1)$ *such that* $\neg\alpha \in body(r_2)$

Note that in [5], the well-formedness condition consists of constraints 1, 2, 4(a); and a more rigid version of constraint 3 with $Pa_\alpha = \emptyset$, which implies that for each probabilistic assumption α, $\mathcal{R}_p$ must contain two rules of the forms $[\alpha : x] \leftarrow$ and $[\neg\alpha : 1 - x] \leftarrow$ (which, according to [5], encode the default/unconditional probability of α). So the *PABA* given in Example 4 below[9] is not well-formed according to [5]. We do not require $Pa_\alpha = \emptyset$ because we want to have *Bayesian* PABA frameworks, defined as follows, to be well-formed.

Definition 3. *An PABA* $\mathcal{P} = (\mathcal{A}_p, \mathcal{R}_p, \mathcal{F})$ *is said to **include** a Bayesian network* $\mathcal{N} = (\mathcal{G}, CPTs), \mathcal{G} = (V, E)$ *where* V *consists of only binary variables, if* $\mathcal{A}_p, \mathcal{R}_p$ *represents the same probabilistic information as* $\mathcal{N}$[10]. *A **Bayesian PABA** framework is an PABA framework that includes a Bayesian network.*

Example 4. Below are a Bayesian *PABA* $\mathcal{P} = (\mathcal{A}_p, \mathcal{R}_p, \mathcal{F})$ and its network.

- $\mathcal{F} = (\mathcal{R}, \mathcal{A}, \bar{})$ where $\mathcal{A} = \{\alpha, \beta, \gamma, \eta\}$ and $\bar\alpha = \neg\alpha$, $\bar\beta = \neg\beta$, $\bar\gamma = \neg\gamma$ and $\bar\delta = \neg\delta$ and $\mathcal{R}$ consists of $r_0 : \neg\alpha \leftarrow \alpha, p_0$ $r_1 : \neg\alpha \leftarrow \beta, p_1$ $r_2 : \neg\beta \leftarrow \alpha, p_2$ $r_3 : \neg\gamma \leftarrow \delta, p_3$ and $r_4 : \neg\delta \leftarrow \gamma, p_4$
- $\mathcal{A}_p = \{p_0, p_1, p_2, p_3, p_4\}$ and $\mathcal{R}_p$ consists of the following probabilistic rules
  $[p_0 : .1] \leftarrow p_2$ $[p_0 : .9] \leftarrow \neg p_2$ $[p_1 : .95] \leftarrow p_4$ $[p_1 : .2] \leftarrow \neg p_4$
  $[p_2 : .3] \leftarrow p_3, p_4$ $[p_2 : .05] \leftarrow p_3, \neg p_4$ $[p_2 : .9] \leftarrow \neg p_3, p_4$
  $[p_2 : .5] \leftarrow \neg p_3, \neg p_4$ $[p_3 : .6] \leftarrow$ $[p_4 : .7] \leftarrow$

---

[8] In examples, we will not list complementary rules to save space.
[9] We will use this framework in running examples from now on.
[10] That is, each pair $\alpha, \neg\alpha$ of probabilistic assumptions of $\mathcal{P}$ corresponds to truth assignments of variable $\alpha \in V$ and vice versa; and each probabilistic rule in $\mathcal{R}_p$ corresponds to one entry of an *CPT* in $\mathcal{N}$ and vice versa.

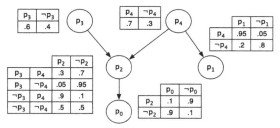

For convenience, let's adopt some notations wrt an $PABA\,\mathcal{P} = (\mathcal{A}_p, \mathcal{R}_p, \mathcal{F})$.

- A **possible world** is a maximal (wrt set inclusion) consistent subset of $\mathcal{A}_p \cup \neg\mathcal{A}_p$. A **partial world** is a subset (not necessarily proper) of a possible world. $\mathcal{W}$ denotes the set of all possible worlds. For each $\omega \in \mathcal{W}$,
  - $ABA\,\mathcal{F}_\omega \triangleq (\mathcal{R}_\omega, \mathcal{A}, \overline{})$ where $\mathcal{R}_\omega \triangleq \mathcal{R} \cup \{p \leftarrow | \ p \in \omega\}$
  - $ABA\,\mathcal{P}_\omega \triangleq (\mathcal{R}_\omega \cup \mathcal{R}_p, \mathcal{A}, \overline{})$.
  - $AA\,\mathcal{P}_\omega$ denotes AA framework $(AR\,\mathcal{P}_\omega, Att\,\mathcal{P}_\omega)$ where $AR\,\mathcal{P}_\omega$ is the set of arguments of $ABA\ \mathcal{P}_\omega$, and $Att\,\mathcal{P}_\omega$ consists of three types of attacks as defined by Definition 4.
- An argument with conclusion being a probabilistic proposition (resp., non-probabilistic proposition) is referred to as a **probabilistic argument** (resp. **non-probabilistic argument**).

Definition 4 [5]. *Let* $A = (Q, \alpha), A' = (Q', \alpha')$ *be arguments in* $AR\ \mathcal{P}_\omega$ *for some possible world* ω. *A* **attacks** *A' if one of three conditions below holds:*

1. *(**type-1 attack**) A is a non-probabilistic argument and α is the contrary of some assumption in Q'.*
2. *(**type-2 attack**) A, A' are probabilistic arguments and A attacks A' by specificity as defined by Definition 5.*
3. *(**type-3 attack**) α is a probabilistic assumption, $A = (\emptyset, \alpha)$ and A' is a probabilistic argument with conclusion of the form $[\neg\alpha : x]$.*

Definition 5 [5]. *Let* $A = (Q, \delta, [\alpha : x])$ *and* $A' = (Q', \delta', [\beta : y])$ *be probabilistic arguments in* $AR\ \mathcal{P}_\omega$ *for some possible world* ω. *Further let* $\delta = S_1, S_2, \ldots, S_m$, $\delta' = S'_1, S'_2, \ldots, S'_n$, *and the rules used to derive S_2 from S_1 and S'_2 from S'_1 are r_1 and r'_1 respectively. A* **attacks** *A' by* **specificity** *if* $body(r'_1) \subset body(r_1)$.

The following definition extends the definition of $Prob_{gr}(.)$ in [5]. Intuitively, it tells how the probabilities of probabilistic assumptions, which are decided by the grounded semantics, propagate to influence the probabilities of accepting other propositions under an arbitrary semantics of argumentation.

Definition 6. *The probability that a proposition π is acceptable wrt semantics sem is* $Prob_{sem}(\pi) \triangleq \displaystyle\sum_{\omega \in \mathcal{W}:ABA\,\mathcal{F}_\omega \vdash_{sem} \pi} P(\omega)$ *where* $P(\omega) \triangleq$

$$\prod_{\alpha \in \omega: AA\,\mathcal{P}_\omega \vdash_{gr} (\_,[\alpha:x])} x$$

For convenience, for a set $\mathcal{S} = \{s_1, s_2, \ldots, s_n\}$ of partial worlds, we use $P(s_1 \vee s_2 \cdots \vee s_n)$ to refer to $\displaystyle\sum_{\omega \in \mathcal{W}, s \in \mathcal{S}: \omega \supseteq s} P(\omega)$.

Example 5 (Cont. Example 4). It is easy to verify that for any $\omega \in \mathcal{W}$:

- $ABA\,\mathcal{F}_\omega \vdash_{cr} \neg\alpha$ iff $\omega \supseteq \{p_1\}$. So $Prob_{cr}(\neg\alpha) = \displaystyle\sum_{\omega \in \mathcal{W}: \omega \supseteq \{p_1\}} P(\omega) = P(\{p_1\})$.
- $ABA\,\mathcal{F}_\omega \vdash_{id} \neg\alpha$ iff $\omega \supseteq s_1$ or $\omega \supseteq s_2$ where $s_1 = \{p_1, \neg p_2\}$ and $s_2 = \{p_0, p_1, p_2\}$. Hence $Prob_{id}(\neg\alpha) = \displaystyle\sum_{\omega \in \mathcal{W}: \omega \supseteq s_1 \text{ or } \omega \supseteq s_2} P(\omega) = P(s_1 \vee s_2)$.

In [5] Dung and Thang show that an PABA framework is *probabilistic coherent* ($\sum_{\omega \in \mathcal{W}} P(\omega) = 1$) if it is *probabilistically acyclic*.

Definition 7. *1. The dependency graph of ABA $\mathcal{F} = (\mathcal{R}, \mathcal{A}, \overline{})$ is a directed graph of which nodes are sentences occurring in $\mathcal{F}$ and there is an edge from node p to node q if and only if*
   (a) $\mathcal{R}$ contains a rule of the form $p \leftarrow \ldots, q, \ldots$, or
   (b) p is an assumption in $\mathcal{A}$ and q is the contrary of p.
2. The dependency graph of PABA $\mathcal{P} = (\mathcal{A}_p, \mathcal{R}_p, \mathcal{F})$ is a directed graph obtained from that of $\mathcal{F}$ by
   (a) first, adding an edge from node p to a node q if $\mathcal{R}_p$ contains a rule of the form $[p : \_] \leftarrow \ldots, q, \ldots$
   (b) then, for each $p \in \mathcal{A}_p$, merging node $\neg p$ with node p.

*An PABA $\mathcal{P}$ is said to be **probabilistically acyclic** if there is no infinite path starting from a probabilistic assumption in the dependency graph of $\mathcal{P}$.*

It turns out that probabilistic acyclicity is also sufficient for probabilistic coherence in our class of PABA frameworks.

Lemma 1. *Let $\mathcal{P}$ be an PABA framework as defined by Definitions 1 and 2. If $\mathcal{P}$ is probabilistically acyclic, then*

1. (Generalizing Lemma 2.1 of [5]) $\sum_{\omega \in \mathcal{W}} P(\omega) = 1$.
2. $0 \leq Prob_{gr}(\pi) \leq Prob_{id}(\pi) \leq Prob_{cr}(\pi) \leq 1$ for any proposition π[11].

From now on, we restrict ourselves to probabilistically acyclic PABA frameworks that satisfy Definitions 1 and 2.

3   Computing PABA Semantics: Theoretical Basis

In this section, we present the theoretical basis for our inference procedures[12].

---

[11] If π does not occur in $\mathcal{P}$, then $Prob_{sem}(\pi) = 0$ for any semantics *sem*.
[12] From now on we assume an arbitrary but fixed $PABA\,\mathcal{P} = (\mathcal{A}_p, \mathcal{R}_p, \mathcal{F})$ with $\mathcal{F} = (\mathcal{R}, \mathcal{A}, \overline{})$ if not explicitly stated otherwise.

Definition 8. *Let **sem** be an argumentation semantics and π be a proposition. A partial world s is said to be **sem-sufficient** for π if $ABA\,\mathcal{F}_\omega \vdash_{sem} \pi$ for any partial world $\omega \supseteq s$.*

Note that if s is sem-sufficient for π then so is any super set of s.

Example 6 (Cont. Example 5). $\{p_1\}$ is cre-sufficient for $\neg\alpha$; while both $\{p_1, \neg p_2\}$ and $\{p_0, p_1, p_2\}$ are ideal-sufficient for $\neg\alpha$.

Definition 9. *Let **sem** be an argumentation semantics and π be a proposition.*

1. *A set S of partial worlds is said to be a **sem-frame** for π if each partial world in S is sem-sufficient for π.*
2. *A sem-frame S for π is said to be **complete** if for each possible world $\omega \in \mathcal{W}$ where $ABA\,\mathcal{F}_\omega \vdash_{sem} \pi$, $\omega \supseteq s$ for some $s \in S$.*

Example 7 (Cont. Example 5). For $\neg\alpha$, $S_1 = \{\{p_1\}\}$ is a complete cre-frame while $S_2 = \{\{p_1, \neg p_2\}, \{p_0, p_1, p_2\}\}$ is a complete ideal-frame.

Note that there are multiple complete sem-frames for the same proposition. For example, cre-frame $\{\{p_1\}, \{p_1, \neg p_2\}\}$ is also complete for $\neg\alpha$.

Theorem 1 below is at the heart of our inference procedures.

Theorem 1. *If $S = \{s_1, s_2, \ldots, s_n\}$ is a complete sem-frame for a proposition π, then $Prob_{sem}(\pi) = P(s_1 \vee s_2 \cdots \vee s_n)$.*

So, continue Example 7, $Prob_{cr}(\neg\alpha) = P(\{p_1\})$; $Prob_{id}(\neg\alpha) = P(\{p_1, \neg p_2\} \vee \{p_0, p_1, p_2\})$.

4   Computing PABA Semantics: Inference Procedures

In this section we present our inference procedure computing PABA's credulous semantics. As Theorem 1 suggests, computing $Prob_{cr}(\pi)$ could be done via two steps: (1) generating a complete cre-frame $S = \{s_1, s_2, \ldots, s_n\}$ for π; and (2) computing $P(s_1 \vee s_2 \cdots \vee s_n)$. To reduce the load in step 2, we would like, in step 1, to arrive at a cre-frame "as small as possible". To this end, we develop the notion of *cre-frame derivation* (Subsect. 4.2), adapting on the notion of AB-dispute derivation of [2,4] for computing ABA's credulous semantics (recalled in Subsect. 4.1), and the notion of base derivation in [9] used to organize search spaces for dispute derivations in AA. In Subsect. 4.3, we shall show that if the given *PABA* framework is Bayesian (see Definition 3), then existing Bayesian network inference algorithms can be used to compute $P(s_1 \vee s_2 \cdots \vee s_n)$.

4.1 AB-Dispute Derivations

AB-dispute derivations [2,4] simulate a dispute between two fictitious players: proponent and opponent. Formally, a AB-dispute derivation is a sequence of tuples $\langle \mathcal{P}_0, \mathcal{O}_0, A_0, C_0 \rangle \ldots \langle \mathcal{P}_i, \mathcal{O}_i, A_i, C_i \rangle \ldots$, where A_i is the set of defense assumptions (consisting of all assumptions occurring in the proponent's arguments) and C_i is the set of culprits (consisting of all opponent's assumptions that the proponent attacks). Multi-set $\mathcal{P}_i$ consists of propositions belonging to any of the proponent's potential arguments. Multi-set $\mathcal{O}_i$ consists of multi-sets of propositions representing the state of all of the opponent's potential arguments.

Definition 10 *(Modified from [2,4]). An AB-dispute derivation in ABA $\mathcal{F} = (\mathcal{R}, \mathcal{A}, \overline{})$ using a selection strategy sl is a (possibly infinite) sequence of tuples $\langle \mathcal{P}_0, \mathcal{O}_0, A_0, C_0 \rangle, \ldots, \langle \mathcal{P}_i, \mathcal{O}_i, A_i, C_i \rangle, \langle \mathcal{P}_{i+1}, \mathcal{O}_{i+1}, A_{i+1}, C_{i+1} \rangle \ldots$ where*

1. *$\mathcal{P}_i$ is a multi-set of propositions, $\mathcal{O}_i$ is a set of finite multi-set of propositions, and A_i, C_i are set of assumptions.*
2. *For each step $i \geq 0$, selection strategy sl selects a proposition $\sigma \in \mathcal{P}_i$ or $\sigma \in S \in \mathcal{O}_i$, and*
   (a) *If $\sigma \in \mathcal{P}_i$ is selected then*
      i. *if σ is an assumption then $\mathcal{P}_{i+1} = \mathcal{P}_i \setminus \{\sigma\}$ and $\mathcal{O}_{i+1} = \mathcal{O}_i \cup \{\{\overline{\sigma}\}\}$[13]*
      ii. *If σ is not an assumption, then there exists some rule $\sigma \leftarrow Bd \in \mathcal{R}$ such that $C_i \cap Bd = \emptyset$ and $\mathcal{P}_{i+1} = \mathcal{P}_i \setminus \{\sigma\} \cup (Bd \setminus A_i)$ and $A_{i+1} = A_i \cup (\mathcal{A} \cap Bd)$*
   (b) *If S is selected in $\mathcal{O}_i$ and σ is selected in S then*
      i. *If σ is an assumption, then*
         A. *either σ is ignored, i.e. $\mathcal{O}_{i+1} = \mathcal{O}_i \setminus \{S\} \cup \{S \setminus \{\sigma\}\}$*
         B. *or $\sigma \notin A_i$ and $\sigma \in C_i$ and $\mathcal{O}_{i+1} = \mathcal{O}_i \setminus \{S\}$*
         C. *or $\sigma \notin A_i$ and $\sigma \notin C_i$ and*
            (C.1) if $\overline{\sigma}$ is an assumption, then $\mathcal{O}_{i+1} = \mathcal{O}_i \setminus \{S\}$ and $A_{i+1} = A_i \cup \{\overline{\sigma}\}$ and $C_{i+1} = C_i \cup \{\sigma\}$
            (C.2) otherwise $\mathcal{P}_{i+1} = \mathcal{P}_i \cup \{\overline{\sigma}\}$ and $\mathcal{O}_{i+1} = \mathcal{O} \setminus \{S\}$ and $C_{i+1} = C_i \cup \{\sigma\}$
      ii. *If σ is not an assumption, then $\mathcal{O}_{i+1} = \mathcal{O}_i \setminus \{S\} \cup \{S \setminus \{\sigma\} \cup Bd \mid \sigma \leftarrow Bd \in \mathcal{R}$ and $Bd \cap C_i = \emptyset\}$*

Definition 11. *1. An AB-dispute derivation **for a proposition** π is such that the first tuple $\langle \mathcal{P}_0, \mathcal{O}_0, A_0, C_0 \rangle = \langle \{\pi\}, \emptyset, \mathcal{A} \cap \{\pi\}, \emptyset \rangle$.*
*2. An AB-dispute derivation is said to be **successful** if it is ended by a tuple $\langle \emptyset, \emptyset, \_, \_ \rangle$.*

Example 8 (Cont. Example 4). Consider ABA $\mathcal{F}'$ obtained from ABA $\mathcal{F}$ by removing all probabilistic assumptions, i.e. $\mathcal{F}'$ contains $r'_0 : \neg\alpha \leftarrow \alpha$ $r'_1 : \neg\alpha \leftarrow \beta$ $r'_2 : \neg\beta \leftarrow \alpha$ $r'_3 : \neg\gamma \leftarrow \delta$ and $r'_4 : \neg\delta \leftarrow \gamma$. The following table shows a successful AB-dispute derivation for proposition $\neg\alpha$ in $\mathcal{F}'$. Note that in step 5 the proponent reuses $r'_1 : \neg\alpha \leftarrow \beta$ but β is not added into $\mathcal{P}_5$.

i	$\mathcal{P}_i$	$\mathcal{O}_i$	A_i	C_i	By rule (of Definition 10)	Remarks
0	$\{\neg\alpha\}$	$\{\}$	$\{\}$	$\{\}$		Proponent claims
1	$\{\beta\}$	$\{\}$	$\{\beta\}$	$\{\}$	2.a.ii	Proponent uses $r'_1 : \neg\alpha \leftarrow \beta$
2	$\{\}$	$\{\{\neg\beta\}\}$	$\{\beta\}$	$\{\}$	2.a.i	Opponent tries to attack β
3	$\{\}$	$\{\{\alpha\}\}$	$\{\beta\}$	$\{\}$	2.b.ii	Opponent uses $r'_2 : \neg\beta \leftarrow \alpha$
4	$\{\neg\alpha\}$	$\{\}$	$\{\beta\}$	$\{\alpha\}$	2.b.i.C1	Proponent selects α as a culprit
5	$\{\}$	$\{\}$	$\{\beta\}$	$\{\alpha\}$	2.a.ii	Proponent reuses $r'_1 : \neg\alpha \leftarrow \beta$

The following theorem states that AB-dispute derivations are sound for credulous acceptance in any *ABA* framework.

Theorem 2 *(Theorem 4.3 in [2]). If $\langle \mathcal{P}_0, \mathcal{O}_0, A_0, C_0\rangle, \ldots, \langle \mathcal{P}_n, \mathcal{O}_n, A_n, C_n\rangle$ is a successful AB-dispute derivation for a proposition π, then A_n is an admissible set of assumptions and supports π.*

In their Theorem 4.4, the authors of [2] show that AB-dispute derivations are not complete in general, but complete for the class of *positively acyclic ABA* frameworks over finite languages. However AB-dispute derivation are indeed complete for a larger class of positively acyclic and *finitary ABA* frameworks.

Definition 12. *Let $\mathcal{F} = (\mathcal{R}, \mathcal{A}, \overline{})$ be an ABA framework.*

1. *$\mathcal{F}$ is said to be **finitary** if for each node in the dependency graph of $\mathcal{F}$, there is a finite number of nodes reachable from it.*
2. *$\mathcal{F}$ is said to be **positively acyclic** if in the dependency graph of $\mathcal{F}$, there is no infinite directed path consisting solely non-assumption nodes.*

Clearly ABA frameworks over finite languages are all finitary but not the reverse. For example, the framework with $\mathcal{R} = \{\neg\alpha_{i+1} \leftarrow \alpha_i \mid i \in \{1, 2, \ldots\}\}$ and $\mathcal{A} = \{\alpha_1, \alpha_2, \ldots\}$, $\overline{\alpha_i} = \neg\alpha_i$, is finitary but has an infinite language.

Theorem 3 *(Generalizing Theorem 4.4 in [2]). Given a positively acyclic and finitary assumption-based framework $\mathcal{F}$.*

1. *If π is supported by an admissible set S of assumptions, then for any selection strategy there is a successful AB-dispute derivation $\langle \mathcal{P}_0, \mathcal{O}_0, A_0, C_0\rangle, \ldots,$ $\langle \mathcal{P}_n, \mathcal{O}_n, A_n, C_n\rangle$ for π where $A_n \subseteq S$.*
2. *There are no infinite AB-dispute derivations for any proposition.*

In non-finitary and/or positively cyclic frameworks, credulously acceptable propositions may not have successful AB-dispute derivations. For example, consider a positively cyclic ABA framework with $\mathcal{R} = \{\neg\alpha \leftarrow \neg\alpha\}$ and $\mathcal{A} = \{\alpha\}$ where $\overline{\alpha} = \neg\alpha$. Clearly α is credulously acceptable but it has no successful AB-dispute derivation. Note that the only AB-dispute derivation for α is

---

[13] Silence about a component means it remains the same as the previous step. In this case 2.a.i, for example, $A_{i+1} = A_i$ and $C_{i+1} = C_i$.

$\langle\{\alpha\},\emptyset,\{\alpha\},\emptyset\rangle, \langle\emptyset,\{\{\neg\alpha\}\},\{\alpha\},\emptyset\rangle, \langle\emptyset,\{\{\neg\alpha\}\},\{\alpha\},\emptyset\rangle,\dots$ which is infinite. Similarly, q is credulously acceptable in a non-finitary ABA framework with $\mathcal{R} = \{q \leftarrow \beta\}\cup\{\neg\beta \leftarrow \alpha_i \mid i \in \{1,2,\dots\}\}\cup\{\neg\alpha_{i+1} \leftarrow \neg\alpha_i \mid i \in \{1,2,\dots\}\}\cup\{\neg\alpha_1 \leftarrow\}$ and $\mathcal{A} = \{\beta,\alpha_1,\alpha_2,\dots\}$ where $\overline{x} = \neg x$ for each $x \in \mathcal{A}$. However there are no successful AB-dispute derivation for q.

To facilitate the presentations of our inference procedures in next sections, let $\mathcal{DS}_{\mathcal{F}}(t, sl)$ refer the set of tuples that can *immediately* follow a tuple t of the form $\langle\mathcal{P},\mathcal{O},A,C\rangle$ in some AB-dispute derivation using selection strategy sl. From Definition 10 part 2, $\mathcal{DS}_{\mathcal{F}}(t, sl)$ can be computed by the following procedure.

(a) If sl selects $\sigma \in \mathcal{P}$, then
 i. if σ is an assumption, then $\mathcal{DS}_{\mathcal{F}}(t, sl) = \{\langle\mathcal{P} \setminus \{\sigma\}, \mathcal{O} \cup \{\{\overline{\sigma}\}\}, A, C\rangle\}$
 ii. if σ is not an assumption, then $\mathcal{DS}_{\mathcal{F}}(t, sl) = \{\langle\mathcal{P}\setminus\{\sigma\}\cup(Bd\setminus A), \mathcal{O}, A\cup(A\cap Bd), C\rangle \mid \sigma \leftarrow Bd \in \mathcal{R}$ and $C\cap Bd = \emptyset\}$
(b) If sl selects $S \in \mathcal{O}$, then $\mathcal{DS}_{\mathcal{F}}(t, sl) = \emptyset$ if $S = \emptyset$. Otherwise, let σ be the sentence selected in S, and
 i. if σ is an assumption, then $\mathcal{DS}_{\mathcal{F}}(t, sl) = \{\langle\mathcal{P}, \mathcal{O}\setminus\{S\}\cup\{S\setminus\{\sigma\}\}, A, C\rangle\}\cup \delta T$ where δT is computed as follows.
   A. if $\sigma \in A$ then $\delta T = \emptyset$.
   B. if $\sigma \notin A$ and $\sigma \in C$, then $\delta T = \{\langle\mathcal{P}, \mathcal{O}\setminus\{S\}, A, C\rangle\}$
   C. if $\sigma \notin A$ and $\sigma \notin C$, then
    (C.1) if $\overline{\sigma} \in A$ then $\delta T = \{\langle\mathcal{P}, \mathcal{O}\setminus\{S\}, A\cup\{\overline{\sigma}\}, C\cup\{\sigma\}\rangle\}$
    (C.2) otherwise, $\delta T = \{\langle\mathcal{P}\cup\{\overline{\sigma}\}, \mathcal{O}\setminus\{S\}, A, C\cup\{\sigma\}\rangle\}$
 ii. if σ is not an assumption, then $\mathcal{DS}_{\mathcal{F}}(t, sl) = \{\langle\mathcal{P}, \mathcal{O}', A, C\rangle\}$ where $\mathcal{O}' = \mathcal{O}\setminus\{S\}\cup\{S\setminus\{\sigma\}\cup Bd \mid \sigma \leftarrow Bd$ is a rule in $\mathcal{R}$ s.t. $Bd\cap C = \emptyset\}$

4.2   Cre-Frame Derivation

The following notion of cre-frame derivations extends the notion of AB-dispute derivations to gradually construct complete cre-frames.

Definition 13. *A **cre-frame derivation** in PABA $\mathcal{P} = (\mathcal{A}_p, \mathcal{R}_p, \mathcal{F})$ using a selection strategy sl is a possibly infinite sequence $\mathcal{T}_0, \mathcal{T}_1,\dots,\mathcal{T}_i,\dots$ where*

1. *$\mathcal{T}_i$ is a set of pairs of the form (t, ω) where t is a tuple $\langle\mathcal{P},\mathcal{O},A,C\rangle$ as defined in Definition 10 and ω is a partial world.*
2. *For each i, sl selects one pair $(t, \omega) \in \mathcal{T}_i$ and a proposition σ from the $\mathcal{P}$ component or $\mathcal{O}$ component of t, and $\mathcal{T}_{i+1} = \mathcal{T}_i \setminus \{(t,\omega)\}\cup\Delta\mathcal{T}$ where*
   (a) *If σ is a probabilistic assumption not occurring in ω[14], then $\Delta\mathcal{T} = \{(t, \omega\cup\{\sigma\}), (t, \omega\cup\{\neg\sigma\})\}$*
   (b) *Otherwise $\Delta\mathcal{T} = \{(t', \omega) \mid t' \in \mathcal{DS}_{\mathcal{F}_\omega}(t, sl)\}$.*

Definition 14. *1. A cre-frame derivation **for a proposition** π is a cre-frame derivation that begins with $\mathcal{T}_0 = \{(\langle\{\pi\},\emptyset,\mathcal{A}\cap\{\pi\},\emptyset\rangle,\emptyset)\}$[15].*

---

[14] That is, neither σ nor its complement are elements of ω.
[15] $\mathcal{A}$ is the set of assumptions in $ABA\,\mathcal{F}$.

2. *A finite cre-frame derivation $\mathcal{T}_0, \ldots, \mathcal{T}_n$ is said to be* **full** *if it can not be extended further, or equivalently $\mathcal{T}_n$ contains only pairs of the form $(\langle\emptyset, \emptyset, \_, \_\rangle, \_)$. The set $\{\omega \mid (\langle\emptyset, \emptyset, \_, \_\rangle, \omega) \in \mathcal{T}_n\}$ is called the* **derived frame**.

Example 9 (Cont. Example 4). A full cre-frame derivation for $\neg\alpha$ is given in the next page. Note that notation $\underline{x}$ means that x is selected by selection strategy.

Theorem 4 below says that cre-frame derivations provide a sound procedure for generating complete cre-frames.

Theorem 4. *If $\mathcal{D}$ is a full cre-frame derivation for a proposition π, then the frame derived by $\mathcal{D}$ is a complete cre-frame for π.*

So, continue Example 9, Theorem 4 says that $\{\{p_1\}, \{p_1, \neg p_2\}\}$ is a complete cre-frame for $\neg\alpha$. Theorem 5 below says that cre-frame derivations provide a terminating procedure for generating complete cre-frames in a general class of PABA frameworks.

Theorem 5. *In an $PABA\ \mathcal{P} = (\mathcal{R}_p, \mathcal{A}_p, \mathcal{F})$ where $\mathcal{F}$ is positively acyclic and finitary, there are no infinite cre-frame derivations for any proposition.*

	(t, ω) — t	ω	By rule
T_0	$\langle\{\neg\alpha\}, \emptyset, \emptyset, \emptyset\rangle$	$\emptyset$	
T_1	$\langle\{\underline{\alpha}, p_0\}, \emptyset, \{\alpha\}, \emptyset\rangle$	$\emptyset$	2.b
	$\langle\{\beta, p_1\}, \emptyset, \{\beta\}, \emptyset\rangle$	$\emptyset$	
T_2	$\langle\{p_0\}, \{\{\underline{\neg\alpha}\}\}, \{\alpha\}, \emptyset\rangle$	$\emptyset$	2.b
	$\langle\{\beta, p_1\}, \emptyset, \{\beta\}, \emptyset\rangle$	$\emptyset$	
T_3	$\langle\{p_0\}, \{\{\underline{\alpha}, p_0\}, \{\beta, p_1\}\}, \{\alpha\}, \emptyset\rangle$	$\emptyset$	2.b
	$\langle\{\beta, p_1\}, \emptyset, \{\beta\}, \emptyset\rangle$	$\emptyset$	
T_4	$\langle\{p_0\}, \{\{p_0\}, \{\beta, p_1\}\}, \{\alpha\}, \emptyset\rangle$	$\emptyset$	2.b
	$\langle\{\beta, p_1\}, \emptyset, \{\beta\}, \emptyset\rangle$	$\emptyset$	
T_5	$\langle\{p_0\}, \{\{p_0\}, \{\beta, p_1\}\}, \{\alpha\}, \emptyset\rangle$	$\{\neg p_0\}$	2.a
	$\langle\{p_0\}, \{\{p_0\}, \{\beta, p_1\}\}, \{\alpha\}, \emptyset\rangle$	$\{p_0\}$	
	$\langle\{\beta, p_1\}, \emptyset, \{\beta\}, \emptyset\rangle$	$\emptyset$	
T_6	$\langle\{p_0\}, \{\{p_0\}, \{\beta, p_1\}\}, \{\alpha\}, \emptyset\rangle$	$\{p_0\}$	2.b
	$\langle\{\beta, p_1\}, \emptyset, \{\beta\}, \emptyset\rangle$	$\emptyset$	
T_7	$\langle\emptyset, \{\{p_0\}, \{\beta, p_1\}\}, \{\alpha\}, \emptyset\rangle$	$\{p_0\}$	2.b
	$\langle\{\beta, p_1\}, \emptyset, \{\beta\}, \emptyset\rangle$	$\emptyset$	
T_8	$\langle\emptyset, \{\{\ \}, \{\beta, p_1\}\}, \{\alpha\}, \emptyset\rangle$	$\{p_0\}$	2.b
	$\langle\{\beta, p_1\}, \emptyset, \{\beta\}, \emptyset\rangle$	$\emptyset$	
T_9	$\langle\{\beta, p_1\}, \emptyset, \{\beta\}, \emptyset\rangle$	$\emptyset$	2.b
T_{10}	$\langle\{\underline{p_1}\}, \{\{\neg\beta\}\}, \{\beta\}, \emptyset\rangle$	$\emptyset$	2.b

	t	ω	By rule
T_{11}	$\langle\{p_1\}, \{\{\neg\beta\}\}, \{\beta\}, \emptyset\rangle$	$\{\neg p_1\}$	2.a
	$\langle\{p_1\}, \{\{\neg\beta\}\}, \{\beta\}, \emptyset\rangle$	$\{p_1\}$	
T_{12}	$\langle\{p_1\}, \{\{\neg\beta\}\}, \{\beta\}, \emptyset\rangle$	$\{p_1\}$	2.b
T_{13}	$\langle\emptyset, \{\{\underline{\neg\beta}\}\}, \{\beta\}, \emptyset\rangle$	$\{p_1\}$	2.b
T_{14}	$\langle\emptyset, \{\{\underline{\alpha}, p_2\}\}, \{\beta\}, \emptyset\rangle$	$\{p_1\}$	2.b
T_{15}	$\langle\emptyset, \{\{p_2\}\}, \{\beta\}, \emptyset\rangle$	$\{p_1\}$	2.b
	$\langle\{\neg\alpha\}, \emptyset, \{\beta\}, \{\alpha\}\rangle$	$\{p_1\}$	
T_{16}	$\langle\emptyset, \{\{\underline{p_2}\}\}, \{\beta\}, \emptyset\rangle$	$\{p_1, \neg p_2\}$	2.a
	$\langle\emptyset, \{\{p_2\}\}, \{\beta\}, \emptyset\rangle$	$\{p_1, p_2\}$	
	$\langle\{\neg\alpha\}, \emptyset, \{\beta\}, \{\alpha\}\rangle$	$\{p_1\}$	
T_{17}	$\langle\emptyset, \emptyset, \{\beta\}, \emptyset\rangle$	$\{p_1, \neg p_2\}$	2.b
	$\langle\emptyset, \{\{\underline{p_2}\}\}, \{\beta\}, \emptyset\rangle$	$\{p_1, p_2\}$	
	$\langle\{\neg\alpha\}, \emptyset, \{\beta\}, \{\alpha\}\rangle$	$\{p_1\}$	
T_{18}	$\langle\emptyset, \emptyset, \{\beta\}, \emptyset\rangle$	$\{p_1, \neg p_2\}$	2.b
	$\langle\emptyset, \{\{\ \}\}, \{\beta\}, \emptyset\rangle$	$\{p_1, p_2\}$	
	$\langle\{\neg\alpha\}, \emptyset, \{\beta\}, \{\alpha\}\rangle$	$\{p_1\}$	
T_{19}	$\langle\emptyset, \emptyset, \{\beta\}, \emptyset\rangle$	$\{p_1, \neg p_2\}$	2.b
	$\langle\{\underline{\neg\alpha}\}, \emptyset, \{\beta\}, \{\alpha\}\rangle$	$\{p_1\}$	
T_{20}	$\langle\emptyset, \emptyset, \{\beta\}, \emptyset\rangle$	$\{p_1, \neg p_2\}$	2.b
	$\langle\{\underline{p_1}\}, \emptyset, \{\beta\}, \{\alpha\}\rangle$	$\{p_1\}$	
T_{21}	$\langle\emptyset, \emptyset, \{\beta\}, \emptyset\rangle$	$\{p_1, \neg p_2\}$	2.b
	$\langle\emptyset, \emptyset, \{\beta\}, \{\alpha\}\rangle$	$\{p_1\}$	

4.3   Computing $P(s_1 \vee \cdots \vee S_n)$

For a set of partial worlds $\mathcal{S} = \{s_1, s_2, \ldots, s_n\}$, let $NDF_\mathcal{S}$ denote to the DNF formula $\bigvee_{i=1}^{n} \bigwedge_{j=1}^{|s_i|} p_{ij}$ where $p_{ij} \in s_i$. For example, if $\mathcal{S} = \{\{p_1\}, \{p_1, \neg p_2\}\}$ then $NDF_\mathcal{S} = p_1 \vee (p_1 \wedge \neg p_2)$.

Lemma 2. *Let $\mathcal{P}$ be an PABA including Bayesian network $\mathcal{N}$. If $\mathcal{S} = \{s_1, \ldots, s_n\}$ is a set of partial worlds of $\mathcal{P}$, then $P(s_1 \vee s_2 \cdots \vee s_n) = Pr_{\mathcal{N}}(NDF_{\mathcal{S}})$ where $Pr_{\mathcal{N}}$ is the probability distribution defined by $\mathcal{N}$.*

Continue Example 4, the lemma says that $P(\{p_1\} \vee \{p_1, \neg p_2\}) = Pr_{\mathcal{N}}(p_1 \vee (p_1 \wedge \neg p_2))$.

So to compute $P(s_1 \vee s_2 \cdots \vee s_n)$, one can compute $Pr_{\mathcal{N}}(NDF_{\mathcal{S}})$ instead using any BN inference algorithms. Doing so on $\mathcal{N}$, one need to translate the query into standard queries on $\mathcal{N}$ using the inclusion-exclusion rule. For example $Pr_{\mathcal{N}}(p_1 \vee (p_1 \wedge \neg p_2)) = Pr_{\mathcal{N}}(p_1) + Pr_{\mathcal{N}}(p_1 \wedge \neg p_2) - Pr_{\mathcal{N}}(p_1 \wedge p_1 \wedge \neg p_2)$. Alternatively, one could construct a new network $\mathcal{N}_{\mathcal{S}}$ by adding to $\mathcal{N}$, for each $i \in \{1, \ldots n\}$, one AND gate to compute $\bigwedge_{j=1}^{|s_i|} p_{ij}$, and one OR gate to compute $\bigvee_{i=1}^{n} \bigwedge_{j=1}^{|s_i|} p_{ij}$ from the output of n AND gates. For example, if $\mathcal{S} = \{\{p_1\} \vee \{p_1, \neg p_2\}\}$, then $\mathcal{N}_{\mathcal{S}}$ contains the following new nodes and edges. Clearly $Pr_{\mathcal{N}}(NDF_{\mathcal{S}})$ equals $Pr_{\mathcal{N}_{\mathcal{S}}}(Q)$, where Q the child of the OR gate.

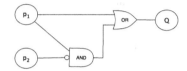

5   Related Work and Conclusions

One of the early known models of Probabilistic Argumentation (PA) is the one of Dung and Thang [5] (DT's PAA) defined as a triple $(\mathcal{F}, \mathcal{W}, P)$ where $\mathcal{F} = (AR, Att)$ is a standard AA framework, $\mathcal{W}$ is a set of possible worlds such that each $\omega \in \mathcal{W}$ defines a set of arguments $AR_\omega \subseteq AR$, and P is a probability distribution over $\mathcal{W}$. DT's PAA defines the grounded probability of argument A by the sum of probabilities of possible worlds in which A is groundedly accepted[16], but, following AA's style, abstracts from: (1) the representation of possible world; (2) the construction of AR_ω for each possible world ω; and (3) the representation and computation of $P(\omega)$. So the authors in the same work [5] combine DT's PAA with Assumption-based Argumentation (ABA [2,4]) to introduce PABA where: (1) a possible world represented by a conjunction of so-called probabilistic assumptions; (2) arguments of AR_ω are constructed from inference rules and assumptions (as in ABA) together with facts representing the occurrences of probabilistic assumptions in ω; and (3) $P(\omega)$ is represented by means of so-called probabilistic rules and computed by grounded semantics. Inference procedures for PABA, however, remain unexplored. To our best knowledge, our work is the first developing PABA inference procedures but there have been many works done on other models of PA, focusing on different computational issues. For example, Li et al. in [8] use a Monte-Carlo simulation to *approximate* the probability of a set of arguments consistent with an

---

[16] That is, $Prob_{gr}(A) \triangleq \sum\limits_{\omega \in \mathcal{W}:(AR_\omega, Att \cap (AR_\omega \times AR_\omega)) \vdash_{gr} A} P(\omega)$.

argumentation semantics in their model of Probabilistic Abstract Argumentation (Li's PAA). The complexity of this problem is recently investigated in [6]. In [1], Doder and Woltran translates Li's PAA frameworks into formulas in a probabilistic logic, so obtaining, as a by-product, a schematic way to compute the above probability precisely using solvers for probabilistic logic. However the authors did not explore this direction further to develop inference procedures for Li's PAA. Interestingly, to use our inference procedures for computing Li's PAA, we just need a simple translator as follows. Consider a Li's PAA framework $(\mathcal{F}, P_{AR}, P_{Att})$. Recall that $\mathcal{F} = (AR, Att)$ is a standard AA framework, $P_{AR} : AR \rightarrow [0,1]$ and $P_{Att} : Att \rightarrow [0,1]$ are probability distributions over AR and Att. $P_{AR}(A)$ is interpreted as the probability of the event that A actually occurs as an argument (denoted $ar(A)$); and $P_{Att}((A, B))$ is interpreted as the probability of the event that A attacks B (denoted $att(A, B)$), conditional to a joint event $ar(A) \wedge ar(B)$. Events in $\{ar(A) \mid A \in AR\}$ are assumed to be pair-wise independent, and $att(A, B)$ is assumed to depend only on $ar(A)$ and $ar(B)$. In other words, the probability distribution over all events $\{ar(A) \mid A \in AR\} \cup \{att(A, B) \mid (A, B) \in Att\}$ is defined by a Bayesian network of the pattern in the next page. A possible world is an AA framework (AR', Att') where $AR' \subseteq AR$ and $Att' \subseteq Att \cap (AR' \times AR')$ with probability of the join event

$$\bigwedge_{A \in AR'} ar(A) \quad \bigwedge_{A \in AR \setminus AR'} \neg ar(A) \quad \bigwedge_{(A,B) \in Att'} att(A, B) \quad \bigwedge_{(A,B) \in Att \setminus Att'} \neg att(A, B).$$

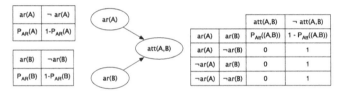

Finally, the semantics of Li's PAA is as follows: the probability that argument A is accepted equals the sum of probabilities of possible worlds in which A is accepted. So, in translating a Li's PAA framework $(\mathcal{F}, P_{AR}, P_{Att})$ into an PABA framework $\mathcal{P} = (\mathcal{A}_p, \mathcal{R}_p, (\mathcal{R}, \mathcal{A}, \overline{}))$, we would like $\mathcal{P}$ to be Bayesian and contains an assumption A, for each argument $A \in AR$, such that the probability of the acceptability of assumption A in $\mathcal{P}$ equals the probability that argument A is accepted according to Li's PAA semantics. The readers can easily verify that $\mathcal{P}$ can be defined as follows: $\mathcal{A} = \{A \mid A \in AR\}$ with $\overline{A} = \neg A$; $\mathcal{R} = \{\neg A \leftarrow \neg ar(A) \mid A \in AR\} \cup \{\neg B \leftarrow A, att(A, B) \mid (A, B) \in Att\}$; $\mathcal{A}_p = \{ar(A) \mid A \in AR\} \cup \{att(A, B) \mid (A, B) \in Att\}$ and $\mathcal{R}_p$ represent the described Bayesian network. Readers can also simplify this translation for subclasses of Li's PAA frameworks such as those in [7] where attacks are all certain given the presences of involved arguments (i.e. $P_{Att}((A, B)) = 1$ for any $(A, B) \in Att$).

Acknowledgment. This work was funded by SIIT Young Researcher Grant under Contract No SIIT-2014-YRG1.

References

1. Doder, D., Woltran, S.: Probabilistic argumentation frameworks – a logical approach. In: Straccia, U., Calì, A. (eds.) SUM 2014. LNCS, vol. 8720, pp. 134–147. Springer, Heidelberg (2014)
2. Dung, P.M., Mancarella, P., Toni, F.: Computing ideal skeptical argumentation. Artif. Intell. **171**(10–15), 642–674 (2007)
3. Dung, P.M.: On the acceptability of arguments and its fundamental role in nonmonotonic reasoning, logic programming and n-person games. Artif. Intell. **77**(2), 321–357 (1995)
4. Dung, P.M., Kowalski, R.A., Toni, F.: Dialectic proof procedures for assumption-based, admissible argumentation. Artif. Intell. **170**(2), 114–159 (2006)
5. Dung, P.M., Thang, P.M.: Towards (probabilistic) argumentation for jury-based dispute resolution. In: COMMA 2010, pp. 171–182 (2010)
6. Fazzinga, B., Flesca, S., Parisi, F.: On the complexity of probabilistic abstract argumentation frameworks. ACM Trans. Comput. Logic **16**(3), 22:1–22:39 (2015)
7. Hunter, A.: A probabilistic approach to modelling uncertain logical arguments. Int. J. Approximate Reasoning **54**(1), 47–81 (2013)
8. Li, H., Oren, N., Norman, T.J.: Probabilistic argumentation frameworks. In: Modgil, S., Oren, N., Toni, F. (eds.) TAFA 2011. LNCS, vol. 7132, pp. 1–16. Springer, Heidelberg (2012)
9. Thang, P.M., Dung, P.M., Hung, N.D.: Toward a common framework for dialectical proof procedure in abstract argumentation. J. Logic Comput. **19**(6), 1071–1109 (2009)

Combining Swarm with Gradient Search for Maximum Margin Matrix Factorization

K.H. Salman[1(✉)], Arun K. Pujari[1,2], Vikas Kumar[1],
and Sowmini Devi Veeramachaneni[1]

[1] School of Computer and Information Sciences, University of Hyderabad,
Hyderabad, India
i.salman.kh@gmail.com, akpcs@uohyd.ernet.in, vikas@uohyd.ac.in,
sowmiveeramachaneni@gmail.com
[2] Central University of Rajasthan, Ajmer, India

Abstract. Maximum Margin Matrix Factorization is one of the very popular techniques of collaborative filtering. The discrete valued rating matrix with a small portion of known ratings is factorized into two latent factors and the unknown ratings are estimated by the resulting product of the factors. The factorization is achieved by optimizing a loss function and the optimization is carried out by gradient descent or its variants. It is observed that any of these algorithms yields near-global optimizing point irrespective of the initial seed point. In this paper, we propose to combine swarm-like search with gradient descent search. Our algorithm starts from multiple initial points and uses gradient information and swarm-search as the search progresses. We show that by this process we get an efficient search scheme to get near optimal point for maximum margin matrix factorization.

1 Introduction

Recommender Systems (RS) are tools and techniques for providing suggestions for items to be used by a user. Recommender systems analyse patterns of users' interest in various products and provide a personalized recommendation to suit the users taste. A good personalized recommendation can provide a new dimension in e-commerce particularly for entertainment products like movies, music, TV shows, and books. Broadly speaking, the techniques of the recommender system are classified into Content Based filtering and Collaborative filtering [9]. Latent factor model is one of the important approaches in Collaborative filtering and it tries to explain the rating of an item by a user in terms of latent factors of users and of items. Separating out the user factors and item factors from the set of ratings is accomplished through matrix factorization (MF), which is a one of the most successful realizations of Collaborative filtering. A user-item rating matrix is factorized into user-latent factor and item-latent factor matrix. The product of these two factor matrices is made to be consistent with the observed ratings and unobserved ratings are estimated. This is achieved computationally by minimizing a loss function [9].

© Springer International Publishing Switzerland 2016
R. Booth and M.-L. Zhang (Eds.): PRICAI 2016, LNAI 9810, pp. 167–179, 2016.
DOI: 10.1007/978-3-319-42911-3_14

Matrix factorization method has recently become very popular for recommender system and among different matrix factorization techniques, Maximum Margin Matrix Factorization (MMMF) [19] is reported to be the most suitable for factorizing a rating matrix. Algorithmically, given few entries of the rating matrix, it is to determine two latent factor matrices U and V such that the product UV^T is optimal with respect to some criterion function for the set of observed ratings. The set of observed ratings is very small proportion of entries of the rating matrix. The matrix factorization in general and MMMF, in particular, are accomplished by solving a nonlinear optimization problem. In the case of MMMF, the criterion function is hinge loss function. Interestingly, though the objective function could not be shown to be convex, it is observed empirically that all variants of gradient descent techniques converge to near-optimal solution for any randomly selected starting point.

Particle Swarm Optimization (PSO) is a swarm intelligence technique where a population of particles moves in a hyper-dimensional solution space to find the global optimum, where the movement is influenced by the social component and the cognitive component of the particle. In other words, PSO is a system which balances between exploration and exploitation by combining local and global information from the neighbours. Over the time, many variants of PSO have been developed like Bare-bone PSO [7], Diversity-based PSO [18], Self-regulating PSO [22], Attraction-Repulsion PSO [15] to name a few. The most characteristic feature of PSO and its variants is that the search trajectory is influenced by the best solutions (local-best and global-best) obtained so far in the search to determine the next solution. These parameters determine the velocity and the velocity determines the next new solution in the search. On the other hand, the most popular and widely used method of search in nonlinear optimization problems is the gradient-based search. Normally, PSO search is preferred over gradient search when the nonlinear objective function is multimodal and there are a large number of local optimizing solutions. In such cases, gradient-based search gets stuck at a local optimizing point whereas population-based techniques search through a broader area ensuring the higher possibility of reaching global optimizing solution. A natural question that arises is whether one can combine gradient information of search direction together with the velocity computed by local/global best solutions to enhance the search. Research on PSO is mostly on devising newer strategies of search in order to obtain global optimizing solution for multi-modal nonlinear objective function. The other direction of research in PSO concentrate on demonstrating the applicability of swarm search for ill-behaved objective functions arising out of various applications.

In this paper, we digress from these directions of research to address a different research problem. The research question that we ask is whether swarm search with its ability to blend exploration with exploitation contributes in any other meaningful ways, besides helping in finding global optimizing point in the presence of several local optimizing solutions. We have taken Maximum Margin Matrix Factorization as our case study. It is observed that even though the gradient search does not yield provably global solution, the search yields almost the similar solutions irrespective of the starting point. Thus, in MMMF, when the search starts with different randomly generated starting point it terminates at different solutions but having almost the same value of the objective function. In other words, there is no apparent advantage of having a population-based search or swarm-search over single-particle search. We show in this paper that there is certainly a definite advantage of swarm-search over single-point search in improving the performance of search in such cases. Let us accept that the first principle of PSO is that the interaction among the particles during the search enabling the search algorithm to evolve to be more efficient than a system which uses independent search sessions of isolated individual particles. In other words, the communications among particles at every step help in selecting a useful *next step* rather than searching without communicating. If we map this hypothesis to gradient-search paradigm, the PSO-like search that is applied concurrently with several initial points that interact among themselves during the search stands to be more efficient than the traditional method that employs a number of repeated gradient search by changing the randomly selected initial points in an isolated manner. The objective of the present work is to demonstrate that even when the traditional gradient search yields a near-optimal solution for a well-behaved objective function (as in the case of MMMF), by employing exploitation-exploration paradigm of PSO we can achieve better search method (faster convergence, in this case). Keeping this objective, we devised a new algorithm of Gradient-based-PSO for MMMF. The contribution of the present is two-fold.

- We devise a new gradient-based swarm search technique for MMMF. We demonstrate that this method is a viable recommender technique.
- Unlike demonstrating the ability to skip local optimizing points of swarm-search, we demonstrate that swarm-search is a more efficient search strategy than the point-search strategy for convergence.

The rest of the paper is organized as follows. In Sect. 2, we briefly review MMMF technique. Section 3 builds the background of PSO. We describe our proposed method in Sect. 4. Section 5 deals with the detailed experimental study to show the efficacy of our approach. We conclude with Sect. 6.

2   Matrix Factorization

In this section, we have given formal definition of matrix factorization and briefly reviewed MMMF. Let Y be a $n \times m$ user-item rating matrix. Each entry $y_{ij} \in \{0, 1, 2, \ldots, R\}$ defines the preference of ith user for j^{th} item. $y_{ij} = 0$ indicates that preference of ith user for j^{th} item is not available (unsampled entry). Given a partially observed rating matrix $Y \in \mathbb{R}^{n \times m}$, matrix factorization aims at determining two matrices $U \in \mathbb{R}^{n \times k}$ and $V \in \mathbb{R}^{m \times k}$ such that $Y \approx UV^T$ where the inner dimension k is called the *numerical rank* of the matrix. Several general purpose matrix factorization techniques have been proposed in the literature such as Nonnegative Matrix Factorization (NMF) [11], Incremental SVD based matrix factorization [20], weighted low-rank approximation [13] to name a few. MMMF [19] is one of the very popular techniques of collaborative filtering.

2.1   Maximum Margin Matrix Factorization

MMMF is one of the most popular latent factor model based approach for Collaborative Filtering [19]. Unlike other latent factor models where the idea is to minimize the rank of the latent factors, MMMF uses regularization on the norms of factor matrices U and V. In addition to U and V matrices, a threshold matrix Θ consisting of $R-1$ thresholds, θ_{ik} ($1 \le k \le R-1$), for every user i is involved to relate a real value to a discrete value. The objective is to determine U, V and Θ such that the following $\mathcal{J}$ is minimized.

$$J(U, V, \Theta) = \sum_{ij \in O} \sum_{k=1}^{R-1} h(T_{ij}^k(\theta_{ik} - u_i v_j^T)) + \frac{\lambda}{2}(\|U\|_F^2 + \|V\|_F^2) \tag{1}$$

where

$$T_{ij}^k = \begin{cases} +1 & \text{for } k \ge y_{ij} \\ -1 & \text{for } k < y_{ij} \end{cases}$$

O is the set of observed values, $\lambda > 0$ is the regularization parameter, $\|.\|_F$ denotes the Frobenius norm and $h(.)$ is smooth hinge loss function defined as follows.

$$h(z) = \begin{cases} 0 & \text{if } z \ge 1 \\ \frac{1}{2}(1-z)^2 & \text{if } 0 < z < 1 \\ \frac{1}{2} - z & \text{otherwise} \end{cases}$$

The gradients of the variables to be optimized are given as follows.

$$\frac{\partial J}{\partial U_{ia}} = U_{ia} - \sum_{k=1}^{R-1} \sum_{j | ij \in O} T_{ij}^k h'(T_{ij}^k(\theta_{ik} - U_i V_j^T))V_{ja}$$

$$\frac{\partial J}{\partial V_{jb}} = V_{jb} - \sum_{k=1}^{R-1} \sum_{j | ij \in O} T_{ij}^k h'(T_{ij}^k(\theta_{ik} - U_i V_j^T))U_{jb}$$

$$\frac{\partial J}{\partial \theta_{ik}} = \sum_{j | ij \in O} T_{ij}^k h'(T_{ij}^k(\theta_{ik} - U_i V_j^T))$$

Finally, the latent factor matrices U and V and threshold matrix Θ are iteratively updated as.

$$U^{t+1} = U^t - c\frac{\partial \mathcal{J}}{\partial U}$$

$$V^{t+1} = V^t - c\frac{\partial \mathcal{J}}{\partial V} \tag{2}$$

$$\Theta^{t+1} = \Theta^t - c\frac{\partial \mathcal{J}}{\partial \theta}$$

where c is a trade-off parameter.

3   Particle Swarm Optimization

Particle Swarm Optimization (PSO) [3] uses a population of a particle to move around in a search space to find the optimal solution. For each particle p_i has a position vector x_i, a velocity vector v_i a personal best pb_i which keeps track of best fitness location of the particle encountered so far and neighbourhood best lb_i. The neighbourhood best vector is the position of best fitness found so far among the neighbourhood N_i of particle i. Where the neighbourhood is $N_i \subseteq P$, $P = \{p_1, p_2, ..., p_n\}$. The velocity and position of each particle is updated as follows.

$$v_i^{t+1} = v_i^t + c_1 r_1 (pb_i^t - x_i^t) + c_2 r_2 (lb_i^t - x_i^t)$$
$$x_i^{t+1} = x_i^t + v_i^{t+1}$$

where c_1, c_2 are acceleration coefficients and $r_1, r_2 \in [0, 1]$ are uniformly distributed random numbers.

Different improved versions of PSO have been proposed [15, 18, 22, 25]. The concept of neighbourhood topologies is about how one particle is connected to the other. Different topologies have been developed for PSO and empirically studied, each affecting the performance drastically [8].

3.1   PSO Topologies

Variants of swarm search make use different social structures and the interaction of individual particles with other particles selected through such social structures. A vast amount of research has been done on sociometrics of PSO [8, 10, 12, 16, 17, 23]. It has been shown that isolated particle swarm individuals perform very poorly [5]. It is the interactions between particles the gives exploration property for each individual. We outline below some basics of social structures or topologies.

gbest: In gbest/global each individual particle is attracted by the best solution found by any member in the neighbourhood (Fig. 1(b)). This structure is equivalent to the fully connected social network. Every particle can compare the performance of every other member of population, imitating the very best [5].

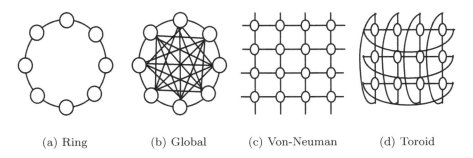

| (a) Ring | (b) Global | (c) Von-Neuman | (d) Toroid |

Fig. 1. Different topologies

lbest: In lbest/ring each individual particle is affected by k immediate neighbours in population. One common lbest case is $k = 2$ (a regular ring lattice), each individual is affected by only its 2 adjacent neighbours (Fig. 1(a)) [5].

Von Neuman: The other kind of popular topology is Von Neuman/mesh topology. Each particle is connected to its left, right, top and bottom neighbours forming a two dimensional lattice (Fig. 1(c)). Some empirical has shown significance of this topology, which performs better in a large number of problems [8].

Toroidal: Toroidal topology is similar to mesh topology, except that all particles in the population have degree of connectivity $k = 4$ (Fig. 1(d)) [12].

In this paper, we primarily focus on two main neighbourhood topologies (ring/lbest and global/gbest). The connections between the particles are static i.e., they do not vary over the course of trail, undirected and unweighted. These two topologies lie in the both extremes of communication spectrum of static topology. *lbest* has minimum number of neighbours whereas *gbest* has connection with all the particles in the population.

4   Proposed Method

In this section, we discuss the method of blending swarm-search with gradient-search to solve the matrix factorization problem. Use of PSO as an optimization technique for MMMF is first introduced in [2]. Many variants of gradient-based PSO exists in literature [1,14,21,24,26]. In [21] gradient search is initiated after PSO terminates, whereas others combine a gradient factor with search direction computed by personal best and global best.

Unlike PSO, in the proposed algorithm the velocity component is updated in every iteration with gradient direction along with the social influence. The search direction of the particle is a combination of gradient direction, and the direction of global best and the objective function is defined as follows.

$$J = \sum_{(i,j)\in O, 1\leq k\leq R-1} K_{ijk} + \frac{\lambda}{2}(\|U\|_F^2 + \|V\|_F^2) \qquad (3)$$

where K_{ijk} is three-dimensional matrix defined as.

$$K_{ijk} = \begin{cases} 1 - \theta_{ik} + x_{ij} & \text{if } k \geq y_{ij} \wedge \theta_{ik} \leq x_{ij} + 1 \\ 1 - x_{ij} + \theta_{ik} & \text{if } k \leq y_{ij} \wedge \theta_{ik} \geq x_{ij} - 1 \end{cases} \qquad (4)$$

The partial derivatives of J comes as following

$$\frac{\partial J}{\partial U} = \lambda U - H_k V^T$$
$$\frac{\partial J}{\partial V} = \lambda V - U^T H_k \qquad (5)$$
$$\frac{\partial J}{\partial \theta} = H_j$$

where $H_k = \sum_k H_{ijk}$ and $H_j = \sum_j H_{ijk}$ where,

$$H_{ijk} = \begin{cases} -1 & \text{if } k \geq y_{ij} \wedge \theta_{ik} \leq x_{ij} + 1 \\ 1 & \text{if } k \leq y_{ij} \wedge \theta_{ik} \geq x_{ij} - 1 \end{cases} \tag{6}$$

Finally, the position of particle p is updated as.

$$U_p^{t+1} = U_p^t - c(\delta \frac{\partial J}{\partial U_{ij}} - (1 - \delta)(U_{best}^t - U_p^t))$$

$$V_p^{t+1} = V_p^t - c(\delta \frac{\partial J}{\partial V_{ij}} - (1 - \delta)(V_{best}^t - V_p^t)) \tag{7}$$

$$\Theta_p^{t+1} = \Theta_p^t - c(\delta \frac{\partial J}{\partial \Theta_{ij}} - (1 - \delta)(\Theta_{best}^t - \Theta_p^t))$$

It is necessary to normalize U and V after every update so that the ith column of U has the same L2 length as the ith column of V [2].

$$U_{norm} = U^{t+1} \sqrt{\frac{||V^{t+1}||^2}{||U^{t+1}||^2}} \qquad V_{norm} = V^{t+1} \sqrt{\frac{||U^{t+1}||^2}{||V^{t+1}||^2}} \tag{8}$$

For notational convenience, we drop subscript from U_{norm}, V_{norm} without loss of generality and refer as U and V in the subsequent discussion.

Algorithm 1. MMMF with PSO-global topology (PSO-global)

---

input : U, V, Θ, swarm size n
output: $U^{best}, V^{best}, \Theta^{best}$

$t \leftarrow 0$;
Initialise $U_p^t, V_p^t, \Theta_p^t, \forall 1 \leq p \leq n$;
while *Stopping criteria met* **do**
  Compute loss function value for all the particles:
  $J_p = \sum_{(i,j) \in O, 1 \leq k \leq r-1} K_{ijk} + \frac{\lambda}{2}(||U_p^t||_F^2 + ||V_p^t||_F^2), \forall 1 \leq p \leq n$;
  $best \leftarrow \underset{p}{\arg\min} \ J_p$;
  Update $U^{best}, V^{best}, \Theta^{best}$;
  Update $U_p^{t+1}, V_p^{t+1}, \Theta_p^{t+1}$ using Eq. 7;
  Normalize U_p^{t+1}, V_p^{t+1} using Eq. 8;
  $t \leftarrow t + 1$;
end

The previous investigation in the particle swarm paradigm finds that population topology has an effect [4,6]. Kennedy [6] observes that populations with fewer connections might perform better on highly multi-modal problems while a highly interconnected network is better for unimodal problems. This is same concern that underlies simulated annealing [7]. All these reasons motivated us

Algorithm 2. MMMF with PSO-ring topology (PSO-ring)

---

input : U, V, Θ, swarm size n
output: $U^{best}, V^{best}, \Theta^{best}$

$t \leftarrow 0$;
Initialise $U_p^t, V_p^t, \Theta_p^t, \forall 1 \le p \le n$;
Compute loss function value for all the particles:
$J_p = \sum_{(i,j) \in O, 1 \le k \le r-1} K_{ijk} + \frac{\lambda}{2}(\|U_p^t\|_F^2 + \|V_p^t\|_F^2), \forall 1 \le p \le n$;
while *Stopping criteria met* **do**
    for $p = 1$ *to* n **do**
        $J_p = \sum_{(i,j) \in O, 1 \le k \le r-1} K_{ijk} + \frac{\lambda}{2}(\|U_p^t\|_F^2 + \|V_p^t\|_F^2)$;
        $best \leftarrow \arg\min_l J_l, l \in \{left, p, right\}$;
        Update $U_p^{best}, V_p^{best}, \Theta_p^{best}$;
        Update $U_p^{t+1}, V_p^{t+1}, \Theta_p^{t+1}$ using Eq. 7;
        Normalize U_p^{t+1}, V_p^{t+1} using Eq. 8;
    end
    $t \leftarrow t + 1$;
end
$J_p = \sum_{(i,j) \in O, 1 \le k \le r-1} K_{ijk} + \frac{\lambda}{2}(\|U_p^{best}\|_F^2 + \|V_p^{best}\|_F^2), \forall 1 \le p \le n$;
$best \leftarrow \arg\min_p J_p$;
Return $U^{best}, V^{best}, \Theta^{best}$;

---

to address the question of way social structures of PSO influence matrix factorization methods. We use two different social structures of PSO namely *global topology* and *ring topology* as mentioned in the previous section. Algorithm 1 describes the detailed procedure related to applicability of PSO-global topology on MMMF. Algorithm 2 describes the detailed procedure related to applicability of PSO-ring topology on MMMF.

5   Experimental Results

In this section, we demonstrate the experiments carried out to evaluate the effectiveness and efficiency of our proposed algorithms using real data set (Movie-Lens) that has been used as benchmark in prior works. In our experiments, we select 80 % of the observed matrix randomly to train the algorithms and use remaining 20 % to test. All the results reported here are the average of 5 such runs.

In our first experiment, we show that MMMF algorithm does not guarantee optimal path all the time. Figure 2 shows that MMMF is following different paths for different initial points. We also observe that algorithm converges faster for some initial points than the others. This observation gives us the motivation to try PSO based search to see how communication between the particles helps to reduce the search time.

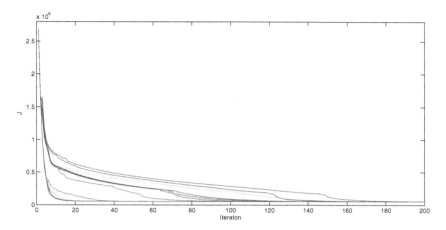

Fig. 2. MMMF finding optimum via different paths

In the second experiment, we employ gradient based PSO to show the communication between the particles help to take optimal path irrespective of different starting points. Figure 3 shows the results related to our second experiment. It can be seen from the figure that particles are started at different initial points yet all are taking same path (optimal). We use global topology for this experiment.

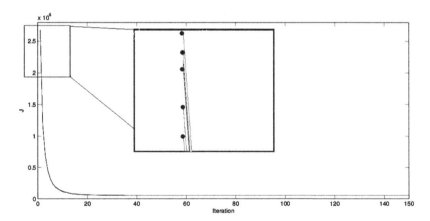

Fig. 3. PSO based MMMF. Particles initiated at different points moves to the optimal path.

In the third experiment, we compare the efficiency of the gradient based PSO with MMMF in terms of its convergence speed. We use both *Global* and *Ring* topologies for this experiment. We refer MMMF with PSO-global topology as

Table 1. The MAE value reached by MMMF for different number of iterations is reached earlier by PSO-based MMMF. Number of particles $= 15$

MAE	Loss			No. of iterations		
	MMMF	PSO-Global	PSO-Ring	MMMF	PSO-Global	PSO-Ring
0.8195	192596	80938	80655	50	14.4	15.6
0.7303	86590	58837	58697	100	25.6	26.4
0.7015	55410	54328	54153	150	50.8	55.4
0.6942	53878	53709	53501	200	65.6	74.8
0.6923	53628	53609	53453	250	70.4	77.8
0.6903	52895	52852	52859	300	88.6	107.7

Table 2. Iteration number for global and ring topology took to reach the MAE. The figures shown are the average value of experiments conducted multiple times.

No. of particles	PSO topology	No. of iterations					
		50	100	150	200	250	300
5	Global	17.2	33.4	58.2	59.0	78.6	79.0
	Ring	18.2	33.8	63.2	64.0	84.4	95.0
10	Global	15.8	26.6	38.0	69.8	79.6	81.0
	Ring	16.1	27.0	39.2	80.2	100.2	124.6
15	Global	14.4	25.6	50.8	65.6	70.4	88.6
	Ring	15.6	26.4	55.4	74.8	77.8	107.7
20	Global	14.8	15.0	47.8	59.2	74.4	87.0
	Ring	15.0	16.1	49.8	67.0	90.2	123.0

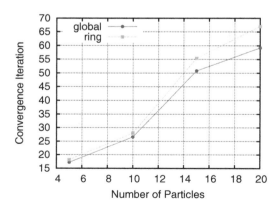

Fig. 4. The performance gap is high as number of particles and iteration increases. (Color figure online)

PSO-global and MMMF with PSO-ring topology as *PSO-ring* in the subsequent discussion. We first run *MMMF* algorithm using gradient descent for fixed number of iterations and observe MAE. We fix the same MAE a target for gradient based PSO. Table 1 reports the number of iterations taken by different search techniques. It can be seen from the table that PSO with global topology and PSO with ring topology are taking less number of iterations to converge than simple gradient search. We also analyse the performance of PSO by varying number of particles and number of iterations. Table 2 shows the number of iterations taken by the PSO to reach target MAE[1]. For example, MMMF with gradient descent takes 50 iterations to reach a particular MAE, PSO with global topology takes 17.2 iterations and PSO with ring topology takes 18.2 iterations to reach same MAE. Another observation we made from the table is PSO-global performing better than PSO-ring. Figure 4 dipicts the number of iterations taken by *PSO-global* versus *PSO-ring* by varying number of particles. We observe from the graph that *PSO-global* is performing better than *PSO-ring*, especially at higher number of particles.

6   Conclusions

In this paper, we combine swarm-like search with gradient descent to provide faster convergence of the optimization problem associated with maximum margin matrix factorization. We show that the search trajectory of gradient descent search does not guarantee the optimal path always. We demonstrate here that when search particles interact during the search, the exploration process helps in faster and better optimizing point. Our experiments on benchmark dataset show that combining swarm-like search with gradient descent improves the search to reach near optimal point. As a result, we get an efficient and novel collaborative filtering method.

Acknowledgements. Part of this work is carried out at Central University of Rajasthan. Authors acknowledge Central University of Rajasthan for providing facilities.

References

1. Borowska, B., Nadolski, S.: Particle swarm optimization: the gradient correction (2009)
2. Devi, V.S., Rao, K.V., Pujari, A.K., Padmanabhan, V.: Collaborative filtering by pso-based mmmf. In: IEEE International Conference on Systems, Man and Cybernetics, pp. 5–8, October 2014
3. Eberhart, R.C., Kennedy, J.: A new optimizer using particle swarm theory. In: Proceedings of the Sixth International Symposium on Micro Machine and Human Science, vol. 1, pp. 39–43 (1995)

---

[1] The MAE that gradient based MMMF could reach in fixed number of iterations.

4. Figueiredo, E.M., Ludermir, T.B.: Effect of the PSO topologies on the performance of the PSO-ELM. In: Neural Networks Brazilian Symposium (SBRN), pp. 178–183. IEEE, October 2012
5. Kennedy, Y.S.J., Eberhart, R.C.: Swarm Intelligence. Morgan Kaufmann, San Francisco (2001)
6. Kennedy, J.: Small world and mega-minds: effects of neighborhood topologies onparticle swarm performance. In: Congress on Evolutionary Computation, pp. 1931–1938 (1999)
7. Kennedy, J.: Bare bones particle swarms. In: IEEE Swarm Intelligence Symposium, pp. 80–87 (2003)
8. Kennedy, J., Mendes, R.: Population structure and particle swarm performance. In: Proceedings IEEE Congress on Evolutionary Computation, vol. 2, pp. 1671–1676 (2002)
9. Koren, Y., Bell, R., Volinsky, C.: Matrix factorization techniques for recommender systems. IEEE Comput. **42**(8), 30–37 (2009)
10. Lane, J., Engelbrecht, A., Gain, J.:Particle swarm optimization with spatially meaningful neighbours. In: Swarm Intelligence Symposium, SIS 2008, pp. 1–8. IEEE (2008)
11. Lee, D.D., Seung, H.S.: Learning the parts of objects by non-negative matrix factorization. Nature **401**(6755), 788–791 (1999)
12. Millie, P., Thangaraj, R., Abraham, A.: Particle swarm optimization: performance tuning and empirical analysis. Found. Comput. Intell. **3**, 101–128 (2009)
13. Nati, N.S., Jaakkola, T.: Weighted low-rank approximations. In: 20th International Conference on Machine Learning, pp. 720–727. AAAI Press (2003)
14. Noel, M.M., Jannett, T.C.: Simulation of a new hybrid particle swarm optimization algorithm. In: Proceedings of the Thirty-Sixth Southeastern Symposium on IEEE System Theory, pp. 150–153 (2004)
15. Pant, M., Radha, T.. Singh, V.: A simple diversity guided particle swarm optimization. In: Proceedings of IEEE Congress Evolutionary Computation, pp. 3294–3299 (2007)
16. Peer, E.S., van den Bergh, F., Engelbrecht, A.P.: Using neighbourhoods with the guaranteed convergence PSO. In: Proceedings of the 2003 IEEE Swarm Intelligence Symposium, SIS 2003, pp. 235–242. IEEE (2003)
17. Ni, J.D.Q.: A new logistic dynamic particle swarm optimization algorithm based on random topology. Sci. World J. (2013)
18. Radha, T., Pant, M., Abraham, A.: A new diversity guided particle swarm optimization with mutation. In: Nature and Biologically Inspired, Computing, pp. 294–299 (2009)
19. Rennie, J.D., Srebro, N.: Fast maximum margin matrix factorization for collaborative prediction. In: Proceedings of the 22nd International Conference on Machine Learning, pp. 713–719. ACM (2005)
20. Sarwar, B., Karypis, G., Konstan, J., Riedl, J.: Incremental singular value decomposition algorithms for highly scalable recommender systems. In: Fifth International Conference on Computer and Information Science, pp. 27–28. Citeseer (2002)
21. Szabo, D.B.: A study of gradient based particle swarm optimisers. Master's thesis, Faculty of Engineering, Built Environment and In- formation Technology University of Pretoria, Pretoria, South Africa (2010)
22. Tanweer, M., Suresh, S., Sundararajan, N.: Self regulating particle swarm optimization algorithm. Inf. Sci. **294**, 182–202 (2015)

23. Toscano-Pulido, G., Reyes-Medina, A.J., Ramírez-Torres, J.G.: A statistical study of the effects of neighborhood topologies in particle swarm optimization. In: Madani, K., Correia, A.D., Rosa, A., Filipe, J. (eds.) Computational Intelligence. SCI, vol. 343, pp. 179–192. Springer, Heidelberg (2011)
24. Vesterstrom, J.S., Riget, J., Krink, T.: Division of labor in particle swarm optimisation. In: WCCI, pp. 1570–1575. IEEE (2002)
25. Yao, J., Han, D.: Improved barebones particle swarm optimization with neighborhood search and its application on ship design. Math. Probl. Eng., 1–12 (2013)
26. Zhang, R., Zhang, W., Zhang, X.: A new hybrid gradient-based particle swarm optimization algorithm and its applications to control of polarization mode dispersion compensation in optical fiber communication systems. In: International Joint Conference on Computational Sciences and Optimization, CSO 2009, vol. 2, pp. 1031–1033. IEEE (2009)

Incorporating an Implicit and Explicit Similarity Network for User-Level Sentiment Classification of Microblogging

Yongyos Kaewpitakkun[(✉)] and Kiyoaki Shirai

Japan Advanced Institute of Science and Technology, Nomi, Japan
{s1320203,kshirai}@jaist.ac.jp

Abstract. In Twitter, the sentiments of individual tweets are difficult to classify, but the overall opinion of a user can be determined by considering their related tweets and their social relations. It would be better to consider not only the textual information in the tweets, but also the relationships between the users. Previous approaches that incorporate network information into the classifier have mainly focussed on "a link" defined by the explicitly connected network, such as, follow, mention, or retweet. However, the presence of explicit link structures in some social networks is limited. In this paper, we propose a framework that takes into consideration the "implicit connections" between users. An implicit connection refers to the relations of users who share similar topics of interest, as extracted from their historical tweet corpus, which contains much data for analysis. The results of experiments show that our method is effective and improves the performance compared to the baselines.

Keywords: Sentiment analysis · Factor-graph model · Topic modeling · Machine learning · Microblogging

1 Introduction

Recently, microblogging services such as Twitter have become popular data sources in the domain of sentiment analysis because of their efficiency and low-cost for development. However, analysing sentiments on tweets is still difficult because tweets are very short and contain slang, informal expressions, emoticons, and mistyping. This causes the data to be sparse and decreases the performance of sentiment classifiers. Existing approaches to sentiment analysis mainly focus on classification at the message level, and ignore information from network relations. The individual tweets of the users are difficult to classify, but their overall opinion can be determined by considering their related tweets and their social relations, which can be of benefit for many opinion mining systems. Unlike traditional previous approaches, not only the textual information in the tweets but also the relationship between the users should be taken into account. Two social science theories (Charu 2011) indicate important phenomena that can apply to social networks:

ⓒ Springer International Publishing Switzerland 2016
R. Booth and M.-L. Zhang (Eds.): PRICAI 2016, LNAI 9810, pp. 180–192, 2016.
DOI: 10.1007/978-3-319-42911-3_15

Homophily: When a link between individuals (such as friendship or other social connection) is correlated with those individuals being similar in nature. For example, friends often tend to be similar in characteristics like age, social background, and educational level.

Co-citation regularity: A related concept, which holds when similar individuals tend to refer or connect to the same things. For example, when two people tweet messages with similar topics, they probably have similar tastes in other things or have other common interests.

Previous approaches that have incorporated network information into a classifier have mainly focussed on the first phenomenon, "Homophily", and define "a link" by the explicitly connected network. Tan et al. (2011) used a friendship network such as that from 'follow' and from the 'mention' graph to perform a user-level sentiment analysis. Pozzi et al. (2013) used the approval relation based on the retweet graph to solve the same problem, and got satisfying results. In some social networks, however, the presence of explicit link structures is limited. The statistics for Twitter in 2009[1] indicate that 55.50 % of the users were not following anyone, 52.71 % had no follower, and only 1.44 % of the tweets are retweets. Therefore, in real-life situations, a large part of the social network does not contain explicit links, and so the current opinion mining systems do not derive any benefit from the network information. In order to overcome this limitation, we propose a framework that incorporates the "implicit connections", based on similarities between users. Following the "co-citation regularity theorem", we will take implicit connections to refer to the relations between users who share similar interests in topics, as extracted from their historical tweet corpus. This will enable us to use more data for sentiment classification. The hypothesis behind this research follows.

Users who have similar interests and often post messages on microblogging containing similar topics tend to have similar opinions in some areas.

In sum, the goal of this research is to develop a method of classifying the overall sentiments (positive or negative) of users about a certain topic by using textual information as well as both explicit and implicit relationships between users in the social network. We also propose an improved method to discover latent topics in the tweets via an enhanced pooling scheme with the conventional Latent Dirichlet Allocation (LDA), called the Hashtag-PMI pooling scheme. In addition, the whole process does not require any human intervention, such as annotation of labeled data. This enables us to apply our method to the sentiment analysis of various targets.

2   Background and Related Work

In this section, we briefly summarize the related research from the two perspectives: topic modeling for short texts and user-level sentiment analysis.

---

[1] http://www.webpronews.com/wonder-what-percentage-of-tweets-are-retweets-2009-06/.

2.1 Topic Modeling for Short Texts

Probabilistic topic models, like Latent Dirichlet Allocation (LDA) (Blei et al. 2003), are widely used and give a successful results for discovering the hidden topics from a large collection of documents. However, previous research has found that the standard topic modeling techniques do not work well with the short and ambiguous form of tweets (Zhao et al. 2011). To overcome this problem, a number of extensions of LDA have been proposed, in two directions. One is to modify the LDA mechanism to deal with short texts, such as Labeled LDA (Ramage et al. 2010) or Twitter-LDA (Zhao et al. 2011). The other, which is simpler and more popular, is to aggregate the tweet messages into more lengthy documents before applying the standard LDA model. This is called the 'message pooling scheme'. The popular message pooling schemes are to merge all tweets under the same author (Hong et al. 2010; Weng et al. 2010), the tweets published in a similar time period, or the tweets with the same hashtag (Mehrotra et al. 2013). Regarding the pooling schemes, Mehrotra et al. (2013) have reported that merging the tweets sharing the same hashtag into one document performed better than other pooling schemes. On the other hand, Wang et al. (2011) reported that only 14.6 % of the tweets contain a hashtag, that is, the remaining un-hashtagged tweets were not used effectively in the hashtag-pooling method. Mehrotra et al. (2013) and Schinas et al. (2014) tackled this problem by merging messages without a hashtag to the most similar document, and found that this method improved the performance of the conventional LDA and achieved the best performance. They used cosine similarity with TF-IDF to measure the textual similarity. In the present paper, we propose an alternative way to extract the potential relationships between tweets that should belong to the same topic. The difference between our approach and previous approaches (Mehrotra et al. 2013; Schinas et al. 2014) is that instead of assigning the un-hashtagged tweets to the document with the highest textual similarity, we consider the co-occurrence between a hashtag and a term, based on Point-wise Mutual Information (PMI), which explicitly captures the relation between them.

2.2 User-Level Sentiment Analysis

Although most previous work on sentiment analysis in Twitter has mainly focused on understanding the sentiments of individual messages (Go et al. 2009; Kaewpitakkun et al. 2014), there have been several attempts to identify the sentiment of the users. They were based on the assumption that the overall sentiment of the users can be estimated by aggregating the sentiments of the individual tweets in their history corpus (Kaewpitakkun et al. 2016; Smith et al. 2013). However, many individual tweets are difficult to classify, due to tweets' being short and ambiguous. Moreover, the simple aggregation of the sentiment of the tweets may cause a lot of noise and errors. To overcome this problem, some researchers have proposed solutions that incorporate the network relation data into their model, such as 'follow' and 'mention' (Tan et al. 2011), and 'retweet' (Pozzi et al. 2013; Nozza et al. 2014). Unfortunately, in some social networks, there is only a limited presence of such explicit link structures. In our approach, we incorporate not only explicit links but also implicit links, which will be extracted from users' historical tweet corpora, in the social network. This enables us to effectively use more of the information in the social network.

3   Proposed Method

This section presents the proposed approach. The system accepts a set of users and a certain topic as input, and the classifies the sentiment (positive or negative) of the users for the given topic. An overview of the framework of the system is shown in Fig. 1. This system is divided into three main parts. First, the implicit relationship between the users is extracted using the LDA with the proposed enhanced pooling scheme. Second, the sentiment of the on-target tweets is classified by a target-dependent sentiment analysis, incorporating target specific knowledge. After that, the information about the implicit relationships and the explicit relationship based on the retweet network and the textual information are incorporated into a heterogeneous factor-graph model. Finally, a loopy belief propagation is applied to predict the sentiment of the users.

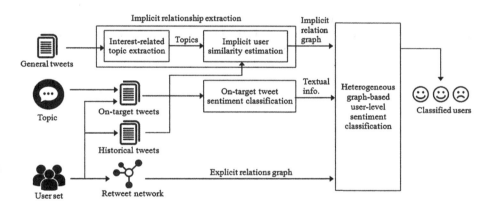

Fig. 1. System framework

3.1   Implicit Relationship Extraction

In this module, we would like to extract the implicit relationship between the users in the social network. As discussed above, the implicit connection refers to the relations of the users who share the similar interested topics extracted from their historical tweet corpus. This module carries out two sub tasks. In the first task, the interest-related topics have been identified by LDA with the enhanced pooling schema. In the second task, the similarity between the users have been estimated based on the cosine similarity in TF-IDF-like vector space.

Interest-Related Topic Extraction. We present an alternative way to discover the latent topics in a general tweet corpus using the conventional LDA, called "Hashtag-PMI" pooling scheme, which constructs a document set by aggregating the tweets that likely to express the same topics into the same documents to create better training data for LDA. First, the tweets including the same hashtag are merged as a single document. The pooled tweets very likely represent the same topic, since the hashtag can be considered as the topic labeled by the user. The tweets with multiple hashtags are assigned to the multiple document and the tweets without hashtag are left unchanged and unmerged. Finally, the LDA is applied on the set of the aggregated documents to infer the latent topics on the tweet corpus.

An additional procedure is performed for pooling the tweets of the same topic. First, correlation between the hashtag H and the candidate terms C (noun and proper noun) is calculated by PMI on the general tweet corpus as shown in Eq. (1). Then, the candidate terms are selected as an extended-hashtag if their PMI value is greater than a threshold T_PMI. The extended-hashtag is the list of synonyms or terms that usually appear together with a given hashtag. Finally, the hashtag is added to the tweets containing one of the terms in the extended-hashtag list. We believe that the potential relations between the tweets can be captured by the terms (the extended-hashtag) that are highly correlated with a hashtag, even when the terms in those tweets are totally different. In the experimental results presented in Sect. 4, the number of LDA topics is set at 100, a value which was decided on after some preliminary experiments.

$$PMI(H, C) = \log \frac{p(H, C)}{p(H)p(C)} = \log \frac{p(H|C)}{p(H)} \tag{1}$$

Implicit User Similarity Estimation. The list of the topics inferred from the previous step are used for estimating the implicit similarity between the users. First, the topic with the highest probability estimated by LDA is assigned to each historical tweet of the user. Only the tweets that have a probability greater than a threshold, T_IMP1, are selected. This screening is applied for filtering out the interest-unrelated or daily chat tweets. After that, the implicit similarities between each pair of users $\left(SIM(user_1, user_2)\right)$ are estimated by cosine similarity in the topic vector space with modified TF-IDF weighting, where the TF refers to the frequency of the 'topic' and IDF refers to the inverse frequency of the 'user'. Only the connections between the users whose implicit similarity is higher than a certain threshold, T_IMP2, are preserved. In the experiments, we empirically set T_IMP1 to 0.01 after some preliminary experimentation, and we varied T_IMP2 from 0 to 1.

3.2   On-Target Tweet Sentiment Classification

This module classifies the sentiments (positive, negative or neutral) of the on-target tweets; they will be used as the textual information in the next step (explained in Subsect. 3.3). We apply the method proposed in Kaewpitakkun et al. (2016), where several techniques are used to improve the performance of target dependent sentiment classification. First, not general but target-dependent training data is constructed for learning the sentiment classifier. It is automatically created from unlabeled tweets by a lexicon-based method and several heuristics. Second, a target-specific add-on lexicon is automatically constructed. A public sentiment lexicon is insufficient for target-specific sentiment analysis, since the words used to express an opinion of the target are often not compiled in it. Finally, an extended target list and competitor list are introduced into the model. The former is the list of synonyms of the target. The latter is a list of the competitors of a given target (e.g. a product). For more detail, see Kaewpitakkun et al. (2016).

3.3 Heterogeneous Graph-Based User-Level Sentiment Classification

Starting with the definition of the user-level sentiment analysis task, the proposed heterogeneous factor-graph model will be described. Then, the inference and prediction algorithm on the graph will be explained.

Social Similarity Factor Graph Model. Given a topic q and a set of users $V_q = \{v_1, v_2, \cdots, v_n\}$ who have tweeted about q, the goal is to infer the sentiment polarities $y = \{y_1, y_2, \cdots, y_n\}$ of the users in V_q, where $y_i \in \{pos, neg\}$. For each user $v_i \in V_q$, we have the set of the tweets of v_i about q, $TW_{v_i,q}$, and the explicit relations of the user v_i's retweeting a message of another user v_j. We also have the users' implicit relationship representing that they tend to tweet about similar topics. We incorporate both textual information and the social similarity network (explicit and implicit relationships) into a single heterogeneous factor graph.

For a given topic q, a *Social Similarity Factor Graph*, denoted by $SG_q = \{V_q, TW_{v_i,q}, E_{tw}, E_{ex}, E_{im}\}$ is constructed as in Fig. 2.

In SG_q, a node is a user in $v_i \in V_q$ or a set of tweets $tw_{v_i} \in TW_{v_i,q}$. There are three types of edges: a user–tweet edge E_{tw} that connects v_i with tw_{v_i}, explicit similarity edges E_{ex}, and implicit similarity edges E_{im} that connect users. E_{tw}, E_{ex} and E_{im} are weighted by $f(v_i)$, $g(v_i, v_j)$ and $h(v_i, v_j)$, respectively, which will be defined later.

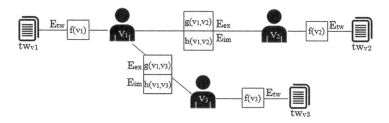

Fig. 2. An example of a social similarity factor graph

Given the social similarity factor graph SG_q, we would like to classify those users v_i with a given sentiment y_i. Based on the Markov assumption, the sentiment $y_i \in \{pos, neg\}$ of the user v_i is influenced by the sentiment labels of the on-target tweets tw_{v_i} and the sentiment labels of the neighboring users $N(v_i)$. This assumption leads us to adapt the concept of factor graph model defined in Tan et al. (2011) and Pozzi et al. (2013) to combine the user's tweet with the explicit and implicit user–user relationships, as shown in Eq. (2).

$$
\log(P(y|SG)) = \left(\sum_{v_i \in V} \left[\log\left(f(y_i|tw_{v_i})\right) \right. \right.
$$
$$
\left. \left. + \sum_{v_j \in N(v_i)} \left[\log\left(g(y_i, y_j|v_i, v_j)\right) + \log\left(h(y_i, y_j|v_i, v_j)\right)\right]/2 \right] \right)
$$
$$
- \log Z
$$
(2)

The first line corresponds to the user–tweet factor and the second line refers to the inclusion of explicit and implicit user–user factors. Z is a normalization factor. We define the feature functions as follows.

The user–tweet factor. This function takes into account the sentiment of the user v_i by analysing his/her on-target tweets. The polarity of a tweet $tw \in tw_{v_i}$ is classified by the on-target tweet-level classifier described in Subsect. 3.2. The probabilities of a positive and negative class, denoted as $P_{pos}(tw)$ and $P_{neg}(tw)$, are also estimated by the classifier. Note that the neutral tweets are discarded since they represent no sentiment. The user–tweet function is defined as follows:

$$
f(y_i|tw_{v_i}) = \sum_{tw \in tw_{v_i}} P_{y_i}(tw)
$$
(3)

The user–user explicit factor. This function takes into account the sentiment of the neighboring users connected by retweet relations. The user–user explicit function is defined as follows:

$$
g(y_i, y_j|v_i, v_j) = \frac{\#retweet_{i \to j}}{\sum\limits_{v_k \in N(v_i)} \#retweet_{i \to k}} . \delta_{y_i, y_j}
$$
(4)

where δ_{y_i, y_j} is the Kronecker's delta (1 when $y_i = y_j$ and 0 otherwise), and $\#retweet_{i \to j}$ denotes the number of times that v_i retweets v_j's posts.

The user–user implicit factor. This function takes into account the sentiment of the neighboring users by analysing the implicit similarity between v_i and their neighbors. The user–user implicit function is defined as follows:

$$
h(y_i, y_j|v_i, v_j) = \frac{SIM(v_i, v_j)}{\sum\limits_{v_k \in N(v_i)} SIM(v_i, v_k)} . \delta_{y_i, y_j}
$$
(5)

where $SIM(v_i, v_j)$ denotes the implicit similarity between user v_i and v_j described in Subsect. 3.1.

Finally, our objective is to maximize the following function with respect to the appropriate sentiment labels.

$$\hat{y} = \arg\max_{\mathbf{y}} \ \log(P(\mathbf{y}|\mathbf{SG})) \tag{6}$$

The Inference and Prediction Algorithm. We adapt the loopy belief propagation (LBP) defined in Wang et al. (2011) to perform the inference and prediction for a given model, i.e., to approximately maximize the function given in (6). LBP is a message passing algorithm for performing inference on graphical models and it has been shown to be a useful approximate algorithm on general graphs. Algorithm 1 shows the pseudo code of LBP. First, the initial labels of the users are assigned through the user–tweet factor function defined in Eq. (3). Messages $m_{i\rightarrow j}(y)$, which represent the degree of influence on the sentiment class y from the node i to j, are inferred by an iterative process. In each iteration, the user–user explicit and implicit factor functions defined in Eqs. (4) and (5) are applied to the sentiment messages from the user v_i to v_j. These messages are continuously updated until they are convergent. Lastly, the final sentiment labels of the users are computed based on the value of their neighbors' converged messages, as shown in the last loop in Algorithm 1.

4   Evaluation

4.1   Dataset

In order to evaluate our proposed system, we used the "Obama Retweet" dataset published by Pozzi et al. (2013), which contains (1) a set of users and their sentiment labels about the topic "Obama", (2) a collection of the tweets posted by users about the topic "Obama" and their sentiment labels (called on-target tweets), and (3) the users' retweet network information, consisting of 252 retweet connections. In order to extract the users' implicit relationship, we further downloaded the last 3,200 (as a maximum) tweets of the users through TwitterAPI. Note that all users and posts in the "Obama Retweet" dataset have been manually labeled with their polarity (positive or negative), but we did not use this for either training or classification. These gold sentiment labels were used only for evaluation.

4.2   Evaluation of the Graph-Based User-Level Sentiment Classification

We conducted several experiments to evaluate the effectiveness of our proposed method. Accuracy is used as the evaluation criterion. The performance of the following methods were measured.

Algorithm 1. Loopy belief propagation

Input: Social Similarity Factor Graph SG
Output: Sentiment label of users V
for $(v_i, v_j) \in E_{ex}, E_{im}$ **do**
  for $y \in \{pos, neg\}$ **do**
    $m_{i \rightarrow j}(y) = 1$
    $m_{j \rightarrow i}(y) = 1$
  end
end
do
  for $v_i \in V$ **do**
    for $v_j \in N(v_i)$ **do**
      for $y_j \in \{pos, neg\}$ **do**
        $m_{i \rightarrow j}(y_j) =$
        $\displaystyle\sum_{y_i \in \{pos, neg\}} ((g(y_i, y_j) + h(y_i, y_j))/2) . f(y_i) . \sum_{v_k \in N(v_i) \backslash v_j} m_{k \rightarrow i}(y_i)$
      end
    end
  end
while *all* $m_{i \rightarrow j}(y_i)$ *stop changing* $\|$ *reach maximum iteration*;
for $v_i \in V$ **do**
  $\hat{y}_i = \displaystyle\arg\max_{y \in \{pos, neg\}} f(y) . \sum_{v_j \in N(v_i)} m_{j \rightarrow i}(y)$
end

Text-only approach (Text-only): The sentiments of the users is computed by a simple majority voting strategy among the labels of their on-target tweets. The on-target tweet-level sentiment classifier described in Subsect. 3.2 is used as the classification tool.

Social similarity factor graph with explicit relations (SG-Exp): The sentiments of the users are inferred by loopy belief propagation on the factor-graph model with the textual information and explicit user relations.

Social similarity factor graph with implicit relations (SG-Imp): The sentiments of the users are inferred by LBP on the factor-graph model using the textual information and implicit user relations.

Social similarity factor graph with explicit and implicit relations (SG-ALL): The sentiments of the users are inferred by LBP on the full factor-graph model described in this paper.

Results on the Full Retweet Dataset. In this experiment, we did the experiment on the full "Obama Retweet" dataset, described in Subsect. 4.1, which contains 62 users and 252 retweet connections. All users had at least one retweet connection. We varied the threshold T_IMP2, which controlled the number of implicit edges in the graph, from 0 to 1. Note that $T\_IMP2 = 1$ means no implicit relation was incorporated in the model. Figure 3(a) shows the results of the graph-based user-level sentiment classification on the full "Obama Retweet" dataset. It shows that the social similarity factor graph with explicit relations (SG-Exp) achieved the best performance, 69.35 % accuracy. The social similarity factor graph with implicit relations (SG-Imp) was effective and improved the performance compared to the baseline (Text-only), especially when the implicit similarity threshold (T_IMP2) was greater than 0.4. It reached a highest accuracy of 64.52 % (a 3.23 % improvement over the text-only method). The combination of explicit and implicit relations (SG-ALL) did not improve the performance compared to SG-Exp because this dataset was designed for retweet network experiments. That is, since the number of the explicit links is much higher than the implicit links, especially when T_IMP2 is high, the implicit relations could not contribute to improve the performance much.

In sum, we found that the information from the explicit links, like retweet, was more effective than the implicit links. This is not surprising because the explicit links were intentionally created by the users while the implicit links might contain some noise because they were estimated from each user's tweet corpus. These noises can be occurred in case that user A and user B have the same interest in some topics, such as soccer or music, and usually post the tweets about it, but they may have different opinions in Obama. However, as we discussed earlier, the presence of such explicit links in real-life situations is limited. Therefore, we conducted another experiment under a more realistic situation.

Results on the Reduced Retweet Dataset. In this experiment, we randomly divided the retweet links in the "Obama Retweet" into 9 parts and constructed 9 datasets, including the retweet connection in only one part. Each dataset contained about 30 retweet links. These datasets are more consistent with a real-life situation in Twitter. Figure 3(b) shows the average of the accuracy

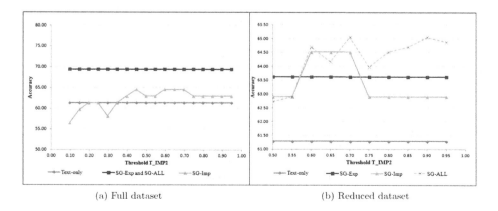

(a) Full dataset   (b) Reduced dataset

Fig. 3. Result of graph-based user-level sentiment classification

of the graph-based user-level sentiment classification over the 9 reduced retweet datasets. It indicates that both SG-Exp and SG-Imp were effective and improved the accuracy compared to the baseline (Text-only) for various T_IMP2 values between 0.5 and 0.95. The combination of explicit and implicit relations (SG-ALL) further improved the accuracy and achieved the best performance compared to other methods when T_IMP2 was set between 0.7 to 0.95. These results indicate that our hypothesis in Sect. 1 is correct. The implicit relations, extracted from users' tweet corpora, are useful, especially when there are few explicit relations.

4.3   Performance of Pooling Methods

We now present the performance of the different pooling schemas for LDA topic extraction. Our proposed hashtag-PMI method was compared to the various baselines:

Unpooled: Represent each tweet as a single document.

Author-based: Merge tweets from the same user into one document.

Hashtag-based: Merge tweets that contain the same hashtag into one document. Tweets containing several hashtags are assigned to several documents, and tweets without a hashtag are left unchanged and unmerged.

Auto Hashtag Labeling (Mehrotra et al. 2013): First, aggregate tweets by using a hashtag-based method. Then, assign the tweets without a hashtag to the document that has the highest textual similarity. In this method, the cosine similarity of the word vector weighted by the term frequency (TF) is used as the similarity measure.

Hashtag-PMI: Our proposed method, presented in Subsect. 3.1. T_PMI was set to 1 based on empirical observations.

We constructed the dataset from StanfordTwitter7[2], a tweet collection posted in June 2009. We chose 14 keywords from "Twitter Suggestion Categories" (Hong et al. 2010), such as 'politics', 'technology', and 'music'. A subset of StandfordTwitter7 was obtained by searching for tweets with these keywords, one by one. In this experiment, the performance of clustering will be measured to evaluate enhanced pooling schemas, but there was no category or topic label in this dataset. We used the hashtags of the keywords, i.e. #politics, as the gold label of the topic cluster. We divided the dataset into two parts. The tweets that contained the keyword hashtag were used as the test data and the remaining tweets were used as the training data. The tweets in the training dataset were merged into a single document according to the different pooling schemas, then the topics were identified by LDA. For evaluation, the topic with the highest probability estimated by LDA was assigned to each tweet, then the tweets with the same topic were merged into one cluster. Note that we removed the hashtag

---

[2] http://snap.stanford.edu/data/twitter7.html.

Table 1. Evaluation of the topic extraction methods

No.	Pooling method	No. of docs	Purity	NMI
1	Unpooled	581,105	0.3705	0.1608
2	Author-based	325,027	0.3781	0.1679
3	Hashtag-based	533,482	0.3721	0.1615
4	Auto Hashtag Labeling (Mehrotra et al. 2013)	490,299	0.3885	0.1810
5	Hashtag-PMI	55,318	**0.4205**	**0.2252**

used as the gold label from the test dataset before clustering. Purity and Normalize Mutual Information (NMI) were used as the evaluation metrics. Purity is the number of correctly assigned documents divided by the total number of documents. NMI is the mutual information between the set of output clusters and the labeled classes of the documents. Table 1 shows the number of documents obtained by pooling, the purity score obtained by each method, as well as its NMI. It indicates that our proposed method was effective and improved the purity and NMI from the strongest baseline (Auto Hashtag Labeling), by 3.2 % and 4.4 %, respectively.

5   Conclusions

This paper presented a novel graph-based method that incorporates the information of both textual information, as well as the explicit and implicit relationships between the users, into a heterogeneous factor graph for the sentiment analysis of the tweets at the user level. The implicit relationships are guessed by the LDA with our proposed enhanced pooling scheme. Loopy belief propagation is applied on the graph to predict the final sentiment of the users. The results of the experiments indicate that our proposed method is effective and improves the classification accuracy compared to the baseline methods that consider only textual information or explicit links. Moreover, we have proposed a new enhanced pooling method, "Hashtag-PMI", to more precisely infer the latent topics by the conventional LDA from the tweet corpus. It outperformed the other state-of-the-art pooling schemas.

One drawback of LDA is that the number of the topics must be defined beforehand. Therefore, we plan to find an effective method that determines automatically the optimal number of topics. The implicit user similarity threshold (T_IMP2) is another important parameter for which we plan to explore a sophisticated method to find an optimal value. In addition, we plan to conduct more experiments with a larger dataset and various topics.

References

Blei, D.M., Ng, A.Y., Jordan, M.I.: Latent dirichlet allocation. J. Mach. Learn. Res. **3**, 993–1022 (2003)

Charu, A.C.: Social Network Data Analytics. Springer, New York (2011)

Go, A., Bhayani, R., Huang, L.: Twitter sentiment classification using distant supervision. CS224N Project Report, Stanford University (2009)

Hong, L., Davison, B.D.: Empirical study of topic modeling in twitter. In: Proceedings of the First Workshop on Social Media Analytics. ACM (2010)

Kaewpitakkun, Y., Shirai, K., Mohd, M.: Sentiment lexicon interpolation and polarity estimation of objective and out-of-vocabulary words to improve sentiment classification on microblogging. In: The 29th Pacific Asia Conference on Language, Information and Computation, pp. 204–213 (2014)

Kaewpitakkun, Y., Shirai, K.: Incorporation of target specific knowledge for sentiment analysis on microblogging. IEICE Trans. Inf. Syst. **E99–D**(4), 959–968 (2016)

Mehrotra, R., Sanner, S., Buntine, W., Xie, L.: Improving lda topic models for microblogs via tweet pooling and automatic labeling. In: Proceedings of the 36th International ACM SIGIR Conference on Research and Development in Information Retrieval, pp. 889–892. ACM (2013)

Nozza, D., Maccagnola, D., Guigue, V., Messina, E., Gallinari, P.: A latent representation model for sentiment analysis in heterogeneous social networks. In: Canal, C., Idani, A. (eds.) SEFM 2014 Workshops. LNCS, vol. 8938, pp. 201–213. Springer, Heidelberg (2015)

Pozzi, F.A., Maccagnola, D., Fersini, E., Messina, E.: Enhance user-level sentiment analysis on microblogs with approval relations. In: Baldoni, M., Baroglio, C., Boella, G., Micalizio, R. (eds.) AI*IA 2013. LNCS, vol. 8249, pp. 133–144. Springer, Heidelberg (2013)

Ramage, D., Dumais, S.T., Liebling, D.J.: Characterizing microblogs with topic models. In: ICWSM 2010, pp. 1–1 (2010)

Schinas, M., Papadopoulos, S., Kompatsiaris, Y., Mitkas, P.A.: StreamGrid: summarization of large scale events using topic modelling and temporal analysis. In: SoMuS@ICMR (2014)

Smith, L.M., Zhu, L., Lerman, K., Kozareva, Z.: The role of social media in the discussion of controversial topics. In: International Conference on Social Computing (SocialCom), pp. 236–243. IEEE (2013)

Tan, C., Lee, L., Tang, J., Jiang, L., Zhou, M., Li, P.: User-level sentiment analysis incorporating social networks. In: Proceedings of the 17th ACM SIGKDD International Conference on Knowledge Discovery and Data Mining, pp. 1397–1405. ACM (2011)

Wang, X., Wei, F., Liu, X., Zhou, M., Zhang, M.: Topic sentiment analysis in twitter: a graph-based hashtag sentiment classification approach. In: Proceedings of the 20th ACM International Conference on Information and Knowledge Management, pp. 1031–1040. ACM (2011)

Weng, J., Lim, E.P., Jiang, J., He, Q.: Twitterrank: finding topic-sensitive influential twitterers. In: Proceedings of the Third ACM International Conference on Web Search and Data Mining, pp. 261–270. ACM (2010)

Zhao, W.X., Jiang, J., Weng, J., He, J., Lim, E.-P., Yan, H., Li, X.: Comparing twitter and traditional media using topic models. In: Clough, P., Foley, C., Gurrin, C., Jones, G.J.F., Kraaij, W., Lee, H., Mudoch, V. (eds.) ECIR 2011. LNCS, vol. 6611, pp. 338–349. Springer, Heidelberg (2011)

Threshold-Based Direct Computation of Skyline Objects for Database with Uncertain Preferences

Venkateswara Rao Kagita[1]($\boxtimes$), Arun K. Pujari[1,2], Vineet Padmanabhan[1], Vikas Kumar[1], and Sandeep Kumar Sahu[1]

[1] School of Computer and Information Sciences,
University of Hyderabad, Hyderabad, India
585venkat@gmail.com, {akpcs,vineetcs}@uohyd.ernet.in,
{vikas,uusandeepsahu}@uohyd.ac.in
[2] Central University of Rajasthan, Ajmer, India

Abstract. Skyline queries aim at finding a set of skyline objects from the given database. For categorical data, the notion of preferences is used to determine skyline objects. There are many real world applications where the preference can be uncertain. In such contexts, it is relevant to determine the probability that an object is a skyline object in a database with uncertain pairwise preferences. Skyline query is to determine a set of objects having skyline probability greater than a threshold. In this paper, we address this problem. To the best of our knowledge, there has not been any technique which handles this problem directly. There have been proposals to compute skyline probability of individual objects but applying these for skyline query is computationally expensive. In this paper, we propose a holistic algorithm that determines the set of skyline objects for a given threshold and a database of uncertain preferences. We establish the relationship between skyline probability and the probability of the union of events. We guide our search to prune objects which are unlikely to be skyline objects. We report extensive experimental analysis to justify the efficiency of our algorithm.

1 Introduction

Skyline queries aim at finding a set of skyline objects from the given database. A skyline object is an object that is not dominated by any other object − an object p dominates another object q if p is no worse than q on all dimensions and better than q on at least one dimension. Numerous algorithms have been proposed in the literature to determine skyline objects for continuous data [4,7–9,14]. In case of numerical data, preferences are implicit and certain. For instance, if the attribute is price, low price is preferred to high price always. In case of categorical data, preferences are required to be specified explicitly. For example, if the attribute is a genre of a movie, it is necessary to state that whether comedy is preferred to action or action is preferred to comedy.

Zhang et al. [13] introduce a method to compute skyline probability of a given object with uncertain preferences. This method relies on *inclusion-exclusion*

© Springer International Publishing Switzerland 2016
R. Booth and M.-L. Zhang (Eds.): PRICAI 2016, LNAI 9810, pp. 193–205, 2016.
DOI: 10.1007/978-3-319-42911-3_16

principle and computes joint probabilities of all elements of the power set. It essentially involves an exponential number of terms and hence, skyline objects cannot be determined even for a database of moderate size. An approximation scheme using Monte Carlo estimation is proposed to overcome performance shortfall [13]. Pujari et al. [10,11] propose an efficient algorithm to find the skyline probability as the basic step to handle skyline query. A concept of *zero-contributing set* [10] is introduced to avoid redundant computation in an exponential search space. It is further extended to an efficient characterization of the zero-contributing set [11]. This method, namely *Usky-base* algorithm, is the best-known algorithm to compute skyline probability of individual object in a database when uncertainity is represented in terms of pairwise preference probabilities of attribute values. Skyline query is handled by computing skyline probabilities of all objects and by identifying objects with probability exceeding a user specified threshold.

In this paper, we address the problem of determining a set of objects having skyline probability greater than a given threshold without having to compute the skyline probability of all individual objects. To accomplish this, we make use of lower and upper bounds of the probability of the union of events which helps us to prune the search space at early stages. We show that our algorithm saves substantial computational time in determining a set of skyline objects. Theoretical analysis is developed to justify the correctness of our proposed method. We compare our method with *Usky-base* algorithm and show that our method outperforms the best-known method.

The rest of the paper is organized as follows. In Sect. 2, We define the problem and discuss the preliminary concepts that are necessary for the discussion. In Sect. 3 we briefly review the best-known algorithm. We review the various bounds and develop theoretical background necessary for bounds with absorption in Sect. 4. We discuss the proposed heuristic in Sect. 5. Experimental analyses of the proposed algorithm are reported in Sect. 6.

2   Problem Definition

Let $\Theta = \{Q^1, Q^2, \ldots, Q^n\}$ be a set of n d-dimensional objects and O be another d-dimensional object not in Θ. Q_j^i (O_j) denotes the value of Q^i, (O, respectively) at dimension j. Given two distinct attribute values α and β, we introduce notation $\alpha \prec \beta$ to denote that α is *preferred* to β. We say Q^i *dominates* O ($Q^i \preceq O$) iff Q^i is *preferred or equal to* O on any dimension and strictly preferred on at least one dimension. Object O is said to be a *skyline object* with respect to set of objects Θ, if O is not dominated by any object in Θ. The power set of Θ is denoted as 2^Θ. Consider the power set lattice L with subsets of Θ as its elements and set inclusion as the partial order. For a lattice with unique infimum and unique supremum, we use the pair of minimal and maximal elements to represent the lattice. In this sense, the power set lattice is denoted as $L = [\emptyset, \Theta]$. A lattice $[\alpha, \beta]$ is said to be *trivial* if $\alpha = \beta$ and is non-trivial, otherwise. A *trivial* lattice contains a single element, and non-trivial lattice contains an even number of elements. We denote L_k to represent a family of subsets of Θ with cardinality k. For notational convenience, we list the indices instead of the objects for a subset of Θ.

Define $distinct(O; S, j)$ for any $S \in 2^{\Theta}$ as set of distinct attribute values in dimension j of all objects in S except $O's$ attribute values and can be written formally as follows.

$$distinct(O; S, j) = \{a \mid a = Q_j^i \; for \; some \; Q^i \in S \; and \; a \neq O_j\}$$

and $distinct(O; S) = \cup_j distinct(O; S, j)$.

Given two distinct attribute values α and β, the probabilistic model to describe the uncertain preferences between them is the following.

$$Pr(\alpha \prec \beta) + Pr(\beta \prec \alpha) \leq 1$$

We have $Pr(\alpha \preceq \alpha) = 1$. Let e_i denote the event $Q^i \preceq O$. The probability of e_i is a joint probability of attribute value preferences:

$$Pr(e_i) = Pr(\bigcap_{j=1}^{d}(Q_j^i \preceq O_j))$$

If we assume that attribute value preferences of different dimensions are mutually independent then we get

$$Pr(e_i) = \prod_{j=1}^{d} Pr(Q_j^i \preceq O_j)$$

The skyline probability [13] of an object O, denoted as $sky(O)$, is defined as the probability that object O is not dominated by others.

$$sky(O) = Pr(\bigcap_{i=1}^{n} \overline{e_i}) = 1 - Pr(\bigcup_{i=1}^{n} e_i) \tag{1}$$

Let for $J \in 2^{\Theta}$, $E_J = \bigcap_{Q^i \in J} e_i$. Using inclusion exclusion principle, we get

$$sky(O) = 1 + \sum_{k=1}^{n}(-1)^k \sum_{J \in 2^{\Theta},|J|=k} Pr(E_J), \tag{2}$$

where

$$Pr(E_J) = \prod_{j=1}^{d} \prod_{\nu \in distinct(O;J,j)} Pr(\nu \prec O_j) \tag{3}$$

Skyline query filters a given database and returns a set of skyline objects. Hence, in a setting where we have uncertain preferences, the skyline query is desired to return a set of most probable objects. In other words, it returns a set of objects having skyline probability greater than a user specified threshold τ.

There are methods to compute skyline probability of individual objects with uncertain preferences [10,11,13]. To the best of our knowledge, there is no method which is directed towards identifying a set of objects whose skyline probability exceeds a certain threshold.

The following problem is addressed in the present work. Given Θ, τ and preference probabilities between pairs of attribute values of every dimension, the problem is to determine the set of objects that has skyline probability exceeding the threshold τ, i.e., $\{O \mid sky(O) > \tau, \forall O \in \Theta\}$.

2.1   Running Example

Consider a database of 6 objects (Table 1) with four attributes A, B, C and D. We treat this as a running example throughout the paper to illustrate different concepts. Pairwise preference probabilities for values of attributes A, B, C and D are assumed to be 0.5. For instance, preference probabilities of values of A are $Pr(a1 \prec a2) = Pr(a2 \prec a1) = 0.5$. Skyline probability of $O, sky(O)$, with respect to $\{Q^1, Q^2, Q^3, Q^4, Q^5\}$ [11] is given as follows.

$$sky(O) = 1 - \sum_{i=1}^{5} Pr(e_i) + \sum_{\substack{i,j=1 \\ i<j}}^{5} Pr(e_i \cap e_j) - \sum_{\substack{i,j,k=1 \\ i<j<k}}^{5} Pr(e_i \cap e_j \cap e_k)$$

$$+ \sum_{\substack{i,j,k,l=1 \\ i<j<k<l}}^{5} Pr(e_i \cap e_j \cap e_k \cap e_l) - [Pr(e_1 \cap e_2 \cap e_3 \cap e_4 \cap e_5)].$$

where e_i denote the event $Q^i \preceq O$. Making use of the assumption that preferences are mutually independent, we get $Pr(e_1) = Pr(a1 \prec a2) \times Pr(d2 \prec d1) = 0.25$. Some of the joint probability values are calculated and shown below.

$Pr(e_2) = Pr(b1 \prec b2) \times Pr(c2 \prec c1) = 0.25$
$Pr(e_1 \cap e_2) = Pr(a1 \prec a2) \times Pr(b1 \prec b2) \times Pr(c2 \prec c1) \times Pr(d2 \prec d1) = 0.0625$

Thus $sky(O) = 1 - [1.125] + [1.0625] - [0.75] + [0.3125] - [0.0625] = 0.4375$.

Table 1. Running example

Objects	A	B	C	D
O	a2	b2	c1	d1
Q^1	a1	b2	c1	d2
Q^2	a2	b1	c2	d1
Q^3	a1	b1	c1	d1
Q^4	a1	b2	c2	d1
Q^5	a1	b1	c2	d2

3   Computing Skyline Probability of Individual Objects

There are essentially three major algorithms to compute skyline probability of an object in a database with uncertain preferences [10,11,13]. *Usky-base* Algorithm, proposed in [11], can be viewed as the most efficient among these and having most general concepts of these three algorithms. In this section, these concepts are reviewed briefly. We adopt the same notations given in [11]. Readers are directed to the original paper for details.

When two sets have same set of distinct attribute values, the corresponding joint probabilities are equal. That is, if $distinct(O; S, j) = distinct(O; S', j)$, $1 \leq j \leq d$, then $Pr(S) = Pr(S')$ [11]. For any non-empty set $S \subsetneq \Theta$, if $distinct(O; S) = distinct(O; S \cup \{x\})$ for some $x \notin S$ then $distinct(O; S') = distinct(O; S' \cup \{x\})$ for every S' containing S. The terms corresponding to S' and $S' \cup \{x\}$ in Eq. (2) appear in opposite signs and if $distinct(O; S') = distinct(O; S' \cup \{x\})$ then they have no contribution when considered together. The elements of the sub-lattice $[S, \Theta]$ can be viewed as a collection of pairs of the form S' and $S' \cup \{x\}$ and if $distinct(O; S) = distinct(O; S \cup \{x\})$ then the overall effect of all terms in this lattice is zero while computing $sky(O)$. The sub-lattice $[S, \Theta]$ is said to be *zero-contributing* set [11]. Note that if S appears in k^{th} level the $S \cup \{x\}$ appears in $(k + 1)^{th}$ level.

For any non-empty set $S \subsetneq \Theta, |S| = k - 1$, if $distinct(O; S \cup \{y\}) = distinct(O; S \cup \{x, y\})$ for $x, y \notin S$, then $S \cup \{x\}$ is said to absorb $S \cup \{y\}$ with respect to O at level k. Let $S \subsetneq \Theta$, $S = \{i_1, i_2, \ldots, i_{k-1}\}$ and $i_1 < i_2 < \ldots < i_{k-1}$. If $distinct(O; S \cup \{y\}) = distinct(O; S \cup \{x, y\})$ and $y, x > i_{k-1}$, then x is said to absorb y with respect to O at level k with prefix S.

When x absorbs y at level k with prefix S, the corresponding zero-contributing sub-lattice is $[S \cup \{y\}, S \cup \Theta_S]$, where Θ_S is $\{i_{(k-1)} + 1, i_{(k-1)} + 2, \ldots, n\}$. In other words, all elements of this zero-contributing set have S prefix. Usky-base algorithm employs prefix-based k-level absorption as it proceeds from the bottom to the top in the powerset lattice. With multiple instances of absorption, multiple zero-contributing sub-lattices are generated which are generally overlapping. It is seen that intersection of two zero-contributing lattices is a zero-contributing lattice, if the intersection is non-degenerate. It is shown in [11] that the zero-contributing sub-lattices arising out of prefix-based k-level absorption cannot have a degenerate intersection.

4   Bounds on the Skyline Probability

The problem of finding the exact value for the probability of the union of events involves calculation of 2^n joint probabilities spread across n levels wherein level k contains the terms of size k, $1 \leq k \leq n$. The probability of the union of events can be rewritten as follows.

$$Pr(\bigcup_{i=1}^{n} e_i) = \sum_{k=1}^{n} (-1)^{k-1} S_k \tag{4}$$

where
$$S_k = \sum_{1 \leq i_1 < ... < i_k \leq n} Pr(e_{i_1} \cap e_{i_2} \cap ... \cap e_{i_k})$$

Computation of $Pr(\bigcup_{i=1}^n e_i)$ requires 2^n terms. The joint probabilities of higher degree may not be easy to compute, and hence, researchers have investigated upper and lower bounds for $Pr(\bigcup_{i=1}^n e_i)$ when k is restricted to m, $m << n$. If m is small, relatively smaller number of joint probabilities would be used to compute the bound. Some of the major bounds are discussed here. We use these bounds in our proposed algorithm.

Lower and upper bounds for $m=2$ [3,6,12]:

$$Pr(\bigcup_{i=1}^n e_i) \geq \frac{2}{h+1}S_1 - \frac{2}{h(h+1)}S_2,$$

where $h = 1 + \left\lfloor \frac{2S_2}{S_1} \right\rfloor$.

$$Pr(\bigcup_{i=1}^n e_i) \leq min\{S_1 - \frac{2}{n}S_2, 1\}$$

Lower and upper bounds for $m=3$ [2,5]:

$$Pr(\bigcup_{i=1}^n e_i) \geq \frac{h+2n-1}{(h+1)n}S_1 - \frac{2(2h+n-2)}{h(h+1)n}S_2 + \frac{6}{h(h+1)n}S_3,$$

where $h = 1 + \left\lfloor \frac{-6S_3+2(n-2)S_2}{-2S_2+(n-1)S_1} \right\rfloor$.

$$Pr(\bigcup_{i=1}^n e_i) \leq min\left(S_1 - \frac{2(2h-1)}{h(h+1)}S_2 + \frac{6}{h(h+1)}S_3, 1\right),$$

where $h = 2 + \left\lfloor \frac{3S_3}{S_2} \right\rfloor$.

Upper bound for $m=4$ [2]:

$$Pr(\bigcup_{i=1}^n e_i) \leq min\left(S_1 - \frac{2((h-1)(h-2)+(2h-1)n)}{h(h+1)n}S_2 + \frac{6(2h+n-4)}{h(h+1)n}S_3 - \frac{24}{h(h+1)n}S_4, 1\right),$$

where $h = 1 + \left\lfloor \frac{(n-2)S_2+3(n-4)S_3-12S_4}{(n-2)S_2-3S_3} \right\rfloor$.

For higher degree, the most general bound is the family of Bonferroni bounds [1] and is as follows.

$$Pr(\cup_{i=1}^n e_i) \begin{cases} \geq \sum_{k=1}^m (-1)^{k-1}S_k & \text{if } m \text{ is even} \\ \leq \sum_{k=1}^m (-1)^{k-1}S_k & \text{Otherwise.} \end{cases}$$

Any lower bound on the probability of the union of events become an upper bound for skyline probability and vice versa. Bounds for $sky(O)$ using Bonferroni bounds are given by

$$Sky(O) \begin{cases} \leq 1 - \sum_{k=1}^{m}(-1)^k S_k & \text{if } m \text{ is even} \\ \geq 1 - \sum_{k=1}^{m}(-1)^k S_k & \text{Otherwise} \end{cases}$$

4.1   Bounds on the Skyline Probability with Absorption

In this section, we discuss the effects of above bounds. Both lower and upper, when computed in conjuction with level-wise absorption. Employing prefix-based k-level absorption at every level absorbs some terms at every level. As a result, while generating terms of the next level, many terms will be removed. We denote the sum of residual terms at level k as S_k'. We show here that all the bounds described above with respect to S_k, for different values of k, are tighter for S_k'. We present below the results to this effect for Bonferroni bounds. Similar results hold also for other bounds and can be proved trivially.

Lemma 1. $1 - \sum_{k=1}^{m}(-1)^k S_k \leq 1 - \sum_{k=1}^{m}(-1)^k S_k' \leq Sky(O)$, when m is odd.

Proof. We know that $sky(O) = 1 - \sum_{k=1}^{n}(-1)^k S_k = 1 - \sum_{k=1}^{n}(-1)^k S_k'$. This implies that $sky(O) = 1 - \sum_{k=1}^{m}(-1)^k S_k' + X$, where $X = [\sum_{k=m+1}^{n}(-1)^k S_k']$. $X \geq 0$ when m is odd, because when m is odd X starts a positive value in the inclusion exclusion. Hence, $1 - \sum_{k=1}^{m}(-1)^k S_k' \leq sky(O)$. Prefix-based k-level absorption identifies the zero contributing set as pairs of several zero contributing sets. That is, if absorption takes place at any level $k \leq m$, the terms that have no zero effect when we consider the lattice till level m are lies only at level m. Therefore, $1 - \sum_{k=1}^{m-1}(-1)^k S_k = 1 - \sum_{k=1}^{m-1}(-1)^k S_k'$. When m is odd, S_m' get negative sign in the formula. The terms which are supposed to have been subtracted are not getting subtracted. Hence $-S_m' \geq -S_m$. Therefore, $1 - \sum_{k=1}^{m}(-1)^k S_k \leq 1 - \sum_{k=1}^{m}(-1)^k S_k'$.

Lemma 2. $1 - \sum_{k=1}^{m}(-1)^k S_k \geq 1 - \sum_{k=1}^{m}(-1)^k S_k' \geq Sky(O)$, when m is even.

Proof. We know that $sky(O) = 1 - \sum_{k=1}^{n}(-1)^k S_k = 1 - \sum_{k=1}^{n}(-1)^k S_k'$. This implies that $sky(O) = 1 - \sum_{k=1}^{m}(-1)^k S_k' + X$, where $X = [\sum_{k=m+1}^{n}(-1)^{k-1} S_k']$. $X \leq 0$ when m is even, because when m is even X starts a negative value in the inclusion exclusion. Hence $1 - \sum_{k=1}^{m}(-1)^k S_k' \geq sky(O)$. Prefix-based k-level absorption identifies the zero contributing set as pairs of several zero contributing sets. That is, if absorption takes place at any level $k \leq m$, the terms that have no zero effect when we consider the lattice till level m are lies at only level m. Therefore, $1 - \sum_{k=1}^{m-1}(-1)^k S_k = 1 - \sum_{k=1}^{m-1}(-1)^k S_k'$. When m is even, S_m' get positive sign in the formula. The terms which are supposedly being added are not getting added. Hence $S_m \geq S_m'$. Therefore, $1 - \sum_{k=1}^{m}(-1)^k S_k \geq 1 - \sum_{k=1}^{m}(-1)^k S_k'$.

For running example, upper Bonferroni bound at level 2 without absorption is 0.875 and with absorption is 0.5. Upper bound [6,12] at level 2 is 0.60416 and with absorption it is 0.5. Similarly, lower Bonferroni bound at level 3 without absorption is 0.1875, upper bound [6,12] is 0.3875 and with absorption, it is 0.4375.

Algorithm 1. Proposed Algorithm

Input: O, Θ, τ
Output: Determine whether O is a skyline object or not
Initialize L_1 to be all singleton elements of Θ:
$L_1 \leftarrow \{Q^1, Q^2, \ldots, Q^n\}$;
Prefix-based-absorption(L_1);
$sky(O) \leftarrow 1$;
$S'_1 = \sum_{J \in L_1} Pr(E_J)$;
Determine $sky_1^L(O)$ using S'_1;
if $sky_1^L(O) > \tau$ **then** Declare O is a skyline object; **exit**;
$sky(O) \leftarrow Sky(O) - S'_1$;
for $k = 2$ to n **do**
$\quad$ $L_k \leftarrow$ GenerateCandidates(L_{k-1});
$\quad$ **if** $L_k = \emptyset$ **then**
$\quad\quad$ **if** $sky(O) > \tau$ **then** Declare O as skyline object;
$\quad\quad$ **else** Declare O is not a skyline object;
$\quad\quad$ **break**;
$\quad$ **end**
$\quad$ $L_k \leftarrow$ Prefix-based-absorption(L_k, O);
$\quad$ $S'_k = \sum_{J \in L_k} Pr(E_J)$;
$\quad$ Determine $sky_k^L(O)$ or $sky_k^U(O)$ or both using $S'_1, S'_2, \ldots, S'_k$;
$\quad$ **if** $sky_k^L(O) > \tau$ **then** Declare O is a skyline object; **exit**;
$\quad$ **else if** $sky_k^U(O) \leq \tau$ **then** Declare O is not a skyline object; **exit**;
$\quad$ $sky(O) \leftarrow Sky(O) + (-1^k)S'_k$;
end

---

5   Proposed Approach

Using the theory from the previous sections, we propose an algorithm to determine a set of skyline objects. Our algorithm takes advantage of both prefix-based k-level absorption and bounds on the skyline probability. Our algorithm traverses the powerset lattice from bottom to top. At each level k, it employs prefix-based k-level absorption. Using the residual terms available till that level, we compute either lower bound ($sky_k^L(O)$) or upper bound ($sky_k^U(O)$) or both. If $sky_k^L(O) > \tau$ then we conclude that $sky(O) > \tau$. Hence, we declare an object

O as a skyline object. The algorithm terminates for the current O and moves to examine another object in the database. If $sky_k^U(O) \leq \tau$ then it can be inferred that $sky(O) \leq \tau$. Hence, O cannot be a skyline object and the algorithm terminates for the current O. At any level k, depending whether k is odd or even, only one of upper-bound and lower-bound condition is employed. If this test fails, that is, we cannot determine O as non-skyline or cannot accept definitely as skyline, then the algorithm moves to the next level to generate candidate sets and to invoke prefix-based absorption. Detailed procedure is given in Algorithm 1. Algorithms for candidate generation and prefix-based absorption as described in [11].

We illustrate the working of our algorithm with the help of running example. We assume threshold to be 0.15. At level 1 all the singletons are there. We employ prefix-based k-level absorption, 1 absorbs 5. Residual terms are 2, 3, 4 and 5. We find lower bound using S_1', which is 0 and is not greater than 0.15. We continue to generate next level candidates. Candidates at second level are 12, 13, 14, 23, 24, 34. We again employ prefix-based k-level absorption, 13 absorbs 12 and 23 absorbs 24. We find lower bound using S_1' and S_2'. We obtain 0.166, which is greater than the threshold. Hence, we declare O to be a skyline object and exit.

6   Empirical Analysis

In this section, we provide empirical analysis to show the efficiency of the proposed method. We evaluate our method using real and synthetic datasets. We use three real datasets (*nursery, flare* and *car*), having categorical attributes without missing values, from UCI ML repository[1]. Our experimental study also examines synthetic data wherein datasets of different sizes (number of objects in the range [100, 10000]) with dimensions varying in the range [5, 12] are randomly generated with uniform distribution for each dimension and assuming mutual independence of dimensions. Details about size and dimensionality of datasets can be found in Table 2. We compare proposed method with best-known algorithm, *Usky-base*.

We evaluate our method considering three different parameters, 1. Time taken by the algorithm, 2. Number of terms used by the algorithm to determine skyline objects, and 3. Termination level. In the first set of experiments, we study the performance of our algorithm as compared to *Usky-base* for different values of τ. We choose τ values starting from the value where most of the objects are declared as a skyline object and ends with the value where most of the objects are declared as non-skyline objects. All the results reported here are the average of hundred randomly selected objects. Figure 1 compares execution time of *proposed* and *Usky-base* methods for different values of *tau*. It can be seen from the figure that proposed algorithm always exhibiting better performance than *Usky-base* algorithm. Another observation is that at higher values of τ our algorithm saves substantial time. In general, we set the threshold value

---

[1] https://archive.ics.uci.edu/ml/datasets.html.

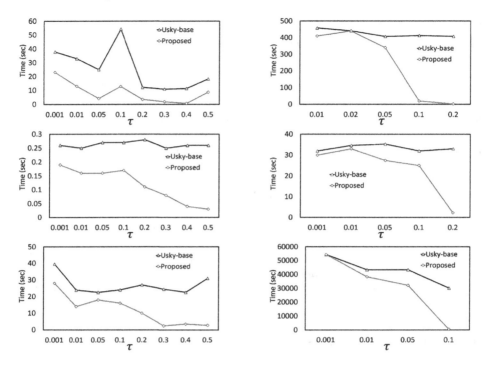

Fig. 1. Comparison of the time for different datasets by varying τ (left to right and top to bottom, flare, car, synthetic (100, 500, 1000, 10000)).

to be high to get smaller number of objects in the skyline set. We could not determine set of skyline objects for *Nursery* dataset using *Usky-base* algorithm with the available resources. The average time taken by the proposed approach for a threshold value 0.01 is 151118 (s). Similarly, for $\tau = 0.05 - 151146$ (s), for $\tau = 0.1 - 272$ (s), and for $\tau = 0.2 - 12.77$ (s).

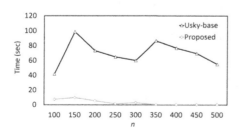

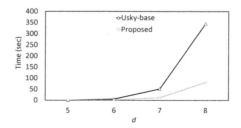

Fig. 2. Time comparison by varying n, $d = 4$, $\tau = 0.3$.

Fig. 3. Time comparison by varying d, $n = 100$, $\tau = 0.3$.

Table 2. Efficiency of our algorithm over number of terms and termination level. "-" denotes this could not be computed with available resources.

Dataset	n	d	τ	No. of terms		Termination level	
				Usky-base	Proposed	Usky-base	Proposed
Nursery	12926	8	0.01	-	354521	-	10
			0.05	-	354521	-	10
			0.1	-	190	-	2
Flare	1329	10	0.01	3848	1951	8.8	4.4
			0.01	1765	858	3.3	1.6
			0.05	2127	620	4.2	1.8
			0.1	2897	763	3.7	1.4
			0.5	1634	289	3.5	1.2
Car	1728	6	0.01	32752	30369	15	10.3
			0.05	32752	23209	15	8
			0.1	32752	1549	15	2.4
			0.2	32752	120	16	2
Synthetic	100	5	0.01	519	341	7.9	4.7
			0.05	519	378	7.9	5.0
			0.1	519	350	7.9	4.7
			0.5	519	58	7.9	2
Synthetic	500	5	0.01	8945	7230	12.9	9.2
			0.05	9450	7002	13.2	7.5
			0.1	9830	1621	13.4	3.2
Synthetic	1000	12	0.01	3848	1951	8.8	4.4
			0.05	3641	2592	8.8	4.4
			0.1	3567	2006	7.4	3.4
			0.5	5089	234	10.6	2
Synthetic	10000	5	0.001	262125	249527	19	12
			0.01	65519	58650	17	10
			0.1	196589	170	18	2

We also analyse average number of terms (joint probabilities) used by the algorithm to determine set of skyline objects, and termination level of the algorithm. Table 2 shows efficiency of our algorithm over number of terms and termination level. It is observed from the table that *proposed* approach is using less number of terms to determine skyline objects as compared to *Usky-base* algorithm, and also our algorithm is terminating at early stage as compared to *Usky-base*.

In the second set of experiments, we analyse the performance of our algorithm by varying number of objects and dimensionality. Figure 2 depicts the execution time comparison by varying n. It can be seen from the figure that proposed approach is performing well over *Usky-base* in terms of time for any value of n. Figure 3 depicts the execution time comparison by varying d. We observe that proposed approach is performing well for higher values of d and our approach is constantly giving good performance over *Usky-base*.

7   Conclusion

In this paper, we propose a novel way of determining a set of skyline objects without having to compute skyline probability of individual object. Our algorithm efficiently identifies whether an object is likely to be a skyline object or not at very early stage. We combine the process of absorption with bound calculation. Our theoretical analysis shows that bounds with absorption are sharper than bounds without absorption. Experimental results based on real and synthetic datasets demonstrates the efficiency of our algorithm.

Acknowledgements. Part of this work is carried out at Central University of Rajasthan. Authors acknowledge Central University of Rajasthan for providing facilities.

References

1. Bonferroni, C.E.: Teoria statistica delle classi e calcolo delle probabilita. Libreria internazionale Seeber, Florence (1936)
2. Bonos, E., Prekopa, A.: Closed form two-sided bounds for probabilities that exactly r and atleast r out of n events occur. Math. Oper. Res. **14**, 317–342 (1989)
3. Dawson, D.A., Sankoff, D.: An inequality for probabilities. Proc. Am. Math. Soc. **18**(3), 504–507 (1967)
4. Han, X., Li, J., Yang, D., Wang, J.: Efficient skyline computation on big data. IEEE Trans. Knowl. Data Eng. **25**(11), 2521–2535 (2013)
5. Kwerel, S.M.: Bounds on the probability of the union and intersection of m events. Adv. Appl. Probab. **7**, 431–448 (1975)
6. Kwerel, S.M.: Most stringent bounds on aggregated probabilities of partially specified dependent probability systems. J. Am. Stat. Assoc. **70**(350), 472–479 (1975)
7. Morse, M., Patel, J.M., Grosky, W.I.: Inf. Sci. Efficient continuous skyline computation **177**(17), 3411–3437 (2007)
8. Papadias, D., Tao, Y., Greg, F., Seeger, B.: Progressive skyline computation in database systems. ACM Trans. Database Syst. (TODS) **30**(1), 41–82 (2005)
9. Papapetrou, O., Garofalakis, M.: Continuous fragmented skylines over distributed streams. In: 2014 IEEE 30th International Conference on Data Engineering (ICDE), pp. 124–135. IEEE (2014)
10. Pujari, A.K., Kagita, V.R., Garg, A., Padmanabhan, V.: Bi-directional search for skyline probability. In: Ganguly, S., Krishnamurti, R. (eds.) CALDAM 2015. LNCS, vol. 8959, pp. 250–261. Springer, Heidelberg (2015)

11. Pujari, A.K., Kagita, V.R., Garg, A., Padmanabhan, V.: Efficient computation for probabilistic skyline over uncertain preferences. Inf. Sci. **324**, 146–162 (2015)
12. Sathe, Y.S., Pradhan, M., Shah, S.P.: Inequalities for the probability of the occurrence of at least m out of n events. J. Appl. Prob. **17**, 1127–1132 (1980)
13. Zhang, Q., Ye, P., Lin, X., Zhang, Y.: Skyline probability over uncertain preferences. In: EDBT/ICDT, pp. 395–405 (2013)
14. Zhang, S., Mamoulis, N., Cheung, D.W.: Scalable skyline computation using object-based space partitioning. In: Proceedings of the 2009 ACM SIGMOD International Conference on Management of data, pp. 483–494. ACM (2009)

Learning from Numerous Untailored Summaries

Yuta Kikuchi[✉], Akihiko Watanabe, Sasano Ryohei, Hiroya Takamura,
and Manabu Okumura

Tokyo Institute of Technology, Yokohama, Japan
{kikuchi,watanabe,sasano,takamura,oku}@lr.pi.titech.ac.jp

Abstract. We present an attempt to use a large amount of summaries contained in the New York Times Annotated Corpus (NYTAC). We introduce five methods inspired by domain adaptation techniques in other research areas to train our supervised summarization system and evaluate them on three test sets. Among the five methods, the one that is trained on the NYTAC followed by fine-tuning on the target data (i.e. the three test sets; DUC2002, RSTDTB$_{long}$ and RSTDTB$_{short}$) performs the best for all the test sets. We also propose an instance selection method according to the faithfulness of the extractive oracle summary to the reference summary and empirically show that it improves summarization performance.

1 Introduction

Machine learning is one of the most important elements for many tasks of natural language processing. Text summarization is no exception. Many recent text summarization studies involved machine learning as a part of their systems [1,11,16,24,27,29,32]. However, one critical problem in the text summarization context is the lack of training data. The official training data for single document summarization used in the Document Understanding Conference (DUC) 2001 consist of around 300 pairs of a document and its summary generated by human annotators (corresponding to approximately 4,000 sentences) and the Ziff-Davis corpus consists of around 7,000 pairs [9,21][1], while the Hansard corpus [5], a rather conventional dataset for machine translation, contains approximately 1.3 million sentence pairs. This situation prevents summarization methods from using a large number of features.

A promising dataset in this problematic situation is the New York Times Annotated Corpus (NYTAC) [26], which is a collection of newspaper articles published as a part of New York Times between January 1, 1987 and June 19, 2007. Out of over 1.8 million articles in the NYTAC, approximately 650,000 articles are accompanied with human-written (single document) summaries. While there have been studies that have used the human summaries in the NYTAC [13,31], there have been no studies that directly involved the NYTAC as training data. One difficulty in using NYTAC for training lies in the fact that

---

[1] The current datasets for multi-document summarization are also small.

R. Booth and M.-L. Zhang (Eds.): PRICAI 2016, LNAI 9810, pp. 206–219, 2016.
DOI: 10.1007/978-3-319-42911-3_17

the summaries in it are not tailored to a specific summarization approach. There are a number of different types of summaries in NYTAC including extractive and abstractive summaries.

We propose an effective way for using NYTAC as a large training resource for single document summarization. Due to the untailored summaries of the NYTAC, the problem we tackle in this paper can be regarded as a special type of domain adaptation [18], and to the best of our knowledge this work is the first for text summarization. In the area of domain adaptation, there are simple but strong methods, such as linear interpolation of parameters or use of the output of the classifier trained with out-domain data as a feature for training with in-domain data [8]. In addition, instance selection and weighting are also known as effective ways for domain adaptation [2,30]. Hence, we prepare some methods based on ideas borrowed from standard domain adaptation techniques and investigate the best way to use NYTAC.

We empirically show that training the summarization system on NYTAC followed by fine-tuning on the target data performs the best for all the test sets. We also propose an instance selection method according to the faithfulness of the extractive oracle summary to the reference summary[2]. The faithfulness of an instance is defined to be the ROUGE-2 score of the extractive oracle summary calculated against the reference summary written by human annotators.

The rest of this paper is organized as follows. Section 2 discuss related work that uses machine learning techniques for text summarization. Section 3 introduces the NYTAC and the human-written summaries in the corpus. In Sect. 4, we present our five methods that rely on the NYTAC and the instance selection method as a preprocessing for those five methods. We then present the experiments on three test sets in Sect. 5. Finally, we conclude the paper and discuss future work in Sect. 6.

2 Related Work

Recently, machine learning has become one of the most important elements in text summarization. Certain current methods learn the importance scores of unigrams, bigrams or dependency arcs [1,13,16,32] to estimate the benefits of extraction units. Others estimate the importance of the entire sentence [11,22]. Some use a structure-learning method to optimize the parameters [24,27,29].

Some efforts to use the NYTAC for summarization have been done prior to our work. Hong and Nenkova [13,15] used the language model trained on the summaries of the NYTAC. Li and Nenkova [17] measured sentence specificity, and Yang and Nenkova [31] identified the informative parts of text. Note that these studies involved the NYTAC as side information but not as training data for summarization, as opposed to our work, in which the NYTAC is directly used as training data of a summarization model.

---

[2] In this paper, *the (extractive) oracle summary* is defined to be the best possible summary that can be generated by sentence extraction, and *the reference summary* is defined to be the original human-written summary in NYTAC.

Although we focus on the NYTAC in this paper, it is not the only summary dataset that our method can be applied to. For example, the dataset of story highlights used by Svore et al. [28][3] is extracted from CNN.com, and can be a target of our method.

3 New York Times Annotated Corpus

NYTAC contains over 1.8 million articles written and published by New York Times, and approximately 650,000 of those articles are accompanied with human-written summaries. While this corpus is the largest dataset for single document summarization studies, its summaries are very diverse. Some summaries in NYTAC can be generated through a sentence-extractive approach. Some require sentence compression or sentence fusion in addition to extraction. Others can be generated only through an abstractive approach. Moreover, there are summaries that are written from a meta-viewpoint. The presence of various types of summaries in the dataset may be beneficial to the development of different techniques in summarization, such as sentence extraction, compression, fusion and abstractive generation [14, 23].

However, different techniques may require different types of training instances. For example, when we use a sentence-extractive summarization method, summaries generated in an abstractive manner should not be appropriate as training instances. Therefore, we investigate effective ways for selecting training instances from NYTAC according to the faithfulness of the extractive oracle summary to the reference summary since we use an extractive approach (Sect. 4.4), which is the most standard approach in the current summarization community.

Table 1 shows three examples from the NYTAC[4]. For each example, we show the human-written summary (*abs*) and its source document (*doc*), and marked their aligned text segments with the same number and colored underline. Each value in brackets next to document ID (*id*) is the ROUGE-2 score of the sentence-extractive oracle summary described in Sect. 4.2. By analyzing *abs* in the NYTAC, we notice that human-summarizers do not only extract text segments in source documents, but they often shorten them by several techniques: deletion of function words including articles, deletion of temporal adverbs ([e] in the table), deletion of symbols including quotations ([b, c] in the table), periods ([d] in the table) and hyphens, or paraphrasing ([a] in the table). Moreover, some summaries in NYTAC do not describe the content of the source document, but describe "what the document is about" from a meta-viewpoint.

4 Training Summarization System on NYTAC

We now describe our five summarization methods that rely on the NYTAC. We first describe the summarization system that we used. We then describe

---

[3] Although only 1,365 articles were used in their work [28], there are potentially many more articles in CNN.com.

[4] ©2016 The New York Times Annotated Corpus, used with permission.

Table 1. Three examples from the NYTAC. For each example, *id*, *abs* and *doc* indicate document id, human-written summary and its source document, respectively. We clip *doc*s to keep space. The clipped sentences are indicated as ([...]), and we show the number of sentences in the entire *doc*. Each value in brackets next to document id is the ROUGE-2 score of the sentence-extractive oracle summary described in Sect. 4.2. For each example, we marked aligned text segments between *abs* and *doc* with the same number and colored underline. The words or phrases in bold with superscripts ([a–f]) are mentioned in Sect. 3.

id	1996/07/28/0868107 (0.111)
abs	Neil Strauss recommends hearing reunion of German progressive-rock group Cluster at the Knitting Factory
doc	Cluster, the Knitting Factory, 74 Leonard Street, TriBeCa, (212) 219-3006. What can make reunions by obscure European groups like Cluster more exciting than classic-rock reunions is that the former often haven't performed in the United States before. [...] (7 sentences)
id	1988/03/23/0129961 (0.366)
abs	Gov James E McGreevey, whose insistence on staying in office until Nov 15 set off political, legal and public relations challenges, needs only to remain in office until midnight[1] Sept 3 to outlast state deadline that would automatically force special election to choose his successor[2]; his announcement Aug 12 that he was stepping down because of extramarital affair with[3] **unidentified man**[a] led to coup attempt from within his own party, which he has managed to fend off; Richard J Codey, Senate president and fellow Democrat, will complete final 14 months of McGreevey's term[4] under provisions of New Jersey's Constitution; photos
doc	After revealing on national television that he is gay, seeing his name become the punch line of countless late-night comedy gags, and fending off a coup attempt from within his own party, Gov. James E. McGreevey needs only to remain in office until midnight[1] Friday to outlast the state deadline that would automatically force a special election to choose his successor.[2] Mr. McGreevey announced on Aug. 12 that he was stepping down because he had had an extramarital affair with[3] **a man he did not identify**[a], but his insistence on remaining in office until Nov. 15 set off political, legal and public relations challenges. [...] Mr. McGreevey will pass that deadline as of Saturday, allowing the Senate president, Richard J. Codey, a fellow Democrat, to complete the final 14 months of his term[4]. [...] (27 sentencces)
id	1996/08/30/0874273 (0.692)
abs	CBS announces that it has signed Diane English, executive producer of **Murphy Brown**[b], to take over as executive producer of its troubled new comedy, **Ink**[c], which stars Ted Danson and Mary Steenburgen[5]; Ink will have delayed debut of Oct 21[d]
doc	CBS announced **yesterday**[e] that it had signed Diane English, the longtime executive producer of **"Murphy Brown"**[b] to take over as executive producer of its troubled new comedy, **"Ink,"**[c] which stars Ted Danson and Mary Steenburgen. CBS also said the show, whose first four episodes fell so short of expectations that they were shelved, would come back with new episodes created by Ms. English starting on **Oct. 21**[d]. [...] (6 sentences)

five different ways for using the NYTAC during training. Finally, we describe a method for selecting training instances according to the faithfulness of the extractive oracle summary to the reference summary as a preprocessing for those five methods.

4.1   Sentence Extraction as Knapsack Problem

We use a sentence-extractive summarization method formulated as a knapsack problem (KP) where sentences are items and sentence lengths (e.g., the number of words in the sentence) are the costs of items. KP is the most standard and reasonable model for single document summarization. Actually, many recent summarization models are based on an extension of KP [12,24]. In supervised training of a KP, the benefit of item s is often represented as the dot product of a weight vector and a feature vector[5]: $\mathbf{w} \cdot \phi(s)$. We can formulate a KP as

$$\text{max.} \quad \sum_i^n b_i x_i$$
$$\text{s.t.} \quad \sum_i^n c_i x_i \leq L; \tag{1}$$
$$x_i \in \{0, 1\}; \quad \forall i, \tag{2}$$

where b_i is the benefit of the i-th sentence; x_i is 1 if the i-th sentence is selected as a part of the summary, 0 otherwise; L is the length limit of the summary; and c_i is the length of the i-th sentence.

For estimating weight vector $\mathbf{w}$, we use a supervised online learning method. Specifically, we use the passive aggressive (PA)-II, a variation of the online PA algorithm [6] formulated as

$$\mathbf{w}_{t+1} = \underset{\mathbf{w}}{\text{argmin}} \quad \frac{1}{2}\|\mathbf{w} - \mathbf{w}_t\|^2 + C\xi^2 \tag{3}$$
$$\text{s.t.} \quad loss(\mathbf{w}) \leq \xi. \tag{4}$$

In more detail, we use a structure learning setting of PA-II [6,7]. As we described in Sect. 2, there are various approaches to use machine learning for text summarization. While binary classifiers or regression models estimate a local decision of whether a word or a sentence should be in a summary or not, the structure learning estimates whether the whole summary is good or not. Recent studies empirically show the effectiveness of the structure learning for summarization [24,27,29].

For updating $\mathbf{w}$, we use the closed-form solution:

$$\mathbf{w}_{t+1} = \mathbf{w}_t + \tau_t(\phi(o_t) - \phi(s_t)), \tag{5}$$
$$\tau_t = \frac{loss(o_t, s_t)}{\|\phi(o_t) - \phi(s_t)\| + \frac{1}{2C}}, \tag{6}$$
$$loss(o_t, s_t) = 1 - ROUGE(o_t, s_t), \tag{7}$$

---

[5] When the benefit of each sentence is represented as the dot product of a weight vector and a feature vector, the benefit can be negative. The optimization problem with negative benefits cannot be regarded as a KP. However, such cases are very rare and can be ignored in practice.

where s_t is the summary generated with the current model $\mathbf{w}$ by solving the 0-1 KP with the cost limit L; o_t is the sentence-extractive oracle summary which we describe in the subsequent section; $\phi(\cdot)$ is the feature vector of a given document; $ROUGE(o, s)$ is the ROUGE-1 [19] score of system summary s for oracle summary o.

We use the following as features in $\phi(\cdot)$ for calculating $b_i = \mathbf{w} \cdot \phi(\cdot)$:
(a) the normalized number of content words,
(b) bag-of-words with part-of-speech tags from the vocabulary consisting of the most frequent 10,000 words,
(c) the logarithm of the length of the sentence; $\log(\#$ *of words in the sentence* $+ 1)$,
(d) the relative length of the sentence in its source document; *# of words in the sentence/# of words in the source document,*
(e) the absolute position of the sentence in its source document,
(f) the relative position of the sentence in its source document,
(g) 1 if the sentence occurs in the first 20 % of the source document; else 0,
(h) the unigram coverage of the sentence for the source document,
(i) the sentence benefit from *NytOnly* (this feature is used only in *Featurize* discussed in Sect. 4.3).

In addition, we parallelize the training process following Zhao et al. [33].

4.2 Sentence Extractive Oracle Summary

An oracle summary is a summary with the highest evaluation metric. We use ROUGE [19] as an evaluation metric. In other words, we select the best possible set of sentences as a summary according to ROUGE-2 score from sentences in the source document. The score of an oracle summary is thus the upper-bound score that a sentence extractive summarization system can achieve.

We use oracle summaries of the summary-article pairs in the NYTAC for two purposes:

- One is to use the selected instances for a perceptron-based structure learning method, where each instance needs to be converted to a feature vector via its extractive oracle summary; we can extract features such as the absolute or relative positions only from extractive summaries.
- The other is to select training instances from the NYTAC according to the ROUGE-2 scores, which in this paper we call *the faithfulness of the extractive oracle summary to the reference summary.* The values of faithfulness can be calculated only with oracle summaries. With this selection strategy, we are able to collect instances that are suitable for extractive summarizers. The instance selection process is described in detail in Sect. 4.4.

We obtain oracle summaries by solving the following budgeted maximum coverage problem:

$$\text{max.} \quad \sum_j^m z_j$$

$$\text{s.t.} \quad \sum_i^n c_i x_i \le L; \tag{8}$$

$$\sum_i^n a_{ij} x_i \ge z_j; \; \forall j \tag{9}$$

$$x_i \in \{0, 1\}; \quad \forall i \tag{10}$$

$$z_j \in \{0, 1\}; \quad \forall j, \tag{11}$$

where z_j is 1 if the j-th bigram appears in both the reference and resulting oracle summaries, 0 otherwise; and a_{ij} is 1 if j-th bigram appears in the i-th sentence. We performed the same preprocessing as the official evaluation script of ROUGE with options "-m -r 2", i.e., eliminating all non-alphabetical and non-numerical words and stemming all the remaining words. Note that Marcu [21] first attempted to create extractive summaries from human-written summaries, but in a different manner and with a different criterion.

4.3   Use of NYTAC for Training

We describe five methods for using the NYTAC during training and use the 5-fold cross validation technique when using the target data for training. As a first baseline, we prepare the *TrgtOnly* method, which is simply trained only on the test data by 5-fold cross validation.

NytOnly is trained only on the NYTAC. This method does not use any information of the target data. Hence, its score purely shows the impact of a large number of training examples.

Mixture is trained on the union of NYTAC data and the training part of the target data for each fold. While this method enlarges the size of the training data, the following three methods use the trained weight vector of *NytOnly*.

LinInter linearly interpolates the weight vectors of *TrgtOnly* and *NytOnly*. We select the interpolation parameter using development data for each fold.

Featurize uses the output of *NytOnly* as an additional feature for predicting sentence benefit with test data.

FineTune initializes a weight vector with the vector of pre-trained *NytOnly* and then performs fine-tuning on the target data (i.e., training on the target data). We assume that the weight vector of *NytOnly* as the initial vector helps in the search for a better weight vector than the zero vector as the initial vector, which is the standard procedure to initialize the parameters in the PA algorithm.

Table 2. Number of summary-article pairs for each threshold *thr* value

thr	0.0	0.1	0.2	0.3	0.4	0.5	0.6	0.7	0.8	0.9	1.0
datasize	524,216	403,222	285,628	203,813	134,068	74,595	31,181	10,472	2,430	299	71

4.4   Instance Selection According to the Faithfulness

As we already described in Sect. 3, there are many types of summaries in NYTAC. Some summaries are extractive, i.e., they can be generated by selecting a set of sentences from the source document, but some are abstractive due to, for example, paraphrasing or generalization of expressions. Since we use a summarization system based on sentence-extraction, we assume that the extractive summaries help in the training of this system, but abstractive summaries do not. Generally speaking, we assume that, if a given set of instances were generated under an uncontrolled situation, we should select appropriate training instances from the set depending on the characterization of the summarization system (e.g., extractive summarization). This concept is related to instance selection developed in domain adaptation [3, 25].

As the criterion for instance selection, we use the ROUGE-2 score calculated against the corresponding sentence-extractive oracle summary described in Sect. 4.2. When the ROUGE-2 score of an oracle summary is high enough, it means that we can generate a summary that is similar to the reference summary by selecting sentences. If the score is low, in contrast, even the best possible sentence-extractive summary should be very dissimilar to the reference summary. In such a case, we would need more complex operations, such as sentence compression, fusion, and paraphrasing. To select instances, we first calculate the oracle summaries of all the documents in the NYTAC and then collect the documents whose oracle summaries yield a ROUGE-2 score higher than a given threshold thr. Table 2 lists the numbers of selected instances for each value of thr. There are only 71 documents that can completely reproduce their reference summaries by sentence extraction. We selected the best thr that yields the best result on the development set through experiments. The examples in Table 1 are selected for different value of thr, i.e. 0.1, 0.3 and 0.6, to show the difference of summaries with respect to the threshold. A summary with a higher threshold is more extractive.

5   Experiments

5.1   Datasets

We used three test sets for evaluation: DUC2002, RSTDTB$_{long}$ and RSTDTB$_{short}$. DUC2002 and RSTDTB$_{long}$ consist of informative summaries, while RSTDTB$_{short}$ consists of indicative summaries (see Sect. 5.4 for detail). Their statistics are listed in Table 3 along with the NYTAC. The DUC2002 [10] consists of 567 document-summary pairs with the summary length limit being 100 words.[6] The RSTDTB$_{long}$ and RSTDTB$_{short}$ are included in the Rhetorical Structure Theory Discourse Treebank (RSTDTB, LDC2002T07) [4], which consists of 385 Wall Street Journal articles from the Penn Treebank manually

---

[6] Some reference summaries contain more than 100 words. Such summaries as well as system summaries were truncated to 100 words during the evaluation.

Table 3. Statistics of each corpus. Bottom three rows are statistics of test sets used in our experiments.

	Corpus size	Summary		Document		Avg. comp. rate
		Avg. # of sentences	Avg. # of words	Avg. # of sentences	Avg. # of words	
NYTAC	524,216[a]	2.42	38.79	29.36	607.62	0.128
DUC2002	533	5.62	101.09	27.40	549.61	0.363
RSTDTB$_{long}$	30	9.57	186.10	116.77	930.23	0.246
RSTDTB$_{short}$	30	3.5	39.43	116.77	930.23	0.093

[a] Although NYTAC contains 658,874 summary-article pairs, we deleted noisy instances, such as that whose text body is empty, and ended up with 524,216 instances.

annotated with a discourse structure in the RST framework and 30 of those articles are associated with their manually generated summaries. We used two types of summaries in the RSTDTB; RSTDTB$_{long}$ is a set of long summaries with the length of 25 % of the source documents and RSTDTB$_{short}$ is a set of short summaries consisting of 2 to 3 sentences. Following the previous work on RSTDTB, we used the elementary discourse units (EDU), which are the minimum units of RST and roughly correspond to clauses, as extraction units. We set the summary length limit L to 100 for DUC2002 and to the number of words in each reference summary for RSTDTB$_{long}$ and RSTDTB$_{short}$.

We evaluated each method by ROUGE-1 (with stemmed and stop-word removed)[7], which has been shown to correlate well with human judgments in DUC2002 test data [19] and is used as a loss function in this paper. We also added ROUGE-2 (with stemmed)[8] to the tables for reference.

5.2   Results: DUC2002

We show the results on DUC2002 in Table 4 with four blocks separated by horizontal lines.

First, the difference between the first and second blocks shows the effect of using the NYTAC for training. All of our methods (in the second block), except *Featurize*, significantly[9] outperformed *TrgtOnly* on ROUGE-1 score, although their ROUGE-2 scores are comparable. Interestingly, although *NytOnly* does not use any information from the test data, its score significantly differed from that of *TrgtOnly*. This indicates the importance of the size of training data. The differences within the second block show that *FineTune* was the most effective for using the NYTAC.

The third block shows the results when we performed the instance selection process. Although there was no significant gain derived from instance selection, the scores improved, except that of *Featurize*, and *FineTune$_{slct}$* significantly outperformed all the other methods. The selected *thr* values for each fold were 0.2, 0.3, 0.3, 0.3 and 0.3, respectively.

---

[7] With options "-a -x -n 1 -m -s" on version 1.5.5 of the official ROUGE script.

[8] With options "-a -x -n 2 -m" on version 1.5.5 of the official ROUGE script.

[9] For the statistical significance test, we used Wilcoxon signed-rank test ($p \leq 0.05$).

Table 4. ROUGE score of each method on DUC2002. Table is separated into four blocks by horizontal lines. Bold scores are best ones in each block.

	ROUGE-1	ROUGE-2
$TrgtOnly$	**0.400**	**0.218**
$NytOnly$	0.409	0.217
$Mixture$	0.411	0.219
$Featurize$	0.404	0.220
$LinInter$	0.409	0.217
$FineTune$	**0.415**	**0.221**
$NytOnly_{slct}$	0.411	0.224
$Mixture_{slct}$	0.412	0.224
$Featurize_{slct}$	0.403	0.219
$LinInter_{slct}$	0.411	0.224
$FineTune_{slct}$	**0.418**	**0.225**
LEAD	0.413	0.224
TextRank [22]	0.409	0.206
s21	0.416	0.223
s27	0.401	0.212
s28	**0.428**	**0.228**
s29	0.400	0.213
s31	0.393	0.203

Next, let us compare $FineTune_{slct}$ with other current methods (fourth block). The **LEAD** method simply extracts the first 100 words from the source document. Despite being very simple, the LEAD method is well known as a very powerful baseline in single document summarization. **TextRank** is a graph-based ranking method inspired by the PageRank algorithm and is also well known in single document summarization [22]. The **s21–31** are the methods submitted to the single document summarization task in DUC2002. We only show the scores of the top five methods. While the score of $FineTune_{slct}$ significantly outperformed LEAD, s27, s29 and s31, the score of s28 was significantly better than $FineTune_{slct}$.

5.3   Result: RSTDTB$_{long}$

Table 5 lists the results on RSTDTB$_{long}$. We can see that $FineTune$ outperformed $TrgtOnly$. This difference is statistically significant. In fact, $FineTune$ performed significantly better than $NytOnly$, $Mixture$, $Featurize$, and $LinInter$.

The difference between the second and third blocks shows that the instance selection is more effective for RSTDTB$_{long}$ than for DUC2002. It suggests that

Table 5. ROUGE score of each method on RSTDTB$_{long}$. Table is separated into four blocks by horizontal lines. Bold scores are best ones in each block.

	ROUGE-1	ROUGE-2
$TrgtOnly$	**0.324**	**0.121**
$NytOnly$	0.313	0.128
$Mixture$	0.320	0.132
$Featurize$	0.359	0.150
$LinInter$	0.313	0.128
$FineTune$	**0.385**	**0.158**
$NytOnly_{slct}$	0.320	0.134
$Mixture_{slct}$	0.323	0.133
$Featurize_{slct}$	0.376	0.156
$LinInter_{slct}$	0.320	0.134
$FineTune_{slct}$	**0.408**	**0.173**
LEAD$_{EDU}$	0.323	0.128
Marcu [20]	0.362	0.155
Hirao [12]	**0.405**	**0.161**

Table 6. ROUGE score of each method on RSTDTB$_{short}$. Table is separated into four blocks by horizontal lines. Bold scores are best ones in each block.

	ROUGE-1	ROUGE-2
$TrgtOnly$	**0.258**	**0.086**
$NytOnly$	0.272	0.085
$Mixture$	0.272	0.085
$Featurize$	0.264	0.087
$LinInter$	0.272	0.085
$FineTune$	**0.283**	**0.094**
$NytOnly_{slct}$	0.265	0.087
$Mixture_{slct}$	0.264	0.088
$Featurize_{slct}$	0.246	0.075
$LinInter_{slct}$	0.265	0.087
$FineTune_{slct}$	**0.286**	**0.092**
LEAD$_{EDU}$	0.240	0.082
Marcu [20]	0.272	0.094
Hirao [12]	**0.321**	**0.106**

the strategy of generating reference summaries, i.e. how human summarizers generate their summaries, is different between RSTDTB$_{long}$ and DUC2002.

We also compared our methods with well-known current summarization methods. The $\mathbf{LEAD}_{EDU}$ method extracts the first K textual units from a source document until the summary length reaches L. **Marcu** proposed an EDU ranking-based method based on the rhetorical tree structure of a document [20]. The method developed by Hirao et al. is the state-of-the-art method of the EDU extraction-based summarization method [12]. They regard the summarization problem as a tree KP from a dependency-based RST tree. Note that the two methods by Marcu and Hirao et al. rely on the rhetorical structure of documents annotated by human annotators. However, it is worth noting that even without using any information of the rhetorical structure of documents, $FineTune_{slct}$[10] outperformed these methods including the state-of-the-art method.

5.4   Results: RSTDTB$_{short}$

The summaries in DUC2002 and RSTDTB$_{long}$ we have discussed are *informative*, while those in RSTDTB$_{short}$ are *indicative*[11], as written in their annotation instructions or descriptions [4,10]. Although our summarization method as well as most existing methods are designed for generating informative summaries, it is interesting to examine how they perform in the task of generating indicative summaries.

We show the results on RSTDTB$_{short}$ in Table 6. The ROUGE score increased when we used the NYTAC. *FineTune* yielded the best score among all the methods except Hirao [12], which uses the rhetorical structure of documents annotated by human annotators. On the other hand, the effect of instance selection[12] cannot be confirmed in this experiment. It might be reasonable because the summaries in RSTDTB$_{short}$ are short, and thus sentence compression or paraphrasing occur more frequently than in the other two test sets.

6   Conclusion

We presented five methods that use the NYTAC for training a single document summarization system. We also proposed an instance selection method according to the faithfulness of the extractive oracle summary to the reference summary for the purpose of selecting appropriate training instances for extractive summarization systems. The experimental results showed that *FineTune* performed better than the other methods in all the experiments. In addition, our instance selection scheme improved summarization performance and achieved the state-of-the-art ROUGE score on RSTDTB$_{long}$.

---

[10] The selected values of *thr* for each fold were 0.1, 0.1, 0.1, 0.1 and 0.1, respectively.

[11] Explanation of these two types of summaries can be found in the book written by Nenkova and McKeown [23]. We quote the relevant part of the book: *A summary that enables the reader to determine about-ness has often been called an indicative summary, while one that can be read in place of the document has been called an informative summary.*

[12] The selected values of *thr* for each fold are 0.3, 0.6, 0.3, 0.3 and 0.6, respectively.

For future work, we will explore more effective and possibly complicated features that can be trained only with a large amount of training data since we now have access to a large amount of training data. We will also attempt to test the effectiveness of using the NYTAC in other types of summarization systems.

Acknowledgement. This work was supported by JSPS KAKENHI Grant Number JP26280080.

References

1. Almeida, M., Martins, A.: Fast and robust compressive summarization with dual decomposition and multi-task learning. In: Proceedings of ACL 2013, pp. 196–206 (2013)
2. Axelrod, A., He, X., Gao, J.: Domain adaptation via pseudo in-domain data selection. In: Proceedings of EMNLP 2011, pp. 355–362 (2011)
3. Biçici, E.: Domain adaptation for machine translation with instance selection. Prague Bull. Math. Linguist. **103**, 5–20 (2015)
4. Carlson, L., Marcu, D., Okurowski, M.E.: RST discourse treebank. In: Linguistic Data Consortium (2002). https://catalog.ldc.upenn.edu/LDC2002T07
5. Consortium, L.D: Hansard corpus of parallel english and french. In: Linguistic Data Consortium (1997). http://www.ldc.upenn.edu/
6. Crammer, K., Dekel, O., Keshet, J., Shalev-Shwartz, S., Singer, Y.: Online passive-aggressive algorithms. J. Mach. Learn. Res. **7**, 551–585 (2006)
7. Crammer, K., McDonald, R., Pereira, F.: Scalable large-margin online learning for structured classification. In: Proceedings of NIPS05 Workshop on Learning With Structured Outputs (2005)
8. Daumé III., H.: Frustratingly easy domain adaptation. In: Proceedings of ACL 2007, pp. 256–263 (2007)
9. Daumé, H., Marcu, D.: Induction of word and phrase alignments for automatic document summarization. Comput. Linguist. **31**(4), 505–530 (2005)
10. DUC: Document understanding conference. In: ACL Workshop on Automatic Summarization (2002)
11. Hirao, T., Isozaki, H., Maeda, E., Matsumoto, Y.: Extracting important sentences with support vector machines. In: Proceedings of COLING 2002, vol. 1, pp. 1–7 (2002)
12. Hirao, T., Yoshida, Y., Nishino, M., Yasuda, N., Nagata, M.: Single-document summarization as a tree knapsack problem. In: Proceedings of EMNLP 2013, pp. 1515–1520 (2013)
13. Hong, K., Nenkova, A.: Improving the estimation of word importance for news multi-document summarization. In: Proceedings of EACL 2014, pp. 712–721 (2014)
14. Jing, H., McKeown, K.R.: Cut and paste based text summarization. In: Proceedings of NAACL 2000, pp. 178–185 (2000)
15. Li, C., Liu, Y., Zhao, L.: Using external resources and joint learning for bigram weighting in ILP-based multi-document summarization. In: Proceedings of NAACL 2015, pp. 778–787 (2015)
16. Li, C., Qian, X., Liu, Y.: Using supervised bigram-based ILP for extractive summarization. In: Proceedings of ACL 2013, pp. 1004–1013 (2013)
17. Li, J.J., Nenkova, A.: Fast and accurate prediction of sentence specificity. In: Proceedings of AAA 2015, pp. 2281–2287 (2015)

18. Li, Q.: Literature survey: domain adaptation algorithms for natural language processing. Technical report, Department of Computer Science. The Graduate Center, The City University of New York (2012)
19. Lin, C.Y.: ROUGE: a package for automatic evaluation of summaries. In: Proceedings of the Workshop on Text Summarization Branches Out, pp. 74–81 (2004)
20. Marcu, D.: Improving summarization through rhetorical parsing tuning. In: Proceedings of Sixth Workshop on Very Large Corpora, pp. 206–215 (1998)
21. Marcu, D.: The automatic construction of large-scale corpora for summarization research. In: Proceedings of SIGIR99, pp. 137–144 (1999)
22. Mihalcea, R., Tarau, P.: Textrank: bringing order into texts. In: Proceedings of EMNLP 2004, pp. 404–411 (2004)
23. Nenkova, A., McKeown, K.: Automatic summarization. Found. Trends. Inf. Retrieval **2–3**, 103–233 (2011)
24. Nishikawa, H., Arita, K., Tanaka, K., Hirao, T., Makino, T., Matsuo, Y.: Learning to generate coherent summary with discriminative hidden semi-Markov model. In: Proceedings of COLING 2014, pp. 1648–1659 (2014)
25. Remus, R.: Domain adaptation using domain similarity- and domain complexity-based instance selection for cross-domain sentiment analysis. In: Proceedings of ICDMW 2012) Workshop on SENTIRE, pp. 717–723 (2012)
26. Sandhaus, E.: The New York Times annotated corpus. In: Linguistic Data Consortium (2008). https://catalog.ldc.upenn.edu/LDC2008T19
27. Sipos, R., Shivaswamy, P., Joachims, T.: Large-margin learning of submodular summarization models. In: Proceedings of EACL 2012, pp. 224–233 (2012)
28. Svore, K., Vanderwende, L., Burges, C.: Enhancing single-document summarization by combining RankNet and third-party sources. In: Proceedings of EMNLP-CoNLL 2007, Association for Computational Linguistics, Prague, Czech Republic, pp. 448–457. http://www.aclweb.org/anthology/D/D07/D07-1047
29. Takamura, H., Okumura, M.: Learning to generate summary as structured output. In: Proceedings of CIKM 2010, pp. 1437–1440 (2010)
30. Xia, R., Zong, C., Hu, X., Cambria, E.: Feature ensemble plus sample selection: domain adaptation for sentiment classification. IEEE Intell. Syst. **28**(3), 10–18 (2013)
31. Yang, Y., Nenkova, A.: Detecting information-dense texts in multiple news domains. In: Proceedings of AAAI 2014, pp. 1650–1656 (2014)
32. Yih, W.T., Goodman, J., Vanderwende, L., Suzuki, H.: Multi-document summarization by maximizing informative content-words. In: Proceedings of IJCAI 2007, pp. 1776–1782 (2007)
33. Zhao, J., Qiu, X., Liu, Z., Huang, X.: Online distributed passive-aggressive algorithm for structured learning. In: Sun, M., Zhang, M., Lin, D., Wang, H. (eds.) CCL and NLP-NABD 2013. LNCS, vol. 8202, pp. 120–130. Springer, Heidelberg (2013)

Selecting Training Data for Unsupervised Domain Adaptation in Word Sense Disambiguation

Kanako Komiya[1]([⊠]), Minoru Sasaki[1], Hiroyuki Shinnou[1], Yoshiyuki Kotani[2], and Manabu Okumura[3]

[1] Ibaraki University, 4-12-1 Nakanarusawa, Hitachi-shi, Ibaraki 316-8511, Japan
{kanako.komiya.nlp,minoru.sasaki.01,
hiroyuki.shinnou.0828}@vc.ibaraki.ac.jp
[2] Tokyo University of Agriculture and Thechnology,
2-24-16 Naka-cho, Koganei, Tokyo 184-8588, Japan
kotani@cc.tuat.ac.jp
[3] Tokyo Institute of Technology,
4259 Nagatuta, Midori-ku, Yokohama 226-8503, Japan
oku@pi.titech.ac.jp

Abstract. This paper describes a method of domain adaptation, which involves adapting a classifier developed from source to target data. We automatically select the training data set that is suitable for the target data from the whole source data of multiple domains. This is unsupervised domain adaptation for Japanese word sense disambiguation (WSD). Experiments revealed that the accuracies of WSD improved when we automatically selected the training data set using two criteria, the degree of confidence and the leave-one-out (LOO)-bound score, compared with when the classifier was trained with all the data.

Keywords: Domain adaptation · Word sense disambiguation · Data selection

1 Introduction

Domain adaptation involves adapting the classifier that has been trained from data in a domain (source domain) to data in another domain (target domain). This has been studied intensively (see Sect. 2).

However, the optimal training data varied according to the properties of the source and target data when domain adaptation for word sense disambiguation (WSD) was carried out [12]. Therefore, if we could determine the optimal training data for each source and target data, the performance of WSD would improve. However, the existing methods to determine the suitable domain adaptation method for each source and target data were used for supervised domain adaptation, and therefore, they require some labeled target data. This paper therefore proposes automatic domain adaptation in an unsupervised manner

© Springer International Publishing Switzerland 2016
R. Booth and M.-L. Zhang (Eds.): PRICAI 2016, LNAI 9810, pp. 220–232, 2016.
DOI: 10.1007/978-3-319-42911-3_18

based on a comparison of multiple classifiers when Japanese WSD is performed (see Sect. 3). Our experiments (see Sects. 4 and 5) revealed that the average accuracy of WSD when the training data set that was automatically determined was used was higher than that when all the data were used collectively (see Sect. 6). In addition, some existing methods determined the optimal domain adaptation method for each instance and the others did for each word type. Therefore, we investigate which is better for domain adaptation, to determine the optimal training data set according to the word type or the instance. We discuss the results in Sect. 7 and conclude this paper in Sect. 8.

2   Related Work

The domain adaptation problem can be categorized into three types depending on the information for learning, i.e., that in supervised, semi-supervised, and unsupervised approaches. According to Daumé [9], a classifier in a supervised approach is developed from a large amount of labeled source data and a small amount of labeled target data. A classifier in a semi-supervised approach is developed from a large amount of labeled source data, a small amount of labeled and a large amount of unlabeled target data. Finally, a classifier is developed from a large amount of labeled source data and unlabeled target data in an unsupervised approach.

Since there are some cases where there have been no manually annotated data and the tagging of corpora is time-consuming in the cases, we focused on the unsupervised domain adaptation for Japanese WSD in the research reported in this paper.

Many researchers have investigated domain adaptation within or outside the area of natural language processing.

Chan et al. carried out the domain adaptation of WSD by estimating class priors using an EM algorithm, which were unsupervised domain adaptation and supervised domain adaptation using active learning [5,6]. Daumé augmented an input space and made triple length features that were general, source-specific, and target-specific for supervised domain adaptation [8]. Daumé et al. extended the earlier work to semi-supervised domain adaptation [9]. Agirre et al. applied singular value decomposition (SVD) to a matrix of unlabeled target data and a large amount of unlabeled source data, and trained a classifier with them [1,2]. Kunii et al. proposed combined use of topic models on unsupervised domain adaptation for WSD [16]. Jiang et al. demonstrated that the performance increased as examples were weighted when domain adaptation was applied [11]. Shinnou et al. reported active learning to remove source instances for domain adaptation for WSD [24]. Shinnou et al. also proposed learning under covariate shift for domain adaptation for WSD [23]. Kouno et al. performed unsupervised domain adaptation for WSD using stacked denoising autoencoder [15].

Zhong et al. proposed an adaptive kernel approach that mapped the marginal distribution of source and target data into a common kernel space [27]. They also conducted sample selection to make the conditional probabilities between the two domains closer.

Raina et al. proposed self-taught learning that utilized sparse coding to construct higher level features from unlabeled data collected from the Web. This method was based on unsupervised learning [22].

Tur proposed a co-adaptation algorithm where both co-training and domain adaptation techniques were used to improve the performance of the model [25]. The research by Blitzer et al. involved work on semi-supervised domain adaptation, where they calculated the weight of words around the pivot features (words that frequently appeared both in source and target data and behaved similarly in both) to model some words in one domain that behaved similarly in another [4]. They applied SVD to the matrix of the weights, generated a new feature space, and used the new features with the original features.

McClosky et al. focused on the problem where the best model for each document is not obvious when parsing a document collection of heterogeneous domains [18]. They studied it as a new task of *multiple source parser adaptation*. They proposed a method of parsing a sentence that first predicts accuracies for various parsing models using a regression model, and then uses the parsing model with the highest predicted accuracy. The main difference is that their work was about parsing but ours discussed here is about Japanese WSD and we determined the best training data with only comparison of some scores of classifiers.

Van Asch et al. reported that the performance in domain adaptation could be predicted depending on the similarity between source and target data using automatically annotated corpus in parsing [3]. They focused on how corpora were selected for use as source data according to the distance between domains, but here we have focused on how to select a training data set depending on the degrees of confidence of multiple classifiers and LOO-bound score.

Komiya and Okumura determined an optimal method of domain adaptation using decision tree learning given a triple of the target word type of WSD, source data, and target data [12,14]. They discussed what features affected how the best method was determined.

Finally, the closest work to ours is that by Komiya and Okumura who determined the optimal method, i.e., the optimal training data set, for each instance using the degree of confidence, which was also used in this paper, for supervised domain adaptation in WSD [13]. We found that the method that they proposed was also effective for unsupervised domain adaptation. In addition, we investigate which is better for domain adaptation, to determine the optimal training data set according to the word type like [12,14] or the instance like [13].

3   Automatic Selection of Training Data

We assume that the labels, i.e., the word senses, of the target data are fully unknown and we have the source data of the multiple domains. On these assumption, we would like to automatically select the subset of the training data that is suitable for the target data from the whole set of the source data. Komiya and Okumura assumed that the optimal training data set would vary according to each instance [13]. However, they also determined an optimal method of domain adaptation for each word type [12,14]. Therefore, we investigate which is better for domain adaptation, to determine the optimal training data set according to the word type or the instance[1]. Here, the instances are the instance vectors for machine learning and they are generated for each word token in the corpora, and multiple classifiers are developed for each word type. When we determine the optimal training data for each instance, we select it for each instance vector of the target data. On the other hand, when we do for each word type, we select it for each instance set, i.e., all the instances of a word type of the target data.

The training data set is automatically determined for each word type or each instance in four steps:

(1) Select some instances randomly from the whole source data and create multiple training data sets,

(2) Train multiple classifiers based on the training data sets (in (1)) and apply them to the target data,

(3) Compare the scores of multiple classifiers (in (2)) for each word type or each instance, and

(4) Predict the label of the word (token) using the classifier whose score is the highest. The classifier is selected for each word type or each instance.

It should be noted that we determine the optimal training data set in step (4) by selecting a classifier developed from it. In addition, we predict the label of each word token, i.e., each instance, not only when we select the classifier for each instance but also when we do for each word type.

We use three types of scores for classifiers and compare them:

– *Confidence* [13],
– *LOO*: LOO-bound score, which is the score based on the LOO-bound [26],
– *Confidence*LOO*: The product of the two scores above.

As Komiya and Okumura [13] reported, the degrees of confidence are the predicted values that indicate how confident classification is and these are often used to select instances to be labeled in active-learning. Since the classifier outputs the degree of confidence per instance, we use the average for all the instances in the target data set of each word type when we determine the optimal training data for each word type. In other words, the score is an averaged value that indicates how confidently a classifier classifies the whole target data of each word type. We use the degree of confidence per instance directly when we determine the optimal training data for each instance. Komiya and Okumura [13] carried out ensemble learning by comparing the degrees of confidence focusing on the fact that these degrees were probabilities output from classifiers. We use the same method for unsupervised domain adaptation.

---

[1] However, it must be noted that their scenario were supervised but our experiments are carried out in an unsupervised manner.

The LOO-bound score is the upper bound of error for the leave-one-out estimation of SVM and is calculated as:

$$LOO_{original,i} = \frac{SV_i}{TR_i},\tag{1}$$

where $LOO_{original,i}$ denotes the original LOO-bound score of each (i_{th}) classifier, and SV_i and TR_i denote the number of support vectors and training data of each classifier, respectively. However, when we select the suitable classifier from multiple classifiers, it is necessary to take into account the numbers of training data of each classifier because they vary a great deal. Therefore, we use LOO-bound of a certain classifier weighted by the number of training data of each classifier.

$$LOO_{selecting,i} = \frac{TR_1}{TR_i}LOO_{original,1} = \frac{SV_1}{TR_i},\tag{2}$$

where $LOO_{selecting,i}$ denotes the LOO-bound for selecting the training data set, TR_1 and SV_1 denote the number of training data and support vectors of a certain classifier in multiple classifiers developed in step (2) respectively, and TR_i denotes the number of training data of each classifier. Since SV_1 is constant for every classifier, $LOO_{selecting,i}$ weights the number of training data of each classifier. In addition, the weight is based on SV_1, i.e., the number of support vectors of a certain classifier.[2] We use $1 - LOO_{selecting,i}$ instead of $LOO_{selecting,i}$ for the score of the classifiers since the LOO-bound is the error rate. Finally, we use the following equation to avoid illegal division because the number of training data could be zero when there is only a single word sense in the training data set; the instances are randomly selected.

$$LOO_i = 1 - \frac{SV_1 + 0.5}{TR_i + 0.5},\tag{3}$$

where LOO_i denotes the LOO-bound score of each classifier.

We are able to automatically determine the best training data set using ensemble learning based on the classifier score for each word type or each instance. Therefore, we expect the average accuracy of WSD, when the training data set that is automatically determined is used for each word type or each instance, to be higher than when the whole training data are collectively used. Navigli [20] introduced this method as an ensemble approach to WSD and called it a *probability mixture*. We used the *probability mixture* assuming that each classifier is trained for each training data set, rather than for each method of domain adaptation like reported in Komiya [13] or each method of WSD like introduced in Navigli [20].

4   Experiment

Libsvm [7], which supports multi-class classification, was used as the classifier for WSD. We used the -b option of libsvm to train a model to estimate probability

---

[2] Maybe we can use SV_{total} instead of SV_1 but it did not affect our results so much.

for the degree of confidence. We trained 100 classifiers for a word type of a domain of source data and employed the classifier with the highest degree of confidence for each word type or each instance. We randomly selected the number of instances in one training data set, from one to the number of all the training data we could use, for each word type[3]. Since the experiments were greatly affected by the randomness of the setting of each experiment, we performed the experiments 10 times and evaluated the averaged accuracies. A linear kernel was used according to the results obtained from preliminary experiments. Twenty features were introduced to train the classifier.

- Morphological features
  - Bag-of-words
  - Part-of-speech (POS)
  - Finer subcategory of POS
- Syntactic features
  - If the POS of a target word is a noun, the verb that the target word modifies is used.
  - If the POS of a target word is a verb, the case element of 'ヲ' (wo, objective) for the verb is used.
- Semantic feature
  - Semantic classification code

Morphological features and a semantic feature were extracted from the surrounding words (two words to the right and left) of the target word and the target word itself. POS and the finer subcategory of POS could be obtained by using a morphological analyzer. We used ChaSen[4] as a morphological analyzer, the Bunruigoihyo thesaurus [19] for semantic classification codes (e.g. The code of the 'program' was 1.3162.), and CaboCha[5] as a syntactic parser. Five-fold cross validation was used in the experiments.

5   Data

Three data that were the same as those utilized by Komiya and Okumura [12, 13] were used for the experiments: (1) the sub-corpus of white papers in the Balanced Corpus of Contemporary Japanese (BCCWJ) [17], (2) the sub-corpus of documents from a Q & A site on the WWW in BCCWJ, and (3) Real World Computing (RWC) text databases (newspaper articles) [10]. Domain adaptation was conducted in three directions according to different source and target data, i.e., one data in three was used for the target data and the other two data

---

[3] How to generate candidate training sets is our future work.

[4] http://sourceforge.net/projects/masayu-a/.

[5] http://sourceforge.net/projects/cabocha/.

Table 1. Minimum, maximum, and average number of instances of each word type for each corpus

Genre	Min	Max	Avg
White papers	58	7,610	2,240.14
Q & A site	130	13,976	2,741.95
Newspaper	56	374	183.36

Table 2. The number of instances of WSD for each corpus

Target data	No. of instances
White paper	49,283
Q & A site	60,323
Newspaper	4,034
Total	232,116

were used for the source data in one setting[6]. Word senses were annotated in these corpora according to a Japanese dictionary, i.e., the Iwanami Kokugo Jiten [21]. It has three levels for sense IDs, and we used the fine-level sense in the experiments. Multi-sense words that appeared equal or more than 50 times in all the data were selected as the target words in the experiment. Twenty-two word types were used in the experiments. Table 1 lists the minimum, maximum, and average number of instances of each word type for each corpus and Table 2 summarizes the total number of the instances of the 22 word types for WSD found in each corpus. Table 3 summarizes the list of target word types. "No. of senses" in the first column is the number of the senses of each word type in the dictionary. For example, the word type "場合 (case)" has two senses and "手 (hand)" has 22 senses in the dictionary. Please note that there is no guarantee that all the senses in the dictionary appear in the corpora.

Komiya and Okumura [12,14] found that the optimal method of domain adaptation varied depending on each 'case' (i.e., a triple of the target word type of WSD, the source data, and the target data). They also assumed that it varied according to each instance [13]. Here, we investigate which is better for domain adaptation, to determine the optimal training data set according to the word type or the instance.

6   Results

Table 4 lists the micro- and macro-averaged accuracies of WSD for the whole data set according to the methods of domain adaptation and Table 5 summarizes the micro- and macro-averaged accuracies of WSD according to the corpora and methods of domain adaptation. Micro-average is the average over word tokens and macro-average is that over word types.

---

[6] Komiya and Okumura [12,13] conducted domain adaptation in six directions with the source data from one domain and the target data from another domain. They used multi-sense words that appeared equal or more than 50 times in both the source and target data, whereas we used multi-sense words that appeared equal or more than 50 times in the two source data and the target data. Therefore, we used fewer target word types than they did.

Table 3. List of target words types

No. of senses	Target words (in Japanese)	Sense example in English	No. of senses	Target words (in Japanese)	Sense example in English
2	場合	case	6	関係	connection
	自分	self		時間	time
3	事業	project		一般	general
	情報	information		現在	present
	地方	area	7	今	now
	社会	society	8	前	before
	思う	suppose	10	持つ	have
	子供	child	12	見る	see
4	考える	think	14	入る	enter
5	含む	contain	16	言う	say
	技術	technique	22	手	hand

We tested *Self*, which is standard supervised learning, in other words, unadapted classifier, with the whole target data by five-fold cross validation, assuming that fully annotated data were obtained and could be used for learning, *MFS*, which is the most frequent sense of the target corpus, *Averaged*, which is the averaged accuracies of the supervised learning with the source data of each domain, *Bigger*, which is standard supervised learning with the bigger source data in two domains of the source data, and *All*, which is standard supervised learning with all the source data as references.

When the target data were Q & A sites and the source data were white papers and newspapers, for example, *Averaged* would be the averaged accuracy of the two accuracies of the Q & A sites: those of the classifier trained with all the white papers and with all the newspapers. *Bigger* would be the accuracy of the classifier trained with all the newspapers because the number of instances in the newspapers was greater than that in the white papers. Finally, *All* would be the accuracy of the Q & A sites whose classifier was trained with all the source data, i.e., both all the white papers and all the newspapers.

Self was an upper bound and *Averaged*, *Bigger*, and *All* were baselines. *Confidence (ty)* and *Confidence (in)* determined the optimal training data set using *Confidence* score for each word type and each instance, respectively. *Confidence*LOO (ty)* and *Confidence*LOO (in)* determined the optimal training data set using *Confidence*LOO* score for each word type and each instance, respectively. *LOO* determined the optimal training data set using *LOO* score for only each word type because the system output *LOO* for each word type. The highest accuracies except for *Self* have been written in bold for each corpus in Tables 4 and 5. The asterisk means the difference between accuracies of the best proposed method (*Confidence*LOO (ty)* or *Confidence*LOO (in)*) and each method is statistically significant according to a chi-square test. The level of significance in the test was 0.05.

Table 4. Average accuracies of WSD for whole data set

Method	Micro avg	Macro avg
Self	93.29 %*	85.97 %
MFS	77.32 %*	73.44 %
Averaged	76.92 %*	71.20 %
Bigger	81.99 %*	74.25 %
All	81.76 %*	75.86 %
Confidence (ty)	75.07 %*	72.16 %
Confidence (in)	75.10 %*	72.66 %
*Confidence*LOO (ty)*	82.15 %*	**76.38 %**
*Confidence*LOO (in)*	**82.25 %**	76.35 %
LOO	81.83 %*	7 6.34 %

Table 5. Average accuracies of WSD according to corpora and methods of domain adaptation

Target data	Micro avg.			Macro avg.		
	White papers	newspapers	Q & A sites	White papers	newspapers	Q & A sites
Self	96.07 %*	79.57 %*	91.93 %*	91.53 %	78.59 %	87.80 %
MFS	78.74 %*	68.59 %*	76.74 %*	77.58 %	69.81 %	72.93 %
Averaged	73.54 %*	72.94 %*	79.95 %*	70.80 %	71.23 %	71.57 %
Bigger	80.72 %*	74.86 %*	**83.50 %***	75.64 %	74.39 %	72.73 %
All	81.80 %	75.95 %	82.11 %*	76.91 %	74.91 %	75.76 %
Confidence (ty)	74.62 %*	73.64 %*	75.53 %*	71.95 %	72.59 %	71.93 %
Confidence (in)	74.01 %*	74.54 %*	76.03 %*	72.12 %	73.14 %	72.73 %
*Confidence*LOO (ty)*	**82.43 %**	76.10 %	82.33 %*	**77.68 %**	75.17 %	76.29 %
*Confidence*LOO (in)*	81.92 %	**76.49 %**	82.91 %	77.31 %	**75.51 %**	76.23 %
LOO	82.15 %	76.06 %	81.95 %*	77.54 %	75.17 %	**76.30 %**

7 Discussion

First, Table 4 indicates that the micro-averaged accuracy of *Bigger* is higher than that of *All* and Table 5 shows the same when the target data were the Q & A sites, which means that the biggest training data did not always provide the highest accuracy.

Second, the same tables reveal that the accuracies of *Confidence (ty)* and *Confidence (in)* are lower than those of the three baselines. This is different from the results obtained by Komiya and Okumura [13] who found that it was more effective to select the method of domain adaptation, i.e., the training data set, using the degree of confidence for each instance. We think this is because the correctness of the degree of confidence decreased when there were few instances of the training data set. Since we randomly determined the number of the instances of the training data set, the classifiers were sometimes trained with a small number of instances and this affected the decline in accuracies. When the training data set included only one instance, for example, the degree of confidence was one, which was the highest value, because there was no other alternative word sense in the training data set. However, it is difficult to deem that the best classifier is that trained with only one training instance, which means that the degrees of confidence are not particularly trustworthy when there are few instances of the training data set.

Moreover, Tables 4 and 5 reveal that the accuracies of *LOO* outperformed those of the three baselines except for the micro-averaged accuracies of the whole data set and those when the target data were the Q & A sites. We think that the LOO-bound score was effective for selecting the classifier because it weighted the number of the training data and therefore the score indicated how trustworthy the classifier was. However, the micro-averaged accuracy for the whole data set of *LOO* could not outperform that of *Bigger* because the micro-averaged accuracy of *Bigger* was higher than that of *LOO* when the target data were Q & A sites.

The micro- and macro-averaged accuracies of *Confidence*LOO (ty)* or *Confidence*LOO (in)*, on the other hand, were the best except for *Self*, i.e., the upper bound, for the whole data set, although the micro-averaged accuracy could not outperform that of *Bigger* when the target data were Q & A sites. Although the difference between the accuracies of *Confidence*LOO (ty)* and *Bigger* was not statistically significant, the difference between the micro-averaged accuracies of *Confidence*LOO (in)* and *Bigger*, that of *Confidence*LOO (ty)* and *All*, and that of *Confidence*LOO (in)* and *All* were statistically significant according to a chi-square test. The level of significance in the test was 0.05. We think that *Confidence*LOO (ty)* and *Confidence*LOO (in)* were the best because it selected the most suitable classifier for each target data. *LOO* returned the same score for any target data if the training data were the same, but *Confidence* returned the score for each combination of the training data set and each instance of the target data. Therefore, the product of two criteria leaded to the better result.

Moreover, the macro-average accuracies of *Confidence*LOO (ty)*, *Confidence*LOO (in)*, and *LOO* outperformed those of the three baselines although the differences were not statistically significant because there were few samples for these. This indicated that these criteria could be used to select a better training data set for various word types.

Next, we investigate which is better for domain adaptation, to determine the optimal training data set according to the word type or the instance. Tables 4 and 5 revealed that the best method of domain adaptation of each target corpus was *Confidence*LOO (ty)* or *Confidence*LOO (in)* except for the micro- and macro-averaged accuracies of Q & A sites; the best methods were *Bigger* and *LOO* for the micro- and macro-averaged accuracy of Q & A sites, respectively. The same tables also show that the differences among accuracies of *Confidence*LOO (ty)*, *Confidence*LOO (in)*, and *LOO* are not so big. This indicates that the effect of *Confidence*LOO* mainly comes from *LOO* and the unit for selecting the optimal training data set, i.e., the word type or the instance, does not affect the results so much although the effect of *Confidence* did improved the accuracies.

However, although the differences are not so big, Table 5 indicates that the best methods vary according to the target corpus; the best method is *Confidence*LOO (ty)* for white papers and it is *Confidence*LOO (in)* for newspapers. We think that it is associated with the balance of the senses in the target corpus. As Table 5 shows, the ratio of MFS for white papers is highest and that for newspapers is the lowest. Therefore, we think that *Confidence*LOO (ty)* is the best for white papers because it is better than *Confidence*LOO (in)* when the senses are biased. Likewise, we think that *Confidence*LOO (in)* is the best for newspapers because it is better than *Confidence*LOO (ty)* when the senses are balanced. When the senses of a data set of a word type are biased, the correct senses tend to be the same for the data set and if not, they will vary. Therefore, the system should determine the optimal training data set according to the word type when the senses of the target corpus are biased and it should determine them according to the instance when the senses of the target corpus are balanced.

Finally, Table 4 demonstrates that *Confidence*LOO (in)* is the best for micro-averaged accuracy and *Confidence*LOO (ty)* is the best for macro-averaged accuracy for the whole data set. We think this is because the method that selects training data set for each word type, i.e., *Confidence*LOO (ty)*, improves macro-averaged accuracy, i.e., averaged accuracy over word types and the method that selects training data set for each instance, i.e., *Confidence*LOO (in)*, improves micro-averaged accuracy, i.e., averaged accuracy over instances.

8   Conclusion

This paper described how to automatically select the training data set by using two criteria, the degree of confidence and the LOO-bound score when there were multiple domain source data for unsupervised domain adaptation in WSD. We selected a suitable training data set using the degree of confidence, the score based on a LOO-bound, and their product. The method with the product of two criteria demonstrated the best micro- and macro-averaged accuracies. We also investigated which was better for domain adaptation, to determine the optimal training data set according to the word type or the instance. Although the differences between accuracies of the units for selecting the training data set, i.e., the word type or the instance, were not so big, the experimental results indicated that the system should determine the optimal training data set according to the word type when the senses of the target corpus are biased and it should determine them according to the instance when the senses of the target corpus are balanced. Finally, the differences between the micro-averaged accuracies of the proposed method, i.e., the method with the product of two criteria for the instances, and three baselines were significant according to a chi-square test.

References

1. Agirre, E., de Lacalle, O.L.: On robustness and domain adaptation using SVD for word sense disambiguation. In: Proceedings of the 22nd International Conference on Computational Linguistics, pp. 17–24 (2008)
2. Agirre, E., de Lacalle, O.L.: Supervised domain adaption for WSD. In: Proceedings of the 12th Conference of the European Chapter of the Association of Computational Linguistics, pp. 42–50 (2009)
3. van Asch, V., Daelemans, W.: Using domain similarity for performance estimation. In: Proceedings of the 2010 Workshop on Domain Adaptation for Natural Language Processing, ACL 2010, pp. 31–36 (2010)
4. Blitzer, J., McDonald, R., Pereira, F.: Domain adaptation with structural coppespondence learning. In: Proceedings of the 2006 Conference on Empirical Methods in Natural Language Processing, pp. 120–128 (2006)
5. Chan, Y.S., Ng, H.T.: Estimating class priors in domain adaptation for word sense disambiguation. In: Proceedings of the 21st International Conference on Computational Linguistics and 44th Annual Meeting of the Association for Computational Linguistics, pp. 89–96 (2006)
6. Chan, Y.S., Ng, H.T.: Domain adaptation with active learning for word sense disambiguation. In: Proceedings of the 45th Annual Meeting of the Association of Computational Linguistics, pp. 49–56 (2007)
7. Chang, C.C., Lin, C.J.: LIBSVM: a library for support vector machines (2001). Software available at cjlin/libsvm http://www.csie.ntu.edu.tw/
8. Daumé III., H.: Frustratingly easy domain adaptation. In: Proceedings of the 45th Annual Meeting of the Association of Computational Linguistics, pp. 256–263 (2007)
9. Daumé III., H., Kumar, A., Saha, A.: Frustratingly easy semi-supervised domain adaptation. In: Proceedings of the 2010 Workshop on Domain Adaptation for Natural Language Processing, ACL 2010, pp. 23–59 (2010)
10. Hashida, K., Isahara, H., Tokunaga, T., Hashimoto, M., Ogino, S., Kashino, W.: The RWC text databases. In: Proceedings of the First International Conference on Language Resource and Evaluation, pp. 457–461 (1998)
11. Jiang, J., Zhai, C.: Instance weighting for domain adaptation in NLP. In: Proceedings of the 45th Annual Meeting of the Association of Computational Linguistics, pp. 264–271 (2007)
12. Komiya, K., Okumura, M.: Automatic determination of a domain adaptation method for word sense disambiguation using decision tree learning. In: Proceedings of the 5th International Joint Conference on Natural Language Processing, IJCNLP 2011, pp. 1107–1115 (2011)
13. Komiya, K., Okumura, M.: Automatic domain adaptation for word sense disambiguation based on comparison of multiple classifiers. PACLIC **2012**, 77–85 (2012)
14. Komiya, K., Okumura, M.: Automatic selection of domain adaptation method for WSD using decision tree learning. J. NLP **19**(3), 143–166 (2012). (In Japanese)
15. Kouno, K., Shinnou, H., Sasaki, M., Komiya, K.: Unsupervised domain adaptation for word sense disambiguation using stacked denoising autoencoder. In: Proceedings of PACLIC-29, pp. 224–231 (2015)
16. Kunii, S., Shinnou, H.: Combined use of topic models on unsupervised domain adaptation for word sense disambiguation. In: Proceedings of PACLIC-27, pp. 224–231 (2013)

17. Maekawa, K.: Balanced corpus of contemporary written Japanese. In: Proceedings of the 6th Workshop on Asian Language Resources (ALR), pp. 101–102 (2008)
18. McClosky, D., Charniak, E., Johnson, M.: Automatic domain adaptation for parsing. In: Proceedings of the 2010 Annual Conference of the North American Chapter of the Association for Computational Linguistics, pp. 28–36 (2010)
19. National Institute for Japanese Language: Linguistics: Bunruigoihyo. Shuuei Shuppan, In Japanese (1964)
20. Navigli, R.: Word sense disambiguation: a survey. ACM Comput. Surv. **41**(2), 1–69 (2009)
21. Nishio, M., Iwabuchi, E., Mizutani, S.: Iwanami Kokugo Jiten Dai Go Han. Iwanami Publisher, In Japanese (1994)
22. Raina, R., Battle, A., Lee, H., Packer, B., Ng, A.Y.: Self-taught learning: transfer learning from unlabeled data. In: ICML 2007, Proceedings of the 24th International Conference on Machine Learning, pp. 759–766 (2007)
23. Shinnou, H., Onodera, Y., Sasaki, M., Komiya, K.: Active learning to remove source instances for domain adaptation for word sense disambiguation. In: Proceedings of PACLING-2015, pp. 224–231 (2015)
24. Shinnou, H., Sasaki, M., Komiya, K.: Learning under covariate shift for domain adaptation for word sense disambiguation. In: Proceedings of PACLIC-29, pp. 215–223 (2015)
25. Tur, G.: Co-adaptation: Adaptive co-training for semi-supervised learning. In: Proceedings of the IEEE International Conference on Acoustics, Speech and Signal Processing, ICASSP 2009, pp. 3721–3724 (2009)
26. Vapnik, V., Chapelle, O.: Bounds on error expectation for support vector machines. Neural Comput. **12**(9), 2013–2036 (2000)
27. Zhong, E., Fan, W., Peng, J., Zhang, K., Ren, J., Turaga, D., Verscheure, O.: Cross domain distribution adaptation via kernel mapping. In: Proceedings of the 15th ACM SIGKDD International Conference on Knowledge discovery and data mining, pp. 1027–1036 (2009)

ALLIANCE-ROS: A Software Architecture on ROS for Fault-Tolerant Cooperative Multi-robot Systems

Minglong Li, Zhongxuan Cai, Xiaodong Yi$^{(\boxtimes)}$, Zhiyuan Wang, Yanzhen Wang, Yongjun Zhang, and Xuejun Yang

State Key Laboratory of High Performance Computing,
College of Computer, National University of Defense Technology,
Changsha 410073, People's Republic of China
yixiaodong@nudt.edu.cn
https://micros.trustie.net

Abstract. Programming multi-robot systems is a complicated and time-consuming work, due to two challenges, i.e., the distributed multi-robot cooperation and the robot software reusability. ALLIANCE [1] is a fully distributed, fault-tolerant and behavior-based model. ROS (Robot Operating System) provides abundant robot software modules. In this paper, by combining both, we propose a software architecture named ALLIANCE-ROS for developing fault-tolerant cooperative multi-robot systems with a lot of software resources available. We encapsulate the ROS mechanisms and Python libraries to construct the basic function units of the ALLIANCE model. One may inherit them to construct the ALLIANCE-model-application with all ROS algorithms, modules and resources available. This work is demonstrated by an experiment of multi-robot patrol in both the simulated and the real environments.

Keywords: Multi-robot · Fault-tolerant · Cooperative · ALLIANCE-ROS

1 Introduction

Multi-robot systems are becoming one of the most important robotics research areas. Programming robots is complicated and time-consuming. Working with multiple and distributed robot systems is further more complicated [2].

Researchers were committed to the multi-task allocation algorithm of a robot team for a long time [3–5]. In multi-agent theory, multiple agents can be organized together to build an intelligent system (reactive [6], deliberative reasoning [7], layered/hybrid architectures [8] etc.). Researchers have applied them to multi-robot systems since an early time [9–11]. But the attempts are conceptual and do not suit for real robots.

Reactive architectures have a major highlight of the overall behavior emergence [12]. Subsumption model [13] is an extremely classical reactive architecture. See Fig. 1 for an example with detailed explanation. ALLIANCE model [1]

© Springer International Publishing Switzerland 2016
R. Booth and M.-L. Zhang (Eds.): PRICAI 2016, LNAI 9810, pp. 233–242, 2016.
DOI: 10.1007/978-3-319-42911-3_19

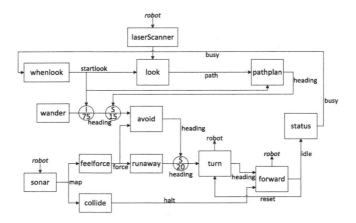

Fig. 1. The three-layer control architecture of an autonomous wander robot based on the subsumption model. The rectangles are the function modules which are loosely coupled to form layers. The 'I' and 'S' circles are used for message inhibiting and suppressing. The upper layers can either suppress or subsume some function units of the lower layers, and the lower layers can send feedback to the upper layers. This increasingly leads to complex intelligence of a robot.

is a natural extension of the subsumption model to multiple robots. This model is a fully distributed, behavior-based architecture that incorporates the use of mathematically-modeled motivations within each robot to achieve fault-tolerant adaptive action selection. See Fig. 2 for an example and its details. For practical multi-robot applications, however, there still needs a developing platform with algorithms, modules and other resources available.

In recent years, some robot software architectures (ROS [14], YARP [15], OPRoS [16], OpenRTM-Aist [17], MRS [18] etc.) were proposed to achieve the modularity and reusability. ROS is becoming the de facto standard. Previously, Li [19] introduced a template based on ROS for implementing the subsumption model, such that one may develop a control system for a single mobile robot by leveraging ROS-provided software resources. COROS [20] is an illuminating work, which presents a high-level framework for cooperative multi-robot systems. It is a generic software architecture implemented on ROS, but COROS provides no model for developing fault-tolerant cooperative multi-robot systems.

In this paper, we designed a software architecture named ALLIANCE-ROS for developing fault-tolerant and cooperative multi-robot applications, which obeys the ALLIANCE model and is implemented on ROS. We encapsulate the ROS mechanisms (ROS node, ROS communication) and Python libraries (Python multiprocessing) to construct the basic function units needed by ALLIANCE. By using ALLIANCE-ROS APIs, one may benefit from: (1) behavior description and coordination templates; (2) fault-tolerant methods for multi-robot task allocation; (3) abundant ROS modules; and (4) Python programming frameworks.

Finally, a multi-robot security patrol application is constructed by using ALLIANCE-ROS. Three robots patrol in a house with three rooms, one for each room. When an intruder irrupts the house, someone of the three should detect and follows it immediately and alarm to notice the owner. At the same time, the other two robots should take the responsibility of the above one to patrol its room to keep the whole house secure. The whole system works well in both the simulated and real environment.

This paper is organized as follows. Section 2 is an overview of ALLIANCE-ROS framework and its APIs. Section 3 illustrates the internals of ALLIANCE-ROS. By presenting an instance of a multi-robot patrol application. Section 4 exemplifies the processes using ALLIANCE-ROS APIs to build a fault-tolerant cooperative multi-robot system. Section 5 is the experimental results. Finally, Sect. 6 concludes the paper.

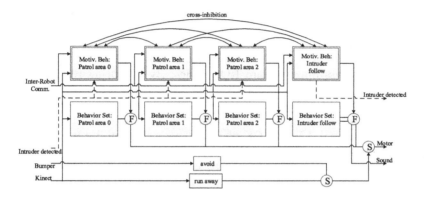

Fig. 2. The ALLIANCE model graph of the multi-robot patrol application. Each single robot should obey a same ALLIANCE model. The model is composed of several behavior sets and behavior layers, which are both implemented by following the subsumption model. The behavior sets correspond to some high-level task-achieving functions. The module named 'motivational behavior' can activate the behavior sets to select a task adaptively. The 'F' circles connect the output of each motivational behavior with the output of its corresponding behavior set, indicating that a motivational behavior either allows all or none of the outputs of its behavior set to pass through to the robot's actuators. Note that the motivational behavior needs to receive the sensor information. Besides, the output of some behavior sets can also be used as the sensor information, which is called 'virtual sensors'.

2   ALLIANCE-ROS Framework

2.1   Framework Overview

ALLIANCE-ROS framework provides abstract classes corresponding to ALLIANCE model elements (see Fig. 3) based on ROS. ROS relies on the concept of computational graph, which represents the network of ROS processes.

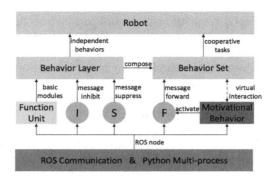

Fig. 3. ALLIANCE-ROS framework.

Each ROS process is a ROS node. ROS nodes communicate with each other by using the ROS-provided message passing mechanisms, for example the 'publish/subscribe' method.

At the bottom layer of Fig. 3, we encapsulate the ROS communication and Python multi-process libraries into ROS nodes. By inheriting ROS nodes, one may implement Function Units and connect them (i.e. establish message passing channels among them) to construct Behavior Layers quite conveniently. To do that, one could focus only on application logic, leaving others to the framework. Besides, several ALLIANCE elements including Inhibitors, Suppressors and Forwarders (i.e. the 'I', 'S' and 'F' circles) are all available, which can be directly put between Behavior Layers to construct Behavior Sets. A Motivational Behavior template is also available for developers. Once again, one could focus only on the parameters of the Motivational Behavior, others including the following three aspects have already been handled by the framework: (1) the fault-tolerant adaptive multi-task selection algorithm that incorporates the use of mathematically-modeled motivations; (2) the interconnected communication among all motivational behaviors in the team robots by using ROS broadcasting mechanism; (3) By specifying the forwarder name as a parameter, the interaction (imaginary lines in Fig. 3) with the behavior set can be built automatically. Finally, Behavior Layers and Behavior Sets can aggregate the Robots.

To sum up, in ALLIANCE-ROS, the Function Unit, the Behavior Layer, the Behavior Set and the Robot construct the application logic. The Inhibitor, the Suppressor, the Forwarder and the Motivational Behavior are encapsulated as APIs, which are presented in next part. The ROS communication and Python multi-process mechanism provide the support of distributed environment.

2.2 The ALLIANCE-ROS APIs

The ALLIANCE-ROS APIs are presented below:

Inhibitor(name, topicIn, msgType1, topicOut, inhibitingTopic, msgType2)
Suppressor(name, topicIn, msgType2, topicOut, topicSupress, msgType2)

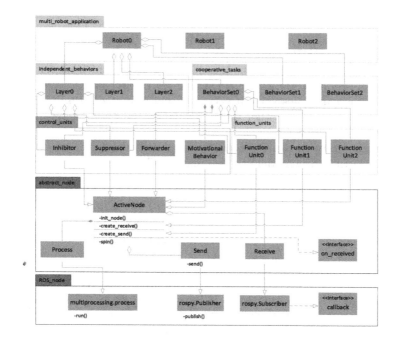

Fig. 4. The UML class diagram of the ALLIANCE-ROS architecture.

Forwarder(name, topicIn1, msgType1, topicOut1, topicIn2, msgType2, topicOut2)

MotivationalBehavior(name, robotID, behaviorSetID, forwarderName)

The interaction of the ALLIANCE elements and function modules is indicated by the 'topic' parameters.

3 ALLIANCE-ROS Implementation

Figure 4 illustrates the UML Class Diagram of ALLIANCE-ROS. The bottom 'ROS_node' layer encapsulates the Python multi-process and the ROS message passing mechanisms. The function units, inhibitors, supressors and forwarders of ALLIANCE model are all fundamentally ROS nodes. By using the Python 'multiprocessing.process' class and rewrite the *run()* method, new processes could be created for constructing ROS nodes. The 'rospy.Publisher' and 'rospy.Subscriber' classes provide an asynchronous message passing mechanism. The *publish()* method in the 'Publisher' class can be used to send messages to other ROS nodes. The 'callback' interface is responsible for handling the incoming messages.

Furthermore, the 'abstract_node' layer encapsulates all things above into an 'ActiveNode', which is the basis of all ALLIANCE elements. To do that, several classes, i.e. 'Process', 'Send' and 'Receive', are constructed first. In 'ActiveN-ode' class, the *init_node()* method is responsible for the node initialization, a 'node_name' parameter should be passed into it. The *create_receive()* method creates a listener for monitoring the incoming messages, which has an interface called *on_received()* in it. Programmers should rewrite the *on_received()* method to handle the incoming messages. If necessary, call the *create_send()* method to send the handled results to other units. The *spin()* method keeps the process running until this node is stopped.

By inheriting the 'ActiveNode' class, different control elements and function units of ALLIANCE model can be created. The message-control and task-coordination algorithms of the ALLIANCE elements are encapsulated, which can be used as ALLIANCE-ROS APIs.

'MotivationalBehavior' is the core class of ALLIANCE-ROS, which encapsulates the adaptive multi-task selection algorithm. And a type of ROS message named 'HeartBeat' is created for it indicating the robot ID, the behavior set ID and whether some behavior set is active. By each motivational behavior broadcasting and receiving this message, the interconnection among all the motivational behaviors in the robot team is built. Besides, the connection with the attached forwarder is encapsulated in it by using the ROS 'publish/subscribe' mechanism.

By inheriting and aggregating the control_unit and the function_unit classes, one may construct independent behavior to implement the lower layers and cooperative tasks to implement the upper behavior sets of the ALLIANCE model, respectively.

Finally, the independent behavior classes and cooperative task classes can be aggregated together to implement a multi-robot application. This layer contains different 'Robot' classes, each of them has a *start()* method responsible for starting the robot.

The different abstract layers are depicted and interpreted as logic containers. At the file system level, we use Python packages to organize them which are represented by separate folders. Physically, classes are embedded in their respective Python modules in different folders.

4   A Multi-robot Patrol Application Using ALLIANCE-ROS

In what follows, we present a multi-robot patrol application, and exemplify how to use ALLIANCE-ROS APIs to build a multi-robot system.

4.1   Application Scenario Description

We assume a scenario that there are three rooms in a house and each robot patrols a room according to several pre-selected goals. If another intruder robot moves into the house, one of the three patrol robots can detect it, follow it and alarm. But at the same time, the room the robot patrols before is empty. So the other two patrol robots will compensate for this fault and patrol different rooms alternately.

4.2 The Design Phase

As shown in Fig. 2, we draw an ALLIANCE model graph of the multi-robot patrol application. Firstly, we determine the independent behavior of a single robot. A robot should have the self-protection ability. When another moving object is approaching, it should sense the object and run away. This is the 'layer 0' behavior. Besides, a robot should have the obstacle-avoid ability. When it is blocked by an obstacle, it should avoid it proactively. This is the 'layer 1' behavior. Secondly, we determine the cooperative tasks (behavior sets) of each robot, including 'patrol area 0', 'patrol area 1', 'patrol area 2' and 'intruder follow' behavior. Thirdly, we determine the function units, the behavior layers and the behavior sets. For example, to construct the 'intruder follow' behavior set, we need three function units including 'detect', 'follow' and 'alarm'. And then, we insert the ALLIANCE control elements (inhibitor, suppressor, forwarder and motivational behavior) in the proper positions of the ALLIANCE model. Finally, we draw an ALLIANCE model graph as shown in Fig. 2.

4.3 The Implementation Phase

Firstly, we create messages that the function units need and recompile the files. For example, the 'goal' message is used by the 'patrol' unit to communicate with the 'navigation' unit. Secondly, we create the function unit classes including 'patrol', 'navigation', 'detect', 'follow', 'avoid', 'run away' and 'alarm'. We overwrite the *on_received()* method of the function units to handle the received messages. For instance, the 'detect' unit receives the image data to recognize the intruder, so the *on_received()* function should be overwritten to handle the image data. Besides, the 'navigation' unit is created by reusing the ROS-provided software resources (navigation stack). Then we use the ALLIANCE-ROS APIs to leverage the created function units. At the same time, the behavior layer classes and the behavior set classes are created. Next, we aggregate the different behavior layer classes and the behavior set classes to construct the different robot classes. Finally, we instantiate the robot classes and call the *start()* method of the instantiated robot objects in the team robots separately. The cooperative multi-robot patrol system runs as a whole.

5   Experimental Result

In the simulated environment, as shown in Fig. 5, we used keyboard to control the red robot to invade the rooms. Figure 7 presents the multi-task selection of the robot team. At first, the three robots were patrolling the three rooms. At about 125 seconds, the intruder robot invaded and the task selection line of the yellow robot moved to 'intruder follow' in Fig. 7, the yellow robot followed it

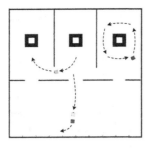

Fig. 5. Experimental result in the simulated environment. The dotted arrows indicate the robots' trajectory. (Color figure online)

Fig. 6. Experimental result in the real environment. The black arrows indicate the robots' trajectory.

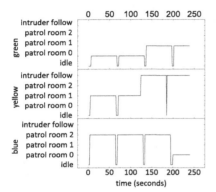

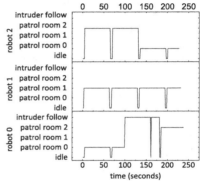

Fig. 7. Task selection in the simulated environment.

Fig. 8. Task selection in the real environment.

immediately with alarming. The green and blue robot patroled the remaining rooms alternately.

In the real environment, as shown in Fig. 6, we controlled an intruder turtlebot with a face photo on it to invade the rooms. Figure 8 presents the multi-task selection of the robot team. At first, the three robots were patrolling the three rooms. At about 100 s, the intruder turtlebot invaded and the task selection line of robot 0 moved to 'intruder follow' task in Fig. 7, robot 0 followed the intruder immediately with alarming. The other two turtlebots would patrol the remaining two rooms. When the intruder left, the robot 0 returned to patrol room 2.

The experimental videos can be downloaded on the micROS website of our research team.

(https://www.trustie.net/organizations/61)

6 Conclusions

This paper combines ALLIANCE model and ROS together to provide developers with a software architecture named ALLIANCE-ROS for building fault-tolerant cooperative multi-robot systems. In future work, we aim to optimize the bottom multi-process and communication mechanism to increase the efficiency. We also try to develop GUI tools to facilitate the system design. Besides, we attempt to add learning mechanism to ALLIANCE-ROS to enhance the self-adaption.

Acknowledgments. This work is supported by Research on Foundations of Major Applications, Research Programs of NUDT under Grant No. ZDYYJCYJ20140601 and the National Science Foundation of China under Grant No. 61221491, 61303185 and 61303068.

References

1. Parker, L.E.: ALLIANCE: an architecture for fault tolerant multirobot cooperation. IEEE Trans. Robot. Autom. **14**(2), 220–240 (1998)
2. Gerkey, B., Vaughan, R.T., Howard, A.: The player/stage project: tools for multi-robot and distributed sensor systems. In: Proceedings of the 11th International Conference on Advanced Robotics, vol. 1, pp. 317–323 (2003)
3. Capitan, J., Spaan, M.T.J., Merino, L., Ollero, A.: Decentralized multi-robot cooperation with auctioned POMDPs. Int. J. Robot. Res. **32**(6), 650–671 (2013)
4. Yuan, M., Ye, Z., Cheng, S., Jiang, Y.: Multi-robot cooperation handling based on immune algorithm in the known environment. In: Proceedings of 2013 Chinese Intelligent Automation Conference, pp. 273–279 (2013)
5. Huang, G., Kaess, M., Leonard, J.: Consistent unscented incremental smoothing for multi-robot cooperative target tracking. Robot. Auton. Syst. **69**, 52–67 (2015)
6. Brooks, R.A.: Intelligence without representation. Artif. Intell. **47**(1), 139–159 (1991)
7. Rao, A.S., Georgeff, M.P., et al.: BDI agents: from theory to practice. In: ICMAS, vol. 95, pp. 312–319 (1995)
8. Connell, J.H.: SSS: a hybrid architecture applied to robot navigation. Robot. Autom. 2719–2724 (1992)
9. Dudek, G., Jenkin, M.R.M., Milios, E., Wilkes, D.: A taxonomy for multi-agent robotics. Auton. Robots. **3**(4), 375–397 (1996)
10. Vail, D., Veloso, M.: Multi-robot dynamic role assignment and coordination through shared potential fields. Multi-Robot Syst., 87–89 (2003)
11. Burgard, W., Moors, M., Fox, D., Simmons, R., Thrun, S.: Collaborative multi-robot exploration. In: Proceedings of IEEE International Conference on Robotics and Automation, ICRA 2000, pp. 476–481 (2000)
12. Jennings, N.R., Sycara, K., Wooldridge, M.: A roadmap of agent research and development. Auton. Agent. Multi-Agent Syst. **1**(1), 7–38 (1998)
13. Brooks, R.A.: A robust layered control system for a mobile robot. Robot. Autom. **2**(1), 14–23 (1986)
14. Quigley, M., Conley, K., Gerkey, B., Faust, J., Foote, T., Leibs, J., Wheeler, R., Ng, A.Y.: ROS: an open-source robot operating system. In: ICRA Workshop on Open Source Software, vol. 3, no. 5 (2009)

15. Metta, G., Fitzpatrick, P., Natale, L.: YARP: yet another robot platform. Int. J. Adv. Robot. Syst. **3**(1), 43–48 (2006)
16. Jang, C., Lee, S., Jung, S.-W., Song, B., Kim, R., Kim, S., Lee, C.-H.: OPRoS: a new component-based robot software platform. ETRI J. **32**(5), 646–656 (2010)
17. Ando, N., Suehiro, T., Kitagaki, K., Kotoku, T., Yoon, W.-K.: RT-middleware: distributed component middleware for RT (robot technology). In: IROS, pp. 3933–3938 (2005)
18. Jackson, J.: Microsoft robotics studio: a technical introduction. Robot. Autom. Mag. **14**(4), 82–87 (2007)
19. Li, M., Yi, X., Wang, Y., Cai, Z., Zhang, Y.: Subsumption model implemented on ROS for mobile robots. In: SysCon (Systems Conference) (2016, to appear)
20. Koubâa, A., Sriti, M.-F., Bennaceur, H., Ammar, A., Javed, Y., Alajlan, M., Al-Elaiwi, N., Tounsi, M., Shakshuki, E.: COROS: a multi-agent software architecture for cooperative and autonomous service robots. In: Koubâa, A., Martínez-de Dios, J.R. (eds.) Cooperative Robots and Sensor Networks 2015. SCI, pp. 3–30. Springer, Heidelberg (2015)

Distributed B-SDLM: Accelerating the Training Convergence of Deep Neural Networks Through Parallelism

Shan Sung Liew[1(✉)], Mohamed Khalil-Hani[1], and Rabia Bakhteri[2]

[1] VeCAD Research Laboratory, Faculty of Electrical Engineering,
Universiti Teknologi Malaysia, 81310 Skudai, Johor, Malaysia
`ssliew2@live.utm.my, khalil@fke.utm.my`
[2] Machine Learning Developer Group, Sightline Innovation,
#202, 435 Ellice Avenue, Winnipeg, MB R3B 1Y6, Canada
`rbakhteri@sightlineinnovation.com`

Abstract. This paper proposes an efficient asynchronous stochastic second order learning algorithm for distributed learning of neural networks (NNs). The proposed algorithm, named distributed bounded stochastic diagonal Levenberg-Marquardt (distributed B-SDLM), is based on the B-SDLM algorithm that converges fast and requires only minimal computational overhead than the stochastic gradient descent (SGD) method. The proposed algorithm is implemented based on the parameter server thread model in the MPICH implementation. Experiments on the MNIST dataset have shown that training using the distributed B-SDLM on a 16-core CPU cluster allows the convolutional neural network (CNN) model to reach the convergence state very fast, with speedups of 6.03× and 12.28× to reach 0.01 training and 0.08 testing loss values, respectively. This also results in significantly less time taken to reach a certain classification accuracy (5.67× and 8.72× faster to reach 99 % training and 98 % testing accuracies on the MNIST dataset, respectively).

Keywords: Deep learning · Distributed machine learning · Stochastic diagonal Levenberg-Marquardt · Convolutional neural network

1 Introduction

Deep learning (DL) is a branch of machine learning (ML) algorithms that learn deeper abstractions of meaningful features by constructing a hierarchical model that perform nonlinear transformations [2]. However, training such complex models is extremely computationally expensive and difficult. This motivates the development of distributed ML techniques that aim to accelerate the training process through parallelism.

The concept of distributed ML is to distribute the training process to multiple processing units or machines in a parallel or distributed computing platform [3]. Distributed versions of the learning algorithms have been developed to

© Springer International Publishing Switzerland 2016
R. Booth and M.-L. Zhang (Eds.): PRICAI 2016, LNAI 9810, pp. 243–250, 2016.
DOI: 10.1007/978-3-319-42911-3_20

train the DL models in the distributed ML environment. Common distributed learning algorithms are usually derived from conventional first order methods (particularly SGD) [3]. However, first order learning algorithms are known to be inefficient because of their slow convergence. Second order algorithms can converge much faster than first order algorithms [6]. Research reported in [1,3] have applied second order learning algorithms for distributed ML in batch learning mode; however, in most cases, they did not outperform the distributed SGD. Some distributed learning algorithms, like those proposed in [3,8] are effective in training deep models, but they are too computationally expensive.

Therefore, this paper aims to improve on the existing distributed learning algorithms by proposing an efficient distributed second order learning algorithm to achieve fast training speedup. The proposed distributed learning algorithm supports asynchronous weight updates for faster performance speedup. The paper is organized as follows. Section 2 presents the proposed algorithm. Section 3 describes the experimental design. Section 4 presents the results and discussions. The final section concludes the work and suggests possible future work.

2   Proposed Distributed Learning Algorithm

2.1   Bounded Stochastic Diagonal Levenberg-Marquardt (B-SDLM)

B-SDLM is an improved and optimized version of the SDLM algorithm. It requires fewer computations than existing second order learning algorithms, yet can achieve fast convergence. Readers may refer to the work in [7] for more details on the B-SDLM algorithm.

An NN training procedure using the B-SDLM algorithm consists of four main stages. Table 1 lists down all the main tasks in the training procedure. The algorithm is represented as a directed cyclic graph (DCG) that consists of 21 nodes (as shown in Fig. 1(a)). The DCG has four feedback loops in total.

2.2   Proposed Algorithm: Distributed B-SDLM

This paper proposes a novel asynchronous stochastic second order learning algorithm for distributed ML of CNN models. The proposed algorithm is derived from the B-SDLM algorithm, and is mapped for implementation on a parallel computing framework by adapting the systematic flow in [4].

Parallelization. A new epoch in the training procedure starts only after the current epoch has completed. Hence, the feedback loop (4) is removed from Fig. 1(a) for simplicity purposes.

Initialization Stage. Weight initialization is performed only once before the training procedure starts, which applies the normalized initialization method. Dataset shuffling is performed at the beginning of every new training epoch.

Table 1. List of tasks in the training procedure with the B-SDLM algorithm.

Stage	Node	Task
Initialization	0	Calculate fan-in and fan-out
	1	Initialize weights
	2	Shuffle dataset
Hessian estimation	3	Fetch data for Hessian estimation
	4	Forward propagation
	5	Second order backward propagation
	6	Accumulate Hessian
	7	Calculate average Hessian
	8	Calculate learning rates
Training	9	Fetch data for gradient computation
	10	Forward propagation
	11	Calculate error
	12	Accumulate misclassification error
	13	First order backward propagation
	14	Update weights
	15	Calculate accuracy
Testing	16	Fetch data for testing
	17	Forward propagation
	18	Calculate error
	19	Accumulate misclassification error
	20	Calculate accuracy

Hessian Estimation Stage. This stage reveals the inherent parallelism of processing all samples concurrently due to the data independency. Therefore, the NN model can be replicated into several model replicas to process the data batches iteratively, and accumulate all results in the end of the stage. This presents a reasonable resource consumption, while still achieving scalable parallelism speedup.

Training Stage. Extracting parallelism in the training stage imposes a greater challenge due to the stochastic learning mode. A more recent learning concept named asynchronous weight update has been proposed to perform training on several machines concurrently [3]. Firstly, nodes 9 to 14 are duplicated to represent multiple model replicas operating on different data. A separate group of tasks can be created to process the gradients (see Table 2). All model replicas fetch the updated weights from a single processing entity (i.e. a parameter server), process the gradients, send the gradients to the parameter server, and repeat the process for all training samples. At the same time, the parameter server receives gradients from any model replicas and updates the weights.

Table 2. Additional tasks for parallel gradient computation with a parameter server.

Node	Operation (Task)
21	Send gradients to parameter server
22	Fetch weights for gradient computation

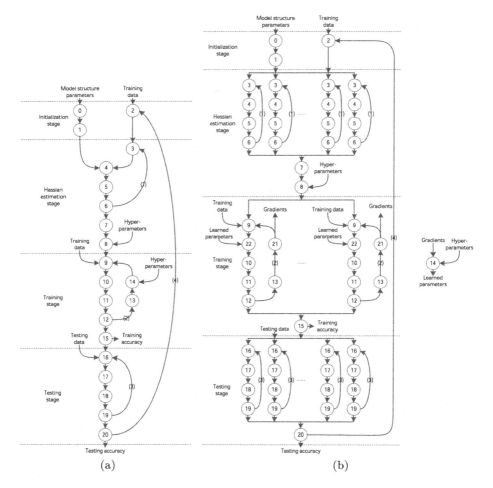

Fig. 1. DCGs of (a) the B-SDLM algorithm, and (b) the proposed distributed B-SDLM algorithm.

Testing Stage. Similar to the Hessian estimation stage, several model replicas are created to evaluate the testing samples in batches. Figure 1(b) depicts the DCG of the proposed distributed training procedure with the B-SDLM learning algorithm based on the parameter server approach.

Scheduling and Synchronization. Figure 2 shows the sequence diagram based on the DCG in Fig. 1(b). Based on the sequence diagram, all the critical sections that require synchronizations are labeled and numbered. The green line represents a synchronization point, the red bounding box is where a writing process occurs, and the blue bounding box denotes a reading process.

The MPI_Barrier() routine is utilized to implement the synchronization points. A writing process requires an exclusive memory window lock by issuing the MPI_Win_lock() with the lock type set as exclusive. MPI_Put() is used to write the data to the remote memory, while new values can be accumulated to the remote memory by calling MPI_Accumulate(). A reading process on the remote memory can be implemented using MPI_Get() and MPI_Win_lock() routines with a shared lock.

3 Experimental Design

The MNIST database consists of 60000 training and 10000 testing images of handwritten digits [5]. The only preprocessing techniques to be applied on the images is z-score normalization.

The CNN model in this work is derived from the baseline LeNet-5 CNN model [5]. Readers may refer to the work in [7] for more details on the CNN model. It is trained based on the training procedure as described in [7] as well, with an additional warm-starting procedure as suggested in [3].

The distributed versions of the learning algorithms are implemented as an MPI program based on the parameter server distributed ML framework using the MPICH library, which is referred to as the MPICH implementation. It runs on a Beowulf CPU cluster consisting of a total of four identical computing platforms with a 4.5 GHz Intel Core i7 4790K CPU each, and are interconnected with a Gigabit network switch.

4 Results and Discussions

4.1 Parallelism Speedup

The performance speedup achieved in the MPICH implementation is compared to the Pthreads implementation that utilizes the Pthreads library (readers may refer to [7] for more details). By referring to Fig. 3(a), the MPICH implementation is slightly slower than the Pthreads implementation. Better fine-grained parallelism can be achieved by implementing Pthreads for the multi-core processors in a single computing platform, and MPICH for the CPU cluster.

This work also studies the impact of the communication overhead to the overall performance of the proposed algorithm. By referring to Fig. 3(b), running in the stochastic learning mode seems unable to benefit from adding more workers, which signifies that it may suffer from the communication bottleneck problem. On the contrary, increasing the mini-batch size during the training alleviates the communication bottleneck issue, where a smooth decrease of the

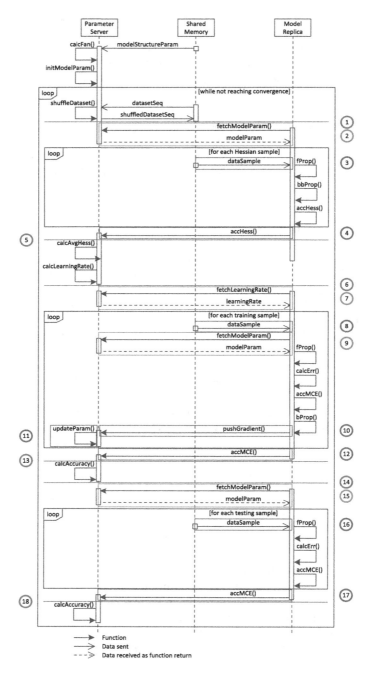

Fig. 2. Critical sections of the sequence diagram for the distributed B-SDLM algorithm. (Color figure online)

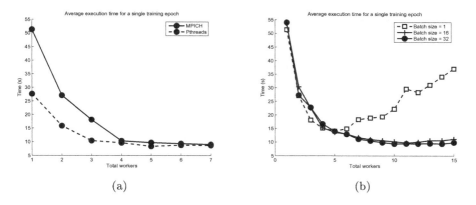

(a)   (b)

Fig. 3. Average execution time per training epoch for (a) Pthreads and MPICH implementations on a single computing platform, and (b) MPICH implementation with different mini-batch sizes.

execution time is observed in Fig. 3(b). Further analysis indicates that adding more workers (i.e. >10) does not produce significant performance speedup. This is most likely due to the limitation of having only a single parameter server to process the incoming gradients from all the workers.

4.2 Impact of Parallelism to Convergence Rate

Figure 4(a) illustrates the time required to reach the convergence state with a certain loss value on the MNIST dataset. Compared to a single worker, training the CNN model using more workers results in a very steep convergence rate, i.e. 6.03× faster to reach 0.01 training loss value, and 12.28× to reach 0.08 testing loss value. This leads to significantly faster training time (5.67× and 8.72× faster to reach 99 % training and 98 % testing accuracies, respectively).

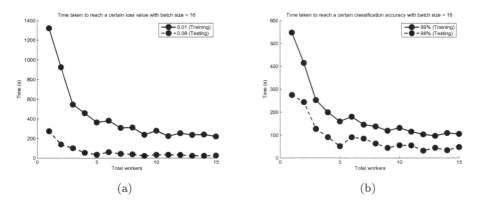

(a)   (b)

Fig. 4. Time taken to reach a certain (a) loss value and (b) classification accuracy on the MNIST dataset when training with batch size = 16 in the MPICH implementation.

5   Conclusion and Future Works

A distributed second order stochastic learning algorithm, i.e. distributed B-SDLM, is proposed to achieve fast training speedup. It is implemented based on the parameter server thread model, and is evaluated on the MPICH implementation. Training using multiple workers with the mini-batch size of 16 allows the CNN model to reach the convergence state very fast, with speedups of 6.03× and 12.28× to reach 0.01 training and 0.08 testing loss values, respectively. This also results in significantly less time taken to reach a certain classification accuracy (5.67× faster to reach 99 % training accuracy, and 8.72× faster to reach 98 % testing accuracy on the MNIST dataset).

Future work of this research involves improving the framework by supporting distributed weight storing and update among multiple parameter servers to reduce the communication bottleneck among the workers and these parameter servers. Parallel implementation of the mini-batch learning mode is viable to achieve the parallelism speedup of processing multiple samples simultaneously, which can be realized by hardware acceleration using GPUs.

Acknowledgements. This work is supported by Universiti Teknologi Malaysia (UTM) and the Ministry of Science, Technology and Innovation of Malaysia (MOSTI) under the Science Fund Grant No. 4S116.

References

1. Agarwal, A., Chapelle, O., Dudík, M., Langford, J.: A reliable effective terascale linear learning system. J. Mach. Learn. Res. **15**(1), 1111–1133 (2014)
2. Chen, Y., Yang, X., Zhong, B., Pan, S., Chen, D., Zhang, H.: CNNTracker: online discriminative object tracking via deep convolutional neural network. Appl. Soft Comput. **38**, 1088–1098 (2016)
3. Dean, J., Corrado, G., Monga, R., Chen, K., Devin, M., Mao, M., Ranzato, M., Senior, A., Tucker, P., Yang, K., Le, Q.V., Ng, A.Y.: Largescale distributed deep networks, pp. 1223–1231. Curran Associates, Inc. (2012)
4. Gebali, F.: Algorithms and Parallel Computing, vol. 84. Wiley, New York (2011)
5. Lecun, Y., Bottou, L., Bengio, Y., Haffner, P.: Gradient-based learning applied to document recognition. Proc. IEEE **86**(11), 2278–2324 (1998)
6. LeCun, Y.A., Bottou, L., Orr, G.B., Müller, K.-R.: Efficient BackProp. In: Montavon, G., Orr, G.B., Müller, K.-R. (eds.) Neural Networks: Tricks of the Trade, 2nd edn. LNCS, vol. 7700, pp. 9–48. Springer, Heidelberg (2012)
7. Liew, S.S., Khalil-Hani, M., Bakhteri, R.: An optimized second order stochastic learning algorithm for neural network training. Neurocomputing **186**, 74–89 (2016)
8. Zeiler, M.D.: ADADELTA: an adaptive learning rate method (2012). arXiv:1212.5701

Single Image Super-Resolution Based on Nonlocal Sparse and Low-Rank Regularization

Chunhong Liu, Faming Fang, Yingying Xu, and Chaomin Shen[✉]

Shanghai Key Laboratory of Multidimensional Information Processing,
East China Normal University, Shanghai 200241, China
cmshen@cs.ecnu.edu.cn

Abstract. Image super resolution (SR) is an active research topic to obtain an high resolution (HR) image from the low resolution (LR) observation. Many results of existing methods may be corrupted by some artifacts. In this paper, we propose an SR reconstruction method for single image based on nonlocal sparse and low-rank regularization. We form a matrix for each patch with its vectorized similar patches to utilize the redundancy of similar patches in natural images. This matrix can be decomposed as the low rank component and sparse part, where the low rank component depictures the similarity and the sparse part depictures the fine differences and outliers. The SR result is achieved by the iterative method and corroborated by experimental results, showing that our method outperforms other prevalent methods.

Keywords: Super resolution · Low-rank · Sparsity · Nonlocal self-similarity

1 Introduction

High resolution (HR) images, compared with low resolution (LR) images, provide more details. However, it is often difficult to acquire HR images due to limitations such as digital imaging systems or imaging environments [1]. Therefore, image super resolution (SR) reconstruction, a technique to reconstruct the HR image from one or several LR images, has become an active topic in many fields such as image processing and computer vision.

For single image SR reconstruction, the simplest type of method is the interpolation-based method, such as bilinear, bicubic and other resampling methods [2,3], in which an interpolation kernel was applied to estimate the missing pixels in the HR image. This tends to produce blur edges with ringing and jagged artifacts [1]. In this regards, many structure priors and nonlocal methods were proposed as regularizers to improve results. One popular prior is the total variation (TV) based method [4], in which the ℓ_1 norm of the image derivative were applied.

© Springer International Publishing Switzerland 2016
R. Booth and M.-L. Zhang (Eds.): PRICAI 2016, LNAI 9810, pp. 251–261, 2016.
DOI: 10.1007/978-3-319-42911-3_21

In recent years, learning based methods have become prevalent for image SR. This type of methods recovers lost high frequency details with the help of the trained LR and HR image patch pairs. For instance, [5] proposed an example based method, and the relationship between the patch pairs is estimated by the Markov random field. Inspired by the locally linear embedding (LLE) methods from manifold learning [6], the assumption that the LR and HR local patch pairs have similar local geometries in two distinct feature spaces was established [7]. Then, the local geometry from the LR patch and its k-neighbors was learned, and mapped to the HR space to reconstruct the SR image. [8,9] proposed the sparse coding based SR methods, which assumed that the sparse representation is consistent between the LR and HR patch pairs.

Later, the nonlocal self-similarity [10], a property for nature images, was introduced into the image processing. [1,11] extended the nonlocal means method to SR reconstruction. In order to obtain more details for SR images, a local target patch was reconstructed by the weighted average from its similar patches. To alleviate the defect that the TV regularization method tends to produce piecewise constant images, nonlocal TV was proposed with the benefit of image nonlocal patch redundancy in [12], where the sparse representation and nonlocal similarity prior were incorporated.

Recently, the low-rank matrix recovery technique is frequently used in image SR problem, e.g., in [13], a structural low-rank regularization method for single image SR has been proposed, and the SR image is reconstructed by fusing multiple reconstructions which are outputs of several simple training methods with small training set. In addition, the nonlocal low-rank regularization of the similar patches matrix and sparse representation of image patches have been also adopted in [14], where the sharp edges and fine structures can be preserved and the visual artifacts also be repressed for image SR.

Motivated by the nonlocal self-similarity of image patches, we integrate the nonlocal low-rank and sparse regularization into SR reconstruction framework. One underlying key observation is that the matrix formed by vectorized similar patches should live in a low dimension space, and the outliers produced in the image reconstruction are sparse with respect to image size [15]. Therefore, similar to [15] for optical flow estimation, we decompose the matrix into the low-rank component and sparse part, and formulate our SR model as the recovery of a low-rank and sparse matrix from missing and corrupted observations. In practice, we first obtain an initial estimate of the SR image by using the bicubic interpolation method on the LR observation. Then, based on the initial reconstructions, we iteratively perform the patch grouping for each patch and solve the SR image by minimizing our nonlocal sparse and low-rank regularization model. By combining the observation model and the regularization term, our model can not only guarantee the similarity between the estimated HR image and the LR observation, but also preserve sharp edges and fine structures and suppress random perturbation and outliers.

2 Proposed Method

In this section, we describe our SR algorithm for single image. Denoting the observed LR image by Y, and the desired HR image by X, the degradation model is:

$$Y = DHX + N, \tag{1}$$

where D and H are the downsampling and the blurring operators, respectively, and N is the additive Gaussian white noise. In order to make the SR problem well-posed, some prior knowledges, denoted by $R(X)$, should be added. Therefore, the SR problem is:

$$\min_X \frac{1}{2}\|DHX - Y\|_F^2 + c_1 R(X), \tag{2}$$

where $\|\cdot\|_F$ denotes the Frobenius norm, and c_1 is the constant to balance the first fidelity and the second regularization terms.

In this paper, we focus on constructing a suitable regularization term.

2.1 The Regularization Term

For each pixel i in X, we define a patch centered at i with size $n \times n$. Vectorizing this patch and denoting the vectorized patch by x_i, its similar vectorized patches from the whole image X can be found by calculating the distance between x_i and other patches. Then, x_i and its $m - 1$ nearest similar patches constitute a matrix $O_i = [x_{i1}, x_{i2}, \cdots, x_{im}] \in \mathbb{R}^{n^2 \times m}$, where $x_{i1} = x_i$ and x_{ij} $(j = 2, \cdots, m)$ is its similar patch.

By definition, the matrix O_i should be low-rank in ideal case. Considering the situation that these similar patches may have small differences and the image may be corrupted by some sparse outliers [16], O_i should have some sparse components.

The above characteristics of O_i, together with the Robust Principal Component Analysis (RPCA) [17], inspires us that O_i can be decomposed as:

$$O_i = L_i + S_i + N_i, \tag{3}$$

where L_i is the low-rank component, S_i the sparse component, and N_i the Gaussian white noise.

The low-rank and the sparse components can be recovered by solving the following objective function:

$$\min_{L_i, S_i} \ \text{rank}(L_i) + c_2\|S_i\|_0, \quad \text{s.t.} \ \ \|O_i - L_i - S_i\|_F^2 \leq \sigma_n^2, \tag{4}$$

where c_2 is a balancing parameter, and σ_n is a preset standard deviation of the Gaussian white noise.

2.2   The Total Energy Formulation

Using the Lagrangian form of (4) and setting it as $R(x)$ in (2), we can rewrite (2) as:

$$E(X, L_i, S_i) = \frac{1}{2}\|DHX - Y\|_F^2$$
$$+ \sum_{i=1}^{P}(\frac{\alpha}{2}\|R_iX - L_i - S_i\|_F^2 + \mu \cdot \text{rank}(L_i) + \lambda\|S_i\|_0), \tag{5}$$

where P is the number of patches in the image, α, μ and λ are positive parameters, and R_i is an operator that extracts O_i for x_i, i.e., $R_iX = O_i$. Note that we allow the overlap between the adjacent patches, thus in the later part of this paper we will introduce a parameter called overlap to measure the overlap phenomenon.

Since the rank-minimization and the ℓ_0 norm problem in (5) are both NP-hard, in order to avoid the difficulty, we use the nuclear norm $\|\cdot\|_*$ to approximate the rank matrix, and ℓ_1 norm to approximate the ℓ_0 norm [15]. So the model (5) can be rewritten as:

$$E(X, L_i, S_i) = \frac{1}{2}\|DHX - Y\|_F^2$$
$$+ \sum_{i=1}^{P}(\frac{\alpha}{2}\|R_iX - L_i - S_i\|_F^2 + \mu\|L_i\|_* + \lambda\|S_i\|_1), \tag{6}$$

which is our final model.

3   Algorithm of the Proposed Method

In our algorithm, the first step is to group the similar patches to form the matrix O_i for each patch in image. Then, we need to solve the low-rank matrix L_i, the sparse matrix S_i, and the SR image X in (6).

These three variables can be solved alternately as follows:

(1) With fixed X and low-rank matrix L_i, the solution of S_i can be formulated as:

$$\hat{S}_i = \arg\min_{S_i} \frac{\alpha}{2}\|R_iX - L_i - S_i\|_F^2 + \lambda\|S_i\|_1, \tag{7}$$

then we can obtain the solution of Eq. (7) as:

$$\hat{S}_i = D_\tau(R_iX - L_i), \tag{8}$$

where D_τ is the soft-thresholding operator with the threshold $\tau = \lambda/\alpha$.

(2) With fixed X and S_i, the solution of L_i can be formulated as:

$$\hat{L}_i = \arg\min_{L_i} \frac{\alpha}{2}\|R_iX - L_i - S_i\|_F^2 + \mu\|L_i\|_*, \tag{9}$$

which can be easily solved via the singular value thresholding algorithm [18]:

$$\hat{L}_i = UD_{\hat{\tau}}(\Sigma)V^T, \quad D_{\hat{\tau}}(\Sigma) = \text{diag}((\sigma_i - \hat{\tau})_+), \tag{10}$$

where $(\sigma_i - \hat{\tau})_+ = \max(0, \sigma_i - \hat{\tau})$, σ_i is the singular value of the matrix $R_i X - S_i$, U and V are the left and right singular vectors of singular values, and $\hat{\tau} = \mu/\alpha$.

(3) For fixed L_i and S_i, we update the SR image X by:

$$\hat{X} = \arg\min_X \frac{1}{2}\|Y - DHX\|_F^2 + \sum_{i=1}^{P} \frac{\alpha}{2}\|R_i X - L_i - S_i\|_F^2. \tag{11}$$

In order to separately solve the fidelity and regularization terms in (11), we introduce an intermediate variable Z which is close to X. Thus Eq. (11) becomes:

$$\hat{X} = \arg\min_X \frac{1}{2}\|Y - DHZ\|_F^2$$
$$+ \sum_{i=1}^{P} \frac{\alpha}{2}\|R_i X - L_i - S_i\|_F^2 + \langle W, X - Z\rangle + \frac{\rho}{2}\|X - Z\|_F^2, \tag{12}$$

where $\langle \cdot, \cdot \rangle$ denotes the inner product, and W is the matrix of the Lagrange multiplier. Then, we apply the alternating direction method of multipliers (ADMM) to Eq. (12), and the three subproblems are solved as:

$$\hat{X} = \arg\min_X \sum_{i=1}^{P} \frac{\alpha}{2}\|R_i X - L_i - S_i\|_F^2 + \langle W, X - Z\rangle + \frac{\rho}{2}\|X - Z\|_F^2, \tag{13}$$

$$\hat{W} = \arg\min_W \langle W, X - Z\rangle, \tag{14}$$

$$\hat{Z} = \arg\min_Z \frac{1}{2}\|Y - DHZ\|_F^2 + \langle W, X - Z\rangle + \frac{\rho}{2}\|X - Z\|_F^2. \tag{15}$$

In summary, the algorithm for optimizing (6) is listed in Algorithm 1. In the algorithm, $\gamma < 1$ is a pre-determined constant, K is the maximum number of iterations, and P is the number of patches in the image.

4   Experimental Results

In this section, we test our proposed method on various test images which are also used in [9,11], and compare it with other prevalent methods.

In order to quantitatively evaluate and compare the results, we use an original image as the "groundtruth" and regard it as the HR image, and the LR image is generated by downsampling the given HR images via bicubic interpolation. Our algorithm is only performed on the illuminance component for color

Algorithm 1: SR based on Low-Rank and Nonlocal Sparsity

Input:
  –The LR image Y;
  –Magnification factor
Output:
  –Super-resolution image X^*
Initialization:
  Initialize the SR image X_0 via bicubic interpolation from Y;
  For $k = 0, 1, 2, \cdots, K - 1$
    1. Inner loop: for $i = 1, 2, \cdots, P$
    For patch x_i, select the first m most similar patches in X_k to form the matrix O_i. Decompose the matric O_i into the low-rank component L_i and sparse component S_i.
      (a) Calculate the sparse matrix S_i via (8);
      (b) Calculate the low-rank matrix L_i via (10);
    2. Reconstruct the SR image X
      (a) Update the image X by solving (13);
      (b) Update the Lagrange multiplier W by (14);
      (c) Update the intermediate variable Z by (15);
      (d) $\alpha = \gamma\alpha$; $X_{k+1} = \hat{X}$.
  End;
    $X^* = X_K$

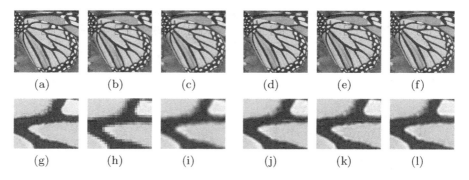

(a)   (b)   (c)   (d)   (e)   (f)

(g)   (h)   (i)   (j)   (k)   (l)

Fig. 1. SR results of the "Buttery" image with magnification factor 2. (a) Original image (groundtruth); (b) LR image; (c) bicubic interpolation; (d) SC method; (e) SCBP method; (f) our method; The images in (g)–(l) are the close-ups of the small red rectangles in the corresponding examples. (Color figure online)

images, as humans are more sensitive to the illuminance change. The initial estimated HR image is obtained by upscaling the LR image to the desired size via bicubic interpolation. The peak signal-to-noise ratio (PSNR) and the structural similarity (SSIM) [19] are used as the objective measures for evaluation.

To demonstrate the effectiveness of our algorithm, we have compared our experiments with the bicubic interpolation method, the sparse coding-based method (SC) [9] and its enhanced version called SCBP [9] which applied the iter-

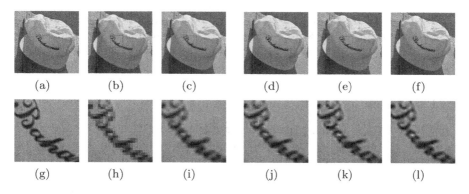

Fig. 2. SR results of the "Hat" image with magnification factor 2. (a) Original image (groundtruth); (b) LR image; (c) bicubic interpolation; (d) SC method; (e) SCBP method; (f) our method; The images in (g)–(l) are the close-ups of the small red rectangles in the corresponding examples. (Color figure online)

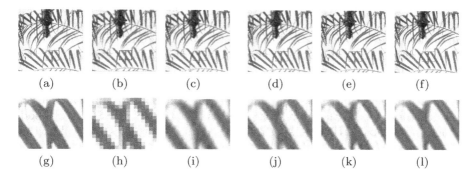

Fig. 3. SR results of the "Leaves" image with magnification factor 3. (a) Original image (groundtruth); (b) LR image; (c) bicubic interpolation; (d) SC method; (e) SCBP method; (f) our method; The images in (g)–(l) are the close-ups of the small red rectangles in the corresponding examples. (Color figure online)

ative back-projection (IBP) [20] algorithm to enhance the sparse coding based method.

4.1 Experimental Settings

In our experiments, we set the number of similar patches $m = 30$, and the patch size as 5×5, so matrix $O_i \in \mathbb{R}^{25 \times 30}$. In order to enhance consistency, we set the overlap as 2 pixels between the adjacent patches. From experience, the convergence results are satisfactory when the iteration number K is set as 30 in Algorithm 1. The other parameters are set to $\alpha = \mu = 1$, $\lambda = 0.45$, $\rho = 0.1$ and $\gamma = 0.9$.

In the experiments of SC and SCBP, 100, 000 patch pairs are randomly sampled for training the coupled dictionaries [21]. The patch sizes are the same as ours, i.e., 5 × 5 with overlap of 2 pixels, and the coupled dictionary sizes are 1024.

4.2   Experimental Results and Analysis

We qualitatively and quantitatively compare our model with the SC and SCBP algorithms. For displaying the results in detail, we choose an area from the images with a red rectangle and illustrated them on the second row in Figs. 1, 2 and 3. The results for the "Butterfly" and "Hat" images with magnifying factor 2 are shown in Figs. 1 and 2, while the results with magnification factor 3 on the image "Leaves" are shown in Fig. 3.

In Figs. 1 and 2, our algorithm can achieve better results by visual inspection. In the image details of Fig. 1, our result has sharper edges and preserves finer structure than those of SC and SCBP. Especially in the pattern edge of the red rectangle in the butterfly, the result of the edge via bicubic interpolation is a little blurry. Although the SC and SCBP methods obtained some high frequency components, they are still a little inferior to ours. Figure 2 shows that some artifacts and outliers around the letters are produced in all cases, but they are effectively suppressed in our case.

Figure 3 is the comparative results with magnification factor 3. When the magnification factor increases, our method can still obtain more satisfactory results than others. Furthermore, Fig. 3 shows that the SC and SCBP methods produce some ghost artifacts along the leaf edges. Nevertheless, our algorithm not only produces fewer artifacts, but also can generate better structure and more details for the SR image.

Table 1. PSNR and SSIM with magnification factor 2

Images	Bicubic	SC	SCBP	Ours
Butterfly	27.4568	29.4886	30.4611	**32.1689**
	0.898	0.9278	0.9373	**0.9531**
Hat	31.7267	33.0280	33.6281	**34.4013**
	0.8768	0.8928	0.9058	**0.9146**
Flower	39.8765	40.2063	40.6245	**41.3082**
	0.9364	0.9365	0.9326	**0.9405**
Lena	32.7947	33.9595	34.5194	**35.5566**
	0.8915	**0.9189**	0.9087	0.9188
Leaves	27.4438	29.3093	30.4306	**32.8808**
	0.9185	0.9468	0.9572	**0.9726**
Parrot	31.3752	32.4906	33.4228	**34.3722**
	0.9278	0.9315	0.9390	**0.9463**

Table 2. PSNR and SSIM with magnification factor 3

Images	Bicubic	SC	SCBP	Ours
Butterfly	24.0528	24.5870	25.5044	**27.1100**
	0.7922	0.8114	0.8292	**0.8853**
Hat	29.1968	29.6534	30.2517	**30.9291**
	0.8003	0.8029	0.8195	**0.8411**
Flower	37.2007	37.0430	37.7643	**38.8482**
	0.8890	0.8786	0.8894	**0.9023**
Lena	30.0986	30.2076	31.0173	**32.0953**
	0.8138	0.8031	0.8253	**0.8545**
Leaves	23.4523	23.4709	24.8520	**26.7961**
	0.7915	0.7997	0.8410	**0.9038**
Parrot	28.0962	28.2647	29.3148	**30.1213**
	0.8715	0.8671	0.8835	**0.9016**

The above observations have also been verified by the quantitative measure in Tables 1 and 2, obtained by each algorithm performed on a set of 6 images. The bicubic interpolation method produces the lowest PSNR and SSIM values. Although the SC and SCBP methods perform better than the bicubic interpolation, their results are still inferior to ours. In the comparison of the Tables 1 and 2, the PSNR and SSIM of all algorithms decrease when the magnification factor increases from 2 to 3, but our algorithm can still obtain better values than the SC and SCBP methods. This indicates that, in terms of the PSNR and SSIM, our reconstructed SR image is closer to the groundtruth than those via other stat-of-the-art algorithms.

Thus, with the above analysis, we conclude that our model is effective for the SR problem.

5   Conclusion

In this paper, inspired by the nonlocal self-similarity of image patches, we have proposed an SR method by incorporating nonlocal sparse and low-rank regularization. Based on the initial estimated SR image, we group similar patches to form a matrix for each patch, and then decompose it into the low-rank and sparse components. Low-rank component is used for achieving the similar structure between similar patches, and sparse component is for repressing outliers. Then the obtained low-rank and sparse matrix are used to improve the SR estimation. We iterate such process until convergence and obtain the optimal SR image. Experimental results on image SR demonstrate that our model can significantly improve the reconstruction quality.

Acknowledgement. This work is supported by the NSFC (No. 61273298), and Science and Technology Commission of Shanghai Municipality (No. 14DZ2260800).

References

1. Zhang, K., Gao, X., Tao, D., Li, X.: Single image super-resolution with non-local means and steering kernel regression. IEEE Trans. Image Process. **21**(11), 4544–4556 (2012)
2. Zhang, L., Wu, X.: An edge-guided image interpolation algorithm via directional filtering and data fusion. IEEE Trans. Image Process. **15**(8), 2226–2238 (2006)
3. Li, M., Nguyen, T.Q.: Markov random field model-based edge-directed image interpolation. IEEE Trans. Image Process. **17**(7), 1121–1128 (2008)
4. Marquina, A., Osher, S.J.: Image super-resolution by TV-regularization and bregman iteration. J. Sci. Comput. **37**(3), 367–382 (2008)
5. Freeman, W.T., Jones, T.R., Pasztor, E.C.: Example-based super-resolution. Comput. Graphics Appl. **22**(2), 56–65 (2002)
6. Roweis, S.T., Saul, L.K.: Nonlinear dimensionality reduction by locally linear embedding. Science **290**(5500), 2323–2326 (2000)
7. Chang, H., Yeung, D., Xiong, Y.: Super-resolution through neighor embedding. In: IEEE Computer Society Conference on Computer Vision and Pattern Recognition, vol. 4, pp. 275–282 (2004)
8. Yang, J., Wright, J., Huang, T., Ma, Y.: Image super-resolution as sparse representation of raw image patches. In: IEEE Computer Society Conference on Computer Vision and Pattern Recognition, pp. 1–8 (2008)
9. Yang, J., Wright, J., Huang, T., Ma, Y.: Image super-resolution via sparse representation. IEEE Trans. Image Process. **19**(11), 2861–2873 (2010)
10. Buades, A., Coll, B., Morel, J.-M.: A non-local algorithm for image denoising. In: IEEE Computer Society Conference on Computer Vision and Pattern Recognition, pp. 60–65 (2005)
11. Dong, W., Zhang, L., Shi, G., Wu, X.: Image deblurring and super-resolution by adaptive sparse domain selection and adaptive regularization. IEEE Trans. Image Process. **20**(7), 1836–1857 (2011)
12. Peyré, G., Bougleux, S., Cohen, L.: Non-local regularization of inverse problems. In: 10th European Conference on Computer Vision, pp. 57–68 (2008)
13. Peng, J., Hon, B.Y., Kong, D.: A structural low rank regularization method for single image super-resolution. Mach. Vis. Appl. **26**(7–8), 991–1005 (2015)
14. Li, Q., Lu, Z., Sun, C., Li, H., Li, W.: Single image super-resolution based on nonlocal similarity and sparse representation. In: 2015 IEEE China Summit and International Conference on Signal and Information Processing (ChinaSIP), pp. 156–160 (2015)
15. Dong, W., Shi, G., Hu, X., Ma, Y.: Nonlocal sparse and low-rank regularization for optical flow estimation. IEEE Trans. Image Process. **23**(10), 4527–4538 (2014)
16. Huang, C., Ding, X., Fang, C., Wen, D.: Robust image restoration via adaptive low-rank approximation and joint kernel regression. IEEE Trans. Image Process. **23**(7), 3085–3098 (2014)
17. Candes, E.J., Li, X., Ma, Y., Wright, J.: Robust principal component analysis? J. ACM **58**(3), 11:1–11:37 (2011)
18. Yang, J., Wright, J., Huang, T., Ma, Y.: A singular value thresholding algorithm for matrix completion. Soc. Ind. Appl. Math. **20**(4), 1956–1982 (2008)

19. Wang, Z., Bovik, A.C., Sheikh, H.R., Simoncelli, E.P.: Image quality assessment: from error visibility to structural similarity. IEEE Trans. Image Process. **13**(4), 600–612 (2004)
20. Irani, M., Peleg, S.: Motion analysis for image enhancement: resolution, occlusion and transparency. J. Vis. Commun. Image Represent. **4**(4), 324–335 (1993)
21. Yang, J., Wang, Z., Lin, Z., Cohen, S., Huang, T.: Coupled dictionary training for image super-resolution. IEEE Trans. Image Process. **21**(8), 3467–3478 (2011)

Generating Covering Arrays
with Pseudo-Boolean Constraint Solving
and Balancing Heuristic

Hai Liu[2], Feifei Ma[1($\boxtimes$)], and Jian Zhang[1]

[1] State Key Laboratory of Computer Science, Institute of Software,
Chinese Academy of Sciences, Beijing, China
{maff,zj}@ios.ac.cn
[2] Beijing Information Science and Technology University, Beijing, China
stuliuhai@gmail.com

Abstract. Covering arrays (CAs) are interesting objects in combinatorics and they also play an important role in software testing. It is a challenging task to generate small CAs automatically and efficiently. In this paper, we propose a new approach which generates a CA column by column. A kind of balancing heuristic is adopted to guide the searching procedure. At each step (column extension), some pseudo Boolean constraints are generated and solved by a PBO solver. A prototype tool is implemented, which turns out to be able to find smaller CAs than other tools, for some cases.

1 Introduction

In complex software systems, faults usually arise from the interaction of a few components/factors. It is reported that up to 90 % of the faults are caused by interactions of at most 3 factors, among which 70 % are caused by pairwise interactions [6]. Combinatorial testing is a useful black-box testing technique to reveal such faults. It usually uses the concept of Covering Arrays (CAs) [13]. Such arrays are important objects in combinatorics. A covering array of strength t is an array with property that each ordered combination of t values from different columns appears at least once in the rows. Each row corresponds to a test case.

Given parameters and coverage criteria, we would like to obtain CAs with the least number of rows, which correspond to the smallest test suites for the System Under Test (SUT). However, generating such CAs is a challenging task, which has been proved to be NP-complete. Therefore, most works on CA generation are based on heuristic search, so as to obtain a solution in reasonable time. The heuristics adopted in the search algorithms are vital to the optimality of the solutions and the efficiency of the algorithms.

This work has been supported in part by the National 973 Program under grant No. 2014CB340701, and by the Youth Innovation Promotion Association, CAS (grant No. 2016104).

R. Booth and M.-L. Zhang (Eds.): PRICAI 2016, LNAI 9810, pp. 262–270, 2016.
DOI: 10.1007/978-3-319-42911-3_22

In general, search algorithms for CAs using problem-specific heuristics fall into two categories: the one-test-at-a-time strategy and its variants, and the In-Parameter-Order (IPO) family [7,8]. The one-test-at-a-time strategy is widely employed in many CA generation algorithms [1]. The basic idea is to generate test cases one by one in a greedy manner until the coverage requirement is met. During the process, each test case covers as many uncovered target combinations as possible, so as to minimize the number of test cases in the test suite. The IPO algorithm expands a CA both horizontally and vertically. It firstly initializes a small sub-array by enumerating all value combinations of the first t parameters, then adds an additional column so that as many target combinations are covered as possible, and then adds rows to cover the remaining uncovered combinations. The horizontal extension stage and the vertical extension stage is repeated alternatively until a CA is completed.

Both of the above strategies are greedy approaches, trying to cover as many target combinations as possible when expanding the array. In this paper, we propose a different heuristic called the balancing heuristic, which only imposes constraints upon each individual column. With this heuristic, our main algorithm generates CAs in the horizontal way: Each time it derives pseudo-Boolean constraints for the next column, and then employs a pseudo-Boolean constraint solver to obtain the solution of the new column. Simple as it is, our algorithm demonstrates much advantage over the state-of-the-art CA generators for instances of strength 2. It can generate smaller CAs, while the execution time is comparable to that of other solvers.

2   Preliminaries

Definition 1. *A **covering array** $CA(N, d_1 d_2 \cdots d_k, t)$ of strength t is an $N \times k$ array having the following two properties:*
1. There are exactly d_i symbols in each column i ($1 \leq i \leq k$);
2. In every $N \times t$ sub-array, each ordered combination of symbols from the t columns appears at least once.

Each column of the CA corresponds to a factor or parameter p_i ($1 \leq i \leq k$), and d_i is called the *level* of p_i.

The parameters in a CA can be combined when their levels are the same. If every parameter has the same level, the array can be denoted by $CA(N, d^k, t)$.

$$
\begin{array}{cccc}
0 & 0 & 0 & 0 \\
0 & 1 & 1 & 1 \\
1 & 0 & 1 & 1 \\
1 & 1 & 0 & 1 \\
1 & 1 & 1 & 0
\end{array}
\qquad\qquad
\begin{array}{ccc}
0 & 1 & 1 \\
0 & 0 & 0 \\
1 & 1 & 0 \\
1 & 0 & 1 \\
2 & 1 & 1 \\
2 & 0 & 0
\end{array}
$$

Fig. 1. $CA(5, 2^4, 2)$ **Fig. 2.** $MCA(6, 3^1.2^2, 2)$

Otherwise, it is called a *mixed level covering array* (MCA) [2]. For a quick example, Fig. 1 shows an instance of $CA(5, 2^4, 2)$. In any two columns, each ordered pair of symbols occurs at least once. Similarly, an instance of $CA(6, 3^1 2^2, 2)$ is given in Fig. 2.

3   Encoding Column Restrictions as Pseudo-Boolean Constraints

As mentioned before, our algorithm constructs a CA column by column. As an initial step, the algorithm randomly generates $t - 1$ columns. Assume that we have constructed m columns ($m \geq t - 1$), we now discuss how to generate pseudo-Boolean constraints for column $m + 1$.

A pseudo-Boolean (PB) constraint is an equation or inequality between polynomials in 0-1 variables. A linear PB clause has the form: $\sum c_i \cdot L_i \sim d$, where $c_i, d \in \mathbb{Z}$, $\sim \in \{=, <, \leq, >, \geq\}$, and L_is are literals.

Let us denote the variable at the ith entry of column $m + 1$ by V_i. For each entry i and each value v ($0 \leq v < d_{m+1}$), we introduce a Boolean variable $P_{i,v}$ such that $P_{i,v} \equiv (V_i = v)$.

The first class of constraints guarantees that each entry can only take one value. For the ith entry of column $m + 1$, we have:

$$\sum_{0 \leq v < d_{m+1}} P_{i,v} = 1$$

Now consider the covering property of the array. Given an array A, suppose we extract $t - 1$ columns (denoted by $C_{i_1}, C_{i_2}, \ldots, C_{i_{t-1}}$) from A and denote the sub-array by A_s. The p-set corresponding to a row vector v is the set of row indices i such that i_{th} row of A_s is v. Apparently, there are $d_{i_1} \times \cdots \times d_{i_{t-1}}$ mutually exclusive p-sets induced by the sub-array A_s.

Example 1. Figure 3 illustrates the p-sets from the $CA(5, 2^4, 2)$ in Fig. 1. Each column induces 2 p-sets. More specifically, the p-set $\{1, 2\}$ is induced by the sub-array formed by column 1 because row 1 and row 2 in the sub-array share the same row vector $\langle 0 \rangle$.

Column	P-set
1	{1,2} {3,4,5}
2	{1,3} {2,4,5}
3	{1,4} {2,3,5}
4	{1,5} {2,3,4}

Fig. 3. p-sets from CA(5,2^4,2)

```
0 0 0 1 0
0 0 0 0 1
1 0 0 0 0
0 1 0 0 0
0 0 1 0 0
1 1 0 1 1
1 0 1 1 1
0 1 1 1 1
1 1 1 1 0
1 1 1 0 1
```

Fig. 4. Balanced CA(10, 2^5, 3)

Theorem 1. *An $N \times (m+1)$ array is a $CA(N, d_1 \ldots d_{m+1}, t)$ **iff** the array formed by its first m columns is a $CA(N, d_1 \ldots d_m, t)$, and for each $N \times (t-1)$ sub-array of the first m columns, the d_{m+1} symbols in column $m+1$ all appear within the rows indexed by each p-set induced by the sub-array.*

Proof. For the **if** way of the implication: Suppose an $N \times (m+1)$ array M satisfies the latter condition and denote the array formed by the first m columns of M by M'. For any t columns extracted from M, we are to show all combinations of symbols from these columns are covered in rows. If column $m+1$ isn't chosen, the conclusion obviously holds since all the t columns are contained in M' and M' is a $CA(N, d_1 \ldots d_m, t)$. Otherwise, denote the other $t-1$ columns from M' by $C_{i_1}, \ldots, C_{i_{t-1}}$, and label an arbitrary combination of symbols from these columns and column $m+1$ as $\langle v_{i_1}, \ldots, v_{i_{t-1}}, v_{m+1} \rangle$. Since M' is also a $CA(N, d_1 \ldots d_m, t-1)$, it covers $\langle v_{i_1}, \ldots, v_{i_{t-1}} \rangle$. Among all the p-sets induced by the sub-array formed by the columns $C_{i_1}, \ldots, C_{i_{t-1}}$, we denote the one corresponding to the vector $\langle v_{i_1}, \ldots, v_{i_{t-1}} \rangle$ by T. From the presumption we know that v_{m+1} of column $m+1$ appears within the rows indexed by T. Hence the vector $\langle v_{i_1}, \ldots, v_{i_{t-1}}, v_{m+1} \rangle$ is covered in M. By definition, M is a $CA(N, d_1 \ldots d_{m+1}, t)$.

For the **only if** way of implication: Suppose the array M is a CA $(N, d_1 \ldots d_{m+1}, t)$. Let the array formed by the first m columns be M'. Obviously M' is a CA $(N, d_1 \ldots d_m, t)$. Now suppose there exists a p-set T in the rows of which the symbol from column $m+1$, namely v_{m+1} does not appear. We denote the symbol combination corresponding to T by $\langle v_{i_1}, \ldots, v_{i_{t-1}} \rangle$. Then the symbol combination $\langle v_{i_1}, \ldots, v_{i_{t-1}}, v_{m+1} \rangle$ is not covered in M, contradicting the presumption that M is a CA of strength t. $\square$

According to Theorem 1, firstly we should calculate all p-sets induced by all $N \times (t-1)$ sub-arrays from the first m columns, then the constraints for the covering property of column $m+1$ can be obtained directly. In practice, we may add the p-sets incrementally to a stack as the columns expand, so that re-computation can be avoided.

Once all the p-sets are computed, it's easy to translate the constraints to pseudo-Boolean constraints. For an arbitrary p-set T, each of the d_{m+1} symbols from column $m+1$ should appear at least once in the rows indexed by T. The constraint of a p-set is naturally represented by the following pseudo-Boolean clauses:

$$\bigwedge_{0 \leq v < d_{m+1}} \sum_{i \in T} P_{i,v} \geq 1$$

4   The Balancing Heuristic

By observing many CA instances, we find that the CAs of the optimal sizes are likely to have the following property:

Definition 2. *A CA is **balanced** if in each column, all symbols occur nearly equally often. Formally, for a $CA(N, d_1 d_2 \cdots d_k, t)$, denote the number of occurrences of symbol v in column j by $O_j(v)$. If for any pair $\langle v, j \rangle$, $\left\lfloor \frac{N}{d_j} \right\rfloor \leq O_j(v) \leq \left\lceil \frac{N}{d_j} \right\rceil$, then the $CA(N, d_1 d_2 \cdots d_k, t)$ is balanced.*

The balancing property indicates that the difference of the numbers of occurrences of any two symbols within the same column is no larger than 1. For example, the $CA(10, 2^5, 3)$ in Fig. 4 is balanced.

Our algorithm employs the balancing heuristic, searching for balanced CAs so as to enhance the probability of finding optimal CAs. Hence in addition to the constraints encoding the CA properties, there are constraints encoding the balancing property. For each value v ($1 \leq v < d_{m+1}$), we have:

$$\left\lfloor \frac{N}{d_{m+1}} \right\rfloor \leq \sum_{1 \leq i \leq N} P_{i,v} \leq \left\lceil \frac{N}{d_{m+1}} \right\rceil$$

Currently we are unable to prove the rationality of the balancing heuristic, nevertheless we can provide some explanation which may shed light on this issue. Unlike the greedy strategies, which aim to locally optimize the current CA solution, the balancing heuristic is concerned with the expansibility of the current solution. The symbol distribution in column $m + 1$ will influence the column that follows (if any). Take a $CA(N, d^k, 2)$ for example. According to Theorem 1, the d symbols in column $m + 2$ must all appear within each p-set induced by column $m + 1$. Intuitively the smallest p-set is the most restrictive one. Since the sum of the sizes of all these p-sets equals to N, it is better if all these p-sets are nearly of the same size, which means column $m + 1$ is balanced.

5   Implementation and Experimental Evaluation

The column generation approach requires that the size of a CA is specified beforehand. To overcome this limitation, we employ the binary search strategy to determine the optimal size of the CA. Our tool CAB (CA searcher with Balancing heuristic) was implemented in the C++ programming language and integrated with the pseudo-Boolean constraint Solver clasp v3.1.1 [16].

For comparison, we selected three state-of-the-art CA generators, PICT [3], ACTS [9] and CASA [4]. PICT is a widely used test case generator developed by Jacek Czerwonka at Microsoft Corporation. The core generation algorithm of PICT adopts the one-test-at-a-time strategy. ACTS is a powerful test generation tool which implements the IPO algorithms. CASA is a CA generator based on Simulated Annealing. We compared CAB with these tools on various benchmarks. The experiments were conducted on an Intel 1.7 GHZ Core Duo i5-4210U PC with virtual Linux2.6 OS. Timeout (TO) means more than one hour.

The experimental results on pure-level CAs of strength 2 with level d ranging from 2 to 4 are illustrated in Tables 1, 2 and 3. Given the number of parameters (denoted by k), the number of rows (denoted by N) generated by each tool and the running time (denoted by T, measured in milliseconds) are listed. It can be seen that in most cases, CAB produces the best results, covering the pairwise interactions of k parameters with the least number of rows. The running times of CAB are also reasonable. In particular, for strength $t = 2$ and level $d = 2$, CAB is able to obtain the best results in dramatically shorter time. The results

Table 1. CA(N,2^k,2)

K	CAB		CASA		PICT		ACTS	
	N	T	N	T	N	T	N	T
3	4	2	4	80	4	84	4	72
4	5	2	5	50	5	25	6	0
10	6	2	6	140	9	20	10	4
15	7	2	7	420	10	37	10	4
35	8	2	9	1010	12	29	14	8
56	9	2	10	2720	14	28	14	8
126	10	2	11	19130	16	185	16	28
210	11	2	12	50670	18	376	18	8
462	12	2	14	303030	20	872	20	624
792	13	4	15	1785650	22	2816	22	2965
1716	14	5	-	TO	24	16221	24	39564
3003	15	9	-	TO	26	59832	26	190143
6435	16	15	-	TO	28	436264	28	1391365
11440	17	37	-	TO	-	crash	-	crach
24310	18	71	-	TO	-	crash	-	crash
43758	19	124	-	TO	-	crash	-	crash
92378	20	251	-	TO	-	crash	-	crash
167960	21	474	-	TO	-	crash	-	crash
352716	22	1036	-	TO	-	crash	-	crash
646646	23	2248	-	TO	-	crash	-	crash
1352078	24	4457	-	TO	-	crash	-	crash
2496144	25	9293	-	TO	-	crash	-	crash
5200300	26	23993	-	TO	-	crash	-	crash
9657700	27	45162	-	TO	-	crash	-	crash

Table 2. CA(N,3^k,2)

K	CAB		CASA		PICT		ACTS	
	N	T	N	T	N	T	N	T
4	9	140	9	29	13	16	9	0
5	12	90	11	100	12	21	15	4
6	13	70	12	360	14	16	15	0
9	14	120	15	340	17	16	15	4
12	15	150	16	1070	19	16	19	0
18	17	230	17	12350	22	29	21	0
24	18	350	19	5080	23	17	24	0
30	19	1190	21	3600	25	21	25	4
39	20	1560	21	38540	27	23	26	4
52	21	2830	22	111140	29	36	28	4
64	22	4600	23	148610	30	40	29	12
83	23	10620	25	108110	33	109	31	16
117	24	20330	26	1376720	34	112	33	32
137	25	39280	27	855600	35	152	34	48
170	26	84420	28	2733730	37	268	36	84
248	27	167000	-	TO	39	537	37	184
289	28	513110	-	TO	39	800	39	264
361	29	996630	-	TO	40	1201	40	456
476	30	919390	-	TO	42	2336	42	888

Table 3. CA(N,4^k,2)

K	CAB		CASA		PICT		ACTS	
	N	T	N	T	N	T	N	T
3	16	160	16	140	17	17	16	0
5	19	80	16	190	22	41	24	0
6	21	160	19	1980	25	36	24	0
7	23	260	22	1660	27	24	32	0
8	24	380	24	4330	28	27	32	0
9	25	760	27	820	30	23	32	0
10	26	6350	26	5290	31	28	33	0
12	27	5360	28	2720	34	25	33	0
15	28	13830	29	41270	35	28	36	0
16	29	10630	29	90910	36	35	38	0
18	30	151950	30	105930	36	48	41	4
21	31	169020	31	457320	39	57	41	4
26	32	313420	33	492560	42	68	43	4

Table 4. CA(N,2^k,3)

K	CAB		CASA		PICT		ACTS	
	N	T	N	T	N	T	N	T
4	8	40	8	140	8	20	8	0
5	10	50	10	90	13	28	12	0
8	12	90	12	450	16	24	18	0
10	16	120	12	1700	18	28	20	0
12	18	310	16	2410	19	28	22	0
14	22	600	18	6890	22	36	25	4
15	22	540	19	1890	23	29	26	4
17	26	940	20	5640	23	20	26	0
20	26	1210	21	24040	26	32	27	4
24	28	1860	23	150800	29	44	29	8

obtained by CASA are quite close to that of CAB, but it often takes much more running times. ACTS and PICT are the fastest solvers in general. However, the CAs found by them are usually larger.

Table 5. Mixed Covering Array

Model	Description	CAB		CASA		PICT		ACTS	
		N	T	N	T	N	T	N	T
Apache	$CA(2^{158}3^84^45^16^1;2)$	**30**	3490	33	44230	32	151	33	172
Bugzilla	$CA(2^{49}3^14^2;2)$	**16**	360	**16**	2660	17	24	18	4
gcc	$CA(2^{189}3^{10};2)$	**15**	9780	17	71710	20	133	20	76
SpinS	$CA(2^{13}4^5;2)$	19	320	**16**	2480	23	13	24	0
SpinV	$CA(2^{42}3^24^{11};2)$	**27**	6460	28	7010	32	28	33	8
Banking1	$CA(3^44^1;2)$	13	70	**12**	130	16	17	15	0
Banking2	$CA(2^{14}4^1;2)$	**10**	180	**10**	210	12	12	**10**	0
CommProtocol	$CA(2^{10}7^1;2)$	**14**	40	**14**	320	16	15	**14**	0
Concurrency	$CA(2^5;2)$	**6**	0	**6**	60	7	11	**6**	0
Healthcare1	$CA(2^63^25^16^1;2)$	**30**	40	**30**	220	**30**	16	**30**	0
Healthcare2	$CA(2^53^64^1;2)$	**15**	140	**15**	210	18	11	**15**	0
Healthcare3	$CA(2^{16}3^64^55^16^1;2)$	**30**	250	32	1590	35	19	32	4
Healthcare4	$CA(2^{13}3^{12}4^65^26^17^1;2)$	**42**	320	**42**	9150	47	27	44	8
Insurance	$CA(2^63^15^16^211^113^117^131^1;2)$	**527**	26420	**527**	215930	**527**	92	**527**	4
NetworkMgmt	$CA(2^24^15^310^211^1;2)$	**110**	28610	**110**	98160	118	28	**110**	0
ProcessorComm1	$CA(2^33^64^6;2)$	**21**	270	22	5390	26	16	26	4
ProcessorComm2	$CA(2^33^{12}4^85^2;2)$	**28**	2690	29	32810	36	21	37	4
Services	$CA(2^33^45^28^210^2;2)$	**100**	350	102	5810	101	31	102	4
Storage1	$CA(2^13^14^15^1;2)$	**20**	10	**20**	100	**20**	28	**20**	71
Storage2	$CA(3^46^1;2)$	**18**	20	**18**	170	19	32	**18**	0
Storage3	$CA(2^93^15^36^18^1;2)$	**48**	80	**48**	610	52	53	51	0
Storage4	$CA(2^53^74^15^26^27^110^113^1;2)$	**130**	2330	132	7770	**130**	68	134	4
Storage5	$CA(2^53^85^36^28^19^110^211^1;2)$	**113**	521870	**113**	243140	123	84	119	4
SystemMgmt	$CA(2^53^45^1;2)$	**15**	50	**15**	310	19	2	16	0
Telecom	$CA(2^53^14^25^16^1;2)$	**30**	40	**30**	4550	**30**	23	**30**	4

Table 4 demonstrates the results on CAs of strength 3 and level 2. CASA produces the smallest CAs in all cases, although its execution times are significantly longer than those of the other tools. Compared with PICT and ACTS, CAB usually obtains smaller CAs in longer times.

We also performed experiments on a number of MCAs, as listed in Table 5. These MCAs are derived from some benchmark SUT models in previous papers on combinatorial testing, see [12] for example. CAB outperforms CASA in both the quality of results and the execution times. PICT and ACTS are very fast, and both fail to produce the smallest CAs in most occasions. Interestingly, for many cases in Table 5, the smallest size (in bold type) happens to be the lower bound of the optimal size in theory (the product of t largest levels).

6   Related Work

The computational methods for covering array generation have been extensively studied in literature. Besides the aforementioned one-test-a-time strategy and IPO strategy, there are also some metaheuristic search and evolutionary algorithms applied to the automatic generation of CAs, including simulated annealing, which is the core algorithm of CASA, genetic algorithms, and particle swarm optimization. For a detailed review, one can refer to [11]. Recently, an efficient local search algorithm has been proposed for generating CAs with constraints [14].

Constraint Solving techniques are also used for automatic generation of CAs. Hinch et al. [5] developed constraint programming models which exploited global constraints and symmetry breaking constraints. They also studied the local search algorithm for a SAT-encoding of the model. Yan and Zhang developed another backtrack search tool for finding CAs [10]. A kind of balancing heuristic is applied to value-tuples in the CA, so as to prune the search space. In particular, pseudo-Boolean constraint solving has also been employed to generate orthogonal arrays (OA), which can be viewed as a special class of CA [15].

7   Conclusion

In this paper, we propose a non-backtracking algorithm which generates covering arrays column by column. It integrates a pseudo-Boolean constraint solver for column generation, and adopts a new heuristic named the balancing heuristic to guide the searching procedure. The principle of balancing heuristic is very different from the greedy strategies such as one-test-at-a-time and IPO, which have been widely employed by CA generators. The balancing heuristic suggests that, rather than locally optimizing the current CA solution, the algorithm should generate balanced columns so that the current solution is more likely to be horizontally expanded. Simple as it is, our algorithm demonstrates advantage over the state-of-the-art CA generators for instances of strength 2. It can generate smaller CAs in reasonable time. In the future, we will study how to improve the performance of our tool CAB on CAs with higher strengths.

References

1. Cohen, D.M., Dalal, S.R., Fredman, M.L., Patton, G.C.: The AETG system: an approach to testing based on combinatorial design. IEEE Trans. Software Eng. **23**(7), 437–444 (1997)
2. Cohen, M.B., Gibbons, P.B., Mugridge, W.B., Colbourn, C.J.: Constructing test suites for interaction testing. In: Proceedings of ICSE 2003, pp. 38–48 (2003)
3. Czerwonka, J.: Pairwise testing in the real world. In: Proceedings of the 24th Pacific Northwest Software Quality Conference (PNSQC 2006), pp. 419–430 (2006)
4. Garvin, B.J., Cohen, M.B., Dwyer, M.B.: An improved meta-heuristic search for constrained interaction testing. In: Proceedings of the 1st International Symposium on Search Based Software Engineering (SBSE 2009), pp. 13–22 (2009)
5. Hnich, B., Prestwich, S.D., Selensky, E., Smith, B.M.: Constraint models for the covering test problem. Constraints **11**(2–3), 199–219 (2006)
6. Kuhn, D.R., Michael, J.R.: An investigation of the applicability of design of experiments to software testing. In: Proceedings of the 27th Annual NASA Goddard/IEEE Software Engineering Workshop (2002)

7. Lei, Y., Tai, K.C.: In-parameter-order: a test generation strategy for pair-wise testing. In: Proceedings of the 3rd IEEE International Symposium on High-Assurance Systems Engineering (HASE 1998), pp. 254–261. IEEE Computer Society (1998)
8. Lei, Y., Kacker, R., Kuhn, D.R., Okun, V., Lawrence, J.: IPOG: a general strategy for T-way software testing. In: Proceedings of the 14th Annual IEEE International Conference and Workshops on the Engineering of Computer-Based Systems (ECBS 2007), pp. 549–556. IEEE (2007)
9. Lei, Y., Kacker, R., Kuhn, D.R., Okun, V., Lawrence, J.: IPOG/IPOG-D: efficient test generation for multi-way combinatorial testing. Softw. Test. Verification Reliab. **18**(3), 125–148 (2008)
10. Yan, J., Zhang, J.: A backtracking search tool for constructing combinatorial test suites. J. Syst. Softw. **81**(10), 1681–1693 (2008)
11. Zhang, J., Zhang, Z., Ma, F.: Automatic Generation of Combinatorial Test Data. Springer Briefs in Computer Science. Springer, Heidelberg (2014). ISBN 978-3-662-43428-4
12. Zhang, Z., Yan, J., Zhao, Y., Zhang, J.: Generating combinatorial test suite using combinatorial optimization. J. Syst. Softw. **98**, 191–207 (2014)
13. Hartman, A., Raskin, L.: Problems and algorithms for covering arrays. Discrete Math. **284**, 149–156 (2004)
14. Lin, J., Luo, C., Cai, S., Su, K., Hao, D., Zhang, L.: TCA: an efficient two-mode meta-heuristic algorithm for combinatorial test generation. In: Proceedings of ASE 2015, pp. 494–505 (2015)
15. Ma, F., Zhang, J.: Finding orthogonal arrays using satisfiability checkers and symmetry breaking constraints. In: Ho, T.-B., Zhou, Z.-H. (eds.) PRICAI 2008. LNCS (LNAI), vol. 5351, pp. 247–259. Springer, Heidelberg (2008)
16. The clasp webpage. http://potassco.sourceforge.net/

A Microblog Hot Topic Detection Algorithm Based on Discrete Particle Swarm Optimization

Huifang Ma$^{(\boxtimes)}$, Yugang Ji, Xiaohong Li, and Runan Zhou

College of Computer Science and Engineering, Northwest Normal University,
Lanzhou, China
mahuifang@yeah.net

Abstract. Traditional hot topic detection algorithms cannot show its optimal performance on microblogs for their inherent flaws in constructing short-text representation model, implementing the core algorithm in large corpus with short time and evaluating the algorithms' qualities during the process of detecting hot topics. In this paper, a novel method for detecting hot topics in microblogs is presented. This approach takes advantage of a probabilistic correlation-based representation measure in order to ensure a dense and low-dimension microblog representation matrix. Besides, we take the clustering as an optimization problem and introduce a discrete particle swarm optimization (DPSO) to simplify the clustering process to detect topics. Furthermore, the clustering quality evaluation criteria is adopted as the optimization objective function for topic detection which can evaluate the algorithms' qualities after each iteration. Experimental results with corpora containing more than 148,000 twitters show that our algorithm is an effective hot topic detection method for microblog.

Keywords: Microblog · Hot topic detection · Probabilistic Correlation-based representation · Short-Text representation model · Discrete particle swarm optimization

1 Introduction

Over the past several years, real-time social networks, such as microblogs, have become a powerful platform where people can vent mood, share opinions, and even disseminate emerging news. Many scholars and researchers are attracted to detect underlying information as hot topics upon these online communication platforms for their perfect interactivity in users' daily life.

As for microblogs, a topic is usually defined as an event on which people focus publicly, and the hot topics are the highly condensed summary of enormous microblogs [1]. It is a challenging problem to detect microblog hot topics because microblog posts are generally much shorter. Thus, traditional topical clustering techniques based on lexical overlap are undoubtedly weak and it is too difficult to construct an effective microblog representation model by using traditional methods like Vector Space Model (VSM) [2], Latent Semantic Analysis (LSA) [3] or Latent Dirichlet Allocation (LDA) [4]. There are

© Springer International Publishing Switzerland 2016
R. Booth and M.-L. Zhang (Eds.): PRICAI 2016, LNAI 9810, pp. 271–282, 2016.
DOI: 10.1007/978-3-319-42911-3_23

many studies attempting to overcome the deficiencies of VSM [5, 6]. Tang et al. [7] devise an approach that enriches microblog representation by employing machine translation to increase the number of features from different languages. Cheng et al. [8] propose a coupled term-term relation model for text representation, which considers both the intra-relation and inter-relation between a pair of words, yet it is obvious that the process of calculating the relationships between words is too complex. A probabilistic correlation-based similarity measure [9] can be introduced to avoid the complex calculation.

Most of researchers prefer to use clustering method when detecting hot topics in microblogs. Since proposed by Kennedy et al. [10], the Particle Swarm Optimization (PSO) has been used in solving many clustering problems. Zhao et al. [11] discover that the PSO has the incomparable superiority in both operation time and complexity. Omran et al. [12] use PSO algorithm in image clustering. There are two types of clustering evaluation validation, one is external validation and the other is internal validation. Traditional external validations [13], such as F-measure and information entropy, are often adopted to evaluate the quality of final cluster while internal validations, the Global Silhouette (GS) coefficient and the Expected Density Measure (EDM), not only can evaluate a cluster's quality effectively, but also have the possibility to optimize the clustering result in process for their feedback characteristic. On the basis of the previous research, Leticia et al. [14, 15] presents an efficient discrete PSO approach to cluster short texts and find that GS is more suitable to be the fitness function than EDM in that algorithm.

Inspired by the observation mentioned above, we present a hot topic detection approach for microblogs based on discrete PSO. First of all, we construct a new representation model for microblogs. And then, to reduce time cost of discrete PSO, a useful method is proposed to check whether particles of different forms are the same clustering results. GS, which is set as the fitness function to optimize the clustering for a better result, can be improved by the probabilistic correlation-based similarity measure mentioned above. Finally, we compare our algorithm with other classical methods and prove its superiority.

The remainder of the paper is organized as follows. In Sect. 2, we describe our representation model for microblogs. Section 3 introduces the discrete PSO in this work. In Sect. 4, experiments are conducted to show the effectiveness of this work in different angles. Section 5 concludes and discusses future work.

2   Representation Model for Microblogs

Though detecting microblog hot topics is a new domain of computer science research, it can be viewed as an instance of mining information from numerous short texts. How to construct an effective microblog representation model is one of the most significant steps for clustering because the characteristics of microblogs may hinder the application of conventional text mining algorithms. In this section, a probabilistic correlation-based similarity measure is adopted to calculate the similarity between words. Furthermore, the intra-correlation and inter-correlation are utilized to construct the representation model and calculate the similarity value between microblogs.

2.1 Probabilistic Correlation Definition

Bag of words (BOW) is a classical model to map documents to a matrix which makes an assumption that words in texts are independent of each other, and the correlations between words are ignored. In practice, word correlations do exit and shouldn't be ignored while detecting hot topics. Considering the conditional probability of word co-occurrence, a probabilistic term correlation model is then developed.

At first, we deem that the words occurring in a microblog have latent relationships and the conditional probability adopted to model the probability of these latent relationships is defined as follows

$$Pr(w_i|w_j) = \frac{Pr(w_iw_j)}{Pr(w_j)} = \frac{df(w_iw_j)/N}{df(w_j)/N}, \tag{1}$$

where $Pr(w_iw_j)$ and $df(w_iw_j)$ respectively denote the probability and the number that words w_i and w_j occur in the same microblog while $Pr(w_i)$ and $df(w_i)$ respectively mean the probability and the number that words w_i appears in the a microblog, and N is the total number of microblogs in the corpora.

To ensure the similarity between microblogs symmetric, the probabilistic correlation of words is described as

$$cor(w_i, w_j) = Pr(w_i|w_j) \cdot Pr(w_j|w_i). \tag{2}$$

The value of $cor(w_i, w_j)$ in range [0, 1] is proportional to the co-occurrence frequency of words w_i and w_j. When w_i and w_j appear in all microblogs that contain one of them, we have $cor(w_i, w_j) = 1$.

2.2 The Representation Model for Microblogs

To construct the representation model for microblogs, a microblog-word matrix $S_{N \times M}$ must be initialized at first where M is the number of different words that occur in the corpora and N denotes the total number of these microblogs. Different from traditional initialization method such as *tf-idf*, we present a novel approach by capturing both the intra-relation (explicit) and inter-relation (implicit).

Definition 1 (intra-relation). Given a microblogs mb_i with words $\{w_{i1}, w_{i2}, \cdots, w_{iM}\}$, each word in mb_i have the intra-relation with other words in the same microblog, i.e., as is shown in Fig. 1, *love* and *flowers* are of intra-relation in mb_1.

Definition 2 (inter-relation). Given two microblogs mb_i with words $\{w_{i1}, w_{i2}, \cdots, w_{iM}\}$ and mb_n with words $\{w_{n1}, w_{n2}, \cdots, w_{nM}\}$, if there are one or more words appear both mb_i and mb_n, these words are called linking words and each word in mb_i have inter-relation with the words in mb_n. I.e., *flowers* and *chocolates* have the inter-relation and *love* is the link word in Fig. 1.

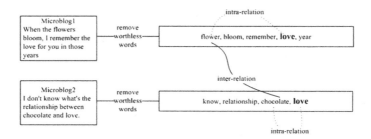

Fig. 1. One specific example of intra-relation and inter-relation, *love* is the link word.

The intra-relation between two words in a microblog is given by:

$$IaR(w_i, w_k) = \begin{cases} 1, & i = k \\ \frac{cor(w_i, w_k)}{\sum_{j=1, j \neq k}^{m} cor(w_j, w_k)}, & i \neq k \end{cases}, \tag{3}$$

where m is the number of different words in the corpora, and it is easy to derive that $\sum_{i=1, i \neq k}^{m} IaR(w_i, w_k) = 1$ when $i \neq k$. Note that all words mentioned in the context belong to the dictionary based on the corpora, in other words, these words are different from each other which ensures that each word can be identified by its subscript.

However, Eq. 3 can only capture the co-occurrence frequency between w_i and w_k, while many words are also closely related though they don't co-occur in the same microblog. Therefore, we define inter-relation as

$$IeR(w_i, w_j) = \begin{cases} 0, & i = j \\ \frac{\sum_{\forall w_k \in L} \min\{IaR(w_i, w_k), IaR(w_j, w_k)\}}{|L|}, & i \neq j \end{cases}, \tag{4}$$

where $|L|$ is the amount of linked words set L, i.e., $|L| = 1$ and $L = \{love\}$ in Fig. 1. We let $IeR(w_i, w_j) = 0$ when $i = j$ since it is worth nothing to calculate the inter-relation between a word itself.

Taking both the intra-relation and inter-relation into consideration, the correlation between w_i and w_j can be defined as

$$WR(w_i, w_j) = \begin{cases} 1, & i = j \\ \alpha \times IeR(w_i, w_j) + (1 - \alpha) \times IaR(w_i, w_j), & i \neq j \end{cases}, \tag{5}$$

where WR is the abbreviation for *word relation*, and $\alpha \in [0, 1]$ determines the importance of intra-relation and inter-relation between w_i and w_j. It is easy to prove that $WR(w_i, w_j) = 1$ if $i = j$.

Instead of term frequency (*tf*) with *weight*$_i$ = 1 in most cases, a new local weighting scheme of words in a microblog is proposed in this approach, namely correlation weight.

Definition 3 (correlation weight). Given a microblog mb_n with an initial $weight_{ni}$ of each word w_{ni}, the correlation weight of w_{ni} in mb_n is defined as

$$cow(w_{ni}) = weight_{ni} + \frac{\sum_{w_{nj} \in mb_n} weight_{nj} \cdot WR(w_{ni}, w_{nj})}{|mb_n|}, \tag{6}$$

where $WR(w_{ni}, w_{nj})$ is the word relation between w_{ni} and w_{nj} in this microblog, $|mb_n|$ is the total number of words in mb_n.

Finally, a new representation model for microblog is proposed and each element rm_{ij} in the matrix $\mathbf{RM}_{N \times M}$ is defined as

$$rm_{ij} = cow(w_{ij}) \cdot idf(w_{ij}), \tag{7}$$

where $i \in \{1, 2, \cdots, N\}$ is the subscript of microblogs in the corpora, $j \in \{1, 2, \cdots, M\}$ is the subscript of words in the dictionary, $idf(w_{ij}) = \log\left(\frac{N}{df(w_{ij})}\right)$, N is the amount of microblogs and M is the length of words in this dictionary.

With the advantages of the conditional probability between words and inner/inter relation, the new matrix $\mathbf{RM}_{N \times M}$ can reflect the relationship between any two words, and is proved denser, lower-dimensional than tradition models in the experimental section. Taking advantages of the inter-relation and inner-relation between words, the similarity function of microblogs is defined as follows [10]:

$$sim(mb_x, mb_y) = \frac{\sum_{(w_i, w_j) \in \mathbf{D}} weight_i weight_j cow(w_i, w_j)}{\|mb_x \oplus \mathbf{D}\| \cdot \|mb_y \oplus \mathbf{D}\|}, \tag{8}$$

where $\mathbf{D}$ denotes the words correlation of mb_x and mb_y, and $\|mb_x \oplus \mathbf{D}\|$, $\|mb_y \oplus \mathbf{D}\|$ denote the sizes of mb_x and mb_y so as to normalize the similarity value.

$$\|mb_x \oplus \mathbf{D}\| = \sqrt{\sum_{(w_i, w_j) \in \mathbf{D}} \left(weight_i^2 cow(w_i, w_j)\right) + \sum_{w_i \in mb_x \setminus mb_y} weight_i^2}, \tag{9}$$

and $\|mb_y \oplus \mathbf{D}\|$ can be calculate in a similar way.

Thus, $sim(mb_x, mb_y) \in [0, 1]$.

3 The DPSO Algorithm for Microblog Hot Topic Detection

Particle Swarm Optimization (PSO) is a population-based search algorithm inspired by the behavior of biological communities that exhibit both individual and social behavior; examples of these communities are flocks of birds, swarms of bees. In PSO, each solution to the problem at hand is called a particle and per particle represents a real vector within the search space, corresponding to a solution of the mazy problem.

However, traditional PSO is originally developed for continuous space but many problems are defined for discrete valued spaces where the domain of the variables is finite. We propose a discrete PSO approach and fit it to clustering microblogs for detecting hot topics in this section. The fitness function of PSO is redefined by GS and the correlation similarity of microblogs. Finally, a valid method to reduce time cost of the algorithm by dealing with particles is also shown then.

3.1   DPSO Algorithm

The basic PSO algorithm contains a swarm of particles in which each particle includes a potential solution. The particles fly through a multi-dimensional search space where the position of each particle is adjusted according to its own experience and the experience of its neighbors during per iteration. In each iteration, the velocity and the position of every particle are calculated by Eqs. 10 and 11. Besides, the current global best position (*gbest*) and individual history best position (*pbest*) are all recorded in the corresponding matrices.

$$v_{id} \leftarrow \omega(v_{id} + \gamma_1(pbest_{id} - par_{id}) + \gamma_2(gbest_d - par_{id})), \tag{10}$$

$$par_{id} \leftarrow (par_{id} + v_{id}), \tag{11}$$

where ω is the inertia factor whose goal is to balance global exploration and local exploitation, γ_1 is the personal learning factor, γ_2 is the social learning factor, par_{id}, $pbest_{id}$, and v_{id} are the position, history best position, and velocity of ith particle in d^{th} dimension respectively, and $gbest_d$ is the best position of the whole swarm in d^{th} dimension. In our context, these concepts are redefined in Definitions 4, 5 and 6 to make it easier to understand and calculate.

DPSO (Discrete Particle Swarm Optimization) is a discrete version of the basic PSO algorithm. In this method, each particle represents a clustering result of the corpora during the process of hot topics detection as the final aim of our algorithm is to get an excellent clustering result.

To detail the process of DPSO, we define four matrices as follows

Definition 4 (particles cluster result matrix). The particles cluster result matrix, $\mathbf{P}_{K \times N}$, is defined to store the current position of each particle and each element, p_{id}, is the cluster of ith particle in d^{th} dimension. i.e., $p_{12} = $ cluster1 means the first particle of the swarm in second dimension is clustered to cluster1. In addition, K is the number of particles in this paper and is set manually.

Definition 5 (particles cluster quality matrix). The particles cluster quality matrix, $\mathbf{PAR}_{K \times N}$, is proposed to record the quality of $\mathbf{P}$, and each element, par_{id} is calculated by the fitness function mentioned below. By comparing the value of par_{id} and $pbest_{id}$ and $gbest_d$, we can judge whether it is worth putting the corresponding element in $\mathbf{P}$ to a new cluster.

Definition 6 (particles cluster velocity matrix). The particles cluster velocity matrix, $\mathbf{V}_{K \times N}$, is provided to represent the probability of choosing the corresponding particle to optimize. The value of per element, v_{id}, can be changed during each iteration of optimization by Eq. 8.

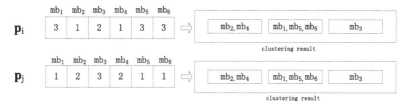

Fig. 2. The situation that different particles represent same clustering result.

In DPSO algorithm, Eq. 11 is modified as

$$par_{id} \leftarrow pbest_{id}, \tag{12}$$

During the initialization of **P**, there is a high probability that different particles may refer to the same clustering result as shown in Fig. 2.

In order to avoid unnecessary time cost of the iterative process, we propose an effective method as follows.

Program.1. The algorithm for checking repeated clustering results.

Input: $P = \bigcup_{i=1}^{k}(p_{i1}, p_{i2}, \ldots, p_{iN})$.
Output: the updated $P = \bigcup_{i=1}^{k}(p'_{i1}, p'_{i2}, \ldots, p'_{iN})$.
1. An integer set clusters $= \bigcup_{i=1}^{l} i$, the preset clusters's amount l,
    $Q_{k \times m} = \bigcup_{i=1}^{k}(q_{i1}, q_{i2}, \ldots, q_{iN}) = \text{null}, count = 0.$
2. for every p_{id} in **P**
3.   if $p_{id} \notin \bigcup_{r=1}^{d-1} p_{ir}$
4. $q_{id} = \text{clusters[count]};$
5.   count++;
6.   else
7.   catch $t \in [1, d-1]$ to ensure $p_{id} = p_{it};$
8. $q_{id} = q_{it};$
9.   end if
10. end for
11. for eyery Q_i in Q
12.   if $Q_i = Q_j$ and $i < j$
13. $P_j = P_i;$
14.   end if
15. end for
16. Output the updated **P**.

By using Program 1, P_i and P_j in Fig. 2 are in the same form of $\{1,2,3,2,1,1\}$ or $\{a, b,c,b,a,a\}$ etc.

3.2   The Improved Fitness Function

The fitness function in PSO is usually used to measure the quality of particles. In other words, it is a method to judge whether the corresponding cluster result is a better one during clustering.

As an external validation used to deal with the corpora of unknown structure and evaluate the clustering quality based on the corpora only, the Global Silhouette coefficient (GS) is a good choice to be the fitness function since it provides a succinct graphical representation of how well each object lies within its cluster. Assuming the corpora have been clustered into l clusters. For each microblog mb_i, let $a(mb_i)$ be the average dissimilarity between mb_i and all other microblogs within the same cluster, which can be interpreted as how well mb_i is assigned to its cluster. Let $b(mb_i)$ be the average dissimilarity between mb_i and the microblogs in the neighboring cluster of mb_i. The formula to calculate the GS value is given by

$$GS(mb_i) = \frac{b(mb_i) - a(mb_i)}{\max\{a(mb_i), b(mb_i)\}}, \tag{13}$$

with $-1 \leq GS(mb_i) \leq 1$. From this formula it can be observed that negative values for this measure are undesirable and that for this coefficient values as close to 1 as possible are desirable.

Given a mb_i belonging to cluster C_a, and the neighboring cluster C_b, the function to compute $a(mb_i)$ is defined as follows

$$a(mb_i) = \frac{\sum_{mb_j \in C_a} (1 - sim(mb_i, mb_j))}{|C_a|}, \tag{14}$$

and $b(mb_i)$ is defined as

$$b(mb_i) = \frac{\sum_{mb_j \in C_b} (1 - sim(mb_i, mb_j))}{|C_b|}, \tag{15}$$

where $|C_a|$ and $|C_b|$ are the amount of microblogs in cluster C_a and C_b respectively.

3.3   The Algorithm's Framework and Details

Combined with correlations (intra/inter relation) of words, a clustering algorithm based on DPSO is proposed to detect microblog hot topics. The algorithm is mainly divided into three steps: constructing a microblog representation model by the conditional probability of word co-occurrence and correlations of words; using DPSO algorithm to initialize and optimize the cluster result of microblogs; judging whether a cluster result need to be optimized and whether the algorithm should end by calculating the corresponding GS (Fig. 3).

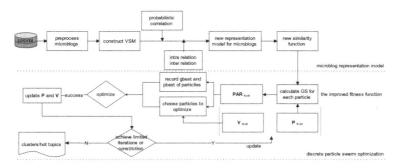

Fig. 3. The algorithm's framework.

Note that V is initialized by a random value in range [0,1] where each element indicates the probability of choosing the corresponding particles to optimize, and **P** is initialized by the random cluster subscripts at the beginning of DPSO.

4   Experiments

In this section, we report our experimental results. Section 4.1 introduces the datasets, parameter settings and effectiveness criteria in the approach. Section 4.2 evaluates the performance of our approach with various parameters and compare our method with existing approaches such as traditional k-means, and MicroBlog Hierarchical Dirichlet Process (MB-HDP) [16] and a topic detection method based on microblog weight named Weighted LDA (W-LDA) [17].

4.1   Data Sets, Parameter Settings and Effectiveness Criteria

Datasets.   We grab the first one hundred search results by Twitter API from March 1st 2015 to Oct. 31st 2015. After removing those invalid twitters and those too short twitters, meanwhile merging the same content, the remaining data set is 148090 microblogs in total.

Parameter Settings.   In our experiments, 50 independent runs are performed, with 10,000 iterations per run, the swarm size $K = 100$ particles, dimensions of each particle N = number of twitters, inertia factor $\omega = 0.9$, personal and social learning factors $\gamma1$, $\gamma2$ are set to 1.0, and the clusters number $l = 10$.

Effectiveness Criteria.   Two evaluative metrics, the Purity and F1-measure are used to evaluate experiments performance. Purity is a simple and transparent evaluation measure for cluster quality while F1-measure, is a weighted harmonic mean of Recall and Precision (R, P).

4.2   Experimental Results and Analysis

In this section, experimental results concerning our algorithm are discussed, such as the influence of regulatory factor α to determine the importance of intra-relation and inter-relation between words, the size of microblogs to deal with, the number of iterations of the algorithm and the effectiveness of the new similarity function. Since our algorithm has some parameters to be tuned, all the involved parameters are carefully tuned and the parameters with best performance are used to report the final results. And then, we make the comparison among our algorithm and other hot topic detection methods such as W-LDA, traditional DPSO and MB-HD.

Figure 4(1) shows *F1-measure* values, *purity* values and GS values with different regulatory factor α. Obviously, our algorithm shows its best performance while α is in the vicinity of 0.5 which indicates the need for a balance between the intra-relation and inter-relation.

Figure 4(2) shows *F1-measure* values, *purity* values and GS values with the increasing size of iterations for optimizing particles. when the number of iteration is more than 10000, the performance of our algorithm becomes stable.

Figure 4(3) shows *F1-measure* values, *purity* values and GS values with different particles' number. Each horizontal coordinate denotes the corresponding times the number of microblogs in the corpora.

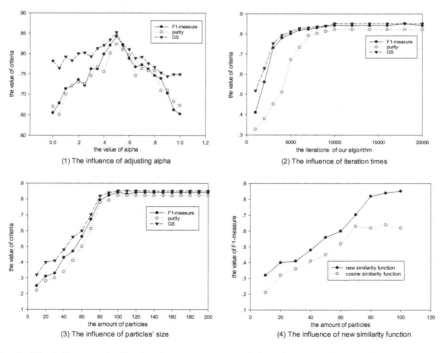

Fig. 4. The influence of adjusting key parameters and the advantage of using the new similarity function based on intra/inter relation.

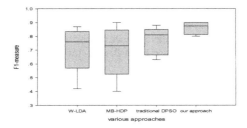

Fig. 5. The comparison of various topic detection approaches.

Figure 4(4) shows *F1-measure* values with the new similarity function and cosine similarity function respectively. The new similarity function shows a better performance for considering the correlation and simplifying the steps of calculating GS which can reduce the algorithm's time cost.

In summary, the above experimental results demonstrate that intra/inter relation, the similarity function play significant roles in this algorithm, which indicates that these aspects should not be ignored.

For effectiveness, we then compare our methods with several existing methods. As we can see, overall, our method outperform all the compared methods. The box plots in Fig. 5 represent that our approach reach the best performance with a median value close to 0.86 and without dispersion except for some outliers. This aspect is important because it shows that our approach is able to find very similar F-measure values in all the runs. The worst one is MB-HDP because its median value is the lowest and its box plot shows the highest dispersion. Both of MB-HDP and W-LDA are based on LDA, which cannot be well generalized. Compared with traditional DPSO, our algorithm present a representational model for microblogs and improve the fitness function which ensure a novel optimization.

5   Conclusion

Hot topic detection for microblogs is a challenging research problem in data mining domain. Different from traditional documents or texts, microblogs are often very short, sparse and spreading rapidly online. In this work, we propose an effective algorithm based on discrete particle swarm optimization. A representation model for microblogs considering correlations and intra/inter relations between words by calculating conditional probability of word-occurrences is constructed to capture the semantic associations between words. In addition, a new method to compute the similarity between microblogs is proposed so as to improve the fitness function, GS. Furthermore, DPSO algorithm is developed to obtain the hot topics in the corpora. Besides, an effective program to overcome the problem that various particles denotes the same clustering result. Finally, experiments have demonstrated its effectiveness in mining microblogs. The future research can be targeted at particle dimension reduction, a more suitable fitness function selection.

Acknowledgement. This work is supported by the National Natural Science Foundation of China (No.61363058), Youth Science and technology support program of Gansu Province (145RJZA232, 145RJYA259), 2016 undergraduate innovation capacity enhancement program and 2016 annual public record open space Fund Project 1505JTCA007.

References

1. Ding, Z.Y., Jia, Y., Zhou, B.: Survey of data mining for micro-blogs. J. Comput. Res. Dev. **04**, 691–706 (2014)
2. Salton, G., Wong, A., Yang, C.S.: A vector space model for automatic indexing. Commun. ACM **18**(11), 613–620 (1975)
3. Deerwester, S., Dumais, S.T., Furnas, G.W., et al.: Indexing by latent semantic analysis. J. Am. Soc. Inf. Sci. **41**(6), 391 (1990)
4. Blei, D.M., Ng, A.Y., Jordan, M.I.: Latent dirichlet allocation. the. J. Mach. Learn. Res. **3**, 993–1022 (2003)
5. Ma, H.F., Zhao, W.Z., Shi, Z.Z.: A nonnegative matrix factorization framework for semi-supervised document clustering with dual constraints. Knowl. Inf. Syst. **36**(3), 629–651 (2013)
6. Ma, H., Jia, M., Xie, M., Lin, X.: A microblog recommendation algorithm based on multi-tag correlation. In: Zhang, S., et al. (eds.) KSEM 2015. LNCS, vol. 9403, pp. 483–488. Springer, Heidelberg (2015). doi:10.1007/978-3-319-25159-2_43
7. Tang, J., Wang, X., Gao, H., et al.: Enriching short text representation in microblog for clustering. Front. Comput. Sci. **6**(1), 88–101 (2012)
8. Cheng, X., Miao, D., Wang, C., et al.: Coupled term-term relation analysis for document clustering. In: The 2013 International Joint Conference on Neural Networks (IJCNN), pp. 1–8. IEEE (2013)
9. Song, S., Zhu, H., Chen, L.: Probabilistic correlation-based similarity measure on text records. Inf. Sci. **289**, 8–24 (2014)
10. Kenndy, J., Eberhart, R.C.: Particle swarm optimization. In: Proceedings of IEEE International Conference on Neural Networks, vol. 4, pp. 1942–1948 (1995)
11. Zhao, X.C., Liu, G.L., Liu, H.Q., et al.: Particle swarm optimization algorithm based on non-uniform mutation and multiple stages perturbation. Chin. J. Comput. **9**, 2058–2070 (2014)
12. Omran, M., Engelbrecht, A.P., Salman, A.: Particle swarm optimization method for image clustering. Int. J. Pattern Recogn. Artif. Intell. **19**(03), 297–321 (2005)
13. Zhang, W.J., Liu, C.H., Li, F.Y.: Method of quality evaluation for clustering. J. Comput. Eng. **31**(20), 10–12 (2005)
14. Cagnina, L.C., Errecalde, M.L., Ingaramo, D.A., et al.: An efficient particle swarm optimization approach to cluster short texts. Inf. Sci. **265**, 36–49 (2014)
15. Cagnina, L.C., Errecalde, M.L., Ingaramo, D.A., et al.: A discrete particle swarm optimizer for clustering short-text corpora. In: Proceedings of the Bioinspired Optimization Methods and their Applications, BIOMA-2008, Ljubljana, Slovenia (2008)
16. Liu, S.P., Yin, J., Ouyang, J., et al.: Topic mining from microblogs based on MB-HDP model. Chin. J. Comput. **7**(008), 1408–1419 (2015)
17. Guo, K., Shi, L.: A topic detection method based on microblog weight. In: 2015 International Conference on Cyber-Enabled Distributed Computing and Knowledge Discovery (CyberC), pp. 209–212. IEEE (2015)

Local Search with Noisy Strategy for Minimum Vertex Cover in Massive Graphs

Zongjie Ma[1]([✉]), Yi Fan[1], Kaile Su[1], Chengqian Li[2], and Abdul Sattar[1]

[1] Institute for Integrated and Intelligent Systems,
Griffith University, Brisbane, Australia
zongjie.ma@griffithuni.edu.au
[2] Department of Computer Science, Sun Yat-sen University, Guangzhou, China

Abstract. Finding minimum vertex covers (MinVC) for simple undirected graphs is a well-known NP-hard problem. In the literature there have been many heuristics for obtaining good vertex covers. However, most of them focus on solving this problem in relatively small graphs. Recently, a local search solver called FastVC is designed to solve the MinVC problem on real-world massive graphs. Since the traditional best-picking heuristic was believed to be of high complexity, FastVC replaces it with an approximate best-picking strategy. However, since best-picking has been proved to be powerful for a wide range of problems, abandoning it may be a great sacrifice. In this paper we have developed a local search MinVC solver which utilizes best-picking with noise to remove vertices. Experiments conducted on a broad range of real-world massive graphs show that our proposed method finds better vertex covers than state-of-the-art local search algorithms on many graphs.

Keywords: Minimum vertex cover · Heuristic search · Massive graphs · Combinatorial optimization · Social networks

1 Introduction

The rapid growth of the Internet, widespread deployment of sensors and other fields produced huge quantity of massive data sets, which has generated a series of computational challenges to existing algorithms. Hence, new algorithms need to be designed to deal with these data sets. Many of these data can be modeled as graphs, and the interest in real-world massive graphs, also known as complex networks [21], is growing significantly over recent decades.

The Minimum Vertex Cover (MinVC) problem is a fundamental NP-hard problem in computer science. Given a simple undirected graph G, a vertex cover S is a subset of vertices s.t. every edge in G has at least one endpoint in S. The objective of MinVC is to find a vertex cover of the minimum size. MinVC is one of the well-known optimization problems with many real-world applications, such as network security, scheduling, VLSI design and industrial machine assignment [5,16]. Also the MinVC problem is closely related to the Maximum

© Springer International Publishing Switzerland 2016
R. Booth and M.-L. Zhang (Eds.): PRICAI 2016, LNAI 9810, pp. 283–294, 2016.
DOI: 10.1007/978-3-319-42911-3_24

Independent Set (MIS) and Maximum Clique (MC) problems, in that algorithms for MinVC can be directly applied to solve the MIS and MC problems. The applications of these three problems involve computer version, information retrieval, signal transmission, aligning DNA and protein sequences [10,11,15], etc.

1.1   Previous Heuristics and Motivations

Due to the great importance to many real-world applications, a large number of algorithms for solving MinVC (MIS, MC) have been proposed during the past decades. Practical algorithms for them can be roughly grouped into two categories, i.e., exact algorithms and heuristic algorithms. The exact ones, mainly based on the general branch-and-bound framework [14,20], confirm the optimality of the solutions they find. However, for large and hard instances, exact methods may become ineffective and fail to return a solution within reasonable time.

On the other hand, heuristic methods are able to find near-optimal solutions within reasonable time for large and hard instances. Local search is a popular strategy among the heuristic approaches, such as [5,16] for MinVC, [1,2] for MIS, and [13] for MC.

The evaluation of existing local search approaches for MinVC are mainly based on standard benchmarks from academic community, such as the DIMACS [12] and BHOSLIB[1] benchmarks [5,16]. In order to improve the performance on these benchmarks, a number of heuristics combined with local search for MinVC have been proposed in the literature recently. COVER [16] introduces edge weighting to MinVC, and is an iterative best improvement approach through updating edge weights at each step to guide a local search. EWLS [4] also exploits the edge weighting strategy, but it only updates the edge weights when being stuck in local optima. EWCC [6] introduces the configuration checking (CC) heuristic into EWLS, and CC is a strategy for handling the cycling problem in local search. Especially, NuMVC [5] introduces two strategies, named two-stage exchange and edge weighting with forgetting, and makes a significant improvement in MinVC solving. Since the benchmarks graphs used by these previous algorithms are not large (usually with less than five thousand vertices), the impact of the complexity of heuristics on the performance is not significant.

In this work, we focus on studying the local search for MinVC in massive real world graphs. Many of these real world graphs have millions of vertices and dozens of millions of edges [17]. The complexity of most previous heuristics is not sufficiently small, and they suffer from these massive graphs with millions of vertices. Thus massive graphs call for new heuristics and algorithms. Recently, an algorithm called FastVC [3] takes a first step towards solving the MinVC problem for real-world massive graphs. FastVC outperforms other existing local search algorithms on finding vertex covers in massive graphs. It is designed by withdrawing or modifying some techniques with high computational cost in NuMVC [5]. Specifically, FastVC replaces the best-picking heuristic in NuMVC

---

[1] http://www.nlsde.buaa.edu.cn/~kexu/benchmarks/graph-benchmarks.htm.

with a low-complexity heuristic named Best from Multiple Selection (BMS). BMS approximates the best-picking heuristic very well.

In local search phase, FastVC abandons the best-picking heuristic in NuMVC. However, this best-picking heuristic guides the search towards very promising areas with a suitable criterion, and is thus widely used in local search algorithms. Besides, when finding a k-vertex cover, FastVC exploits traditional best-picking strategy to remove a vertex with minimum loss to generate a $(k-1)$-candidate solution. However, our experiments show that this process happens very frequently, which can be time-consuming due to the $O(|V|)$ complexity.

1.2   Contribution and Paper Organization

In this work, we propose a new algorithm called NoiseVC, which is dedicated to solve the MinVC problem in massive graphs. Instead of abandoning the best-picking heuristic, we exploited an efficient data structure named min-heap for best-picking in this work. The procedure of min-heap [8] runs in $O(|\lg V|)$ to maintain its property for best-picking, while the complexity of traditional best-picking heuristic in NuMVC is believed to be $O(|V|)$ which is very time-consuming for massive graphs.

Since best-picking can easily be trapped by local minima, we design a noisy strategy to help it escape from local minima. Given a candidate vertex set C, *with probability p, remove a vertex with the minimum loss, breaking ties in favor of the oldest one; with probability $1 - p$, remove a vertex randomly.*

Besides, when finding a k-vertex cover, min-heap is also used to remove a vertex with the minimum loss to generate a $(k-1)$-candidate solution, which lowers the complexity and helps to save time in updating the best solution.

We conduct experiments to compare NoiseVC with FastVC on a wide range of real-world massive graphs. Experimental results show that for all the 12 classes of instances in this benchmark, NoiseVC significantly outperforms FastVC on solution quality for 8 classes. More specifically, over the 136 graphs we tested, NoiseVC finds smaller vertex cover on 23 graphs.

Using the 23 graphs above, we tested FastVC with a cutoff of 100,000 s. Even within such a large cutoff the solutions obtained by FastVC are still worse than those obtained by NoiseVC within 1000 s. This means that NoiseVC is *at least 100 times as efficient as* FastVC on these graphs. It rarely happens in literature to find a better solution. The existing MinVC algorithms often obtain the same quality solutions, and concern on comparing the success rate of finding a solution of such a quality.

The rest of paper is organized as follows. Section 2 gives some necessary definitions and notations, and shows the framework of the local search for MinVC problem. Then, we describe our new algorithm for MinVC on massive graphs in Sect. 3. In Sect. 4, we carry out extensive experiments to evaluate NoiseVC. Finally, we conclude our work in Sect. 5.

2 Preliminaries

2.1 Definitions and Notation

Given an undirected graph $G = (V, E)$, where $V = \{v_1, v_2, ...v_n\}$ is a vertex set and $E \subseteq V \times V$ is an edge set. Each edge is a 2-element subset of V. For an edge $e = \{u, v\}$, the vertices u and v are called the endpoints of edge e. Two vertices, such as u and v, are neighbors if and only if there exists an edge between them. The neighborhood of v is defined as $N(v) = \{u \in V | \{u, v\} \in E\}$. The degree of v is defined as $deg(v) = |N(v)|$, which is equal to the number of its neighbors. An edge $e \in E$ is covered by a vertex set $S \subseteq V$ if at least one endpoint of e is in S; otherwise, e is uncovered by S.

For a graph $G = (V, E)$, a vertex cover of G is a subset of V which contains one or two endpoints of each edge in E. The complementary graph of G is denoted as $\bar{G} = (V, \bar{E})$, where $\bar{E} = \{(u, v) | (u, v) \notin E\}$. Then for a subset $S \subseteq V$, there are three equivalent statements [22]: S is a vertex cover of G, $V \backslash S$ is an independent set of G and $V \backslash S$ is a clique of $\bar{G}$.

2.2 Local Search for MinVC

We use C to denote the current candidate solution, which is a set of vertices selected for covering. Algorithm 1 shows a general framework of local search for MinVC.

Algorithm 1. Local Search Framework for MinVC

---

1 construct C until it becomes a vertex cover;
2 **while** *not reach terminate condition* **do**
3   if C *covers all edges* **then**
4          $C^* \leftarrow C$;
5   remove a vertex from C;
6   exchange a pair of vertices;
7 **return** C^*

There are two phases in Algorithm 1: a construction stage (Line 1) and a local search phase (Lines 2 to 6). A vertex cover is constructed at the first phase, and such a vertex cover is also called the *starting vertex cover* throughout this paper. In the local search phase, whenever the algorithm finds out a k-sized cover (Line 3), one vertex is removed from C (Line 5) and continues to search for a $(k-1)$-sized cover, until some termination condition is reached (Line 2).

The move to a neighboring candidate solution consists of an exchange of two vertices (Line 6): a vertex $u \in C$ is removed from C and a vertex $v \notin C$ is added into C. This step is also called an exchanging step. Thus the local search moves step by step in the search space to find a smaller vertex cover. After the algorithm terminates, it outputs the smallest vertex cover that has been found.

For a vertex $v \in C$, the *loss* of v, denoted as $loss(v)$, is defined as the number of covered edges that will become uncovered by removing v from C. For a vertex $v \notin C$, the *gain* of v, denoted as $gain(v)$, is defined as the number of uncovered edges that will become covered by adding v into C. Both *loss* and *gain* stand for *scoring properties* of vertices. In any step, a vertex v has two possible states: inside C and outside C. We use $age(v)$ to denote the number of steps since its state was last changed.

3 NoiseVC for MinVC in Massive Graphs

3.1 The Top-Level Algorithm of NoiseVC

We describe the NoiseVC algorithm on a top level in this subsection and the details of the proposed functions will be given and analyzed in the next subsection.

Algorithm 2. NoiseVC

---

    input : a graph $G = (V, E)$, the cutoff time
    output: a vertex cover of G

1 $C \leftarrow ConstructVC()$;
2 **while** *elapsed time < cutoff* **do**
3   **if** C *covers all edges* **then**
4        $C^* \leftarrow C$;
5   remove a vertex with the minimum loss;
6   continue;
7     $u \leftarrow$ BestPickingWithNoise(C);
8   remove u from C;
9     $e \leftarrow$ a random uncovered edge;
10    $v \leftarrow$ the endpoint of e with greater *gain*, breaking ties in favor of the older one;
11   add v into C;
12 **return** C^*;

NoiseVC outlined in Algorithm 2 adopts the local search framework for MinVC in Algorithm 1. In the construction phase (Line 1), a starting vertex cover is constructed by a function named $ConstructVC()$ which is introduced in FastVC. In the local search phase (Lines 2 to 11), whenever NoiseVC obtains a vertex cover, a vertex with minimum *loss* is removed based on the min-heap data structure. In the exchanging step (Lines 7 to 11), NoiseVC first chooses a vertex in C to remove, which is accomplished by the BestPickingWithNoise() function. Then the algorithm selects an uncovered edge randomly and chooses the endpoint of this edge with greater *gain* (breaking ties in favor of the older one) to add it into C. It should be noted that along with removing or adding a vertex, the algorithm will update the *loss* and *gain* values of the vertex and its neighbors.

3.2   Min-heap for Best-Picking in Massive Graphs

Many of these real world graphs have millions of vertices and dozens of millions of edges. The traditional best-picking heuristic, which aims at choosing the best element according to some criterion, is very time-consuming for these massive data sets. Therefore, in FastVC best-picking heuristic is replaced by BMS, which approximates best-picking very well. BMS works as follows: Choose k vertex randomly with replacement from C, and then return a vertex with the minimum loss, breaking ties in favor of the oldest one. Since k is set to 50 in FastVC, BMS chooses a vertex whose *loss* value is among the best 10 % in C with probability of 99.48 %. However, the best 10 % can be a very large number in massive graphs.

Instead of abandoning best-picking heuristic, we exploit min-heap [8] for best-picking in this work. The vertices in C are used to build a min-heap according to their *loss* value, which is used for choosing the vertex with the minimum loss. The vertex with the minimum *loss* is returned by `heap-minimum`(C) in our min-heap. Throughout this paper, we use `heap-minimum`(C) to denote the result of best-picking by min-heap. To the best of our knowledge, it is the first time that a heap is used for best-picking in MinVC solving.

Complexity Analysis. Along with remove or add a vertex v, we have to update the *loss* value of its neighbors. Therefore the worst complexity of our best-picking is $O(d_{max}|\lg V|)$, where d_{max} is the maximum degree in a graph. Since most massive real world graphs are sparse graphs [7,9], the value of d_{max} is usually not large. On the other hand, the complexity of traditional best-picking heuristic in NuMVC is $O(|V|)$ which is very time-consuming for massive graphs.

3.3   Best-Picking with Noisy Strategy

During the exchanging step, best-picking can easily be trapped by local minima. In this work, we introduce noise to help it escape from local minima. When choosing a vertex to remove, `BestPickingWithNoise`() is shown in Algorithm 3.

Algorithm 3. BestPickingWithNoise

---

input : a vertex set C, a noise parameter p
output: a vertex $v \in C$

1 With probability p: $v \leftarrow$ a vertex with the minimum *loss*, breaking ties in favor of the oldest one;
2 With probability $1 - p$: $v \leftarrow$ a random vertex in C;
3 **return** v;

Throughout this paper, the parameter p is fixed to 0.4 in advance for all the experiments. This means that the parameter p in this study is instance-independent. Different p values are only used to test parameter sensitivity.

There are two modes in Algorithm 3: the greedy mode (Line 1) and the random mode (Line 2). `BestPickingWithNoise()` switches between the two modes at a certain probability. In the greedy mode, a vertex with minimum *loss* is returned by `heap-minimum`(C). In the random mode, a vertex is selected randomly to avoid local optima.

Relationship with Other Heuristics. The proposed best-picking with noisy strategy is a new kind of random walk, which is an efficient and effective method with very low time complexity to improve local search. The existing random walk strategy focused on choosing a variable (vertex) from a random unsatisfied constraint (unsatisfied clause or uncovered edge). Our random walk strategy considers all satisfied constraints, and it chooses a vertex from C where all the incident edges are covered.

4   Experimental Results and Analysis

This section compares NoiseVC with FastVC using real-world massive graphs, since FastVC outperforms other existing local search algorithms on finding vertex covers in massive graphs.

4.1   Benchmark

To verify the effectiveness of our method, we conducted comparative experiments on a broad range of real-world massive graphs. All 139 instances are downloaded[2], which were originally online[3]. We excluded three extremely large ones, since they are out of memory for both solvers here. Many of these real-world massive graphs contain millions of vertices and dozens of millions of edges. Recently, some of these graphs are used to test parallel algorithms for Maximum Clique [19] and Coloring problems [18].

The benchmark used in our experiments can be divided into 12 classes: biological networks, collaboration networks, Facebook networks, interaction networks, infrastructure networks, Amazon recommend networks, retweet networks, scientific computation networks, social networks, technological networks, web lint networks, and temporal reachability networks.

4.2   Setup

NoiseVC was implemented in C++[4], and FastVC was also implemented in C++[5]. Both solvers were compiled by g++ 4.6.3 with the option $-O3$. The experiments were conducted on a cluster equipped with a number of Intel(R)

---

[2] http://lcs.ios.ac.cn/~caisw/Resource/realworld%20graphs.tar.gz.
[3] http://www.graphrepository.com/networks.php.
[4] https://github.com/math6068/NoiseVC.
[5] http://lcs.ios.ac.cn/~caisw/Code/FastVCv2015.11.zip.

Xeon(R) central processing units (CPUs) X5650 @2.67 GHz with 8 GB RAM, running Red Hat Santiago OS. In our experiments, the probability parameter p is fixed to be 0.4. For FastVC, we adopt the parameter setting reported in [3]. Both solvers are run 10 times on each instance with a time limit of 1000 s if not mentioned explicitly. For each solver on each instance, we report the minimum size ("C_{min}") and averaged size ("C_{avg}") of vertex covers found by the solver. To clarify the comparisons, we report the difference ("Δ") between the minimum sizes of vertex cover found by NoiseVC and FastVC. A positive (negative) means NoiseVC (FastVC) finds a smaller vertex cover.

Table 1. Experimental results on real-world massive graphs. A positive Δ means NoiseVC finds a smaller vertex cover, while a negative Δ means FastVC finds a smaller vertex cover. For $\Delta \neq 0$, we bold the smaller value of minimum size (C_{min}) between the two algorithms, and for $\Delta = 0$, we bold the smaller value of average size (C_{avg})

Graph	$|V|$	$|E|$	FastVC $C_{min}(C_{avg})$	NoiseVC $C_{min}(C_{avg})$	Δ
socfb-A-anon	3097165	23667394	**375231**(375232.8)	375233(375233)	-2
socfb-B-anon	2937612	20959854	**303048**(303048.8)	303049(303049)	-1
socfb-Berkeley13	22900	852419	17210(17212.8)	17210(**17212.1**)	0
socfb-CMU	6621	249959	4986(4986.5)	4986(**4986**)	0
socfb-Duke14	9885	506437	7683(7683.1)	7683(**7683**)	0
socfb-Indiana	29732	1305757	23315(23317.3)	**23314**(23316.3)	1
socfb-OR	63392	816886	36548(36549.2)	**36547**(36547.8)	1
socfb-Penn94	41536	1362220	31162(31164.8)	**31161**(31164)	1
socfb-Stanford3	11586	568309	8518(8518)	**8517**(8517.6)	1
socfb-Texas84	36364	1590651	28167(28171.4)	**28166**(28170.3)	1
socfb-UCLA	20453	747604	15223(15224.3)	**15222**(15223.8)	1
socfb-UConn	17206	604867	13230(13231.6)	13230(**13231.2**)	0
socfb-UCSB37	14917	482215	11261(11263.1)	11261(**11261.1**)	0
socfb-UF	35111	1465654	27306(27309.1)	**27305**(27308.3)	1
socfb-UIllinois	30795	1264421	24091(**24092.6**)	24091(24093.60)	0
socfb-Wisconsin87	23831	835946	18383(18385.1)	18383(**18383.8**)	0
ia-infect-dublin	410	2765	293(293.5)	293(**293**)	0
inf-roadNet-CA	1957027	2760388	1001273(1001310.9)	**1001269**(1001302)	4
inf-roadNet-PA	1087562	1541514	555220(555242.8)	**555191**(555226.4)	29
rec-amazon	91813	125704	47606(47606)	**47605**(47605.5)	1
rt-retweet-crawl	1112702	2278852	81048(81048)	**81046**(81046.9)	2
sc-ldoor	952203	20770807	**856755**(856757.4)	856756(856758.2)	-1
sc-nasasrb	54870	1311227	51244(51247.4)	**51242**(51245.2)	2
sc-pkustk11	87804	2565054	83911(83912.5)	83911(**83911.8**)	0
sc-pkustk13	94893	3260967	**89217**(89220.6)	89241(89248.5)	-24
sc-pwtk	217891	5653221	207716(207719.9)	**207707**(207715.2)	9
sc-shipsec1	140385	1707759	117318(117338.4)	**117246**(117273.6)	72
sc-shipsec5	179104	2200076	147140(147175)	**147115**(147142.8)	25
soc-buzznet	101163	2763066	30625(30625)	**30618**(30622.3)	7
soc-delicious	536108	1365961	85686(85696.4)	**85527**(85596.9)	159
soc-digg	770799	5907132	103244(103245.3)	103244(**103245**)	0
soc-flickr	513969	3190452	153272(153272)	**153271**(153272)	1
soc-FourSquare	639014	3214986	90109(90109.3)	**90108**(90109.3)	1
soc-gowalla	196591	950327	84222(84222.3)	84222(**84222.2**)	0
soc-livejournal	4033137	27933062	1869045(1869053.7)	**1869036**(1869048.2)	9
soc-pokec	1632803	22301964	**843422**(843434.8)	843426(843433.6)	-4
tech-as-skitter	1694616	11094209	**527185**(527196)	527253(527274.7)	-68
tech-RL-caida	190914	607610	74930(74938.9)	**74863**(74883.5)	67
scc_infect-dublin	10972	175573	9104(9104)	**9103**(9103)	1
web-arabic-2005	163598	1747269	**114426**(114427.2)	114427(114428.2)	-1
web-BerkStan	12305	19500	**5384**(5384)	5385(5385)	-1
web-it-2004	509338	7178413	414671(**414676.3**)	414671(414676.8)	0
web-spam	4767	37375	2298(2298)	**2297**(2297)	1
web-wikipedia2009	1864433	4507315	648317(**648321.7**)	648317(648323.5)	0

Table 2. Comparative performances on instances where both solvers return the same C_{min} and C_{avg} values. We bold the better value between two solvers

Graph	time		#step/ms	
	FastVC	NoiseVC	FastVC	NoiseVC
bio-celegans	<0.01	<0.01	992	**1999**
bio-diseasome	<0.01	<0.01	794	**2351**
bio-dmela	**<0.01**	0.012	929	**1515**
bio-yeast	<0.01	<0.01	903	**2766**
ca-AstroPh	0.038	**0.035**	683	**735**
ca-citeseer	1.503	**0.562**	334	**811**
ca-coauthors-dblp	**14.344**	22.297	**258**	135
ca-CondMat	0.030	**0.021**	678	**1090**
ca-CSphd	<0.01	<0.01	919	**4135**
ca-dblp-2010	2.197	**0.749**	381	**911**
ca-dblp-2012	5.625	**1.799**	316	**809**
ca-Erdos992	<0.01	<0.01	1091	**2687**
ca-GrQc	<0.01	<0.01	701	**1575**
ca-HepPh	0.02	0.02	**650**	580
ca-hollywood-2009	**25.814**	50.033	**220**	104
ca-MathSciNet	5.296	**4.159**	376	**689**
ca-netscience	<0.01	<0.01	759	**2520**
ia-email-EU	<0.01	<0.01	940	**1006**
ia-email-univ	<0.01	<0.01	942	**1562**
ia-enron-large	0.072	**0.054**	585	**1150**
ia-fb-messages	<0.01	<0.01	1108	**1649**
ia-infect-hyper	<0.01	<0.01	**1041**	653
ia-reality	<0.01	<0.01	1153	**1368**
ia-wiki-Talk	0.159	**0.112**	601	**913**
inf-power	0.013	**<0.01**	850	**2889**
rt-retweet	<0.01	<0.01	962	**4351**
rt-twitter-copen	<0.01	<0.01	863	**3312**
sc-msdoor	**18.180**	37.031	**312**	143
socfb-MIT	72.995	**6.442**	**952**	386
soc-BlogCatalog	**0.323**	0.679	**541**	228
soc-brightkite	0.281	**0.219**	648	**1106**
soc-dolphins	<0.01	<0.01	1036	**2427**
soc-douban	**0.012**	0.014	964	**1002**
soc-epinions	0.231	**0.176**	619	**1418**
soc-flixter	3.398	**1.924**	**438**	430
soc-lastfm	1.637	**1.604**	**617**	493
soc-LiveMocha	26.357	**21.853**	**483**	310
soc-slashdot	**0.287**	0.301	547	**818**
soc-twitter-follows	**0.037**	0.044	**1004**	211
soc-wiki-Vote	<0.01	<0.01	951	**2380**
soc-youtube	9.200	**7.350**	318	**644**
soc-youtube-snap	26.553	**12.115**	198	**592**
tech-as-caida2007	<0.01	<0.01	746	**1707**
tech-internet-as	0.026	**0.022**	495	**1409**
tech-p2p-gnutella	0.023	**0.014**	746	**1147**
tech-routers-rf	<0.01	<0.01	914	**2138**
tech-WHOIS	<0.01	<0.01	940	**964**
scc_enron-only	<0.01	<0.01	**813**	191
scc_fb-forum	<0.01	<0.01	**719**	164
scc_fb-messages	<0.01	<0.01	**509**	75
scc_infect-hyper	<0.01	<0.01	**850**	163
scc_reality	0.035	**0.023**	**323**	24
scc_retweet	<0.01	<0.01	**914**	264
scc_retweet-crawl	0.060	**0.029**	513	**1848**
scc_rt_alwefaq	<0.01	<0.01	842	**2121**
scc_rt_assad	<0.01	<0.01	849	**2700**
scc_rt_bahrain	<0.01	<0.01	846	**4403**
scc_rt_barackobama	<0.01	<0.01	963	**3600**
scc_rt_damascus	<0.01	<0.01	1105	**5219**
scc_rt_dash	<0.01	<0.01	780	**5205**
scc_rt_gmanews	<0.01	<0.01	947	**1186**
scc_rt_gop	<0.01	<0.01	972	**7795**
scc_rt_http	<0.01	<0.01	1442	**7897**
scc_rt_israel	<0.01	<0.01	1107	**7918**
scc_rt_justinbieber	<0.01	<0.01	1091	**1452**
scc_rt_ksa	<0.01	<0.01	1058	**4544**
scc_rt_lebanon	<0.01	<0.01	1338	**8566**
scc_rt_libya	<0.01	<0.01	901	**6338**
scc_rt_lolgop	<0.01	<0.01	**1156**	820
scc_rt_mittromney	<0.01	<0.01	966	**5342**
scc_rt_obama	<0.01	<0.01	1328	**9077**
scc_rt_occupy	<0.01	<0.01	777	**4513**
scc_rt_occupywallstnyc	<0.01	<0.01	907	**1285**
scc_rt_oman	<0.01	<0.01	801	**6134**
scc_rt_onedirection	<0.01	<0.01	**867**	864
scc_rt_p2	<0.01	<0.01	1019	**7554**
scc_rt_qatif	<0.01	<0.01	882	**7185**
scc_rt_saudi	<0.01	<0.01	1004	**2793**
scc_rt_tcot	<0.01	<0.01	956	**7273**
scc_rt_tlot	<0.01	<0.01	1010	**7216**
scc_rt_uae	<0.01	<0.01	946	**6493**
scc_rt_voteonedirection	<0.01	<0.01	1145	**7006**
scc_twitter-copen	**<0.01**	0.01	**790**	110
web-edu	<0.01	<0.01	981	**2237**
web-google	<0.01	<0.01	754	**1849**
web-indochina-2004	**0.148**	0.172	629	**1214**
web-polblogs	<0.01	<0.01	902	**1942**
web-sk-2005	15.511	**6.455**	424	**864**
web-uk-2005	0.045	**0.038**	**360**	60
web-webbase-2001	0.024	**<0.01**	695	**1801**

4.3   Experimental Results

The main experimental results are shown in Table 1. We tested all the 139 instances, and the results on graphs where NoiseVC and FastVC return precisely return solutions with both the same C_{min} and C_{avg} are not reported in Table 1.

According to the results in Table 1, we observe that:

1. Out of 44 graphs, NoiseVC finds better and worse vertex covers than FastVC in 23 and 8 graphs, respectively.
2. NoiseVC finds the same minimum vertex cover as FastVC in 13 graphs, among which NoiseVC obtains smaller average size of vertex cover for 10 graphs (bold value in Table 1).
3. NoiseVC outperforms FastVC for 8 classes of instances listed in Table 1, except for technological networks and web link networks.

Table 2 shows the performances on those instances where two solvers return both the same C_{min} and C_{avg} values. We also present the averaged number of steps to locate a solution, and the number of steps executed in each millisecond. The time columns show that NoiseVC outperforms FastVC on 24 instances, while FastVC is faster than NoiseVC on 10 instances. The last two columns reveal that the complexity per step in NoiseVC is significantly lower than that in FastVC.

Speed Improvements. Over a half of the 23 graphs where we found smaller covers, NoiseVC makes a substantially large progress. Now we show how great the progress is. We enlarged the cutoff to be 100 times as large as before (i.e., **100,000 s**), and tested FastVC over such graphs. The results are shown in Table 3. Also we present the respective results of NoiseVC within **1,000 s** in this table.

As is shown in Table 3, even within such a large cutoff, FastVC does not get the same solution quality as NoiseVC does with a cutoff of 1,000 s for any of

Table 3. Results on the 12 graphs on which NoiseVC makes a substantially large progress

Graph	$|V|$	$|E|$	FastVC×100 $C_{min}(C_{avg})$	NoiseVC $C_{min}(C_{avg})$	Δ
inf-roadNet-CA	1957027	2760388	1001272(1001306.3)	**1001269**(1001302)	3
inf-roadNet-PA	1087562	1541514	555220(555242)	**555191**(555226.4)	29
rec-amazon	91813	125704	47606(47606)	**47605**(47605.5)	1
sc-pwtk	217891	5653221	207712(207717.2)	**207707**(207715.2)	5
sc-shipsec1	140385	1707759	117298(117313.8)	**117246**(117273.6)	52
sc-shipsec5	179104	2200076	147130(147171.3)	**147115**(147142.8)	15
soc-buzznet	101163	2763066	30625(30625)	**30618**(30622.3)	7
soc-delicious	536108	1365961	85685(85695.5)	**85527**(85596.9)	158
soc-flickr	513969	3190452	153272(153272)	**153271**(153272)	1
tech-RL-caida	190914	607610	74930(74938.9)	**74863**(74883.5)	67
scc_infect-dublin	10972	175573	9104(9104)	**9103**(9103)	1
web-spam	4767	37375	2298(2298)	**2297**(2297)	1

Table 4. Experimental results on different values of p over 10 runs. $\bar{\Delta}$ means the average Δ on all 136 graphs

p	0.1	0.2	0.3	0.4	0.5	0.6	0.7	0.8	0.9
$\bar{\Delta}$	5.8	4	2.7	2.2	1.2	1	-0.1	-0.7	-0.8

these 12 graphs. That is, our solver is *at least 100 times as efficient as* FastVC on these graphs.

Parameter Testing. Table 4 shows the results of NoiseVC with different parameter p ranging from 0.1 to 0.9. We make a comparison with FastVC on average Δ over 136 instances. NoiseVC outperforms FastVC significantly on a wide range of parameter settings, from 0.1 to 0.6.

5 Conclusions

In this work, we propose a new algorithm named NoiseVC for MinVC in massive real-world graphs. A heap is exploited for best-picking in the local search phase, and a new noisy strategy is combined with best-picking to avoid local minima. The experimental results indicate that NoiseVC significantly outperforms FastVC on finding smaller vertex cover in most of the massive graphs of the benchmarks. In the future, we would like to design heap-based best-picking heuristic for other combinatorial optimization problems.

Acknowledgment. This work is supported by ARC Grant FT0991785, NSF Grant No. 61463044 and Grant No. [2014]7421 from the Joint Fund of the NSF of Guizhou province of China.

References

1. Andrade, D.V., Resende, M.G.C., Werneck, R.F.F.: Fast local search for the maximum independent set problem. J. Heuristics **18**(4), 525–547 (2012)
2. Barbosa, V.C., Campos, L.C.D.: A novel evolutionary formulation of the maximum independent set problem. J. Comb. Optim. **8**(4), 419–437 (2004)
3. Cai, S.: Balance between complexity and quality: local search for minimum vertex cover in massive graphs. In: Proceedings of the Twenty-Fourth International Joint Conference on Artificial Intelligence, IJCAI, pp. 25–31 (2015)
4. Cai, S., Su, K., Chen, Q.: EWLS: a new local search for minimum vertex cover. In: AAAI (2010)
5. Cai, S., Su, K., Luo, C., Sattar, A.: NuMVC: an efficient local search algorithm for minimum vertex cover. J. Artif. Intell. Res. **46**, 687–716 (2013)
6. Cai, S., Su, K., Sattar, A.: Local search with edge weighting and configuration checking heuristics for minimum vertex cover. Artif. Intell. **175**(9), 1672–1696 (2011)

7. Chung Graham, F., Lu, L.: Complex graphs and networks american mathematical society (2006)
8. Cormen, T.H., Leiserson, C.E., Rivest, R.L., Stein, C.: Introduction to Algorithms, vol. 6. MIT Press, Cambridge (2001)
9. Eubank, S., Kumar, V., Marathe, M.V., Srinivasan, A., Wang, N.: Structural and algorithmic aspects of massive social networks. In: Proceedings of the Fifteenth Annual ACM-SIAM Symposium on Discrete Algorithms, pp. 718–727. Society for Industrial and Applied Mathematics (2004)
10. Ji, Y., Xu, X., Stormo, G.D.: A graph theoretical approach for predicting common RNA secondary structure motifs including pseudoknots in unaligned sequences. Bioinformatics 20(10), 1603–1611 (2004)
11. Jin, Y., Hao, J.: General swap-based multiple neighborhood tabu search for the maximum independent set problem. Eng. Appl. AI 37, 20–33 (2015)
12. Johnson, D.S., Trick, M.A.: Cliques, Coloring, and Satisfiability: Second DIMACS Implementation Challenge, vol. 26. American Mathematical Society, Providence (1996)
13. Katayama, K., Hamamoto, A., Narihisa, H.: An effective local search for the maximum clique problem. Inf. Process. Lett. 95(5), 503–511 (2005)
14. Li, C.M., Quan, Z.: An efficient branch-and-bound algorithm based on maxsat for the maximum clique problem. In: Proceedings of the Twenty-Fourth AAAI Conference on Artificial Intelligence, AAAI 2010, vol. 10, pp. 128–133. AAAI Press, Atlanta (2010)
15. Pullan, W.J., Hoos, H.H.: Dynamic local search for the maximum clique problem. J. Artif. Intell. Res. (JAIR) 25, 159–185 (2006)
16. Richter, S., Helmert, M., Gretton, C.: A stochastic local search approach to vertex cover. In: Hertzberg, J., Beetz, M., Englert, R. (eds.) KI 2007. LNCS (LNAI), vol. 4667, pp. 412–426. Springer, Heidelberg (2007)
17. Rossi, R., Ahmed, N.: The network data repository with interactive graph analytics and visualization. In: AAAI, pp. 4292–4293 (2015)
18. Rossi, R.A., Ahmed, N.K.: Coloring large complex networks. Soc. Netw. Anal. Min. 4(1), 1–37 (2014)
19. Rossi, R.A., Gleich, D.F., Gebremedhin, A.H., Patwary, M.M.A.: Fast maximum clique algorithms for large graphs. In: Proceedings of the Companion Publication of the 23rd International Conference on World Wide Web Companion, pp. 365–366. International World Wide Web Conferences Steering Committee (2014)
20. Segundo, P.S., Rodríguez-Losada, D., Jiménez, A.: An exact bit-parallel algorithm for the maximum clique problem. Comput. OR 38(2), 571–581 (2011)
21. Traud, A.L., Mucha, P.J., Porter, M.A.: Social structure of facebook networks. Phys. A Stat. Mech. Appl. 391(16), 4165–4180 (2012)
22. Wu, Q., Hao, J.K.: A review on algorithms for maximum clique problems. Eur. J. Oper. Res. 242(3), 693–709 (2015)

An FAQ Search Method Using a Document Classifier Trained with Automatically Generated Training Data

Takuya Makino[✉], Tomoya Noro[✉], and Tomoya Iwakura[✉]

Fujitsu Laboratories Ltd., Sunnyvale, USA
{makino.takuya,t.noro,iwakura.tomoya}@jp.fujitsu.com

Abstract. We propose an FAQ (Frequently Asked Question) search method that uses classification results of input queries. FAQs aim at covering frequently asked topics and users usually search topics in FAQs with queries represented by bag-of-words or natural language sentences. However, there is a problem that each question in FAQs is not usually sufficient enough to cover variety of queries that have the similar meaning but different surface expressions, such as synonyms, paraphrase and causal relations due to each topic usually consists of a representative question and its answer. As a result, users who cannot find their answers in FAQs ask a call center operator. To consider similarity of meaning among different surface expressions, we use a document classifier that classifies each query into topics of FAQs. A document classifier is trained with not only FAQs but also corresponding histories of operators for covering variety of queries. However, corresponding histories do not include links to FAQs, we use a method for generating training data from the corresponding histories with FAQs. To generate training data correctly, the method takes advantage of a characteristic that many answers in corresponding histories related to FAQs are created by quoting corresponding FAQs. Our method uses a surface similarity between answers in corresponding histories and the answer part of each topic in FAQs for automatically generating training data. Experimental results show that our method outperforms an FAQ search based method using word matching in terms of Mean Reciprocal Rank and Precision@N.

1 Introduction

Call centers are managed to respond their users' question from not only customers of companies but also employees of large companies. To reduce the cost of operators of call centers, FAQs are prepared, which are a set of a question and its answer. Development of FAQs have the following benefits.

1. If users can find answers from FAQs by themselves, the cost of operators is reduced.
2. When users ask call center operators and their answers exist in FAQs, the cost of operators is reduced because operators can answer questions from users by quoting FAQs.

© Springer International Publishing Switzerland 2016
R. Booth and M.-L. Zhang (Eds.): PRICAI 2016, LNAI 9810, pp. 295–305, 2016.
DOI: 10.1007/978-3-319-42911-3_25

Traditional FAQ search methods use a surface similarity between a query and an FAQ for ranking FAQs [2,8]. However, this approach cannot handle queries that have the same meaning but different surface expressions such as synonyms, paraphrase and causal relations. Let us consider the following example.

query: I lost my credit card.
the question part of an FAQ: How do I request a new / replacement card?
the answer part of an FAQ: You can request a replacement card by signing
    in to the online banking and going to the information tab of your account.

In the above example, only "I" and "card" are matched with the words in the question part of an FAQ, therefore, the score for the answer given by a search based method would not be high.

One of the solutions to this problem is to create knowledge such as a causal relation between "lose" and "replacement". However, the cost of constructing these knowledge should be high because we have to prepare domain specific knowledge in many cases. Some previous works proposed to use a word alignment model for finding semantically similar questions [10,11,13]. They regarded pairs of similar questions or pairs of question and its answer as parallel corpus. Even though their models can learn the probability of a relatedness between two words, their model cannot learn how important a probability of two words for searching correct FAQs.

We propose the use of a document classifier that predicts whether or not a given query corresponds to each FAQ. We need certain amount of training data for training a document classifier that covers variety of surface patterns, however, we usually have only FAQs that are made by grouping past same topic questions and only a representative question of each topic is reserved. To train a document classifier that covers variety of surface patterns, we use training data automatically generated from corresponding histories and FAQs. Corresponding histories do not include links to FAQs, therefore, we take advantage of characteristic that many answers in corresponding histories related to an FAQ are created by quoting the answer part of the FAQ. Our method uses a surface similarity between answers in corresponding histories and the answer part of the topic for automatically generating training data. Then, we train a document classifier with FAQs and the generated training data for predicting corresponding FAQs of a given query. The trained document classifier is used for augmenting features of a learning to rank-based search. Classification results of given queries by the classifier are used features for ranking FAQs.

We evaluate our method with an in-company FAQs and corresponding histories. The experimental results show that our method outperforms a search-based method only uses surface similarities and a word alignment model-based method in terms of MRR, Precision@N.

2   Proposed Method

This section describes the problem definition for a learning to rank of this paper and our proposed method. The proposed method consists of the following three steps. At first, our method generates training data automatically. Then, our method trains a document classifier that predicts whether a given question corresponds to each FAQ or not. Finally, our model trains a ranking model to assign higher score for a correct FAQ than an incorrect FAQ for a question.

2.1   Preliminaries

We employ a learning to rank approach for searching FAQs with a given query. The learning objective is to induce a model that assigns higher score to the correct FAQ than the all incorrect FAQs of each given question. The features are not only surface-based ones but also classification results obtained with a document classifier that classifies queries into FAQs. We train a document classifier and a ranking model with FAQs and corresponding histories. FAQs are a set of a question Q and its answer A: $D_1 = \{(Q_1, A_1), ..., (Q_M, A_M)\}$. Corresponding histories are a set of pairs of a user's question I and its answer R: $D_2 = \{(I_1, R_1), ..., (I_N, R_N)\}$. Figure 1 shows examples of FAQs and corresponding histories.

2.2   Generating Training Data Automatically

We have corresponding histories and FAQs, however, answers in corresponding histories do not include links to FAQs. Annotating such data manually is very

FAQs

ID		
FAQ1	question	How do I request a new / replacement card?
	answer	You can request an replacement card by signing to the online banking and going to the information tab of your account.
FAQ2	question	When will I receive my credit card?
	answer	You will receive your card within 10 business days.

Corresponding histories

ID		
Log1	question	I lost my credit card.
	answer	You can request by signing to the online banking
Log2	question	I forgot the password of my credit card.
	answer	Select 'Forgot your password?' link on the log in screen.
Log3	question	When can I get my credit card?
	answer	It is about four weeks.

Fig. 1. Examples of FAQs and corresponding histories.

expensive, therefore, we employ an automatic generation method of training data. To know characteristics of corresponding histories, we first asked operators how they answer for a question. The answers are the following. When an operator receives a question from a user, the operator searches FAQs based on the question. If the operator can find its answer from FAQs, they answer by quoting the answer part of an FAQ. Based on such characteristics, we generate training data automatically by calculating a surface similarity between the answer records in corresponding histories and the answer part of an FAQ.

Our method generates training data automatically by following the previous work [7] that calculates a surface similarity between answer records of corresponding histories. To generate training data automatically, we use FAQs and corresponding histories. FAQs are a set of a question Q and its answer A: $D_1 = \{(Q_1, A_1), ..., (Q_M, A_M)\}$. Corresponding histories are a set of pairs of a user's question I and its answer R: $D_2 = \{(I_1, R_1), ..., (I_N, R_N)\}$.

We use harmonic mean of reciprocal ranks (hrank) that is defined in Eq. 1 as a similarity between the answer records in corresponding histories and the answer part of an FAQ. To handle a large size of corresponding histories, we use the score of a full text search engine Elasticsearch[1] that calculates the tfidf similarity between an answer records in corresponding histories and the answer part of an FAQ. $rank_{A_i}$ indicates the rank of the answer part of an FAQ A_i when searching FAQs by an answer record R_j. $rank_{R_j}$ indicates the rank of an answer record R_j when searching answer records in corresponding histories by the answer part of an FAQ A_i. When $hrank(A_i, R_j)$ is larger than a threshold t for generating training data, our method links an FAQ A_i and its question Q_i to the question I_j of an answer record R_j. Our method generates $D_3 = \{(Q_i, A_i, I_j)|1 \leq m \leq M, 1 \leq n \leq N\}$ as training data.

$$hrank(A_i, R_j) = \frac{1}{2}(\frac{1}{rank_{A_i}} + \frac{1}{rank_{R_j}}) \tag{1}$$

We describe an example of generating training data with FAQs and corresponding histories in Fig. 1. Our method searches corresponding histories with FAQs. Then, our method searches FAQs with corresponding histories. For example, when our method searches answers in corresponding histories with the answer part of FAQ1 in FAQs, the answer of Log1 in corresponding histories is ranked at the first place. Next, our method searches answer parts of FAQs with the answers in corresponding histories. Here, the answer part of FAQ1 in FAQs is ranked at the first place. In this situation, the hrank of FAQ1 and Log1 takes 1 and our method appends FAQ1 and the question of Log1 (("How do I request a new / replacement card?", "You can request an replacement card by signing to online banking and going to information tab for your account."), "I lost my credit card.") to training data D_3.

---

[1] https://www.elastic.co/jp/.

2.3   Training a Document Classifier

We train a document classifier with the automatically generated training data in addition to FAQs. Since the number of FAQs is more than two, we employ a multi class classification method based a one-versus-the-rest method. A binary document classifier for each FAQ is trained with the questions that are linked with the FAQ as positive examples, the questions that are linked with the other FAQs as negative examples. Question parts of FAQs are also used for training. When our model trains a binary document classifier for an FAQ (Q_i, A_i), questions that are linked to (Q_i, A_i) in D_3 and Q_i are used as positive examples, and questions that are linked to (Q_k, A_k) in D_3 $(1 \leq k \leq M \text{and} k \neq i)$ is used as negative examples.

For training binary classifiers, we use Adaptive Regularization of Weight Vectors [5] with the following features. We use unigrams of content words in base forms, word bigrams, pairs of content words in base forms that has a dependency relation as features. Except for word bigrams, we convert a word to a base form.

For example, when training a binary classifier for the FAQ "How can I request a replacement card" and positive example is "I lost my credit card.". Our method extracts the following text representations as features from a given question:

Unigrams of content words in base forms: "lose", "credit" and "card"
Bigrams of words: "(I, lost)", "(lost, my)", ..., "(credit, card)"
A pair of content words that have a dependency relation: "lose→card" and "lose→credit"

2.4   Learning a Ranking Model

To train a ranking model, we use the automatically generated training data in addition FAQs as in the document classifier training. Our approach is based on a pairwise learning algorithm that learns a ranker to assign higher score for a correct FAQ than an incorrect FAQ.

We show our learning to rank procedure in Algorithm 1. For a question in the automatically generated training data, GetFeatVec($\hat{Q}, \hat{A}, I$) extracts a feature vector from a question and a correct FAQ. GetRndFalsePair(I, D_1) randomly samples an incorrect FAQ and GetFeatVec(Q_k, A_k, I) extracts a feature vector from a question and an incorrect FAQ. UpdateWeight($\mathbf{w_r}, \mathbf{x}$) updates a weight vector with difference of these two feature vectors. ϕ_r indicates a feature vector extracted from input question I, both of the question part Q and the answer part A of an FAQ. We use Adaptive Regularization of Weight Vectors [5] for updating weight vector and set K to 10.

A GetFeatureVec(Q_i, A_i, I) function in Algorithm 1 extracts the following features:

– **cos-q, cos-a:** A cosine similarity between a given question and the question part of an FAQ and a cosine similarity between a given question and the answer part of an FAQ

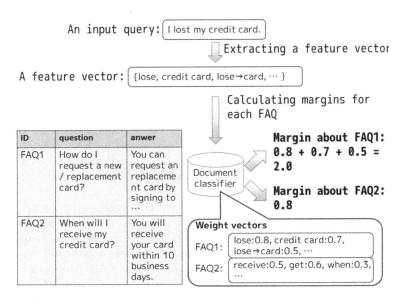

Fig. 2. The example of calculating margins for each FAQ by using a document classifier

---

Algorithm 1. pairwise learning to rank

1: $\mathbf{w_r} \leftarrow \mathbf{0}$
2: **for** $(\hat{Q}, \hat{A}, I) \in D_3$ **do**
3:     $\mathbf{x}_p \leftarrow \text{GetFeatVec}(\hat{Q}, \hat{A}, I)$
4:     **for** k **do**1...K
5:         $(Q_k, A_k, I) \leftarrow \text{GetRndFalsePair}(I, D_1)$
6:         $\mathbf{x}_n \leftarrow \text{GetFeatVec}(Q_k, A_k, I)$
7:         $\mathbf{x} \leftarrow \mathbf{x}_p - \mathbf{x}_n$
8:         $\mathbf{w_r} \leftarrow \text{UpdateWeight}(\mathbf{w_r}, \mathbf{x})$
9:     **end for**
10: **end for**

- **dep**: The number of pairs of words each of which has a dependency relation.
- **np**: The number of noun phrases that each of which occurs in both a given question and the question part of an FAQ divided by the number of noun phrases that occur either a given question and the question part of an FAQ.
- **syn**: Whether the same synset occurs in the both of question and question part of FAQ. A synset is a group of word senses that has similar meaning in WordNet.
- **faq-cat**: Whether the category of an FAQ exists in predicted top-5 categories of a given question. Since an FAQ has categories, we used category information of an FAQ. We train a document classifier that predicts the category for a given question because an FAQ has categories but a given question does not have

categories. We train a category classifier by using bag-of-words as features and employs Adaptive Regularization of Weight Vectors [5] for learning.

– **faq-scorer**: We use the margin of a binary classifier for an FAQ (Q_i, A_i) with a sigmoid fitting as a feature.

In a training phase of a ranker, our model uses the difference of feature vectors between true FAQ and false FAQ for learning $\mathbf{w}_r$. In a test phase a ranker, our model extracts a feature vector for each FAQ, and ranks FAQs by sorting scores of FAQs assigned by the trained ranker.

3   Experiments

3.1   Experimental Setup

We used an in-company FAQs and corresponding histories for experiments. For creating test data, we annotated randomly sampled 286 questions from corresponding histories with their correct FAQs.

For generating training data, we set the threshold for automatically generating training data to 0.6. If the answer part of an FAQ is too short, linked questions in corresponding histories may include noise. Therefore, we discarded generated training data for FAQs those number of characters in answer part is less than 10. Our method generated $27,040$ pairs of questions and FAQs. We removed questions that exists in the both of the training data and the test data for training our model.

We used a Japanese morphological analyzer MeCab[2] for word segmentation and a Japanese dependency parser CaboCha[3] for dependency parsing. We used Mean Reciprocal Rank (MRR) and Precision@N (P@N) as evaluation metrics. MRR is the average of reciprocal ranks of correct FAQs and it is going to 1 when correct FAQs ranked higher than incorrect FAQs. P@N is the ratio of a correct FAQ that are ranked higher than Nth and it is going to 1 when correct FAQs are ranked higher than Nth.

We compare our proposed method with baselines that are a full text search engine Elasticsearch and a word alignment model-based one [7]. A word alignment model-based model is formalized as follows:

$$P(Q|I) = \prod_{w \in Q} P(w|I), \tag{2}$$

where $P(w|I)$ is calculated as

$$P(w|I) = (1 - \lambda) \sum_{t \in Q} (P_{tr}(w|t) P_{ml}(t|I)) + \lambda P_{ml}(w|C). \tag{3}$$

We estimated $P_{tr}(w|t)$ with the automatically generated training data by using GIZA++[4]. Following Jeon et al. [7], we set $P_{tr}(w|w) = 1$. We set λ that maximizes MRR of test data.

---

[2] https://taku910.github.io/mecab/.
[3] https://taku910.github.io/cabocha/.
[4] http://www.statmt.org/moses/giza/GIZA++.html.

3.2   Experimental Results

Evaluation of Automatically Generated Training Data: We randomly sampled 50 pairs from automatically generated training data and evaluated those pairs by a human annotator. We show the evaluation result in Table 1.

Almost half of pairs are correct. When the answer part of an FAQ is short, paired questions are noisy.

Table 1. Evaluation of automatically generated pairs

label	number
true	24
false	26

Evaluation of an FAQ Category Classifier: The in-company FAQ used in our experiment has hierarchical categories. We used categories at depth two and number of categories is 107. We conducted 10-fold cross validation using FAQ. Since FAQ has categories which is assigned in advance. We expected that FAQ category classifier can predict that category-level similarity between question and FAQ. We used Adaptive Weight Regularization of Weight Vectors [5] for updating a parameter vector. We show the results of FAQ category classification in Table 2.

Table 2. P@N for FAQ category classification

	P@N
P@1	0.758
P@2	0.839
P@3	0.872
P@4	0.889
P@5	0.898

Results of FAQ Ranking: We show ranking results in Table 3. For full text search, we use search queries consist of content words which are joined with OR. Our proposed method outperformed full text search and word alignment-based model in terms of MRR, P@1, P@5, P@10.

We conducted ablation tests and Table 4 shows the evaluation results. We can see that the contribution of faq-scorer is the largest in our feature set.

Figure 3 shows an MRR learning curve of our proposed method. To plot MRR, we selected 1,000 training data and incrementally learned our model at each step. MRR of our proposed method is improved by increasing the number of training data.

Table 3. Comparison with baselineValues with † significantly differ from our proposed method. We conducted paired t-test in terms of MRR, P@1, P@5, P@10.

method	MRR	P@1	P@5	P@10
Proposed Method	0.478	0.367	0.605	0.727
word alignment-based model	0.315†	0.238†	0.402†	0.476†
full text search	0.276†	0.174†	0.388†	0.483†

Table 4. Ablation tests

method	MRR	P@1	P@5	P@10
Proposed method	0.478	0.367	0.605	0.727
w/o syn	0.478	0.367	0.601	0.727
w/o dep	0.478	0.363	0.612	0.731
w/o np	0.476	0.360	0.605	0.717
w/o faq cat	0.469	0.357	0.598	0.710
w/o cos-{q,a}	0.397	0.311	0.486	0.605
w/o faq scorer	0.346	0.220	0.486	0.601

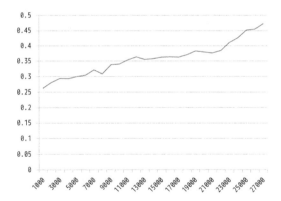

Fig. 3. Learning curve of MRR of our proposed method

Analysis of FAQ Scorer: Table 5 shows features that have larger weight for FAQ "How can I request a new / replacement card?". For example, this FAQ can correspond to a question that has "lose→card", "magnetic failure" or "wallet", which are not included in the original question of the FAQ.

Error Analysis: Main reason of failure of our method is that some FAQs do not have paired questions. If the answer part of an FAQ is similar with the answer part of the other FAQ, our method tends to fail because the questions that are linked automatically to those FAQs have similar contents.

Table 5. Features that have positive weight

Feature name	Feature
dependency	lose→card
noun phrase	replacement application
word bigram	magnetic failure
dependency	card→stole
word unigram	lose
word unigram	wallet

4 Related Works

For searching FAQs or a community QA site such as Yahoo! Answers, some previous research used WordNet or Wikipedia [2,14] for using synonym dictionaries. Since WordNet and Wikipedia do not cover the topic of FAQs, it's difficult to use knowledge such as synonyms.

There are some researches propose the use of IBM models [1] that learn a probability of two words with a statistical machine translation algorithm [7,10, 11,13]. These models learn an alignment probability of two words. It is important to align two words in machine translation, however, an alignment probability does not indicate an importance for finding correct FAQs.

Cao et al. [3,4] proposed to use category information for language model. Their models calculate a probability that a given question belongs to the category of a candidate question and assign it to an alignment probability. Their methods are similar to our model in terms of using a document classifier, but our document classifier predicts directly whether a given question corresponds to each FAQs or not directly.

Ko et al. [9], Surdeanu et al. [12] and Higashinaka and Isozaki [6] used learning to rank for question answering. Their models are similar to our model in terms of learning to rank but features of their model are based on a surface similarity, an alignment probability, a dictionary based similarity and a query log based similarity. Our model uses the output of a document classifier that predicts whether a given question corresponds to FAQs or not.

5 Conclusion

We proposed an FAQ search method that uses classification results of input queries. For training a document classifier, our method generates training data automatically. By utilizing a document classifier that predicts whether a given question corresponds to each FAQ or not, our method outperformed baselines, which are a full text search and a word alignment model-based one. In the future work, we need to improve the generating method of training data.

References

1. Brown, P.F., Pietra, V.J.D., Pietra, S.A.D., Mercer, R.L.: The mathematics of statistical machine translation: parameter estimation. Comput. Linguist. (1993)
2. Burke, R., Hammond, K., Kulyukin, V., Lytinen, S., Tomuro, N., Schoenberg, S.: Natural language processing in the FAQ finder system: results and prospects. In: Working Notes from AAAI Spring Symposium on NLP on the WWW (1997)
3. Cao, X., Cong, G., Cui, B., Jensen, C.S.: A generalized framework of exploring category information for question retrieval in community question answer archives. In: Proceedings of the WWW (2010)
4. Cao, X., Cong, G., Cui, B., Jensen, C.S., Zhang, C.: The use of categorization information in language models for question retrieval. In: Proceedings of CIKM (2009)
5. Crammer, K., Kulesza, A., Dredze, M.: Adaptive regularization of weight vectors. In: Proceedings of NIPS (2010)
6. Higashinaka, R., Isozaki, H.: Corpus-based question answering for why-questions. In: Proceedings of IJCNLP (2008)
7. Jeon, J., Croft, W.B., Lee, J.H.: Finding similar questions in large question and answer archives. In: Proceedings of CIKM (2005)
8. Jijkoun, V., de Rijke, M.: Retrieving answers from frequently asked questions pages on the web. In: Proceedings of CIKM (2005)
9. Ko, J., Mitamura, T., Nyberg, E.: Language-independent probabilistic answer ranking for question answering. In: Proceedings of ACL (2007)
10. Riezler, S., Vasserman, A., Tsochantaridis, I., Mittal, V., Liu, Y.: Statistical machine translation for query expansion in answer retrieval. In: Proceedings of ACL (2007)
11. Soricut, R., Brill, E.: Automatic question answering using the web: beyond the factoid. Inf. Retr. **9**, 191–206 (2006)
12. Surdeanu, M., Ciaramita, M., Zaragoza, H.: Learning to rank answers on large online QA collections. In: Proceedings of ACL (2008)
13. Xue, X., Jeon, J., Croft, W.B.: Retrieval models for question and answer archives. In: Proceedings of SIGIR (2008)
14. Zhou, G., Liu, Y., Liu, F., Zeng, D., Zhao, J.: Improving question retrieval in community question answering using world knowledge. In: Proceedings of IJCAI (2013)

An Analysis of Influential Users for Predicting the Popularity of News Tweets

Krissada Maleewong[✉]

School of Information Technology, Shinawatra University, Pathumthani, Thailand
krissada@siu.ac.th

Abstract. Twitter plays an important role in today social network. Its key mechanism is retweet that disseminates information to broad audiences within a very short time and help increases the popularity of the social content. Therefore, an effective model for predicting the popularity of tweets is required in various domains such as news propagation, viral marketing, personalized message recommendation, and trend analysis. Although many studies have been extensively researched on predicting the popularity of tweets, they mainly focus on the content-based and the author-based features, while retweeter-based features are less concerned. This paper aims to study the impact of influential users who retweet tweets, also called retweeters, and presents simple yet effective measures for predicting the influence of retweeters on the popularity of online news tweets. By analyzing the popularity of news tweets and the impact of the retweeters, a number of useful measures are defined to evaluate influence of users in the retweeter network, and used to establish the prediction model. The experimental results show that the application of the retweeter-based features is highly effective and enhances the performance of the prediction model with high accuracy.

Keywords: Twitter · Retweet · Influential user · Active user · Popular user · News tweet · Social network

1 Introduction

Nowadays, Twitter is considered as the most prominent micro-blogging service available on the Web. It allows people to publish 140-character short messages known as tweets, which can also contain images, videos, or URLs that link to the original online sources. In the Twitter network, users (a.k.a. *followers*) can follow other notable users (a.k.a. *followees*) to gain real-time updates on news and statuses. When a user finds an interesting tweet written by another user and wants to share it with his/her followers, the user (a.k.a. *retweeter*) can retweet such a tweet by either using a retweet button or manually editing the original tweet and adding a text indicator (e.g., RT *@user* or via *@user*) to mention that the original tweet came from the specified user. Therefore, retweeting mechanism is an important technique for information diffusion in Twitter and utilized in several applications such as breaking news detection, personalized message recommendation, viral marketing, trend analysis, and Twitter-based early warning systems. By focusing on online news, Twitter is adopted as a new medium for

© Springer International Publishing Switzerland 2016
R. Booth and M.-L. Zhang (Eds.): PRICAI 2016, LNAI 9810, pp. 306–318, 2016.
DOI: 10.1007/978-3-319-42911-3_26

disseminating news from their websites to the readers. Several news sources such as BBC News, CNN, and Bangkok Post distribute latest news to readers using their Twitter, while readers read the articles and might follow the URLs attached in the news tweets to the original news sources for further reading. A reader who finds an interesting news tweet can retweet the news tweets to his/her followers. This mechanism results in not only the popularity of news tweets but also the popularity of the root of news pages published on the news websites. Accordingly, the retweeting mechanism can help people to access the news faster and empowers the online news channels to widespread their content in social network.

One important factor that affects the popularity of news tweets is users. In this paper, a user who retweets tweets is called a *retweeter*. Since different retweeters have unequally influence, an analysis of the impact of influential retweeters on the popularity of tweets is required. In addition, understanding how a tweet becomes popular can help to gain a better insight into how the information is dispersed over the social network. However, predicting the popularity of news tweet is a challenging task comparing to other kinds of tweet due to theirs short life time. Therefore, this paper aims to study important features that impact the popularity of news tweets based on retweeting mechanism and retweeter-based features. By analyzing the influence of different types of retweeters, a number of useful measures are proposed to evaluate the influence of retweeters. Based on the proposed measures, a number of retweeters-based features are introduced for predicting the popularity of news tweets.

The organization of this paper is as follows: Sect. 2 discusses the related work. Section 3 analyzes the popularity of news tweets and studies the impact of influential retweeters. Section 4 presents the prediction model. Section 5 performs an experiment and reports the experiment results. Section 6 draws conclusions and future research direction.

2 Related Work

Many studies have been researched on predicting the popularity of tweets by proposing a variety of features such as content-based features, contextual or author-based features, network structural features, and temporal features. These features describe the characteristics and the past evolution of tweets, as well as their social interaction. By applying a generalized linear model, the content-based features (i.e., numbers of URLs and hashtags), and the author-based features (i.e., number of followers and followees, and age of the account) are investigated in order to calculate the retweetability of tweets (Suh et al. 2010). Later, various types of features are introduced including the content-based and temporal information of tweets, metadata of tweets and users, as well as structural properties of the users in social network for estimating the number of future retweets using binary and multi-class classification models (Hong et al. 2011). By concerning a set of social features (i.e., number of followers, friends, statuses, favorites, and number of times the user was listed), the propagation of tweets is predicted based on the passive-aggressive algorithm (Petrovic et al. 2011). In addition, the evolution of retweets is predicted based on the size of the retweeter network and the depth from the source of tweets using a probabilistic model (Zaman et al. 2014). By investigating the network of SinaWeibo, the biggest microblogging system in China, the *structural characteristics* (Bao et al. 2013) incorporates the early popularity with the link density and the diffusion depth of early adopters for predicting the popularity of short messages. To predict the popularity of newly emerging hashtags in Twitter (Ma et al. 2013), a set of content-based and author-based features is applied to five standard classification models (i.e., Naïve bayes, k-nearest neighbors, decision trees, support vector machines, and logistic regression). The results show that the logistic regression model performs the best but the experiment relaxed the problem and predicted the range of popularity instead of the exact value of the popularity.

With emphasis on analyzing the influence of users in social network, various measures have been introduced. *FollowerRank* (Nagmoti et al. 2010) and *Structural Advantage* (Cappelletti and Sastry 2012) adapt the traditional *in-degree measure* (Hajian and White 2011) for determining the popularity of a user. *Follower-Followee ratio* (Bigonha et al. 2012) and *Paradoxical Discounted* (Gayo-Avello 2013) consider numbers of followers and followees for detecting spammers (users with many followees and few followers). By concerning influential users as celebrities, *StarRank* (Khrabrov and Cybenko 2010) applies PageRank algorithm for determining the acceleration of mentions over time, while *Acquaintance Score* (Srinivasan et al. 2013) utilizes numbers of mentions, replies, and retweets. In order to rank users based on their activities, *TweetRank* (Nagmoti et al. 2010) counts the number of tweets of the user, and *TweetCount-Score* (Neves et al. 2015) counts the number of tweets of a user plus the number of retweets. Based on time concerning, *Effective readers* (Lee et al. 2010) measures the speed of a user to tweet about a new topics, while *ActivityScore* (Yuan et al. 2013) counts the number of followers, followees, and tweets of each user during a period of time. *IP Influence* (Romero et al. 2011) evaluates the influence of the users and their passivity using metrics of retweets, followers, and followees. However, most researches have focused on the content-based features of the tweets and the author-based features of the authors who post the tweets, while the the retweeter-based features of users who help disseminate the tweets are less concerned in the prediction models.

3   Analysis of Influential Users in Popular News Tweets

This section describes the dataset and presents data analysis of the characteristics of popular news tweets as well as the impact of influential retweeters.

3.1   Dataset and Popularity of News Tweets

The dataset was collected from BBC News Twitter using Twitter Streaming API. It provides various topics including business, entertainment, science, sport, and weather, etc. The dataset contains 192,821 retweet data collected from 3,336 news tweets posted before December 2015 to make sure that there were likely to be no more retweet occurring.

In most of the recent works, the popularity of tweets is evaluated by the number of retweets because it is the most effective measure to disseminate messages comparing to other metrics such as the number of favorites or replies. In this paper, the *popularity* of news tweets, therefore, refers to the total number of retweets that the tweets receive. Due to problems with identifying the connection between the root tweet and the subsequent manual retweets in the dataset retrieved from Twitter's API, and the previous studies report that retweet graphs typically have most vertices at depth one, suggesting that root tweets (posted by the authors) get retweeted much more often than the retweets get retweeted (Kwak et al. 2010; Goel et al. 2012). This study, therefore, considers only the retweets made using the retweet button and the reduced dataset contains 163,312 retweets collected from 2,545 news tweets. This whole dataset was randomly separated into two smaller sets with a 70:30 ratio for training and testing the models, respectively. Figure 1 shows the frequency distribution of the popularity of the reduced dataset. The histogram demonstrates the skewed right distribution indicating that a few number of news tweets (1 %) receive high popularity (greater than 1,000 retweets), while about 90 % of news tweets gain the popularity between 0–150 retweets. The most popular tweet gains 2,865 retweets, and the median and the average of the popularity are 50 and 75 retweets, respectively.

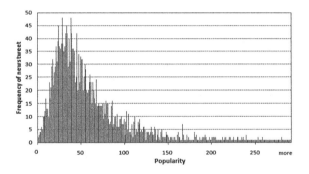

Fig. 1. Frequency of the popularity of news tweets.

To study the evolution of the popularity, Fig. 2 illustrates the correlation of the retweet times and the cumulative number of retweets for the four popularity levels: minimum (2 retweets), median (50 retweets), average (75 retweets), and maximum (2,865 retweets). In Fig. 2, the numbers of retweets for all popularity levels rapidly increase in the early stage after the tweets were posted (within 1,000–10,000 s, which is difference from other kinds of tweets that ranged from many hours to many days (Ma et al. 2013; Szabo and Huberman 2010) and then are almost stable. This makes the prediction model for news tweets more challenging. The time for the final retweet to occur for all levels ranged from 1.5 h to 6 days. In order to identify the prediction time with the highest performance of the prediction model, the time to reach 50 % of number of retweets for the median popularity (50 retweets) is adopted, which is 550 s. Note that the prediction time can be defined differently in which the time close to the time of final retweet yields higher accuracy but it might be too late for enhancing the popularity of the news tweets, while the less time gains lower accuracy (e.g., most of tweets may receive similar number of retweets at 10 s after the tweets are posted) but have enough time to disseminate the tweets to the right target group or influential users. This temporal analysis is considered as an interesting open research question.

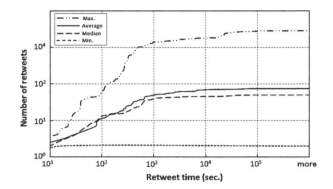

Fig. 2. Correlation of the retweet time and the cumulative number of retweets for the four popularity levels.

3.2   Impact of the Influential Users

The popularity of news tweets relies on many factors and an important factor is users. This Subsection, therefore, studies the impact of the influential retweeters on the popularity of news tweets. The retweeters are distinguished into two types including (*i*) *active user* and (*ii*) *popular user* as described following.

3.2.1   Active User

An *active user* is a person who has stable and frequent participation in the social network during a period of time (Fabián 2015; Yin and Zhang 2012). However, the activity rate of a user often fluctuates with time and it is costly to calculate the activity rate at any given time point. Thus, this study defines an active user based on user's participation during his/her Twitter's account lifetime. Let U be the set of users. For a user u U, his/her activity rate, denoted by $ActivityRate(u)$, is defined to specify the participation of user u and is calculated by counting the number of tweets plus the number of retweets posted and shared by the user throughout his/her Twitter's account lifetime, which could then be formally defined as follows:

$$ActivityRate(u) = \frac{T_u + RT_u}{accountTime(u)}, \qquad (1)$$

where

- u is a retweeter,
- T_u is the number of tweets posted by user u,

- RT_u is the number of retweets shared by user u, and
- *accountTime u* is the Twitter's account lifetime of user u in day unit.

A user who participates in posting high number of tweets and retweeting many tweets during his/her account lifetime is considered as an active user. Note that the number of tweets is considered in order to avoid users who have high number of retweets without posting any tweet, whose tend to be spammers or bots (Messias et al. 2013).

3.2.2   Popular User

A *popular user*, also called a celebrity, is a person who is recognized or followed by many other users on the network (Fabián 2015). Many recent researches evaluate the popularity of a user by concerning his/her number of followers and/or followees. A user who has high number of followers or high ratio of followers and followees is considered as a popular user. On the other hand, this research determines the popularity of a user by focusing on the number of *list*. In twitter, a user creates a list to organize and keep a closer eye on his/her followees into a group related to a specific topic such as celebrities, technical leaders, or news lists. Hence, a popular user, evaluated based on the list, is an expert or an influencer in a specific field considered based on the community point of view. For a user u U, his/her popularity score, denoted by *PopularityScore(u)*, is defined to specify the recognition of a user in the Twitter's network. Intuitively, it is computed by counting the number of lists that user u is listed, which could be defined as follows:

$$PopularityScore(u) = L_u, \tag{2}$$

where

- u is a retweeter, and
- L_u is the number of lists that user u is listed.

Thus, a user who is listed in many lists is considered as a popular user.

Figure 3(a) shows the correlation between the *ActivityRate* of retweeters and the popularity of news tweets, while Fig. 3(b) shows the correlation between the *PopularityScore* of retweeters and the popularity of news tweets. The figures plot the averages of the

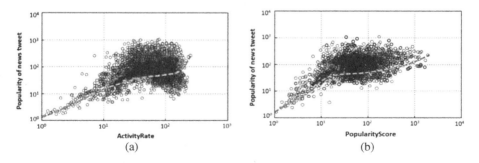

Fig. 3.   Correlation between (a) *ActivityRate* and (b) *PopularityScore* and the popularity of news tweets.

ActivityRate and *PopularityScore* of retweeters for each news tweet against the popularity of the tweets retweeted by the retweeters, while the medians of the *ActivityRate* and the *PopularityScore* per bin are drawn in the dashed line. Obviously, highly popular tweets are mainly retweeted by users who have high activity rate and high *PopularityScore*. The average of the *ActivityRate* and the *PopularityScore* of retweeters are above the medians, indicating that there are many retweeters who have very high activity rate and concerned as popular users retweet these popular tweets. Therefore, a tweet retweeted by retweeters who have high *ActivityRate* and *PopularityScore* is likely to be a popular tweet. According to the linear relationships, the *ActivityRate* and the *PopularityScore* features are introduced for predicting the popularity of news tweets (Fig. 3).

4   The Prediction Model

This section presents the feature-based approach for predicting the popularity of the news tweets by means of a multiple linear regression. The features of the retweeters, the content, and the authors are described.

Table 1. Features for the popularity prediction

Name	Description
Retweeter-based features	
Number of Retweeters	Number of all retweeters who retweet a tweet
ActivityRate	Average of *ActivityRate* of all retweeters who retweet a tweet
TweetCountScore	Average of *TweetCountScore* of all retweeters who retweet a tweet
PopuralityScore	Average of *PopuralityScore* of all retweeters who retweet a tweet
FollowerRank	Average of *FollowerRank* of all retweeters who retweet a tweet
Content-based features	
Hashtag	Number of hashtags in a tweet
Mention	Number of usernames mentioned in a tweet (excluding "RT @username")
URL	Number of URLs found in a tweet
Media	Number of attached photos or videos
Tweet period	Period of tweet time (e.g., after midnight, morning, afternoon, evening)
Tweet day	Day of tweet's time (e.g., weekday or weekend)
Author-based features	
Follower	Number of followers of the author who posts a tweet
Followee	Number of followees of the author who posts a tweet
Status	Number of tweets and retweets made by the author who posts a tweet

4.1   Features

Table 1 summarizes the features for predicting the popularity of the news tweets, divided into three groups including: (*i*) *retweeter-based*, (*ii*) *content-based*, and (*iii*) *author-based features*.

(i) **Retweeter-based features** describe the characteristics of retweeters.
   - *Number of Retweeters*: Number of all retweeters who retweet a tweet.
   - *ActivityRate*: This feature finds the average of *ActivityRate* of retweeters who retweet a tweet.
   - *TweetCountScore*: By applying the *TweetCountScore* (Noro et al. 2012), this feature calculates the average of *TweetCountScore* of all retweeters. The *TweetCountScore* counts the number of tweets plus the number of retweets shared by a retweeter u, and is divided by the *TweetCountScore* of another retweeter u' who post the maximum number of tweets and retweets, which can be computed as follows:

$$TweetCountScore(u) = \frac{T_u + RT_u}{\max(TweetCountScore(u'))}.$$

   A retweeter who posts and retweets many tweets is considered as an active user.
   - *PopularityScore*: This feature computes the average of *PopularityScore* of all retweeters who retweet a tweet.
   - *FollowerRank*: By applying the *FollowerRank* (Nagmoti et al. 2010), this feature finds the average of *FollowerRank* of all retweeters who retweet a tweet. The *FollowerRank* evaluates the ratio of followers and followees of a retweeter u, which can be evaluated as follows:

$$FollowerRank(u) = \frac{\#followers(u)}{\#followers(u) + \#followees(u)}.$$

   A retweeter who has high number of followers and less number of followees is concerned as a popular user.

(ii) **Content-based features** explain the content attributes of a tweet.
   - *Hashtag*: Number of hashtags found in a tweet. In Twitter, a hashtag is frequently used to identify a topical keyword (e.g., #mufc identifies that the tweet is classified in mufc topic).
   - *Mention*: Number of usernames mentioned in a tweet. Usually, an ampersand symbol is used for mentioning to another user (e.g., @POTUS refers to the President Barack Obama).
   - *URL*: Number of URLs found in a tweet.
   - *Media*: Number of photos or videos attached in a tweet.
   - *Tweet period*: Period of tweet's time in day, which can be divided into four slots including *after midnight*, *morning*, *afternoon*, and *evening*.
   - *Tweet day*: Day of tweet's time, which can be classified into two groups including *weekday* and *weekend*.

(iii) **Author-based features** identify social characteristics of an author who posts a tweet.
   - *Follower*: Number of followers of the author who posts a tweet.
   - *Followee*: Number of followees of the author who posts a tweet.

- *Status*: Number of tweets and retweets made by the author who posts a tweet. Intuitively, it specifies the participations of the author in Twitter's network.

4.2 Modeling the Prediction of Popular Tweets

Based on the features formally defined in the previous section, the prediction model has been developed using a multiple linear regression to predict the popularity of news tweets. In previous researches (Suh et al. 2010; Hong et al. 2011; Ma et al. 2013), the prediction problem was relaxed by predicting the *range* of the popularity. In contrast, this research aims at predicting the *exact value* of the popularity which can be achieved using the multiple linear regression. Several key assumptions of the multiple linear regression analysis were investigated and summarized (Due to limitation of space, the statistical results are omitted). Firstly, the linear relationships between the independent and dependent variables were examined using scatterplots and the results show that all retweeter-based features, author-based features, and media feature have different level of strong linear relationship, while most of the content-based features present a roughly linear relationship. Secondly, the normal distribution was investigated using a histogram. The results show that the data is normally distributed. Thirdly, there is no serious multicollinearity found in the dataset and the highest collinearity is *mention* and *PopularityScore* features with an acceptable value of 0.48. In addition, the homoscedasticity was investigated by creating a scatterplot of the residuals against the independent variables. The distributions of residuals show that there is no serious heteroscedasticity. Hence, the abovementioned features were applied to the regression model for predicting the popularity of news tweets, while the prediction time was set to 550 s (time to reach 50 % of the median popularity). Let NT be the set of news tweets, the popularity of news tweet $nt \in NT$, denoted by $P(nt)$ could then be calculated as follows:

$$P(nt) = \beta_0 + \beta_1 x_1 + \beta_2 x_2 + \cdots + \beta_i x_i,$$

where

- β_0 is the intercept term,
- $\beta_1, \beta_2, \ldots, \beta_i$ are the regression coefficient of features $x_1, x_2, \ldots, x_i$ respectively.

5 Experiment

This section presents an experimental setup including the evaluation metrics and the popularity predicting methods used to compare the experimental results. In addition, the experimental results are reported and discussed.

5.1 Evaluation Metrics and Methods Compared

Three standard performance metrics are applied as described below:

- *Root Mean Squared Error* (*RMSE*) is the square root of the average squared distance of a data point from the fitted line. However, one important limitation of squared errors is that it often places emphasis on the effect of outliers.

- *Mean Absolute Error (MAE)* is a quantity used to measure how close forecasts or predictions are to the eventual outcomes.
- *Adjusted R^2* is the most widely used and reported measure of error and goodness-of-fit of a regression model. Based on R^2, *adjusted R^2* calculates the proportion of the variation in the dependent variable accounted by the explanatory variables.

The performance metrics compare the efficiency of the methods that applied different features as depicted in Table 2.

Table 2. Evaluation methods.

Features	Method							
	Baseline	I	II	III	IV	V	VI	VII
Retweeter-based features								
Number of Retweeters		√						
ActivityRate			√				√	
TweetCountScore				√				√
PopularityScore					√		√	
FollowerRank						√		√
Content-based features	√	√	√	√	√	√	√	√
Author-based features	√	√	√	√	√	√	√	√

In Table 2, Baseline method evaluates the popularity of news tweets based on the *content-based* and the *author-based features* only, while other methods apply different *retweeter-based features* together with the *content-based* and the *author-based features*. Method I concerns that all retweeters have identical influence on the popularity and then uses the *number of retweeters* as a retweeter-based feature. By concerning the influence of active retweeters, Method II and Method III apply *ActivityRate* and *TweetCountScore*, respectively. By focusing on the influence of popular users, Method IV and Method V adapt *PopularityScore* and *FollowerRank*, respectively. Method VI determines the popularity using the two proposed measures, *ActivityRate* and *PopularityScore*, while Method VII applies *TweetCountScore* and *FollowerRank*.

5.2   Experimental Results and Discussion

In this Subsection, the experimental results are reported and discussed. Table 3 summarizes the accuracy of Baseline and the seven methods for predicting the popularity.

Table 3 shows that the accuracy of Baseline method was significantly lower than that of other methods which implies that the popularity prediction built based on the content-based and author-based features yields high error and needs to be improved using other effective features. By applying a retweeter-based feature, the accuracies of all the seven methods were increased. Obviously, the accuracy of Method VI was significantly higher than that of other methods with small values of RSME and MAE (6.304 and 3.833, respectively) suggesting that the *ActivityRate* and *PopularityScore* are considered as important and effective features in order to enhance the performance of the prediction model. Notice that the values of RSME and MAE might be larger than

Table 3. Accuracy of baseline and the seven compared methods.

Method	RMSE	MAE	Adjusted R^2
Baseline method	27.617	23.763	0.453
Method I (number of retweeters)	19.336	13.621	0.551
Method II (*ActivityRate*)	8.103	4.174	0.745
Method III (*TweetCountScore*)	15.801	9.135	0.673
Method IV (*PopularityScore*)	9.657	5.209	0.710
Method V (*FollowerRank*)	13.285	8.478	0.655
Method VI (*ActivityRate* and *PopularityScore*)	6.304	3.833	0.815
Method VII (*TweetCountScore* and *FollowerRank*)	15.021	9.013	0.705

that of the other related works because this experiment predicted the popularity as an exact value, while the other works estimated the popularity in term of a popularity range (e.g., popular or not popular).

By focusing on the influence of active user, the accuracy of Method II was higher than that of Method III indicating that the *ActivityRate* is an effective measure for determining active users. An important reason is that the participation of a retweeter during his/her account lifetime is considered by the *ActivityRate*, while the account lifetime is ignored by the *TweetCountScore*. Thus, an active retweeter measured by the *TweetCountScore* may be very active in a specific period and vary vastly during his/her account lifetime that causes a higher error of the predicting result. By considering the influence of popular user as a retweeter-based feature, the accuracy of Method IV was higher than that of Method V suggesting that the *PopularityScore* is a potential measure for evaluating the influence of popular users. One reason is that the *PopularityScore* concerns only the expertise or the popularity of a user based on the other users' point of view but the *FollowerRank* takes account of not only the number of followers but also the number of followees (or friends). Hence, a retweeter who has large number of followers and knows or follows a lot of friends is not considered as a popular user when evaluated by the *TweetCountScore*.

In addition, the prediction model of Method VI generated based on the multiple linear regression resulted that the retweeter-based features are strongly predictive of the popularity when comparing to the content-based and author-based features. The *ActivityRate* and *PopularityScore* obtained high importance with the significant ($p < 0.000$) and positive coefficient (β) values (1.437 and 1.105, respectively). According to the regression coefficient values, active users are considered as influential users who affect the popularity of news tweets, while popular users have lower influence. The result is intuitive since active users frequently participate in posting and sharing vast amount of social media content, as well as retweeting news tweets. An interesting issue is that how popular users affects the popularity. To investigate this issue, the retweet network of news tweets retweeted by top ten popular users (retweeters who obtain highest values of *PopularityScore*) was analyzed. The result revealed that the popular users mainly retweeted a tweet in the early state (about 10–20 min after a tweet is posted) and before many active users. By using Twitter REST API, the relationship between the popular users and the active users was examined. The result shown that there was a number of active users who follow the popular users retweeted the same tweet after the popular users, while such active users do not follow the news source. The active users who later retweeted the same tweet may receive the news tweets from the popular users. Thus, the popular users also play important role in disseminating news tweets and help increase the popularity. Accordingly, a tweet retweeted by many active retweeters who have high *ActivityRate* and many popular users who gain high *PopularityScore* is likely to receive high number of retweets and considered as a popular tweet.

Furthermore, most of the author-based features obtained higher coefficient values than that of the content-based features which similar to the results reported in the previous researches (Hong et al. 2011; Ma et al. 2013). Interestingly, the media feature (a content-based feature) had regression coefficient (0.453) close to the author-based features. One possible reason is that the photos or videos attached in a tweet give informative data and attract large number of retweeters.

6 Conclusions and Future Work

This paper presents *retweeter-based features* for predicting the popularity of news tweets using a multiple linear regression model. In order to understand how news tweets are disseminated over the social network, the characteristics and the evolution of popular news tweets are analyzed. The result reveals that the popularity of news tweets is dramatically increased in a few minutes after the tweets are posted and then almost stable. By analyzing the impact of the two types of retweeters, *active users* and *popular users*, the results of the analysis shows strong linear relationships and suggests that a tweet retweeted by very active and most popular users is likely to receive large number of retweets. The *ActivityRate* and the *PopularityScore* are proposed for measuring the influence of retweeters and used as retweeter-based features for modeling the prediction of the popularity of news tweets. The experimental results demonstrate that the application of the *ActivityRate* and the *PopularityScore* (method VI) enhances the performance of the prediction model with high accuracy when comparing to other retweeter-based features as well as the content-based and the author-based features. The result of this research can be applied in online news applications to help promoting news content and enhancing a news recommendation system.

The future work, therefore, includes the development of social media analytic such as breaking news detection application and news recommendation system. For example, a news article retweeted by popular users in the early stage should be recommended to other similar popular users or to active users as many as possible in order to increase the number of views of the news article. In addition, an intensive research on the temporal analysis is required in order to identify the prediction time and investigate more time evolving features for the prediction model.

References

Bao, P., Shen, H.-W., Huang, J., Cheng, X.-Q.: Popularity prediction in microblogging network: a case study on Sina Weibo. In: WWW 2013 (2013)

Bigonha, C.A., Cardoso, T.N., Moro, M.M., Goncalves, M.A., Almeida, V.: Sentiment-based influence detection on Twitter. J. Braz. Comp. Soc. **18**(3), 169–183 (2012)

Cappelletti, R., Sastry, N.: IARank: ranking users on Twitter in near real-time, based on their information amplification potential. In: Social Informatics 2012, Washington, DC, USA, pp. 70–77 (2012)

Fabián, R.: Measuring user influence on Twitter: a survey (2015). arXiv:1508.07951

Gayo-Avello, D.: Nepotistic relationships in Twitter and their impact on rank prestige algorithms. Inf. Process. Manag. **49**(6), 1250–1280 (2013)

Goel, S., Watts, D.J., Goldstein, D.G.: The structure of online diffusion networks. In: EC 2012, New York, USA (2012)

Hajian, B., White, T.: Modelling influence in a social network: metrics and evaluation. In: PASSAT/SocialCom 2011, Boston, MA, USA (2011)

Hong, L., Dan, O., Davison, B.D.: Predicting popular messages in twitter. In: WWW 2011 (2011)

Khrabrov, A., Cybenko, G.: Discovering influence in communication networks using dynamic graph analysis. In: PASSAT 2010, Minneapolis, Minnesota, USA (2010)

Kwak, H., Lee, C., Park, H., Moon, S.: What is Twitter, a social network or a news media? In: WWW 2010, New York, USA (2010)

Lee, C., Kwak, H., Park, H., Moon, S.B.: Finding influentials based on the temporal order of information adoption in Twitter. In: WWW 2010, Raleigh, North Carolina, USA (2010)

Ma, Z., Sun, A., Cong, G.: On predicting the popularity of newly emerging hashtags in Twitter. J. Am. Soc. Inf. Sci. Technol. **64**(7), 641399–641410 (2013)

Messias, J., Schmidt, L., Oliveira, R., Benevenuto, F.: You followed my bot! Transforming robots into influential users in Twitter. First Monday **18**(7) (2013)

Nagmoti, R., Teredesai, A., Cock, M.D.: Ranking approaches for microblog search. In: WI 2010, Toronto, Canada (2010)

Neves, A., Vieira, R., Mourao, F., Rocha, L.: Quantifying complementary among strategies for influeners' detection on Twitter. In: ICCS 2015 (2015)

Noro, T., Ru, F., Xiao, F., Tokuda, T.: Twitter user rank using keyword search. In: 22nd European-Japanese Conference on Information Modelling, pp. 31–48. IOS Press, Prague (2012)

Petrovic, S., Osborne, M., Lavrenko, V.: RT to Win! Predicting message propagation in Twitter. In: ICWSM 2011 (2011)

Romero, D.M., Galuba, W., Asur, S., Huberman, B.A.: Influence and passivity in social media. In: Gunopulos, D., Hofmann, T., Malerba, D., Vazirgiannis, M. (eds.) ECML PKDD 2011, Part III. LNCS, vol. 6913, pp. 18–33. Springer, Heidelberg (2011)

Srinivasan, M.S., Srinivasa, S., Thulasidasan, S.: Exploring celebrity dynamics on Twitter. In: I-CARE 2013, Hyderabad, India (2013)

Suh, B., Hong, L., Pirolli, P., Chi, E.H.: Want to be retweeted? Large scale analytics on factors impacting retweet in Twitter network. In: SOCIALCOM 2010 (2010)

Szabo, G., Huberman, B.A.: Predicting the popularity of online content. Commun. ACM **53**(8), 80–88 (2010)

Yin, Z., Zhang, Y.: Measuring pair-wise social influence in microblog. In: ASE/IEEE International conference on Social Computing and 2012 ASE/IEEE International Conference on Privacy, Security, Risk and Trust (2012)

Yuan, J., Li, L., Huang, L.L.: Topology-based algorithm for users' influence on specific topics in micro-blog. J. Inf. Comput. Sci. **10**(8), 2247–2259 (2013)

Zaman, T., Fox, E.B., Bradlow, E.T.: A Bayesian Approach For Predicting The Popularity Of Tweets. MIT, Cambridge (2014)

Learning with Additional Distributions

Sanparith Marukatat[✉]

IMG Lab, NECTEC, 112 Thailand Science Park, Pathumthani, Thailand
sanparith.marukatat@nectec.or.th

Abstract. This paper studies the problem of learning with distributions. In this work, we do not focus on the distribution that represents each data point. Instead, we consider the distribution that is an additional information around each data point. The proposed method yields a new kernel that is similar to an existing one. The main difference is that our kernel requires an integration in the kernel space. Theoretically, the proposed method yields a better generalization compared to normal SVM.

Keywords: SVM · SMM · Kernel for distributions

1 Introduction

Several data analysis methods such as classification or regression rely on data in vectorial form. These techniques can be extended using kernel trick to cover other representations such as strings (Lodhi et al. 2002), trees (Moschitti 2006), graphs (Harchaoui and Bach 2007) and distributions (Dalal and Triggs 2005; Vedaldi and Zisserman 2012). The latter is often used for describing the content of an image such as color (Chapelle et al. 1999), gradient information (Dalal and Triggs 2005; Maji et al. 2008), or visual words obtained from local descriptors (Sivic and Zisserman 2006). These data representation can be plugged into an SVM using an appropriate kernel function. Recently Jebara et al. (2004), Hein and Bousquet (2005), and Muandet et al. (2012) proposed new kernel families that combine Mercer kernel with distributions. These kernels exploit the advantages of probabilistic model within discriminative framework of SVM.

From previous works, there are two types of distributions that can be distinguished namely the *distribution that represents each data point* and the *distribution around each data point*. Terms histogram in text classification, histogram of visual words in object recognition or histogram of colors in image processing belong to the first type of distribution. The second type of distribution can be assigned to data represented in any format. For example, given a set of vectors, we can compute a set of neighbor points and derive a Gaussian distribution around each vector. If the data were already represented as distributions such as histograms of colors, we can still assign additional distribution such as Dirichlet on top of them. This Dirichlet distribution describes, indeed, how the histograms are distributed around a specific histogram.

© Springer International Publishing Switzerland 2016
R. Booth and M.-L. Zhang (Eds.): PRICAI 2016, LNAI 9810, pp. 319–326, 2016.
DOI: 10.1007/978-3-319-42911-3_27

The additional distribution, called *type-2 distribution* hereafter, allows introducing prior knowledge that could be useful especially when dealing with small dataset or uncertainty. In previous works, the distinction between the two types of distributions is not clear. For example, the structural kernel proposed by Hein and Bousquet or the level-2 kernel proposed by Muandet et al. involve type-2 distribution. Nonetheless, some experiments reported therein were based on data represented as type-1 distributions without additional type-2 distributions. This could explain why in some cases, these kernels were outperformed by kernels that work directly on type-1 distributions.

This paper studies the classical learning algorithm when provided with type-2 distributions. Section 2 presents the learning framework that leads to the definition of a new composite kernel in Sect. 2.3. Section 3 discusses the density estimation in the kernel space required by the proposed kernel. Then Sect. 4 studies the generalization of linear classifier using additional distributions. Sections 5 and 6 presents experiments and conclusions respectively.

2   Enhanced Linear Classifier and Composite Kernel

2.1   Enhanced Classification Function

Traditional supervised learning search for the function f that minimizes the risk:

$$\mathcal{R}(f) = \int_{\mathcal{X} \times \mathcal{Y}} l(y, f(x)) dP(x, y) \tag{1}$$

where l is a loss function and P is a fixed but unknown distribution of example (x, y) with input data $x \in \mathcal{X}$ and the desired output $y \in \mathcal{Y}$. Suppose that we are also given distribution P_x around each input x. With this additional knowledge, we may consider the following enhanced risk instead:

$$\mathcal{R}_d(f) = \iint l(y, f(z)) dP_x(z) dP(x, y) = \int \mathbb{E}_{z \sim P_x}[l(y, f(z))] dP(x, y). \tag{2}$$

This is a special case of the risk considered earlier by Muandet et al. (2012). The main difference is that we enforce our view on local distribution that depends on each data point. However, the theoretical result reported therein could be applied to our case as well. In particular, the Theorem 3 of Muandet et al. (2012) says that $l(y, \mathbb{E}_{z \sim P_x}[f(z)])$ is a close estimation of $\mathbb{E}_{z \sim P_x}[l(y, f(z))]$ with high probability. The estimation error can be, indeed, upper bounded as a function of the maximum variance of all distribution P_x. This result implies that one should consider the following enhanced classification function, denoted f_d, instead of f:

$$f_d(x) = \mathbb{E}_{z \sim P_x}[f(z)] = \int f(z) dP_x(z). \tag{3}$$

2.2   Enhanced Linear Classifier

Consider linear classification function

$$f(x) = \langle w, x \rangle. \tag{4}$$

If we enhance it with distribution P_x, then we obtain

$$f_d(x) = \int \langle w, z \rangle dP_x(z) = \langle w, \int z dP_x(z) \rangle. \tag{5}$$

It is interesting noted that if $x = \int z dP_x(z)$, then the additional distribution does not bring any contribution to the new classification rule.

Let $\bar{x} = \int z dP_x(z)$, we have

$$f_d(x) = \langle w, \bar{x} \rangle. \tag{6}$$

Given a training data $(x_1, y_1), ..., (x_n, y_n)$ and the associated distribution $P_{x_t}, t = 1, ..., n$ the optimal separating hyperplane is of the form

$$w = \sum_{t=1}^{n} \alpha_t y_t \bar{x}_t \tag{7}$$

where coefficients $\alpha_1, ..., \alpha_n$ are obtained from the training procedure. By injecting this quantity into the enhanced classification rule (Eq. (5)) we obtain

$$f_d(x) = \sum_t \alpha_t y_t \langle \bar{x}_t, \bar{x} \rangle \tag{8}$$

$$= \sum_t \alpha_t y_t \left\langle \int z dP_{x_t}(z), \int z' dP_x(z') \right\rangle \tag{9}$$

$$= \sum_t \alpha_t y_t \iint \langle z, z' \rangle dP_{x_t}(z) dP_x(z'). \tag{10}$$

This means that the enhanced classification rule can be considered as a linear classifier in a feature space induced by the following kernel:

$$\kappa(x, x') = \iint \langle z, z' \rangle dP_x(z) dP_{x'}(z'). \tag{11}$$

2.3   Composite Kernel

Previous section shows that enhanced linear classifier in the input space can be considered as linear classifier working in the RKHS induced by the kernel described in the Eq. (11). In this section, we consider the application of the enhanced linear classifier in the RKHS induced by an additional *base kernel k*. Let ϕ be the the the mapping function of k. In this RKHS, the kernel in the Eq. (11) becomes:

$$\kappa(\phi(x), \phi(x')) = \iint \langle \phi(z), \phi(z') \rangle dP_{\phi(x)}(\phi(z)) dP_{\phi(x')}(\phi(z')). \tag{12}$$

From the above equation, we define the following *composite kernel*:

$$\kappa(x, x') = \iint k(z, z')\, dP_{\phi(x)}(\phi(z))dP_{\phi(x')}(\phi(z')). \tag{13}$$

The above equation is similar to the *structural kernel I* proposed by Hein and Bousquet (2005) or the *level-2 kernel* proposed recently by Muandet et al. (2012):

$$\kappa(x, x') = \iint k(z, z')\, dP_x(z)dP_{x'}(z'). \tag{14}$$

The main difference is that the latter integration runs over input space, whereas in our case, it is done in the RKHS induced by the base kernel k. Previous works justified this integration as a mechanism to incorporate structural information about the probability space into the kernel function. It was proved to yield valid positive definite kernel. However, it does not correspond to the enhanced classification rule in the kernel space. We believe that the composite kernel in the Eq. (13) can better capture the structural information contained in the kernel space using additional distribution.

3   Density Estimation in Kernel Space

The composite kernel proposed in previous section requires an integration using local density in kernel space. To render this calculation flexible in practice, we consider the use of non-parametric density estimation to model the distribution around each point. In addition, we will consider a discrete integration over set of neighbors around each point instead of numerical integration.

Given a set of input data $x_1, ..., x_n$ and a kernel k. We first compute the set $\mathrm{NN}(x_t), t = 1, ..., n$ that contains neighbors of $\phi(x_t)$ under the Euclidean distance in the kernel space

$$\|\phi(x) - \phi(x')\|_k^2 = k(x, x) + k(x', x') - 2k(x, x'). \tag{15}$$

We consider m neighbors including x_t itself.

For each point $z \in \mathrm{NN}(x)$, the density at $\phi(z)$ is estimated non-parametrically as follows

$$P_{\phi(x)}(\phi(z)) = \frac{1}{Z} \sum_{z' \in \mathrm{NN}(x)} \exp\left(-\frac{\|\phi(z) - \phi(z')\|^2}{b}\right) \tag{16}$$

where b is the bandwidth and Z is the normalizing constant. The bandwidth b is heuristically set to $\hat{\sigma}$ that is the average of sample local variances computed as

$$\hat{\sigma} = \frac{1}{n} \sum_t \hat{\sigma}_t \tag{17}$$

$$\hat{\sigma}_t^2 = \frac{1}{m} \sum_{z \in \mathrm{NN}(x_t)} \|\phi(z) - \overline{\phi(z)}\|^2 \tag{18}$$

where $\overline{\phi(z)} = (1/m) \sum_z \phi(z)$ is the mean of $NN(x)$. The distance to this mean can be computed using kernel trick as follows

$$\|\phi(z) - \overline{\phi(z)}\|^2 = k(z, z) - \frac{2}{m} \sum_{z'} k(z, z') + \frac{1}{m^2} \sum_{z'} \sum_{z''} k(z', z''). \qquad (19)$$

4   Note on Generalization of the Enhanced Linear Classifier

The VC-dimension of a canonical linear classifier w is upper bounded by $\Lambda^2 R^2$ where $\|w\| \leq \Lambda$ and $R = \max_t \|x_t\|$ (Scholkopf and Smola 2001). For enhanced linear classifier, by following similar reasoning we obtain that its VC-dimension, h, can be bounded as follows

$$h \leq \Lambda^2 \overline{R}^2 \qquad (20)$$

where $\overline{R} = \max_t \|\bar{x}_t\|$. Noted that we have

$$Var(P_{x_t}) = \int \|z - \bar{x}_t\|^2 dP_{x_t}(z) \qquad (21)$$

$$= \int \|z\|^2 dP_{x_t}(z) - \|\bar{x}_t\|^2. \qquad (22)$$

If we adopt the strategy of integrating over set of neighbors, then the first term of the right hand side is upper bounded by R^2. Thus, one obtain $\overline{R}^2 \leq R^2 - \sigma_{\min}^2$ where $\sigma_{\min}^2$ is the minimum variance of the distributions $P_{x_t}, t = 1, ..., n$. This leads to the following inequality:

$$h \leq \Lambda^2 (R^2 - \sigma_{\min}^2). \qquad (23)$$

This is similar to the normal bound, with an additional term that tighten it up. Thus with additional distribution, a better generalization can be expected.

5   Experiments

The proposed method was evaluated against the normal SVM as well as Support Measure Machine (SMM) that is SVM with the level-2 kernel (Eq. (14)) using various base kernels. In the following, we denoted the proposed method that is SVM with additional distributions by *SVMd*. Our SVMd and SMM implementation is based on LIBSVM[1] with 1-vs-1 strategy for multi-class classification.

It is worthy noted that the composite kernel as well as the level-2 kernel provide natural frameworks to incorporate un-labeled data in supervised learning. Indeed, we can use un-labeled data to learn local distribution around labeled examples. This leads to novel semi-supervised learning techniques. Hence, in the following we shall evaluate the proposed method in semi-supervised setting as

---

[1] https://www.csie.ntu.edu.tw/~cjlin/libsvm/.

well. To this end, the classical handwritten digit images from MNIST dataset[2] was used. The dataset contains 60,000 training examples and 10,000 test examples. In semi-supervised evaluation, the whole training set (60,000 examples), without labels, were used to construct the neighbor set for both training examples and test examples. However, only part of the training set will be used to train SVMd and SMM. We consider neighbor set of 5 elements. This already leads to 25-fold more calculation compared to the evaluation of simple base kernel.

5.1   Experiment 1: Vectorial Data

Figures 1(a) and (c) show the test accuracy obtained from the three methods when trained with 5,000, 10,000, 20,000 first training examples as well as the results obtained when trained with the whole training set. The base kernels used in this experiment were RBF kernel with $\gamma = 0.00015$ and polynomial kernel degree 2. Figures 1(b) and (d) show the number of support vectors obtained from these tests. From these figures, one can see that SVMd achieved better results compared to normal SVM especially when the number of training examples is limited. The SMM also yielded similar performance but slightly lower than the proposed method. However, the number of support vectors obtained from our method is lower than SMM and much lower than the normal SVM.

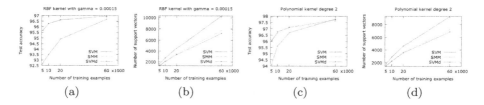

(a)   (b)   (c)   (d)

Fig. 1. Comparative results from normal SVM, SMM and SVMd. Figures (a) and (c) show the test accuracies using different base kernels whereas Figures (b) and (d) show the obtained number of support vectors. (Color figure online)

5.2   Experiment 2: Histogram Data

For the second experiment, we investigate another data representation namely the histogram. To this end, we consider the handwritten digit bitmap from previous experiment as a distribution of intensities. Indeed, the 28×28 bitmap was discretized into 49 bins using 7×7 grid. We summed up the intensities in each grid. The counts were then normalized so that the sum of all bins is one. Four base kernels evaluated in this work are: the *Hellinger's* kernel, the *Chi-square* (χ^2) kernel, the *Histogram intersection (HI)* kernel, and the *Jensen-Shannon's* (JS) kernel (see Vedaldi and Zisserman 2012 for more detail on these kernels).

---

[2] http://yann.lecun.com/exdb/mnist/.

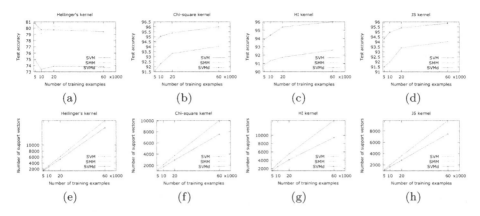

Fig. 2. Comparative results from normal SVM, SMM and SVMd. The top row (Figures (a), (b), (c), and (d)) shows the test accuracies using different base kernels whereas the bottom row (Figures (e), (f), (g), and (h)) shows the obtained number of support vectors. (Color figure online)

It should be noted that the normal Euclidean distance is not a natural distance measure for histogram representation used in this experiment. Indeed, the classical measure between histograms is the histogram intersection that is also one of the base kernel considered in this work. Hence, the Euclidean distance in HI kernel space will be used to compute the nearest neighbors in SMM. Consequently, with HI as base kernel, the level-2 SMM kernel is exactly the same as the proposed composite kernel.

From these figures, one can see that SVMd and SMM produce clearly better results compared to normal SVM. SMM and SVMd performance are almost the same. We believe this is due to the similarity between the distance in the RKHS induced by the considered kernels. Indeed, if the distances in the two spaces are highly correlated, then we may simply perform the integration in the input space instead of the RKHS; in which case SVMd becomes SMM.

6   Conclusion

In this work, we consider the problem of learning with additional distributions. Indeed, the integration of distributions in a kernel machine has been done previously in form of kernel between distributions. However, in these works, the distributions are added mainly in order to take advantages of probabilistic model. In this work, the proposed composite kernel arises naturally within the framework of linear classification with additional distributions. We have shown that, theoretically, it can lead to a tighter confidence interval; hence a better generalization capacity. The proposed method yields similar performance as SMM. However, we believe that the advantage of the proposed method should become clearer if the distance in input space and that in the RKHS are not correlated.

Compared to normal SVM, the proposed method outperforms in several cases, with lower number of support vectors.

The main disadvantage of the proposed method is in the evaluation of the composite kernel. This is also true for SMM. Nonetheless, as SMM consider the integration in the input space, it is possible to derive analytic forms for several choices of base kernel and additional distributions. In our case, we consider the integration in the kernel space. Deriving similar analytic forms is a challenging and opening problem for future works.

References

Chapelle, O., Haffner, P., Vapnik, V.: Support vector machines for histogram-based image classification. IEEE Trans. Neural Netw. **10**(5), 1055–1064 (1999)

Dalal, N., Triggs, B.: Histograms of oriented gradients for human detection. In: 2005 IEEE Computer Society Conference on Computer Vision and Pattern Recognition (CVPR 2005), pp. 886–893. IEEE Computer Society (2005)

Harchaoui, Z., Bach, F.R.: Image classification with segmentation graph kernels. In: 2007 IEEE Computer Society Conference on Computer Vision and Pattern Recognition (CVPR 2007). IEEE Computer Society (2007)

Hein, M., Bousquet, O.: Hilbertian metrics and positive definite kernels on probability measures. In: Cowell, R.G., Ghahramani, Z. (eds.) Proceedings of the Tenth International Workshop on Artificial Intelligence and Statistics, AISTATS. Society for Artificial Intelligence and Statistics (2005)

Jebara, T., Kondor, R., Howard, A.: Probability product kernels. J. Mach. Learn. Res. **5**, 819–844 (2004)

Lodhi, H., Saunders, C., Shawe-Taylor, J., Cristianini, N., Watkins, J.C.H.C.: Text classification using string kernels. J. Mach. Learn. Res. **2**, 419–444 (2002)

Maji, S., Berg, A.C., Malik, J.: Classification using intersection kernel support vector machines is efficient. In: 2008 IEEE Computer Society Conference on Computer Vision and Pattern Recognition (CVPR 2008). IEEE Computer Society (2008)

Moschitti, A.: Making tree kernels practical for natural language learning. In: McCarthy, D., Wintner, S. (eds.) EACL 2006, 11st Conference of the European Chapter of the Association for Computational Linguistics, pp. 113–120. The Association for Computer Linguistics (2006)

Muandet, K., Fukumizu, K., Dinuzzo, F., Schölkopf, B.: Learning from distributions via support measure machines. Adv. Neural Inf. Process. Syst. **25**, 10–18 (2012)

Scholkopf, B., Smola, A.J.: Learning with Kernels: Support Vector Machines, Regularization, Optimization, and Beyond. MIT Press, Cambridge (2001)

Sivic, J., Zisserman, A.: Video google: efficient visual search of videos. In: Ponce, J., Hebert, M., Schmid, C., Zisserman, A. (eds.) Toward Category-Level Object Recognition. LNCS, vol. 4170, pp. 127–144. Springer, Heidelberg (2006)

Vedaldi, A., Zisserman, A.: Efficient additive kernels via explicit feature maps. IEEE Trans. Pattern Anal. Mach. Intell. **34**(3), 480–492 (2012)

Acquiring Activities of People Engaged in Certain Occupations

Miho Matsunagi[(✉)], Ryohei Sasano, Hiroya Takamura, and Manabu Okumura

Tokyo Institute of Technology, Yokohama, Japan
matsunag@lr.pi.titech.ac.jp, {sasano,takamura,oku}@pi.titech.ac.jp

Abstract. We present a system to acquire knowledge on the *activities* of people engaged in certain occupations. While most of the previous studies acquire phrases related to the occupation, our system acquires pairs of a verb and one of its arguments, which we call activities. Our system acquires activities from sentences written by people engaged in the target occupations as well as from sentences whose subjects are the target occupations. Through experiments, we show that the activities collected from each resource have different characteristics and the system based on the two resources would perform robustly for various occupations.

Keywords: Activity acquisition · Occupational knowledge · Social media

1 Introduction

Knowledge about people engaged in certain occupations is useful in many situations. For instance, it would be valuable for those who want to become medical doctors to know that doctors often read academic papers as well as perform surgery. As another example, e-commerce companies can recommend e-book readers to news reporters if they know that news reporters frequently take bullet trains and read books on the train.

There are several studies on acquiring knowledge about people with certain attributes (e.g., [1,11]). These studies aim to extract phrases related to target attributes. For example, *movie* and *IMAX camera* are related to film directors if there are many sentences like

(1) The film director shot a movie with an IMAX camera.

However, such phrases do not always capture the characteristics of the target attribute. For example, film directors and actors are both related to *movie* but a film director *shoots* a movie, while an actor *appears in* a movie. This difference can be helpful for people looking for occupations related to *movie*. To capture such differences, we acquire usual *activities* related to the occupation, such as *shoot a movie*, where an activity is defined as a pair of a verb and one of its arguments except its subject.

© Springer International Publishing Switzerland 2016
R. Booth and M.-L. Zhang (Eds.): PRICAI 2016, LNAI 9810, pp. 327–339, 2016.
DOI: 10.1007/978-3-319-42911-3_28

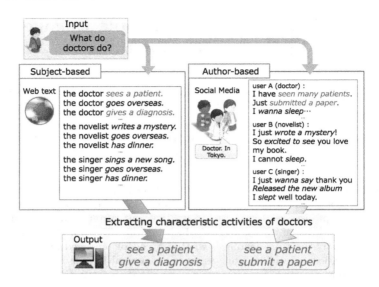

Fig. 1. Overview of our system.

In this paper, we present a system that acquires activities from two types of resources. The first type is sentences whose grammatical subjects are occupational titles. Example (1) is an instance of this type, from which we can extract *shoot a movie* and *shoot with an IMAX camera* as activities of film directors. This approach is based on the assumption that the grammatical subject of a sentence mostly denotes the agent of activities described in the sentence. Therefore, we can acquire activities related to the occupation by this approach. Most of these activities are, however, typical activities of the occupation, i.e. most of them are obviously associated with the occupation.

It is also valuable to know activities that are not obviously associated with the occupation but related to it as well as typical activities for those who are interested in the target occupations as their future occupation. For example, the medical doctor is known as an occupation that sees a patient and gives a diagnosis, but medical doctors often write and submit academic papers as researchers in their field. However, such activities as *submit a paper* cannot be collected from the first type of the sentences, because they rarely appear in these sentences. Thus, we also extract activities from sentences written in the first person by people engaged in target occupations, which we regard as the second type of sentences. If many medical doctors write "I just submitted a paper." on their blogs, we can extract *submit a paper* as an activity of medical doctors. This approach is based on the assumption that the sentences written in the first person often contain activities that rarely appear in the first type of sentences. We assume that activities with different characteristics can be collected by using both of these two resources.

Figure 1 shows an overview of our system. It consists of a subject-based component and an author-based component. The subject-based component collects sentences of the first type mentioned above, i.e., sentences whose grammatical subjects are occupational titles, and extracts activities from them. The author-based component collects social media users engaged in target occupations and extracts these users' own activities from their posts.

2 Related Work

There are several studies on acquiring knowledge about certain attributes. Bergsma and Durme [1] acquired knowledge on the properties of gender by calculating the pointwise mutual information between a gender and each phrase. Sap et al. [11] acquired knowledge on the properties of age and gender by using linear multivariate regression and classification models. These studies, however, focused on attributes with a limited number of classes, where it is easy to prepare labeled data and take a supervised approach. On the other hand, the number of occupations is much larger than that of age or gender and is usually not fixed. Therefore, it is impractical to prepare labeled data for all occupations in advance. We thus explore an unsupervised approach to acquire activities related to an input occupation.

The subject-based component of our system adopts a similar approach to unsupervised keyphrase extraction. Although most of the existing studies use various techniques such as language modeling [12] or graph-based ranking [9,13], they are basically based on co-occurrence information between the target domain and keyphrases. We follow these studies in the subject-based component that uses co-occurrence information between the target occupation and an activity.

Our system also uses an author-based component that leverages social media text written by people engaged in target occupations. Although this component extracts authors' activities by looking for the first person, first person pronouns are often omitted in social media texts, especially those in Japanese, which is our focus. Kanouchi et al. [4] addressed a similar problem and built a supervised classifier to predict subjects for diseases/symptoms mentioned in sentences. On the other hand, we use several rules to extract authors' own activities in the first person sentences to remedy this problem.

Filatova and Prager [2] and Kozareva [7] also extracted activities from text. Filatova and Prager automatically extracted activities in the documents about a target person and classified them into occupation-specific or others. Kozareva acquired activities by which one could answer the question such as "What are the duties of a medical doctor?" for various entities, including persons, organizations, and other objects. However, these studies focus on only typical activities such as *see a patient* by doctors. On the other hand, we focus on not only typical activities but also non-typical activities such as *submit a paper* by doctors, and thus our target activities are more diverse than those of their studies.

3   Our System

As shown in Fig. 1, our system acquires activities of an input occupation by using a subject-based component and an author-based component.

3.1   Subject-Based Component

This component acquires activities from sentences whose grammatical subjects are occupational titles, because such sentences often contain activities of the target occupation. For example, both *"the doctor sees a patient"* and *"the doctor goes overseas"* in Fig. 1 contain an activity of a medical doctor. Such activities, however, are not always specific to the target occupation. In Fig. 1, while *see a patient* is specific to doctors, *go overseas* is not. Thus, we calculate the chi-square score χ^2 [10] between the activity and the target occupational title to measure how specific the activity is to the occupation.

Table 1. Notation for frequencies regarding with a target activity and occupation for calculating the chi-square score. x and y in $N_{x,y}$ denote a target occupation and activity, respectively.

Type	Description
$N_{1,1}$	# of times that the target activity is performed by people with the target occupation
$N_{1,0}$	# of times that activities other than the target activity are performed by people with the target occupation
$N_{0,1}$	# of times that the target activity is performed by people with occupations other than the target occupation
$N_{0,0}$	# of times that activities other than the target activity are performed by people with occupations other than the target occupation

Table 1 shows the notation related to frequencies $N_{i,j}$ for calculating χ^2. For example, $N_{0,1}$ of a doctor and *see a patient* denotes the frequency of *see a patient* by people other than doctors. χ^2 is calculated using $N_{i,j}$ by the following equations:

$$E_{i,j} = \sum_{i'} N_{i',j} \sum_{j'} N_{i,j'} / \sum_{i',j'} N_{i',j'},$$

$$\chi^2 = \sum_{i,j} (N_{i,j} - E_{i,j})^2 / E_{i,j}.$$

χ^2 compares the observed frequency of co-occurrence between the activity and the target occupation with the expected frequency of co-occurrence when the activity and target occupation are assumed to be independent of each other. We consider that one activity is related to the target occupation if its χ^2 score is large.

The process of this component is summarized as follows. It first collects pairs of a subject and an activity from parse trees. To avoid using incorrect parts of parse trees, it applies Kawahara and Kurohashi [5]'s method, which extracts unambiguous parts of the parse tree, and uses only reliable parts. It then calculates the χ^2 score between the input occupation and each activity and outputs activities with large χ^2 scores.

Table 2. Rules for extracting the authors' own activities. Only activities that satisfy these constraints are extracted.

Name	Description
Subject	The grammatical subject is the first person "I" or omitted
Object	The grammatical object is not the author
Modification	The verb representing the target activity does not modify a noun
Modality	The modality of the sentence is not interrogatory, imperative, subjunctive, injunctive, or potential; these modalities suggest that the target activity might not actually be performed

3.2 Author-Based Component

This component acquires activities from social media texts in three steps. Since most social media posts are about the daily lives of users, we can collect activities related to the target occupation from their posts.

The component first collects users engaged in the target occupation from the profiles of social media users. Since some users describe their occupations in their profiles, we collect users whose profiles contain the target occupation. However, not all such users are actually engaged in the target occupation. Some users may mention an occupation that they want to have. Therefore, we use several rules to filter out users who are actually not engaged in the target occupation. When the target occupation is doctor, users with profile (2) are collected and users with profile (3) are not.

(2) I'm a doctor.

(3) My dream: to be a doctor.

Secondly, the component extracts users' activities from their posts. Since activities mentioned in users' posts are not always performed by the author of the post, we use several rules listed in Table 2 to select the authors' own activities[1]. As a result, the component extracts the underlined activities in (4) and (5), and filters out the activities in (6), (7) and (8).

---

[1] The author-based component extracts the authors' own activities with the accuracy of 65.0 % by using our rules. This accuracy does not directly affect the performance of the system, because specific activities are finally selected on the basis of χ^2 scores.

(4) I just <u>had dinner</u>.

(5) <u>Arrived at Tokyo</u>.

(6) <u>Call me</u> when you get home.

(7) I saw a <u>running dog</u>.

(8) I should <u>go to hospital</u>.

In these examples, (6) is filtered because the object of call in (6) is the author, (7) is filtered because "running" in (7) modifies "dog.", and (8) is filtered because this sentence implies that "I" actually do not yet "go to hospital."

Lastly, it calculates χ^2 scores in the same manner as in the subject-based component and outputs activities with large χ^2 scores as specific to the target occupation.

4 Experiment

4.1 Experimental Setting

We experimentally applied our system to the well-known occupations that satisfied all of the following constraints.

Table 3. Populations and accuracies of collecting users engaged in target occupations. Numbers in **bold** are less than 100 in population, and accuracies in *italic* are less than 60 %.

Occupation	Acc.	(Pop.)	Occupation	Acc.	(Pop.)
Announcer	69.0 %	(206)	Babysitter	84.5 %	(2,819)
Novelist	79.5 %	(3,186)	Homemaker	97.0 %	(23,556)
Photographer	85.5 %	(1,664)	Lawyer	90.0 %	(428)
Carpenter	*54.5 %*	(625)	Musician	90.5 %	(694)
Cook	66.5 %	(367)	Nurse	90.0 %	(2,819)
Counselor	91.5 %	(618)	Painter	88.3 %	(552)
Curator	89.6 %	(**91**)	Pharmacist	92.5 %	(1,030)
Detective	*11.1 %*	(**9**)	Pilot	14.0 %	(290)
Nutritionist	84.5 %	(1,410)	Civil-servant	80.0 %	(1,502)
Doctor	*59.0 %*	(383)	Singer	*60.0 %*	(1,348)
Editor	96.5 %	(1,373)	Station staff	*50.0 %*	(**2**)
Engineer	93.5 %	(5,843)	Teacher	71.0 %	(1,664)
Guard	*44.5 %*	(1,528)	Actor	88.5 %	(239)
Beautician	85.5 %	(3,527)	News reporter	94.0 %	(396)

1. The occupation is registered as a noun in the dictionary of Japanese morphological analyzer JUMAN[2].
2. The occupation is listed in the Japanese Wikipedia's occupation list.
3. The occupational name appears more than 10,000 times in a Japanese Web corpus consisting of approximate 10 billion sentences.

Table 3 shows 28 occupations that were fed into our system. In the subject-based component, we used predicate-argument pairs extracted from approximately 6.5 billion parse trees[3]. In the author-based component, we crawled Twitter users who tweeted in Japanese in 2013 and their tweets by using Twitter API[4]. Consequently, we collected approximately 11,287,300 Japanese users. We finally extracted approximately 32,000 users tied to the target occupations in the author-based component. We obtained syntactic structures from tweets by using a Japanese parser KNP [8].

As a preliminary experiment, we checked whether the author-based component correctly collected Twitter users engaged in the target occupation. Two human annotators examined 100 user profiles per occupation[5] and judged whether the automatically estimated occupations matched the occupations that the annotators considered the users have according to their profiles. Table 3 shows the population of collected users and accuracies. Although our author-based component failed to accurately collect users in some occupations that rarely appeared in social media, it collected the users with an accuracy of 80 % for 22 occupations out of the 28 target occupations.

As mentioned in Sect. 1, we assume that the activities collected from each resource have different characteristics. We thus conducted an evaluation with a crowdsourcing service to confirm this assumption. We presented an activity with a target occupation to the crowd-workers in the Japanese crowdsourcing service Lancers[6], and asked them to judge which of three categories the activity and occupation match.

1. **Obvious:** The presented activity is obviously associated with the presented occupation.
2. **Non-obvious:** The presented activity is not obviously associated with the presented occupation, but related to it in some way.
3. **Irrelevant:** The presented activity is not associated with the presented occupation.

In order to ensure the quality of the results, we selected only the workers who correctly answered quality control questions, which were very easy to answer if

---

[2] http://nlp.ist.i.kyoto-u.ac.jp/EN/index.php?JUMAN.
[3] We used predicate-argument pairs provided by Kawahara and Kurohashi. For details of the method of extracting predicate-argument pairs from a Web corpus, please see Kawahara and Kurohashi [6].
[4] https://dev.twitter.com/overview/documentation.
[5] Less than 100 users are collected for curator, detective, and station staff. Annotators examined all users for them.
[6] http://www.lancers.jp.

workers actually read the questions, such as *see a patient* by doctors. Each pair of an activity and an occupation was evaluated by five workers, and the score was calculated as the number of activities to which at least three out of five workers answered 1 or 2.

We evaluated occupations that had more than 200 activities in each component, because the quality of the acquired activities is not necessarily high if the number of acquired activities is too small. Both the subject-based and author-based components were evaluated for 13 out of 28 occupations, and only the author-based component was evaluated for 11 out of 28 occupations. Neither of these components were evaluated for remaining four occupations.

4.2 Comparisons of Two Types of Activities

Table 4 shows the scores of each component. The subject-based component had accuracies of over 65 % for 10 out of 13 evaluated occupations, though it failed in some occupations. In particular, the accuracies of homemakers and teachers were lower than 40 %. We manually examined the acquired activities of these occupations and found that the system often contain irrelevant activities that frequently appeared in advertisements on the Web. Most of these activities, such as *earn with a blog* by housemakers, do not match their real lives, and thus the accuracies for these occupations were low.

In the author-based component, the accuracies of collecting users engaged in the target occupations were correlated with the accuracies of acquiring activities in most cases. The component failed to acquire activities related to civil-servants and nutritionists, though their accuracies for collecting users were high, whose accuracies were 6 % and 36 %, respectively, because people with those occupations hardly wrote posts about their occupational activities. On the other hand, it successfully acquired activities related to singers though their accuracy of collecting users was not high, because they frequently announced the activities related to their occupations in the social media.

Next, we compared the activities for the 13 occupations, for which the two components were evaluated. Figure 2 compares the two components' accuracies in the scatter plot for 13 occupations. From Fig. 2, we can see that both of the two components acquired activities for 13 occupations with the average accuracies of approximately 70 %. However, the intersection of activities acquired by each component is very small; the average number of the common activities in the two components for 13 occupations is only 2.92 out of 100. This fact suggests that the system can acquire diverse activities by using both of the two components.

We then turn to the performance for each occupation. We found that the two components compensated for each other's weaknesses. For example, the subject-based component succeeded in acquiring activities of doctors and cooks, though the author-based component failed for them. Likewise, the author-based component succeeded in acquiring activities of homemakers and teachers, though the subject-based component failed for them. As a result, our system achieved an accuracy of at least 59 % for each occupation by one of the components. Therefore, the system would become robust by combining the two components.

Table 4. Scores of each component. The numbers in second and third columns denote the number of related activities out of 100 activities. For reference, we also show in the rightmost column the accuracies of collecting users by the author-based component in Table 3.

Occupation	The accuracy of activities acquisition (subject-based)	The accuracy of activities acquisition (author-based)	The accuracy of user collection
Announcer	**74 %**	56 %	69.0 %
Novelist	**74 %**	78 %	79.5 %
Photographer	**85 %**	95 %	85.5 %
Cook	**91 %**	48 %	66.5 %
Counselor	**75 %**	53 %	91.5 %
Doctor	**66 %**	19 %	59.0 %
Engineer	**78 %**	84 %	93.5 %
Homemaker	33 %	92 %	97.0 %
Lawyer	**70 %**	64 %	90.0 %
Nurse	**76 %**	78 %	90.0 %
Singer	**85 %**	81 %	60.0 %
Teacher	38 %	81 %	71.0 %
News reporter	59 %	55 %	94.0 %
Carpenter	-	17 %	54.5 %
Nutritionist	-	36 %	84.5 %
Editor	-	85 %	96.5 %
Guard	-	10 %	44.5 %
Beautician	-	68 %	85.5 %
Babysitter	-	77 %	84.5 %
Musician	-	95 %	90.5 %
Painter	-	96 %	88.3 %
Pharmacist	-	84 %	92.5 %
Civil-servant	-	6 %	80.0 %
Actor	-	63 %	88.5 %
Average	69.5 %	63.4 %	-

We further investigated the characteristics of the activities that were correctly acquired by the two components. As described above, we regarded the activities as correct if they were evaluated as "obvious" or "non-obvious" by at least three out of five crowd-workers. These activities were classified into the following three groups on the basis of the breakdown of the evaluation.

1. **Obvious:** The number of workers who evaluated the activity as "obvious" is larger than the number of workers who evaluated the activity as "non-obvious".

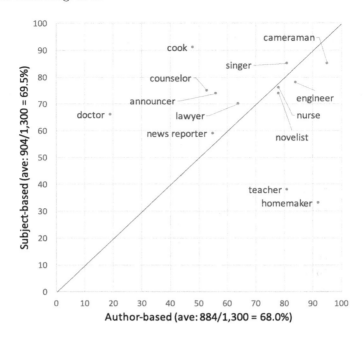

Fig. 2. Comparison of activities acquired by the two components. The x axis denotes the score of a author-based component, and the y axis denotes the score of a subject-based component.

2. **Non-obvious:** The number of workers who evaluated the activity as "non-obvious" is larger than the number of workers who evaluated the activity as "obvious".

3. **Other:** The number of workers who evaluated the activity as "non-obvious" is equal to the number of workers who evaluated the activity as "obvious".

Table 5 shows the classified result. In this table, N_{total} denotes the total number of correctly acquired activities, and N_{ob} (obvious), N_{non-ob} (non-obvious), and N_{other} (other) denote the number of activities the respective groups. The author-based component acquired more "non-obvious" activities than the subject-based component did. This difference was significant according to Fisher's exact test [3] at a significance level 0.01. This result supports our assumption that the activities collected from each resource have different characteristics.

We also investigated the relation between χ^2 rank and an accuracy. We first sort activities in descending order of χ^2 scores, and calculated the accuracy of top-N ($N = 10, 20, ..., 100$) activities. Figure 3 shows the result. From this figure, we found that the accuracy did not change regardless of χ^2 scores for the subject-based component, while the accuracy decreased monotonically in accordance with the increase of the activities for the author-based component. We think it is because the subject-based component occasionally gives high χ^2 scores to peculiar activities that were performed by only few people engaged in

Table 5. The characteristics of correctly acquired activities. N_{total} is the sum of N_{ob} (obvious), $N_{non\text{-}ob}$ (non-obvious), and N_{other} (other) and it corresponds to the total score in Fig. 2.

Component	N_{total}	N_{ob}	$N_{non\text{-}ob}$	N_{other}	$N_{non\text{-}ob} / (N_{ob} + N_{non\text{-}ob})$
Author-based	884	491	344	49	**41.2 %**
Subject-based	904	577	281	46	32.8 %

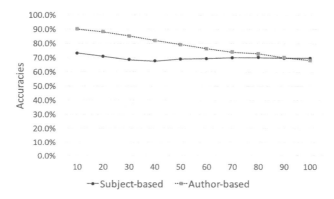

Fig. 3. The change of accuracies associated with the number of activities.

the target occupations. These activities often appear as phrases such as those in book titles and headlines, and thus they appear more frequently than the actual situations. However, note that when we manually investigated the activities with quite low χ^2 scores in the subject-based component, these activities were hardly related to the target occupations. Therefore, χ^2 score is effective for filtering out unrelated activities, while it fails to filter out some peculiar activities.

Table 6 shows some examples from our system. The author-based component often acquired activities that were related to the target occupation and performed in the daily life, though it sometimes acquired activities that were not specific to the target occupation. For example, *prepare a brief* by lawyers is related to them and is likely to be performed in their daily lives, while *look down* and *pay tax* by doctors are performed by everyone. On the other hand, the subject-based component often acquired activities that were frequently mentioned by others, though it sometimes acquired peculiar activities that were not actually performed by the target occupation. For example, *accept a consultation* and *establish a defense counsel* are indeed typical activities of lawyers, while *deceive a patient* by doctors is actually unlikely to be performed by them. We think the subject-based component incorrectly acquired *deceive a patient* because it is often mentioned in the book titles and headlines as a doctors' activity.

Table 6. Examples of activities acquired in each component. Numbers in brackets denote the number of crowd-workers who evaluated the activity as (obvious, non-obvious, irrelevant), respectively.

Occupation	Subject-based	Author-based
Lawyer	Accept a consultation (3,2,0)	Write a brief (0,5,0)
	Be in charge of the defense (5,0,0)	Recruit a lawyer (2,2,1)
	Establish a defense counsel (5,0,0)	Appear in the office (0,4,1)
Doctor	Deceive a patient (2,1,2)	Go to an academic meeting (1,4,0)
	Charge for a treatment (5,0,0)	Look down (0,1,4)
	Write a medical certificate (5,0,0)	Pay tax (0,0,5)
Homemaker	Succeed in business (0,0,5)	Hang out the laundry (5,0,0)
	Earn with a blog (1,2,2)	Prepare a lunch box (5,0,0)
	Try for pocket money (2,1,2)	Take a daughter out (3,2,0)

5   Conclusion

We presented a system that had two components to acquire knowledge about the activities of people engaged in certain occupations. The subject-based component acquires activities from the sentences whose grammatical subjects are the target occupational title, while the author-based component acquires activities from text written by people engaged in the target occupation. In the evaluation with a crowdsourcing service, the subject-based component and author-based component acquired activities for 13 occupations with the average accuracies of 69.5 % and 68.0 %, respectively. As a whole, our system achieved an accuracy of at least 59 % for each occupation by one of the components. We also showed that the activities acquired with each component have different characteristics. The author-based component acquired more activities that were not obviously associated with the occupation but related to it than the subject-based component did. For future work, we plan to explore the strategy for combining the two components that robustly acquires activities for various occupations.

Acknowledgement. We would like to acknowledge Prof. Kurohashi and Prof. Kawahara for providing us with the data of predicate-argument pairs used in the experiments. This work was supported by JSPS KAKENHI Grant Number JP26280080 and the Center of Innovation Program from Japan Science and Technology Agency, JST.

References

1. Bergsma, S., Van Durme, B.: Using conceptual class attributes to characterize social media users. In: Proceedings of the 51st Annual Meeting of the Association for Computational Linguistics (ACL), pp. 710–720 (2013)

2. Filatova, E., Prager, J.: Tell me what you do and I'll tell you what you are: learning occupation-related activities for biographies. In: Proceedings of the Conference on Human Language Technology and Empirical Methods in Natural Language Processing (HLT/EMNLP), pp. 113–120 (2005)
3. Fisher, R.A.: On the interpretation of χ^2 from contingency tables, and the calculation of P. J. Roy. Stat. Soc. **85**(1), 87–94 (1922)
4. Kanouchi, S., Komachi, M., Okazaki, N., Aramaki, E., Ishikawa, H.: Who caught a cold? - Identifying the subject of a symptom. In: Proceedings of the 53rd Annual Meeting of the Association for Computational Linguistics and the 7th International Joint Conference on Natural Language Processing (ACL-IJCNLP), pp. 1660–1670 (2015)
5. Kawahara, D., Kurohashi, S.: Fertilization of case frame dictionary for robust Japanese case analysis. In: Proceedings of the 19th International Conference on Computational linguistics (COLING), pp. 425–431 (2002)
6. Kawahara, D., Kurohashi, S.: Case frame compilation from the web using high-performance computing. In: Proceedings of the 5th International Conference on Language Resources and Evaluation (LREC), pp. 1344–1347 (2006)
7. Kozareva, Z.: Learning verbs on the fly. In: Proceedings of the 24th International Conference on Computational Linguistics (COLING), pp. 599–610 (2012)
8. Kurohashi, S., Nagao, M.: A syntactic analysis method of long Japanese sentences based on the detection of conjunctive structures. Comput. Linguist. **20**(4), 507–534 (1994)
9. Mihalcea, R., Tarau, P.: TextRank: bringing order into texts. In: Proceedings of the 2004 Conference on Empirical Methods in Natural Language Processing (EMNLP), pp. 404–411 (2004)
10. Miller, R., Siegmund, D.: Maximally selected chi square statistics. Biometrics **38**(4), 1011–1016 (1982)
11. Sap, M., Park, G., Eichstaedt, J., Kern, M., Stillwell, D., Kosinski, M., Ungar, L., Schwartz, H.A.: Developing age and gender predictive lexica over social media. In: Proceedings of the 2014 Conference on Empirical Methods in Natural Language Processing (EMNLP), pp. 1146–1151 (2014)
12. Tomokiyo, T., Hurst, M.: A language model approach to keyphrase extraction. In: Proceedings of the ACL 2003 Workshop on Multiword Expressions: Analysis, Acquisition and Treatment, pp. 33–40 (2003)
13. Zha, H.: Generic summarization and keyphrase extraction using mutual reinforcement principle and sentence clustering. In: Proceedings of the 25th Annual International ACM SIGIR Conference on Research and Development in Information Retrieval, pp. 113–120 (2002)

Motion Primitive Forests for Human Activity Recognition Using Wearable Sensors

Nguyen Ngoc Diep[1,2(✉)], Cuong Pham[1,2], and Tu Minh Phuong[1,2]

[1] Computer Science Department, Posts and Telecommunications Institute
of Technology, Hanoi, Vietnam
{diepnguyenngoc,cuongpv,phuongtm}@ptit.edu.vn
[2] Machine Learning and Applications Lab, Posts and Telecommunications Institute
of Technology, Hanoi, Vietnam

Abstract. Human activity recognition is important in many applications such as fitness logging, pervasive healthcare, near-emergency warning, and social networking. Using body-worn sensors, these applications detect activities of the users to understand the context and provide them appropriate assistance. For accurate recognition, it is crucial to design appropriate feature representation of sensor data. In this paper, we propose a new type of motion features: *motion primitive forests*, which are randomized ensembles of decision trees that act on original local features by clustering them to form motion primitives (or words). The bags of these features, which accumulate histograms of the resulting motion primitives over each data frame, are then used to build activity models. We experimentally validated the effectiveness of the proposed method on accelerometer data on three benchmark datasets. On all three datasets, the proposed motion primitive forests provided substantially higher accuracy than existing state-of-the-art methods, and were much faster in both training and prediction, compared with k-means feature learning. In addition, the method showed stable results over different types of original local features, indicating the ability of random forests in selecting relevant local features.

Keywords: Human activity recognition · Wearable sensors · Motion primitive forests · Random forests · Bag of features

1 Introduction

Human activity recognition (HAR) is an active research field as it has a broad range of practical applications such as situated services for supporting peoples lives [16], pervasive healthcare, fitness logging, and social networking [10]. In general, there are two ways to capture data of human behaviors: installing sensors inside fabric object surroundings and wearing sensors on the different parts of human body. The former can allow users to perform their activities in natural manner while the latter often acquires fewer sensors and more mobilized than the former. As activity recognition is a time series problem that needs

© Springer International Publishing Switzerland 2016
R. Booth and M.-L. Zhang (Eds.): PRICAI 2016, LNAI 9810, pp. 340–353, 2016.
DOI: 10.1007/978-3-319-42911-3_29

to classify what and when an activity is being performed by a user, analyzing captured data in activity recognition task often includes 4 subtasks: signal processing, data segmentation, feature extraction, and real-time classification. In which, processing signals involves low and high pass filtering and data resampling if necessary. Segmenting data is the subtask to segment continuously data streams into length-fixed sliding windows. Features, which can be either statistical features or frequency-domain features, are extracted from processed data within sliding windows to form feature vectors. These feature vectors are used for training the activity models which are then used for recognizing human activities in real-time.

Some recent works show that multilevel features provide good recognition performance [11,24]. These features are computed by quantizing local features or local descriptions extracted from small fragments of each data frame. Existing works use unsupervised learning algorithms such as k-means clustering, Gaussian mixture model (GMM), or topic modeling [12] to generate such high level features by grouping local descriptions to form so-called motion primitives or motion words. The use of such motion primitives, combined with the bag-of-features, has resulted in significant advances in HAR. Although high recognition accuracies could be achieved, the above methods are computationally expensive in both training and prediction due to the cost of assigning local descriptions to motion primitives. Moreover, GMM and topic modeling are generative methods, while discriminative features often lead to higher accuracy.

In this paper, we propose an alternative method to create motion primitive vocabulary and perform description assignment by using a small ensemble of decision trees, which we call *motion primitive forests* (MPF). MPF works as clustering trees by grouping similar local descriptions in leaf nodes. This process is guided by activity class labels and is much faster than k-means. The advantage of using decision tree ensembles in learning visual codebook has been shown in image processing [15], but this is the first time it is applied to sensor-based activity recognition, to the best of our knowledge. Our second contribution is that we introduce new simple local features to use with MPF. These are original values of data points or their pairwise sums and differences, which are fast to compute while providing the same recognition performance as more complex local descriptions. We conducted experiments with three public datasets widely used in activity recognition research and the results show that our proposed method outperforms k-means motion learning as well as two other state-of-the-art methods in [18,22].

2   Related Work

As activity recognition is a time series problem, several activity recognition works exploited non-linear time series based algorithms on sensor data (after being pre-processed) such as dynamic time warping [17] or time-delay embeddings [7]. Usually, these algorithms extract frequency-domain features such as Fourier Frequency Transform (FFT) features from sensor signals and then directly applied to the time series matching schemas. An advantage of time series based activity recognition is that it is light-weight and therefore suitable for real-time applications on devices with limited resources such as mobile phones or wearable computers. In reverse, a challenge for any matching procedure is to deal with the variants of activity patterns (which often occur when collecting datasets under real-world settings).

Most existing activity recognition methods rely on either pre-defined statistical features or frequency domain features extracted from fixed-length sliding windows (or frames) [1]. Features can be local [24], global [8,9,12,18,22], or both [25] which are combined into feature vectors and used for training human activity classification models. For example, work by [23] proposed a hierarchical feature selection schema for enhancing recognition performance. A few other works utilize feature learning methods that are able to automatically learn features extracted from sensor data using Principle Component Analysis (PCA) [9] or Restricted Boltzmann Machine (RBM) [18]. In addition, an effective feature learning method based on PCA-based Empirical Cumulative Distribution Function (ECDF) is also proposed by [18], and Convolutional Neural Networks (CNN) is proposed by [22]. These works yielded significant results and considerably improve the performance of activity recognition systems.

Bag-of-features (BoF) feature selection framework for activity recognition is used in [12,25], which originally used in text categorization and image classification. In their study [12], BoF is used for the classification of a complex set of 34 activities with large variations. Moreover, when combined with a topic model (Latent Dirichlet Allocation), BoF has proven the effectiveness of discovering up to 10 high-level daily activities (routines) with 73 % accuracy despite of there are a number of short and similar tasks in the dataset.

In brief, BoF models for activity recognition are often constructed in three steps: feature extraction, vocabulary or codebook learning, and histogram-of-words/motions construction. Features extracted are often local features which are computed from small segments of sliding windows. Vocabulary can be learnt by using unsupervised clustering algorithms such as k-means [11,24], predictive clustering trees [6] or GMM [8,24] to group features into clusters. Each cluster center forms a word (code vector), and a collection of words over a dataset constitutes the vocabulary. Frequency of words in a vocabulary forms the histogram of words. Most of previous works based on BoF assumed small segments (from which local features are extracted) are independent and non-overlapped. This tends to ignore the relationships between adjacent segments within sliding windows. In addition, using k-means or GMM to build vocabulary can have limitations such as over-fitting, the complexity of high dimension data and speed. To deal with these issues, we propose *motion primitive forests* which leverage random forests [2], ensembles of decision trees, to construct the vocabulary for improving recognition accuracies and speed.

3 Methods

The overview of our method is illustrated in Fig. 1. The system takes as input tri-axial continuous sensor data streams from accelerometers attached to different positions on the user's body. Using a sliding window, we segment continuous sensor data streams into equal-size frames, whose lengths are longer than the durations of any activities. Each frame is then divided into equal *slices* (possibly with overlaps) so that each slice is much smaller than the frame itself (Fig. 1(a)). From each slice we extract features to form a *local feature vector*. These can be statistical features like mean, variance, or complex features such as skewness, kurtosis, etc. (Fig. 1(b)). In the training phase, if a frame contains an activity then the activity's label is assigned to whole frame as well as all slices belonging to it. In prediction, the problem is to predict activity class label for each frame.

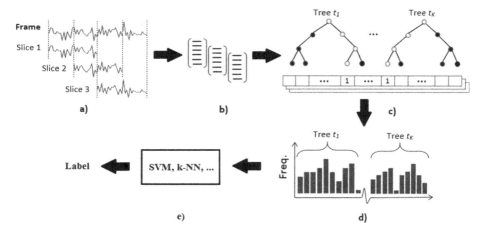

Fig. 1. (a) 3 slices segmented from an activity frame with overlap of 50 %, (b) Features extracted from each slice, (c) MPF as motion vocabulary, (d) Motion primitive histogram, (e) Classification.

Resulting local feature vectors from all training activity (and background) are then pooled together and quantized to form the motion primitive vocabulary or codebook. This is the process of assigning a discrete index to each local feature vector so that similar feature vectors have the same index with a high probability. The quality of motion primitive construction is crucial for recognition accuracy. Previous approaches for activity recognition used unsupervised learning methods such as k-means clustering for this important step with index assignment using expensive nearest neighbor search. In this work, we use random forests - randomized ensembles of decision trees - that act on local feature vectors and cluster them so that similar vectors belong to the same leaf (with a high probability) (Fig. 1(c)). These random forests, which we call *motion primitive forests* (MPF), are faster than k-means and nearest neighbor assignment, and produce motion primitives that are highly discriminative despite the large number of background slices. This is achieved mostly because activity labels are used to guide the construction of decision trees.

In the next step, we combine the motion primitives learned by MPF with the bag-of-features model (Fig. 1(d)). We compute the histogram of leaf node indices for each frame and use this histogram as input features for classification algorithms such as k-NN or SVM (Fig. 1(e)).

In the rest of this section, we present details of MPF and other components of our method.

3.1   Random Forests

First we give a brief review of random forests [2]. A random forest is a collection of T decision trees as shown in Fig. 1(c). Data samples traverse a decision tree by recursively branching left or right depending on split functions learned on features until reaching leaf nodes. The final assignment of labels is decided by averaging over all T trees to achieve robust and accurate classification. Random forests have been shown to be effective in both classification and clustering tasks [15, 20]. In this study, we use random forests as a means for clustering and mapping local feature vector to motion primitives.

Constructing Random Forests. Given a dataset N, each element represented by D features $A = \{a_1, ..., a_D\}$ and has the same activity label as the containing frame. Each tree is trained on a subset N', which we sample at random from N. The tree is built recursively top-down. The training data N_p of node p is split into two disjoint subsets left N_l and right N_r in accordance with a threshold c of some split function f of feature vector a:

$$N_l = \{n \in N_p | f(a_n) < c\} \tag{1}$$

$$N_r = N_p \backslash N_r \tag{2}$$

At each split node, we generate at random a group of size m of candidate attributes for function f and search over all possible values c and choose the one that maximize the gain in Gini index:

$$\Delta I_G(N_p) = I_G(N_p) - \frac{|N_l|}{|N_p|} I_G(N_l) - \frac{|N_r|}{|N_p|} I_G(N_r) \tag{3}$$

where $I_G(N_p)$ is the Gini index for the training data N_p. During the learning process, the training data is split recursively at each node p until all elements of N_p are of the same class or less than a given number.

3.2   Motion Primitive Forests

Motion Primitive Forests are random forests used for clustering and mapping local feature vectors to motion primitives (Fig. 1(c)). In standard random forests,

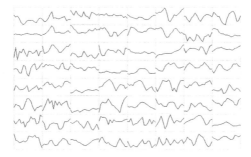

Fig. 2. Examples of motion primitives. Each cell shows the average signal stream of slices assigned to a leaf node.

voting over the ensemble of trees is used to reduce class label variance. Here, we do not use voting but treat each leaf node from each tree as a separate motion primitive. In other words, the leaf nodes define a partitioning, in which each leaf corresponds to a cluster of similar local feature vectors. As an illustration of learned motion primitives, Fig. 2 shows examples of 64 clusters corresponding to 64 leaf nodes. Each cell describes an acceleration signal stream (only signal on X axis are shown), which is the average of all signal streams in the training slices assigned to the corresponding leaf nodes. Features learned include motion primitives characterized by different movement intensities and different directions in the 3-D space.

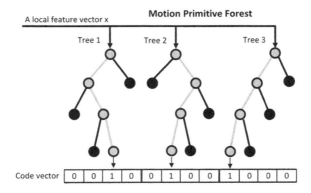

Fig. 3. The code vector created from a random forest with three trees for a local feature vector. Paths to the resulting leafs are shown in yellow. (Color figure online)

For each input local feature vector, the MPF returns a set of leaf indices, one for each tree. We use these leaf indices to form the code vector as illustrated in Fig. 3, where a "1" means the input vector has been mapped to the corresponding motion primitive. Then, following the bag-of-features approach, we sum over the code vectors of all slices that belong to a frame to form the histogram of motion primitives for this frame (Fig. 1(d)). We use the normalized histogram as the feature representation of the frame in the final classifier. Thus, it is not the forests, but the final classifier being responsible for voting.

In our method, the vocabulary size is equal to the total number of leaf nodes from all trees in the forest. To control the vocabulary size, we control the size of trees by pruning them until reaching the desired number of leaf nodes.

3.3 Local Features

Different types of local features have been proposed for activity recognition, including signal-based features, which are further divided into statistical features and physical features [24], body model features, event-based features [3]. Among these, statistical and physical features used by Zhang and Sawchuk [24] are most suitable to represent small slices and thus are used in our experiments.

All the above features require some computation over the data points of a slice. To further simplify the computation, we introduce new local features, which require very little or no computation because they are original values of data points or pair-wise sums and differences of two data points. Specifically, we use the following three types of simple features: (i) the value $p_{i,k}$ of data point with index i in axis k (*unary feature*), (ii) the sum $p_{i_1,k_1} + p_{i_2,k_2}$ (*sum feature*), and (iii) the difference $p_{i_1,k_1} - p_{i_2,k_2}$ (*difference feature*) of a pair of data points from axes k_1 and k_2, (k_1 and k_2 are possibly different), and $i_1 - i_2 = 1$ (sum and difference of only neighbor points). As shown by experiments, such simple features, when combined with MPF, provide accuracy comparable with more complex features, while being much cheaper to compute.

When k-means clustering is used to quantize local feature vectors, it is important to select appropriate local features. Irrelevant features will result in incorrect distances between feature vectors, which lead to bad clustering. Here, we conjecture that MPF is able to select relevant local features while learning split function. To verify this hypothesis, we group features in three sets as shown in Table 1, and build MPF using each of them. The first feature set consists of five simple but popular statistical features in time series data while the second feature set is more complex because it uses physical features derived from the physical parameters of human motion [23]. The third feature set comprises the proposed simple features.

3.4 Classifier

A wide range of classifiers have been used in activity recognition including decision trees, Bayesian methods, k-NN, neural networks, Markov models and classifier ensembles [14]. In experiments, we selected two classifiers: 1-NN for its popularity in activity recognition; and SVM because it has proved its superior prediction accuracy in a wide range of application domains including activity recognition.

Table 1. Local features.

Feature set	Features
1	Mean, standard deviation, root mean square, averaged derivatives, mean crossing rate
2	Mean and variance of movement intensity, normalized signal magnitude area, eigenvalues of dominant directions, averaged velocity along heading direction and averaged acceleration energy
3	Unary, sum and difference

4   Experiments and Results

4.1   Datasets

We conducted experiments on three public datasets widely used in activity recognition research. These datasets contain data streams from tri-axial accelerometers worn on subjects, who performed various activities in different contexts. The activity labels were manually assigned and provided with the datasets. We used a sliding window with a size of 64 sample points with 50 % overlap to segment the data streams into frames.

Activity Prediction (AP) [13]. This is a close-dataset (without background activity), which contains accelerometer data for six daily activities such as "jogging", "walking", "ascending stairs", etc., performed by 36 subjects. The data was collected from cell phones in the users' pockets under laboratory settings. Sensor readings were sampled at 20 Hz, resulting in around 29,000 frames.

Opportunity (OP) [19]. This is an open dataset (including unknown activities), containing data from multiple sensors worn on subjects or embedded on objects manipulated by them. The subjects performed different activities in a kitchen environment, from which 11 activities such as "cleaning", "drinking coffee", "open door", etc. were annotated. We used an excerpt of the dataset, which consists of data from a sensor worn on the right hand of a single subject. The same excerpt was also used by [18,22]. The data was sampled at 64 Hz resulting in about 4,200 frames.

Skoda (SK) [21] is an open dataset collected from multiple accelerometers worn on an assembly-line worker in a car production environment. The SK dataset contains 46 activities such as "open hood", "close left hand door", "check steering wheel", etc. We restrict our experiments to a single sensor worn on the right arm and 10 activities plus unknown activities. With sampling rate of 48 Hz, the dataset consists of 7,500 frames.

4.2   Experimental Settings

The performance metrics used in this study is in structure of overall mean accuracy $\pm$ standard deviation to evaluate the proposed method and compare with other approaches. The accuracy is computed as the ratio of number of frames correctly classified over the total number of frames (i.e. true positives + true negatives). This is the performance metrics widely used in activity recognition as it provides convenient comparison of different methods when multiple activity classes are present [3,18,24].

All methods and settings were evaluated with 10-fold cross validation for AP and OP datasets. For SK dataset, because the samples are distributed unequally across different classes, we performed only 4-fold cross validation. For all experiments we report average accuracies over folds. For each fold, we held out 10 % of the training set (via stratified sampling) as the validation set for tuning parameters and used these parameters for the test set (unless mentioned otherwise). We used LibSVM [4] with radial basis kernel (RBF) and our own implementation of 1-NN for the experiments. Parameters C and $gamma$ of SVM were selected using a grid search procedure on the validation set. MPF was implemented by modifying the random forest implementation provided by Breiman, keeping the number of features considered at each split as default, i.e. square root of the feature number. All experiments were run on a PC running Windows with 2.8 GHz dual-core processor and 8 GB RAM.

4.3   Comparison with k-means Feature Learning

The first experiment was designed to compare MPF with k-means feature learning, in which k-means clustering was used to build vocabulary and nearest neighbor search for motion primitive assignment [11,24]. Because these two methods under comparison use the same framework with only k-means clustering and nearest neighbor assignment replaced by MPF, the comparison results would show how much improvement could be achieved by using MPF. For a fair comparison, we kept most settings similar to ones used by [11,24]. Specifically, we used feature set 1 from Table 1, which is the same as statistical feature set used by Zhang and Sawchuk in [24] and SVM for classification. We also used slices of the same size and no overlap as described in their works. Other parameters were optimized separately for each method by using the validation set.

Figure 4 shows the accuracy of MPF and k-means for different vocabulary sizes (i.e. the numbers of clusters and leaf nodes for k-means and MPF respectively) on each dataset. As shown, the best vocabulary size varies across datasets but MPF outperforms k-means in all cases. The differences in accuracy between the two methods are substantially large for OP dataset, and for some cases of SK dataset. It is interesting to note that MPF tends to favor larger vocabulary size than k-means. For example, on the SK dataset, the accuracy of k-means decreases when the size exceeds 100, while the accuracy of MPF continues to increase.

It was impossible to directly compare the running times of two methods because MPF is written in C++ and k-means is from Matlab, and thus are not reported in detail here. However, we noticed that while MPF has almost the same running time as vocabulary size increases, the running time of k-means

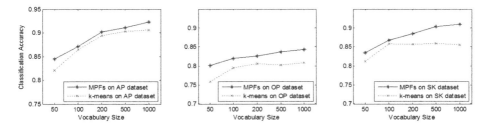

Fig. 4. Comparison of MPF and k-means based vocabulary construction algorithms on three datasets.

increases linearly. At the size of 200, the running time of k-means is already an order of magnitude longer than that of MPF (580 s vs. 54 s, SK dataset).

4.4   Influence of Local Features

The next experiment was designed to evaluate the influence of different local feature types. To show the general applicability of MPF, we tuned parameters only for SK dataset using a hold-out validation set, and used these parameters for all three datasets. The tuned parameters are: slice size = 8 with 87.5 % overlap, number of trees = 5, number of leaf nodes per tree = 500. Cross-validation was run for each feature set from Table 1 in turn with SVM as the classifier.

Figure 5 (left) summarizes the accuracies achieved when using each feature set on three datasets. The results show that the different feature types provide comparable accuracies with very small fluctuations across datasets. This result is important as it suggests two implications. First, MPF is less sensitive to the choice of local features, possibly because trees are able to select relevant features to consider at nodes. This is an advantage over unsupervised motive primitive learning such as k-means, which requires preselection of appropriate local features [24]. Second, simple features from set 3 perform as well as more complex statistical and physical features, while being faster to compute. Combined with MPF, using such simple features further reduces computational cost, which is crucial for real-time recognition.

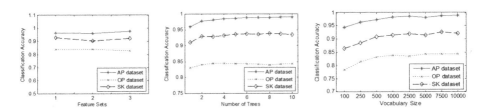

Fig. 5. Influence of local features (left), number of trees (middle) and vocabulary size (right) on accuracy.

4.5   Number of Trees and Vocabulary Size

In the next experiment, we investigated the influence of two important parameters - number of trees and vocabulary size - on the accuracy. We used feature set 3 and kept other parameters the same as in the previous experiment while varying the number of trees and vocabulary size. Figure 5 (middle) shows the classification accuracies for different numbers of trees. As seen, the accuracy increases when the number of trees increases from 1 to 5 on AP and SK datasets, and from 1 to 4 on OP dataset. Afterward, the accuracy tends to be steady and using more trees does not lead to considerable accuracy improvements. Thus, we conclude that a small ensemble of as few as five tree is sufficient for reducing the variance and achieving good recognition accuracy.

To investigate the influence of vocabulary size on classification performance, we varied the number of leaf nodes when training MPF. As illustrated in Fig. 5 (right), the accuracies over three datasets increases if the vocabulary size increases up to 2,500 and then goes steady. Note that, this size is much larger than previously reported when using k-means or GMM clustering [11,24]. The above results also indicate that MPF is not very sensitive to number of trees and vocabulary size, provided they are large enough.

4.6   Comparison with Other Methods

In the last experiment, we compared the recognition accuracy of MPF when using simple local features with two state-of-the-art methods. The first method for comparison is feature learning using PCA and ECDF (PCA + ECDF for short) combined with 1-NN classifier by [18]. The second method uses CNN with partial weight sharing for both feature learning and classification [22]. We re-implemented the methods in Matlab and kept the best parameter values as reported in their papers. MPF was run with both 1-NN and SVM.

Table 2 summarizes the accuracies of three methods MPF, PCA + ECDF and CNN partial weight sharing, represented in structure of mean accuracy standard deviation. As can be seen, all methods achieved high accuracies on AP dataset (more than 95 %) while OP dataset proves to be the most challenging. Both variants of MPF, with 1-NN and SVM, consistently achieved higher accuracies than PCA + ECDF and CNN in all three datasets. SVM provides better results

Table 2. Classification accuracy comparisons.

Method	Dataset		
	AP	OP	SK
MPF + 1-NN	97.93 % ± 0.4 %	84.17 % ± 1.6 %	92.58 % ± 0.9 %
MPF + SVM	**98.48 % ± 0.6 %**	**85.68 % ± 1.8 %**	**95.08 % ± 0.4 %**
PCA + ECDF	95.75 % ± 0.5 %	79.39 % ± 1.5 %	90.48 % ± 0.7 %
CNN partial weight sharing	96.88 % ± 2.2 %	76.83 % ± 0.4 %	88.19 % ± 0.8 %

than 1-NN, which is not surprising given the high dimension of input histograms and 1-NN's sensitivity to irrelevant features. The accuracy improvements of MPF over PCA + ECDF and CNN are especially noticeable for OP and SK datasets (almost 10 % difference between MPF and CNN). The results also show that there is no clear winner between CNN and PCA + ECDF.

It can be observed that the variation level of activity patterns in OP dataset would be high. In fact, activities addressed by [5] are quite complex as for each activity of daily living (ADL) consisting of a number of sub-activities which might not always be repeated in all time of performing the activity. Nevertheless, MPF had made substantial accuracy improvement on OP with almost 5 % difference compared to the second best method. This also demonstrates that MPF deals well on the activity set with high level of variation.

5   Conclusion

We have presented motion primitive forests as effective motion features for human activity recognition. These forests quantize local feature vectors of small data segments in a supervised manner to produce motion primitive vocabulary. By leveraging labeled data, the method can generate highly discriminative motion primitives. In addition, it is able to automatically select relevant local features, making it less sensitive to heuristics used in selecting local features. This makes it possible to apply MPF directly on very simple features such as original values of data points without loss of recognition accuracy. Combined with the bag-of-features approach, the proposed motion primitive forests advance the state-of-the-art in recognition accuracy when tested on three benchmark datasets. The results indicate the potential of the proposed method for activity recognition in mobile and ubiquitous computing as well as time series analysis in general.

References

1. Bao, L., Intille, S.S.: Activity recognition from user-annotated acceleration data. In: Ferscha, A., Mattern, F. (eds.) PERVASIVE 2004. LNCS, vol. 3001, pp. 1–17. Springer, Heidelberg (2004)
2. Breiman, L.: Random forests. Mach. Learn. 45(1), 5–32 (2001)
3. Bulling, A., Blanke, U., Schiele, B.: A tutorial on human activity recognition using body-worn inertial sensors. ACM Comput. Surv. 46(3), 33 (2014)
4. Chang, C.C., Lin, C.J.: LIBSVM: a library for support vector machines. ACM Trans. Intell. Syst. Technol. (TIST) 2(3), 27 (2011)
5. Chavarriaga, R., Sagha, H., Calatroni, A., Digumarti, S.T., Tröster, G., Millán, J.D.R., Roggen, D.: The opportunity challenge: a benchmark database for on-body sensor-based activity recognition. Pattern Recogn. Lett. 34(15), 2033–2042 (2013)

6. Dimitrovski, I., Kocev, D., Loskovska, S., Džeroski, S.: Fast and efficient visual codebook construction for multi-label annotation using predictive clustering trees. Pattern Recogn. Lett. **38**, 38–45 (2014)
7. Frank, J., Mannor, S., Precup, D.: Activity recognition with time-delay embeddings. In: 2011 AAAI Spring Symposium Series (2011)
8. Ghasemzadeh, H., Loseu, V., Jafari, R.: Collaborative signal processing for action recognition in body sensor networks: a distributed classification algorithm using motion transcripts. In: Proceedings of the 9th IPSN, pp. 244–255. ACM (2010)
9. He, Z., Jin, L.: Activity recognition from acceleration data based on discrete consine transform and svm. In: IEEE International Conference on Systems, Man and Cybernetics, SMC 2009, pp. 5041–5044. IEEE (2009)
10. Hoey, J., Plötz, T., Jackson, D., Monk, A., Pham, C., Olivier, P.: Rapid specification and automated generation of prompting systems to assist people with dementia. Pervasive Mob. Comput. **7**(3), 299–318 (2011)
11. Huỳnh, T., Blanke, U., Schiele, B.: Scalable recognition of daily activities with wearable sensors. In: Hightower, J., Schiele, B., Strang, T. (eds.) LoCA 2007. LNCS, vol. 4718, pp. 50–67. Springer, Heidelberg (2007)
12. Huỳnh, T., Fritz, M., Schiele, B.: Discovery of activity patterns using topic models. In: Proceedings of the 10th International Conference on Ubiquitous Computing, pp. 10–19. ACM (2008)
13. Kwapisz, J.R., Weiss, G.M., Moore, S.A.: Activity recognition using cell phone accelerometers. ACM SIGKDD Explor. Newsl. **12**(2), 74–82 (2011)
14. Lara, O.D., Labrador, M.A.: A survey on human activity recognition using wearable sensors. IEEE Commun. Surv. Tutorials **15**(3), 1192–1209 (2013)
15. Moosmann, F., Triggs, B., Jurie, F.: Fast discriminative visual codebooks using randomized clustering forests. In: NIPS 2006, pp. 985–992. MIT Press (2007)
16. Pham, C., Hooper, C., Lindsay, S., Jackson, D., Shearer, J., Wagner, J., Ladha, C., Ladha, K., Plötz, T., Olivier, P., et al.: The ambient kitchen: a pervasive sensing environment for situated services. In: DIS 2012. ACM (2012)
17. Pham, C., Plötz, T., Olivier, P.: A dynamic time warping approach to real-time activity recognition for food preparation. In: de Ruyter, B., Wichert, R., Keyson, D.V., Markopoulos, P., Streitz, N., Divitini, M., Georgantas, N., Mana Gomez, A. (eds.) AmI 2010. LNCS, vol. 6439, pp. 21–30. Springer, Heidelberg (2010)
18. Plötz, T., Hammerla, N.Y., Olivier, P.: Feature learning for activity recognition in ubiquitous computing. In: IJCAI 2011, vol. 22, p. 1729 (2011)
19. Roggen, D., Calatroni, A., Rossi, M., Holleczek, T., Forster, K., Troster, G., Lukowicz, P., Bannach, D., Pirkl, G., Ferscha, A., et al.: Collecting complex activity datasets in highly rich networked sensor environments. In: 2010 Seventh International Conference on Networked Sensing Systems (INSS), pp. 233–240. IEEE (2010)
20. Shotton, J., Johnson, M., Cipolla, R.: Semantic texton forests for image categorization and segmentation. In: IEEE Conference on CVPR 2008, pp. 1–8. IEEE (2008)
21. Zappi, P., Lombriser, C., Stiefmeier, T., Farella, E., Roggen, D., Benini, L., Tröster, G.: Activity recognition from on-body sensors: accuracy-power trade-off by dynamic sensor selection. In: Verdone, R. (ed.) EWSN 2008. LNCS, vol. 4913, pp. 17–33. Springer, Heidelberg (2008)
22. Zeng, M., Nguyen, L.T., Yu, B., Mengshoel, O.J., Zhu, J., Wu, P., Zhang, J.: Convolutional neural networks for human activity recognition using mobile sensors. In: 6th International Conference on Mobile Computing, Applications and Services (2014)

23. Zhang, M., Sawchuk, A.A.: A feature selection-based framework for human activity recognition using wearable multimodal sensors. In: Proceedings of the 6th International Conference on Body Area Networks, pp. 92–98. ICST (2011)
24. Zhang, M., Sawchuk, A.A.: Motion primitive-based human activity recognition using a bag-of-features approach. In: 2nd ACM SIGHIT, pp. 631–640. ACM (2012)
25. Zheng, Y., Wong, W.k., Guan, X., Trost, S.: Physical activity recognition from accelerometer data using a multi-scale ensemble method. In: Proceedings of the 25th Innovative Applications of Artificial Intelligence Conference, pp. 1575–1581 (2013)

An Orientation Histogram Based Approach for Fall Detection Using Wearable Sensors

Nguyen Ngoc Diep[1,2]([✉]), Cuong Pham[1,2], and Tu Minh Phuong[1,2]

[1] Computer Science Department, Posts and Telecommunications Institute of Technology, Hanoi, Vietnam
{diepnguyenngoc,cuongpv,phuongtm}@ptit.edu.vn
[2] Machine Learning and Applications Lab, Posts and Telecommunications Institute of Technology, Hanoi, Vietnam

Abstract. Histogram features are extracted by calculating the distribution of orientations of small fragments or quanta of sliding windows on the sensors continuously acceleration data stream. Bins of the histogram is automatically computed based on clusters of similar orientations of quanta, making it less sensitive to parameters used in selection of bins than a heuristic approach. We also present a finer representation of the sliding window by applying the above extraction method to extract local feature vectors of small data segments instead of calculating features from the whole sliding window. Extracted features are used with support vector machines trained to classify frames of data streams into containing falls or non-falls. We evaluated the proposed method on three public datasets with acceleration data including falls and other activities of daily living. On all three datasets, performance of the proposed method is substantially higher than two other fall detection methods.

Keywords: Fall detection · Wearable sensors · SVM · Feature extraction

1 Introduction

Falls are one of the leading causes of morbidity and mortality among the older adults. One of three adults age 65 and older falls each year. Up to 30 % of those who fall suffer moderate to severe injuries, such as hip fractures and head traumas, and can increase the risk of premature death. Many elderly people experience falls and they cannot stand up, even from non-injurious falls, and as result remains on ground for even longer than an hour [24]. Furthermore, about 50 % of those elderly who has been in such circumstances lose their lives within six months, even when they suffer from non serious injury [27]. Therefore, it is critical to detect falls as early as possible because it might help to reduce the time between the fall and the arrival of medical care. In this context, a fall detection system to send alerts automatically to the caregivers when patients fall would be an extreme need.

© Springer International Publishing Switzerland 2016
R. Booth and M.-L. Zhang (Eds.): PRICAI 2016, LNAI 9810, pp. 354–366, 2016.
DOI: 10.1007/978-3-319-42911-3_30

Common approaches to automatic fall detection are computer vision [14,15,25], ambient sensing [2,11,30] and wearable sensing [11,19,23,28,29]. Computer vision analyzes video sequences of motion, provided by digital cameras equipped in the environment, to recognize the events of an activity (including fall). Ambient sensing approach acquires context data of the users using multiple sensors installed in the surrounding environments (e.g. pressure sensors on the floor) to analyze and determine if a fall occurred. Both of these approaches are typically accurate but lack of flexibility as they often require pre-settings of the environments so that fall detection is strictly limited to the area equipped with sensors. Wearable sensing approach, in contrast, uses a single or several sensors such as accelerometers or gyroscopes worn on different parts of user's body to acquire posture and/or motion information for detecting falls. Users can be more mobilized and are not limited to a room or a flat. Moreover, compared to computer vision and ambient sensing, wearable sensing are low cost while providing reasonable detection accuracy. Thus, in this work, we focus on wearable sensing for fall detection, particularly the systems that learn features extracted from sensor data stream to detect falls.

Fall detection techniques used in wearable systems can be mainly divided into two types [16]: threshold-based and machine learning. Several existing systems used threshold-based techniques, the machine learning approach, however, has attracted to researchers. In threshold-based methods, a fall event is triggered when one or several features such as acceleration peaks, valleys or other shape features reach predefined thresholds. Even though these methods are low-complexity and simple for implementation, the rate of fall alarms is a critical issue. Moreover, they are proven to be ineffective in many situations (such as [3]). Instead of using such fixed thresholds, machine learning approach applies classifiers like support vector machines (SVM) to distinguish falls from other activities. This approach often has better detection rates [16]. A key for the success of these fall detection methods is to design appropriate feature representation of sensor data. Ideally, to achieve higher accurate detection, features should be designed to clearly separate between falls and fall-like activities and work robustly across different people.

In this work, we propose a novel feature extraction method for automatic fall detection using wearable sensors. Our method bases on the distribution of differences of angles and orientations of data points within quanta (small fragments) segmented from sliding windows to form the histogram feature vectors. Instead of using heuristics to decide bins of the histogram, we applied another approach using unsupervised clustering methods to automatically calculate bins and it results in good performance. When using with SVM [7], experimental results shown that these features can lead to significant accuracy improvements. In addition, we will show that detection accuracy can be improved if we apply the proposed algorithm to local segments of the frame to form local feature vectors, then aggregating these local feature vectors to create the global feature vector. This approach can alleviate the problem of false alarm, caused by some activities of daily life (ADL) which have similar acceleration signals to falls (such as stand-to-sit and sit-to-lie). By extracting local features, we can keep characterization of a fall in different phases (pre-fall, critical, post fall and recovery) [22]. The performance of our proposed algorithm is benchmarked on three public datasets containing falls and ADL. We also compare our proposed algorithm to two recent and high-performance fall detection algorithms. To the best of our knowledge, this is one of the few works which performs comparison of fall detection algorithms on public datasets.

The remaining of the paper is organized as follows. Section 2 is a literature review of existing approaches to fall detection. Section 3 describes our proposed method for automatic fall detection. Experimental results are presented in Sect. 4 and the paper ends with the conclusion.

2 Related Work

A broad range of wearable fall detection systems used threshold-based algorithms in continuous measurements. These algorithms are low-complexity and achieve high detection rates. For example, Lindeman et al. [20] used accelerometers which were integrated into a hearing-aid housing fixed behind the ear and detected falls by considering the spatial direction of the head, the velocity right before the initial contact with the ground and the impact. Chen et al. [9] considered the impact and the change in orientation to detect fall and location of the user. Kangas et al. [18] proposed four thresholds comprising of total sum vector, dynamic sum vector, vertical acceleration and difference between the maximum and minimum acceleration values to detect fall. Bourke et al. proposed fall algorithms separately based on thresholds on both signals from a tri-axial accelerometer [6] and a biaxial gyroscope [5]. However, because of using a set of fixed thresholds for features extracted from sensor data streams but testing on individuals with different factors such as mass, age, clinical history and diseases [3], it results in the sensor data could be affected by those factor. Thus, it makes these algorithms to be ineffective.

Recent wearable fall detection systems use classification methods with higher reliability. Zhang et al. [29] proposed temporal and magnitude features extracted from acceleration signal and used a 1-class SVM classifier to detect fall. Doukas et al. [11] presented an SVM based method to classify acceleration data in three axis into fall, walk and run. Jantaraprim et al. [17] also used SVM with short time min-max features to detect falls. Pham et al. in [23] used a set of simple but effective statistical features with Hidden Markov Model (HMM) to detect fall and recognize some ADL. These methods reported with high detection rates but only tested on their own datasets. In this work, we propose an SVM based method with new features for fall detection. Through a rigorous evaluation process over several public datasets, these features are proved to be more effective in separating falls and ADL.

3   Methods

In our proposed algorithm, falls are detected by classifying a window of a signal stream into "fall" and "non-fall". Given input as three streams of acceleration data along three axes x, y, z, the algorithm detects falls using the following steps: segmentation, feature extraction and classification. In the segmentation step, the system takes tri-axial continuous sensor data streams from accelerometers attached on the user's body as input. Using a sliding window, we segment continuous sensor data streams into equal-size frames (possibly with overlaps), whose length is longer than duration of any ADL and falls. In the feature extraction step, we extract features from each frame which we use as input for the classification step. In classification step, the system classifies each frame into "fall" and "non-fall" by using an SVM or other classification algorithms with the extracted features.

For the rest of this section, we present details of the feature extraction algorithm and other important aspects in our method.

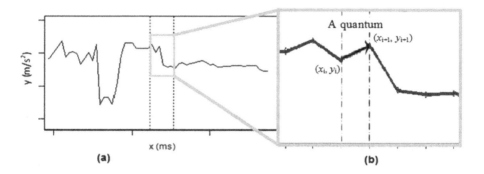

Fig. 1. An example of a quantum and its orientation

3.1   Feature Extraction Algorithm

First, each frame (or window) is divided into small fragments or *quanta* of size *l*. After computing orientation of each quantum, we construct a histogram of these orientations by dividing the entire range of orientations into series of intervals/bin and count the number of quanta having orientation falling into each interval. Bin ranges can be learnt by using clustering methods such as k-means to group orientations of quanta of training frames into clusters, instead of being assigned to some predefined values. The constructed histogram is used as feature for the given frame. Details for the feature extraction process are described with following steps.

Computing Orientations of Quanta. Given a frame with signal on x-axis and its quanta as illustrated in Fig. 1. Notice that the signal stream is a time series represented in xy-coordinates where x is in millisecond and y is in m/s^2. The orientation of a quantum q is defined as the angle between the vector connecting its start (x_i, y_i) and end (x_{i+1}, y_{i+1}) points and x-axis, as shown in Fig. 1(b). It can be calculated using this formula:

$$\theta(q) = \tan^{-1}\left(\frac{y_{i+1} - y_i}{x_{i+1} - x_i}\right) \qquad (1)$$

Orientation values can be in range of (-90, 90°).

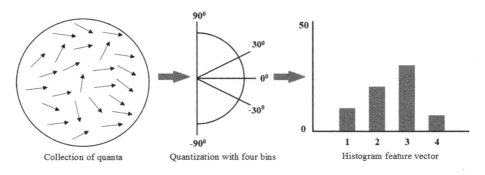

Fig. 2. An example of quantization with four bins and histogram feature vector

Clustering the Orientations of Quanta. To create a histogram, we have to define its bins first. This can be done by breaking 180° of orientation into M bins (possibly with different ranges). Here, a bin is a group of quanta with similar orientations. Figure 2 shows an example of quantization setting with four bins: $(-90$ to $-30°)$, $(-30$ to $0°)$, $(0$ to $30°)$ and $(30$ to $90°)$. Bin ranges can be manually chosen or automatically calculated by using unsupervised clustering methods. In case of using unsupervised clustering methods to automatically calculate bin ranges, to map orientation of a quantum into a bin, we need to perform cluster index assignment using nearest neighbor search. If number of clusters increases, the cost of cluster assignment increases. We can improve this step by calculating bin ranges directly as follows. Except the first bin and the last bin, given a bin i^{th}: the lowest value of the range is calculated by mean value of cluster centers of cluster $(i-1)^{th}$ and cluster i^{th}; the highest value of the range is calculated by mean value of cluster centers of cluster i^{th} and cluster $(i+1)^{th}$. Range of the first bin is from 0° to mean value of the first and second cluster centers. And range of the last bin is from mean value of the cluster centers of cluster $(M-1)^{th}$ and cluster M^{th} to 90°.

Calculating the Orientation Histogram. In this step, we map the orientation of each quantum q into one of these bins. The number of quanta belonging to each of M bins is counted to form a histogram vector of size M, elements of which are the counted numbers (see the example in the right part of Fig. 2).

Then, we concatenate three vectors counted this way for the three signal streams (along the x, y, z axes) to form a vector of size $3 \times M$, which used as the feature vector of the given frame.

3.2 Constructing Finer Representation Using Orientation Histogram Based Features

The intuition behind using this kind of features is that frames corresponding to similar activities or falls would have similar curves of signal streams and thus have similar number of quanta with close orientations. Therefore, these features provide an approximate representation of signal streams, which can be computed efficiently and is suitable for a wide range of classification algorithms. Unfortunately, histograms of quanta orientations do not contain information about relative locations of the quanta within a window. We illustrate this problem by an example below.

Consider two frames of two different activities as described in Fig. 3. Assume that we need to construct a histogram of two 90° bins. Range of bin 1 is from -90 to 0° and range of bin 2 is from 0 to 90°. Using the above proposed algorithm, we can extract features of these two frames, which are two histogram vectors. Both of them have same values of (2, -2), even though they are computed from the two different activities which have different signal streams. But if we divide the frames into two segments first (using the red line in Fig. 3), then extract features from each segments and concatenate the two local histogram vectors, we have two different histogram feature vectors of the two frames. One is (1, 1, 1, 1) and the other is (2, 0, 0, 2). It is shown that we can solve this problem by extracting

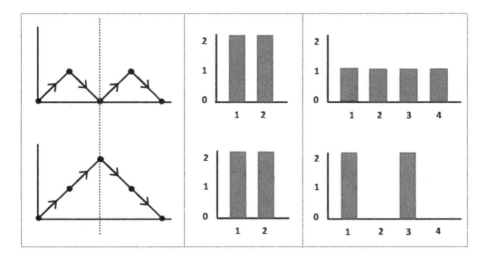

Fig. 3. Same histogram features (in the middle) extracted from two frames (in the left) without segmentation and different features (in the right) extracted from two frames (in the left), each segmented into two segments (by the red line). (Color figure online)

local features of the frame instead of only extracting features of the whole frame. The modified algorithm is described in general as follows.

For each frame, we divide it into K smaller non-overlapping, equal-size segments. Then for each segment of the frame, we extract features using the above proposed feature extraction method to form a local feature vector. The feature vector of the frame is then constructed by concatenating K local feature vectors from K segments within the frame.

In feature extraction process, quantum length l, number of bins M and number of segments K are important parameters. Since the common values are from 50 Hz to 100 Hz, in this work, we only use quanta with length of 1 because length of 2 or higher means the sampling rate may be smaller than 50 Hz which is inadequate for fall detection purpose. Number of bins and number of segments could be decided by experiments or tuning process.

3.3 Classification Using SVM

To separate falls from non-fall activities, we can use the features extracted in the previous step with various classification algorithms. In this study, we employ support vector machines (SVM) as the classifier because SVM can effectively perform classification in a wide range of application domains including automatic fall detection [17,29]. SVM can achieve good generalization on unseen data by relying on two techniques: (i) mapping input features into a new feature space often of higher dimensions by using a kernel function; and (ii) finding in the new feature space a hyperplane with max-margin that separate negative examples from positive ones. For training and testing with SVM, we normalize all calculated feature vectors.

4   Experiments and Results

This section represents experiments for verifying our proposed method. This starts with data pre-processing, then followed by a brief for describing three publicly datasets: DLR [13], MobiFall2 [26] and tFall [21]. The section will also cover the influence of some parameters on detection accuracy and ends up with the comparison of our method with some state-of-the-art fall detection methods using accelerometers.

4.1   Dataset and Preprocessing

We conduct a rigorous experiment on 3 public datasets collected by different research institutions: DLR dataset [10] from the Institute of Communications and Navigation of the German Aerospace Center, MobiFall2 dataset [4] from the Biomedical Informatics & eHealth Laboratory of the Technological Educational Institute of Crete and tFall dataset [12] from the University of Zaragoza. All datasets comprise of various simulated falls (in which younger people mimicked elderly people) and ADL, contexts and highly rich sensor data.

- **DLR:** this dataset contains activity data for falls and six daily activities. Several types of falls and daily activities such as "sitting", "standing", "walking", "running", "jumping" and "lying" are performed by 16 male and female subjects aged between 23 and 50. Each participant worn the Xsens MTx inertial measurement unit (IMU) with a single tracker placed on the belt. The accelerometers sampling frequency is set at 100 Hz.

- **MobiFall2:** this dataset consists of 4 types of falls and 9 daily activities performed by 11 volunteers: 6 males and 5 females, with ages from 22 to 36. Fall patterns comprise of "forward-lying", "front-knees-lying", "sideward-lying" and "back-sitting-lying" and nine daily activities comprise of "standing", "walking", "jogging", "jumping", "stairs up", "stairs down", "sitting on a chair", "step in a car" and "step out a car". Each participant positions a Samsung Galaxy S3 smartphone in his pocket. The frequency of accelerometer in the phones is set at 87 Hz.

- **tFall:** this dataset comprises of 8 types of falls and several real-life activities which are collected by 7 males and 3 females aged from 20 to 42 under real-life conditions. The simulated set of falls consists of "forward", "forward straight", "backward", "lateral left", "lateral right", "sitting on empty air", "syncope" and "forward fall with obstacle". Each participant used Samsung Galaxy Mini phone to acquire ADL data, in which only ADL with acceleration peak over a given threshold (1.5 g) were recorded. In fall data collection process, each participant put their phones in both their two pockets. The accelerometer sampling frequency is set at 50 Hz. After preprocessing, the dataset consists of 7816 ADL and 503 fall records, each has length of 6 seconds. This is the biggest dataset used in the experiment.

Before evaluating the fall detection algorithms, the datasets have been pre-processed using 50 % overlapping, 6 second sliding windows, labeled as ADL or falls with acceleration peak over a threshold of 1.5 g (similar with samples in tFall dataset). The acceleration peak of a frame is the maximum value of acceleration magnitude of a frame. After preprocessing process, we obtained 2771 frames of ADL and 36 frames of falls from DLR dataset, 1832 frames of ADL and 288 frames of falls from MobiFall2 dataset and 7816 frames of ADL and 503 frames of falls from tFall dataset. Summary of all datasets used in this experiment is shown in Table 1.

4.2 Experiment Settings

A simple performance metric often used for effectiveness evaluation on imbalanced classification is overall accuracy, which is the ratio of number of frames correctly classified over the total of frames. Two other metrics often used to evaluate a fall detection algorithm are sensitivity and specificity. The sensitivity is the number of true positive responses divided by the number of actual positive cases. It shows how good the classifier is at detecting a test condition correctly. The specificity is the number of true negatives divided by the number of actual negatives. It shows how good a classifier is at avoiding false alarms.

Table 1. Summary of datasets used in the experiment

	DLR	MobiFall2	tFall
Participants	16 males and females (aged 23–50)	6 males and 5 females (aged 22–36)	7 males and 3 females (aged 20–42)
No. types of falls	Not specified	4	8
No. types of ADL	6	9	Not specified
No. fall frames	36	288	503
No. ADL frames	2771	1832	7816
Acc. frequency	100 Hz	87 Hz	50 Hz
Acc. range	7g	2g	2g
Sensor position	Belt	Pocket	Pocket

Following [1,3], we used accuracy, sensitivity and specificity as the performance metrics for fall detection algorithms.

Because the samples are distributed unequally between ADL and fall data in the benchmark datasets, all methods and settings are evaluated with only 4-fold cross validation. For all experiments we report performance metrics over folds. For each fold, we held out 10 % of the training set (via stratified sampling) as the validation set for tuning parameters and used these parameters for the test set (unless mentioned otherwise).

We used LibSVM [8] with radial basis kernel (RBF) for the experiments. Parameters C and *gamma* of SVM were selected by grid search on the validation set. All experiments were run on a PC running Windows with 2.8 GHz dual-core processor and 8 GB RAM.

4.3   Results

First, we performed an experiment on tFall dataset to examine the influence of different aspects of our algorithms in fall detection accuracy: (a) features extracted from whole frame with manual bin selection and bins having same degrees; (b) features extracted from whole frame with automatic bin selection using k-means clustering method; (c) features extracted after frame segmentation with automatic bin selection using k-means clustering method. To show the general applicability of the proposed algorithm, we tuned parameters only for tFall dataset using a hold-out validation set, and used these parameters for all three datasets. The tuned parameters are: the number of bins $M = 60$ (from a set of $\{10, 30, 60, 90, 120, 180\}$) and the number of segments of a frame $K = 8$ (from a set of $\{4, 8, 16, 32, 64\}$). Cross validation was run for each method in turn with SVM as the classifier. Table 2 summarizes the results obtained on tFall dataset.

Noticed that all methods in Table 2 achieve high accuracies. It can be realized that automatic bin selection using clustering gives a noticeable improvement

Table 2. Comparative fall detection results on tFall dataset

Methods	Accuracy	Sensitivity	Specificity
(a) Extracted from whole frame + manual bin selection	95.1 % ± 0.35 %	97.36 % ± 0.27 %	71.74 % ± 1.25 %
(b) Extracted from whole frame + automatic bin selection	97.47 % ± 0.24 %	98.75 % ± 0.17 %	86.45 % ± 1.15 %
(c) Extracted from segments of frame + automatic bin selection	98.71 % ± 0.32 %	99.40 % ± 0.27 %	93.75 % ± 1.89 %

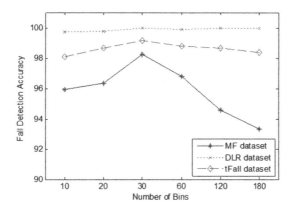

Fig. 4. Fall detection accuracy vs. number of bins

(about 2 % higher than (a)). The result is even better for (c) with about 3.5 % higher than (a). These results can be explained that by using local features, no information about relative locations of the quanta within a frame is lost. Thus, it can keep characterization of a fall (i.e. characterization of different phases of a fall), which contributes to distinguish between falls and fall-like ADL. Even sensitivity of (c) is not improved much, specificity gets remarkable result. This shows that (c) can avoid much more false alarms compared to (a) and (b).

To examine the influence of number of bins M to the feature extraction method (c), we varied the number of bins while keeping number of segments same as in previous experiment. Then, we plot the accuracies of all three datasets against the number M of bins and the number K of segments per frame to see changes in performance in Figs. 4 and 5. As seen in Fig. 4, the accuracy increases when number of bins increases from 10 to 30 and then the accuracy decreases afterward, especially for MF dataset the accuracy drops significantly. Figure 5 shows that the fall detection accuracy increases when number of segments increases from 4 to 6, and then it slowly decreases when number of segments reaches 16 and go down steadily afterward.

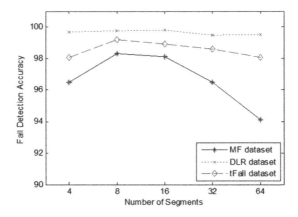

Fig. 5. Fall detection accuracy vs. number of segments per frame

Table 3. Fall detection results on three benchmark datasets

Methods	DLR	MobiFall2	tFall
(1)	98.11 % ± 0.20 %	96.51 % ± 0.41 %	95.62 % ± 0.53 %
(2)	98.83 % ± 0.50 %	96.22 % ± 0.34 %	96.18 % ± 0.46 %
(3)	99.93 % ± 0.08 %	99.20 % ± 0.45 %	98.70 % ± 0.32 %

(1) Pham et al. [23], (2) Jantaraprim et al. [17], (3) Our method.

Next, we compare against two recent and high-performance methods using statistical based features proposed by Pham et al. [23] and Jantaraprim et al. [17] on all three datasets. We re-implemented these methods in Matlab and kept the best parameter values as reported in their papers. The results are summarized in Table 3, represented in the structure of mean accuracy ± standard deviation.

As can be seen, all methods achieved very high accuracies on DLR dataset (over 98 %) while tFall dataset is more challenging. Our method has highest accuracies over all three datasets. The accuracy improvements of our method over (1) and (2) are noticeable on all three datasets while there is no clear winner between (1) and (2).

5   Conclusion

We have presented an effective feature extraction method for automatic fall detection. These features are constructed by aggregating multiple local histogram features which is created by quantizing orientations of quanta of each segment in a frame. The histogram bins are automatically computed based on unsupervised clustering techniques instead of manual selection. This makes the proposed method can be applied for various accelerometer data without loss of detection accuracy due to the influence of different characteristics of datasets such as sampling frequency, acceleration ranges or position of accelerometers.

The proposed features are validated using support vector machines with 4-fold cross validation protocol. The experiment results show that the detection accuracy of our method is greater than two other state-of-the-art fall detection methods [17,23] when tested on three public datasets containing simulated falls and ADL. Even no real data from older people used for the testing, the results indicate the potential of our method for real-world fall detection at large.

References

1. Abbate, S., Avvenuti, M., Bonatesta, F., Cola, G., Corsini, P., Vecchio, A.: A smartphone-based fall detection system. Pervasive Mob. Comput. **8**(6), 883–899 (2012)
2. Alwan, M., Rajendran, P.J., Kell, S., Mack, D., Dalal, S., Wolfe, M., Felder, R.: A smart and passive floor-vibration based fall detector for elderly. In: Information and Communication Technologies, ICTTA 2006, vol. 1, pp. 1003–1007. IEEE (2006)
3. Bagala, F., Becker, C., Cappello, A., Chiari, L., Aminian, K., Hausdorff, J.M., Zijlstra, W., Klenk, J.: Evaluation of accelerometer-based fall detection algorithms on real-world falls. PLoS ONE **7**(5), 1–9 (2012)
4. Lab, B.M.I.: MobiFall2 dataset. http://www.bmi.teicrete.gr/index.php/research/mobifall. Accessed 05 Dec 2015
5. Bourke, A.K., Lyons, G.M.: A threshold-based fall-detection algorithm using a bi-axial gyroscope sensor. Med. Eng. Phys. **30**, 84–90 (2008)
6. Bourke, A.K., OBrien, J.V., Lyons, G.M.: Evaluation of a threshold-based tri-axial accelerometer fall detection algorithm. Gait & Posture **26**(2), 194–199 (2007)
7. Burges, C.J.C.: A tutorial on support vector machines for pattern recognition. Data Min. Knowl. Disc. **2**(2), 121–167 (1998)
8. Chang, C.C., Lin, C.J.: LIBSVM: a library for support vector machines. ACM Trans. Intell. Syst. Technol. (TIST) **2**(3), 27 (2011)
9. Chen, J., Kwong, K., Chang, D., Luk, J., Bajcsy, R.: Wearable sensors for reliable fall detection. In: Engineering in Medicine and Biology Society, pp. 3551–3554 (2005)
10. DLR: DLR dataset. http://www.dlr.de/kn/en/Portaldata/27/Resources/dokumente/04_abteilungen_fs/kooperative_systeme/high_precision_reference_data/Activity_DataSet.zip. Accessed 05 Dec 2015
11. Doukas, C., Maglogiannis, I., Tragas, P., Liapis, D., Yovanof, G.: Patient fall detection using support vector machines. In: Boukis, C., Pnevmatikakis, A., Polymenakos, L. (eds.) Artificial Intelligence and Innovations 2007: From Theory to Applications, pp. 147–156. Springer, Heidelberg (2007)
12. EduQTech Group: tFall dataset. http://eduqtech.unizar.es/fall-adl-data/. Accessed 22 Dec 2015
13. Frank, K., Nadales, M.J.V., Robertson, P., Pfeifer, T.: Bayesian recognition of motion related activities with inertial sensors. In: Proceedings of the 12th ACM International Conference Adjunct Papers on Ubiquitous Computing - Ubicomp 2010, p. 445 (2010)
14. Fu, Z., Culurciello, E., Lichtsteiner, P., Delbruck, T.: Fall detection using an address-event temporal contrast vision sensor. In: IEEE International Symposium on Circuits and Systems, ISCAS 2008, pp. 424–427. IEEE (2008)

15. Hazelhoff, L., Han, J., de With, P.H.N.: Video-based fall detection in the home using principal component analysis. In: Blanc-Talon, J., Bourennane, S., Philips, W., Popescu, D., Scheunders, P. (eds.) ACIVS 2008. LNCS, vol. 5259, pp. 298–309. Springer, Heidelberg (2008)

16. Igual, R., Medrano, C., Plaza, I.: Challenges, issues and trends in fall detection systems. BioMed. Eng. OnLine 12(1), 1 (2013)

17. Jantaraprim, P., Phukpattaranont, P., Limsakul, C., Wongkittisuksa, B.: Fall detection for the elderly using a support vector machine. Int. J. Soft Comput. Eng. 2(1), 484–490 (2012)

18. Kangas, M., Konttila, A., Lindgren, P., Winblad, I., Jämsä, T.: Comparison of low-complexity fall detection algorithms for body attached accelerometers. Gait & posture 28(2), 285–291 (2008)

19. Lai, C.F., Chang, S.Y., Chao, H.C., Huang, Y.M.: Detection of cognitive injured body region using multiple triaxial accelerometers for elderly falling. IEEE Sens. J. 11(3), 763–770 (2011)

20. Lindemann, U., Hock, A., Stuber, M., Keck, W., Becker, C.: Evaluation of a fall detector based on accelerometers: a pilot study. Med. Biol. Eng. Comput. 43(5), 548–551 (2005)

21. Medrano, C., Igual, R., Plaza, I., Castro, M.: Detecting falls as novelties in acceleration patterns acquired with smartphones. PLoS ONE 9(4), e94811 (2014)

22. Noury, N., Rumeau, P., Bourke, A.K., ÓLaighin, G., Lundy, J.E.: A proposal for the classification and evaluation of fall detectors. Irbm 29(6), 340–349 (2008)

23. Pham, C., Phuong, T.M.: Real-time fall detection and activity recognition using low-cost wearable sensors. In: Murgante, B., Misra, S., Carlini, M., Torre, C.M., Nguyen, H.-Q., Taniar, D., Apduhan, B.O., Gervasi, O. (eds.) ICCSA 2013, Part I. LNCS, vol. 7971, pp. 673–682. Springer, Heidelberg (2013)

24. Reece, A.C., Simpson, J.M.: Preparing older people to cope after a fall. Physiotherapy 82(4), 227–235 (1996)

25. Rougier, C., Meunier, J., St-Arnaud, A., Rousseau, J.: Fall detection from human shape and motion history using video surveillance. In: 21st International Conference on Advanced Information Networking and Applications Workshops, AINAW 2007, vol. 2, pp. 875–880. IEEE (2007)

26. Vavoulas, G., Pediaditis, M., Spanakis, E.G., Tsiknakis, M.: The MobiFall dataset: an initial evaluation of fall detection algorithms using smartphones. In: 13th IEEE International Conference on BioInformatics and BioEngineering, pp. 1–4 (2013)

27. Wild, D., Nayak, U.S., Isaacs, B.: How dangerous are falls in old people at home? BMJ 282(6260), 266–268 (1981)

28. Wu, G.E., Xue, S.: Portable preimpact fall detector with inertial sensors. IEEE Trans. Neural Syst. Rehabil. Eng. 16(2), 178–183 (2008)

29. Zhang, T., Wang, J., Xu, L., Liu, P.: Fall detection by wearable sensor and one-class SVM algorithm. In: Huang, D.-S., Li, K., Irwin, G.W. (eds.) Intelligent computing in signal processing and pattern recognition, pp. 858–863. Springer, Heidelberg (2006)

30. Zigel, Y., Litvak, D., Gannot, I.: A method for automatic fall detection of elderly people using floor vibrations and soundproof of concept on human mimicking doll falls. IEEE Trans. Biomed. Eng. 56(12), 2858–2867 (2009)

Learning of Evaluation Functions to Realize Playing Styles in Shogi

Shotaro Omori$^{(\boxtimes)}$ and Tomoyuki Kaneko

Graduate School of Arts and Sciences, The University of Tokyo, Tokyo, Japan
{omori,kaneko}@graco.c.u-tokyo.ac.jp

Abstract. This paper presents a method to give a computer player an intended playing style by the machine learning of an evaluation function. Recent improvements in machine learning techniques have realized the automated tuning of the feature weight vector of an evaluation function. To make a strong player, as many moves as possible of strong players' game records are needed, though the number of available game records decreases when we focus on a specific playing style. To pursue both goals of playing style and playing strength, we present three steps of learning: classifying moves with respect to playing styles, training the weight vector of an evaluation function by using the whole set of game records to maximize its playing strength, and modifying the weight vector carefully so as to improve agreement with the moves of the intended playing style. We applied our method to realize players of defense or attack-oriented style in shogi and tested the players by self-play against the original version. The results confirmed that the presented method successfully adjusted evaluation functions in that the frequency of defensive moves is significantly increased or decreased in accordance with the game records used while keeping the winning ratio at almost 50 %.

1 Introduction

Development of personalized AI systems that are not only useful but also communicate with users in a preferable manner depending on a user is a challenging goal in artificial intelligence research. For example, research in game programming has yielded many strong programs in various kind of games, where some programs have outperformed top human players (Campbell et al. 2002; Buro 2002; Tesauro 2002; Matsubara 2015). This paper focuses on the playing styles of computer programs on top of the strength of programs. Many human players have their own playing styles, and many of us enjoy moves that match their styles in addition to their strengths when watching games. Developing such techniques that create a computer program with an intended playing style is a challenging goal in artificial intelligence research. If a wide variety of computer players are also available, it might bring fun as well as an actual benefit to the training of skills when a user plays against a computer program.

S. Omori—presently with Yahoo Japan Corporation.

R. Booth and M.-L. Zhang (Eds.): PRICAI 2016, LNAI 9810, pp. 367–379, 2016.
DOI: 10.1007/978-3-319-42911-3_31

To develop a computer program having an intended playing style, we present a method based on the machine learning of evaluation functions. We chose shogi, which is a popular chess variant in Japan (Iida et al. 2002), for evaluation of our method because there are established machine learning techniques that adjust forty million parameters in an evaluation function by using the game records of experts (Hoki and Kaneko 2014). The machine learning techniques are so effective that no shogi program with conventional handcrafted evaluation functions has broken into the top five of computer shogi tournaments in more than five years. Also, there are many professional players in shogi, where various playing styles are recognized and discussed (Namai and Ito 2010; Sawa and Ito 2011).

In the learning, the weight vector of an evaluation function is adjusted so that the decisions by the game tree search agree with a given set of game records. Intuitively, if we use a subset of moves consistent with an intended playing style for the learning, the game tree search with the resulted evaluation function is expected to follow the playing style. However, this naive approach has a serious drawback in the playing strength because the number of game records available for teacher data drastically decreases in the naive approach. Empirically, the number of game records for teacher data should be at an order of at least ten thousand (e.g., 40,000) to make a strong shogi program while at most 1000 records are available for a specific player in shogi. The playing strength is actually degraded when we decrease the number of game records, as shown in our experiments.

To pursue both goals of playing style and playing strength, we present three steps of learning: selecting a subset of moves with respect to a playing style to be realized, training the weight vector of an evaluation function by using the whole set of game records to maximize its playing strength, and modifying the weight vector carefully. In the second step, we use the whole set of game records of expert players regardless of their playing styles to maximize its playing strength, and this is the main difference from the naive approach above. Then, the weight vector is adjusted so as to improve agreement with the selected moves while keeping the difference from the weight vector obtained in the second step as small as possible. We applied our method to realize players with a defense or attack-oriented style in shogi and conducted self-play between each player with a learned evaluation function and the original version. We measured how the intended playing style is reproduced as well as the winning ratio. The results confirmed that the presented method successfully adjusted the evaluation functions in that the frequency of defensive moves is significantly increased or decreased in accordance with the game records used while keeping the winning ratio at almost 50 %.

2   Related Work

This section briefly introduces related work on the playing styles and machine learning of evaluation functions.

2.1   Playing Styles

While it is usually assumed that black and white share a common evaluation function in minimax search methods, one can exploit the difference between strategies if he or she knows the opponent's strategy (Carmel and Markovitch 1993; Donkers et al. 2005). After techniques on machine learning of evaluation functions have improved, work in chess was presented that identifies whether a record is played by Kasparov or Kramnic (Levene and Fenner 2009). In that work, evaluation functions are trained via temporal difference learning. Because the playing styles of Kasparov or Kramnic are different in tactical and positional aspects, evaluation functions are expected to be different when they are tuned to fit each player. In the learning, only the fifth to the 35-th moves are considered. We followed up in our work in that only moves in a middlegame are analyzed and used for classification of a game record. Although strong chess programs are obtained through a variant of temporal difference learning (Baxter et al. 2000; Veness et al. 2009), the number of feature weights adjusted was on the order of thousands. In this work, to obtain more accurate evaluation functions for a specific playing style, we used large-scale learning techniques (Hoki and Kaneko 2014) and adjusted forty million parameters in an evaluation function.

Playing styles can be realized by tuning parameters other than those in evaluation functions. In order to develop an interesting player in a real time game, modifying the preferences of actions is proposed so that the final scores of both players become small (Ortiz B. et al. 2010). For players using Monte-Carlo tree search methods, playing styles are shown to be adjusted by setting appropriate prior knowledge in experiments in a game of Dobutsu Shogi (Shimizu and Kaneko 2014). Additionally, it is also reported that playing strength is degraded by inducing prior knowledge oriented to a playing style, although strength can be recovered without losing its style by increasing the number of playouts or by enhancing forward pruning in UCT. In this work, we assume a computer player playing with traditional game tree search with an evaluation function.

2.2   Shogi and Related Work

Shogi is a popular chess variant in Japan, where the rules of shogi, as well as a survey of techniques in computer programs, are described in (Iida et al. 2002). We chose shogi for evaluation of our method, because there is a standard method to tune the weight vector in evaluation functions (Hoki and Kaneko 2014). There are also many professional game players in shogi, and we followed the existing research about the playing styles of professional players.

Shogi players have some common understanding in playing styles, and many popular words exist related to playing styles, e.g., whether pieces are densely built-up or not. However, a study in psychology showed that professional players do not always agree on the playing style of a player, even in the basic classification of an attack-oriented or defense-oriented style (Okamoto and Hashiguchi 1989). Recently, some studies analyzed the relation between statistical properties (e.g., the frequency of each type of piece moved) and playing styles to distinguish the famous player Habu as well as others (Tosaka and Matsubara 2006), or to distinguish an attack-oriented or defense-oriented style (Sawa and Ito 2011). Because the number of game records played by a single player is limited, we chose the notion of attack and defense oriented styles in our work while we used a different definition suitable for machine classification to increase the number of game records available for training.

Preliminary work exists that tried to realize a computer player that has the playing style of a specific player (Namai and Ito 2010). While the selection of opening moves is reported to work well by adjustment of the opening database, there was much room for improvement in the tuning of evaluation functions to adjust moves in the middlegame. The reason is apparently an insufficient number of game records by considering observations in other research (Hoki and Kaneko 2014; Kaneko 2012; Takise and Tanaka 2012). There is another work that tried to improve such a skill of computer programs that moves their king as well as other pieces into the opponent area to win by a declaration rule (Takise and Tanaka 2012). The work reported that it is not sufficient to adjust the weight vector of an evaluation function with a set of game records related to the intended skill. Instead, to obtain improved evaluation functions with respect to the skill, a small number of features specific to the skill were introduced, and only the weights of the features were adjusted by using the selected game records. These observations show the difficulty of realizing a playing style via adjustment of evaluation functions.

2.3 Learning of Evaluation Functions in Bonanza

We used Bonanza (version 6.0) in this work, which is an open source shogi program[1] and which won the world computer shogi championship twice. The details of game-tree search in Bonanza are described in (Hoki and Muramatsu 2012), and the Minimax Tree Optimization method (MMTO) for machine learning of evaluation functions is described in (Hoki and Kaneko 2014). We briefly introduce MMTO here because our work is constructed on top of MMTO. About forty million parameters are used for the feature weight vector $\boldsymbol{w}$ of its evaluation function, and vector $\boldsymbol{w}$ is tuned by MMTO to maximize agreement with a given set of game records. MMTO minimizes the following objective function:

$$J^{\mathcal{P}}(\boldsymbol{w}) = J(\mathcal{P}, \boldsymbol{w}) + J_C(\boldsymbol{w}) + J_R(\boldsymbol{w}), \tag{1}$$

where $J(\mathcal{P}, \boldsymbol{w})$ is the main term presenting how $\boldsymbol{w}$ inconsistent with a set of training positions $\mathcal{P}$ in the given game records, $J_C(\boldsymbol{w})$ is for the constraint of the weights of the pieces, and $J_R(w)$ is a standard l_1 regularizer that governs the weights of positional features other than the pieces. Term $J(\mathcal{P}, \boldsymbol{w})$ is the summation of differences of minimax values between a move played in a game record and another legal move in all positions:

$$J(\mathcal{P}, \boldsymbol{w}) = \sum_{p \in \mathcal{P}} \sum_{m \in \mathcal{M}'_p} T(s(p.d_p, \boldsymbol{w}) - s(p.m, \boldsymbol{w})),$$

---

[1] http://www.geocities.jp/bonanza_shogi/.

where p is a position, d_p is a move played in position p, $\mathcal{M}'_p$ is a set of legal moves in position p except for played move d_p, $s(p, \boldsymbol{w})$ is the minimax value for position p identified by tree search with evaluation function of weight vector $\boldsymbol{w}$, $p.m$ is the position after move m in position p, and $T(x)$ is a horizontally mirrored sigmoid function $1/(1 + \exp(ax))$ where $a = 0.0273$ in the implementation of Bonanza we used. In short, term $J(\mathcal{P}, \boldsymbol{w})$ is minimized when the move played is evaluated better with respect to a player to move than other moves, by game tree search with weight vector $\boldsymbol{w}$ for each position. Weight vector $\boldsymbol{w}$ is iteratively updated following an approximated gradient of objective function $J^{\mathcal{P}}(\boldsymbol{w})$. In each update, small randomness is added in the implementation of Bonanza 6.0.

The evaluation function of Bonanza equips an éxtended piece square table considering the combination of three pieces including the king. Because these features can discriminate small differences in the location of the king and that of another piece, we believe that they can represent the attack or defense-oriented playing style defined in the next section.

3 Learning of Playing Style

We present a method to give a computer player an intended playing style by the machine learning of an evaluation function. We chose a defense or attack-oriented style in shogi for an example of the playing style to be realized. Our method is general and not specific to these playing styles; however, it is important that a large number of game records are available for the training of a style. Let $\mathcal{P}_f$ be a set of training positions, each of which is labeled by a desired move. Typically, $\mathcal{P}_f$ consists of the moves and positions of all available game records played by expert players. Our method consists of three steps of learning:

Selection select a subset of training positions $\mathcal{P}_s \subset \mathcal{P}_f$ so that the desired moves in $\mathcal{P}_s$ are consistent with a playing style to be realized,

Strength-improvement train the weight vector of an evaluation function by using the full set of training positions $\mathcal{P}$ to maximize its playing strength (we call the resulting vector $\boldsymbol{w}_f$), and

Style-adjustment modify the weight vector carefully $\boldsymbol{w}_f$ and obtain $\boldsymbol{w}_s$ so as to improve agreement with the selected moves $\mathcal{P}_s$ prepared in the Selection step while keeping the difference from vector $\boldsymbol{w}_f$ obtained in the Strength-improvement step as small as possible.

In the following subsections, the details of the Selection step and the Style-adjustment step are described. In the Strength-improvement step, standard MMTO briefly reviewed in Sect. 2.3 is used to obtain the weight vector, and we used the implementation in Bonanza.

3.1 Classification of Attack and Defense Records

We call the sequence of all moves in a game record played by a single player (e.g., black or white) a *half record*. Usually, the playing style of black and that of the white player are different. Therefore, we present the selection of a set of half records for the training of an intended playing style.

We classify each move as *attack* or *defense* in accordance with the Manhattan distance between the destination square of the move and each king. If the destination is closer to the player's king, it is treated as a defense move. Otherwise, i.e., if the destination is closer to the opponent's king or at an equal distance from both kings, it is treated as an attack move. Then, we classify the style of a half record as attack or defense in accordance with the relative frequency of attack or defense moves. If the number of attack moves in a half record is more than or equal to (less than) that of the defense moves, the style of the half record is treated as attack (defense). In order to make the classification as consistent with the feeling of human players as possible, the number of attack and defense moves is measured only in a middlegame, where we divided each half record into three stages: opening, middlegame, and endgame, each of which consists of equal move lengths. It is because other aspects than the playing style have relatively high influence on the destination of a move. Most pieces inherently go near their king in the opening by the initial placement of pieces, and in the endgame it is natural that the winning (losing) side plays more attack (defense) moves. Also, we excluded the game records that resulted in a draw in our training because the move length of such records tends to be much different from the average length of other records. This automated classification is effective in keeping the number of half records available as large as possible. The number of attack or defense moves is expected to be similar, and each half record is classified as attack or defense without ties. Therefore, almost half of all records are available for our training. Our classification was basically consistent with the style of players discussed in (Sawa and Ito 2011).

3.2   Learning of Evaluation Functions

Let $\boldsymbol{w}_f$ be the weight vector of an evaluation function trained with the whole set of game records in the Strength-improvement step. Also, we have a set of half records $\mathcal{P}_s$ consisting of an intended playing style, attack or defense. Then, in the Style-adjustment step, we modify the weight vector so that it is consistent with the set $\mathcal{P}_s$.

In order to maintain the playing strength of $\boldsymbol{w}$ throughout adjustments in this step, we introduced an additional constraint term for l_2 regularization to keep $\boldsymbol{w}$ near $\boldsymbol{w}_f$, by following the work by Yano et al. (2009):

$$J_{R_f}(\boldsymbol{w}, \boldsymbol{w}_f) = \frac{C}{2}||\boldsymbol{w} - \boldsymbol{w}_f||^2. \tag{2}$$

Therefore, the objective function to be minimized in this step is slightly modified from Eq. (1):

$$J(\mathcal{P}_s, \boldsymbol{w}) + J_C(\boldsymbol{w}) + J_R(\boldsymbol{w}) + J_{R_f}(\boldsymbol{w}, \boldsymbol{w}_f).$$

We also used the implementation of MMTO in Bonanza for this step with two modifications: incorporation of the new regularization term J_{R_f} and learning

from a set of half records instead of a set of full game records. The same number
of half records are prepared for each combination of (black, white) × (attack,
defense) for stability in the learning process.

4   Experimental Results

To show the effectiveness of our method, we conducted experiments with
Bonanza in shogi. The game records of 295 professional players were used
throughout the experiments and are available in the shogi-kifu-database[2]. Note
that the games ended in draw by repetition or by jishogi are excluded in advance,
as discussed in Sect. 3.1.

4.1   Game Records

We analyzed a playing style for each half record (i.e., each player of each record),
and prepared exclusive sets of 20, 000 half records for each combination of player
(black or white) and style (attack or defense), listed as a, d, e, and h in Table 1(a).
Also, we collected a set of the accompanying half records played by the opponent,
to compare learning with the full record later. A full game record consists of the
half record moves of black and those of white. For example, when we find the
half record of black is in an attack (defense)-oriented style, we classify not only
the half record of black into set a (e), but also the accompanying half record of
white into set b (f).

Table 1. Configurations of game records used for learning of evaluation functions

| full | half | | | name | record type | #half records | |
	black	white				attack	defense
ab	a attack	b ?		attack 4:defense 0	attack (ad)	40, 000	0
cd	c ?	d attack		attack 0:defense 4	defense (eh)	0	40, 000
ef	e defense	f ?		attack 2.2:defense 5.8	(efgh)	22, 131	57, 869
gh	g ?	h defense		attack 7.3:defense 8.7	(abcdefgh)	73, 015	86, 985

(a) Four sets of game records   (b) Combination of sets of half records.
(eight sets of half records).

By using these sets of half records, we designed four configurations of game
records in the learning of evaluation functions, as summarized in Table 1(b). In
the first two configurations "attack 4:defense 0" and "attack 0:defense 4", each
of the two sets of half records are directly used, respectively, to realize attack or
defense-oriented players. The number placed after "attack" or "defense" stands

---

[2] http://kifdatabase.no-ip.org/shogi/.

for ten thousand half records rounded to two digits at most. In these configurations, half records (i.e., moves by a single player of an intended playing style) are used for the learning of evaluation functions, as described in Sect. 3. The last two configurations are for the simulation of learning with full game records instead of half records, where moves of both players in each full record are used for training. The third configuration "attack 2.2:defense 5.8" uses the moves of both players in the corresponding full game records of "attack 0:defense 4". For 40, 000 half records of defense style, the number of attack and defense style of the opponents was 22, 131 and 17, 869, respectively. Similarly, the fourth configuration "attack 7.3:defense 8.7" uses the moves of both players in the corresponding full game records of "attack 0:defense 4" and those of "attack 4:defense 0". These numbers include the half records collected in accordance with the playing styles (i.e., set a, d, e, and h) and the accompanying records (i.e., b, c, f, and g). Note that the playing style of the opponent is not controlled in advance.

4.2   Analysis of Learned Evaluation Functions

We trained the weight vector of the evaluation function of Bonanza for each configuration in Table 1. For the initial value of weight vector w_f in the Style-adjustment step in Sect. 3, we adapted the weight vector distributed along with the source codes by the author of Bonanza. Also, parameter C in Eq. (2) is set to be zero in the experiments in this subsection. The weight vector of evaluation functions were tuned by MMTO by giving command "learn no-ini 32 -1 -1 12 12" to Bonanza.

We evaluated the obtained evaluation functions by two criteria: how the intended playing style is reproduced in game-playing and the playing strength. We conducted a the self-play of Bonanza with the original weight vector $w0$ and with the obtained weight vector. Each program searches at most 1,000,000 nodes for each move, and 300 games are played for each pair of players. Then, we analyzed the game records of self-play and measured the frequency of defense records and the winning rate. We classified each half record as attack or defense, and the frequency of defense records for each player is defined as the number of defense (half) records divided by the number of all records.

Figure 1a shows the frequency of defense records in the vertical axis and iteration in the horizontal axis. Since MMTO is an iterative method, the weight vector is yielded in each iteration, and we analyzed them for each of ten iterations. For "attack 0:defense 4" (shown in blue line) and "attack 4:defense 0" (shown in red line), the frequency quickly increases and decreases, respectively, in the first ten iterations and becomes stable. In contrast, for "attack 2.2:defense 5.8" (shown in cyan line) and "attack 7.3:defense 8.7" (shown in green line), the frequency remains around 50 % in each iteration. Therefore, we can see that the frequency of defense records played by the learned evaluation functions changes in accordance with to the configuration of game records used in learning, and that it is important to select moves in a unit of half records.

Figure 1b shows the winning rate of each configuration against the original version. Although the winning rates are not stable and vary depending on

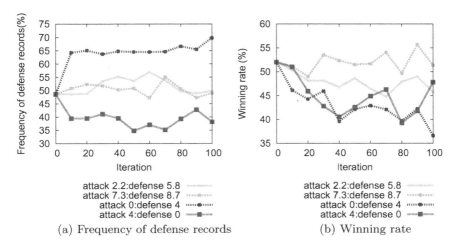

(a) Frequency of defense records   (b) Winning rate

Fig. 1. Analysis of four evaluation functions by self-play (Color figure online)

iterations, most values fall under 50 %, except for those of "attack 7.3:defense 8.7" (shown in green line). The differences in the winning rate can be explained by the differences in the total number of game records used in the learning. For example, in "attack 7.3:defense 8.7", 160,000 half records (i.e., 80,000 game records) were used, which are four times larger than that used in "attack 0:defense 4". Note that in the original MMTO (Hoki and Kaneko 2014), it is reported that 47,566 game records were used in the learning.

To prevent the loss in the playing strength, we introduced a penalty term in the Style-adjustment step in our method. Note that the loss in the playing strength is small and easily recovered by increasing the number of nodes for search, etc., in these cases because the winning rate observed was between 40 % to 50 %. However, the winning rate may fall more if the number of available game records is more limited, depending on the playing style to be realized.

4.3   Relationship Between Playing Style and Strength

To realize a playing style without degrading its playing strength, we introduced a penalty term in the Style-adjustment step in our method. For evaluation functions obtained with various values in parameter C, $5, 1, 0.5, 0.5 \cdot 10^{-1}, 0.5 \cdot 10^{-2}, 0.5 \cdot 10^{-3}, 0.5 \cdot 10^{-4}$, and $0.5 \cdot 10^{-5}$, we measured the winning rate and the frequency as in the previous experiments. Figure 2a shows the frequency of defense records for various C. To focus on the influence of C to the frequency, the frequency plotted for each configuration is the average over those for seven evaluation functions in the 40th, 50th, 60th, 70th, 80th, 90th, and 100th iterations. The frequency of "attack 4:defense 0" (shown with red point) and that of "attack 0:defense 4" (shown with blue point) are about 65 % and 40 %,

respectively, where $0.5 \cdot 10^{-5}$. We can see that the frequencies approach around
50 % as parameter C increases.

Similarly, Fig. 2b shows the winning rate for various C. Although the winning
rate is not so stable, the lowest winning rate is observed for the smallest C at
$0.5 \cdot 10^{-5}$. Also, many points around 50 % are observed for $C \geq 10^{-3}$.

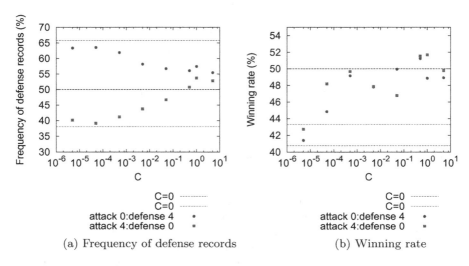

(a) Frequency of defense records   (b) Winning rate

Fig. 2. Frequency of defense records and winning rate for parameter C

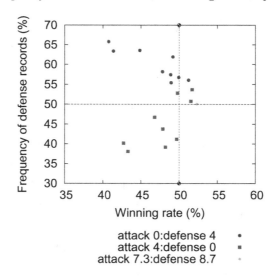

Fig. 3. Scatter plot of evaluation functions with various values of parameter C. The
vertical (horizontal) axis shows the winning rate (frequency of defense records).

Figure 3 shows the same data in a scatter plot where the vertical (horizontal) axis shows the winning rate (frequency of defense records). We can see that parameter C controls the balance between the frequency of defense records and the winning rate against the original version.

Table 2 shows the rate of agreement with moves in test data of defense records which are disjoint from training data. In MMTO, it is assumed that the rate of agreement with experts' moves is important in the playing strength. Therefore, it is natural that many values are around 40 %. In detail, we can see that "attack 0:defense4" ("attack 4:defense 0") achieved the lowest rate of agreement in attack (defense) moves, though the differences are small. It is consistent with the expectation that evaluation functions trained by "attack 0:defense4" ("attack 4:defense 0") have more (less) of a chance to learn defense moves.

Table 2. Rate of agreement with attack or defense moves in middlegame

Configuration	Rate of agreement (%)	
	Attack	Defense
Original	40.32	36.63
attack 4:defense 0	38.94	<u>34.82</u>
attack 0:defense 4	<u>38.25</u>	35.97
attack 2.2:defense 5.8	38.95	36.11
attack 7.3:defense 8.7	39.51	36.37
attack 0:defense 4 ($C = 0.5 \cdot 10^{-3}$)	39.72	36.69
attack 0:defense 4 ($C = 0.5 \cdot 10^{-4}$)	39.05	36.62
attack 0:defense 4 ($C = 0.5 \cdot 10^{-5}$)	38.58	36.38

Observing these results, we argue that our method successfully adjusted the playing style of a computer program with respect to attack or defense-oriented moves while keeping its playing strength compared to the original version.

5   Conclusion

We presented a method to give a computer player an intended playing style by the machine learning of an evaluation function. Our method trains an evaluation function by three steps: (1) selecting a subset of moves with respect to a playing style to be realized, (2) training the weight vector of an evaluation function by using the whole set of game records to maximize its playing strength, and (3) modifying the weight vector carefully so as to improve agreement with the selected moves prepared in step (1) while penalizing the difference from the weight vector obtained in step (2). We applied our method to realize a defense or attack-oriented style in shogi and tested adjusted evaluation functions by self-play against the original version. We measured how frequently an intended playing style is reproduced as well as the winning ratio. The frequency of defensive moves was significantly increased or decreased in accordance with the game records used while keeping the winning ratio around 50 %. We also confirmed that the weight in the penalty term used in step (3) controls the balance between the strength and playing style.

Interesting future work would be to apply our method to various playing styles and also to various games other than shogi.

Acknowledgment. A part of this work was supported by JSPS KAKENHI Grant Numbers 25330432 and 16H02927.

References

Baxter, J., Tridgell, A., Weaver, L.: Learning to play chess using temporal-differences. Mach. Learn. **40**(3), 242–263 (2000)

Buro, M.: Improving heuristic mini-max search by supervised learning. Artif. Intell. **134**(1–2), 85–99 (2002)

Campbell, M., Hoane Jr., A.J., Hsu, F.-H.: Deep blue. Artif. Intell. **134**(1–2), 57–83 (2002)

Carmel, D., Markovitch, S.: Learning models of opponent's strategy in game playing. In: In Proceedings of the AAAI Fall Symposium on Games: Planning and Learning, pp. 140–147. The AAAI Press (1993)

Donkers, H., van den Herik, H., Uiterwijk, J.: Selecting evaluation functions in opponent-model search. Theoret. Comput. Sci. **349**(2), 245–267 (2005)

Hoki, K., Kaneko, T.: Large-scale optimization for evaluation functions with minimax search. J. Artif. Intell. Res. (JAIR) **49**, 527–568 (2014)

Hoki, K., Muramatsu, M.: Efficiency of three forward-pruning techniques in shogi: futility pruning, null-move pruning, and late move reduction (LMR). Entertain. Comput. **3**(3), 51–57 (2012)

Iida, H., Sakuta, M., Rollason, J.: Computer shogi. Artif. Intell. **134**(1–2), 121–144 (2002)

Kaneko, T.: Evaluation functions of computer shogi programs and supervised learning using game records. J. Jpn. Soc. Artif. Intell. **27**(1), 75–82 (2012). (In Japanese)

Levene, M., Fenner, T.I.: A methodology for learning players' styles from game records. In: CoRR abs/0904.2595 (2009)

Matsubara, H.: Declaration of termination of computer shogi project. IPSJ Mag. **56**(11), 1054–1055 (2015). (In Japanese)

Namai, S., Ito, T.: A trial AI system with its suggestion of Kifuu (playing style) in shogi. In: 2010 International Conference on Technologies and Applications of Artificial Intelligence (TAAI), pp. 433–439 (2010). doi:10.1109/TAAI.2010.94

Okamoto, K., Hashiguchi, H.: Psychological Analysis by Rorschach, MDS of 11 professional shogi players' Kifuu. Brain Shuppan. (In Japanese)

Ortiz B., S.E., et al.: An interesting opponent for fighting videogames. In: SIG Technical Reports. GI 4. IPSJ, pp. 1–8 (2010)

Sawa, N., Ito, T.: Statistical analysis of elements of play style in shogi (Japanese Chess). In: SIG Technical Reports. GI 3. IPSJ, pp. 1–8 (2011). (In Japanese)

Shimizu, S., Kaneko, T.: Evaluation, implementation of UCT with prior knowledge for computer's styles of playing two-player games. In: Proceedings of 19th Game Programming Workshop 2014, pp. 188–195 (2014). (In Japanese)

Takise, R., Tanaka, T.: Development of entering-king oriented shogi programs. IPSJ J. **53**(11), 2544–2551 (2012). (In Japanese)

Tesauro, G.: Programming backgammon using self-teaching neural nets. Artif. Intell. **134**(1–2), 181–199 (2002)

Tosaka, K., Matsubara, H.: Feature extraction of players from game records in shogi. In: SIG Technical Reports. 2006-GI-016. IPSJ, pp. 1–8 (2006). (In Japanese)

Veness, J., et al.: Bootstrapping from game tree search. Adv. Neural Inf. Process. Syst. **22**, 1937–1945 (2009)

Yano, Y., et al.: Adaptive learning utilizing parameters of existing evaluation function. In: Proceedings of 14th Game Programming Workshop 2009, pp. 1–8 (2009). (In Japanese)

Offline Text and Non-text Segmentation
for Hand-Drawn Diagrams

Buntita Pravalpruk[✉] and Matthew M. Dailey

Computer Science and Information Management, Asian Institute of Technology,
Klong Luang, Pathumtani, Thailand
buntita@gmail.com, mdailey@ait.asia
http://www.cs.ait.ac.th/

Abstract. Writing and drawing are basic forms of human communication. Handwritten and hand-drawn documents are often used at initial stages of a project. For storage and later usage, handwritten documents are often converted into a digital format with a graphics program. Drawing with a computer in many cases requires skill and more time than less formal handwritten drawings. Even when people have experience in computer drawing and are familiar with the application, it takes time. Automatic conversion of images of hand-drawn diagrams into a digital graphic format file could save time in the design process. One of early critical tasks in hand-drawn diagram interpretation is segmentation of the diagram into text and non-text components. In this paper, we compare two approaches for offline text and non-text segmentation of contours in an image. We describe the feature extraction and classification processes. Our methods obtain 82–86 % accuracy. Future work will explore the application of these techniques in a complete diagram interpretation system.

Keywords: Image recognition · Hand-drawn diagram · Hand-written diagram · Text and non-text classification

1 Introduction

Writing is a simple and natural way for people to take notes and exchange information, exchange thoughts and feelings, express ideas, and plan. Handwritten notes often contain pictures or diagrams in addition to the associated text. Such notes contain a trove of useful information that may take many forms, including doctors' summaries of patient interviews, business analysts' depictions of business processes, engineers' descriptions of computer system designs, architects' sketches of a building design, and so on.

Handwritten and hand-drawn documents are very common at the initial stage of a design or brainstorming process. However, oftentimes, rough initial handwritten documents must be converted into a clean digital format with a graphics program or other tool. Unfortunately, producing formal diagrams with a computer usually requires more skill and more time than drawing less formal

© Springer International Publishing Switzerland 2016
R. Booth and M.-L. Zhang (Eds.): PRICAI 2016, LNAI 9810, pp. 380–392, 2016.
DOI: 10.1007/978-3-319-42911-3_32

diagrams on paper. Even when the person transcribing the document is skilled, the conversion process wastes precious time, making the prospect of automatic conversion attractive. For this reason, we are interested in automatic conversion of hand-drawn documents into digital format after they are scanned or photographed.

There are generally three main steps in automated hand-drawn document interpretation. First, the document should be divided into text and non-text regions, because diagrams and text require different strategies for interpretation. Second, in text regions, each letter or digit can be detected and classified (Freitas et al. 2007; Kara and Stahovich 2007; Lauer et al. 2007; Zhong et al. 2010), then we can form words and numbers from the letters and digits. Meanwhile, in non-text regions, we can detect the lines and curves comprising objects (Hammond and Davis 2006; Stahovich 2004), and if the document is known to be structured (i.e., a representative of a specific type of document such as a UML or electrical circuit diagram), the lines and curves can be given semantic interpretations. In a final step, we can combine the results of the two flows, redraw the document consistently and accurately in terms of the recognized text segments and objects, and provide the semantic interpretation of the diagram to downstream processes.

In this paper, we focus on the first step, namely text and non-text segmentation. There are two general approaches to the problem, depending on the type of input device used to acquire the diagram. The first type of device is a sketchpad or tablet that records the timing and sequence of each pen/finger stroke. There have been many methods proposed to deal with such *online* documents by researchers, focusing on either the segmentation step (Waranusast et al. 2009) or the interpretation step (Costagliola et al. 2006; Hammond and Davis 2007; Kara and Stahovich 2007). One area in which successful methods have been demonstrated is in flowchart interpretation (Bresler et al. 2013; Lemaitre et al. 2013).

However, when working with the second type of devices, scanners and cameras, we are faced with the task of interpreting *offline* documents, simple raster images without stroke or timing information. This problem is more difficult, because letters, digits, lines, and curves must be inferred before they can be classified then recognized. There has been some work in this area, for example, on the use of pattern mapping and statistics to detect shapes in offline flowchart diagrams (Wu et al. 2015). Successful segmentation and interpretation of more complex kinds of diagrams, however, would require more sophisticated methods.

We tackle the problem of offline document segmentation by first detecting contours in the input image then applying sequence analysis methods to the result. In the first stage, we classify individual contours as to whether they are text or non-text objects, and then we analyze relationships between neighboring contours to classify them into the same group. Exploiting relationships between neighboring contours enables improved interpretation.

We have implemented two approaches to the problem of offline text and non-text segmentation and, as a case study, have tested the two methods on a collection of hand-drawn diagrams in the Business Process Model and Notation

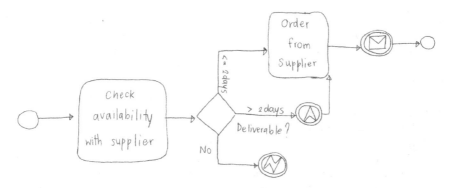

Fig. 1. An example diagram in Business Process Model and Notation (BPMN) format.

(BPMN) format. An example diagram is shown in Fig. 1. In an experiment with 120 diagrams, we obtain 80–84 % accuracy on a per-pixel basis. In the rest of this paper, we present the methodology, experiments, and analysis of the experimental results in more detail.

2   Method

Our method performs the following steps (Fig. 2):

1. Acquire input image I.
2. Binarize I to obtain binary image I_b.
3. Find the connected components $C_1, C_2, \ldots$ in I_b. Let N be the number of connected components found.
4. Convert each connected component C_i to a sequence S_i consisting of the *turns* made along the *outer contour* of C_i. We use turn codes S (straight), L (left), and R (right).
5. Classify each sequence code S_i using two alternative classifiers:
   (a) *Turn ratio classifier*: simple threshold on the ratio of straight elements to turn elements in the sequence.
   (b) *Bayes maximum a posteriori (MAP) classifier*: determine the class (text or non-text) with the highest conditional probability: $P(\text{text} \mid S_i)$ or $P(\text{non-text} \mid S_i)$.
6. Output the result of each classifier (labels y_i^B for the MAP classifier and labels y_i^R for the ratio classifier) to a database.

2.1   Binarization

For binarization, we use the standard Otsu algorithm (Otsu 1979) implemented by OpenCV.

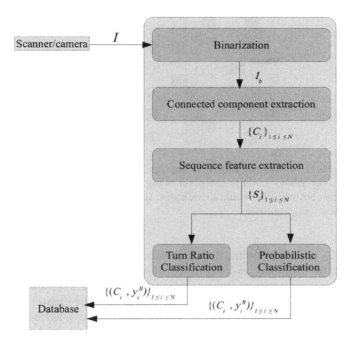

Fig. 2. System overview.

2.2 Contour Conversion

We find the connected components $C_i, 1 \leq i \leq N$ of I_b. Currently, we assume each contour is entirely text or non-text without any overlap. This assumption will be relaxed in future research. We then find the outer contour of each connected component using the contour tracing algorithm of Suzuki and Abe (Suzuki and Abe 1985) as implemented in OpenCV (OpenCV Dev Team 2016).

2.3 Sequence Feature Extraction

Once we find the contour for a given connected component C_i, we trace the contour, converting each pixel into a feature indicating whether, relative to the previous pixel in the chain, the current pixel represents a straight forward motion or a turn to the left or right. For each connected component C_i we obtain a sequence S_i of turn features S (straight), L (left turn), and R (right turn).

See Fig. 3 for the coding of pixels in the contour sequence as S, L, and R. Note that the encoding is ambiguous (the original contour cannot be re-created from the turn feature sequence), but the encoding performs well in classification, as we shall see in the results.

For straight moves, assume we are given the position of the middle point of three contour pixels as (x, y). There are three cases that should be classified

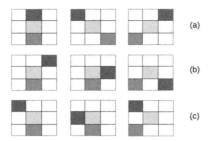

Fig. 3. Examples of turn types. In every diagram, the green pixel is the previous pixel in the contour, the yellow pixel is the pixel being considered, and the red pixel is the next pixel in the contour. (a) Straight examples. (b) Right turn examples. (c) Left turn examples. (Color figure online)

as straight: (1) the three points have the same position in x with all different positions of y, or the same position in y and all different positions in x; (2) the first point has both x position and y position less than the position of the middle point and the last point has both x position and y position greater than those of the middle point, or vice versa; (3) the first point has an x position greater than that of the middle point and a y position less than that of the middle point, while the last point has a x position less than the x position of the middle point and a y position greater than that of the middle point, or vice versa. If the sequence of three consecutive pixels around a target pixel fall into one of these three cases, the pixel is considered straight and assigned a label S.

For right turns, there are four cases: (1) the first point has less position of y than that of the last point and both of them have the same or less position of x than that of the middle point. (2) the first point has greater position of y than that of the last point and both of them have the same or greater position of x than that of the middle point. (3) the first point has less position of x than that of the last point and both of them have the same or less position of y than that of the middle point. (4) the first point has greater position of x than that of the last point and both of them have the same or greater position of y than that of the middle point. If the sequence of three consecutive pixels around a target pixel fall into one of these three cases, the pixel is considered right turn and assigned a label R.

For left turns, there are also four cases: (1) the first point has greater position of y than that of the last point and both of them have the same or less position of x than that of the middle point. (2) the first point has less position of y than that of the last point and both of them have the same or greater position of x than that of the middle point. (3) the first point has greater position of x than that of the last point and both of them have the same or less position of y than that of the middle point. (4) the first point has less position of x than that of the last point and both of them have the same or greater position of y than that of the middle point. If the sequence of three consecutive pixels around a target pixel fall into one of these three cases, the pixel is considered right turn and assigned a label L.

On the hypothesis that second-order relationships between contour turns could be useful for contour classification, we also consider pairs of consecutive turns as features. After converting contour i into turn sequence S_i, we can obtain a sequence of second-order turn features left turn after left turn (LL), right turn after left turn (RL), straight after left turn (SL), left turn after right turn (LR), right turn after right turn (RR), straight after right turn (SR), left turns after straight (LS), right turn after straight (RS) and straight after straight (SS). We use the consecutive turn counts in the probabilistic classifier described below.

2.4 Sequence Feature Classification

Most classification methods require a fixed vector of inputs and produce a binary or multi-class output. For sequential data such as the turn sequences S_i, whose length are not pre-determined, we need to convert the arbitrary-length data sequence into a fixed size vector of summary features.

For the turn ratio classifier, we do this by simply counting the number of turn features in a given sequence, e.g., for left turn features in turn sequence S_i, we find the number of L features in S_i, written $N_i^L = |S_i|_L$, and similarly to obtain N_i^R and N_i^S.

For the probabilistic classifier, as will be seen in the next section, we use a first-order Markov model that requires counts of the consecutive turn features, i.e., N_i^{LL}, N_i^{LR}, N_i^{LS}, N_i^{SL}, N_i^{SR}, N_i^{SS}, N_i^{RL}, N_i^{RR}, and N_i^{RS}. When we sum these features over all sequences in a training set to obtain the total frequency of each type of consecutive turn, we denote the result $N^{SL} = \sum_i N_i^{SL}$, and similarly for the other consecutive turn types.

Turn Ratio Calculation. Based on the hypothesis that contours of non-text elements of a drawing will tend to have longer straight strokes than text elements, we can expect that the ratio of turn sequence elements to straight sequence elements, i.e., $R_i = (N_i^R + N_i^L)/N_i^S$, should be relatively small when C_i is a text element and relatively large when C_i is a non-text element.

Classifying elements as text or non-text based on the feature R_i, then, merely requires a threshold above which we classify the element as text and below which we classify the element as non-text. We use a straightforward optimization over a training set to find the best threshold.

Markov Chain Processing. For the probabilistic classifier, we attempt to find posterior probabilities $P(\text{text} \mid S_i)$ and $P(\text{non-text} \mid S_i)$, and answer with the class having the highest posterior. From Bayes' rule, we have

$$P(\text{text} \mid S_i) = \frac{P(S_i \mid \text{text})P(\text{text})}{P(S_i)}. \tag{1}$$

Expanding the sequence S_i into its tokens and ignoring the constant denominator, we have

$$P(\text{text} \mid S_i) \propto P(S_{i,1}, S_{i,2}, \ldots, S_{i,|S_i|} \mid \text{text})P(\text{text}). \tag{2}$$

We can use the chain rule for the joint conditional probability and the first-order Markov assumption to obtain

$$P(\text{text} \mid S_i) \propto P(\text{text}) \prod_{j=1}^{|S_i|} P(S_{i,j} \mid S_{i,1} \cdots S_{i,j-1}, \text{text}) \tag{3}$$

$$\approx P(\text{text}) \prod_{j=1}^{|S_i|} P(S_{i,j} \mid S_{i,j-1}, \text{text}). \tag{4}$$

The probabilities $P(S_{i,j} \mid S_{i,j-1}, \text{text})$ merely express the frequency with which a turn or straight element of the contour sequence follows a turn or straight element in a text element contour. For each type of element (text and non-text), there are nine such probabilities that are easily estimated based on turn count features. For example, $P(S_{i,j} = \text{S} \mid S_{i,j-1} = \text{L}, \text{text})$ is simply $N^{LS}/(N^{RS} + N^{SS} + N^{LS})$, or more simply, N^{LS}/N^{S}.

As an example, see Fig. 4 for a visualization of the conditional probabilities we observed for text and non-text contours in a training set consisting of BPMN diagram elements. Confirming the basic hypothesis, we can observe that $P(S_{i,j} = \text{S} \mid S_{i,j-1} = \text{S}, \text{non-text}) > P(S_{i,j} = \text{S} \mid S_{i,j-1} = \text{S}, \text{text})$, i.e., $0.852 > 0.645$.

To classify a given sequence, we simply compare the posterior probabilities. If $P(\text{text} \mid S_i) > P(\text{non-text} \mid S_i)$, we conclude that the sequence is text; otherwise we conclude it is non-text.

3   Experimental Design

To empirically evaluate the methods described in Sect. 2, we collected 15 example BPMN 2.0 process specifications in XML format from the OMG Website[1] and imported them into Signavio Process Editor.[2] Signavio could only successfully convert nine of 15 process specifications, so we exported those nine diagrams to PDF then printed them. We had 200 secondary school and undergraduate students in Thailand copy the diagrams, then we scanned each diagram using a desktop scanner at 2400 dpi. The result was 600 hand-drawn BPMN diagrams as raw bitmaps. We selected 120 of these diagrams for ground truth annotation at random.

To annotate the hand-drawn diagrams with ground truth information, we developed a simple application able to binarize and extract contours from the bitmaps. The application highlights each connected component of the diagram

---

[1] http://www.omg.org/spec/BPMN/20100602/.
[2] https://editor.signavio.com/.

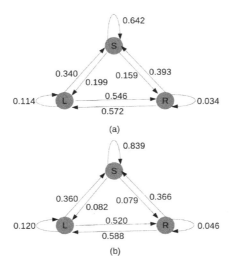

(a)

(b)

Fig. 4. Example Markov chains for turn sequences for text and non-text contours, calculated from a training set of 96 BPMN diagrams. (a) Markov chain for text contours. (b) Markov chain for non-text contours. Non-text contours tend to have more S elements than do text contours.

in turn for the user to annotate. The user identifies each contour as *text*, *non-text*, or *noise*. *Text* contours are those that contain letters or parts of letters without connection to anything other than letters. *Non-text* contours are those that have no connection to any letter, that contain just lines, curves, and other shapes meaningful for the overall diagram. An example of a partial annotation of a diagram is shown in Fig. 5. Contours without any meaning to the diagram are classified as *noise*. We used colors to identify contour types as shown in the figure.

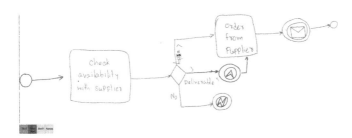

Fig. 5. Example partially annotated hand-drawn BPMN diagram. Text contours are marked green. Non-text contours are marked red. Contours that contain both text and non-text are marked with yellow and split into contours that can be classified into one type. (Color figure online)

When a contour is marked as containing both text and non-text, the annotator is asked to annotate the text and non-text parts of the contour using a paint-like tool. An example is shown in Fig. 6.

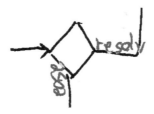

Fig. 6. Example annotation of text and non-text for a contour containing both. Nontext parts are painted black, text parts are painted red, and overlapping parts are painted yellow. (Color figure online)

For convenience, we also preprocess each contour to get the turn counts during the annotation process. (In an online system, of course, the processing would be done online as soon as the diagram is loaded.) As described in Sect. 2, we classify each pixel along the contour of each connected component as either *straight*, a *right turn*, or a *left turn* (refer again to Fig. 3), then we count the number of pixels falling into each type of turn and consecutive turn category (the quantities N_i^S, N_i^L, N_i^R, N_i^{SS}, N_i^{LS}, N_i^{RS}, N_i^{SL}, N_i^{LL}, N_i^{RL}, N_i^{SR}, N_i^{LR}, and N_i^{RR} introduced in Sect. 2). The result of the process is a repository of ground truth files in YAML format, for example as shown in Fig. 7.

```
%YAML:1.0
file: "2"
collectDate: "Wed Jul  8 14:55:30 2015\n"
contours:
    - { type:Noise, xmin:177, ymin:233, xmax:177, ymax:233, leftcount:1,
        rightcount:0, straightcount:0, LR:1, LL:0, LS:0, RL:0, RR:0, RS:0,
        SL:0, SR:0, SS:0, total:1 }
    - { type:Noise, xmin:190, ymin:253, xmax:190, ymax:253, leftcount:1,
        rightcount:0, straightcount:0, LR:1, LL:0, LS:0, RL:0, RR:0, RS:0,
        SL:0, SR:0, SS:0, total:1 }
    - { type:Nontext, xmin:389, ymin:101, xmax:430, ymax:204, leftcount:75,
        rightcount:72, straightcount:150, LR:7, LL:36, LS:32, RL:37, RR:2,
        RS:33, SL:31, SR:34, SS:85, total:297,
        GT:"/home/anita/Desktop/Thesis program/outputfile/2/2_contour91.bmp" }
    - { type:Noise, xmin:463, ymin:520, xmax:469, ymax:521, leftcount:4,
        rightcount:2, straightcount:7, LR:1, LL:2, LS:1, RL:2, RR:0, RS:0,
        SL:1, SR:0, SS:6, total:13 }
```

Fig. 7. Example ground truth data file.

At analysis time, we use the contour and turn counts for classification. To evaluate the two classifiers, we split the 120 images into five partitions for leave-one-out five-fold cross validation. For each of the five splits of the data into test and train sets, we used the training set to find the best threshold for the turn ratio classifier, and we calculated the probabilities $P(S_{i,j} \mid S_{i,j-1}, \text{text})$ over all text contours in the training set and the probabilities $P(S_{i,j} \mid S_{i,j-1}, \text{non-text})$ over all non-text contours in the training set. For example, for the probabilities $P(S_{i,j} \mid S_{i,j-1} = \text{S}, \text{text})$ we simply calculate the ratios N_t^{SS}/N_t^S, N_t^{LS}/N_t^S and N_t^{RS}/N_t^S, where the counts are taken over all text contours in the training set. These probabilities are then used to evalute, for each contour i in the test set, $P(\text{text} \mid S_i)$ and $P(\text{non-text} \mid S_i)$. Repeating for all five splits, we obtain test results over the entire 120-image data set.

4   Experimental Results

After collecting and processing the data as described in the previous section, we obtained the following results for turn ratio classification and maximum a posteriori classification.

4.1   Experiment 1 (Turn Ratio Classification)

The results of turn ratio classification from five-fold cross validation are shown in Table 1.

Table 1. Experiment 1 results. Accuracy of turn ratio classification for training set and test set, for each partition of the data for five-fold cross validation, on a per-pixel and per-contour basis. Column 2 shows the turn ratio threshold that is optimal for per-pixel classification on that partition's training set.

Cross validation partition	Optimal ratio	Train set accuracy per pixel	Test set accuracy per pixel	Train set accuracy per contour	Test set accuracy per contour
1	0.779	83.68 %	86.41 %	71.05 %	71.14 %
2	0.778	84.42 %	82.92 %	71.83 %	67.87 %
3	0.784	83.61 %	86.04 %	70.06 %	73.91 %
4	0.775	84.28 %	83.42 %	71.30 %	71.11 %
5	0.779	85.03 %	81.42 %	70.98 %	71.35 %
Average	0.779	84.204 %	84.042 %	71.044 %	71.076 %
S.D	0.003	0.5218	1.904	0.575	1.918

The turn ratio method is surprisingly accurate. Similar results on the training and test sets indicate that the data set is sufficiently large to provide a stable estimate of the single turn ratio threshold parameter.

4.2   Experiment 2 (Probabilistic Classification)

The results of probabilistic classification from five-fold cross validation are shown in Table 2.

Again surprisingly, the per-pixel accuracy of the MAP classifier is approximately 1.4 % less accurate than the turn ratio classifier. The turn ratio classifier was tuned to obtain the best possible per-pixel accuracy on the training set, whereas the probabilistic classifier is trained to provide an estimate of $P(\text{text} \mid S_i)$ on a per-contour basis. Since the probabilistic method is approximately 1.2 % more accurate than the turn ratio threshold classifier on a per-contour basis, optimizing the threshold on $P(\text{text} \mid S_i)$ on a per-pixel basis would improve results. This would roughly correspond to adjusting the Bayes risk for each contour based on the size of the contour. When we perform this small optimization, we obtain an average test set per-pixel accuracy of 85.6 %.

Table 2. Experiment 2 results. Accuracy of probabilistic classification for training set and test set, for each partition of the data for five-fold cross validation, on a per-pixel and per-contour basis.

Cross validation partition	Train set accuracy by pixels	Test set accuracy by pixels	Train set accuracy by contours	Test set accuracy by contours
1	82.05 %	84.87 %	72.44 %	74.28 %
2	83.38 %	81.31 %	72.66 %	67.68 %
3	82.19 %	84.04 %	72.12 %	74.27 %
4	82.59 %	82.73 %	72.99 %	71.50 %
5	83.65 %	80.23 %	71.99 %	73.89 %
Average	82.772 %	82.636 %	72.44 %	72.324 %
S.D	0.638	1.704	0.362	2.542

4.3   Discussion

The BPMN diagram data set generally contains many short text contours and fewer long non-text contours. The proportion of text pixels overall is approximately 30.16 %, and the proportion of non-text pixels is approximately 69.84 %, whereas the proportion of text contours is 81.48 %, while the proportion of non-text contours is 18.52 %. Turn ratio classification yields an accuracy of 84.0 %, whereas the probabilistic classifier yields an accuracy of 82.6 % or 85.6 % depending on its objective function. Looking more closely at these results, we found that the two classifiers agreed about 94 % of the time, with 81 % of the contours overall classified correctly by both methods and 13 % incorrectly classified by both methods.

Figure 8 shows examples of contours that are correct classified by both methods and examples of contours on which the two classifiers disagree. Non-text contours with long straight lines or smooth curves are generally classified correctly,

Fig. 8. Example contours classification example. (a) Contours that are correctly classified by both methods. (b) Contours that correctly classified by the turn ratio classifier but incorrectly classified by the probabilistic classifier. (c) Contours that are correctly classified by the probabilistic classifier but incorrectly classified by the turn ratio classifier. (d) Contours that are incorrectly classified by both methods.

and non-text contours forming circles or shorter straight lines with turns are likely to be classified as text. Overall, non-text contours contain longer straight lines than do text contours, so the probability of moving to the S state of the Markov chain are higher in the non-text model than in the text model. While this works well overall, it does mean that text contours with longer, straighter segments, such as l, d, p, and H, are likely to be classified as non-text contours.

The data set was collected in Thailand from students who may write English letters differently than native English speakers, so model parameters may need re-estimation for different data sets. Binarization might also require adaptation to different scanners or drawing styles.

5   Conclusion

This paper has presented a method for text and non-text segmentation of static hand-written drawings that works without requiring stroke information. Although the methods are relatively simple, the results on BPMN diagrams are quite accurate, in the 82 % to 86 % range. The method could be improved with a richer feature set and a binary classifier such as the support vector machine (SVM) (Burges 1998; Niu and Suen 2012; Yang et al. 2013). The segmenter described here will make an excellent starting point for a complete diagram digitization system. The fact that such accurate results can be obtained for BPMN diagrams, which describe executable business processes, raises the tantalizing prospect of bootstrapping business automation software development processes with a simple drawing.

References

Burges, C.J.C.: A tutorial on support vector machines for pattern recognition. Data Min. Knowl. Disc. **2**, 121–167 (1998)

Lemaitre, A., Carton, C., Couasnon, B.: Fusion of statistical and structural information for flowchart recognition. In: 2013 12th International Conference on Document Analysis and Recognition, pp. 1210–1214. IEEE (2013)

Costagliola, G., Deufemia, V., Risi, M.: A multi-layer parsing strategy for on-line recognition of hand-drawn diagrams. In: IEEE Symposium on Visual Languages and Human-Centric Computing, VL/HCC 2006, pp. 103–110. IEEE Computer Society, September 2006

Freitas, C.O.A., Oliveira, L.S., Bortolozzi, F., Aires, S.B.K.: Handwritten character recognition using nonsymmetrical perceptual zoning. Int. J. Pattern Recogn. Artif. Intell. **21**(01), 135–155 (2007)

Hammond, T., Davis, R.: Tahuti: a geometrical sketch recognition system for UML class diagrams. In: ACM SIGGRAPH 2006 Courses, SIGGRAPH 2006. ACM (2006)

Hammond, T., Davis, R.: Ladder, a sketching language for user interface developers. In: ACM SIGGRAPH 2007 Courses, SIGGRAPH 2007. ACM (2007)

Kara, L.B., Stahovich, T.F.: Hierarchical parsing and recognition of hand-sketched diagrams. In: ACM SIGGRAPH (2007)

Lauer, F., Suen, C.Y., Bloch, G.: A trainable feature extractor for handwritten digit recognition. Pattern Recogn. **40**(6), 1816–1824 (2007)

Bresler, M., Prua, D., Hlaváč, V.: Modeling flowchart structure recognition as a max-sum problem. In: 2013 12th International Conference on Document Analysis and Recognition, pp. 1215–1219. IEEE (2013)

Niu, X.-X., Suen, C.Y.: A novel hybrid CNN-SVM classifier for recognizing handwritten digits. Pattern Recogn. **45**(4), 1318–1325 (2012)

OpenCV Dev Team. Finding contours in your image (2016). http://docs.opencv.org/2.4/doc/tutorials/imgproc/shapedescriptors/find_contours/find_contours.html

Otsu, N.: A threshold selection method from gray-level histograms. IEEE Trans. Syst. Man Cybern. **9**(1), 62–66 (1979)

Stahovich, T.F.: Segmentation of pen strokes using pen speed. In: Proceedings of the AAAI Fall Symposium on Making Pen-Based Interaction Intelligent and Natural, pp. 21–24 (2004)

Suzuki, S., Abe, K.: Topological structural analysis of digitized binary images by border following. Comput. Vis. Graph. Image Process. **30**(1), 32–46 (1985)

Waranusast, R., Haddawy, P., Dailey, M.: Segmentation of text and non-text in on-line handwritten patient record based on spatio-temporal analysis. In: Combi, C., Shahar, Y., Abu-Hanna, A. (eds.) AIME 2009. LNCS, vol. 5651, pp. 345–354. Springer, Heidelberg (2009)

Wu, J., Wang, C., Zhang, L., Rui, Y.: Offline sketch parsing via shapeness estimation. In: Proceedings of the 24th International Conference on Artificial Intelligence, IJCAI 2015, pp. 1200–1206. AAAI Press (2015)

Yang, X., Qiaozhen, Y., He, L., Guo, T.: The one-against-all partition based binary tree support vector machine algorithms for multi-class classification. Neurocomputing **113**, 1–7 (2013)

Zhong, C., Ding, Y., Fu, J.: Handwritten character recognition based on 13-point feature of skeleton and self-organizing competition network. In: Proceedings of the 2010 International Conference on Intelligent Computation Technology and Automation, ICICTA 2010, vol. 02, pp. 414–417. IEEE Computer Society, Washington, D.C. (2010)

Face Verification Algorithm with Exploiting Feature Distribution

Zhi Qu, Xuan Li, Yong Dou[✉], and Ke Yang

Key Laboratory of Parallel Distribution, National University of Defense Technology,
SanYi Avenue, Changsha, Hunan, China
quziyan@qq.com, yongdou@nudt.edu.cn

Abstract. Deep feature is widely applied in many fields such as image retrieval, image classification, face verification, etc. All the post-processing methods using deep feature make some assumptions about feature distribution. However, in most situations, features do not follow the hypothesised distribution approximately. In this paper, we focus on face-verification applications which also suffer from these problems. We propose an up-sample method called IUSM to alleviate the problems caused by biased samples. Additionally, by analyzing the Joint Bayesian model theoretically and practically, we propose a feature fusion method called LFF which utilizes the distribution properties of Joint Bayesian. Based on IUSM, the face verification accuracy of biased data is improved by 6 % while generalization ability of convolution network is not crippled. On the widely used Labeled Face in the Wild(LFW) dataset, LFF method can slightly improve the accuracy 0.15 % while the baseline accuracy is more than 97.51 %. We also argue that LFF can improve each deep face verification algorithm which uses Joint Bayesian model due to LFF's linear combination of features.

1 Introduction

Face verification using deep convolution network has received much attention in recent years([15–19]) due to the expressiveness and generalization of deep feature. As reported in [5,11,13], deep feature is more effective for extracting high-level visual descriptors than other methods such as SIFT [10], HOG [6], LBP [1], etc. So most recent works [16,18,19] use deep feature with Joint Bayesian model [4] for face verification. Works upon using deep learning achieved extremely high accuracy on LFW [9], the hardest face dataset at present, from 97 % to 99 %, which is exceed the human performance.

Although deep feature works well in most situations, it is reported in [13] the performance gets less at scale. Also we find that its generalization ability is crippled in some feature spaces which is called biased spaces. In this work we chose the dataset proposed in [19] to train our net. And we propose an effective way to find the potential biased feature spaces belong to training set and give an up-sampling method called IUSM for solving the problem to some extent. In this process, we first find the biased feature spaces by analysing each identity's feature

© Springer International Publishing Switzerland 2016
R. Booth and M.-L. Zhang (Eds.): PRICAI 2016, LNAI 9810, pp. 393–405, 2016.
DOI: 10.1007/978-3-319-42911-3_33

distribution in the training set and create a LFW-like dataset called CASTest consisting of validate images belong to the biased identities. Then we do IUSM through aligning an illumination distribution whose statistical parameters are estimated from unbiased data. The accuracy of the net trained by data after IUSM is improved by more than 6 % on CASTest than the net trained by original data. IUSM also slightly improves the performance on LFW.

In this work, we also elaborately analyse the Bayesian model in theory. Further, we propose a practical feature fusion method that can make the face similarity score distributions which under the same and different identities conditions more separate in Fisher's criterion [20] meaning. This method slightly improves the accuracy on LFW about 0.15 % while the baseline accuracy is more than 97.51 %.

The remaining sections are organized as follows. Section 2 introduces the related work. Section 3 presents details of our method. Experimental results are reported in Sect. 4. In the end, we draw our conclusions and make some planning of future work.

2   Related Work

Deep feature has become a mainstream in the field of face verification and their precisions on LFW [9] has achieved extremely high accuracy range from 97 % to 99 % ([15–19]) in recent years. Further, the Joint Bayesian model [4] is reported more effective [4, 16, 18, 19] than other distance metric such as L2, L1, cosine, etc.

2.1   Deep Representation with Bias

In the work [18], they trained a group of deep nets on several face patches respectively, then did PCA on the concatenating feature of nets output. [16] did not only employ a set of nets similar to [18], but also introduced the supervised signal in training stage. [17] utilized multi-layer supervised technology and feature selection method for further improvements. [19] proposed a similar method to [16] except using only one net.

The accuracy on LFW benchmark achieved near perfect results in the works above. However, as reported in [13], once evaluated at scale recognition, the precision of algorithms get less. [21] argued that this problem is caused by the data bias in the training set and cross factors such as pose, occlusion and age variation in reality.

To solve the biased data distribution, [2] proposed a bootstrapping method which handled the imbalance by iteratively selecting a small subset of images, and made use of an entropy-based diversity measure for the initial selection, thereby achieving over a two-fold reduction in human time required. [7] proposed an up-sample method using adaptive synthetic sampling approach (ADASYN) to generate synthetic instances for the minority class. [8] reviewed academic

activities special for the class imbalance problem and investigated various reme-
dies in several different levels according to learning phases. [14] discussed some
of the most frequently used methods that aim to solve the problem of learning
with imbalanced data sets and analyzed their results respectively.

2.2   Joint Bayesian Model

Joint Bayesian Model [4] is widely used for face verification task. As reported
in [3,18,19], Joint Bayesian outperforms other metrics such as L1, L2, cosine,
etc. The similarity under Joint Bayesian can be seen as a quadratic formation
of random vectors, so the ratio is also a random variable. In the face verification
field, there are no further discusses on the mechanism of Joint Bayesian model
so far. But in statistics field, [12] has analyzed the numerical characteristics of
this form variables. This paper will utilize the conclusion in [12] to improve the
performance of Joint Bayesian model.

3   Method

This section will introduce our experiment methods. We first do pre-processing
on the dataset, including face alignment and up-sampling for less-image identities. Then we introduce our training method which uses both identification and
verification signals. Finally, we propose the method for face verification by utilizing the distribution properties of Joint Bayesian model. Subsections 3.1 and 3.2
are the pre-processing steps. Subsections 3.3 and 3.4 show the details reference to
net training. Subsection 3.5 introduces our method to improve the performance
of joint Bayesian model. Works in Subsections 3.1, 3.3 and 3.4 mainly follow the
work introduced in [19] with some modifications.

3.1   Face Alignment

Face has its own structural information, same identity's faces have the similar
texture information, different faces' texture information varies a lot. Face alignment can exploit this structural information, and we find it makes easy for deep
model to converge. So we adapt the two point face alignment method introduced
in [19]. The two-point face alignment image size is 100×100.

3.2   Up-Sampling Mechanism

In our experiments, we observe features fetched from the network which trained
on the original CASIA-WebFace [19] data have some distribution biases on the
small samples identities (see Fig. 1).

   We create a data set(called **CASTest**) on these biased identities' data which
belongs to validation set. The structure of this data set is same to LFW face
verification set, which contains 2200 training pairs and 6000 testing pairs.

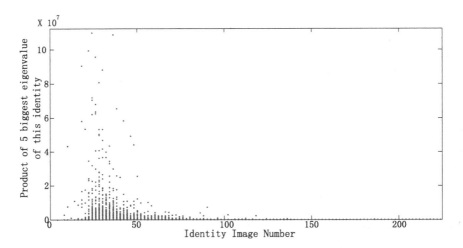

Fig. 1. X-axis represents the image number of one identity, y-axis is the product of first five eigenvalues of this identity's covariance matrix.

We say these data being **trained bias**, or not being fully trained, because of distributions and their inner structures. So we propose a method called **IUSM** to enrich samples belong to the small classes by aligning illumination factor distribution. We will see in Sect. 4 that precisions both on **CASTest** and **LFW** are improved using this up-sample strategy.

The **illumination factor** is defined as

$$I_{s,t} = \frac{1}{m \times n} \sum_{i=1}^{m} \sum_{j=1}^{n} M_{s,t}(i,j) \tag{1}$$

where $M_{s,t}(i,j)$ represents gray value on pixel(i,j) which belongs to the identity s's image t. m and n are the width and height respectively.

We model the distribution of $I_{s,t}$ as a normal distribution if s and t are randomly selected.

$$\varepsilon = \frac{I_{s,t} - \bar{I}_s}{\bar{I}_s} \sim N(0, \sigma) \\ \bar{I}_s = \frac{1}{n_s} \sum_t I_{s,t} \tag{2}$$

where n_s is the image number of identity s.

Then we estimate the σ in (2) from images in the identities which are trained unbiased (we select identity s under criterion $n_s \geq 200$ empirically).

For an identity s has less than L images, we desire to increase its samples. Exploiting the observation in Fig. 1, we set L as 200. First, we generate a number set SN whose size is $L - n_s$ from the distribution of (2). Then we select the image i whose illumination factor $I_{s,i}$ is median number of identity s. For each number num in set SN, we randomly select an image m of identity s whose illumination factor is $I_{s,m}$ and generate a new image m_{new} using the Eq. 3.

$$m_{new} = m \otimes (num - \frac{I_{s,m} - I_{s,i}}{I_{s,i}}) \frac{I_{s,i}}{I_{s,m}} + 1 \tag{3}$$

where $m \otimes t$ means each pixel in m product the number t.

3.3 The Structure of Deep Convolution Network

The deep convolution network structure we select is similar to the one proposed in [19]. However, there are two main differences: 1. The input size is modified to 96×96 which adapt to our training method. 2. Output dimension of $Conv52$ is 160, which is verified by us that have little effect to the final precision.

3.4 Training Methodology

We follow most of the training methods proposed in [19] with two modifications. The loss function is formulated as

$$L(\theta) = L_{soft\max}(\theta) + \lambda L_{verification}(\theta) \tag{4}$$

Due to the up-sample procedure which increase the number of less-sample identities, we adjust the λ from 6e-4 to 2.3e-3, which a little smaller than the one proposed in [19]. And do verification task on each model to select the best. Based on the best model, we continue training to reinforce this state. Another modification is we use a 96×96 sliding window upon each aligned image to randomly generate training samples in each batch.

3.5 Face Verification

As demonstrated in [3,4,16,18,19], verification method using Bayesian model [4] is more effective than Euclid distance and cosine distance, etc. So we dig into Bayesian model for some improvements.

As introduced in [4], a face is represented by the sum of two independent Gaussian variables:

$$x = \mu + \varepsilon \tag{5}$$

where x is the observed face with the mean of all faces subtracted, μ represents its identity, ε is the face variation (e.g., lightings, poses, and expressions) within the same identity. the latent variable μ and ε follow two Gaussian distributions

$$\mu \sim N(0, S_\mu) \\ \varepsilon \sim N(0, S_\varepsilon) \tag{6}$$

The similarity of two faces can be expressed as:

$$r(x_1, x_2) = \log \frac{P(x_1, x_2 | H_I)}{P(x_1, x_2 | H_E)} = x_1^T A x_1 + x_2^T A x_2 - 2x_1^T G x_2 \tag{7}$$

where

$$A = (S_\mu + S_\varepsilon)^{-1} - (F + G) \tag{8}$$

$$\begin{pmatrix} F+G & G \\ G & F+G \end{pmatrix} = \begin{pmatrix} S_\mu + S_\varepsilon & S_\mu \\ S_\mu & S_\mu + S_\varepsilon \end{pmatrix}^{-1} \tag{9}$$

There are four conclusions we can get: 1. Both matrix A and G are negative semi-definite matrices. 2. Easy to prove matrix A-G is positive semi-definite matrix. 3. The negative log likelihood ratio will degrade to Mahalanobis distance if A = G. 4. The log likelihood ratio metric is invariant to any full rank linear transform of the feature.

The best model is selected under the criterion:

$$m_{opt} = \arg \min_m (\min_{\rho_m}(\int_{-\infty}^{\rho_m} f(r|H_I) + \int_{\rho_m}^{+\infty} f(r|H_E))) \tag{10}$$

where m_{opt} is the best model, ρ_m represents the threshold select in one model, $f(r|H_I)$ and $f(r|H_E)$ is the similarity probability density function conditioned with H_I and H_E repectively.

Because of the formulation upon is not derivable, also the distribution of r is the sum of quadratic forms of Gaussian random variables, whose probability density function is too complex, normally experimenters select the best model using grid search on training set.

We further propose a feature fusion method which can improve the performance of Bayesian-face model by utilizing the distribution property of r and optimizing a loss function inspired by LDA. The distributions of $r|H_I$ and $r|H_E$ are illustrated in Fig. 2.

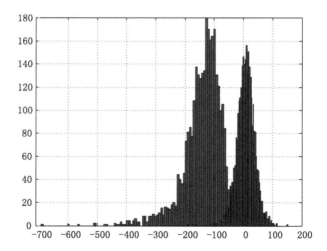

Fig. 2. The distribution of $r|H_I$ and $r|H_E$ in our baseline experiment. We can see that they can be distinguished by a threshold that minimize the α and β mistake in statistical meaning.

Under the condition of different identities(H_E)

$$x_1 = \bar{\mu} + \mu_1 + \varepsilon_1$$
$$x_2 = \bar{\mu} + \mu_2 + \varepsilon_2 \tag{11}$$

where $\mu_1 \sim N(0, S_\mu)$, $\mu_2 \sim N(0, S_\mu)$ and $\varepsilon \sim N(0, S_\varepsilon)$. $\mu_1, \mu_2, \varepsilon_1, \varepsilon_2$ are independent with each other. The whole bias on the feature distribution is $\bar{\mu}$. Then, we can get the deduction:

$$
\begin{aligned}
r|H_E =& 2\bar{\mu}^T(A - G)\bar{\mu} + 2\bar{\mu}^T(A - G)\mu + 2\bar{\mu}^T(A - G)\varepsilon_1 + 2\bar{\mu}^T(A - G)\varepsilon_2 \\
& + 2\mu^T(A - G)\mu + 2\mu^T(A - G)\varepsilon_1 + 2\mu^T(A - G)\varepsilon_2 + \varepsilon_1{}^T A\varepsilon_1 \\
& + \varepsilon_2{}^T A\varepsilon_2 - 2\varepsilon_1{}^T G\varepsilon_2
\end{aligned} \tag{12}
$$

Further, we get the mean and variance of r under H_E condition:

$$
\begin{aligned}
E(r|H_E) =& 2\bar{\mu}^T(A - G)\bar{\mu} + 2tr(AS_\mu) + 2tr(AS_\varepsilon) \\
D(r|H_I) =& 8\bar{\mu}^T(A - G)S_\mu(A - G)^T\bar{\mu} + 8\bar{\mu}^T(A - G)S_\varepsilon(A - G)^T\bar{\mu} \\
& + 4tr(S_\mu AS_\mu A) + 4tr(S_\varepsilon AS_\varepsilon A) + 8tr(S_\mu AS_\varepsilon A) + 4tr(S_\mu GS_\mu G) \\
& + 4tr(S_\varepsilon GS_\varepsilon G) + 8tr(S_\mu GS_\varepsilon G)
\end{aligned} \tag{13}
$$

where $tr(A)$ represents the trace of A. Readers can refer to supplementary materials [12] for deduction details.

Under the condition of same identity(H_I)

$$x_1 = \bar{\mu} + \mu_1 + \varepsilon_1$$
$$x_2 = \bar{\mu} + \mu_1 + \varepsilon_2 \tag{14}$$

where variables under the same distributions as Eq. 11, but $\mu_1 = \mu_2$. And similar to $r|H_E$,

$$
\begin{aligned}
r|H_I =& 2\bar{\mu}^T(A - G)\bar{\mu} + 4\bar{\mu}^T(A - G)\mu + 2\bar{\mu}^T(A - G)\varepsilon_1 + 2\bar{\mu}^T(A - G)\varepsilon_2 \\
& + 2\mu^T(A - G)\mu + 2\mu^T(A - G)\varepsilon_1 + 2\mu^T(A - G)\varepsilon_2 + \varepsilon_1{}^T A\varepsilon_1 \\
& + \varepsilon_2{}^T A\varepsilon_2 - 2\varepsilon_1{}^T G\varepsilon_2
\end{aligned} \tag{15}
$$

The mean and variance of r under H_I condition are:

$$
\begin{aligned}
E(r|H_I) =& 2\bar{\mu}^T(A - G)\bar{\mu} + 2tr((A - G)S_\mu) + 2tr(AS_\varepsilon) \\
D(r|H_I) =& 16\bar{\mu}^T(A - G)S_\mu(A - G)^T\bar{\mu} + 8\bar{\mu}^T(A - G)S_\varepsilon(A - G)^T\bar{\mu} \\
& + 8tr(S_\mu(A - G)S_\mu(A - G)) + 8tr(S_\mu(A - G)S_\varepsilon(A - G)) \\
& + 4tr(S_\varepsilon AS_\varepsilon A) + 4tr(S_\varepsilon GS_\varepsilon G)
\end{aligned} \tag{16}
$$

Feature Fusion Strategy. During experiments, we find that if we apply the point in multi-features' polygon to the Joint Bayesian model selected under the criterion Eq. 10, the final precision outperform which use the frontal face feature only.

We select 18 features to fuse, first 1~9 are fetched from sliding windows in original face image, 10~18 are from flip face. Each sliding window has the same size. The 5th, 14th sliding windows are at the center of a original and flip face respectively.

The ith sliding window has offsets on horizontal and vertical direction relative to the center window:

$$dh = 2(floor(((i-1)\%9)/3) - 1)$$
$$dv = \begin{cases} -2 \ if(i\%3 == 1) \\ 0 \ if(i\%3 == 2) \\ 2 \ if(i\%3 == 0) \end{cases} \quad (17)$$

where dh is the horizontal offset relative to center, dv is the vertical offset relative to center, $\%$ is the operation of mod.

Then the fusion strategy uses a linear combination of features fetched from some sliding windows over original face and its flip face, which we call it **LFF** briefly:

$$f_{fusion} = \sum_{i=1}^{k} \alpha_i f_i$$
$$s.t. : \begin{cases} \sum_{i=1}^{k} \alpha_i = 1 \\ \forall \alpha_i > 0 \end{cases} \quad (18)$$

where f_{fusion} is the fused feature, k is the number of features being fused, f_i is the ith feature and α_i is its weight. Our goal is to find the α making Eq. 19 minimum.

$$\alpha = \arg\min_{\alpha, \rho_m} (\int_{-\infty}^{\rho_m} f(r|H_I)dr + \int_{\rho_m}^{+\infty} f(r|H_E)dr) \quad (19)$$

where α is a vector consisting of feature weights. Other parameters are same to which in Eq. 10.

As introduced in Sect. 3.5 upon, the possibility density function of r is too complex. So we loosen the target we want by using the Fisher's criterion [9] to evaluate the model. An observation we get is weights around $\frac{1}{k}$ always perform better. But it's not best to set all weights to $\frac{1}{k}$. We argue that points around center of these features have better generalization ability for face representation. And then we introduce a penalty term corresponding to this observation. The full objective function is:

$$\begin{cases} f(\beta) = \dfrac{D(r|H_I) + D(r|H_E)}{(E(r|H_I) - E(r|H_E))^2} + \lambda_1 \left(\sum_{i=1}^{k} \beta_i^2 - 1 \right)^2 + \lambda_2 \left(\sum_{i=1}^{k} (\beta_i^2 - \frac{1}{k}) \right)^2 \\ \alpha_i = \beta_i^2 \end{cases}$$
$$(20)$$

The introduction of β eliminates the positive constraint of α. We need to optimize the objective function $f(\beta)$, and then get the value of α as feature weights.

Optimizing Method. The derivative of objective function Eq. 20 is a polynomial that has 4th degree of α, which makes the computation complexity be $O(k^5)$, k is the parameter number. To deal with the complexity of computation, we iteratively optimize each 2 parameters in an epoch and rearrange the optimization order in each epoch. In experiments, we set λ_1 to 1e0 and λ_2 to 3e-2. In Sect. 4 we will illustrate the algorithm converge quickly.

4 Experiments

We evaluate our method on face verification tasks of two datasets, including **Labelled Faces in the Wild** and **CASTest**. CASTest is created by collecting the validate images of bias identities introduced in Sect. 3.2. Its components are similar to LFW, having 2200 training pairs and 6000 testing pairs. In CASTest task, we compare the precision using our up-sample method with not using it. In the LFW task, we first do the comparison same to CASTest. Then we evaluate our linear feature fusion method(LFF). The fusion weights are trained both on the CASWebface training images and LFW training images. Finally, comparison between LFF and concatenation of multi features(CFF) is delivered.

4.1 Results of Up-Sample Hard Identities

All images in the LFW and CASTest datasets are processed by the similar pipleline as [19], and normalized to 100×100 which left 4×4 margin adapt to 96×96 input size of the convolution network. Because of the CASTest images are selected from CASWebface [19], we evaluate on this data set without Bayesian face model transfer learning. The hyper-parameters are tuned on the training set of CASTest like View1 of LFW. According to the protocols of LFW, we both evaluate the accuracy with and without transfer learning of Bayesian face model. The accuracies are listed in Tables 1 and 2 respectively.

Table 1. The performance on CASTest.

Method	Accuracy±SE
Without IUSM	75.07±0.54
With IUSM	81.57±0.52

By inspecting the results shown in Tables 1 and 2, we can see that our up-sampling method can significantly improve the accuracy on CASTest, which is composed of biased data in CASWebface dataset. It slightly improves the performance on LFW. So we can say that this method dose not weaken the deep net's generalization ability.

Table 2. The performance on LFW.

Method	Accuracy±SE	Protocol
Without IUSM	97.48±0.51	unsupervised
With IUSM	97.51±0.54	unsupervised
Without IUSM & JB	97.62±0.53	supervised
With IUSM & JB	97.66±0.58	supervised

4.2 Results of Feature Fusion

In this section we first compare Bayesian face method introduced in [3, 4, 19] with our Linear Feature Fusion(**LFF**) method. Then we compare our method with **CFF** method which concatenates features and does PCA for reducing dimension to 160. All experiments are based on the convolution network trained using IUSM.

In the first step, the following experiments are conducted:

- A1: Single feature + **JB** on **CW**;
- A2: Single feature + Transfer **JB** on **LFW**;
- B1: 6 LFF(Trained on **CW**) + **JB** on **CW**;
- B2: 6 LFF(Trained on **LFW**) + Transfer **JB** on **LFW**;
- C1: 10 LFF(Trained on **CW**) + **JB** on **CW**;
- C2: 10 LFF(Trained on **LFW**) + Transfer **JB** on **LFW**;
- D1: 18 LFF(Trained on **CW**) + **JB** on **CW**;
- D2: 18 LFF(Trained on **LFW**) + Transfer **JB** on **LFW**;

where **JB** represent Joint Bayesian, **CW** is CASIA-WebFace dataset, **LFW** means LFW training set, α LFF(Trained on **DATASET**) means fuse α features using LFF method whose weights are trained on training set of $DATASET$. Corresponding results are listed in Table 3.

By exploiting the results showed in the left 3 columns of Table 3, we can draw three conclusions: (1) Using the fusion method can improve accuracy slightly. (2) The more features fused, the more accuracy increased. (3) The fusion method affects the stability of face verification algorithm slightly(standard deviation rises).

Next we compare our method with CFF. In CFF training step, we first concatenate a fixed number features as a single feature, then do PCA on the feature space and reduce dimensions to 160. Note that we only compare the accuracy without Joint Bayesian transfer learning. The following experiments are conducted. For clearness, we list 3 contrast experiments in Table 3 repeatedly:

- B1: 6 LFF(Trained on **CW**) + **JB** on **CW**;
- E: 6 CFF + **JB** on **CW**;
- C1: 10 LFF(Trained on **CW**) + **JB** on **CW**;
- F: 10 CFF + **JB** on **CW**;

Table 3. Left 3 columns: Accuracies of normal method and different strategies of our method. Right 3 columns: Comparison between LFF and CFF.

Method	Accuracy±SE	Protocol	Method	Accuracy±SE	Protocol
A1	97.51±0.54	unsupervised	B1	97.58±0.88	unsupervised
B1	97.58±0.88	unsupervised	E	97.50±0.74	unsupervised
C1	97.61±0.84	unsupervised	C1	97.61±0.84	unsupervised
D1	97.65±0.86	unsupervised	F	97.55±0.82	unsupervised
A2	97.66±0.58	supervised	D1	97.65±0.86	unsupervised
B2	97.75±0.77	supervised	G	97.65±0.70	unsupervised
C2	97.75±0.73	supervised			
D2	97.80±0.76	supervised			

- D1: 18 LFF(Trained on **CW**) + **JB** on **CW**;
- G: 18 CFF + **JB** on **CW**;

Corresponding results are listed in the right 3 columns of Table 3. We can see that the performance using our method is higher than or equal to CFF, but the stability is a little lower than CFF.

4.3   Effectiveness and Efficiency of Feature Fusion Learning Method

In this sub-section, we evaluate the effectiveness and efficiency of our learning method.

Effectiveness. First we randomly search the 18 weights 20000 times to get 20000 weights-groups. By using each group, we linear fuse 18 features fetched from LFW training set and calculate the similarity of each pair. We get the hyper parameter from LFW training set and use it to get the accuracy on LFW Testing set. By descending the accuracies of these groups, we get the sub-groups corresponding to the highest 10 % accuracies. Then mean and standard deviation of these accuracies are shown in the top of Fig. 3. To be comparable, the weights learned by our method on LFW training data is shown in the bottom of Fig. 3.

Top figure in Fig. 4 shows the Fisher's Criterion Factor of LFW training set varying with the weights learning procedure. We can see the distributions of $r|H_i$ and $r|H_e$ are more distinguishable than original under the Fisher's Criterion.

Efficiency. We randomly generate several group initial values which are constrained by $\sum_{i=1}^{k} \alpha_i = 1$, and then we do our training procedure, all results of these experiments are similar. One sample is shown in the bottom figure of Fig. 4, which we can see the values converge quickly.

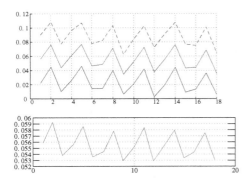

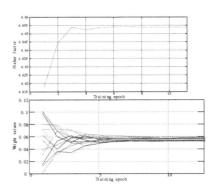

Fig. 3. (TOP)Random search in weight space, select top 10 best weights. Red line is mean of each weight, blue line is one standard deviate away from mean. (BOTTOM)The 18 weights learned by our algorithm. (Color figure online)

Fig. 4. (TOP)Fisher's factor rise with training epoch. (BOTTOM)Weights converge fast with training epoch.

5   Conclusion

This work find biased data in training set and propose a method called **IUSM** to relieve the potential effects these data caused. IUSM dose not weaken the generalization ability of convolution network. This work also introduce a feature fusion(**LFF**) method theoretically and practically based on Joint Bayesian model. LFF is slightly effective on improving the accuracy of face verification task in our experiments. We also argue that this method can improve each of the deep face verification algorithm that use Joint Bayesian model due to the linear combination. Further work will be done in three directions: (1) training more effective convolution network under limited dataset by using up-sample and down-sample methods; (2) improving the face verification accuracy by utilizing the statistical properties of feature; (3) conducting further research on feature fusion methods which are more stable for face verification task.

References

1. Ahonen, T., Hadid, A., Pietikainen, M.: Face description with local binary patterns: application to face recognition. IEEE Trans. Pattern Anal. Mach. Intell. **28**(12), 2037–2041 (2006)
2. Berry, J., Fasel, I., Fadiga, L., Archangeli, D.: Training deep nets with imbalanced and unlabeled data. In: INTERSPEECH (2012)
3. Cao, X., Wipf, D., Wen, F., Duan, G., Sun, J.: A practical transfer learning algorithm for face verification. In: 2013 IEEE International Conference on Computer Vision (ICCV), pp. 3208–3215. IEEE (2013)

4. Chen, D., Cao, X., Wang, L., Wen, F., Sun, J.: Bayesian face revisited: a joint formulation. In: Fitzgibbon, A., Lazebnik, S., Perona, P., Sato, Y., Schmid, C. (eds.) ECCV 2012, Part III. LNCS, vol. 7574, pp. 566–579. Springer, Heidelberg (2012)

5. Ciresan, D., Meier, U., Schmidhuber, J.: Multi-column deep neural networks for image classification. In: 2012 IEEE Conference on Computer Vision and Pattern Recognition (CVPR), pp. 3642–3649. IEEE (2012)

6. Dalal, N., Triggs, B.: Histograms of oriented gradients for human detection. In: 2005 IEEE Computer Society Conference on Computer Vision and Pattern Recognition, CVPR 2005, vol. 1, pp. 886–893. IEEE (2005)

7. Guan, S., Chen, M., Ha, H.Y., Chen, S.C., Shyu, M.L., Zhang, C.: Deep learning with MCA-based instance selection and bootstrapping for imbalanced data classification. In: First IEEE International Conference on Collaboration and Internet Computing (CIC) (2015)

8. Guo, X., Yin, Y., Dong, C., Yang, G., Zhou, G.: On the class imbalance problem. In: 2008 Fourth International Conference on Natural Computation, ICNC 2008, vol. 4, pp. 192–201. IEEE (2008)

9. Huang, G.B., Ramesh, M., Berg, T., Learned-Miller, E.: Labeled faces in the wild: a database for studying face recognition in unconstrained environments. Technical report 07–49, University of Massachusetts, Amherst (2007)

10. Ke, Y., Sukthankar, R.: PCA-SIFT: A more distinctive representation for local image descriptors. In: Proceedings of the 2004 IEEE Computer Society Conference on Computer Vision and Pattern Recognition, CVPR 2004, vol. 2, pp. II-506–II-513. IEEE (2004)

11. Krizhevsky, A., Sutskever, I., Hinton, G.E.: Image net classification with deep convolutional neural networks. In: Advances in Neural Information Processing Systems, pp. 1097–1105 (2012)

12. Mathai, A.M., Provost, S.B.: Quadratic forms in random variables: theory and applications. J. Am. Stat. Assoc. (1992)

13. Miller, D., Kemelmacher-Shlizerman, I., Seitz, S.M.: Megaface: a million faces for recognition at scale (2015). arXiv:1505.02108

14. Monard, M.C., Batista, G.E.: Learnng with skewed class distrihutions. In: Advances in Logic, Artificial Intelligence, and Robotics: LAPTEC 2002, vol. 85, p. 173 (2002)

15. Schroff, F., Kalenichenko, D., Philbin, J.: Facenet: a unified embedding for face recognition and clustering (2015). arXiv:1503.03832

16. Sun, Y., Chen, Y., Wang, X., Tang, X.: Deep learning face representation by joint identification-verification. In: Advances in Neural Information Processing Systems, pp. 1988–1996 (2014)

17. Sun, Y., Liang, D., Wang, X., Tang, X.: Deepid3: face recognition with very deep neural networks (2015). arXiv:1502.00873

18. Sun, Y., Wang, X., Tang, X.: Deep learning face representation from predicting 10,000 classes. In: 2014 IEEE Conference on Computer Vision and Pattern Recognition (CVPR), pp. 1891–1898. IEEE (2014)

19. Yi, D., Lei, Z., Liao, S., Li, S.Z.: Learning face representation from scratch (2014). arXiv:1411.7923

20. Yu, H., Yang, J.: A direct lda algorithm for high-dimensional data with application to face recognition. Pattern Recogn. $34(10)$, 2067–2070 (2001)

21. Zhou, E., Cao, Z., Yin, Q.: Naive-deep face recognition: touching the limit of LFW benchmark or not? (2015). arXiv:1501.04690

A Multi-memory Multi-population Memetic Algorithm for Dynamic Shortest Path Routing in Mobile Ad-hoc Networks

Nasser R. Sabar[✉], Ayad Turky, and Andy Song

School of Computer Science and I.T., RMIT University, Melbourne, Australia
{nasser.sabar,ayad.turky,andy.song}@rmit.edu.au

Abstract. This study investigates the dynamic shortest path routing (DSPR) problem in mobile ad-hoc networks. The goal is to find the shortest possible path that connects a source node with the destination node while effectively handling dynamic changes occurring on the ad-hoc networks. The key challenge in DSPR is how to simultaneously keep track changes and search for the global optima. A multi-memory based multi-population memetic algorithm is proposed for DSPR in this paper. The proposed algorithm combines the strength of three different strategies, multi-memory, multi-population and memetic algorithm, aiming to effectively explore and exploit the search space. It divides the search space by multiple populations. The distribution of solutions in each population is kept in the associated memory. The multi-memory multi-population approach is to capture dynamic changes and maintain search diversity. The memetic component, which is a hybrid Genetic Algorithm (GA) and local search, is to find high quality solutions. The performance of the proposed algorithm is evaluated on benchmark DSPR instances under both cyclic and acyclic environments. Our method obtained better results when compared with existing methods in the literatures, showing the effectiveness of the proposed algorithm in handling dynamic optimisation.

Keywords: Dynamic shortest path routing · Memetic algorithms · Dynamic optimisation · Evolutionary algorithm

1 Introduction

This study is to establish a new method for solving the dynamic shortest path routing (DSPR) problem under mobile ad-hoc networks (MANET) environments where the topological structure of network keeps changing. MANET is made of an arbitrary group of mobile devices such as mobile phones. Nodes may be appearing or disappearing on the network due to factors like flat battery, poor reception, interrupted services and so on [12]. Unfortunately optimisation algorithms that have been proposed to solve static shortest path routing (SPR) problem for MANETs are not directly suitable for DSPR [2,11,13]. Because dealing with dynamic environments requires tracking changes and searching for

© Springer International Publishing Switzerland 2016
R. Booth and M.-L. Zhang (Eds.): PRICAI 2016, LNAI 9810, pp. 406–418, 2016.
DOI: 10.1007/978-3-319-42911-3_34

optimal solutions simultaneously. One remedy to this issue is through maintaining the search diversity so the search process can cope with problem changes more effectively.

A multi-memory multi-population memetic algorithm (M-MMA) is therefore proposed for DSPR. It is built upon a recently established memetic algorithm for DSPR. Memetic algorithm incorporates local search algorithm with Genetic Algorithm (GA) so local exploitation can be combined with exploration [5,10]. Our algorithm introduces two extra components, multi-memory and multi-population to further improve the search process. Multi-population is to divide a population of solutions into several sub-populations [9]. Each sub-population occupies a different region of the search space. An area which was bad but becomes good may be quickly identified by the search. The second component multi-memory is to maintain the solution distribution of each sub-population. The improvement of solutions in each sub-population is done through memetic algorithm (MA). A well-known DSPR simulator proposed by Yang et. al is used in our study [12]. Experiments show that the proposed algorithm can achieve better performance compared to state-of-the-art algorithms in the literature.

2   Problem Description

DSPR problems can be represented as an undirected connected graph in which there are a set of nodes and a set of edges, $G(V, E)$. Each node v ($v \in V$) represents a mobile device or a wireless router. All nodes are connected by edges that link adjacent nodes. Each edge is associated with a weight or a cost that represents distance or the cost of communication between v_i to v_j. On a MANET two nodes will be connected if they can reach each other for packet transmission. Hence any two nodes within the radio transmission range of each other and operating on the same channel will be connected. For each connection or edge, a transmission delay is also added. Due to the dynamic nature of MANET, the topology may change over time. An initial network G_0 may change to G_1, G_2 to G_n.

The formal notation for DSPR in MANETs is presented in Table 1. This notion is from [3]. The main goal to find the shortest possible path between the source node s and the destination node t. The generated path is considered feasible if it contains no loops (loop-free) and the total communication delay is within the upper bound. When a change occurs in a MANET, meaning devices joining or leaving the network, a DSPR algorithm should still be able to find a feasible and shortest path to reconnect. Thus an effective DSPR algorithm should response to a change very quickly regardless the nature of the change. The objective functions of DSPR can be formulated as follow:

$$D(P_i) = \sum_{l \in P_i(s,t)} d_i \leq DELAY \tag{1}$$

$$C(P_i) = \min_{P_i \in G_i} \sum_{l \in P_i(s,t)} c_l \tag{2}$$

Table 1. Notation of DSPR for MANETs

$G_0(V_0, E_0)$	A graph representing the initial MANET
$G_i(V_i, E_i)$	Graph of the MANET after the ith change
s	Source node
t	Destination node
$P_i(s, t)$	Path from node s to r on graph G_i
l	A link connecting two nodes
d_l	Transmission delay on link l
c_l	Communication cost of link l
$D(P_i)$	Total transmission delay on path P_i
$C(P_i)$	Total cost of path P_i

where $DELAY$ is the delay upper bound. The total delay $D(P_i)$ along the transmission path from s to t should not exceed delay upper bound $DELAY$ and the total cost $C(P_i)$ of the transmission should be minimum.

3   Methodology

Our proposed M-MMA method combines the strengths of three strategies: (1) multi-memory, (2) multi-population, (3) memetic algorithm. This approach aims to search for good solutions while effectively respond to dynamic changes occurred during the optimisation process. These three strategies are applied in a sequence on a given problem instance as follows. Firstly, the multi-population component divides the entire population into several sub-populations. Secondly, the solutions of each sub-population is improved by memetic algorithm through evolutionary operators including selection, crossover and mutation, and a local search process. Thirdly, the memory mechanism is called to update the solutions of each sub-population.

The flowchart of the algorithm is shown in Fig. 1. It first sets the parameters, randomly create an initial population of solutions and evaluate their fitness value. Next, it divides the population into m sub-populations. The aforementioned three strategies are then applied on each sub-population separately. Once a change in the environment is detected, all solutions are merged into one big population to be re-partitioned again. This process is repeated until the stopping criteria is met. The details are discussed in the following subsections.

3.1   Set Parameters

The proposed algorithm has six parameters: the maximum number of iterations ($MaxIt$), population size (Ps), memory size (M_s), crossover rate (CR), mutation rate (MR) and the number of sub-populations (m). The value of each parameter is set based on preliminary tests which are discussed in Sect. 4.3.

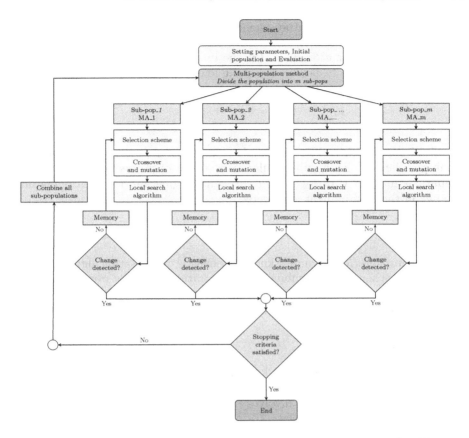

Fig. 1. Flowchart of our Multi-memory Multi-population memetic algorithm

3.2   Initial Population

The population of solutions with size Ps are randomly generated. Each solution is represented by one chromosome that is a one-dimensional array. Each gene contains an integer number which represents the ID of a node on the MANET $G(V, E)$. The first and last genes are the source and destination nodes respectively.

3.3   Evaluation

The fitness values of solutions, including the initial solutions, are calculated using the following equation:

$$f(s) = \left[\sum_{l \in p(s,t)} C_l \right]^{-1} \tag{3}$$

where $f(Ch_i)$ represents the fitness value of chromosome (Ch_i), s and t represent the source and destination nodes respectively, $p(s, t)$ is the path between the source node and destination node, l is the link between nodes in $p(s, t)$ and C is the cost of lth link. It can be seen that low cost leads to high fitness value, hence having better chance to be selected for reproducing the next generation.

3.4 Multi-population

The multi-population component randomly divides the whole population into m sub-populations. Each sub-population is to be optimised by MA. All sub-populations interact with each other through merging and re-partitioning once a change in the environment is detected.

3.5 Memetic Algorithm

Memetic algorithm (MA) is a well-known stochastic optimisation algorithm [5,6]. It is a hybrid scheme that combines the exploration aspect of population based evolutionary search and the exploitation aspect of local search [8]. The MA in this study hybridises genetic algorithm (GA) with local search.

3.6 Genetic Algorithm

Genetic algorithm (GA) is a well-known problem solving method inspired by survival-of-the-fittest principle in nature [3]. In our method, each sub-population has its own GA process which involves the following major steps.

- **Selection**: Selection picks up two solutions from the population for reproduction [3]. A pair-wise without replacement tournament selection scheme is used here [4,12]. Solutions in a population are randomly paired. The better one from each pair will be considered as winner for that tournament.
- **Crossover**: Crossover exchanges the genes of two selected solutions to generate new solutions of the next generation [3]. Single-point crossover operator is used here [1,4,12]. In our method that single point selected as the crossover point is always an intermediate node between the source node and the destination node. Hence the new solutions generated by crossover have same source and destination. The probability of performing crossover is determined by the crossover rate CR.
- **Mutation**: Mutation complements crossover in allowing the search to get out from the local optima point [3]. A one-point mutation operator is used [1,7]. It first randomly chooses a point as the mutation point and then randomly changes the values of all points behind that mutation point.

– **Repair procedure**: Both crossover and mutation may generate infeasible solutions which contain loops in the path. The repair procedure is to fix infeasible solutions and make feasible. It removes loops by eliminating duplicated node and reconnecting the path with neighboured one [7,12].

3.7 Local Search

Local search aims to improve the convergence of the search process. Simple descent is used in this work. It iteratively explores the neighbourhood area of a given solution, seeking for a better alternative. In each iteration, a neighbourhood solution is generating by modifying the current solution using a replace operator. This operator randomly select one node and then replace it with one of its neighbouring nodes. The updated solution will be accepted if it is better than the original one in term of the fitness measure. This iterative process continues until the termination condition of the local search is met. In our M-MMA the local search will stop if there is no improvement after a predefined number of iterations (see Sect. 4.3). To reduce computational cost, the local search is only triggered if the fitness value of the new solution is no better than the worst one in the population.

3.8 Change Detection

This part checks whether there is a significant difference in the environment, meaning G_n is different with G_{n+1}. If that is positive then the search process of all sub-populations will terminate. Otherwise, a new generation will start for every sub-population.

3.9 Multi-memory

The main role of memory is to ensure that the solutions of each sub-population are well scattered over the landscape. So changes can be dealt with more promptly. At each generation, redundant solutions in the population are removed and replaced with solutions stored in the memory. In this study three different types of memories are introduced. Each type stores a set of solutions described below. Each MA process is randomly assigned with one type of memory:

1. M1: a set of random solutions.
2. M2: the best solutions from the previous generations.
3. M3: a set of solutions generated by modifying existing best solutions.

3.10   The Stopping Criteria

M-MMA terminates if the maximum number of generations is reached. Otherwise, when a change is detected all sub-populations will be combined to form a large population which will be again divided to multi-population search for the new MANET topology.

4   Experiments

This section describes our experiments including the simulation of dynamic MANET, the performance evaluation metric and the parameter settings of M-MMA.

4.1   The Simulation of Dynamic Environments

The dynamic simulator for MANET and the network topology instances are introduced by [12]. The simulator first generates a square region of 200×200, where the x axis coordinate and y axis coordinate are set between $[0, 200)$. Next, it randomly places 100 nodes and establish links between them. Nodes are linked if the Euclidean distance is less than the given radio transmission *Range*, where, *Range* is set to 50. Each link is randomly assigned a cost and delay. The delay upper bound is twice the minimum end-to-end delay, same as that in [12]. These steps will continue until all connections are created and the network is established.

To simulate a dynamic aspect, the initial topology will be changed over the time by modifying a number of selected nodes. Two parameters, R and M are to control the dynamic environment. R represents the number of generations between consecutive changes, while M represents the severity of change. For instance, if R is set to 5 then the topology will be changed for every 5 generations during evolution. If M is set to 2 this mean at each change 2 nodes will be randomly selected and changed. Each node must be either in active or sleep mode. If a selected node is active, its status will be changed into inactive. Similarly if a node is inactive, then it will be activated to join the network.

The simulator instances consists of four series of different characteristics. These sets are namely series #1, series #2, series #3, and series #4. Two different dynamic environments are considered in this paper: acyclic dynamic environment (series #2, #3 and #4) where there is no repeat of topology and cyclic dynamic environment (represented by series #1) where the repeat of topology is permitted. In series #2, #3 and #4, M is set to 2, 3 and 4, respectively. In series #1, M is equal to 2. Network topology 1 is set the same as the last one, topology 21.

4.2   Performance Evaluation Metric

In this paper, we use the overall off-line performance $(\overline{F}_{OFF})$ to evaluate the performance of the proposed M-MMA. This measurement is also used by others for algorithm comparisons [12]. It can be calculated as follows:

$$\overline{F}_{OFF} = \frac{1}{Max_{gen}} \sum_{i=1}^{Max_{gen}} (\frac{1}{N_{run}} \sum_{j=1}^{N_{run}} \sum_{l \in p(s,t)} C_{l,i,j}^{best}) \tag{4}$$

where Max_{gen} and N_{run} represent the maximum number of generations and the total number of runs respectively. $C_{l,i,j}^{best}$ is the cost of a link on the path of the best solution at the ith generation during the jth run. The lower the $\overline{F}_{OFF}$ value, the better the performance. Note this measurement is different with the objective function shown in Formula 3. The objective function is to guide the search.

4.3   Parameter Settings

The proposed algorithm has seven different parameters that need to be set by user. To tune these parameters, a series of preliminary experiments were conducted to find out the most appropriate value for each of these parameters. We tested the proposed M-MMA 30 independent runs with different parameters combinations. The best parameter values are listed in Table 2.

Table 2. The parameter settings

M-MMA Parameter	Tested values	Suggested value
Population size (PS)	10, 20, 30, 40, 50, 60, 70	30
Crossover rate (CR)	0.3, 0.5, 0.7, 0.9	0.7
Mutation rate (MR)	0.1, 0.3, 0.6 , 0.9	0.1
Number of sub-population (m)	3, 5, 7 , 10	5
Consecutive non-improvement iterations	5, 10, 15 , 20	5
Memory size	1–20	8

5   Results and Discussions

The experimental results are presented here and compared with other methods. M-MMA was tested on four series of network instances for cyclic dynamic environments and the acyclic dynamic environments.

5.1   Results Under Cyclic Dynamic Environment

M-MMA is compared with the following three algorithms taken from the litera-ture for cyclic dynamic environment:

1. **MEGA**:Genetic algorithm with memory scheme [12].
2. **MRIGA**: Memory and random immigrants GA [12].
3. **MIGA**: Memory based immigrants GA [12].

Series #1 instances are used here. Parameter M is set to 2, while R value, time for change, is set to 5, 10, and 15 respectively. Series #1 contains 101 topologies. The maximum number of generations for $R=5$, $R=10$ and $R=15$ is 505, 1010 and 1515, respectively.

Table 3 shows the results of M-MMA. Results from other three methods, MEGA, MRIGA and MIGA on series #1 are also listed. The best results are highlighted in bold. Observing these results we see the superb performance of our proposed M-MMA comparing with MEGA, MRIGA and MIGA. M-MMA outperformed these algorithms on all instances.

Table 3. Results under cyclic dynamic environment

Algorithm	Series #1		
	$R=5$	$R=10$	$R=15$
M-MMA	**432.427**	**418.686**	**416.811**
MEGA	464.935	427.622	442.997
MRIGA	437.23	440.984	430.033
MIGA	480.446	461.834	445.168

To examine the search behaviour of our M-MMA as well as MEGA, MRIGA and MIGA, the search progress over 500 generations for $R=5$ is plotted in Fig. 2. As can be seen from the figure, M-MMA is consistently at the bottom of the figure. It is most stable one during the whole the search process. In comparison other methods fluctuate when a change occur. This illustrates that our M-MMA is an effective solution method for handling DSPR problem with cyclic changes.

5.2   Results Under Acyclic Dynamic Environments

In this set of experiments under acyclic dynamic environment, our M-MMA is compared with another three methods proposed for this situation. These meth-ods are:

1. **EIGA**: Elitism based immigrants genetic algorithm [12].
2. **RIGA**: Random immigrants genetic algorithm [12].
3. **HIGA**: Hybrid immigrants genetic algorithm [12].

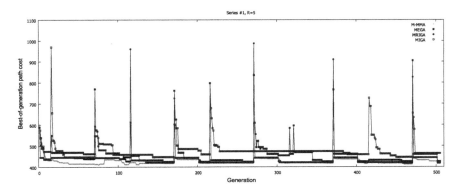

Fig. 2. Search progress of four algorithms on Series #1, $R = 5$

Series #2, Series #3, and Series #4 are used here. Each series involves 21 different network topologies. The M values are set to 2, 3 and 4. The R value are set to be 5, 10 and 15 respectively. The maximum number of generations for these three values are 105, 210 and 315, respectively. So all 21 topologies can be included in the experiments.

Table 4 shows the results from these four methods. The best average result on one row is highlighted in bold. As can be seen from the table, our M-MMA outperformed EIGA, RIGA and HIGA on all series. This good performance is mainly the result of multi-memory and multi-population which can preserve the diversity during the search process while removing redundant solutions.

The search progress of the four algorithms are plotted in Fig. 3 on Series #2, Series #3, and Series #4 when $R=5$. As can be seen from the figure our M-MMA again exhibits its stability under these dynamic environments. M-MMA was the lowest curve on these plots and stayed consistently low when changes occur. These changes on the network caused high path cost, or a spike on the figure, when using other three methods. This comparison shows that the proposed M-MMA can quickly adjust itself to the changes under acyclic dynamic environment.

Table 4. Results under acyclic dynamic environments

	Series #2			Series #3			Series #4		
	$R=5$	$R=10$	$R=15$	$R=5$	$R=10$	$R=15$	$R=5$	$R=10$	$R=15$
M-MMA	**427.142**	**421.01**	**417.643**	**460.124**	**429.75**	**424.749**	**446.732**	**434.839**	**431.69**
EIGA	447.743	446.371	436.565	462.619	485.471	448.524	489.762	468.438	462.937
RIGA	433.838	435.886	440.587	461.19	450.352	445.263	475.543	461.819	464.737
HIGA	445.381	467.776	455.333	490.962	497.276	452.422	506.962	487.052	498.724

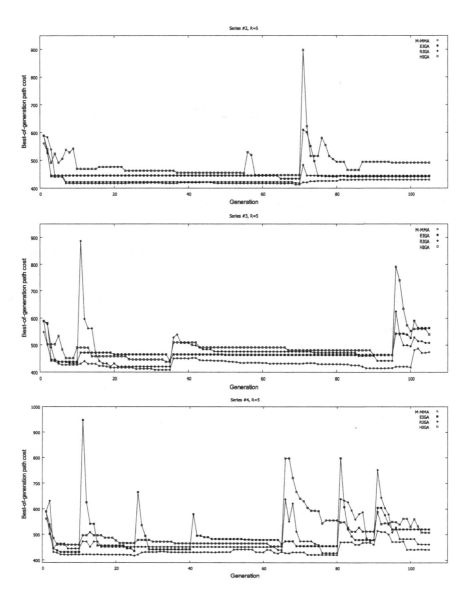

Fig. 3. Search progress of four algorithms on Series #2, #3, #4 ($R = 5$)

6   Conclusion

This study proposed a multi-memory multi-population memetic algorithm for dynamic shortest path routing problems in mobile ad-hoc networks. The proposed algorithm divides a population of solutions into several sub-populations to perform search separately over different parts of the search space. It use a multi-memory mechanism to store solutions for each sub-population so good solutions can be preserved to cope with future changes. Memetic algorithm, which hybridises genetic algorithm and local search, is performed on each sub-population to find high quality solutions for that sub-population.

The proposed method has been evaluated on four series of shortest path routing problems under different dynamic environment. Six different state-of-the-art methods were introduced for comparison. The results shown that our method can handle both cyclic and acyclic dynamic changes without modifications. More importantly the shortest paths found by the proposed method are better than paths found by other methods. Further analysis shows that the search performance of the proposed M-MMA is very stable. It can quickly adjust itself to fit with the new environment. The search is consistent yet efficient in terms of coping with dynamic changes. We conclude that the proposed multi-memory multi-population memetic algorithm is an effective and competitive approach in solving dynamic shortest path routing problems. It can accommodate changes well while performing search. It is a good candidate for dynamic MANETs.

In our future study we will examine the exact contribution of the memory and multi-population components. So the performance may be further improved, the computational cost may be reduced. In addition more instances will be introduced to facilitate further validation and extension.

References

1. Ahn, C.W., Ramakrishna, R.S.: A genetic algorithm for shortest path routing problem and the sizing of populations. IEEE Trans. Evol. Comput. **6**(6), 566–579 (2002)
2. Branke, J.: Evolutionary Optimization in Dynamic Environments, vol. 3. Springer Science & Business Media, New York (2012)
3. Holland, J.H.: Adaptation in Natural and Artificial Systems: An Introductory Analysis with Applications to Biology, Control, and Artificial Intelligence. MIT press, Cambridge (1992)
4. Lee, S., Soak, S., Kim, K., Park, H., Jeon, M.: Statistical properties analysis of real world tournament selection in genetic algorithms. Appl. Intell. **28**(2), 195–205 (2008)
5. Moscato, P.: On evolution, search, optimization, genetic algorithms and martial arts: towards memetic algorithms. Caltech Concurrent Comput. Program, C3P Rep. **826**, 1989 (1989)
6. Neri, F., Cotta, C.: Memetic algorithms and memetic computing optimization: a literature review. Swarm Evol. Comput. **2**, 1–14 (2012)

7. Oh, S., Ahn, C.W., Ramakrishna, R.S.: A genetic-inspired multicast routing optimization algorithm with bandwidth and end-to-end delay constraints. In: King, I., Wang, J., Chan, L.-W., Wang, D.L. (eds.) ICONIP 2006. LNCS, vol. 4234, pp. 807–816. Springer, Heidelberg (2006)

8. Sabar, N.R., Song, A.: Dual population genetic algorithm for the cardinality constrained portfolio selection problem. In: Dick, G., Browne, W.N., Whigham, P., Zhang, M., Bui, L.T., Ishibuchi, H., Jin, Y., Li, X., Shi, Y., Singh, P., Tan, K.C., Tang, K. (eds.) SEAL 2014. LNCS, vol. 8886, pp. 703–712. Springer, Heidelberg (2014)

9. Sabar, N.R., Song, A., Tari, Z., Yi, X., Zomaya, A.: A memetic algorithm for dynamic shortest path routing on mobile ad-hoc networks. In: 2015 IEEE 21st International Conference on Parallel and Distributed Systems (ICPADS), pp. 60–67. IEEE (2015)

10. Sabar, N.R., Song, A., Zhang, M.: A variable local search based memetic algorithm for the load balancing problem in cloud computing. In: Squillero, G., Burelli, P. (eds.) EvoApplications 2016. LNCS, vol. 9597, pp. 267–282. Springer, Heidelberg (2016). doi:10.1007/978-3-319-31204-0_18

11. Turky, A.M., Abdullah, S., Sabar, N.R.: A hybrid harmony search algorithm for solving dynamic optimisation problems. Procedia Comput. Sci. **29**, 1926–1936 (2014)

12. Yang, S., Cheng, H., Wang, F.: Genetic algorithms with immigrants and memory schemes for dynamic shortest path routing problems in mobile ad hoc networks. IEEE Trans. Syst. Man Cybern. Part C: Appl. Rev. **40**(1), 52–63 (2010)

13. Yang, S., Yao, X.: Population-based incremental learning with associative memory for dynamic environments. IEEE Trans. Evol. Comput. **12**(5), 542–561 (2008)

Detecting Critical Links in Complex Network to Maintain Information Flow/Reachability

Kazumi Saito[1]([✉]), Masahiro Kimura[2], Kouzou Ohara[3], and Hiroshi Motoda[4,5]

[1] School of Administration and Informatics, University of Shizuoka, Shizuoka, Japan
k-saito@u-shizuoka-ken.ac.jp
[2] Department of Electronics and Informatics, Ryukoku University, Otsu, Japan
kimura@rins.ryukoku.ac.jp
[3] Department of Integrated Information Technology, Aoyama Gakuin University,
Sagamihara, Japan
ohara@it.aoyama.ac.jp
[4] Institute of Scientific and Industrial Research, Osaka University, Ibaraki, Japan
motoda@ar.sanken.osaka-u.ac.jp
[5] School of Computing and Information Systems, University of Tasmania,
Hobart, Australia

Abstract. We address the problem of efficiently detecting critical links in a large network. Critical links are such links that their deletion exerts substantial effects on the network performance. Here in this paper, we define the performance as being the average node reachability. This problem is computationally very expensive because the number of links is an order of magnitude larger even for a sparse network. We tackle this problem by using bottom-k sketch algorithm and further by employing two new acceleration techniques: marginal-link updating (MLU) and redundant-link skipping (RLS). We tested the effectiveness of the proposed method using two real-world large networks and two synthetic large networks and showed that the new method can compute the performance degradation by link removal about an order of magnitude faster than the baseline method in which bottom-k sketch algorithm is applied directly. Further, we confirmed that the measures easily composed by well known existing centralities, e.g. in/out-degree, betweenness, PageRank, authority/hub, are not able to detect critical links. Those links detected by these measures do not reduce the average reachability at all, i.e. not critical at all.

Keywords: Social networks · Link deletion · Critical links · Node reachability

1 Introduction

Studies of the structure and functions of large complex networks have attracted a great deal of attention in many different fields such as sociology, biology, physics and computer science [23]. It has been recognized that developing new

© Springer International Publishing Switzerland 2016
R. Booth and M.-L. Zhang (Eds.): PRICAI 2016, LNAI 9810, pp. 419–432, 2016.
DOI: 10.1007/978-3-319-42911-3_35

methods/tools that enable us to quantify the importance of each individual node and link in a network is crucially important in pursuing fundamental network analysis. Networks mediate the spread of information, and it sometimes happens that a small initial seed cascades to affect large portions of networks [27]. Such information cascade phenomena are observed in many situations: for example, cascading failures can occur in power grids (e.g., the August 10, 1996 accident in the western US power grid), diseases can spread over networks of contacts between individuals, innovations and rumors can propagate through social networks, and large grass-roots social movements can begin in the absence of centralized control (e.g., the Arab Spring). These problems have mostly been studied from the view point of identifying influential nodes under some assumed information diffusion model. There are other studies on identifying influential links to prevent the spread of undesirable things. See Sect. 2 for related work.

We study this problem from a slightly different angle in a more general setting. Which links are most critical in maintaining a desired network performance? For example, when the desired performance is to minimize contamination, the problem is reduced to detecting critical links to remove or block. When the desired performance is to maximize evacuation or minimize isolation, the problem is to detect critical links that reduce the overall performance if these links do not function. This problem is mathematically formulated as an optimization problem when a network structure is given and a performance measure is defined. In this paper we define the performance as being the average node reachability with respect to a link deletion, i.e. average number of nodes that are reachable from every single node when a particular link is deleted. The problem is to rank the links in accordance with the performance and identify the most critical link(s).

Since the core of the computation is to estimate reachability, an efficient method of counting reachable nodes is needed. We borrow the idea of bottom-k sketch [11,12] which can estimate the number of reachable nodes quite efficiently by sampling a small number of nodes. Although it is very efficient, it still is computationally heavy when applied to our problem because we have to compute reachability from every single node for a particular link deletion and repeat this for all nodes and take the average. We repeat this for all the links and rank the results. To cope with this difficulty, we introduce two acceleration techniques called marginal-link updating (MLU) and redundant-link skipping (RLS). These are designed to improve the computational efficiency of bottom-k sketch.

We have tested our method using two real-world benchmark networks taken from Stanford Network Analysis Project and two synthetic networks which we designed to control the structural properties. We confirmed that about an order of magnitude reduction of computation time is obtained by use of these two acceleration techniques over a baseline method in which no acceleration techniques are used and bottom-k sketch algorithm is applied directly. We also analyzed which acceleration technique works better in which situations. We further investigated whether other measures which can easily be composed by the well known existing centralities can detect critical links. We composed four measures each computed by degree centrality, betweenness centrality, PageRank centrality and authority/hub centrality, respectively. These four measures rank the links very differently and those identified critical according to these measures do not reduce the performance at all, i.e. they are not critical by no means. This series of experiments confirm that the proposed method is unique and can efficiently detect critical links.

The paper is organized as follows. Section 2 briefly explains studies related to this paper. Section 3 revisits bottom-k sketch algorithm and introduces two new acceleration techniques. Section 4 reports four datasets used and the experimental results: computational efficiency and comparison with other measures. Section 5 summarizes the main achievement and future plans.

2   Related Work

Finding critical links in a network is closely related to the problem of efficiently preventing the spread of undesirable things such as contamination and malicious rumors by blocking links. An effective method of blocking a limited number of links in a social network was presented to solve the contamination minimization problem under a fundamental information diffusion model such as the independent cascade and the linear threshold models [17]. Many studies were also made on exploring effective strategies for reducing the spread of infection by removing nodes in a network [1,5,6,22]. Moreover, we note that the contamination minimization problem can be converse to the influence maximization problem, which has recently attracted much interest in the field of social network mining [3,8,9,15,16,18,21,26,28].

To find critical links in a network, we consider quantifying how influential each link is. It is closely related to quantifying how influential each node is in the network. To this end, several node-centrality measures have been presented in the field of social network analysis. Representative node-centrality measures include degree centrality [14], HITS (hub and authority) centrality [7], PageRank centrality [4] and betweenness centrality [14]. Here, note that for some node-centrality measures such as betweenness centrality, their computation becomes harder as the network size increases, since it needs to take the global network structure into account. Thus, several researchers presented methods of approximating such node-centralities [2,10,24]. Moreover, given an information diffusion model on a social network, influence degree centrality can be defined by evaluating the influence of each node. Unlike node-centrality measures derived only from network topology, influence degree centrality exploits a dynamical process on the network as well. An efficient method of simultaneously estimating the influence degrees of all the nodes was presented under the SIR model setting [20]. We note that influence degree centrality can also be employed for identifying super-mediators of information diffusion in the social network [25]. In this paper, we propose a method of efficiently evaluating how critical each link is in the network (i.e., calculating our new link-centrality measure). Since conventional node-centrality measures can naturally derive link-centrality measures, we also compare the proposed link-centrality measure with those link-centrality measures (see Sect. 4.3).

A bottom-k sketch [11,12] used in this paper is a summary of a set of nodes, which is obtained by associating with each node in a network an independent random rank value drawn from a probability distribution. The bottom-k estimator includes the k smallest rank values, and the kth smallest one is used for the estimation. This estimate has a Coefficient of Variation (CV), which is the ratio of the standard deviation to the mean, that is never more than $1/\sqrt{k-2}$ and is well concentrated [11]. We can quite efficiently calculate the bottom-k sketch of each node in the network by orderly assigning the rank values from the smallest one to those nodes reachable by reversely following links over the network. Based on this framework, a greedy Sketch-based Influence Maximization (SKIM) algorithm has been proposed, and it has been shown that the SKIM algorithm scales to graphs with billions of edges, with one to two orders of magnitude speedup over the best greedy methods [13]. Thus, we also develop our method of detecting critical links under the framework of the bottom-k sketching algorithm.

3   Proposed Method

Let $G = (\mathcal{V}, \mathcal{E})$ be a given simple network without self-loops, where $\mathcal{V} = \{u, v, w, \cdots\}$ and $\mathcal{E} = \{e = (u, v), f, g, \cdots\}$ are sets of nodes and directed links, respectively. Let $\mathcal{R}(v; G)$ and $\mathcal{Q}(v; G)$ be the sets of reachable nodes by forwardly and reversely following links from a node v over G, respectively, where note that $v \in \mathcal{R}(v; G)$ and $v \in \mathcal{Q}(v; G)$. Also, let $\mathcal{R}_1(v; G)$ and $\mathcal{Q}_1(v; G)$ be the sets of those nodes adjacent to v, i.e., $\mathcal{R}_1(v; G) = \{w \in \mathcal{R}(v; G) \mid (v, w) \in \mathcal{E}\}$ and $\mathcal{Q}_1(v; G) = \{u \in \mathcal{Q}(v; G) \mid (u, v) \in \mathcal{E}\}$, respectively. Here, we briefly revisit the bottom-k sketch [11,12] and describe the way to estimate the number of the reachable nodes from each node $v \in \mathcal{V}$, i.e., $|\mathcal{R}(v; G)|$. First, we assign to each node $v \in \mathcal{V}$ a value $r(v)$ uniformly at random in $[0, 1]$. When $|\mathcal{R}(v; G)| \geq k$, let $\mathcal{B}_k(v; G)$ be the subset of the k smallest elements in $\{r(w) \mid w \in \mathcal{R}(v; G)\}$, and $b_k(v; G) = \max \mathcal{B}_k(v; G)$ be the k-th smallest element. Then, we can unbiasedly estimate the number of the reachable nodes from v by $H(v; G) = |\mathcal{B}_k(v; G)|$ if $|\mathcal{B}_k(v; G)| < k$[1]; otherwise $H(v; G) = (k-1)/b_k(v; G)$. Here note that for any $c > 0$, it is enough to set $k = (2 + c)\epsilon^{-2} \log |\mathcal{V}|$ to have a probability of having relative error larger than ϵ bounded by $|\mathcal{V}|^{-c}$ [11,12]. Here, we can efficiently calculate the bottom-k sketch $\mathcal{B}_k(v; G)$ for each node $v \in \mathcal{V}$ by reversely following links $k|\mathcal{E}|$ times. Namely, we first initialize $\mathcal{B}_k(v; G) \leftarrow \emptyset$ and sort the random values as $(r(v_1), \cdots, r(v_i), \cdots, r(v_{|\mathcal{V}|}))$ in ascending order, i.e., $r(v_i) \leq r(v_{i+1})$. Then, from $i = 1$ to $|\mathcal{V}|$, for $w \in \mathcal{Q}(v_i; G)$, we repeatedly insert $r(v_i)$ into $\mathcal{B}_k(w; G)$ by reversely following links from v_i if $|\mathcal{B}_k(w; G)| < k$.

As described earlier, we focus on the problem of detecting a critical link $\hat{e} \in \mathcal{E}$, where the average number of reachable nodes maximally decreases by its removable. Let $G_e = (\mathcal{V}, \mathcal{E} \backslash \{e\})$ be the network obtained by removing a link e, then we can define the following objective function to be minimized with respect to $e \in \mathcal{E}$.

---

[1] $\mathcal{B}_k(v; G)$ can still be defined when $|\mathcal{R}(v; G)| < k$. In this case its cardinality is the number of reachable nodes from v.

$$F_0(G_e) = \frac{1}{|\mathcal{V}|} \sum_{v \in \mathcal{V}} |\mathcal{R}(v; G_e)|. \tag{1}$$

In this paper, by using the estimation based on the bottom-k sketches, we focus on the following objective function.

$$F(G_e) = \frac{1}{|\mathcal{V}|} \sum_{v \in \mathcal{V}} H(v; G_e). \tag{2}$$

Here we can straightforwardly obtain a baseline method which re-calculates the bottom-k sketches, $\mathcal{B}_k(v; G_e)$, with respect to G_e for all nodes from scratch. However, the baseline method generally requires a large amount of computation for large-scale networks. In order to overcome this problem, by borrowing and extending the basic ideas of pruning techniques proposed in [19, 20], below we propose new acceleration techniques called marginal-link updating (MLU) and redundant-link skipping (RLS).

The MLU technique locally updates the bottom-k sketches of some nodes when removing links incident to a node with in-degree 0 or out-degree 0 in the network G. First, let $v \in \mathcal{V}$ be a node with in-degree 0, i.e., $|\mathcal{Q}_1(v; G)| = 0$. Here, note that by removal of a link from v to its child node w, say $e = (v, w)$ and $w \in \mathcal{R}_1(v; G)$, only the bottom-$k$ sketch of node v changes, i.e., $\mathcal{B}_k(u; G) = \mathcal{B}_k(u; G_e)$ for any node $u \neq v$. Namely, we can locally update the bottom-k sketch of node v by computing $\mathcal{B}_k(v; G_e)$ as the k smallest elements in $\cup_{w \in \mathcal{R}_1(v; G_e)} \mathcal{B}_k(w; G)$. On the other hand, let $v \in \mathcal{V}$ be a node with out-degree 0, i.e., $|\mathcal{R}_1(v; G)| = 0$, then the bottom-$k$ sketch of node v is $\mathcal{B}_k(v; G) = \{r(v)\}$. Here, note that by removal of a link to v from its parent node u, say $e = (u, v)$ and $u \in \mathcal{Q}_1(v; G)$, only the bottom-$k$ sketch of node x such that $x \in \mathcal{Q}(v; G) \backslash \mathcal{Q}(v; G_e)$ possibly changes. Thus, by computing the bottom-$(k + 1)$ sketch of any node $u \in \mathcal{V}$, i.e., $\mathcal{B}_{k+1}(u; G)$, in advance, we can locally update the bottom-k sketch of such a node x just by replacing $r(v) \in \mathcal{B}_k(x; G)$ with $b_{k+1}(x; G)$ unless $|\mathcal{B}_{k+1}(x; G)| \leq k$, by reversely following links from v as performed in the bottom-k sketches calculation.

The RLS technique selects each link $e \in \mathcal{E}$ for which $F(G_e) = F(G)$ and prune some subset of such links. Here, we say that a link $e = (v, w) \in \mathcal{E}$ is a *skippable link* if there exist some node $x \in \mathcal{V}$ such that $f = (v, x) \in \mathcal{E}$ and $g = (x, w) \in \mathcal{E}$, i.e., $x \in \mathcal{R}_1(v; G) \cap \mathcal{Q}_1(w; G)$, which means $|\mathcal{R}_1(v; G) \cap \mathcal{Q}_1(w; G)| \geq 1$. Namely, we can skip evaluating $F(G_e)$ for the purpose of solving our problem due to $F(G_e) = F(G)$. Moreover, we say that a link $e = (v, w) \in \mathcal{E}$ is a *prunable link* if $|\mathcal{R}_1(v; G) \cap \mathcal{Q}_1(w; G)| \geq 2$. Namely, we can prune such a link e for our problem by setting $G \leftarrow G_e$ due to $F((G_e)_f) = F(G_f)$ for any link $f \in \mathcal{E}$. For each node $v \in \mathcal{V}$, let $\mathcal{S}(v)$ and $\mathcal{P}(v)$ be sets of skippable and prunable links from v. We can calculate $\mathcal{S}(v)$ and $\mathcal{P}(v)$ as follows: for each child node $w \in \mathcal{R}_1(v; G)$, we first initialize $c(v, w; G) \leftarrow 0$, $\mathcal{S}(v) \leftarrow \emptyset$ and $\mathcal{P}(v) \leftarrow \emptyset$. Then, for each node $x \in \mathcal{R}_1(v; G)$, we repeatedly set $c(v, w; G) \leftarrow c(v, w; G) + 1$ and $\mathcal{S}(v) \leftarrow \mathcal{S}(v) \cup \{(v, w)\}$ if $\{w\} \in \mathcal{R}_1(x; G)$, and set $\mathcal{P}(v) \leftarrow \mathcal{P}(v) \cup \{(v, w)\}$ and $G \leftarrow G_{(v,w)}$ if $c(v, w; G) \geq 2$.

In our proposed method, the RLS technique is applied before the MLU techniques, because it is naturally conceivable that the RLS technique decreases the number of links in our network G. Clearly we can individually incorporate these techniques into the baseline method. Hereafter, we refer to the proposed method without the MLU technique as the RLS method, and the proposed method without the RLS technique as the MLU method. Since it is difficult to analytically examine the effectiveness of these techniques, we empirically evaluate the computational efficiency of these three methods in comparison to the baseline method.

4   Experiments

We evaluated the effectiveness of the proposed method using two benchmark and two synthetic networks.

4.1   Datasets

We employed two benchmark networks obtained from SNAP (Stanford Network Analysis Project)[2]. The first one is a high-energy physics citation network from the e-print arXiv[3], which covers all the citations within a dataset of $34,546$ papers (nodes) with $421,578$ citations (links). If a paper u cites paper v, the network contains a directed link from u to v. The second one is a sequence of snapshots of the Gnutella peer-to-peer file sharing network from August 2002[4]. There are total of 9 snapshots of Gnutella network collected in August 2002. The network consists of $36,682$ nodes and $88,328$ directed links, where nodes represent hosts in the Gnutella network topology and links represent connections between the Gnutella hosts.

In addition, we utilized two synthetic networks with a DAG (Directed Acyclic Graph) property, which were generated by using the DCNN and DBA methods described in [19,20], respectively. For the sake of convenience, we briefly revisit these methods. First, we explain the DCNN method. Here, we say that a pair of nodes $\{v, w\}$ is a potential pair if they are not directly connected, but have at least one common adjacent node. Then, we can summarize the DCNN method as an algorithm which repeats the following steps from a single node and an empty set of links while $|\mathcal{V}| < L$: (1) With probability $1 - \delta$, create a new node $u \in \mathcal{V}$, select a node $v \in \mathcal{V}$ at random, and add a link (u, v) or (v, u) arbitrary; (2) With probability δ, select a potential pair $\{v, w\}$ at random, and add a link (v, w) or (w, v) to be a DAG direction. Clearly, we can easily see that the DCNN method generates a DAG. In our experiments, we set $L = 35,000$ and $\delta = 0.1$ to make sure that the numbers of nodes and links can be roughly equal to $|\mathcal{V}| = 35,000$ and $|\mathcal{E}| = 350,000$, which can be a network with an intermediate size between the above two benchmark networks.

---

[2] https://snap.stanford.edu/.
[3] https://snap.stanford.edu/data/cit-HepPh.html.
[4] https://snap.stanford.edu/data/p2p-Gnutella30.html.

Table 1. Basic statistics of networks.

No.	Name	$|\mathcal{V}|$	$|\mathcal{E}|$	$|\mathcal{I}_0|$	$|\mathcal{O}_0|$	$|\mathcal{S}|$	$|\mathcal{P}|$
1	CIT	34,546	421,578	2,393	6,320	302,248	176,224
2	DBA	35,000	351,317	5,984	5,999	85,815	24,690
3	DCN	35,000	350,807	4,996	8,868	289,398	175,211
4	P2P	36,682	88,328	26,960	229	1,502	29

Next, we explain the DBA method. Here, we say that a node is selected by preferential attachment if its selection probability is proportional to the number of adjacent nodes. Then, we can summarize the DBA method as an algorithm which repeats the following steps from a DAG having M links generated by the DCNN method while $|\mathcal{V}| < L$: (1) With probability $1 - \delta$, create a new node $u \in \mathcal{V}$, select a node $v \in \mathcal{V}$ by preferential attachment, and create a link (u, v) or (v, u) arbitrary. (2) With probability δ, select a node $v \in \mathcal{V}$ at random, select another node $w \in \mathcal{V}$ by preferential attachment, and create a link (v, w) or (w, v) to be a DAG direction. Again, we can easily see that the DBA method generates a DAG. In our experiments, we also set $L = 35,000$, $\delta = 0.1$, and $M = 100$.

In what follows, we refer to these two benchmark networks of citation and pear-to-pear and those generated by the DCNN an DBA methods as CIT, P2P, DCN and DBA networks. Table 1 summarizes the basic statistics of these networks, consisting of the numbers of nodes and links, $|\mathcal{V}|$ and $|\mathcal{E}|$, the numbers of in-degree 0 and out-degree 0 nodes, $|\mathcal{I}_0|$ and $|\mathcal{O}_0|$, and the numbers of skippable and prunable links, $|\mathcal{S}|$ and $|\mathcal{P}|$, where each network is also identified by its data number as shown in Table 1. From this table, we can conjecture that the RLS technique works well for the CIT and DCN networks, while the MNU technique for the P2P networks. Here note that the numbers of skippable and prunable links appearing in the networks generated by the DCNN method inevitably become larger than those generated by the DBA method because the DCNN method has a link creation mechanism between potential pairs.

4.2   Computational Efficiency

First, we evaluated the efficiency of the proposed method which calculates $F(G_e)$ for each link $e \in \mathcal{E}$. We compared the computation time of the baseline (BL), RLS, MLU, and proposed (PM) methods by performing five trials. Here, we used the same random value $r(v)$ assignment for each trial so that the bottom-k sketches of all the nodes are the same for any method, i.e., it is guaranteed that each method can produce the same result. Figure 1 shows the computation times of each method for five trials plotted by dots and the average values over these trials plotted by different markers as indicated in the figure, where we set $k = 64$ for calculation of the bottom-k sketches of all the nodes according to [13]. Figure 1(a) compares the actual processing times of these methods, where our programs implemented in C were executed on a computer system equipped with two Xeon X5690 3.47 GHz

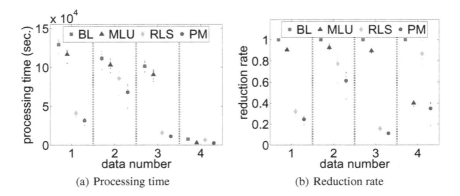

Fig. 1. Computation time comparison.

CPUs and a 192 GB main memory with a single thread within the memory capacity. Figure 1(b) compares the reduction rates of computation times for these methods from the BL method.

From Fig. 1(a), we can see that the computation times were improved largely for the CIT and CNN networks, modestly for the P2P network, and much less modestly for the DBA network, although the computation time of the BL method for the P2P network was smaller than those for the other networks. More specifically, as expected, we consider that the RLS technique worked quite well especially for the CIT and DCN networks, due to large numbers of skippable and prunable links in these networks as shown in Table 1. On the other hand, although the MLU technique is not so remarkably effective, we consider that this technique can steadily improve the reduction rate of computation times especially for the P2P network as shown in Fig. 1(b). In short, we can conjecture that the proposed method combining both the RLS and MLU techniques is more reliable than the other three methods in terms of computation time because it produced the best performance for all of the four networks. Reduction of computation time depends on network structures, but overall we can say that use of both techniques can increase the computational efficiency by about an order of magnitude. These results demonstrate the effectiveness of the proposed method.

4.3   Comparison with Conventional Centralities

As noted earlier, by solving our critical link detection problem that the average number of reachable nodes maximally decreases by link removable, we can obtain the value $F(G_e)$ for each link $e \in \mathcal{E}$ as a measure to evaluate the criticalness of the link e. Thus, we evaluated whether or not our measure $F(G_e)$ can actually provide a novel concept in comparison with some measures derived from conventional centralities.

As conventional centralities, we examined the degree centrality, the betweenness centrality, the PageRank centrality, and the eigenvalue centrality for network G, and straightforwardly extended these centralities so as to evaluate the criticalness of a given link e. The first measure $DGC(e)$ derived from degree centrality for a given link $e = (u, v) \in \mathcal{E}$ is defined as

$$DGC(e) = |\mathcal{Q}_1(u; G)| \times |\mathcal{R}_1(v; G)|, \tag{3}$$

where recall that $\mathcal{Q}_1(u; G)$ and $\mathcal{R}_1(v; G)$ are in-degree of node u and out-degree of node v, respectively. Namely, the first measure $DGC(e)$ highly evaluate a link from a node with high in-degree to a node with high out-degree. Next, for a given link $e = (u, v) \in \mathcal{E}$, the second measure $BWC(e)$ derived from the betweenness centrality is defined as

$$BWC(e) = btw(u) \times btw(v), \quad btw(v) = \sum_{w \in \mathcal{V}} \sum_{x \in \mathcal{V}} \frac{nsp_{w,x}^G(v)}{nsp_{w,x}^G}, \tag{4}$$

where $btw(v)$ stands for the betweenness of a node $v \in \mathcal{V}$ and $nsp_{w,x}^G$ is the total number of the shortest paths between node w and node x in G and $nsp_{w,x}^G(v)$ is the number of the shortest paths between node w and node x in G that passes through node v. Namely, the second measure $BWC(e)$ highly evaluate a link between nodes with high betweenness centrality scores. Going on next, for a given link $e = (u, v) \in \mathcal{E}$, the third measure $PRK(e)$ derived from the PageRank centrality is defined as

$$PRK(e) = prk(u) \times prk(v), \tag{5}$$

where $prk(v)$ stands for the PageRank score of a node $v \in \mathcal{V}$, which is provided by applying the PageRank algorithm with random jump factor 0.15 [4]. Namely, the third measure $PRK(e)$ highly evaluates a link between nodes with high PageRank scores. Finally, for a given link $e = (u, v) \in \mathcal{E}$, the fourth measure $EIG(e)$ derived from eigenvector centrality is defined as

$$EIG(e) = auth(u) \times hub(v), \tag{6}$$

where $auth(u)$ and $hub(v)$ respectively stand for the authority and the hub scores of nodes $u, v \in \mathcal{V}$, which is provided by applying the HITS algorithm [7]. Namely, the fourth measure $EIG(e)$ highly evaluates a link from a node with a high hub score to a node with high authority score.

First, we examined how each of highly ranked links by these standard centralities, $i.e.$, DGC, BWC, PRK, and EIC, can decrease the average number of reachable nodes by its removal. Here, we measured the performance by the following gain $F(G) - F(G_e)$

$$F(G) - F(G_e) = \frac{1}{|\mathcal{V}|} \sum_{v \in \mathcal{V}} H(v; G) - \frac{1}{|\mathcal{V}|} \sum_{v \in \mathcal{V}} H(v; G_e). \tag{7}$$

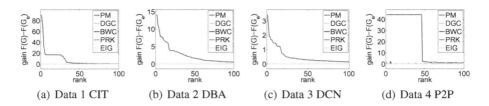

(a) Data 1 CIT   (b) Data 2 DBA   (c) Data 3 DCN   (d) Data 4 P2P

Fig. 2. Gain comparison of extracted links. (Color figure online)

Figure 2 shows our experimental results, where the vertical and horizontal axes stand for the rank until top-100 and the gain, respectively, and Fig. 2(a), (b), (c) and (d) correspond to the CIT, DBA, DCN, and P2P networks. We can see that our proposed method (denoted by PM) detected critical links having substantial amount of gains, where the curves for the top-100 links are somewhat different from each other depending on the network datasets. On the other hand, the gains for all of the top-100 links by those measures defined by the standard centralities were almost zeros for any dataset. In addition, we should emphasize that the gain curves shown in Fig. 2 could uncover some characteristics of these networks, i.e., the DCN network was relatively robust to a single link removal, and so on.

Next, we examined the similarity between our ranking based on $F(G_e)$ and the other ranking, $i.e.$, the DGC, BWC, PRK, and EIC ranking. Here, we measured the similarity between the top j links for our ranking method, denoted as a set $\mathcal{A}_j$, and those for the other ranking method, denoted as a set $\mathcal{A}'_j$, by the precision $Prec(j)$ defined by

$$Prec(j) = \frac{|\mathcal{A}_j \cap \mathcal{A}'_j|}{j}. \tag{8}$$

Figure 3 shows our experimental results, where the vertical and horizontal axes stand for normalized rank for all links and the precision, respectively, and Fig. 3(a), (b), (c) and (d) are those of the CIT, DBA, DCN, and P2P networks. We can see that the results were quite similar to each other regardless of any pair of the centrality measures and the networks although a slightly better precision curve swelling in the upper left corner was obtained especially for the DBA network.

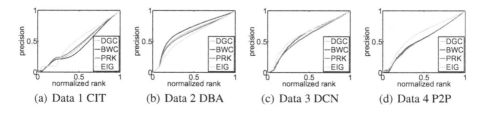

(a) Data 1 CIT   (b) Data 2 DBA   (c) Data 3 DCN   (d) Data 4 P2P

Fig. 3. Precision comparison of extracted links. (Color figure online)

Table 2. Ranks by conventional centralities to top-3 links by proposed measure.

Rank	CIT				DBA			
	DGC	BWC	PRK	EIG	DGC	BWC	PRK	EIG
1	233,638	296,250	161,504	405,661	188,686	309,435	251,492	218,665
2	189,768	424	26,626	402,823	301,228	335,210	309,249	289,465
3	67,333	106,045	2,158	141,559	276,667	320,951	288,338	217,370
Rank	DCN				P2P			
	DGC	BWC	PRK	EIG	DGC	BWC	PRK	EIG
1	284,501	269,488	252,472	168,049	42,915	88,077	82,028	81,162
2	302,753	296,190	293,284	312,283	44,008	79,543	75,022	75,128
3	311,040	311,040	338,940	186,370	74,710	74,947	85,107	65,655

Finally, Table 2 shows ranks by conventional centralities to top-3 links obtained by proposed measure. These results indicate that the rankings by these convention measures were substantially different from those of our measure. Namely, these experimental results suggest that our measure $F(G_e)$ could actually provide a novel concept in comparison with some measures derived from conventional centralities.

5   Conclusion

In this paper we have proposed a novel computational method that can detect critical links quite efficiently for a large network. The problem is reduced to finding a link that reduces the network performance substantially with respect to its removal. Such a link is considered critical in maintaining the good performance. There are many problems that can be mapped to this critical link detection problem, e.g. contamination minimization be it physical or virtual, evacuation trouble minimization, road maintenance prioritization, etc.

Network performance varies with specific problem, but in general it is represented by the reachability performance, i.e. how many nodes are reachable from a node in the network on the average. This brings in computational issue because reachability must be estimated for all the nodes for a particular link removal and to find critical links this has to be repeated for all the links. The number of links is generally an order of magnitude larger than the number of nodes even for a sparse network that is encountered in actual practice. We used bottom-k sketch algorithm as a basis to count reachable nodes, which only uses k-samples to estimate the reachable nodes from a selected node. It has a sound theoretical background and been shown quite efficient and accurate for a k which is far smaller than the number of nodes in the network. Our contribution is to introduce two new acceleration techniques to further reduce the bottom-k sketch computation by clever local update and redundant computations pruning. The first technique MLU (marginal-link updating) locally updates the bottom-k sketches of some nodes when removing links incident to a node with in-degree 0 or out-degree 0 in the network. The second technique RLS (redundant-link skipping) selects each link that does not affect the performance with respect to its removal and prune some subset of such links.

We have tested the performance of the proposed method using four networks with about 35,000 nodes and 90,000 to 420,000 links. Two were taken from Stanford Network Analysis Project and the other two were artificially generated to control the network structure. We verified that the acceleration techniques indeed work for all these four networks of different characteristics and can reduce the computation time by about an order of magnitude. MLU works better for networks with many nodes with in-degree 0 or out-degree 0. RLS works better for networks with many prunable links. We have further evaluated how other measures based on conventional centralities work in estimating our performance measure, i.e. the average number of reachable nodes by a link removal. We have composed four measures, each based on degree centrality, betweenness centrality, PageRank centrality and authority/hub centrality, respectively. All of these measures are not able to detect critical links that were detected by the proposed method. Links detected by these measures do not show any performance degradation, i.e. not critical at all. We can conclude that no existing measure can find critical links.

There are many things to do. Reachability computation is a basic operation and is a basis for many applications. We continue to explore techniques to further reduce computation time. Our immediate future plan is to apply our method to a real world application and show that it can solve a difficult problem efficiently, e.g. identifying important hot spots in transportation network or evacuation network.

Acknowledgments. This material is based upon work supported by the Air Force Office of Scientific Research, Asian Office of Aerospace Research and Development (AOARD) under award number FA2386-16-1-4032, and JSPS Grant-in-Aid for Scientific Research (C) (No. 26330261).

References

1. Albert, R., Jeong, H., Barabási, A.L.: Error and attack tolerance of complex networks. Nature **406**, 378–382 (2000)
2. Boldi, P., Vigna, S.: In-core computation of geometric centralities with hyperball: a hunderd billion nodes and beyond. In: Proceedings of the 2013 IEEE 13th International Conference on Data Mining Workshops (ICDMW 2013), pp. 621–628 (2013)
3. Borgs, C., Brautbar, M., Chayes, J., Lucier, B.: Maximizing social influence in nearly optimal time. In: Proceedings of the 25th Annual ACM-SIAM Symposium on Discrete Algorithms (SODA 2014), pp. 946–957 (2014)
4. Brin, S., Page, L.: The anatomy of a large-scale hypertextual web search engine. Comput. Netw. ISDN Syst. **30**, 107–117 (1998)
5. Broder, A., Kumar, R., Maghoul, F., Raghavan, P., Rajagopalan, S., Stata, R., Tomkins, A., Wiener, J.: Graph structure in the web. In: Proceedings of the 9th International World Wide Web Conference, pp. 309–320 (2000)

6. Callaway, D.S., Newman, M.E.J., Strogatz, S.H., Watts, D.J.: Network robustness and fragility: percolation on random graphs. Phys. Rev. Lett. **85**, 5468–5471 (2000)
7. Chakrabarti, S., Dom, B., Kumar, R., Raghavan, P., Rajagopalan, S., Tomkins, A., Gibson, D., Kleinberg, J.: Mining the web's link structure. IEEE Comput. **32**, 60–67 (1999)
8. Chen, W., Wang, Y., Yang, S.: Efficient influence maximization in social networks. In: Proceedings of the 15th ACM SIGKDD International Conference on Knowledge Discovery and Data Mining (KDD 2009), pp. 199–208 (2009)
9. Chen, W., Yuan, Y., Zhang, L.: Scalable influence maximization in social networks under the linear threshold model. In: Proceedings of the 10th IEEE International Conference on Data Mining (ICDM 2010), pp. 88–97 (2010)
10. Chierichetti, F., Epasto, A., Kumar, R., Lattanzi, S., Mirrokni, V.: Efficient algorithms for public-private social networks. In: Proceedings of the 21st ACM SIGKDD International Conference on Knowledge Discovery and Data Mining (KDD 2015), pp. 139–148 (2015)
11. Cohen, E.: Size-estimation framework with applications to transitive closure and reachability. J. Comput. Syst. Sci. **55**, 441–453 (1997)
12. Cohen, E.: All-distances sketches, revisited: HIP estimators for massive graphs analysis. In: Proceedings of the 33rd ACM SIGMOD-SIGACT-SIGART Symposium on Principles of Database Systems, pp. 88–99 (2015)
13. Cohen, E., Delling, D., Pajor, T., Werneck, R.F.: Sketch-based influence maximization and computation: scaling up with guarantees. In: Proceedings of the 23rd ACM International Conference on Conference on Information and Knowledge Management, pp. 629–638 (2014)
14. Freeman, L.: Centrality in social networks: conceptual clarification. Soc. Netw. **1**, 215–239 (1979)
15. Goyal, A., Bonchi, F., Lakshmanan, L.: A data-based approach to social influence maximization. Proc. VLDB Endowment **5**(1), 73–84 (2011)
16. Kempe, D., Kleinberg, J., Tardos, E.: Maximizing the spread of influence through a social network. In: Proceedings of the 9th ACM SIGKDD International Conference on Knowledge Discovery and Data Mining (KDD 2003), pp. 137–146 (2003)
17. Kimura, M., Saito, K., Motoda, H.: Blocking links to minimize contamination spread in a social network. ACM Trans. Knowl. Disc. Data **3**, 9:1–9:23 (2009)
18. Kimura, M., Saito, K., Nakano, R.: Extracting influential nodes for information diffusion on a social network. In: Proceedings of the 22nd AAAI Conference on Artificial Intelligence (AAAI 2007), pp. 1371–1376 (2007)
19. Kimura, M., Saito, K., Ohara, K., Motoda, H.: Efficient analysis of node influence based on sir model over huge complex networks. In: Proceedings of the 2014 International Conference on Data Science and Advanced Analytics (DSAA 2014), pp. 216–222 (2014)
20. Kimura, M., Saito, K., Ohara, K., Motoda, H.: Speeding-up node influence computation for huge social networks. Int. J. Data Sci. Anal. **1**, 1–14 (2016)
21. Leskovec, J., Krause, A., Guestrin, C., Faloutsos, C., VanBriesen, J., Glance, N.: Cost-effective outbreak detection in networks. In: Proceedings of the 13th ACM SIGKDD International Conference on Knowledge Discovery and Data Mining (KDD 2007), pp. 420–429 (2007)
22. Newman, M.E.J., Forrest, S., Balthrop, J.: Email networks and the spread of computer viruses. Phys. Rev. E **66**, 035101 (2002)
23. Newman, M.: The structure and function of complex networks. SIAM Rev. **45**, 167–256 (2003)

24. Ohara, K., Saito, K., Kimura, M., Motoda, H.: Resampling-based framework for estimating node centrality of large social network. In: Džeroski, S., Panov, P., Kocev, D., Todorovski, L. (eds.) DS 2014. LNCS, vol. 8777, pp. 228–239. Springer, Heidelberg (2014)
25. Saito, K., Kimura, M., Ohara, K., Motoda, H.: Super mediator - a new centrality measure of node importance for information diffusion over social network. Inf. Sci. **329**, 985–1000 (2016)
26. Song, G., Zhou, X., Wang, Y., Xie, K.: Influence maximization on large-scale mobile social network: a divide-and-conquer method. IEEE Trans. Parallel Distrib. Syst. **26**, 1379–1392 (2015)
27. Watts, D.: A simple model of global cascades on random networks. Proc. Natl. Acad. Sci. USA. **99**, 5766–5771 (2002)
28. Zhou, C., Zhang, P., Zang, W., Guo, L.: On the upper bounds of spread for greedy algorithms in social network influence maximization. IEEE Trans. Knowl. Data Eng. **27**, 2770–2783 (2015)

Using Canonical Correlation Analysis for Parallelized Attribute Reduction

Ping Li, Mengting Xu, Jianyang Wu, and Lin Shang[✉]

State Key Laboratory for Novel Software Technology, Department of Computer
Science and Technology, Nanjing University, Nanjing 210046, China
lipingnju@gmail.com, xumtpark@gmail.com, wu.wujy@163.com,
shanglin@nju.edu.cn

Abstract. Attribute reduction in rough sets theory has been widely
used in classification. Classical attribute reduction algorithm only considers correlation between condition attributes and decision attributes,
which ignores the relationship among condition attributes themselves.
Moreover, when faced with large-scale data, running time of classical
attribute reduction algorithm has been increasing. Aiming to solve these
two problems, a parallelized reduction algorithm called $P - CCARough$
$Reduction$ is proposed in this paper. The algorithm employs canonical correlation analysis named $CCAFusion$ and parallelized attribute
reduction algorithm named $P - RoughReduction$. $CCAFusion$ divides
the original set of attributes into two subsets randomly. Then the correlations of these two subsets of features are analyzed. After that, the
attributes are fused into one collection according to the derived correlations. $P - RoughReduction$ algorithm is based on a distributed framework $MapReduce$ which parallelizes the classical attribute reduction
algorithm according to the attribute importance in rough sets theory.
It is shown that $P - CCARoughReduction$ algorithm through experiments on 50000 samples not only performs well on time, the classification
accuracy has also been significantly improved.

Keywords: Canonical correlation analysis · Rough sets theory ·
Attribute reduction · $MapReduce$

1 Introduction

Rough sets theory [1] has been widely used in processing classification [2,3],
attribute reduction [4–6] and other basic issues of data mining [7–9]. Attribute
reduction algorithm has been the focus of attention of rough sets. Lots of effort
has been put into it by many researchers and amounts of contributions have
been achieved. Varieties of attribute reduction algorithms based on rough sets
have been proposed. They are based on attribute significance [10–12], discernibility matrix [13–15], entropy [16–18], and so on. However, with the Internet
development, data increased in an explosive speed. Therefore how to achieve the

© Springer International Publishing Switzerland 2016
R. Booth and M.-L. Zhang (Eds.): PRICAI 2016, LNAI 9810, pp. 433–445, 2016.
DOI: 10.1007/978-3-319-42911-3_36

attribute reduction for massive data in a fast speed has become an intractable issue. This is the focus of the paper.

Hadoop [26] developed by Apache organization is an open source distributed computing platform for big data. It implements a distributed file system - *HDFS* [27] and a distributed computing framework - *MapReduce* [28–30]. The main idea of *MapReduce* is conquer and divide. Our work is running on hadoop.

Inspired by *MapReduce* that the data is generally cut crosswise into slices, we can also cut data lengthwise into slices. Meanwhile, we find that classical attribute reduction algorithms based rough sets only care about the relationship between condition attributes and decision attributes, neglecting the relationships among condition attributes themselves [1,3]. If we can make analysis of the relationship among condition attributes, it would be a new way to reduce attributes. Canonical correlation analysis algorithm (*CCA*) [19,21] is a kind of feature fusion technology [24,25]. It will fuse the features in multi-views into one single view according to the degree of correlation within these features. Moreover, the information of multi-views will not be lost as much as possible due to the reduced attributes. After performed by *CCA*, the feature dimensionality will be reduced. According to the idea, we can divide the features of the original data set into many sub-property collections. The original set is one single view and these sub-property collections are multi-views. Then *CCA* can be adopted to fuse these features in the multi-views to reduce dimensionality.

In this paper, we put forward a parallelized attribute reduction algorithm based on canonical correlation analysis and *MapReduce* named $P - CCARoughReduction$. The algorithm combines the two methods we will propose: *CCAFusion* algorithm and $P-RoughReduction$ algorithm. *CCAFusion* algorithm will reduce the redundant attributes among the condition attributes themselves and the process cuts the original attribute collection equally into two sub-attribute collections. Each sub-attribute collection will be regarded as a sub-view of the original view. Then we will employ *CCA* to analyze the correlations of these different sub-views. After the obtained correlations of features in sub-views are obtained, these sub-views will be fused into one view again by the correlations in descending order. After the process, we have reduced those features with high correlation. In $P - RoughReduction$ algorithm, we will define a simplified metric for computing attribute importance. Then we will parallelize the classical attribute reduction algorithm based on attribute importance on *MapReduce*.

Experimental results show that $P - CCARoughReduction$ runs two times faster than $P-RoughReduction$. The classification accuracy improves more than 10 % compared with $P - RoughReduction$. Thus $P - CCARoughReduction$ not only performs well in time cost, but also significantly improves the classification performance.

The structure of this paper is as follows. In the first part, we will introduce related background knowledge. In the second part, we will demonstrate $P - CCARoughReduction$ algorithm we proposed in detail. In the third part, we will carry out experiments and make analysis on the experimental results. In the end, we will make a conclusion about the work we have done and point out the future work.

2 Preliminaries

In this section, some background knowledge about rough sets theory, *MapReduce* and canonical correlation analysis algorithm will be introduced.

2.1 Rough Sets Theory

Definition 1. A decision table is defined as $S = \langle U, C, D, f \rangle$, where U is the domain. $A = C \cup D$ is the attribute set, among which C is the condition attribute set and D is the decision attribute set, at the same time, $C \cap D = \phi$. $V = \bigcup_{\alpha \in U} V_\alpha$ is the set of attribute values. $f : U \times (C \cup D) \rightarrow V$ is a function, which gives attribute a its value.

Definition 2. The significance of attribute a, $a \in C$, is defined by

$$Sig_a = r_{P \cup \{a\}}(Q) - r_{P - \{a\}}(Q), \tag{1}$$

$$r_P(Q) = |POS_P(Q)| / |U|, \tag{2}$$

where $P, Q \subseteq (C \cup D)$, U is the domain, $r_P(Q)$ is the dependency of attribute P related to Q and $POS_P(Q)$ is the positive region.

2.2 MapReduce: Distributed Computing Framework

MapReduce based on *Hadoop* platform is a programming framework for developers to handle large-scaled data. The programming process is as follows. Firstly, define *Map* method and *Reduce* method. *Map* process will read data in the form of $< key, value >$ pair and generate intermediate results in the same form. Then *Reduce* process will merge the intermediate results which have the same key values and produce the final results in the form of $< key, value >$ pair.

2.3 Canonical Correlation Analysis

Canonical Correlation Analysis algorithm proposed by H. Hotelling [22] has played an important role in multivariate analysis field and has been widely used in feature fusion. It is a statistical analysis method to analyze the relevance of two groups of variables. The algorithm aims to find out two groups of linear combination to maximize the correlation between the two groups of variables.

Definition 3. Suppose X and Y are two feature vectors, $X \in R^n, Y \in R^m, X = \{x_1, x_2, ..., x_n\}, Y = \{y_1, y_2, ..., y_m\}$. The purpose of CCA is to seek out two groups of linear combination a^T and b^T which maximize the relevance of random variables $u = a^T X$ and $v = b^T Y$. The correlation of u and v is defined as follows.

$$Corr(u, v) = Corr(a^T X, b^T Y) = \frac{a^T \sum^{12} b}{\sqrt{a^T \sum^{11} a}\sqrt{b^T \sum^{22} b}}, \tag{3}$$

where $\sum^{11}$ is the convariance matrix of X, $\sum^{22}$ is the convariance matrix of Y, and $\sum^{12}$ is the convariance matrix of X and Y.

We maximize $Corr(u, v)$, which becomes an optimization problem. We make the denominator in the formula above be constant 1. Then formula 3 can be transferred into formula 4.

$$\begin{aligned} Maximize : a^T \sum^{12} b \\ Subject\ to : a^T \sum^{11} a = 1, b^T \sum^{22} b = 1 \end{aligned} \tag{4}$$

Then we will construct Lagrange's equation to solve the optimization problem.

$$L = a^T \sum^{12} b - \frac{\lambda}{2}(a^T \sum^{11} a - 1) - \frac{\theta}{2}(b^T \sum^{22} b - 1), \tag{5}$$

Next, it is computed by derivation and the derivation is equated zero, like the following equations show:

$$\begin{aligned} \frac{\partial L}{\partial a} = \sum^{12} b - \lambda \sum^{11} a = 0. \\ \frac{\partial L}{\partial b} = \sum^{21} a - \theta \sum^{22} b = 0. \end{aligned} \tag{6}$$

The first equation multiplies a^T on the left, and the second equation multiplies b^T on the left, according to the subject condition of Equation (4), we can get

$$\lambda = \theta = a^T \sum^{12} b \tag{7}$$

So the problem of maximizing $Corr(u, v)$ can be converted to maximizing the value of λ. The above solution is based on the condition of $\sum_{11}, \sum_{22}$ is invertible. However, if the two matrices are irreversible, the resolution can be conducted by [23].

3 $P - CCARoughReduction$: One Parallelized Attribute Reduction Algorithm Based on CCA

To obtain the attribute reduction of large-scale data quickly, we put forward an algorithm $P - CCARoughReduction$ to perform the parallel attribute reduction algorithm based on rough sets and CCA. In the algorithm, CCA is employed to fuse condition attributes to achieve the preliminary attribute reduction.

Aiming to implement $P - CCARoughReduction$ algorithm, we propose two other algorithms: $CCAFusion$ and $P - RoughReduction$. Before we introduce $P - CCARoughReduction$ algorithm, we will illustrate $CCAFusion$ and $P - RoughReduction$ algorithms first.

3.1 $CCAFusion$ Algorithm

The basic idea of feature fusion is that we will fuse those high relevant features through computing the relevance among them. In this paper, we will divide the original feature set into two subsets. Among them, each subset represents a description of the original set and it shows the effect of multi-views. Then we can follow the classical canonical correlation analysis algorithm to merge the generated multi-views into one single view to achieve the preliminary effect of attribute reduction. Figure 1 shows the framework of $CCAFusion$ and Algorithm 1 shows the details of $CCAFusion$ algorithm.

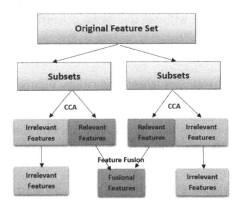

Fig. 1. Framework of $CCAFusion$

$CCAFusion$ can reduce the dimensionality while keep the information of original features as much as possible. Among them, the Fusional Features stage means we will throw away the most relevant features based on different granularity. In the experiment stage, we set the fusional granularity are 5 to 45, respectively. Adjusting the granularity of fusion which represents the number of the most relevant features, so we can control the information loss caused by reduced dimension.

3.2 $P - RoughReduction$ Algorithm

When faced with massive data, parallelization is the method to solve the efficiency problem. Therefore, we put forward a parallelized algorithm in this

---

Algorithm 1. CCAFusion

Require: Origin data set A
Ensure: Data set Res after $CCAFusion$ algorithm
 1: Divide the data set A into two subsets A_1 and A_2 , denoted as X and Y respectively. The dimension both of them is W.
 2: For X and Y, calculate the variance matrix $\sum^{11}$ and $\sum^{22}$.
 3: For X and Y, calculate the covariance matrix $\sum^{12}$ and $\sum^{21}$.
 4: Calculate the value of matrix P, $P = \sum_{11}^{-1} \sum^{12} \sum_{22}^{-1} \sum^{21}$ and matrix Q, $Q = \sum_{22}^{-1} \sum^{21} \sum_{11}^{-1} \sum^{12}$.
 5: Calculate the feature vector p of P and q of Q.
 6: Update $X = p{'}X$, $Y = q{'}Y$. Return to (2) step, and iterate M times.
 7: If the fusion granularity is n, $n < W$. Replace the top n dimension of X and Y, then get the fusion attributes Res.

---

paper to improve the running efficiency. $P - RoughReduction$ algorithm utilizes $MapReduce$ to parallelize classical attribute reduction algorithm based on attribute importance. We put forward a new definition of attribute importance based on Definition 2. We suppose $P = a$, then the definition will be transformed as follows in Formula 6. We will use the forward heuristic algorithm to achieve the final reduction.

$$Sig_a = |POS_{\{a\}}(Q)|/|U|, \qquad (8)$$

where $Q \subseteq D$.

Now, we will present our parallelized algorithm. Figure 2 shows the data flow of attribute reduction on $MapReduce$. The master node will cut the original data into data blocks and submit blocks of data and reduction task to each Mapper node. Each Mapper node conducts attribute reduction task independently and obtain intermediate reduction results. These results will be shuffled and delivered to Reducer nodes. Then Reducer nodes integrate all the results into the final reduction result.

The time complexity of our algorithm is $O(\frac{1}{K} \times |C|^2 \times |U|)$. $|C|$ is the number of attributes, $|U|$ is the number of training samples and $|K|$ is the number of cluster nodes. When the number of cluster nodes is large enough, namely $K \geqslant |C|$, the time complexity will be decreased to $O(|C| \times |U|)$. However, in the practical application, it is difficult to make it equal between cluster scale and attribute dimensionality. In this paper, we keep the cluster scale and attribute dimensionality at the same order of magnitude, which basically make the time complexity of our parallelized algorithm be $O(|C| \times |U|)$.

3.3 $P - CCARoughReduction$ Algorithm

Although the parallelized algorithm based on $MapReduce$ decreases the runtime significantly, the reduction number is unsatisfactory. This is because rough sets reduction algorithm only cares about the relationship between condition attributes and decision attributes, which neglects the relevance among condition attributes themselves. Thus in this paper, we take the issue into consideration.

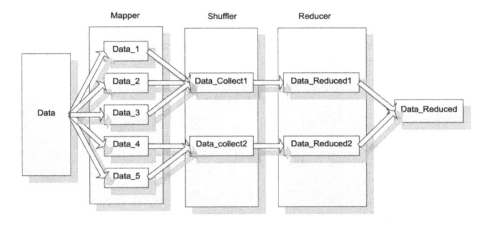

Fig. 2. Data flow of attribute reduction on *MapReduce*

We use *CCAFusion* algorithm to merge those high relevant features firstly. Then we reduce the attributes with classical reduction algorithm. In that case, we can achieve better reduction results.

We combine *CCAFusion* and *P − RoughReduction* algorithms in Sects. 3.1 and 3.2 to put forward a parallelized attribute reduction algorithm based on canonical correlation analysis and *MapReduce* which is called *P − CCARoughReduction*. Figure 3 shows the framework of the algorithm.

We make the results of *CCAFusion* algorithm as the input of *P − RoughReduction* algorithm. Then *P − RoughReduction* output the final results. *P − CCARoughReduction* makes use of the high-efficiency pretreatment of *CCAFusion* and the well-performed operational efficiency of *P − RoughReduction* to obtain a good reduction result at a high rate of speed.

4   Experiments

In this section, we will show how our proposed method works by the experiments and analysis.

4.1   Experiments Setup

In the experiments, we use two data sets: *Y earP redictionM SD_transformed* abbreviated *Y ear_T* and *mushrooms* as representative examples. *Y ear PredictionM SD_transformed* is generated from original data set *Y ear PredictionM SD* in *UCI* repository [31]. *Y earP redictionM SD* data set has 90 condition attributes and has no missing data. The number of samples are up to 50000. Thus it is a non-sparse and large data set for experiments. However, *Y earP redictionM SD* is used for regression to predict the year of samples, which has more than one hundred decision attribute values. Therefore, we firstly discretize the decision attribute values. We flag those values from 1900 to 1909 as 0, from 1910 to 1919 as 1 and the rest can be done at the same manner. After that s so that it can be used for classification. Data set $mushrooms$ also comes from UCI repository. The detail of data sets can be seen from Table 1.

Table 1. Data Sets

Data Set	SampleNum	AtributeNum	ClassNum
$Year\_T$	50000	90	12
$mushrooms$	8124	112	2

The experimental configuration is as follows.

- For $CCAFusion$ algorithm, the experiments are conducted on $matlab2011b$ and $windows7$.
- For $P - RoughReduction$ algorithm, the experimental environment is $Hadoop-1.0.4$ and the operation system is $Ubuntu12.04$. The cluster contains 20 nodes totally. Each node is an independent PC machine. These machines are connected with each other.

4.2 Experimental Results and Analysis

We conduct experiments to compare $P - CCARoughReduction$ algorithm and $P - RoughReduction$ algorithm on the computational time and classification performance.

4.2.1 Computational Time

Three groups of comparative experiments are carried out to measure the time efficiency of $P-CCARoughReduction$. We compare it with $P-RoughReduction$ algorithm based on the runtime when the sample number is 5000, 10000 and 20000. The results can be seen from Fig. 4.

- In Fig. 4., x-axis represents the number of cluster nodes. We set up the number equal to 1, 2, 4, 8 and 16. Y-axis represents algorithms' runtime. The result of $P - RoughReduction$ is flagged in red while that of $P - CCARoughReduction$ is flagged in green. From the figure, $P-CCARoughReduction$ runs faster than $P - RoughReduction$. This is can be explained by the process of $CCAFusion$ for the preliminary reduction.
- Note that with the increase of the number of cluster nodes, the running time does not gain linear growth. Especially, when the number becomes 16, the running time increases slowly. In fact, due to the burden of network and the limitation of experimental machines' performance, we cannot improve the running time by adding cluster nodes unlimitedly.

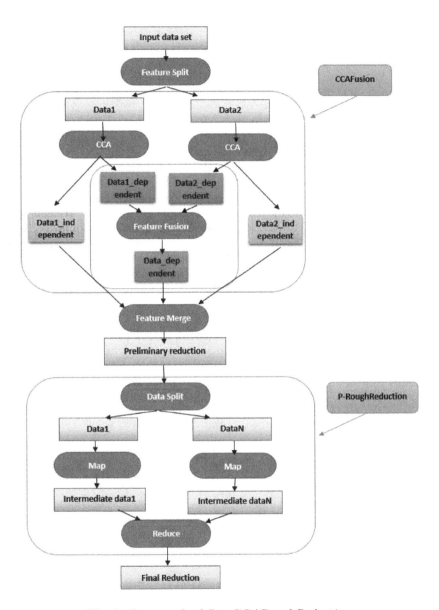

Fig. 3. Framework of $P - CCARoughReduction$

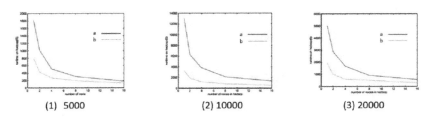

| (1) 5000 | (2) 10000 | (3) 20000 |

Fig. 4. Comparison between $P-RoughReduction$(a) and $P-CCARoughReduction$(b) (Color figure online)

4.2.2   Classification Performance

To test the classification performance of $P - CCARoughReduction$ algorithm, we conducted several experiments on data sets $Year\_T$ and $mushrooms$. We have compared the results at different fusional granularity. Fusional granularity represents the number of the most relevant features in two views. We can adjust the number in CCA algorithm. $CCA5$ means the fusional granularity is five, that is five most relevant features in two views will be fused. $origin$ means the data set skips $CCAFusion$ step. The standard SVM [32,33] is employed as the classifier. Table 2 shows the classification results on data set $Year\_T$. The number of training samples is 5000 and the testing number is 1000.

Table 2. Classification results on $Year\_T$

DataSet	Original AttriNum	Reduced AttriNum	Accuracy
origin	90	23	73.8
CCA5	85	19	85.5
CCA10	80	14	83.3
CCA15	75	15	87.2
CCA20	70	11	**91.0**
CCA25	65	12	89.1
CCA30	60	10	87.2
CCA35	55	18	90.1
CCA40	50	11	86.1
CCA45	45	8	88.2

Table 3 shows the classification results on data set $mushrooms$. The number of training samples is still 5000 and the testing number is still 1000.

– From Tables 2 and 3., we can see that for the original data set, the classification accuracy of SVM is 73.8 % and 81.5 %. After handling by $P - CCARoughReduction$, the highest classification accuracy of $Year\_T$ is 91.0 % and the highest classification accuracy of $mushrooms$ is 92.9 %.

Table 3. Classification results on *mushrooms*

DataSet	Original AttriNum	Reduced AttriNum	Accuracy
origin	112	37	81.5
CCA5	106	12	86.3
CCA10	98	22	88.1
CCA15	90	19	**92.9**
CCA20	82	25	92.8
CCA25	74	20	91.1
CCA30	66	17	92.2
CCA35	56	15	87.6

It improves more than 15 % and 10 % relatively. With the granularity of fusion increasing, the classification accuracy first rises then falls. It is because there are many redundant features that may lead to wrong decisions. $P - CCARoughReduction$ algorithm firstly reduce the redundant features and improve the classification accuracy. However, when the number of reduced attributes is beyond a certain limit, it will cause information loss which brings the decline of classification accuracy.

- We can also obtain that the reduction results are satisfying. The dimensionality of attributes by $P - CCARoughReduction$ is lower than that by $P - RoughReduction$. For the first data set, the lowest dimensionality is 8, while it is 12 in the second data set. Thus $P - CCARoughReduction$ algorithm is not only beneficial to decrease dimensionality but also improves the classification accuracy.

5   Conclusion and Future Work

Attribute reduction is one of the core contents of rough sets theory. In order to solve the low efficiency of classical attribute reduction algorithm facing large-scale data, we come up with $P - CCARoughReduction$, a parallelized attribute reduction algorithm based on canonical correlation analysis and $MapReduce$. The classical attribute reduction algorithms based rough sets only care about the relationship between condition attributes and decision attributes, neglecting the relationships among condition attributes themselves, so we firstly adopts $CCAFusion$ algorithm to pretreat data sets and reduct the most correlation features based on different granularity to obtain preliminary reduction results. Based on the results, we make use of $P - RoughReduction$ to concurrently reduce the attributes. Through the analysis of experiments, we find that the running efficiency of $P - CCARoughReduction$ algorithm is absolutely satisfying, which greatly improves reduction effectiveness and classification accuracy.

However, when it comes across data of large sparse degree, the classification accuracy will be low in most cases. In the future, we will strengthen the ability to process sparse data and transform data format. Apart from that, we will try more divisions and fuse more views to get better results.

Acknowledgements. We would like to acknowledge the support for this work from the National Natural Science Foundation of China (Grant No. 61403200).

References

1. Pawlak, Z.: Rough sets. Int. J. Comput. Inform. Sci. **11**(5), 341–356 (1982)
2. Pawlak, Z.: Rough classification. Int. J. Man Mach. Stud. **20**(5), 469–483 (1984)
3. Tsai, Y., Cheng, C., Chang, J.: Entropy-based fuzzy rough classification approach for extracting classification rules. Expert Syst. Appl. **31**(2), 436–443 (2006)
4. Qian, Y., Liang, J., Pedrycz, W., et al.: Positive approximation: an accelerator for attribute reduction in rough set theory. Artif. Intell. **174**(9), 597–618 (2010)
5. Jensen, R., Shen, Q.: Fuzzy crough attribute reduction with application to web categorization. Fuzzy Sets Syst. **141**(3), 469–485 (2004)
6. Yao, Y., Zhao, Y.: Attribute reduction in decision-theoretic rough set models. Inform. Sci. **178**(17), 3356–3373 (2008)
7. Han, J., Kamber, M., Pei, J.: Data Mining: Concepts and Techniques: Concepts and Techniques. Elsevier, Waltham (2011)
8. WBerkhin, P.: A survey of clustering data mining techniques. In: Kogan, J., Nicholas, C., Teboulle, M. (eds.) Grouping Multidimensional Data, pp. 25–71. Springer, Heidelberg (2006)
9. Hand, D., Mannila, H., Smyth, P.: Principles of data Mining. MIT Press, Cambridge (2001)
10. Deng, D., Yan, D., Wang, J.: Parallel reducts based on attribute significance. In: Yu, J., Greco, S., Lingras, P., Wang, G., Skowron, A. (eds.) RSKT 2010. LNCS, vol. 6401, pp. 336–343. Springer, Heidelberg (2010)
11. Wu, J., Zou, H.: Attribute reduction algorithm based on importance of attribute value. Comput. Appl. Softw. **27**(2), 255–257 (2010)
12. Kong, L., Mai, J., Mei, S., Fan, Y.: An improved attribute importance degree algorithm based on rough set. In: Proceedings of IEEE International Conference on Progress in Informatics and Computing (PIC), pp. 122–126 (2010)
13. Skowron, A., Rauszer, C.: The discernibility matrices and functions in information systems. In: Słowiński, R. (ed.) Intelligent Decision Support, pp. 331–362. Springer, Netherlands (1992)
14. Qian, J., Miao, D., Zhang, Z., et al.: Hybrid approaches to attribute reduction based on indiscernibility and discernibility relation. Int. J. Approximate Reasoning **52**(2), 212–230 (2011)
15. Xu, Z., Zhang, C., Zhang, S., Song, W., Yang, B.: Efficient attribute reduction based on discernibility matrix. In: Yao, J.T., Lingras, P., Wu, W.-Z., Szczuka, M.S., Cercone, N.J., Ślęzak, D. (eds.) RSKT 2007. LNCS (LNAI), vol. 4481, pp. 13–21. Springer, Heidelberg (2007)
16. Wu, S., Gou, P.: Attribute reduction algorithm on rough set and information entropy and its application. Comput. Eng. **37**(7), 56–61 (2011)
17. Guoyin, W., Hong, Y., Dachun, Y.: Decision table reduction based on conditional information entropy. J. Comput. **25**(7), 759–766 (2002)
18. Tsai, Y., Cheng, C., Chang, J.: Entropy-based fuzzy rough classification approach for extracting classification rules. Expert Syst. Appl. **31**(2), 436–443 (2006)

19. Thompson, B.: Canonical correlation analysis. In: Encyclopedia of Statistics in Behavioral Science (2005)
20. Hsing, T., Eubank, R.: Canonical correlation analysis. In: Theoretical Foundations of Functional Data Analysis, with an Introduction to Linear Operators, pp. 265–304 (2015)
21. Bin, G., Gao, X., Yan, Z., et al.: An online multi-channel SSVEP-based brain computer interface using a canonical correlation analysis method. J. Neural Eng. **6**(4), 046002 (2009)
22. Hotelling, H.: Relations between two sets of variates. Biometrika **28**, 321–377 (1936)
23. Borga, M.: Canonical correlation: a tutorial. On line tutorial (2001). http://people.imt.liu.se/magnus/cca. 4:5
24. Yang, J., Zhang, D., et al.: Feature fusion: parallel strategy vs. serial strategy. Pattern Recogn. **36**(6), 1369–1381 (2003)
25. Sun, Q., Zeng, S., Liu, Y., et al.: A new method of feature fusion and its application in image recognition. Pattern Recogn. **38**(12), 2437–2448 (2005)
26. White, T.: Hadoop: The Definitive Guide. O'Reilly Media Inc., Sebastopol (2012)
27. Shvachko, K., Kuang, H., Radia, S., et al.: The hadoop distributed file system. In: IEEE 26th Symposium on Mass Storage Systems and Technologies (MSST), pp. 1–10. IEEE (2010)
28. Dean, J., Ghemawat, S.: MapReduce: simplified data processing on large clusters. Commun. ACM **51**(1), 107–113 (2008)
29. Dean, J., Ghemawat, S.: MapReduce: a flexible data processing tool. Commun. ACM **53**(1), 72–77 (2010)
30. Condie, T., Conway, N., Alvaro, P., et al.: MapReduce Online. NSDI **10**(4), 20 (2010)
31. UCI Machine Learning Repository. http://archive.ics.uci.edu/ml/
32. Suykens, J., Vandewalle, J.: Least squares support vector machine classifiers. Neural Proc. Lett. **9**(3), 293–300 (1999)
33. Wang, K., Liang, C., Liu, J., et al.: Prediction of piRNAs using transposon interaction and a support vector machine. BMC Bioinform. **15**(1), 419 (2014)

A Novel Isolation-Based Outlier Detection Method

Yanhui Shen, Huawen Liu, Yanxia Wang[(✉)], Zhongyu Chen,
and Guanghua Sun

Department of Computer Science, Zhejiang Normal University,
Jinhua 321004, China
wangyx@zjnu.cn

Abstract. Outlier detection is one of the most important tasks in data analysis. It refers to the process of recognizing unusual characteristics which may provide useful insights in helping us to understand the behaviors of data. In the paper, an isolation-based outlier detection method, called Entropy-based Greedy Isolation Tree (EG*i*Tree), is proposed. Unlike other tree-like detection methods, our method exploits a half-baked isolation tree, which is constructed via three entropy-based heuristics, to identify outliers. Specifically, the heuristics are used to guide the selection process of attribute and its split value when constructing the tree. Thus, the outlierness score of each data point is estimated based on the total partition cost of the isolation node in the tree, as well as the path length and complexity of partition. Experiment results on public real-world datasets show that our approach outperforms distanced-based, density-based, subspace-based as well as state-of-the-art isolation-based approaches.

Keywords: Outlier detection · Data mining · Isolation · Isolation tree · Entropy

1 Introduction

An outlier (a.k.a., abnormality, discordant, deviant, or anomaly), refers to a data point which is significantly different from the remaining data [1]. Outlier detection aims to find data objects/points that are considerably dissimilar, exceptional and inconsistent with respect to the majority data [2]. It is a fundamental and important task in data mining and pattern recognition. The reason is that outliers can provide critical and actionable information in various application domains, such as intrusion detection, financial markets, medical diagnosis, credit card fraud and earth science. During the past decades, a lot of outlier detection algorithms have been proposed. According to the techniques adopted, they can be broadly divided into five major categories, i.e., statistical methods, distance-based methods, density-based methods, subspace-based methods and isolation-based methods [1–3].

The assumption of the statistical detection methods is that data points are generated from a distribution or probability model, while outlier points are not conform to the underlying model. Generally, the statistical detection methods can be classified into the parametric and the non-parametric methods [3]. The typical examples include the

© Springer International Publishing Switzerland 2016
R. Booth and M.-L. Zhang (Eds.): PRICAI 2016, LNAI 9810, pp. 446–456, 2016.
DOI: 10.1007/978-3-319-42911-3_37

Gaussian model-based methods [4–7], the histograms-based methods [8–11] and the kernel density function methods [12–14]. Note that the statistical methods are efficient, but the assumption is too strong.

In the distance-based detection methods, the data points, which have larger distances to their neighbors than others, are considered as outliers. Knorr and Ng [15] firstly proposed the notion of distance-based outliers, called DB (k,λ)-Outlier. To address the choice problem of parameter values, they further developed another distance-based method called DB (p,D)-Outlier by virtue of local neighborhood [16, 17]. However, the parameters including λ and D are difficult to be specified. Ramaswamy et al. [18] took the distance of the point to the k-th nearest neighbor as its outlierness score. Zhang et al. [19] measured the outlierness of points by using local distance-based outlier factor (LDOF). To improve the computational efficiency, several techniques, such as sampling-based [20] and indexing-based [21, 22], were also adopted in outlier detection. Even so, the efficiency of the distance-based methods is still a problem, especial for high-dimensional data.

The density-based methods adopt the neighborhood density of a point to represent its outlierness degree, and an outlier has lower density in comparison to normal points. Representative examples include Local Outlier Factor (LOF) [22], Connectivity-based Outlier Factor (COF) [23], Local Correlation Integral (LOCI) [24], Resolution-based Outlier Factor (ROF) [25] and INFLuenced Outlier-ness (INFLO) [26]. In LOF [22], the points with the highest LOF values, which quantify the sparseness of a point to its local neighborhood, are considered as outliers. Unlike others, LOCI uses multi-granularity deviation factor (MDEF) to measure the relative deviation of density of a point's neighborhood [24]. In ROF [25], an outlier is defined as a point which is inconsistent with the majority of the data at different resolutions.

The subspace-based methods assume that normal points are close to each other, while abnormal points are often far from the normal ones in some subspaces. For example, the high contrast subspaces (HiCS) [27] firstly takes the difference between marginal and conditional probable density functions of a subspace as its interest, and then ranks outliers within the most promising subspace. The subspace outlier degree (SOD) algorithm [28] explore axis-parallel subspace spanned by neighbors for each point, and then determine how much the point deviates from the neighbors in the subspace. Although the subspace-based detection methods can identify more interesting outliers in high-dimensional space, searching meaningful subspace is time-consuming.

Unlike the density-based and the distance-based approaches, the isolation-based detection methods consider both scattered and clustered properties when identifying anomalies. Isolation Forest (iForest) [29] is such a detection algorithm. It detects outliers by isolation instance, without relying on any distance or density measure. Given a dataset, iForest builds an ensemble of isolation tree (iTrees). Within the tree, anomalies are those instances which have short average path lengths. iForest has high efficiency if subsampling strategies are used. However, it may fail when there are local clustered anomalies. To address this problem, SCiForest [30] exploits hyper-planes to construct models, and uses heuristic strategies to select split points deterministically. It should be pointed out that both iForest and SCiForest adopt random forest to build models. Thus, they have random nature and their performance heavily relies on the randomness.

In this paper, we propose a novel outlier detection method called EGiTree (Entropy-based Greedy Isolation Tree). Our detection method constructs a half-baked isolation tree instead of a whole isolation tree. Thus it has higher efficiency in outlier detection. In addition, to eliminate the randomness property in *i*Forest, three entropy-based heuristic strategies are exploited to greedily search a proper attribute and its split value of the selected attribute when constructing the tree. In our method, a point is considered as an outlier if its corresponding isolation node has higher partition cost, longer path length and partition complexity. The simulation experiments conducted on public datasets show that the proposed detection method can not only eliminate the randomness of the isolation-based methods, but also be more efficient and effective than the state-of-the-art outlier detection techniques.

The rest of the paper is organized as follows: Sect. 2 introduces three heuristic selection strategies. The construction of isolation tree using the heuristic selection strategies is developed in Sect. 3. Section 4 empirically evaluates the proposed method with the popular ones on public datasets. The final conclusion is provided in Sect. 5.

2   Heuristic Selection Strategies

During the construction process of a tree in random forest, how to select a proper attribute and a split value of attribute is an important issue, which directly determines the performance of the construction tree. Generally speaking, the projection of outlier points maybe belong to following two cases presented in Fig. 1:

Case 1: The projection of an outlier point in one or more attributes is easy to be divided with the projection of other normal points.

Case 2: The projection of an outlier point in any attribute is difficult to be divided with the projection of other normal points.

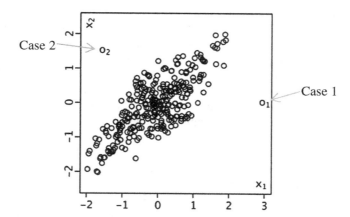

Fig. 1. O_1 and O_2 are two outlier points, where O_1 belongs to case 1 and O_2 belongs to case 2.

For the first case, there is a bigger gap between the minimum projection value and the maximum projection value on the selected attribute, because of the existence of the outlier point. In this case, the projection value of the outlier point does not mix with that of the normal points in the dataset. The outlier point can be divided from the dataset easily. Seeing from the angle of distribution of the projection value, the amount of disorder is lower, or we can say, the entropy of the dataset in the attribute is smaller. Whereas, the projection value of the outlier point mixes with that of the normal points, then there is not a distinct gap and the projection values of the dataset distribute in a nearly uniform fashion and a partition is difficult to carry out. To sum up, the first Heuristic is give as follows:

Heuristic 1: *The attribute with the smallest entropy is selected firstly.*

If the entropy is small enough, then it is mean that there are outlier points like case 1, or there are some clusters with smaller within-class scatter. The biggest gap can be found in the attribute, and the middle point of the gap is selected as a split value. Base on the above analysis, the second heuristic strategy is given as follows:

Heuristic 2: *If the entropy of the dataset in one attribute is smaller than a given threshold, then the biggest gap can be found in the attribute and the partition is generated on the middle point of the gap.*

But when the smallest entropy of the dataset is still bigger than a threshold, the partition becomes more difficult. Under the circumstances, the best strategies may be that a trade-off partition. That is to say, the dataset is divided into two near equal subset and the meaningful partition of using Heuristic 2 is expected in the next partition on the subset. The third heuristic strategy can be described as:

Heuristic 3: *If the entropy of the dataset in one attribute is bigger than a given threshold, then the mean of the projection value of the dataset in the attribute be selected as the split value and two balanced sub-trees are constructed.*

The above three heuristics will be repetitiously used in our method to construct a half-baked isolation tree.

3 EG*i*Tree

As mentioned above, anomalous points often exhibit one of two cases. Thus, we can exploit this property to construct isolation tree. Given a dataset, the proposed method, EG*i*Tree, only builds a half-baked isolation tree (*i*Tree), rather than an ensemble of trees, to detect outliers. To eliminate the random property of selecting the attribute and the split value when constructing the isolation tree, the entropies of the projections in each attribute are computed firstly. The attribute with the smallest entropy is selected to partition data with respect to the heuristic selection strategies in Sect. 2. A half-baked isolation tree, which is a proper binary tree and each node in the tree has exactly zero or two children nodes, is then generated by a recursive partition manner.

Specifically, EG*i*Tree for outlier detection consists of the following steps:

Step 1: Estimating the entropy for each attribute

Let $X = \{x^1, \ldots, x^n\}$ be a dataset consisting of n data points with d-variate attribute, and X^r be the projection of X with respect to the r-th attribute. A nonparametric of the entropy of the r-th attribute is estimated as follows [31]:

$$h_r(X^r, b) = -\sum_{i=1}^{b} p_i ln(p_i) \tag{1}$$

where b is the number of bins, and p_i is the probability of a projection value fallen within the i-th bin. Since the attributes often have different scale, for the sake of simplicity, we fix the number of bins for each attribute when estimating its entropy. After obtaining the entropies, we select the attribute with the smallest entropy to partition the data points according to *Heuristic 1*. If the points conform to a uniform distribution after projected to an attribute, the entropy of the attribute is

$$h_{max}(b) = -b \cdot (\frac{1}{b} \cdot ln(\frac{1}{b})) = -ln(\frac{1}{b}) \tag{2}$$

Step 2: Determining a proper split value

After an attribute is selected, its proper split value should be determined when partitioning the data points. To measure whether a split value is optimal or not, a similar coefficient α'_r is introduced as follows:

$$\alpha'_r = \frac{h_r(X^r, b)}{h_{max}(b)} \tag{3}$$

According to Eq. (3), $\alpha'_r \in (0, 1]$, and the larger the value of α'_r, the probability distribution of the data points projected on the r-th attribute is more close to the uniform distribution.

Based on this efficient, we can determine a split value for the selection attribute. Given a threshold α, there are two situations for the coefficient α'_r. The first one is $\alpha'_r < \alpha$. In this case, we can use Heuristic 2 to determine the split position for the r-th attribute. Firstly, a maximum-gap is searched by the pigeonhole principle firstly. Then the midpoint of the gap is selected as the split point to generate a partition. Thus, the partition cost of the r-th attribute is

$$cost(X) = \frac{max(X^r) - min(X^r) - maxgap(X^r)}{max(X^r) - min(X^r)} = 1 - \frac{maxgap(X^r)}{max(X^r) - min(X^r)} \tag{4}$$

where $max(X^r)$ and $min(X^r)$ are the maximum and minimum value of the projection of the r-th attribute respectively. $maxgap(X^r)$ is the range of the maximum gap.

Another case is $\alpha'_r \geq \alpha$, where Heuristic 3 can be used to determine the split position of the r-th attribute. Specifically, the mean of the projection of X on the r-th attribute is selected as the split position, yielding two balanced subsets of data points. Thus, the partition cost is considered to be the most expensive, i.e., $cost(X) = 1$.

Summarily, the partition cost of the r-th attribute is:

$$\cos t(X) = \begin{cases} 1 - \dfrac{\max gap(X^r)}{\max(X^r) - \min(X^r)}, & \alpha'_r < \alpha \\ \\ 1, & \alpha'_r \geq \alpha \end{cases} \tag{5}$$

Step 3: Constructing a half-baked isolation tree

Based on the above analysis, we can construct a half-baked isolation tree in a recursive manner. Specifically, step 1 and step 2 repeat alternatively, until the number of isolated nodes in the tree reaches the specified requirements. The implement details are given in Algorithm 1.

Algorithm 1: EGiTree(X, b, α, k)

  Input: A dataset $X \in \mathbb{R}^{n \times d}$, a number of bins, a threshold α, a number of isolated points k
  Output: The top k isolated points
  $O = \emptyset, S = \emptyset$;
  $root \leftarrow$ Node $\{index \leftarrow \{1, 2, \cdots, n\}, depth \leftarrow 1, cost \leftarrow 0\}$
  $S = S \cup \{root\}$
  $depth_{max} \leftarrow n$
  while S not empty do
      let s be the first element of S
      if $s.depth > depth_{max}$ then   break;
      if $|s.index| \leq 1$ then
          $O = O \cup \{index \leftarrow s.index, total\ cost \leftarrow s.cost\}$;
          if $|O| \geq k$ then $depth_{max} \leftarrow s.depth$;
      else
          $X' \leftarrow X_{i \in s.index}$;
          $j \leftarrow argmin_{j \in \{1, \cdots, d\}, h_j > 0}\ h_j(X'^j, b)$;
          $z \leftarrow X'^j$;
          $s.cost \leftarrow s.cost + cost(X')$
          $L \leftarrow \{i \in s.index\ and\ X_i^j \leq m\}$
          $R \leftarrow \{i \in s.index\ and\ X_i^j > m\}$
          $s.left \leftarrow$ Node $\{index \leftarrow L, depth \leftarrow s.depth + 1, cost \leftarrow s.cost\}$
          $s.right \leftarrow$ Node $\{index \leftarrow R, depth \leftarrow s.depth + 1, cost \leftarrow s.cost\}$
          $S = S \cup \{s.left\} \cup \{s.right\}$;
      end if
  end while
  Sort elements of O by total cost in ascending order
  return the first k elements of O

In Algorithm 1, a half-baked isolation tree with k isolated points, rather than a whole isolation tree, is generated. Besides, the total cost of each isolated point is estimated. Let S_i is an isolated node and it corresponds to the i-th data point in the dataset. A_i is a node set of traversing the isolation tree form S_i to the root node. The total partition cost of S_i is:

$$cost_t(s_i) = \sum_{a \in A_i} cost(a) \tag{6}$$

Step 4: Estimating outlier score

In the existing isolation-based methods, the path length is often used to measure the susceptible degree of data point to be isolated. However, it has not taken the data points with the equal path length into account. Thus, it cannot discriminate those data points effectively in the isolation tree. In our method, the outlier degree of data point is defined as follows:

$$outlierscore(X_i) = \frac{1}{cost_t(s_i)} \tag{7}$$

where X_i is the i-th data point in X and s_i is an external node corresponding to X_i in the isolation tree. From the definition, we know that the outlier score of X_i takes not only the information of path length but also the partition cost into account. It is noticeable that the smaller the total partition cost of a point, the larger the outlier-ness score.

In our method, there parameters are involved. They are the number of bins b, the entropy threshold α and the number of isolated points k. For the number of bins, if it is too small, the entropy estimated by Eq. (1) may be inaccuracy. However a larger value of b may yield more bias of entropy. Empirically, b is often assigned as 10. α is used to select heuristic strategies to estimate the partition cost for each attribute. If α is too small, the probability of using Heuristic 3 becomes very high. On the contrary, if α is too large, the probability of using Heuristic 2 becomes very high and the outlier belongs to case 2 will be missed. In our experiments, EGiTree achieved better performance as $\alpha = 0.9$. The parameter k controls the depth of the half-baked isolation tree and the running speed of our method. The value of k ranges from 1 to n, which is the size of dataset. Our experimental results show that it is appropriate when k equals to $\lfloor 0.5n \rfloor$.

4 Empirical Evaluation

4.1 Datasets

For validate the effectiveness of the proposed method, we carried out experiments on seven real-world datasets from the UCI Machine Learning Repository [32]. They are *Thyroid* (ANN version), *Wisconsin Diagnostic Breast Cancer* (WDBC), *Ionosphere*, *Pendigits*, *Statlog Shuttle* and *Human Activity*. Their descriptions are briefly shown in Table 1.

Table 1. Brief descriptions of the experimental datasets, where n is the number of points, d is the number of dimensions, and *ratio* is the ratio of outliers.

Dataset	n	d	*ratio*
Ann-Thyroid	3372	6	2.47 %
WDBC	569	30	37.26 %
Ionosphere	352	32	35.90 %
Pendigits	453	16	19.87 %
Statlog Shuttle	14500	9	20.84 %
Human Activity	1860	561	13.55 %

For the *Ann-Thyroid* dataset, there are three classes. Since the first class only covers 93 data points, it has been taken as outliers, while for the WDBC dataset, 212 data points are labeled with '*Malignant*' and the rests are labeled with '*Benign*', therefore we chose '*Malignant*' instances as outliers. For the *Ionosphere* dataset, 32 '*bad*' data were considered as outliers. In *Pendigits*, ten classes cover even number of points. Thus, 90 over 453 data instances from class 1–9 evenly are considered as outliers. Since the second class have 3022 data, approximately 20 %, in the *Statlog Shuttle* dataset, we selected these data points as outliers. For the *Human Activity* dataset, we chose the data instances from the class 4, 5 and 6 as normal points, and samples from the class 1, 2 and 3 as outliers.

4.2   Experimental Results

In the experiments, the proposed method, EGiTree, was compared to iForest [29], LOF [22], LDOF [19] and SOD [28]. We took Aare Under receiver operating characteristic Curve (AUC) as the evaluation criterion. During the whole experiments, the parameters in each comparison algorithms were assigned as default values recommended by the authors. For example, $\psi = 256$ and $t = 100$ in iForest, while $l = 20$, $k = 20$ and $\alpha = 0.8$ in SOD. Similar situation to LOF and LDOF, where $k = 20$. The experimental results are shown in Table 2, where the largest AUC values are highlighted in bold. Since iForest is an ensemble detection algorithm, it was run in 10 times and the average performance was obtained.

Table 2. The performance of AUC of the comparison detection algorithms

Dataset	EGiTree $k = \lfloor 0.5n \rfloor$	EGiTree $k = n$	iForest	LOF	LDOF	SOD
Ann-Thyroid	**0.9896**	**0.9896**	0.9715 ± 0.0022	0.8056	0.7395	0.9105
WDBC	**0.8655**	**0.8801**	0.7677 ± 0.0196	0.5296	0.5352	0.8085
Ionosphere	**0.9303**	**0.9575**	0.8451 ± 0.0045	0.8605	0.8837	0.8445
Pendigits	0.9563	0.9563	**0.9647 ± 0.0029**	0.5107	0.5517	**0.9634**
Statlog Shuttle	0.8169	**0.8538**	0.8197 ± 0.0126	0.5032	0.5186	0.5734
Human Activity	**0.9933**	**0.9933**	0.9707 ± 0.0024	0.4747	0.4891	0.9884

Form the experimental results, one can observe that our method outperformed the comparison algorithms in all datasets except *Pendigits*, where *i*Forest and SOD had better performance. LOF and LDOF achieved relatively lower AUC on five over six datasets. However, their performance on *Ionosphere* was better than SOD and *i*Forest, but still worse than EG*i*Tree. The underlying reason is that the *Ionosphere* dataset includes more outliers, which scatter in the whole space. Additionally, both *i*Forest and SOD adopt subsampling and weighting techniques to detect outliers, respectively.

For EG*i*Tree, the parameter k may bring impacts on the performance of EG*i*Tree. If k is large enough, most of outlier points can be traversed and labeled as isolation nodes in the half-baked isolation tree. The experimental results in Table 2 show that EG*i*Tree with large k had better performance. As $k = n$, EG*i*Tree had the largest AUC values on five datasets, and close to the best one on *Pendigits*. When $k = \lfloor 0.5n \rfloor$, EG*i*Tree also achieved the best performance on four datasets. On the rest datasets, the performance of EG*i*Tree was close to the best ones.

Table 3 presents the runtime of the comparison algorithms. According to the results in the table, we know that EG*i*Tree had relatively better efficient in both $k = n$ and $k = \lfloor 0.5n \rfloor$ cases. For example, the proposed detection method required less time to outlier detection. On *Human Activity*, LOF was the fastest. However, it had relatively poor AUC performance in comparing to others. For *i*Forest, it needs much more time in the experiments, due to the fact that it is an ensemble one. The experimental results demonstrated this fact.

Table 3. The runtime of the comparison detection algorithms (in seconds)

Dataset	EG*i*Tree $k = \lfloor 0.5n \rfloor$	EG*i*Tree $k = n$	*i*Forest	LOF	LDOF	SOD
Ann-Thyroid	**0.04**	**0.05**	1.06	1.10	0.70	5.96
WDBC	**0.03**	**0.04**	0.34	0.07	0.10	0.24
Ionosphere	**0.00**	**0.01**	0.22	0.04	0.03	0.06
Pendigits	**0.01**	**0.01**	0.20	0.03	0.03	**0.01**
Statlog Shuttle	**0.22**	**0.23**	4.74	7.07	5.76	82.19
Human Activity	4.13	5.09	45.58	**3.68**	3.88	4.66

5   Conclusion

In this paper, a novel isolation-based outlier detection method called EG*i*Tree is proposed. The fundamental idea of EG*i*Tree follows that of *i*Forest, that is, anomalies are data points that are more susceptible to isolation. Unlike *i*Forest, which uses the techniques of sampling and ensemble to construct model, the proposed method exploits heuristic strategies to eliminate the uncertain and random effects of random forest. Specifically, we adopt three strategies to select proper attributes and their split values when constructing the isolation tree. Due to the nature of our heuristics preferring to separate the most isolated data points from the rest, the suspicious data points can be revealed in the early stage in constructing isolation tree. Besides, to exactly measure the susceptible degree of data points, a new outlier score involving partition cost and the path length is introduced. Thus EG*i*Tree is capable of identifying the most suspicious data points approximately without estimating outlier scores for all data points. The experimental results show that out approach outperforms distanced-based, density-based, subspace-based as well as the state-of-the-art isolation-based approaches in most cases.

Acknowledgement. This work is partially supported by the NSF of China (No. 61272468, 61272007, 61572443), the NSF of Zhejiang Province (No. LY14F020012, No. LQ13F020026), and the Opening Fund of Top Key Discipline of Computer Software and Theory in Zhejiang Provincial Colleges at Zhejiang Normal University (No. ZSDZZZZXK27).

References

1. Aggarwal, C.C.: Outlier Analysis. Springer, New York (2013)
2. Han, J., Kamber, M.: Data Mining: Concepts and Techniques. Morgan Kaufman Publishers, San Francisco (2000)
3. Zhang, J.: Advancements of outlier detection: a survey. ICST Trans. Scalable Inf. Syst. **13** (1), 1–26 (2013)
4. Barnett, V., Lewis, T.: Outliers in Statistical Data, 3rd edn. Wiley, Chichester (1994)
5. Barnett, V.: The ordering of multivariate data (with discussion). J. Roy. Stat. Soc. Ser. A **139**, 318–354 (1976)
6. Laurikkala, J., Juhola, M., Kentala, E.: Informal identification of outliers in medical data. In: Fifth International Workshop on Intelligent Data Analysis in Medicine and Pharmacology, pp. 20–24 (2000)
7. Solberg, H.E., Lahti, A.: Detection of outliers in reference distributions: performance of horn's algorithm. Clin. Chem. **51**(12), 2326–2332 (2005)
8. Endler, D.: Intrusion detection: applying machine learning to solaris audit data. In: Proceedings of the 14th Annual Computer Security Applications Conference, p. 268 (1998)
9. Fawcett, T., Provost, F.: Activity monitoring: noticing interesting changes in behavior. In: Proceedings of the 5th ACM SIGKDD International Conference on Knowledge Discovery and Data Mining, pp. 53–62 (1999)
10. Kruegel, C., Toth, T., Kirda, E.: Service specific anomaly detection for network intrusion detection. In: Proceedings of the 2002 ACM Symposium on Applied Computing, pp. 201–208 (2002)
11. Yamanishi, K., Takeuchi, J.I.: Discovering outlier filtering rules from unlabeled data: combining a supervised learner with an unsupervised learner. In: Proceedings of the 7th ACM SIGKDD International Conference on Knowledge Discovery and Data Mining, pp. 389–394 (2001)
12. Parzen, E.: On the estimation of a probability density function and mode. Ann. Math. Stat. **33**, 1065–1076 (1962)
13. Bishop, C.: Novelty detection and neural network validation. In: Proceedings of IEEE Vision, Image and Signal Processing, vol. 141, pp. 217–222 (1994)
14. Tarassenko, L.: Novelty detection for the identification of masses in mammograms. In: Proceedings of the 4th IEEE International Conference on Artificial Neural Networks, vol. 4, pp. 442–447, Cambridge, UK (1995)
15. Knorr, E.M., Ng, R.T.: Algorithms for mining distance-based outliers in large dataset. In: Proceedings of 24th International Conference on Very Large Data Bases, VLDB 1998, pp 392–403, New York, NY (1998)

16. Knorr, E.M., Ng, R.T.: Finding intentional knowledge of distance-based outliers. In: Proceedings of 25th International Conference on Very Large Data Bases, VLDB 1999, pp. 211–222, Edinburgh, Scotland (1999)
17. Knorr, E.M., Ng, R.T., Tucakov, V.: Distance-based outliers: algorithms and applications. VLDB J. **8**(3–4), 237–253 (2000)
18. Ramaswamy, S., Rastogi, R., Shim, K.: Efficient algorithms for mining outliers from large data sets. In: Proceedings of the 2000 ACM SIGMOD International Conference on Management of Data, SIGMOD 2000, pp. 427–438. ACM Press, New York (2000)
19. Zhang, K., Hutter, M., Jin, H.: A new local distance-based outlier detection approach for scattered real-world data. In: Theeramunkong, T., Kijsirikul, B., Cercone, N., Ho, T.-B. (eds.) PAKDD 2009. LNCS, vol. 5476, pp. 813–822. Springer, Heidelberg (2009)
20. Sugiyama, M., Borgwardt, K.M.: Rapid distance-based outlier detection via sampling. In: Proceedings of the Twenty-Seventh Annual Conference on Neural Information Processing Systems, Harrahs and Harveys, Lake Tahoe, 5–10 Dec 2013
21. Berchtold, S., Keim, D.A., Kriegel, H.-P.: The X-tree: an index structure for high-dimensional data. In: Proceedings of the 22th International Conference on Very Large Data Bases, pp. 28–39 (1996)
22. Beckmann, N., Kriegel, H.-P., Schneider, R., Seeger, B.: The R*tree: an efficient and robust access method for points and rectangles. In: Proceedings of 1990 ACM SIGMOD International Conference on Management of Data, SIGMOD 1990, pp 322–331, Atlantic City, NJ (1990)
23. Tang, J., Chen, Z., Fu, A.W.-c., Cheung, D.W.: Enhancing effectiveness of outlier detections for low density patterns. In: Chen, M.-S., Yu, P.S., Liu, B. (eds.) PAKDD 2002. LNCS (LNAI), vol. 2336, pp. 535–548. Springer, Heidelberg (2002)
24. Papadimitriou, S., Kitagawa, H., Gibbons, P., Faloutsos, C.: Loci: fast outlier detection using the local correlation integral. In: Proceedings. 19th International Conference on Data Engineering, pp. 315–326 (2003)
25. Fan, H., Zaïane, O.R., Foss, A., Wu, J.: A nonparametric outlier detection for effectively discovering Top-N outliers from engineering data. In: Ng, W.-K., Kitsuregawa, M., Li, J., Chang, K. (eds.) PAKDD 2006. LNCS (LNAI), vol. 3918, pp. 557–566. Springer, Heidelberg (2006)
26. Jin, W., Tung, A.K., Han, J., Wang, W.: Ranking outliers using symmetric neighborhood relationship. In: Ng, W.-K., Kitsuregawa, M., Li, J., Chang, K. (eds.) PAKDD 2006. LNCS (LNAI), vol. 3918, pp. 577–593. Springer, Heidelberg (2006)
27. Keller, F., Muller, E., Bohm, K.: HiCS: high contrast subspaces for density-based outlier ranking. In: IEEE 28th International Conference on Data Engineering, ICDE 2012, pp: 1037–1048 (2012)
28. Kriegel, H.-P., Kröger, P., Schubert, E., Zimek, A.: Outlier detection in axis-parallel subspaces of high dimensional data. In: Theeramunkong, T., Kijsirikul, B., Cercone, N., Ho, T.-B. (eds.) PAKDD 2009. LNCS, vol. 5476, pp. 831–838. Springer, Heidelberg (2009)
29. Liu, F.T., Ting, K.M., Zhou, Z.-H.: Isolation-based anomaly detection. ACM Trans. Knowl. Discov. Data **6**(1), 3:1–3:39 (2012)
30. Liu, F.T., Ting, K.M., Zhou, Z.-H.: On detecting clustered anomalies using SCiForest. In: Balcázar, J.L., Bonchi, F., Gionis, A., Sebag, M. (eds.) ECML PKDD 2010, Part II. LNCS, vol. 6322, pp. 274–290. Springer, Heidelberg (2010)
31. Gyorfi, L., van der Meulen, E.C.: Density-free convergence properties of various estimators of entropy. Comput. Stat. Data Anal. **5**, 425–436 (1987)
32. Lichmanm M.: UCI Machine Learning Repository, Irvine, CA: University of California, School of Information and Computer Science (2013). http://archive.ics.uci.edu/ml

Setting an Effective Pricing Policy for Double Auction Marketplaces

Bing Shi[1(✉)], Yalong Huang[1], Shengwu Xiong[1], and Enrico H. Gerding[2]

[1] School of Computer Science and Technology, Wuhan University of Technology,
Wuhan, China
bingshi@whut.edu.cn
[2] School of Electronics and Computer Science, University of Southampton,
Southampton, UK

Abstract. In this paper, we analyse how double auction marketplaces set an effective pricing policy to determine the transaction prices for matched buyers and sellers. We analyse this problem by considering continuous privately known trader types. Furthermore, we consider two typical pricing policies: *equilibrium k pricing policy* and *discriminatory k pricing policy*. We firstly investigate how to determine the transaction prices to reach the maximal allocative efficiency in an isolated marketplace when the traders adopt Bayes-Nash equilibrium bidding strategies. We find that when the marketplace adopts *discriminatory k pricing policy*, the maximal allocative efficiency is reached by setting $k = 0.41$ or 0.59. We find that *equilibrium k pricing policy* provides higher allocative efficiency than *discriminatory k pricing policy*. We further discuss how different pricing policies can affect traders' Bayes-Nash equilibrium bidding strategies. Furthermore, we extend the analysis to the setting with two marketplaces competing against each other to attract traders. We find that the marketplace using *equilibrium k pricing policy* is more likely to beat the marketplace using *discriminatory k pricing policy*, where all traders converge to the marketplace using *equilibrium k pricing policy* in Bayes-Nash equilibrium. Our analysis can provide meaningful insights for designing an effective pricing policy.

Keywords: Pricing policy · Double auction · Market selection and bidding strategy · Fictitious play · Bayes-Nash equilibrium

1 Introduction

A double auction marketplace allows multiple buyers and sellers to trade goods simultaneously [3]. As a highly efficient market mechanism, it has been widely used by electronic exchanges (such as financial exchanges and Internet Ad exchanges). Now often such electronic exchanges do not exist in isolation, and there are several competing electronic exchanges where traders can participate. These exchanges need effective market policies to compete against each other to attract traders. They need to decide when to clear the markets (i.e. clearing

© Springer International Publishing Switzerland 2016
R. Booth and M.-L. Zhang (Eds.): PRICAI 2016, LNAI 9810, pp. 457–471, 2016.
DOI: 10.1007/978-3-319-42911-3_38

policy) and how to set the transaction prices (i.e. pricing policy) [17]. In this situation, software agents are often used to determine effective market mechanisms because of the speed of trading that is required [8]. Setting transaction prices is one of the major concerns when designing double auction market mechanisms since the prices can significantly affect traders' profits, and thus affect their choices of marketplaces. In this paper, we will analyse how to set effective pricing policies.

Specifically, in this paper, we assume that the double auction marketplaces match traders when all traders have submitted their offers (i.e. clearing house). For the pricing policy, the price associated with any particular transaction must be no more than the buyer's offer (denoted as a bid in the following), and no less than the seller's offer (denoted as an ask). The existing pricing policies can be categorised as two types: discriminatory pricing policies and uniform pricing policies [17]. A discriminatory pricing policy may set different prices to different transactions during a single clear event, while the price of each transaction is bounded only by the offers involved. A typical discriminatory pricing policy is called *discriminatory k pricing policy*. In this policy, the marketplace sorts the bids descendingly and the asks ascendingly, and matches the buyer with v^{th} highest bid with the seller with v^{th} lowest ask if the ask is not greater than the bid. The transaction price is set at the point determined by k in the interval between the matched bid and ask. In contrast, considering fairness for all traders, a uniform pricing policy sets the same price for all transactions, typically using the equilibrium price [16] as the transaction price. The equilibrium price is determined as follows. We define the ask quote as the minimum of the lowest tentatively matchable bid and the lowest unmatchable ask, and define the bid quote as the the maximum of the highest tentatively matchable ask and the highest unmatchable bid. Then the bid quote and ask quote define the range of the equilibrium price (note that the bid quote is not greater than the ask quote). The valid equilibrium price can be at the point determined by a parameter k in the interval between the ask quote and bid quote. In the following, we refer the uniform pricing policy based on the equilibrium price as *equilibrium k pricing policy*. In this paper, we assume that marketplaces adopt either *equilibrium k pricing policy* or *discriminatory k pricing policy*, and can set values to k. Note that $k \in [0, 1]$ and the value of k indicates the marketplace's bias to buyers or sellers when setting transaction prices.[1]

A number of works have analysed the pricing policies for double auction marketplaces. In [12], the authors first proposed a pricing policy to set the transaction price at the point between the buyer and seller's offers. Then a discriminatory pricing policy was proposed in [7] to set different prices for different transactions. A genetic algorithm was used in [9] to find the best price to maximise the market's allocative efficiency and traders' market power when using a discriminatory pricing policy. A uniform pricing policy was proposed in [16] to

---

[1] $k = 0(k = 1)$ is the extreme point where the transaction price is set at the bid(ask) quote for *equilibrium k pricing policy*, and the transaction price is set at the matched ask(bid) for *discriminatory k pricing policy*.

choose the bid/ask quote as the transaction price. In [8], researchers designed pricing policies empirically in the competing environment. In addition, there also exists other analysis on pricing policies for double auctions, such as [2,5,6]. Intuitively, traders' strategies of submitting offers will also affect the transaction prices. Therefore, when determining the pricing policy, the marketplace needs to take into account traders' strategies. Traders' bidding strategies (and market selection strategies in the context of multiple marketplaces) are affected by each other. It is appropriate to use game theory [4] to analyse traders' Bayes-Nash equilibrium strategy (the most widely known solution concept in game theory), where each trading agent makes a best response against the other agents' strategies. However, the existing research did not consider trades' equilibrium bidding strategies when analysing the pricing policy. We analysed the Bayes-Nash equilibrium bidding strategy in double auctions [13], but did not analyse the pricing policy. In this paper, we will analyse how the marketplaces set effective pricing policies when traders use Bayes-Nash equilibrium market selection and bidding strategies.

Specifically, in this paper, we undertake the analysis by considering heterogeneous traders with continuous privately known types (the type is the trader's preference on the goods). We use a computational learning approach (fictitious play) to compute traders' Bayes-Nash equilibrium strategies, and based on this we analyse marketplaces' pricing policies. We consider two settings: isolated marketplaces without competition, and two marketplaces with competition. In the setting of isolated marketplaces, the marketplace intends to maximise the allocative efficiency, which is is one of the most important metrics to measure the performance of marketplaces [14]. In the competing environment, competing marketplaces intend to attract as more traders as they can. We analyse how marketplaces set effective pricing policies in both settings when traders adopt Bayes-Nash equilibrium strategies. In so doing, this is the *first* work on determining the effective pricing policy by taking into account traders' Bayes-Nash equilibrium strategies. In more detail, the contributions of the paper are as follows. In the setting of isolated marketplaces, we find that when the marketplace adopts *discriminatory k pricing policy*, the maximal allocative efficiency is reached by setting $k = 0.41$ or 0.59, in contrast to the intuitive value of 0.5. Furthermore, we find that *equilibrium k pricing policy* provides higher allocative efficiency than *discriminatory k pricing policy*. We also analyse how different pricing policies can affect traders' Bayes-Nash equilibrium bidding strategies. Furthermore, in the competing environment, we find that the marketplace using *equilibrium k pricing policy* can beat the one using *discriminatory k pricing policy*. These results can provide meaningful insights for designing an effective pricing policy.

The structure of the paper is as follows. In Sect. 2, we describe the setting for analysing the pricing policy. In Sect. 3, we introduce the fictitious play algorithm used in our analysis. In Sect. 4, we analyse the pricing policy. Finally, we conclude in Sect. 5.

2   The Framework

In this section, we first introduce the basic setting for analysing our problems. Since we analyse the pricing policy when traders take Bayes-Nash equilibrium strategies, we need to derive the equations to calculate traders' expected utilities, which will be used by the fictitious play algorithm in Sect. 3 to approximate the Bayes-Nash equilibrium market selection and bidding strategy.

2.1   Basic Settings

We assume that there is a set of buyers, $\mathcal{B} = \{1, 2, ...B\}$, and a set of sellers, $\mathcal{S} = \{1, 2, ...S\}$. Each buyer and each seller can only trade one unit of the goods in the marketplace. All goods are identical. Each buyer and seller has a type[2], which is denoted as θ^b and θ^s respectively. We assume that the types of all buyers are i.i.d drawn from the cumulative distribution function F^b, with support $[0, 1]$, and the types of all sellers are i.i.d drawn from the cumulative distribution function F^s, with support $[0, 1]$. The distributions F^b and F^s are assumed to be common knowledge and differentiable. The probability density functions are f^b and f^s respectively. In our setting, the type of each specific trader is not known to the other traders, i.e. *private information*.

In addition, we assume that there exist competing double auction marketplaces $\mathcal{M} = \{1, ..., M\}$, that offer places for trade and determine the transaction prices for the buyers and sellers. The marketplace is cleared when all traders have submitted their offers. The marketplaces use either *equilibrium k pricing policy* or *discriminatory k pricing policy* to set the transaction prices. Traders will incur a small cost ι when they choose any marketplace (for example, the time cost for trading online or travel and time costs for trading in shopping malls). We do this so that they slightly prefer choosing no marketplace than choosing a marketplace but not transacting.

We now describe how traders select marketplaces and submit offers. The allowed offers of buyers and sellers are given by the offer space $\Psi = \{0, \frac{1}{D}, \frac{2}{D}, ..., \frac{D-1}{D}, 1\} \cup \{\ominus\}$, i.e. the offer space comprises $D + 1$ allowable offers from 0 to 1 with step size $1/D$ (D is a natural number), and $\ominus$ means not submitting an offer in the marketplace (i.e. not choosing the marketplace). We refer to a trader choosing a marketplace and bidding as an *action*. Formally, a buyer's action is defined as a tuple $\delta^b = \langle d_1^b, ..., d_M^b \rangle$ where the buyer bids d_m^b in marketplace m if $d_m^b \neq \ominus$, and does not choose marketplace m if $d_m^b = \ominus$. This definition combines the buyer's market selection and submitted offers in the selected marketplaces as a whole. Similarly, a seller's action is given by $\delta^s = \langle d_1^s, ..., d_M^s \rangle$. In our setting, traders are only allowed to bid in one marketplace at a time. Therefore at most one of the bids placed by the action δ is not equal to $\ominus$. The set of all allowed actions constitutes the *action space*, which is

---

[2] The type of a buyer is its *limit price*, the highest price it is willing to buy the item for, and the type of a seller is its *cost price*, the lowest price it is willing to sell the item for.

defined as $\Delta = \{\delta \in \Psi^M : \exists i \in \mathcal{M}, \forall j \neq i, d_i \in \Psi \text{ and } d_j = \ominus\}$. Note that both buyers and sellers have the same action space.

A trader's action choice depends on its type. Hence, a *strategy*, is defined as a mapping from the set of types to the action space. Formally, we use $\sigma^b : [0,1] \rightarrow \Delta$ and $\sigma^s : [0,1] \rightarrow \Delta$ to denote the buyer and the seller trading strategies respectively. Note that the expected utility of a trader is directly dependent on its beliefs about other traders' action choices. Therefore, instead of looking at traders' strategies, in what follows, the expected utility is expressed directly in terms of traders' action distributions. Specifically, we use ω_i^b to denote the probability of action δ_i^b being chosen by a buyer, and use ω_i^s to denote the probability of action δ_i^s being chosen by a seller. Furthermore, we use $\Omega^b = (\omega_1^b, \omega_2^b, ..., \omega_{|\Delta|}^b)$, $\sum_{i=1}^{|\Delta|} \omega_i^b = 1$, to represent the probability distribution of buyers' actions, and $\Omega^s = (\omega_1^s, \omega_2^s, ..., \omega_{|\Delta|}^s)$ for the sellers' action distribution. Note that, given a trader's strategy and the type distribution function, we can derive the probability of a certain action being played by the trader. Specifically, we use $\sigma^{-b}(\delta_i^b) \subseteq [0,1]$ to denote the set of buyer types using action δ_i^b. Then the probability of the action δ_i^b being played is $\omega_i^b = \int_{x \in \sigma^{-b}(\delta_i^b)} f^b(x) dx$. The calculation of the seller action distribution is analogous.

2.2   The Trader's Expected Utility

We now derive equations to calculate expected utilities of traders. In what follows, we derive the expected utility of a buyer, but the seller's is calculated analogously. A buyer's expected utility depends on its type, its own bid, and its beliefs about action choices of other traders. We calculate the expected utility of a buyer with type θ^b when taking action $\delta^b = \langle d_1^b, ..., d_M^b \rangle$ given the other buyers' action distribution Ω^b, the sellers' action distribution Ω^s, and the pricing policy.

For the buyer's action $\delta^b = \langle d_1^b, ..., d_M^b \rangle$, when $d_m^b = \ominus, \forall m \in \mathcal{M}$, i.e. the buyer does not enter any marketplace, the buyer's expected utility is zero. When the buyer chooses to only enter marketplace m (i.e. $d_m^b \neq \ominus$), its expected utility over action δ^b is equal to the expected profit when the buyer bids d_m^b in marketplace m given its pricing policy and the parameter k, which is computed as follows.

Firstly, we need to sort the buyers' bids descendingly and sellers' asks ascendingly in the marketplace. The reason is as follows. When the marketplace adopts *discriminatory k pricing policy*, we need to know the position of the buyer's bid in the marketplace, which determines its matching with sellers. When the marketplace adopts *equilibrium k pricing policy*, we need to sort them to find the ask quote and bid quote, in order to decide the equilibrium price range. Specifically, we use a $|\Delta|$-tuple $\bar{x} = \langle x_1, ...x_{|\Delta|} \rangle \in \mathcal{X}$ to represent the number of buyers choosing different actions, where x_i is the number of buyers choosing action δ_i^b, $\mathcal{X}$ is the set of all such possible tuples and we have $\sum_{i=1}^{|\Delta|} x_i = B - 1$ (note that we need to exclude the buyer for which we are calculating the expected utility). The probability of exactly x_i buyers choosing action δ_i^b is $(\omega_i^b)^{x_i}$, and the probability of this tuple appearing is:

$$\rho^b(\bar{x}) = \begin{pmatrix} B-1 \\ x_1, ..., x_{|\Delta|} \end{pmatrix} \times \prod_{i=1}^{|\Delta|} \left(\omega_i^b\right)^{x_i} \tag{1}$$

Now for a particular $\bar{x}$, we determine the buyer's position as follows. Firstly, we obtain the number of other buyers whose bids are greater than the buyer's bid d_m^b, which is given by:

$$X^>(\bar{x}, d_m^b) = \sum_{\delta_i^b \in \Delta : d_{im}^b > d_m^b} x_i \tag{2}$$

where d_{im}^b is the bid placed by the action δ_i^b in marketplace m. Similarly, we use $X^=(\bar{x}, d_m^b)$ to represent the number of buyers whose bids are equal to the buyer's bid (excluding the buyer itself):

$$X^=(\bar{x}, d_m^b) = \sum_{\delta_i^b \in \Delta : d_{im}^b = d_m^b} x_i \tag{3}$$

Due to having discrete bids and given $X^>(\bar{x}, d_m^b)$ buyers bidding higher than the buyer's bid d_m^b and $X^=(\bar{x}, d_m^b)$ buyers bidding equal to d_m^b, the buyer's position $v_{\bar{x}}$ given $\bar{x}$ in marketplace m could be anywhere from $X^>(\bar{x}, d_m^b) + 1$ to $X^>(\bar{x}, d^b) + X^=(\bar{x}, d_m^b) + 1$, which constitutes the buyer's position range. We then use $\mathcal{V}_{\bar{x}} = \{X^>(\bar{x}, d_m^b) + 1, ..., X^>(\bar{x}, d_m^b) + X^=(\bar{x}, d_m^b) + 1\}$ to denote the position range. Since $X^=(\bar{x}, d_m^b) + 1$ buyers have the same bid, a tie-breaking rule is needed to determine the buyer's position. We adopt a standard rule where each of these possible positions occurs with equal probability, i.e. $1/(X^=(\bar{x}, d_m^b) + 1)$.

The buyer's expected utility also depends on sellers' action choices. Specifically, we use a $|\Delta|$-tuple $\bar{y} = \langle y_1, ... y_{|\Delta|} \rangle \in \mathcal{Y}$ to represent the number of sellers choosing different actions, where y_i is the number of sellers choosing action δ_i^s, and $\mathcal{Y}$ is the set of all such possible tuples and we have $\sum_{i=1}^{|\Delta|} y_i = S$. The probability of this tuple appearing is:

$$\rho^s(\bar{y}) = \begin{pmatrix} S \\ y_1, ..., y_{|\Delta|} \end{pmatrix} \times \prod_{i=1}^{|\Delta|} \left(\omega_i^s\right)^{y_i} \tag{4}$$

Now given the buyer's positions $v_{\bar{x}}$ and the number of sellers choosing different actions $\bar{y}$, we are ready to calculate the buyer's expected utility. When the marketplace chooses *discriminatory k pricing policy*, given the tuple $\bar{y}$, we can sort the asks of the sellers in marketplace m descendingly. Then the ask which is $v_{\bar{x}}^{th}$ highest will be matched with the buyer's bid. We denote this ask as d_m^s. Now the buyer's expected utility can be calculated:

$$U(v_{\bar{x}}, \bar{y}, \theta^b, d_m^b, \Omega^B, \Omega^S, k) = \begin{cases} \theta^b - TP - \iota & \textbf{if } d_m^b \geq d_m^s \\ 0 - \iota & \textbf{if } d_m^b < d_m^s \end{cases} \tag{5}$$

where $TP = d_m^b \times k + d_m^s \times (1 - k)$ is the transaction price and ι is the constant cost.

When the marketplace chooses the *equilibrium k pricing policy*, the calculation is as follows. Given the number of buyers choosing different actions $\bar{x}$ and
the number of sellers choosing different actions $\bar{y}$, we sort the bid descendingly
and asks ascendingly, and then obtain the bid quote and ask quote in marketplace m, which are denoted as Q_m^b and Q_m^s respectively. Then the equilibrium
price of the marketplace is $EP = Q_m^s \times k + Q_m^b \times (1-k)$. Furthermore, according
to the sorted bids and asks, we can get the exact number of transactions that will
be made in the marketplace, which is denoted as T. Now the buyer's expected
utility is given by:

$$U(v_{\bar{x}}, \bar{y}, \theta^b, d_m^b, \Omega^B, \Omega^S, k) = \begin{cases} 0 - \iota & \textbf{if } v_{\bar{x}} > T \\ \theta^b - EP - \iota & \textbf{if } v_{\bar{x}} \leq T \end{cases} \quad (6)$$

Finally, by considering all possible numbers of sellers choosing different
actions, all possible positions and all possible numbers of buyers choosing different actions, the buyer's expected utility is given by:

$$\tilde{U}(\theta^b, \delta^b, \Omega^b, \Omega^s, k) = \sum_{\bar{x} \in \mathcal{X}} \rho^b(\bar{x}) \times \sum_{v_{\bar{x}} \in \mathcal{V}_{\bar{x}}} \frac{1}{X = (\bar{x}, d^b) + 1} \times$$
$$\sum_{\bar{y} \in \mathcal{Y}} \rho^s(\bar{y}) \times U(v_{\bar{x}}, \bar{y}, \theta^b, d^b, \Omega^B, \Omega^S, k) \quad (7)$$

3   Fictitious Play

In this section, we introduce the fictitious play (FP) algorithm to derive the
Bayes-Nash equilibrium strategies of traders. We intend to analyse how to set
effective pricing policies given traders using Bayes-Nash equilibrium strategies.
Therefore, we first need to derive the Bayes-Nash equilibrium trading strategies. In the standard FP algorithm [1], opponents are assumed to play a fixed
mixed strategy. Then by observing relative appearance frequencies of different
actions, the player can estimate their opponents' mixed strategies, and take a
best response. The observed frequencies of opponents' actions are termed *FP
beliefs*. In each round, all players estimate their opponents' mixed strategies and
update their FP beliefs, and play a best response to their FP beliefs. All players
continually iterate this process until it converges to the Nash equilibrium. However, the standard FP algorithm is not suitable for analysing Bayesian games
where the player's type is not known to other players. In such games, a strategy
is a function mapping a player's type to an allowed offer. In the standard FP
algorithm, by observing the frequency of opponents' actions, we cannot know
the actual strategy of a player since we do not know which type submits which
offer. The same as the work done in [13], we use a generalised fictitious play algorithm [10] to derive traders' strategies with continuous types and a finite action
space. Moreover, in reality, it is impossible to run the algorithm to convergence
since it involves an infinite number of iteration rounds. Therefore, it is often

used to approximate the Bayes-Nash equilibrium (i.e. deriving the ϵ-Bayes-Nash equilibrium) by running the fictitious play algorithm for a limited number of rounds.

We first describe how to compute the best response actions against current FP beliefs. Previously, we used Ω^b and Ω^s to denote the probability distributions of buyers' and sellers' offers respectively. In the FP algorithm, we use them to represent FP beliefs about the buyers' and sellers' offers respectively. Then, given their beliefs, we compute the buyers' best response function as follows (mutatis mutandis for sellers):

$$\sigma^{b*}(\theta^b|\Omega^b, \Omega^s) = argmax_{\delta^b \in \Delta}\tilde{U}(\theta^b, \delta^b, \Omega^b, \Omega^s) \qquad (8)$$

where σ^{b*} is the best response action of the buyer with type θ^b given FP beliefs Ω^b and Ω^s. The optimal utility that a buyer with type θ^b can achieve is

$$\tilde{U}^*(\theta^b, \Omega^b, \Omega^s) = max_{\delta^b \in \Delta}\tilde{U}(\theta^b, \delta^b, \Omega^b, \Omega^s) \qquad (9)$$

From the buyer's expected utility equations (see Eqs. 5 and 6) we note that the buyer's expected utility $\tilde{U}(\theta^b, \delta^b, \Omega^b, \Omega^s)$ is linear in its type θ^b for a given action. Given this, and given a finite number of actions, the best response function is the upper envelope of a finite set of linear functions, and thus is piecewise linear.

Now we have computed the best response function and also provided the sets of types corresponding to the best response actions. Based on this, we can calculate the best response action distribution of buyers. Given that the upper envelope is a piece-wise linear function, we know that the set of buyer types corresponding to the best response action δ_i^b is $\sigma^{-b}(\delta_i^b)$. Given this, the probability of action δ_i^b being played by a buyer is $\omega_i^b = \int_{x \in \sigma^{-b}(\delta_i^b)} f^b(x)dx$. By calculating the probability of each action being used, we obtain the current best response action distribution of buyers, denoted by Ω_{br}^b, which is given current FP beliefs. We can then update the FP beliefs of buyers' actions:

$$\Omega_{\tau+1}^b = \frac{\tau}{\tau+1} \times \Omega_\tau^b + \frac{1}{\tau+1} \times \Omega_{br}^b \qquad (10)$$

where $\Omega_{\tau+1}^b$ is the updated FP beliefs of the buyers' actions for the next iteration round $\tau + 1$, Ω_τ^b is the FP beliefs on the current iteration τ, and Ω_{br}^b is the probability distribution of the best response actions against FP beliefs Ω_τ^b. This equation actually gives the FP beliefs at the current round as the average of the best response action distributions of buyers in all previous rounds. The computation of the sellers' best response function and belief updates is analogous. In our setting, we need to update both buyers' and sellers' FP beliefs simultaneously.

We now describe how to check the convergence to a Bayes-Nash equilibrium. If the trader cannot gain more than ϵ by taking a best response action against current best response action distributions, the FP algorithm stops the iteration process and converges. Formally, the measure of convergence is given by:

$$|\tilde{U}^b(\Omega_{br}^b, \Omega_{br}^s) - \tilde{U}_{br}^b(\Omega_{br}^b, \Omega_{br}^s)| \leq \epsilon \ and \ |\tilde{U}^s(\Omega_{br}^b, \Omega_{br}^s) - \tilde{U}_{br}^s(\Omega_{br}^b, \Omega_{br}^s)| \leq \epsilon \quad (11)$$

where $\tilde{U}^b(\Omega^b_{br}, \Omega^s_{br})$ is the expected utility of a buyer in the best response action distributions Ω^b_{br} and Ω^s_{br}:

$$\tilde{U}^b(\Omega^b_{br}, \Omega^s_{br}) = \int_0^1 f^b(x) \times \tilde{U}^b(x, \delta^b, \Omega^b_{br}, \Omega^s_{br})dx \tag{12}$$

where δ^b is the action chosen by the buyer with type x (actually, it is the best response action of this buyer against FP beliefs Ω^b_τ and Ω^s_τ, i.e. $\sigma^{b*}(x|\Omega^b_\tau, \Omega^s_\tau)$). $\tilde{U}^b_{br}(\Omega^b_{br}, \Omega^s_{br})$ is the expected utility of a buyer adopting the best response action against the current best response action distributions Ω^b_{br} and Ω^s_{br}:

$$\tilde{U}^b_{br}(\Omega^b_{br}, \Omega^s_{br}) = \int_0^1 f^b(x) \times \tilde{U}^b(x, \delta^{b*}, \Omega^b_{br}, \Omega^s_{br})dx \tag{13}$$

where $\delta^{b*} = \sigma^{b*}(x|\Omega^b_{br}, \Omega^s_{br})$ is the best response action of the buyer with type x given action distributions Ω^b_{br} and Ω^s_{br}. The equations for sellers are analogous.

When the FP algorithm converges, the current best response actions with corresponding type sets constitute an ϵ-Bayes-Nash equilibrium.

4   Experimental Analysis

We now begin to analyse how to set effective pricing policies when traders adopt Bayes-Nash equilibrium strategies. We firstly use the above FP algorithm to derive the traders' Bayes-Nash equilibrium strategies. Based on the derived Bayes-Nash equilibrium, we go on to analyse how to set effective transaction prices. For illustrative purpose, in the following analysis, we consider 2 marketplaces, 5 buyers and 5 sellers, and 11 discrete bid(ask) levels plus $\ominus$ (denoting the action when the marketplace is not chosen).[3] Furthermore, we assume that both buyer and seller types are independently drawn from a uniform distribution. In addition, we set $\epsilon = 0.00001$ in the ϵ-Bayes-Nash equilibrium, and assume that the small cost for traders to enter a marketplace is set to $\iota = 0.0001$.

4.1   An Isolated Marketplace

We first analyse how to choose an effective pricing policy to achieve the highest allocative efficiency in an isolated marketplace. When the isolated marketplace uses either *discriminatory k pricing policy* or *equilibrium k pricing policy*, we discretise k from 0 to 1 with step size 0.01. Then for each possible k value, we use fictitious play to derive the Bayes-Nash equilibrium bidding strategies. We consider different initial FP beliefs, and find that the algorithm consistently converge to the same Bayes-Nash equilibrium.[4] Now according to the offers placed by

---

[3] We also tried other settings. However, the main insights are similar.

[4] Note that there exists a trivial Nash equilibrium where no traders enter the marketplace. We ignore this equilibrium since there is a zero probability of transactions happening between buyers and sellers. Furthermore, while a pure-strategy Nash equilibrium always exists in this setting [10], in general, the FP algorithm is not guaranteed to converge [15].

traders in equilibrium, we run simulations to obtain the marketplace's allocative efficiency. The allocative efficiency is the ratio of traders' overall actual profits to the theoretical profits. We run 100,000 simulations to obtain the average allocative efficiency. In each run, based on the offers submitted in equilibrium, we can determine the transaction prices. Then each trader's actual profit is the difference between its type and the transaction price. The trader's theoretical profit is made when all traders bid truthfully and an optimal allocation is made. We calculate the average allocative efficiency over 100,000 runs and show the 95 % confidence interval for the significance of the results.

The allocative efficiency of both pricing policies with different k values is shown in Fig. 1. The black line represents the allocative efficiency for *equilibrium k pricing policy*, while the grey line represents the allocative efficiency for *discriminatory k pricing policy*. We can see that both pricing policies can provide high allocative efficiency (more than 90 %). We find that in contrast to [9] where $k = 0.5$ is the best for the marketplace using *discriminatory k pricing policy*, in this setting, the highest allocative efficiency is reached when k is about 0.41 or 0.59. One possible reason is that in [9], the authors assume traders adopt RE bidding strategy [11], while we consider a Bayes-Nash equilibrium bidding strategy.[5] We can see that the allocative efficiency of the marketplace using *discriminatory k pricing policy* changes significantly with respect to the changes of k values. This is because in *discriminatory k pricing policy*, the transaction price is determined by k between the matched bid and ask. When the difference between the matched bid and ask is large, the value of k can significantly affect traders' profits, which results in the changes of the allocative efficiency. Furthermore, we find when *equilibrium k pricing policy* is used, the allocative efficiency does not vary significantly with respect to k values. This is because the allowed equilibrium price range determined by the bid and ask quote is quite small. Therefore, by changing k in *equilibrium k pricing policy*, the actual equilibrium price is not changed too much, and thus the allocative efficiency is not affected significantly by changing k. We also find when $k = 0.5$, the allocative efficiency is the highest. Finally, we can see that the marketplace using *equilibrium k pricing policy* indeed has higher allocative efficiency than the marketplace using *discriminatory k pricing policy*, which may suggest that we need to adopt *equilibrium k pricing policy* (and set $k = 0.5$) in the competing environment.

Furthermore, we also analyse how pricing policy can affect traders' Bayes-Nash equilibrium bidding strategy. For both pricing policies, we show traders' Bayes-Nash equilibrium strategies under typical k values in Fig. 2. The black line represents the buyers' bids in equilibrium for given types and the grey line represents the sellers' asks in equilibrium. We find that traders shade their offers (i.e. buyers bid less than their types and sellers ask more than their types) in equilibrium. Comparing Fig. 2(a) to (b) where both pricing policies provide

---

[5] In [9], the authors intend to find appropriate k value to maximise the combination of allocative efficiency and traders' market power. We also do this in our setting, and still find that when k is equal to 0.41 or 0.59, the combination of allocative efficiency and traders' market power is maximised.

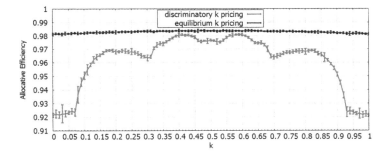

Fig. 1. Allocative efficiency and 95 % confidence interval for different pricing policies with different k values.

the highest allocative efficiency, we find that traders shade their offers more in the marketplace using *discriminatoyy k pricing policy*. For example, buyers with type 1 bid 0.8 in marketplace using *discriminatory k pricing policy*, but bid 0.9 in the marketplace using *equilibrium k pricing policy*. The possible reason is as follows. In *discriminatory k pricing policy*, the transaction price is determined by the matched bid and ask. Therefore, when $k = 0.5$, buyers want to decrease their bids in order to move the transaction price to the sellers' side to make more profits, and similar for sellers. However, in *equilibrium k pricing policy*, the transaction price is uniform for all traders, which is determined by bid quote and ask quote. The individual behavior of shading may not affect the transaction price, but indeed decreases the probability of transacting. Therefore, in *equilibrium k pricing policy*, traders will not shade more. When $k = 0$ in both pricing policies (i.e. setting biased prices to buyers), we find that sellers will shade more than other k values (e.g. comparing Fig. 2(c) to (e), and comparing Fig. 2(d) to (f)) in order to keep profits, and buyers consequently shade less.

4.2   Competing Marketplaces

Now we analyse how to determine an effective pricing policy to attract traders in the competing environment with two marketplaces. We first consider the case of two marketplaces adopting *discriminatory k pricing policy*, but having different k values. We find that depending on the initial FP beliefs, traders may converge to different marketplaces. However, we find that the marketplace with $k = 0.5$, which sets no biased transaction prices to buyers and sellers, is more likely to attract all traders in Nash equilibrium, and how traders bid in equilibrium is the same as that in Fig. 2(e). It even beats the case of $k = 0.41$ or 0.59, which provides slightly higher allocative efficiency. This may suggest that in the competing environment, when both marketplaces provide similar allocative efficiency, setting fair transaction prices is more important. Furthermore, we consider the case of two marketplaces using *equilibrium k pricing policy* with different k values. We find that the marketplace with $k = 0.5$ is more likely to win all traders, and traders' Bayes-Nash equilibrium bidding strategy is the same as that in Fig. 2(b).

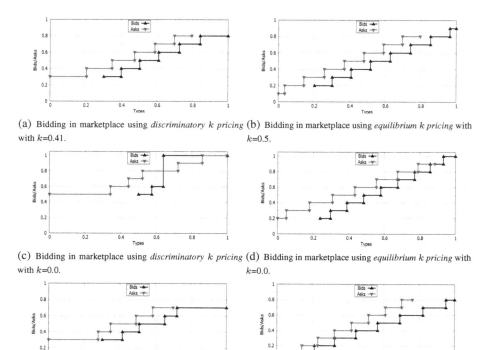

(a) Bidding in marketplace using *discriminatory k pricing* with $k=0.41$.

(b) Bidding in marketplace using *equilibrium k pricing* with $k=0.5$.

(c) Bidding in marketplace using *discriminatory k pricing* with $k=0.0$.

(d) Bidding in marketplace using *equilibrium k pricing* with $k=0.0$.

(e) Bidding in marketplace using *discriminatory k pricing* with $k=0.5$.

(f) Bidding in marketplace using *equilibrium k pricing* with $k=1.0$.

Fig. 2. Bayes-Nash equilibrium bidding strategies in different pricing policies.

Furthermore, we analyse the case that both marketplaces adopt different pricing policies. We assume that marketplace 1 uses *discriminatory k pricing policy* and marketplace 2 uses *equilibrium k pricing policy*. We find that in Bayes-Nash equilibrium, depending on the initial FP beliefs, traders may converge to the marketplace using *discriminatory k pricing policy* or *equilibrium k pricing policy*. Traders are more likely to converge to marketplace 2 which provides higher allocative efficiency. Particularly, when marketplace 1 sets $k=0.41$ and marketplace 2 sets $k=0.5$ (i.e. both marketplaces provide the highest allocative efficiency), we find that marketplace 2 is more likely to beat marketplace 1. We investigate the changes of traders' choices of marketplaces during the learning process. Specifically, when initially each action is equally selected, the dynamic change of traders' probability of choosing each marketplace is shown in Fig. 3.[6] The black dash lines represent the probabilities of traders' actions choosing

---

[6] Note that in this case the traders converge after 1800 rounds. However, we only show the first 200 rounds in order to clearly indicate the dynamic changes of traders choosing marketplaces.

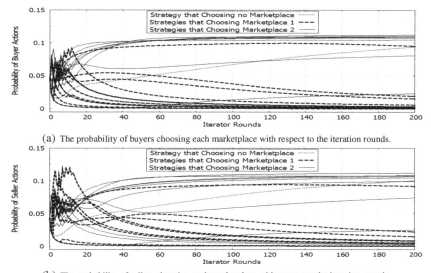

(a) The probability of buyers choosing each marketplace with respect to the iteration rounds.

(b) The probability of sellers choosing each marketplace with respect to the iteration rounds.

Fig. 3. The probability of traders choosing each marketplace with respect to the iteration rounds.

marketplace 1. The black solid lines represent the probabilities of traders' actions choosing marketplace 2. The grey dot line represents the probability of choosing no marketplaces. From Fig. 3, we can see that although marketplace 1 sets a biased transaction price ($k = 0.41$) to attract buyers, as marketplace 2 provides a fair transaction price and higher allocative efficiency, sellers prefer marketplace 2, and buyers consequently move to marketplace 2 as well. This further suggests that in the competing environment, setting a fair transaction price is important.

5   Conclusions

In this paper, we investigate how marketplaces set an effective pricing policy in terms of maximising allocative efficiency in the isolated environment, and attracting traders in the competing environment. We find that the marketplace using *discriminatory k pricing policy* can maximise the allocative efficiency by setting $k = 0.41$ or 0.59 when traders use Bayes-Nash equilibrium bidding strategies. We find that the marketplace using *equilibrium k pricing policy* can provide higher allocative efficiency than that using *discriminatory k pricing policy*. In the competing case, the marketplace using *equilibrium k pricing policy* can beat the one using *discriminatory k pricing policy*. This may suggest that the marketplace should adopt *equilibrium k pricing policy* with $k = 0.5$.

In this paper, we do not consider the dynamic interaction between traders' strategies and the pricing policies. For example, during the learning process, the marketplace using *discriminatory k pricing policy* can adjust its parameter k to keep traders according to the changes of traders' market choices. In the future, we will consider the interaction between trading strategies and pricing policies. Furthermore, in order to obtain an equilibrium solution in a reasonable time, we consider a relatively small number of traders and marketplaces. When the number of participants increases beyond 100 agents, the computational load is intense and it is difficult to get the solution in a reasonable time. A possible way to address this limitation is to consider a higher level of abstraction in which individual agents represent many agents of the same type.

Acknowledgments. This paper was funded by the National Natural Science Foundation of China (No. 61402344), Scientific Research Foundation for the Returned Overseas Chinese Scholars, Ministry of Education of China, and Specialized Research Fund for the Doctoral Program of Higher Education of China (No. 20120092120033).

References

1. Brown, G.W.: Iterative solution of games by fictitious play. Act. Anal. Prod. Allocation **13**(1), 374–376 (1951)
2. Chakraborty, M., Das, S., Peabody, J.: Price evolution in a continuous double auction prediction market with a scoring-rule based market make. In: Proceedings of the 29th AAAI Conference on Artificial Intelligence, pp. 835–841 (2015)
3. Friedman, D., Rust, J.: The Double Auction Market: Institutions, Theories and Evidence, vol. XIV. Perseus Publishing (1993). Santa Fe Institute Studies in the Science of Complexity
4. Fudenberg, D., Tirole, J.: Game Theory. The MIT Press, Cambridge (1991)
5. Ji, M., Li, H.: Exploring price fluctuations in a double auction market. Computational Economics, 1–21 (2015)
6. Li, W., Wang, S., Cheng, X.: Truthful multi-attribute auction with discriminatory pricing in cognitive radio networks. ACM SIGMOBILE Mob. Comput. Commun. Rev. **18**(1), 3–13 (2014)
7. Nicolaisen, J., Petrov, V., Tesfatsion, L.: Market power and efficiency in a computational electricity market with discriminatory double auction pricing. IEEE Trans. Evol. Comput. **5**(5), 504–523 (2001)
8. Niu, J., Cai, K., Parsons, S., McBurney, P., Gerding, E.H.: What the 2007 tac market design game tells us about effective auction mechanisms. Auton. Agents Multiagent Syst. **21**, 172–203 (2010)
9. Phelps, S., McBurney, P., Parsons, S., Sklar, E.: Applying genetic programming to economic mechanism design: evolving a pricing rule for a continuous double auction. In: Proceedings of the 2nd International Conference on Autonomous Agents and Multi-Agent Systems, pp. 1096–1097 (2003)
10. Rabinovich, Z., Naroditskiy, V., Gerding, E.H., Jennings, N.R.: Computing pure bayesian-nash equilibria in games with finite actions and continuous types. Artif. Intell. **195**, 106–139 (2013)
11. Roth, A.E., Erev, I.: Predicting how people play games: reinforcement learning in experimental games with unique, mixed strategy equilibria. Am. Econ. Rev. **88**(4), 848–881 (1998)
12. Satterthwaite, M.A., Williams, S.R.: Bilateral trade with the sealed bid k-double auction: existence and efficiency. J. Econ. Theory **48**(1), 107–133 (1989)

13. Shi, B., Gerding, E.H., Vytelingum, P., Jennings, N.R.: An equilibrium analysis of competing double auction marketplaces using fictitious play. In: Proceedings of the 19th European Conference on Artificial Intelligence, pp. 575–580 (2010)
14. Smith, V.L.: An experimental study of competitive market behavior. J. Polit. Econ. **70**, 111–137 (1962)
15. Vijay, K., Sjöström, T.: On the convergence of fictitious play. Math. Oper. Res. **23**(2), 479–511 (1998)
16. Wurman, P.R., Walsh, W.E., Wellman, M.P.: Flexible double auction for electronic commerce: theory and implementation. Decis. Support Syst. **24**, 17–27 (1998)
17. Wurman, P.R., Wellman, M.P., Walsh, W.E.: Specifying rules for electronic auctions. AI Mag. **23**(3), 15–23 (2002)

An Investigation of Objective Interestingness Measures for Association Rule Mining

Ratchasak Somyanonthanakul[1]([⊠]) and Thanaruk Theeramunkong[2]

[1] Department of Medical Informatics, College of Information
and Communication Technology, Rangsit University, Lak Hok, Thailand
ratchasak.s@rsu.ac.th
[2] Sirindhorn International Institute of Technology,
Thammasat University, Bangkok, Thailand
thanaruk@siit.tu.ac.th

Abstract. While a large number of objective interestingness measures have been proposed to describe an association pattern which encodes meaningful relationship among attributes in a dataset, their characteristics and interrelations are not well explored. In this work, we investigate static and dynamic characteristics of 21 commonly used interestingness measures in order to understand their common and distinct properties. Four systematical methods investigated are (1) trend analysis, (2) fixed-total variable-portion analysis, (3) fixed-total fixed-portion-combination analysis, and (4) imbalance and extreme scenario analysis. A correlation analysis has been made to find interrelation patterns of the measures.

Keywords: Association rules · Interestingness · Measure analysis

1 Introduction

A main objective of association rule mining is to discover relationships among set of items in a transactional database [1], such as market basket analysis [2], protein-DNA binding sequence pattern discovery [3], web usage analysis [4], intrusion detection [5], and so on. So far there have been a number of interestingness measures invented to express various types of information. Among them, some most popular measures are support, confidence, lift, conviction, and so on. Traditionally, support is used to measure the strength of a dataset, while support and confidence are used to measure the strength of an association rule. With a threshold, a measure can be used to investigate different characteristics of an association rule [6].

In [7], 21 measures are proposed and their characteristics are compared under a predefined set of scenarios, which have proven particularly useful for finding association patterns in data sets with random distributions. However, it is unclear whether such predefined set cover all the possibilities or not.

In this work, to grasp usefulness of such different association measures, we present a framework to investigate characteristics of the measures in terms of dynamic and static points of view to contrast them for better understanding of their common and distinct properties. Towards this, a correlation analysis has been made to find interrelation

© Springer International Publishing Switzerland 2016
R. Booth and M.-L. Zhang (Eds.): PRICAI 2016, LNAI 9810, pp. 472–481, 2016.
DOI: 10.1007/978-3-319-42911-3_39

patterns of the measures and a number of suggestions or guidelines for using the measures can be listed.

The outline of this paper is as follows. Section 2 provides definition and background of 21 well-known objective interestingness measures. Section 3 presents an outline of four systematical methods; (1) trend analysis, (2) fixed-total variable-portion analysis, (3) fixed-total fixed-portion-combination analysis, and (4) imbalance and extreme scenario analysis. An analysis result with four viewpoints is described in Sect. 4. Section 5 gives discussion and conclusions.

2 Objective Measures of Interestingness

2.1 Definition of Objective Measures

In contrast to subjective measures where human factor and domain are involved, objective measures, rooted on a data-driven approach, can be used for evaluating the quality of association patterns, independently of domain with minimal labor from the users. Theoretically an objective measure requires a threshold to filter low-quality patterns. Most objective measures are usually computed based on frequency counts tabulated in a contingency table as shown in Table 1.

Table 1. A two-way contingency table for variable A and B.

	B	$\bar{B}$	
A	f_{11}	f_{10}	f_{1+}
$\bar{A}$	f_{01}	f_{00}	f_{0+}
	f_{+1}	f_{+0}	N

In the table, the notation $\bar{A}$ (or $\bar{B}$) indicates that A (or B) is absent from a transaction and each entry f_{ij} denotes a frequency count. f_{11} is the number of transactions that A and B collocate, f_{01} is the number of transactions that consist of A but not B, f_{01} is the number of transactions that are composed of B but not A, and f_{00} is the number of transactions that do not contain either A or B. The row sum f_{1+} ($= f_{10} + f_{11}$) represents the support count for A, while the column sum f_{+1} ($= f_{10} + f_{11}$) represents the support count for B. f_{0+} ($= f_{00} + f_{01}$) and f_{+0} ($= f_{00} + f_{10}$) indicate the number of transactions that do not contain A and B, respectively. N is the total number of transactions, where $(f_{+1} + f_{+0}) = (f_{1+} + f_{0+}) = (f_{00} + f_{01} + f_{10} + f_{11})$.

2.2 Interestingness Measurement

In decades, several different interestingness measures have been developed for representing the strength of association rules. They are various in purpose and usefulness. Traditionally, three most conventional measures are support, confidence and lift (interestingness). Suppose that the association rule considered is $A \rightarrow B$. The support

probability of A $(P(A))$. indicates how many transactions the association rule holds. Two alternative forms of supports are the absolute value and the ratio to the total number of transactions. Table 2 illustrates a summary description for 21 common measures [7].

Table 2. Objective measures for associate pattern [7]

Measures	Definition
φ-coefficient [8]	$\dfrac{P(A,B)-P(A)P(B)}{\sqrt{P(A)P(B)(1-P(A))(1-P(B))}}$
Goodman-Kruskal's [9]	$\dfrac{\sum_j max_k P\left(A_j,B_k\right)+\sum_k max_j P\left(A_j,B_k\right)-max_j P\left(A_j\right)-max_k P(B_k)}{2-max_j P\left(A_j\right)-max_k P(B_k)}$
Odds ratio [10]	$\dfrac{P(A,B)P(\bar{A},\bar{B})}{P(A,\bar{B})P(\bar{A},B)}$
Yule's Q [11]	$\dfrac{P(A,B)P\left(\overline{AB}\right)-P(A,\bar{B})P(\bar{A},B)}{P(A,B)P\left(\overline{AB}\right)+P(A,\bar{B})P(\bar{A},B)}=\dfrac{\sqrt{\alpha}-1}{\sqrt{\alpha}+1}$
Yule's Y [12]	$\dfrac{\sqrt{P(A,B)P\left(\overline{AB}\right)}-\sqrt{P(A,\bar{B})P(\bar{A},B)}}{\sqrt{P(A,B)P\left(\overline{AB}\right)}+\sqrt{P(A,\bar{B})P(\bar{A},B)}}=\dfrac{\sqrt{\alpha}-1}{\sqrt{\alpha}+1}$
Kappa [13]	$\dfrac{P(A,B)+P(\bar{A},\bar{B})-P(A)P(B)-P(\bar{A})P(\bar{B})}{1-P(A)P(B)-P(\bar{A})P(\bar{B})}$
Mutual Information [14]	$\dfrac{\sum_i\sum_j P(A_i,B_j)\log\frac{P(A_i,B_j)}{P(A_i)P(B_j)}}{\min(-\sum_i P(A_i)\log P(A_i),\sum_j P(B_j)\log P(B_j))}$
J-measure (J) [15]	$\max(P(A,B)\log\left(\frac{P(B\mid A)}{P(B)}\right)+P(A\bar{B})\log\left(\frac{P(\bar{B}\mid A)}{P(\bar{B})}\right),$ $P(A,B)\log\left(\frac{P(A\mid B)}{P(A)}\right)+P(\bar{A}B)\log\left(\frac{P(\bar{A}\mid B)}{P(\bar{A})}\right)))$
Gini index [16]	$\max \dfrac{(P(A)[P(B\mid A)^2+P(\bar{B}\mid A)^2]+P(\bar{A})[P(B\mid\bar{A})^2+P(\bar{B}\mid\bar{A})^2]-P(B)^2-P(\bar{B})^2,}{(P(B)[P(A\mid B)^2+P(\bar{A}\mid B)^2]+P(\bar{B})[P(A\mid\bar{B})^2+P(\bar{A}\mid\bar{B})^2]-P(A)^2-P(\bar{A})^2}$
Support [17]	$P(A, B)$
Confidence [18]	$\max (P(B\mid A), P(A\mid B))$
Laplace [19]	$\max\left(\frac{NP(A,B)+1}{NP(A)+2},\frac{NP(A,B)+1}{NP(B)+2}\right)$
Conviction [20]	$\max\left(\frac{P(A)P(\bar{B})}{P(A\bar{B})},\frac{P(B)P(\bar{A})}{P(B\bar{A})}\right)$
Interest [21]	$\dfrac{P(A,B)}{P(A)P(B)}$
Cosine [22]	$\dfrac{P(A,B)}{\sqrt{P(A)P(B)}}$
Piatetsky-Shapio's [21]	$P(A, B) - P(A) P(B)$
Certainty factor [21]	$\max\left(\frac{P(B\mid A)-P(B)}{1-P(B)},\frac{P(A\mid B)-P(A)}{1-P(A)}\right)$
Added Value [21]	$\max(P(B\mid A) - P(B), P(A\mid B) - P(A))$
Collective strength [21]	$\dfrac{P(A,B)+P(\overline{AB})}{P(A)P(B)+P(\bar{A})P(\bar{B})}\times\dfrac{1-P(A)P(B)-P(\bar{A})P(\bar{B})}{1-P(A,B)-P(\overline{AB})}$
Jaccard [22]	$\dfrac{P(A,B)}{P(A)+P(B)-P(A,B)}$
Klosgen [22]	$\sqrt{P(A, B)}\max(P(B\mid A) - P(B), P(A\mid B) - P(A))$

The confidence ($P(B|A)$) indicates the correlation among A and B, that is when an event A occurs, how likely the other event B will occur. Similar to the confidence, the lift shows the ratio of posterior probability of A when B is known ($P(A|B)$) to the prior However, by defining different association measures, it is possible to find different types of association patterns or rules that are appropriate for different types of data and applications. This situation is analogous to that of using different objective functions for measuring the goodness of a set of clusters in order to obtain different types of clustering.

3 Analytical Method

This section describes four systematical methods to investigate static and dynamic characteristics of 21 commonly-used interestingness measures in order to understand their common and distinct properties. They are (1) trend analysis, (2) fixed-total variable-portion analysis, (3) fixed-total fixed-portion-combination analysis, and (4) imbalance and extreme scenario analysis.

3.1 Trend Analysis

In the first analysis, given a fix total number N, three main free variables, $p(A)$, $p(B)$, and $p(AB)$, f_{1+}, f_{+1}, and f_{11} are varied for investigation, each with three possible stages, (1) unchanged (0), (2) increasing (+), and (3) decreasing (−). Based on these three possibilities on the three variables, there are 27 possibilities. The three main free variables can be used to determine the value of other frequencies as shown in Table 3. Based on these all cases, the effect (tendency) on 21 measures are investigated in Sect. 4. The tendency values are considered to assign feasible tendency values to f_{1+}, f_{+1}, and f_{11}.

The assignment tendency values effect to $f_{00}, f_{01}, f_{10}, f_{0+}$ and f_{0+} and shown six possible stages (1) unchanged (0), (2) increasing (+), (3) decreasing (−), (4) unknown (?), (5) more increasing ($+^2$), and (6) more decreasing ($-^2$). The tend analysis result can describe correlation of variables for association rules.

3.2 Fixed-Total Variable-Portion Analysis

As the second analysis, the four main variables are assigned fixed-total values f_{11}, f_{10}, f_{01} and N. The numerical examples are set to specific value ($f_{11} = 100$, $f_{10} = 150$, $f_{01} = 250$, $N = 1000$). These main variables are varied by a constant value that is identified to be a variable-portion. Table 4 shows five example cases and based line is identified in case No. 1.

Table 3. The complete set of 27 cases for trend analysis

	1	2	3	4	5	6	7	8	9	10	11	12	13	14	15	16	17	18	19	20	21	22	23	24	25	26	27
f_{1*}	0	0	0	0	0	0	0	0	0	−	−	−	−	−	−	−	−	−	+	+	+	+	+	+	+	+	+
f_{*1}	0	0	0	−	−	−	+	+	+	0	0	0	−	−	−	+	+	+	0	0	0	−	−	−	+	+	+
f_{11}	0	−	+	0	−	+	0	−	+	0	−	+	0	−	+	0	−	+	0	−	+	0	−	+	0	−	+
f_{00}	0	−	+	−	?	$+^2$	+	$-^2$	?	+	?	$+^2$	$+^2$	?	$+^2$	$+^2$	?	?	−	$-^2$	?	$-^2$	?	?	$-^2$	$-^2$	?
f_{01}	0	+	−	+	?	$-^2$	−	$+^2$	?	0	+	−	−	?	$-^2$	−	$+^2$	?	0	+	−	+	?	$-^2$	+	$+^2$	?
f_{10}	0	+	−	0	+	−	0	+	−	−	?	$-^2$	−	?	$2_,$	+	?	$2_,$	+	$+^2$	?	−	$+^2$	?	+	$+^2$	?
f_{0*}	0	0	0	0	0	0	0	0	0	+	+	+	+	+	+	+	+	+	−	−	−	−	−	−	−	−	−
f_{*0}	0	0	0	−	+	+	+	−	−	?	0	0	+	+	+	−	−	−	?	0	0	+	+	+	−	−	−

Table 4. Fixed-total Variable-portion Analysis

Case	Value	f_{11}	f_{10}	f_{01}	f_{00}	f_{1+}	f_{+1}	f_{0+}	f_{+0}
1	0	100	150	250	500	250	350	750	650
2	−20	100	150	**230**	520	250	330	750	670
3	+20	100	150	**270**	480	250	370	750	630
4	−20	100	**130**	250	520	230	350	770	650
5	−20	100	**130**	**230**	540	230	330	770	670

3.3   Fixed-Total Fixed-Portion-Combination Analysis

In the third analysis, the four main variables are assigned fixed-total values to $f_{11}, f_{10},$ f_{01} and f_{00}. The numerical examples are set to specific value ($f_{11} = 100$, $f_{10} = 150$, $f_{01} = 250$, $f_{00} = 500$). Four main variables are changed value by constant value that is identified to be a fixed-portion-combination portion. Table 5 shows five example cases and case No. 1 is a baseline. Case No. 2–5 shows moving combination values.

Table 5. Fixed-total Fixed-portion-combination

Case	f_{11}	f_{10}	f_{01}	f_{00}	f_{1+}	f_{+1}	f_{0+}	f_{+0}	N
1	100	150	250	500	250	350	750	650	1000
2	100	150	500	250	250	600	750	400	1000
3	100	250	150	500	350	250	650	750	1000
4	100	250	500	150	350	600	650	400	1000
5	100	500	150	250	600	250	400	750	1000

3.4   Imbalance and Extreme Scenario Analysis

The fourth method, an imbalance and extreme data is assigned the smallest and biggest values to main variables. Table 6 shows numerical examples simulating a possible extreme data and fixed total number at 1,000,000,000.

Table 6. Example Dynamic Value Statics Varible

Case	f_{11}	f_{10}	f_{01}	f_{00}	N
1	1	1,999	1,499,999	998,498,001	1,000,000,000
2	750	1,999,250	1,499,250	996,500,750	1,000,000,000
3	200,000	499,800,000	399,800,000	100,200,000	1,000,000,000
4	20,000,000	480,000,000	380,000,000	120,000,000	1,000,000,000
5	375,000	499,625,000	1,125,000	498,875,000	1,000,000,000

4 Analysis Results

4.1 Trend Analysis

Table 7 shows trend analysis of three main variables that consists of three tendency. The result is represented by number i.e. up trend = "+", down trend = "-", and unchanged or fixed trend = "0". The tendency analysis of Sect. 3.1 are calculated and predict tendency of measures.

Table 7. Tendency analysis result

Case	f_{1+}	f_{+1}	f_{11}	1	2	3	4	5	6	7	8	9	10	11	12	13	14	15	16	17	18	19	20	21
				Interestingness Measures																				
1	0	0	−	−	−	−	−	−	−	−	−	−	−	−	−	−	−	0	0	−	−	−	−	+
2	0	0	+	+	+	+	+	+	+	+	+	+	+	+	+	+	+	0	0	+	+	+	+	−
3	0	−	0	0	0	+	−	−	−	−	−	−	−	−	−	−	−	0	0	−	−	−	−	−
4	0	−	−	−	−	−	0	0	−	−	−	−	−	−	−	−	−	0	−	−	−	−	−	+
5	0	−	+	+	+	+	+	+	−	−	−	−	−	−	−	−	0	+	0	−	−	−	−	−
6	0	+	0	0	0	−	−	+	+	+	+	+	+	+	+	+	+	0	0	+	+	+	+	+
7	0	+	−	−	−	−	−	+	+	+	+	+	+	+	+	+	0	−	0	+	+	+	+	+
8	0	+	+	+	+	+	0	0	+	+	+	+	+	+	+	+	+	0	+	+	+	−	−
9	−	0	0	0	0	+	0	+	−	+	+	+	+	+	+	+	+	0	+	+	+	+	+	−
10	−	0	−	−	0	0	−	+	+	+	+	+	+	+	+	+	+	0	+	+	+	+	+	+
11	−	0	+	+	+	+	0	+	+	+	+	+	+	+	+	+	+	0	+	+	+	+	−	−
12	−	−	0	0	0	+	+	0	+	+	+	+	+	+	+	+	+	0	+	+	+	−	+	−
13	−	−	+	+	+	+	+	+	+	+	+	+	+	+	+	+	+	0	+	+	+	+	+	−
14	−	+	0	0	0	+	0	−	+	+	+	+	+	+	+	+	+	0	+	+	+	+	−	+
15	−	+	−	−	−	0	−	−	+	+	+	+	+	+	+	+	+	0	+	+	+	+	−	+
16	−	+	+	+	+	+	+	+	+	+	+	+	+	+	+	+	+	0	+	+	+	+	−	−
17	+	+	+	0	−	−	+	+	+	+	+	+	−	+	+	−	−	0	−	+	+	+	+	−
18	+	0	−	−	−	−	−	0	−	−	−	−	−	−	−	−	−	0	−	−	−	−	−	+
19	+	0	+	+	+	0	0	+	−	−	0	−	−	−	−	−	−	0	−	−	−	+	−	−
20	+	−	0	0	0	−	0	+	−	−	−	−	−	−	−	−	−	−	−	−	−	−	−	−
21	+	−	−	−	−	−	−	−	−	−	−	−	−	−	−	−	−	−	−	−	−	−	−	+
22	+	−	+	+	+	0	+	+	−	−	−	−	−	−	−	−	−	−	−	−	−	−	−	−
23	+	+	0	0	0	−	−	0	+	+	+	+	+	−	+	+	−	−	0	−	+	+	+	+	−
24	+	+	−	−	−	−	−	−	−	−	−	−	−	−	−	−	−	0	−	−	−	−	−	+

4.2 Dynamic Value Static Variable

The result of dynamic value on static variable is shown in Table 8. We define "o" is minimum value and "x" is maximum value and "−" is not identify. The numerical examples of Sect. 3.2 shows the result that case No. 3 is taken majority minimum value of all interestingness measure. However, case No. 5 is taken majority maximum value of interestingness measure.

Table 8. Result of dynamic value static varible

Case	Interestingness Measures																				
	1	2	3	4	5	6	7	8	9	10	11	12	13	14	15	16	17	18	19	20	21
1	x	o	–	–	–	–	–	–	–	–	–	–	–	–	–	o	o	–	–	–	–
2	x	o	x	–	–	–	–	–	–	–	–	–	–	–	x	–	o	–	–	–	o
3	x	o	o	o	o	o	o	o	o	o	o	o	o	o	o	–	o	–	o	o	x
4	x	x	–	–	o	–	–	–	–	–	–	–	–	–	–	x	–	–	–	–	–
5	x	x	x	x	x	x	x	x	x	x	x	x	x	x	x	x	x	x	x	x	–

4.3 Static Value Dynamic Variable

The result of static value dynamic variable shown in Table 9. We define "o" is minimum value and "x" is maximum value and "–" is not identify. The numerical examples of Sect. 3.3 shows the result that case No. 4 is taken majority minimum value of interestingness measure. However, case No. 1 and No. 3 are taken majority maximum value of interestingness measure.

Table 9. Result of static value dynamic varible

Case	Interestingness Measures																				
	1	2	3	4	5	6	7	8	9	10	11	12	13	14	15	16	17	18	19	20	21
1	x	x	–	x	x	x	x	x	x	x	x	x	x	x	o	x	x	x	x	o	o
2	x	x	o	–	–	–	–	–	–	–	–	–	–	o	–	x	–	–	–	–	–
3	x	–	x	x	–	x	x	x	x	x	x	x	x	x	o	x	x	x	x	o	–
4	x	–	o	o	o	o	o	o	o	o	o	o	o	o	o	x	o	o	o	o	x
5	x	o	x	–	–	–	–	–	–	–	–	–	–	o	–	–	–	–	–	–	x

4.4 Imbalance Value Static Variable

The result of imbalance value static variable shown in Table 10. We define "o" is minimum value and "x" is maximum value and "–" is not identify. The numerical examples of Sect. 3.4 shows the result that case No. 4 is taken majority minimum value of interestingness measure. However, case No. 3 is taken majority maximum value of interestingness measure.

Table 10. Result of imbalance value static varible

Case	Interestingness Measures																					
	1	2	3	4	5	6	7	8	9	10	11	12	1	13	14	15	16	17	18	19	20	21
1	o	–	o	–	–	x	x	x	–	o	x	x	o	o	o	–	x	x	o	o	o	o
2	–	o	–	–	–	–	–	–	–	–	x	–	–	–	–	–	o	–	–	–	–	–
3	–	–	–	o	o	o	o	o	o	–	o	o	–	–	x	o	o	o	–	x	–	–
4	x	x	–	–	–	–	–	–	x	–	–	x	x	x	o	–	–	x	–	–	–	–
5	–	–	x	x	x	–	–	–	x	–	–	–	–	–	x	–	–	–	–	x	–	–

5   Discussion and Conclusion

This paper presents four analytic methods and numerical data are systematical simulated. The paper describes 21 interestingness measures that effects to tendency and size of data. We have also investigated the behavior of association pattern interestingness measures over different data sets. In Imbalance data, we observed that all example cases cannot identify pattern association. Comparing the pattern of interestingness measure resulted in imbalance data cannot identify association pattern.

Systematically simulated data are proposed to investigate by four systematically method. Trend Analysis is the first method to analysis three variables with four main interestingness measure. However, we cannot exactly identify tendency interestingness measure. The result of trend analysis shown that tendency depend on $P(A, B)$.

Two method, dynamic value static variable and static value dynamic variable shown pattern of association pattern. Example cases of two method is generated using systematics. The results can shown pattern of association.

The last method, imbalance data is generated based on medical data. The result shown that we cannot identify association pattern. The future work, large and complex data will generate in order to investigate relationship between objective interestingness measure and association pattern.

Acknowledgement. This work has been supported funding by Rangsit University and Sirindhorn International Institute of Technology, Thammasat University.

References

1. Agrawal, R., Imielinski, T., Swami, A.: Mining association rules between sets of items in large databases. In: ACM SIGMOD International Conference on Management of Data, Washington DC, USA, pp. 207–216 (1993)
2. Brin, S., Motwani, R., Ullman, J.D., Tsur, S.: Dynamic itemset counting and implication rules for market basket data. In: ACM SIGMOD International Conference on Management of Data. New York, USA, 255–264 (1997)
3. Leung, K.S., Wong, K.C., Chan, T.M., Wong, M.H., Lee, K.H., Lau, C.K., Tsui, S.K.: Discovering protein–DNA binding sequence patterns using association rule mining. Nucleic Acids Res. 38(19), 6324–6337
4. Srivastava, J., Cooley, R., Deshpande, M., Tan, P.N.: Web usage mining: discovery and applications of usage patterns from web data. ACM SIGKDD Explor. Newsl. 1(2), 12–23 (2000)
5. Lee, W., Stolfo, S.J., Mok, K.W.: A data mining framework for building intrusion detection models. In: Proceedings of the 1999 IEEE Symposium on Security and Privacy, pp. 120–132. IEEE (1999)
6. Piatetsky-Shapiro, G.: Discovery, analysis, and presentation of strong rules. In: Knowledge Discovery in Databases, pp. 229–238 (1991)
7. Tan, P.N., Kumar, V., Srivastava, J.: Selecting the right interestingness measure for association patterns. In: The Eighth ACM SIGKDD International Conference on Knowledge Discovery and Data Mining, Edmonton, Alberta, pp. 32–41 (2002)
8. Agresti, A.: Categorical Data Analysis. Wiley, New York (1990)

9. Goodman, L.A., Kruskal, W.H.: Measures of associationfor cross-classifications. J. Am. Stat. Assoc. **49**, 732–764 (1968)

10. Mosteller, J.: Association and estimation in contingency tables. J. Am. Stat. Assoc. **63**, 1–28 (1968)

11. Yule, G.U.: On the methods of measuring association between two attributes. J. R. Stat. Soc. **75**, 579–642 (1912)

12. Cohen, J.: A coefficient of agreement for nominal scales. Educ. Psychol. Meas. **20**, 37–46 (1960)

13. Cover, T., Thomas, J.: Elements of Information Theory. Wiley, New York (1991)

14. Smyth, P., Goodman, R.M.: Rule induction using information theory. In: Shapiro, G.P., Frawley, W. (eds.) Knowledge Discovery in Databases, pp. 159–176. MIT Press, Cambridge (1991)

15. Breiman, L., Friedman, J., Olshen, R., Stone, C.: Classification and Regression Trees. Chapman & Hall, New York (1984)

16. Clark, P., Boswell, R.: Rule induction with cn2: some recent improvements. In: Proceedings of the European Working Session on Learning EWSL-91, Porto, Portugal, pp. 151–163 (1991)

17. Brin, S., Motwani, R., Ullman, J., Tsur, S.: Dynamic itemset counting and implication rules for market basket data. In: Proceedings of 1997 ACM-SIGMOD International Conference on Management of Data, Montreal, Canada, pp. 255–264 (1997)

18. DuMouchel, W., Pregibon, D.: Empirical bayes screening for multi-item associations. In: The Seventh International Conference on Knowledge Discovery and Data Mining, pp. 67–76 (2001)

19. Shortliffe, E., Buchanan, B.: A model of inexact reasoning in medicine. Math. Biosci. **23**, 351–379 (1975)

20. Tan, P.N., Kumar, V.: Interestingness measures for association patterns: a perspective. In: KDD 2000 Workshop on Post-processing in Machine Learning and Data Mining, Boston, MA, August (2000)

21. van Rijsbergen, C.J.: Information Retrieval, 2nd edn. Butterworths, London (1979)

22. Klosgen, W.: Problems for knowledge discovery in databases and their treatment in the statistics interpreter explora. Int. J. Intell. Syst. **7**(7), 649–673 (1992)

A Novel Multi Stage Cooperative Path Re-planning Method for Multi UAV

Xiao-hong Su$^{(\boxtimes)}$, Ming Zhao$^{(\boxtimes)}$, Ling-ling Zhao, and Yan-hang Zhang

School of Computer Science and Technology, Harbin Institute of Technology,
Harbin 150001, China
sxh@hit.edu.cn, Sequoia00@163.com

Abstract. When the multi-UAVs cooperatively attack multi-tasks, the dynamic changes of environments can lead to a failure of the tasks. So a novel path re-planning algorithm of multiple Q-learning based on cooperative fuzzy C means clustering is proposed. Our approach first reflects the dynamic changes of re-planning space by updating the fuzzy cooperative matrix. Then, the key way-points on the current global paths are used as the initial clustering centers for the cooperative fuzzy C means clustering, which generates the classifications of space points for multi-tasks. Furthermore, we use the classifications as the state space of each task and the fuzzy cooperative matrix as the reward function of the Q-learning. So a multi Q-learning algorithm is presented to synchronously re-plan the paths for multi-UAVs at every step. The simulation results show that the method subtracts the re-planning space of the tasks and improves the search efficiency of the learning algorithm.

Keywords: Multi-UAVs · Cooperative fuzzy C means clustering · Multi Q-learning · Path re-planning

1 Introduction

Multi-Unmanned Aircraft Systems (MUAS) have been widely applied in military and civilian environments because they maintain more reliability, security and flexibility over a single UAV [1]. The cooperative path planning of multi-UAVs plays an essential part in the task planning system of MUAS, which mainly studies how to plan secure and feasible flight paths from their initial positions to the target locations. However, the off-line global paths may fall into some new threat areas because of dynamically changing environments. Therefore, the global paths should be re-planned on the execution of the tasks to ensure flight safety and reduce the probability of failure [2]. There are more difficulties to solve and stronger real-time requirements than the off-line path planning.

The most common approaches directly extend the methods of off-line global planning to dynamic re-planning. For example, the classic mixed integer linear programming (MILP) has been applied to path optimization and collision avoidance in the dynamic environment in [3]. The method of graph-based such

© Springer International Publishing Switzerland 2016
R. Booth and M.-L. Zhang (Eds.): PRICAI 2016, LNAI 9810, pp. 482–495, 2016.
DOI: 10.1007/978-3-319-42911-3_40

as a random road map in [4] and the heuristicalgorithms such as D* and M* in [5,6] are also commonly used for path planning and re-planning. However, these methods suffer from a large amount of calculation, and it's difficult to take the cooperation among the UAVs into account.

Moreover, swarm intelligence algorithms have obvious effectiveness on the cooperative path planning of multi-UAVs, such as genetic algorithms [7] and particle swarm optimization algorithms [8], all have natural advantages in the parallel multi-task planning and collaboration. But those algorithms are too complex with heavy computation costs. The existing research usually shortens the number of iterations and reduces the size of the population to improve the speed of the algorithms, as their results are easy to fall into local optimum. In recent years, the approach of real-time path planning based on dynamic programming has been generally recognized. The Markov Decision Process (MDP) in [9,10] and the Reinforcement Learning (RL) in [11,12] are widely used in the on-line path planning. In this way, the state transition is gradually implemented through regional exploration and prediction, and the ability of the UAVs to deal with complex unknown environments is improved by the feedback of learning.

Previously, a representation method of 3D space based on fuzzy sets have set up in [13], and the global optimal paths for multi-UAVs are generated. On this basis, a multi Q-learning path re-planning method based on cooperative fuzzy C means clustering (CFCM) is proposed in this paper. This method first monitors the current flight states, environmental changes and paths failure. Then it describes the cooperative correlation for the space points, and builds up the fuzzy cooperative matrix based on fuzzy space representation. Furthermore, the fuzzy C means clustering is performed in every space grid surface of re-planning space, while the space points are divided into some state classifications for different tasks according to the task properties and the cooperative correlation. Finally, a multi Q learning algorithm with multi step is used in each task state space to select the optimal action strategy iteratively, which effectively produces the cooperative re-planning paths of multi UAVs.

The rest of this paper is organized as follows. Section 2 describes the re-planning cooperative framework for multi-UAVs. The cooperative correlation coefficient and fuzzy cooperative matrix are given in Sect. 3, and the division of state space for multi-tasking based on the CFCM is also presented. In Sect. 4, we focus on the study of path re-planning based on multiple Q learning with CFCM. Simulation results and analysis are shown in Sect. 5 and the conclusion is given in Sect. 6.

2 The Path Re-planning Model of Multi-UAVs

2.1 The Description of Path Re-planning

In the phase of off-line, a set of feasible global cooperative paths has been planned by the cultural algorithm with fuzzy space representation in [13]. An example of the global path planning is shown in Fig. 1.

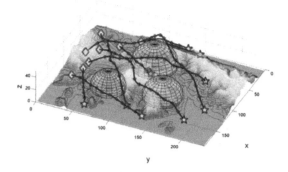

Fig. 1. The scenario of MUCMTA in 3D environment (Color figure online)

In Fig. 1, the diamonds are the starting positions of the UAVs, the stars are the target locations for attack, and the yellow hemispheres are the radiation ranges of the radar threat area. Each optimal path of the UAVs is represented by a blue curve that is connected by a set of key way-points. However, during the UAVs performed tasks, the radar threat in the region can move toward an unknown direction, and some new threats may randomly pop up. Thus, part of the paths will fall into the dynamic threat areas again, which has the risk of being detected by radar. So it is essential to re-plan the failed path section so that the paths in the task execution are secure.

The path re-planning can be described as a multi-step, multi-object cooperative optimization problem that is closely related to time. Each task uses the global path as a reference, and it is constantly adjusting and correcting the key way-points following the time of t, so that the UAVs are able to respond to the changes of the threat area in the environment. The paths of re-planning can be expressed as a function of time t as follows:

$$L'_k = \{s_k, p_{k1}(t), \cdots, p_{ki}(t), t_k\} \quad k \in [1, N] \tag{1}$$

where the key way-point on the path of task k changes its position with time t. It ensures that the UAVs are able to bypass the changed threat areas and maintain cooperation with other UAVs.

The major challenges of the re-planning problem are summarized as the following aspects:

1. How to effectively reduce the planning space to save resources when updating the space itself, and how to keep the cooperative performance of the multi-task functional.

2. When their unexpected environment changes, how to quickly update the information in a limited time frame as well as to immediately inform and give feedback to the re-planning algorithm.

3. Although the re-planning paths are more focused on feasibility, it is still essential for the paths to increase towards and optimal path. Therefore, how to select the most effective algorithm is critical.

2.2   Re-planning Model Based on CFCM and Multi Q Learning

The Q learning algorithm, proposed by Watkins, is an asynchronous dynamic programming method of no-model [14]. It combines the dynamic programming with learning mechanism, in order to solve the optimization problem of multi-step decision making with delayed returns. However, the "Curse of dimensionality" is still unsolved in the Q learning algorithm, especially when the problem is complex and the cooperative demand is high. In order to handle the dynamically changing environment, it is necessary to reduce the search space and to optimize the representation of the cooperative tasks.

According to the previous study, the discrete points on the space grid have been represented as a fuzzy set related to the tasks and the threats, which can effectively enhance the degree of attention. On this basis, a reasonable classification is obtained by using the CFCM to further explore the structure of the multi-dimensional space data, while it is also suitable for dividing the state space during the changes in the environment. Therefore, we use the CFCM to divide the space into the categories of each task with cooperative properties, which can effectively reduce the search space and improve the efficiency of cooperative re-planning. Moreover, taking the spatial fuzzy set as the return matrix of multi Q learning can increase the pertinence of learning and reduce the blindness and randomness in the initial learning. The fundamental structure of combination these two methods is shown in Fig. 2.

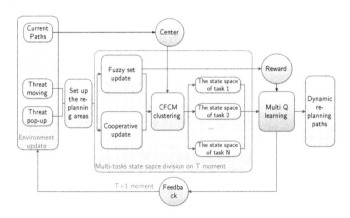

Fig. 2. The model of cooperative path re-planning

3   The Division of State Space Based on the CFCM

3.1   Correlation Coefficient and Fuzzy Cooperative Matrix

In order to deal with the cooperation of multi tasks better, it is not only to measure the membership degree of the space points for all tasks, but also to analyze the reliability that the space point as a key way-point for any one of the tasks. The similarity degree of different categories is expressed by the cooperation coefficient of the clustering method, so the cooperation coefficient can be introduced to describe the cooperative performance of the space point when it is a key way-point for any tasks.

The cooperative correlation is mainly determined by the cooperative relationship of UAV space and time. The space cooperative correlation can be obtained by the similarity degree of membership that the current key way-point performs during different tasks. If the degree of membership for a point to execute a task is much higher than that to execute the other tasks, this point is only suitable to be the key way-point of this task. The time cooperative correlation is obtained by comparing the time difference of which the current key way-point executes different tasks. If the time of a point to perform a task is shorter than to perform other tasks, the point is more suitable to be the key way-point of the short time task. The point therefore has a higher time cooperative correlation for this task.

Based on the above analysis, we introduce the cocorrelation coefficients of space and time, which increases the measurement of the cooperative performance for the key way-points selected. The cocorrelation coefficients are defined as follows:

(1) Co-correlation coefficient of spatial membership degree:

We observe the membership degree set $\{\tilde{\mu}_1, \cdots \tilde{\mu}_i \cdots \tilde{\mu}_j \cdots \tilde{\mu}_k, \quad i \neq j\}$ of a point P for every task. The difference of $|\tilde{\mu}_i - \tilde{\mu}_j|$ can reflect the cooperative correlation that the point executes task i or j. When the difference is large, the point suits for task i, but not for task j. On the contrary, when the difference is small, the point both suit for task i and j, which is not good for cooperation. As the maximum value in the membership degree set has the highest level of attention, the co-correlation coefficient is calculated by the minimum difference between the maximum degree and the other degrees. It is as follows:

$$\beta_1^{k_i} = min\left(\left|max\left(\frac{\tilde{\mu}_{ki}}{\sum\limits_{i=1}^{n} \tilde{\mu}_{ki}}\right) - \left(\frac{\tilde{\mu}_{kj}}{\sum\limits_{i=1}^{n} \tilde{\mu}_{ki}}\right)\right|\right), \quad i,j = 1,2,\cdots,n, i \neq j \quad (2)$$

(2) Co-correlation coefficient of minimum time differences:

Simarlarly to (1), the co-correlation coefficient of time is defined by the minimum difference of execution time where a point executes any tasks. According to the execution time set of the point to each task, the cooperation coefficient of minimum time difference can be described as follows:

$$\beta_2^{k_i} = min\left(\left|max\left(\frac{t_{ki}}{\sum\limits_{i=1}^{n} t_{ki}}\right) - \left(\frac{t_{kj}}{\sum\limits_{i=1}^{n} t_{ki}}\right)\right|\right), \quad i,j = 1,2,\cdots,n, i \neq j \quad (3)$$

When the co-correlation coefficient is defined, the points on the planning space can be expressed as the integrated form of the coordinate position, membership degree of task, and co-correlation coefficient. Such representation makes the information of the points on the planning space more complete. The fuzzy cooperative matrix of the space point set is established as follows:

$$
S = \left\{ \begin{array}{ccc ccccc cc}
\overbrace{\begin{matrix} x_1 & y_1 & z_1 \end{matrix}}^{a1} & \overbrace{\begin{matrix} \tilde{\mu}_1^1 & \tilde{\mu}_2^1 & \cdots & \tilde{\mu}_n^1 \end{matrix}}^{a2} & \overbrace{\begin{matrix} \beta_1^1 & \beta_2^1 \end{matrix}}^{a3} \\
x_2 & y_2 & z_2 & \tilde{\mu}_1^2 & \tilde{\mu}_2^2 & \cdots & \tilde{\mu}_n^2 & \beta_1^2 & \beta_2^2 \\
\vdots & \vdots & \vdots & \vdots & \vdots & \vdots & \vdots & \vdots & \vdots \\
x_k & y_k & z_k & \tilde{\mu}_1^k & \tilde{\mu}_2^k & \cdots & \tilde{\mu}_n^k & \beta_1^k & \beta_2^k
\end{array} \right\} \tag{4}
$$

Where S denotes a set of points on the grid of the planning space, $a1$ is the point coordinate position, $a2$ is the membership degrees for the point to execute each task, and $a3$ is the collection of co-correlation coefficient, which increases more cooperative information for the space points. So if we cluster the planning space according to the fuzzy cooperative matrix, the key way-points for the task can be selected more effectively, and those clusters maintain good cooperation and independent of each other. Therefore, it can quickly plan the reasonable cooperative paths in the cluster space of task.

3.2 The CFCM Based on Fuzzy Cooperative Matrix

Incorporating the measurement of the cooperative relationship into clustering can effectively improve the classification rate [16]. The points on each grid surface of the planning space can be clustered by the fuzzy cooperative matrix and CFCM, which produces the cooperative classifications for different tasks. Furthermore, the data sample is the coordinate set of space points $S(a1)$, and the degree of membership for tasks clustering is normalized by $S(a2)$. So the cooperative target function with the co-correlation coefficient for CFCM is:

$$
J_b(U, v) = \sum_{j=1}^{k} \sum_{i=1}^{N} (\tilde{\mu}_{ji})^b (d_{ji})^2 + \sum_{m=1}^{M} (\beta_m) \sum_{j=1}^{k} \sum_{i=1}^{N} (\tilde{\mu}_{ji} - \tilde{\mu}_{ji'})^b (d_{ji})^2 \tag{5}
$$

The second item in (5) is the sum of the cooperative of space and time. Where d_{ji} is the Euclidean distance. It measures the distance between the sample of x_j and the center of class i, and β is the co-correlation coefficient.

The global paths and the re-planning paths of last time t are known, so if the key way-points on the previous paths are used as the initial clustering center, the space classification should be more reasonable. Moreover, it is also increase the accuracy of the cluster that adopt $S(a2)$ as the initial division matrix U. The update function of the division matrix is defined as follows:

$$
\left\{ \begin{array}{l}
U' = \dfrac{1}{\sum\limits_{i=1}^{N} (\frac{d_{ik}}{d_{i'j}})^{(\frac{2}{b-1})}} \\
U_{ik} = s \cdot U + (1 - s) \cdot U', \qquad 0 \leq s \leq 1
\end{array} \right. \tag{6}
$$

The new division matrix combines the last degree of membership and the updating ones by the proportion factor of s. Then that the method is to update the center of cooperative cluster is as follows:

$$v_{ji} = s' \frac{\sum\limits_{j=1}^{k} (\mu_{ji})^b x_{ji}}{\sum\limits_{j=1}^{k} (\mu_{ji})^b} + (1 - s') \frac{\sum\limits_{j=1}^{k} \sum\limits_{m=1}^{M} (\beta_{mi})^b x_{ji}}{\sum\limits_{j=1}^{k} \sum\limits_{m=1}^{M} (\beta_{mi})^b}, \qquad 0 \le s' \le 1 \qquad (7)$$

The second item in (7) is the effect to cluster the center by the co-correlation coefficient. According to (6) and (7), it modifies the cluster center and the division matrix until the algorithm is convergent. The classification of space points for the re-planning task is shown in Fig. 3.

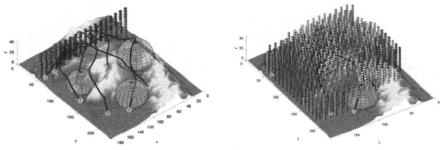

(a) The clustering in one grid surface   (b) The all clustering of task space

Fig. 3. The schematic of state space division for re-planning by CFCM

4   The Re-planning with Multiple Q Learning and CFCM

4.1   Q Learning Algorithm and the Fuzzy Reward Rates of Cooperation

The Q learning algorithm combines the TD learning and extends dynamic programming, which has the merits of no dependence on a model and that the strategy updates on-line. So, it is appropriate for solving the problem of on-line path re-planning. According to the feature of Q learning, we can define the feasible key way-points collection of UAV to be the state space of re-planning as $S = (s_1, s_2, \cdots, s_t)$. The action collection of $A = (a_1, a_2, \cdots, a_t)$ indicates the possible maneuver when UAV passes over a key way-point. The result of this action is a set of alternative points that are selected from the next grid surface. Then the probability state is that transfers from the current state to the next state $T(s_t, a, s_{t+1})$, and $R = (r_1, r_2, \cdots, r_t)$ is a rewarded when action a is executed. The goal of Q learning is to update the Q value by iterations, so that it approaches the optimal strategy selection. The classical update function of Q value [15] is as follows:

$$Q(s_t, a_t) = (1 - \alpha_t)Q(s_t, a_t) + \alpha_t(r(s_t, a_t) + \gamma max_{a_{t+1}}[Q(s_{t+1}, a_{t+1})]) \qquad (8)$$

Where γ is the discount coefficient, and α_t is the learning rate that constantly declines with time t while $0 \leq \alpha \leq 1$. In each iteration, both the state of time $t + 1$ is observed and all the action is updated. Then it also requires the current state-actions to be updated with an estimation value of $Q(s_t, a_t)$, so as to select the optimal strategy for key way-points in the current position.

Although the Q learning algorithm has no model, and it does not need to know the reward function $r(s, a)$, there is great blindness and randomness because the priory knowledge is lacking. Especially in early learning, it is easy to fall into the local optimum. So the reward function based on the priory knowledge of Q learning can effectively reduce the blindness of the search. The membership degree and co-correlation coefficient of point set which corresponds to a task in the space can reflect the level of attention. So it can use them together as a reward of Q learning. Using this reward, when the Q learning algorithm transfers the key way-point in t to the next key way-point $t + 1$, the point with a higher degree of membership and co-correlation coefficient has the greater probability of being chosen. Therefore, it can reduce the blindness of the learning and improve the efficiency of search. Then the reward function can be set up by the cooperative fuzzy matrix, and it is updated with the matrix, as follows:

$$\begin{cases} r(s_t, a_t, s_{t+1}) = [\omega_1 \tilde{\mu}_i^j(t + 1) + \omega_2 \sum_{m=1}^{M} \beta_m^i(t + 1)], & i, j = 1, 2, \cdots, n \\ r(T) = \infty \end{cases} \qquad (9)$$

Where ω_1 and ω_2 are weight coefficients, which the sum is 1. $r(T)$ indicates the reward of task target position, and this reward is set to a greater value

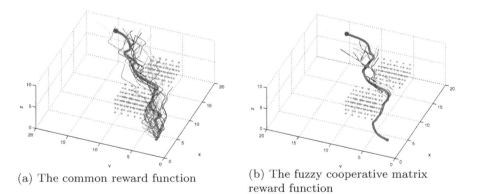

(a) The common reward function

(b) The fuzzy cooperative matrix reward function

Fig. 4. The effect of reward matrix on learning (Color figure online)

or ∞ than any other points, so that the Q learning can eventually find the target. It shows that the effect on re-planning by using the cooperative fuzzy matrix as a reward function in Fig. 4.

In the Fig. 4, the regions of red grid are the threat areas, the color lines are the searched location in the Q learning iteration, and the red solid line is the optimal result. It can be seen from the Fig. 4 that using the fuzzy cooperative matrix as the reward function can effectively optimizes the performance of the search, and reduces the initial blindness.

4.2   The Re-planning Based on CFCM and Multi-Q Learning

The CFCM algorithm can first be used to divide the space based on the membership degree of task and cooperative correlation. Then the space points that are suitable for a same task are divided into a class, while the class is as a state space of Q learning for the task.

Based on the state space classification for multi-tasks, the multi Q learning algorithm can parallel re-planning paths of multi tasks in the same learning process. Multi Q learning only needs to consider the actions $[a_t^1, \cdots, a_t^n]$ for multi tasks that are updated by the Q value at the same time, because the CFCM clustering has divided the state spaces of the multi tasks. So the updated formula of multi Q learning is as follows:

$$
\begin{aligned}
Q(s_t, a_t^1, \cdots, a_t^n) = {} & (1 - \alpha_t)Q(s_t, a_t^1, \cdots, a_t^n) + \alpha_t(r(s_t, a_t^1, \cdots, a_t^n) \\
& + \gamma max_{a_{t+1}}[Q(s_{t+1}, a_{t+1}^1, \cdots, a_{t+1}^n)])
\end{aligned}
\tag{10}
$$

Then the step of multi Q learning based on CFCM clustering is as follows:

Step 1: The algorithm monitors the flight status of UAVs and the environmental information of the current step, sets the re-planning regions by the changed threat areas Tp, and updates the fuzzy cooperative matrix of the space.

Step 2: Using the key way-points of last stage paths on the grid surface as the initial centers, CFCM clustering is performed to generate the point classification for cooperative tasks on each grid surface.

Step 3: It combines the same task classification in every grid surface of the space to form the state space of this task, and updates the reward matrix.

Step 4: To observe the state $[s_t^1, \cdots, s_t^n]$ of each task at the key way-points of the current grid surface, and the action $[a_t^1, \cdots, a_t^n]$ of each task are selected and executed.

Step 5: To observe the state $[s_{t+1}^1, \cdots, s_{t+1}^n]$ of the next step of grid surface corresponding to the action $[a_t^1, \cdots, a_t^n]$, and obtain the reward $[r_t^1, \cdots, r_t^n]$.

Step 6: It updates the value of $Q(s_t, a_t^1, \cdots a_t^n)$ based on $[r_t^1, \cdots, r_t^n]$.

Step 7: Repeat steps 1–6, until all tasks search to their final target.

5   Experimental Results

Based on the global off-line paths planning, the initial input of the algorithm is obtained from previous work [13]. The parameters of clustering iteration number is 200, clustering update interval is 20(s) and the Q learning iteration number is 5000. Moreover, the learning rate is $[1.0 \sim 0.1]$, and the discount coefficient of learning is 0.7. Then several simulation cases are implemented on the software platform of MATLAB (R2015a) in Windows 7, which is based on the PC of a Intel Core i5 3.20 GHz and 16 GB of RAM.

5.1 Exp. 1: Feasibility Experiment for Cooperative Paths Re-planning

We made the experiment to verify the feasibility of path re-planning in a dynamic environment. Figure 5 shows that the results of multi-stages of re-planning paths for single UAV with the approach of CFCM and Q learning.

(a) The first stage of re-planning   (b) The second stage of re-planning   (c) The last stage of re-planning

Fig. 5. The result of multi stages re-planning paths for single UAV (Color figure online)

In Fig. 5, the dark blue lines are the global paths, the yellow lines are the re-planning paths on the first stage, the light blue lines are the re-planning path on the second stage and the red path is the re-planning path on the last stage. The semicircular areas are the moving radar threat areas, where the different colour indicates the position that the radars have passed in each stage. Figure 5(a)–(c) represents the effect of each stage of path re-planning. It can be seen from the Fig. 5 that our approach can re-plan the effective path from the current position of a UAV to the target point with the environment changes, while the global path is gradually adjusted by the state transition at all stages. This approach ensures that the final paths respond to the dynamic of threat areas moving, so that the re-planning paths by dynamic programming are always safe and reliable.

Then the effect of multi-UAVs cooperative path re-planning by the approach of CFCM and multi Q learning in each stage is shown in Fig. 6.

From Fig. 6, we can get conclusion that all re-planning paths have good cooperative with each other and they can effectively avoid the collisions during the different stages, because the paths re-planning are performed respectively by CFCM, and they adapt to the dynamic changes of the environment by multi-Q learning. However, some re-planning paths have large maneuvers at a certain stage in Fig. 6, because they are impacted by the location of the current threat areas. Then the next stage re-planning can fix the maneuver of last stage, which

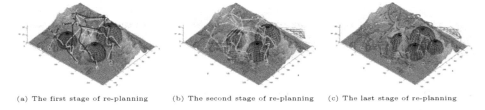

(a) The first stage of re-planning   (b) The second stage of re-planning   (c) The last stage of re-planning

Fig. 6. The result of multi stages re-planning paths for UAVs cooperation

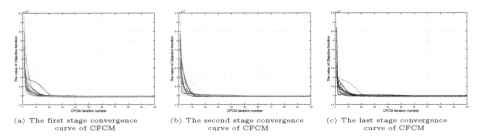

(a) The first stage convergence   (b) The second stage convergence   (c) The last stage convergence
curve of CFCM   curve of CFCM   curve of CFCM

Fig. 7. The convergence curve of objective function for Multi stages CFCM clustering

ensures that the UAVs are always able to fly on the best re-planning paths. From all stages, we can see that the effects of re-planning paths are gradually better, and they are always near the shortest length and highest degree of membership, so it is a feasible approach.

Figure 7 shows that the convergence curve of the CFCM objective function in every stage. The CFCM clustering dynamically divide the task state space so that can reflect the updated current environment. It can be seen from Fig. 7 that the algorithm quickly converges to the optimal clustering center, and the process of convergence are all similar, so the CFCM can better divide the tasks state space in every stage. Table 1 shows the execution time of CFCM.

Table 1. The execution time of re-planning by CFCM for every stage

Stages	Execution time of tasks(s)								
	t1	t2	t3	t4	t5	t6	t7	t8	t9
stage1	7.82	10.41	9.20	6.89	11.78	5.80	10.37	8.52	5.41
stage2	7.21	9.84	8.56	6.90	11.09	5.36	9.79	7.19	5.27
stage3	5.39	7.19	6.61	6.67	10.61	4.88	8.64	7.88	5.27

From Table 1, the execution time shows that the novel approach re-plans the paths very quickly, and the execution time is gradually shorter with the increasing stages because the re-planning spaces are gradually decrease.

5.2   Exp. 2: Compared with Other Methods

To analyse the effect of re-planning for one task in space, we first compared the improved CFCM and Q learning approach with others reward function Q learning and off-line planning methods. Then, the execution time of these methods and the previous off-line global planning are given in Table 2.

Table 2. The execution time comparison of re-planning and offline planning

Methods	Execution time of tasks(s)									
	iteras	t1	t2	t3	t4	t5	t6	t7	t8	t9
Re-planning of standard	5000	8.93	11.53	11.20	6.73	12.82	7.53	11.81	9.33	7.02
Re-planning of global fuzzy	5000	9.10	11.93	12.14	7.22	13.06	7.98	11.96	10.02	7.64
Re-planning of CFCM	5000	7.82	10.41	9.20	6.89	11.78	5.80	10.37	8.52	5.41
Offline-planning of global	2000	84.33	87.42	79.02	83.01	104.21	77.02	78.76	90.65	75.27

From Table 2, we can see that all the re-planning methods have a faster execution speed than the off-line planning, so they can satisfy the requirement of time for re-planning during the on-line phase. As well as our approach has the fastest execution time, because the division of planning space for tasks by CFCM optimizes the state space of task searching and ensures the efficiency of the algorithm.

Then we compared our approach with other dynamic re-planning methods of FPAG in [7] and POMDP in [9], and the result is shown in Table 3.

Table 3. The task cost and execution time comparison of re-planning methods

Task	CFCM-MQL		FPGA		POMDP	
	Cost(km)	Time(s)	Cost(km)	Time(s)	Cost(km)	Time(s)
UAV1	148.38	7.92	140.07	23.21	194.48	11.93
UAV2	221.52	10.31	210.63	32.80	264.73	13.35
UAV3	179.97	9.23	190.32	30.12	182.21	11.20
UAV4	242.33	6.88	220.47	31.55	246.90	6.78
UAV5	301.67	12.01	238.15	38.73	287.19	12.84
UAV6	198.41	6.22	177.82	25.24	223.61	9.34
UAV7	150.16	10.54	149.09	35.06	137.34	15.67
UAV8	249.48	8.47	218.83	28.49	254.38	11.06
UAV9	141.35	5.39	137.59	24.60	140.06	7.64

From Table 3, we can see that the execution time of our approach is better than others. Although our approach and POMDP has longer path cost than FPGA because of they searching the path are not sufficient, those cost are acceptable for re-planning problems.

6   Conclusions

In this paper, we proposed a novel multi-Q learning approach for cooperative path re-planning of UAVs based on the fuzzy C means clustering. This approach first uses the CFCM clustering to divide the planning space. Secondly, the co-correlation coefficient of the space points for tasks is defined to increase the cooperative information of the key way-points. Moreover the fuzzy cooperative matrix of the space points is used as the reward function for Q learning, so that it effectively reduces the blindness of algorithm searching. Finally, the multi Q learning is also optimized by using the parallel update function. The algorithm can deal with multi re-planning tasks in the needed execution time.

In the future, our work will be more focused on the maneuverability optimization for re-planning paths, and deal with the complex re-planning problem when the space dimension and the number of tasks greatly increases.

References

1. Pierre, D.M., Zakaria, N., Pal, A.J.: Self-Organizing Map approach to determining compromised solutions for multi-objective UAV path planning. In: 2012 12th International Conference on Control Automation Robotics & Vision (ICARCV), pp. 995–1000. IEEE (2012)
2. Brezak, M., Petrovic, I.: Real-time approximation of clothoids with bounded error for path planning applications. IEEE Trans. Robot. **30**(2), 507–515 (2014)
3. Berger, J., Boukhtouta, A., Benmoussa, A.: A new mixed-integer linear programming model for rescue path planning in uncertain adversarial environment. Comput. Oper. Res. **39**(12), 3420–3430 (2012)
4. Zucker, M., Kuffner, J., Branicky, M.: Multipartite RRTs for rapid replanning in dynamic environments. In: 2007 IEEE International Conference on Robotics and Automation, pp. 1603–1609. IEEE (2007)
5. Wu, J., Zhang, D.-h.: Path planning for dyanmic target based on kalman filtering algorithm, D* algorithm. Electron. Opt. Control **21**(8), 50–53 (2014)
6. Wagner, G., Choset, H.: M*: A complete multirobot path planning algorithm with performance bounds. In: 2011 IEEE/RSJ International Conference on Intelligent Robots and Systems (IROS), pp. 3260–3267. IEEE (2011)
7. Kok, J., Gonzalez, L.F., Kelson, N.: FPGA implementation of an evolutionary algorithm for autonomous unmanned aerial vehicle on-board path planning. IEEE Trans. Evol. Comput. **17**(2), 272–281 (2013)
8. Lu, L., Gong, D.: Robot path planning in unknown environments using particle swarm optimization. In: Fourth International Conference on Natural Computation, ICNC 2008, vol. 4, pp. 422–426. IEEE (2008)
9. Ragi, S., Chong, E.K.P.: UAV path planning in a dynamic environment via partially observable Markov decision process. IEEE Trans. Aerosp. Electron. Syst. **49**(4), 2397–2412 (2013)

10. Feyzabadi, S., Carpin, S.: Risk-aware path planning using hirerachical constrained Markov decision processes. In: 2014 IEEE International Conference on Automation Science and Engineering (CASE), pp. 297–303. IEEE (2014)
11. Zhang, B., Mao, Z., Liu, W., et al.: Geometric reinforcement learning for path planning of UAVs. J. Intell. Robot. Syst. **77**(2), 391–409 (2015)
12. Wu, Y.-r., Wu, Y.-l., Ding, W., et al.: Dual-Aircraft cooperative path planning based on Q learning. Electron. Opt. Control **21**(8), 15–19 (2014)
13. Zhao, M., Zhao, L., Su, X., Ma, P., Zhang, Y.: A cultural algorithm with spatial fuzzy set to solve multi-UAVs cooperative path planning in a three dimensional environment. J. Harbin Inst. Technol. **47**(10), 29–34 (2015)
14. Watkins, C.J.C.H., Dayan, P.: Q-learning. Mach. Learn. **8**(3–4), 279–292 (1992)
15. Xu, X.: Reinforcement Learning, Approximate Dynamic Programming. Science Press, Beijing (2010)
16. Yin, X.: The Study of Multi-Agent Cooperative Reinforcement Learning Methods. The National University of Defense Technology, Changsha (2003)
17. Tan, X.: Fuzzy Clustering Algorithm Based on Collaborative Research. Changsha University of Science & Technology, Changsha (2013)

Sentiment Analysis for Images on Microblogging by Integrating Textual Information with Multiple Kernel Learning

Junxin Tan[1], Mengting Xu[1], Lin Shang[1(✉)], and Xiuyi Jia[2]

[1] State Key Laboratory for Novel Software Technology,
Department of Computer Science and Technology,
Nanjing University, Nanjing 210023, China
jxtan@smail.nju.edu.cn, xumtpark@gmail.com, shanglin@nju.edu.cn
[2] School of Computer Science and Engineering,
Nanjing University of Science and Technology, Nanjing 210094, China
jiaxy@njust.edu.cn

Abstract. Image is one of the most important means to express users' emotions on microblogging, like Sina Weibo. More and more people post only images on it, due to the fast and convenient nature of image. Taking a post only using images on microblogging has been a new tendency. Most existing studies about sentiment analysis on microblogging focus on the text, or integrate image as an auxiliary information into text, so they are not applicable in this scenario. Although a few methods related to sentiment analysis for image have been proposed, most of them either ignore the semantic gap between low-level visual features and higher-level image sentiments, or require a lot of textual information in the phases of both training and inference. This paper proposes a new sentiment analysis method based on *Simple Multiple Kernel Learning (SimpleMKL)*. Specifically, textual information as a sort of sufficiently emotional source data, we can use it to promote the ability via SimpleMKL to classify images. And once we get the image classifier, none of texts are needed when predicting other unlabelled images. Experimental results show that our proposed method can improve the performance significantly on data we crawled and labelled from Sina Weibo. We find that our method not only outperforms some common methods, like SVM, Naive Bayes, KNN, Random Forest, Adaboost, etc., using the image features of colour, hog, texture, but also outperforms some state-of-the-art methods.

Keywords: Sentiment analysis · Microblogging · Image sentiment · Multiple Kernel Learning

1 Introduction

Microblogging, such as Twitter, Facebook and Sina Weibo, is a popular social media where millions of people express their feelings, emotions, and attitudes every day. Users can post short text messages, images and other type of

© Springer International Publishing Switzerland 2016
R. Booth and M.-L. Zhang (Eds.): PRICAI 2016, LNAI 9810, pp. 496–506, 2016.
DOI: 10.1007/978-3-319-42911-3_41

information (multi-modality) through various application software on different devices, such as laptops, mobile phones and tablets. With the rapid development of the Internet, microblogging, served as platform to connect people with each other, has been an indispensable part of the modern society.

Although mining sentiment information from microblogging had been researched many years and achieved many great breakthroughs, most contributions only focused on texts [1–4]. Most of researchers think that image can be seen as a bag, like text, made up of plentiful visual words. Thus it also contains a lot of sentiment information. More and more researchers try to integrate text and image in order to enhance prediction accuracy of text. Zhang et al. [5]. fused text and image features based on *K-Nearest-Neighbor Algorithm (KNN)* and *Minkowski Distance*, and reported that image could contribute to prediction of emotion for text.

Compared with other forms of contents, image is a much more natural and faster way to express one's emotions, and a growing number of people only post images or some additional very short texts (e.g. ten words or maybe shorter) on microblogging. Taking a post only using images on microblogging has been a new tendency. For text, we can directly extract more sentiment-related information. However, unlike text, image is so obscure in the aspect of sentiment representation that we can not describe its emotion features easily [11]. That is one of the biggest challenges to researchers.

A few methods related to sentiment analysis for image have been proposed. Siersdorfer et al. [6] analyzed and predicted sentiment of images on the social web. They considered each image as bag-of-visual words, and extracted the *SIFT (Scale-Invariant Feature Transform)* features and raw colour distribution of images. Borth et al. [7] constructed a large-scale *Visual Sentiment Ontology (VSO)*, where each of images owned its corresponding textual description with emotion. Then they trained image object detector to predict the sentiment of one certain image based on *VSO*. However, it is very limited to represent emotion only using these raw features of images. And for some intricate images, most of methods based on raw features perform not well. The challenge of this problem lies in the semantic gap between low-level visual features and higher-level image sentiments.

Recently some researchers tended to integrate text and image to exploit relations among visual content and relevant contextual information to bridge the semantic gap in the prediction of image sentiments. Wang et al. [8,10] modelled the interaction between images and textual information systematically so as to support sentiment prediction using both sources of information. Yang et al. [9] considered social relationship between users. They constructed a model based on *Latent Dirichlet Allocation (LDA)* and *Gaussian Mixture Model (GMM)* to extract high-level features of images by integrating comments of each image. Nevertheless, in the process of training and prediction, the textual information for each image that most of these methods consider is required not only to exist, but also to be relatively long. They are not suitable in the circumstance that there are only images, or additional very short texts for each image.

To address the problems above, in this paper, we propose a method based on *Simple Multiple Kernel Learning (SimpleMKL)* [12] that can (1) make use of textual information to bridge the "semantic gap" between low-level visual features and higher-level image sentiments; (2) use textual information to train image classifier, and we need not any textual information in the phase of prediction. Specifically, textual information as a sort of sufficiently emotional source data, we can use it to promote the ability via SimpleMKL to classify images. And once we get the image classifier, none of texts are needed when predicting other unlabelled images. In addition, (3) the data that we use in the experiments is crawled and labelled from Sina Weibo by ourselves.

The rest of this paper is organized as follows. Section 2 describes the related work, including text sentiment analysis using image as supplement, as well as some research work in the field of image sentiment analysis. The implementation details of our method are described in Sect. 3. Section 4 presents the data we crawled and labelled from Sina Weibo, the experimental results, as well as the comparisons with common methods and some state-of-the-art methods. Finally, in Sect. 5 we conclude the paper.

2 Related Work

Image is a sort of important modality on microblogging. Like text, it also contains a lot of sentiment information. The application of the image into sentiment prediction of text draws many researchers' attention. Zhang et al. [5] fused text and image features based on KNN and Minkowski Distance. They firstly computed cosine similarity of text and image for a post in test set with another post in training set on Sina Weibo primarily. Then a two-dimensional space will be constructed, where two axises represented text and image respectively and a point consisted of cosine similarity of text and image. Finally, distance between this point and (1, 1) was seen as the eventual similarity (i.e., more small the distance is, more similar these two posts) and classification result of this post will be got based on KNN. Besides this method, most of methods that fuse text and image after getting the classification results respectively can be categorized into *Late-fusion (or, Post-fusion)*. Late-Fusion does not consider the internal connection between different modalities, and integrates them directly with some sorts of weight, like text length, etc.

Recently, some researchers engaged in the image sentiment analysis work. Yang et al. yang:jia focused on the emotion analysis of image on flickr, an image sharing website. For each image published by one certain user, there were a lot of comments published by other users below that. They considered social relationship between users, and constructed a model based on LDA and GMM to extract high-level features of images by integrating comments of each image. Specifically, they described visual features of images by GMM, and described the comments by LDA. These two parts were integrated together by learning a Bernoulli parameter to model the relationship between the image publisher and comment publishers. Finally, a probabilistic matrix related to image-topic, the new feature matrix of images, would be achieved. Besides this method, most of methods that fuse text and image before getting the classification results respectively (i.e., feature-level fusion) can be categorized into *Pre-fusion*. Unlike the Late-fusion, it considers the internal connection between different modalities, and fuses them in the process of feature extraction.

Nevertheless, for the research work of image sentiment analysis, most of Pre-fusion methods require plenty of textual information. That means they need textural information in the phases of both training and inference. However, nowadays most of posts on microblogging, like Sina Weibo, include only images, and this has been a new tendency. Apparently, aforementioned methods are not suitable for the situation. On the other hand, text and image can be considered as two views for a post. Inspired by that, we try to introduce multi-view learning method into our work. In this paper, our proposed method based on SimpleMKL [12], an algorithm framework for solving Multiple Kernel Learning [13] problem in multi-view learning, belongs to the category of Pre-fusion. It can bridge the "semantic gap" between low-level visual features and higher-level image sentiments. Moreover, it needs textural information only in the process of training.

3   Approach

In this section, we present the details of our proposed approach. In this paper, we focus on the data that crawled by ourselves and labelled into two category, positive and negative. In that, positive images can been considered to own a sort of positive emotion, like happiness, surprise, etc., and negative images are in contrast, like anger, disgust, fear, sadness, etc. Thus we formulate the image sentiment analysis as a binary classification problem.

In what follows, we first briefly describe the framework of SimpleMKL. Then we introduce how to perform training and predicting with SimpleMKL on microblogging data.

3.1   SimpleMKL

Literally, Multiple Kernel Learning (MKL) focuses on two or more kernels in machine learning. In the real world, we often need to solve problems in which the data is combined from different sources or modalities (e.g. texts and images from microblogging). As a result of their different representations or different feature subsets, they have different measures of similarity corresponding to different kernels [19]. That means the solutions to the problems require different kernels. Instead of creating a new kernel, multiple kernel learning can be used to pick or combine kernels already established for each individual data source in linear or nonlinear mode. There are a lot of learning algorithm for MKL based on different approaches, such as functional form, heuristic approaches, bayesian approaches, etc.

SimpleMKL [12], proposed by Rakotomamonjy et al., solves the MKL problem and provides a new insight on MKL algorithms based on mixed-norm regularization, as well as projected gradient method to solve this optimization problem. The reason why we select SimpleMKL instead of other MKL related methods lies in its fast nature (i.e., it needs much less iterations for converging towards a reasonable solution than other MKL methods).

The primal formulation of SimpleMKL is:

$$\underset{w_m, \xi, b, \eta}{\operatorname{argmin}} J(\eta) = \frac{1}{2} \sum_{m=1}^{P} \frac{1}{\eta_m} \|w_m\|^2 + C \sum_{i=1}^{N} \xi_i, \tag{1}$$

$$\text{subject to } y_i \left(\sum_{m=1}^{P} \langle w_m, \Phi(x_i^m) \rangle + b \right) \geq 1 - \xi_i, \quad \forall i$$

$$\sum_{m=1}^{P} \eta_m = 1.$$

where P is the number of kernels, w_m is the vector of weight coefficients assigned to Φ_m (the m_{th} kernel), each η_m controls the squared norm of w_m in the objective function, C is a predefined positive trade-off parameter between model simplicity and classification error, N is the sample size, ξ_i is the vector of slack variables assigned to the n_{th} sample, and b is the bias term of the separating hyperplane.

Due to strong duality, we can also calculate $J(\eta)$ using the dual formulation:

$$\underset{\alpha}{\operatorname{argmax}} \ J(\eta) = \sum_{i=1}^{N} \alpha_i - \frac{1}{2} \sum_{i=1}^{N} \sum_{j=1}^{N} \alpha_i \alpha_j y_i y_j k_\eta(x_i, x_j), \tag{2}$$

$$\text{subject to } k_\eta(x_i, x_j) = \sum_{m=1}^{P} \eta_m k_m(x_i^m, x_j^m),$$

$$\sum_{i=1}^{N} \alpha_i y_i = 0,$$

$$C \geq \alpha_i \geq 0 \ \forall_i.$$

where like η, α_i is also a primal and dual variable (N-dimensional vector) to optimize.

Then, we can update η using the following gradient calculated with α found above.

$$\frac{\partial J(\eta)}{\partial \eta_m} = \frac{1}{2} \sum_{i=1}^{N} \sum_{j=1}^{N} \alpha_i \alpha_j y_i y_j \frac{\partial k_\eta(x_i^m, x_j^m)}{\partial \eta_m} = -\frac{1}{2} \sum_{i=1}^{N} \sum_{j=1}^{N} \alpha_i \alpha_j y_i y_j k_m(x_i^m, x_j^m) \ \forall_m. \tag{3}$$

3.2   Training and Inference

Formulation. We are given a set of posts N (we also describe the size of dataset as N). For each post $n \in N$, we have the image and corresponding text (textual information is used to train an enhanced image classifier), described as f_n and t_n respectively. We use a T dimensional vector x_{t_n} to represent t_n, where each dimension indicates one of t_n's textual features (e.g., top-k TF-IDF words, etc.). And we use a R dimensional vector x_{f_n} to present f_n, where each dimension indicates one of f_n's visual features (e.g., hue, saturation, hog, glcm, tamura textures, etc.). The feature vector of each image can be put behind the feature vector of corresponding text, thus we can achieve a matrix Q (its dimension is N × (T + R)). Each post n has a label (positive +1 or negative −1). We assume that the text and image have the same label in the same post. Thus, for the text and image of each post n, we describe their label as y_{t_n} and y_{f_n} respectively.

Training. During training, we need to compute the α and η. The input is the matrix Q. The algorithm is as follows:

Algorithm 1. SimpleMKL algorithm

1: set $\eta_m = \frac{1}{2}\ for\ m = 1, \ldots, P$
2: **while** stopping criterion not met **do**
3:     compute $J(\eta)$ by using an SVM solver with $K = \sum_m \eta_m^\dagger K_m$
4:     compute $\frac{\partial J}{\partial \eta_m}\ for\ m = 1, \ldots, P$ and descent direction D
5:     set $\mu = \underset{m}{\text{argmax}}\ \eta_m, J^\dagger = 0, J^\dagger = \eta, D^\dagger = D$
6:     **while** $J^\dagger < J(\eta)$ **do**
7:         $\eta = \eta^\dagger, D = D^\dagger$
8:         v= $\underset{\{m|D_m < 0\}}{\text{argmin}}\ -\eta_m/D_m, \gamma_{max} = -\eta_v/D_v$
9:         $\eta^\dagger = \eta + \gamma_{max}D, D_\mu^\dagger = D_\mu - D_v, D_v^\dagger = 0$
10:         compute $J^\dagger$ by using an SVM solver with $K = \sum_m \eta_m^\dagger K_m$
11:     **end while**
12:     line search along D for $\gamma \in [0, \gamma_{max}]$
13:     $\eta \leftarrow \eta + \gamma D$
14: **end while**

Inference. Since MKL is based on SVM, when acquiring optimal α and η, we can get the prediction results according to the target function:

$$f(x) = \sum_{n=1}^{N}(\alpha_n \sum_{m=1}^{P} \eta_m K_m(x, x_n)),$$

where $K_m(.)$ is the $m_t h$ kernel function (e.g. linear, polynomial, gaussian function, etc.). Given a image described as x, if the result of target function is negative, we can consider its label is −1, otherwise, we can consider its label is +1.

4   Experimental Results

4.1   Data Preparation

We perform our experiments on the dataset crawled by ourselves from Sina Weibo[1]. In the dataset, we randomly downloaded more than 1600 posts including texts and corresponding images. We assign a rating score to each post. The post with rating > 3 are labeled positive, and those with rating < 3 are labeled negative. The rest (rating $= 3$) are discarded for their ambiguous polarity. At last 1000 posts are reserved, including 575 positive instances and 425 negative instances. Figure 1 shows two examples of the data.

真的是当你陷入到某个沼泽里如果没有人来拉你一把你会越陷越深。很高兴我有这样一帮伙伴、兄弟，无论你是什么样子无论你犯了什么样的错误他们都会支持你、保护你，这是人生最美的财富。对不起啦兄弟们，让我们一起加油！　◎　苏州·莲堤

(a) A positive post.

有时候看的太透，就失去了爱人的能力。看不透时，只是心里的难过跟伤心，看得透时，那种悲凉确是从骨子里透出来的!

(b) A negative post.

Fig. 1. Two examples of our dataset.

4.2   Feature Extraction

We extract textual features using TF-IDF (we use words of top-k TF-IDF value as feature words). Here k is 1000, that means we have a 1000×1000 feature matrix for text.

We extract visual features with dimension of 284 for a given image. Table 1 describes the visual features we used in this work [20]. According to Osgood dimensional approach [14], saturation and brightness can have direct influence on our pleasure, arousal and dominance. We use the method proposed by Valdez et al. [15] to compute Pleasure, Arousal and Dominance.

$$Pleasure = 0.69Br + 0.22S,$$

$$Arousal = -0.31Br + 0.60S,$$

---

[1] http://weibo.com.

$$Dominance = 0.76Br + 0.32S,$$

where Br denotes Brightness and S denotes Saturation.

Haralock [18] used GLCM method to analyse the emotion of an image. It is a classic method to measure textures. We use GLCM to compute contrast, correlation, energy and homogeneity. Tamura [16] used the Tamura texture features to analyze the emotion of an image. This method is popular among affective image retrieval [17]. We use the first three of the Tamura texture features, i.e., coarseness, contrast and directionality.

Table 1. Summary of visual features utilized in this work.

Visual feature name	Brief description
HSV	Mean and standard deviation value of hue, saturation and brightness for a given image
Pleasure, Arousal, Dominance	One type of affective coordinates calculated by brightness and saturation
HOG	Histogram of oriented gradients for a given image
GLCM	Using grey level cooccurrence matrix to compute contrast, dissimilarity, correlation, energy and homogeneity
Tamura textures	Using the first three of the tamura texture features, i.e., coarseness, contrast, directionality

4.3 Performance Evaluation

Evaluation Measure. We compare the proposed model with alternative methods (**SVM, Naive Bayes (NB), Decision Tree (DT), Random Forest (RF), Adaboost (AD)**, based on image low-level features, and **EL+SVM** method proposed by Yang [9]) in terms of Accuracy, Precision, Recall, and F1-Measure[2]. Table 2 shows performance of our method and alternative methods.

Our Method. We use three types of kernel function, i.e., linear kernel function, polynomial kernel kernel function, gaussian kernel function. And we set maximum iteration to 30. As a matter of convenience, we call it **SMKL**.

Alternative Methods. We conduct 5-fold cross validation. For SVM, we use polynomial kernel function, degree of which is set to 6. And its coefficient is set to 0.05. For Decision Tree and Random Forest, we set maximum depth of tree to 100. For adaboost, we set its learning rate to 0.5. For EL+SVM, we set K = 6, $\alpha = 0.05$, $\beta_0 = 1$, $\beta_1 = 2$, $\gamma = 0.05$, $\tau = 0.05$, and we set maximum iteration to 100.

---

[2] https://en.wikipedia.org/wiki/F1_score.

Table 2. Performance of our method and alternative methods.

Methods	Accuracy	Precision		Recall		F1-score	
		+1	−1	+1	−1	+1	−1
SVM	0.548	0.609	0.450	0.602	0.473	0.603	0.461
NB	0.562	0.594	0.487	0.755	0.288	0.663	0.359
DT	0.639	0.688	0.571	0.678	0.613	0.682	0.589
RF	0.608	0.625	0.601	**0.801**	0.352	**0.701**	0.441
AD	0.544	0.576	0.454	0.718	0.284	0.638	0.342
EL + SVM	0.575	0.623	0.582	0.531	0.502	0.573	0.539
SMKL	**0.659**	**0.736**	**0.718**	0.657	**0.685**	0.694	**0.701**

Analysis and Discussions. As is shown in Table 2, overall, SMKL outperforms other baseline methods. Most of common methods (like, SVM, Naive Bayes, etc.) based on low-level features perform not as better as our method. The reason lies in that they only consider visual features and ignore textual information in the phase of training. However, the low-level visual features can not express the high-level image sentiments well. By contrast, our method can make use of textual information to bridge the "semantic gap" between low-level visual features and high-level image sentiments. Through the proposed method, we can attain a more powerful classifier for image.

Although Yang et al. reported in their work that their method outperforms most of other methods, it performs not well here. We think the reason is that LDA is not suitable to model short text. Specifically, conventional topic models implicitly capture the document-level word cooccurrence patterns to reveal topics, and thus suffer from the severe data sparsity in short documents [21,22]. Moreover, in our method, once we get the image classifier, none of texts are needed when predicting other unlabelled images. EL+SVM can not do that.

5   Conclusion

In this paper, we proposed a new method based on SimpleMKL to improve the performance of image sentiment analysis. We set up a dataset from Sina Weibo to validate our proposed method. In text feature extraction, we use TF-IDF to select the features. In image feature extraction, we extract colour features and text features. Experiments on the dataset demonstrates that our model improves the performance on predicting images' sentiment largely.

Acknowledgments. We would like to acknowledge the support for this work from the National Natural Science Foundation of China (Grant No. 61403200).

References

1. Pang, B., Lee, L.: Opinion mining and sentiment analysis. Found. Trends Inf. Retrieval **2**, 1–135 (2008)
2. Pang, B., Lee, L.: A sentimental education: sentiment analysis using subjectivity summarization based on minimum cuts. In: Proceedings of the 42nd Annual Meeting on Association for Computational Linguistics, pp. 271–282 (2004)
3. Liu, B.: Sentiment analysis and opinion mining. Synthesis Lectures on Human Language Technologies, vol. 5 (2012)
4. McDonald, R., Hannan, K., Neylon, T.: Structured models for fine-to-coarse sentimen analysis. In: Annual Meeting-Association For Computational Linguistics, vol. 45 (2007)
5. Zhang, Y., Shang, L., Jia, X.: Sentiment analysis on microblogging by integrating text and image features. In: Cao, T., Lim, E.-P., Zhou, Z.-H., Ho, T.-B., Cheung, D., Motoda, H. (eds.) PAKDD 2015. LNCS, vol. 9078, pp. 52–63. Springer, Heidelberg (2015)
6. Siersdorfer, S., Hare, J.: Analyzing and predicting sentiment of images on the social web. In: Proceedings of the 18th ACM International Conference on Multimedia, MM 2010, pp. 715–718 (2010)
7. Both, J., Ji, R., Chen, T.: Large-scale visual sentiment ontology and detectors using adjective noun pairs. In: Proceedings of the 21st ACM International Conference on Multimedia, MM 2013, pp. 223–232 (2013)
8. Wang, Y., Wang, S., Tang, J.: Unsupervised sentiment analysis for social media images. In: Proceedings of the 24th International Conference on Artificial Intelligence, pp. 2378–2389 (2015)
9. Yang, Y., Jia, J., Zhang, S.: How do your friends on social media disclose your emotions? In: The 28th Association for the Advancement of Artificial Intelligence, pp. 1–7 (2014)
10. Jia, J., Wu, S., Wang, X., Tang, J.: Can we understand van gogh's mood?: learning into infer affects from images in social networks. In: Proceedings of the 18th ACM International Conference on Multimedia, MM 2012, pp. 857–860 (2012)
11. Shin, Y., Kim, E.: Affective prediction in photographic images using probabilisitic affective model. In: CIVR 2010, pp. 390–397 (2010)
12. Rakotomamonjy, A., Bach, R., Canu, S.: SimpleMKL. J. Mach. Learn. Res. **9**, 2491–2521 (2008)
13. Gönen, M., Alpáydn, E.: Multiple kernel learning algorithms. J. Mach. Learn. Res. **12**, 2211–2268 (2011)
14. Osgood, C.E.: The nature and measurement of meaning. Psychol. Bull. **49**, 197–237 (1957)
15. Valdez, P., Mehrabian, A.: Effects of color on emotions. J. Exp. Psychol.: Gen. **123**(4), 394 (1994)
16. Tamura, H., Mori, S., Yamawaki, T.: Textural features corresponding to visual perception. IEEE Trans. Syst. Man Cybern. **8**(6), 460–473 (1978)
17. Wu, Q., Zhou, C.-L., Wang, C.: Content-based affective image classification and retrieval using support vector machines. In: Tao, J., Tan, T., Picard, R.W. (eds.) ACII 2005. LNCS, vol. 3784, pp. 239–247. Springer, Heidelberg (2005)
18. Haralock, R., Shapiro, L.: Computer and Robot Vision. Addison-Wesley Longman Publishing Co. Inc., Boston (1991)
19. Abramowitz, M., Stegun, I.: Handbook of mathematical functions (1970)

20. Ou, L., Luo, M.: A study of colour emotion and colour preference. Part I: colour emotions for single colours. Color Res. Appl. **29**(3), 232–240 (2004)
21. Yan, X., Guo, J.: A biterm topic model for short texts. In: Proceedings of the 22nd International Conference on World Wide Web, pp. 1445–1456 (2013)
22. Quan, X., Kit, C.: Short and sparse text topic modeling via self-aggregation. In: Proceedings of the 24th International Conference on Artificial Intelligence, pp. 2270–2276 (2015)

Grouped Text Clustering Using Non-Parametric Gaussian Mixture Experts

Yong Tian[✉], Yu Rong, Yuan Yao, Weidong Liu, and Jiaxing Song

Department of Computer Science and Technology, Tsinghua University,
Beijing, China
gnoynait@gmail.com, rongyu9124@hotmail.com, yaoy92@gmail.com,
{liuwd,jxsong}@tsinghua.edu.cn

Abstract. Text clustering has many applications in various areas. Before being clustered, texts often have already been grouped or partially grouped in practise. Texts from the same group are related to each other and concentrate on a few topics. The group information turns out to be valuable for text clustering. In this paper, we propose a model called Non-parametric Gaussian Mixture Experts to get better clustering result through utilizing group information. After converting texts to vectors by semantic embedding, our model can automatically infer proper cluster number for every group and the whole corpus. We develop an online variational inference algorithm which is scalable and can handle incremental datasets. Our algorithm is tested on various text datasets. The results demonstrate our model has significantly better performance in cluster quality than some other classical and recent text clustering methods.

1 Introduction

Text clustering has many applications in various areas, such as information retrieval, document organization, online advertising and social networks mining. Text data are commonly incremental and of huge volumes, which require cluster algorithms to be efficient and scalable. Besides that, we can often find some ways to partition all or part of the text into different groups. For example, web pages can be grouped by their sites or directories, research papers can be grouped by conferences or journals and social network messages can be grouped by users. Texts in the same group are usually similar or related, and have much fewer topics than those of the whole corpus. How these texts are grouped provides us with extra information in addition to the texts themselves.

Group information is valuable for clustering. On one hand, group information can serve as heuristic information to get better clustering result. Clusters consistent with the group partitions are usually preferred. If a clustering result shows no tendency that texts in the same group are assigned to a small number of clusters, it might not be what people want. On the other hand, group information can be used as prior knowledge to help reduce the ambiguity of texts. For example, a web page with a single word *apple* can refer either a fruit or an IT company. However, if we are told that the page exists on a technology web

R. Booth and M.-L. Zhang (Eds.): PRICAI 2016, LNAI 9810, pp. 507–516, 2016.
DOI: 10.1007/978-3-319-42911-3_42

site, we are pretty sure it is about technology. When we try to take advantage of group information to cluster texts, new challenges arise. First of all, each group's cluster number is unknown and it is impractical to set the cluster numbers manually for every groups. The other problem is clusters must be shared among groups. Otherwise, it will lead to many duplicated clusters.

To deal with those problems and make use of group information for better cluster quality, we propose a novel model called Non-Parametric Gaussian Mixture Experts model (NPGME) and develop a scalable online inference algorithm for it. After converting texts to vectors, NPGME can automatically infer proper cluster number for not only the whole corpus but also each groups, and learn clusters shared by all groups.

2   Text Representation

The first step for text clustering is to properly represent them. One way to do that is to transform texts to vectors. Bag-of-words and TF-IDF representations are primary representations. They are simple but high dimensionality and sparsity prevent them from fitting Gaussian models. Topic models, such as Latent Dirichlet Allocation (LDA) [1], are more sophisticated vector representations with dimensionality of a selected number k. However, although their efficient inference methods keep emerging, topic models still require significant time to train. Thus topic model representation could be a choice, but not an economical one.

Semantic embedding method is an un-supervised machine learning technique which learns a continuous vector representation for word or text. The cosine value of two vectors can be used to measure semantic similarity. In the last three years, semantic embedding tools, such as word2vec [2,3], paragraph vector [4] and GloVE [5], have been developed and become widely used in various applications. They are computationally efficient and have great scalability and generality. Particularly, word vectors are quite standard. There are even ready-to-use word vectors on the web that can be downloaded directly. For short texts, such as article titles and short messages, we can use the sum of vectors of words in each text as representation. For long texts, such as paragraph and articles, we resort to paragraph vectors since sum vectors may contain too much noise. We can further normalized these vectors to unit length so that Euclidean distance can be used to approximate their cosine distance and then apply Gaussian distribution to model clusters.

3   Model

Non-parametric Gaussian Mixture Experts (NPGME) is an extension of Dirichlet process Gaussian mixtures [6]. It not only has infinite Gaussian components with weights constructed via Dirichlet process, called global expert, but also has different Dirichlet process weights in each group, called group expert. Global expert gives the weights of each component in the whole dataset and group expert adjust that in its group. If some data belong to no group, no adjustment is needed. This is a significant difference compared with Hierarchical Dirichlet Process [7], which requires all data to be grouped.

In NPGME, Gaussian components' parameters $\theta_t := (\mu_t, \lambda_t^{-1} I)$, are drawn from conjugate distributions,

$$\mu_t \sim \mathcal{N}(\upsilon_t, (\gamma_t I)^{-1}), \quad \lambda_t \sim \text{Gamma}(a_t, b_t), \quad t = 1, 2, \ldots. \tag{1}$$

Each component has a global weight β_t from *stick-breaking process* [8]:

$$\beta_t' \sim \text{Beta}(1, \omega), \quad \beta_t = \beta_t' \prod_{l=1}^{t-1}(1 - \beta_l'). \tag{2}$$

Suppose there are J groups. Each group j has local weights $\pi_{jk}, k = 1, 2, \ldots,$

$$\pi_{jk}' \sim \text{Beta}(1, \alpha), \quad \pi_{jk} = \pi_{jk}' \prod_{l=1}^{k-1}(1 - \pi_{jl}'), \tag{3}$$

All groups share the same global components but with no guarantee that they are in the same order. Therefore, a symmetric multivariate distributed random variable c_{jk} is used to connect local weights and global components. Weight of global component $t = c_{jk}$ in group j is set proportional to $\pi_{jk}\beta_{c_{jk}}$ as adjustment. To get grouped data x_{jn} with the adjusted weight, draw

$$z_{jn}' \sim \text{Mult}(\pi_j), \quad x_{jn} \sim \mathcal{N}(\theta_{z_{jn}}), \quad z_{jn} = c_{jz_{jn}'}, \tag{4}$$

but only keep the sample with probability $\beta_{z_{jn}}$. Ungrouped data x_{0n} can be simply drawn by

$$z_{0n} \sim \text{Mult}(\beta), \quad x_{0n} \sim \mathcal{N}(\theta_{z_{0n}}). \tag{5}$$

Using semantic embedding representation or even topic model representation (as long as the training time is acceptable), we can train NPGME to learn model parameters. After that, we use the model to estimate posterior distribution of latent variable z_{jn} for each text and assign the text to the most probable cluster indicated by z_{jn}. Note that the training and assignment are in different steps. Once the model has been trained, texts can be clustered even if they are new and not in the training set. Thus NPGME can cluster incremental dataset.

4   Inference

In order to process large-scale and incremental dataset efficiently, we develop a scalable variational inference algorithm for NPGME under a framework called *Stochastic Variational Inference* [9]. This method introduces tractable variational distributions to approximate the posterior distribution and cast the inference problem into a stochastic optimization problem.

Table 1. Variational distributions

Variable	Distribution	Parameter	Related expectations
z_{jn}	Multivariate	$\phi_{jn}^{(k)}$	$\mathbb{E}_q[z_{jn}^{(k)}] = \phi_{jn}^{(k)}$
c_{jk}	Multivariate	$\zeta_{jk}^{(t)}$	$\mathbb{E}_q[c_{jk}^{(t)}] = \zeta_{jt}^{(t)}$
β_t'	Beta	$\xi_t^{(1)}, \xi_t^{(2)}$	$\mathbb{E}_q[\ln \beta_t'] = \Psi(\xi_t^{(1)}) - \Psi(\xi_t^{(1)} + \xi_t^{(2)})$,
			$\mathbb{E}_q[\ln(1 - \beta_t')] = \Psi(\xi_t^{(2)}) - \Psi(\xi_t^{(1)} + \xi_t^{(2)})$
π_{jk}'	Beta	$\nu_{jk}^{(1)}, \nu_{jk}^{(2)}$	$\mathbb{E}_q[\ln \pi_{jk}'] = \Psi(\nu_{jk}^{(1)}) - \Psi(\nu_{jk}^{(1)} + \nu_{jk}^{(2)})$,
			$\mathbb{E}_q[\ln(1 - \pi_{jk}')] = \Psi(\nu_{jk}^{(2)}) - \Psi(\nu_{jk}^{(1)} + \nu_{jk}^{(2)})$
λ_t	Gamma	(a_t, b_t)	$\mathbb{E}_q[\lambda_t] = \frac{a_t}{b_t}$
μ_t	Gaussian	(v_t, γ_t)	$\mathbb{E}_q[\ln \mathcal{N}(x_{jn}|\theta_t)] = \frac{D}{2}(\Psi(a_t) - \ln b_t - \ln 2\pi$
			$-\frac{1}{\gamma_t}) - \frac{a_t}{2b_t}(x_{jn} - v_t)^{\mathrm{T}}(x_{jn} - v_t)$

We use *trunking* technique [10] by setting $\beta_T' = 0, \pi_{jK}' = 0$. The variational distribution $q(z, c, \pi, \beta, \theta)$ for posterior distribution $p(z, c, \pi, \beta, \theta|X)$ is fully factored as listed in Table 1. Our target is to maximize the evidence lower bound (ELBO) $\mathbb{E}_q[\ln p(X, \pi', c, z, \beta', \theta)]$ with respect to these variational distribution's parameters, where

$$\ln p(X, \pi', c, z, \beta', \theta) = \ln p(z|\beta) + \ln p(z|\pi') + \ln p(\pi') + \ln p(\beta') +$$

$$\ln p(\theta) + \sum_{j=1}^{J}(\sum_{n,t,k} z_{jn}^{(k)} c_{jk}^{(t)} \ln \beta_t \mathcal{N}(x_{jn}|\theta_t) + \sum_{n,t} z_{0n}^{(t)} \ln \mathcal{N}(x_{jn}|\theta_t)). \tag{6}$$

Our online inference algorithm is presented in Algorithm 1. It maximizes the ELBO by alternately updating each factor's natural parameter individually when other parameters fixed with mini batches.

In line 1 global and group parameters are randomly initialized. Then the algorithm loops until convergence. In iteration τ, learning rate ρ is set to $\tau^{-\kappa}$, where $\kappa \in (0.5, 1]$ to ensure convergence. In each iteration, the algorithm samples a subset of data and splits them by their groups. For ungrouped data x_{0n}, it computes $r_{0n}^{(t)}$, which is the expectation of the probability of that that data is generated by component t. For all data belonging to group j, the algorithm not only computes $r_{jn}^{(t)}$, but also updates group parameters. Even though $r_{jn}^{(t)}$ has the same meaning with $r_{0n}^{(t)}$, it is calculated by a different formula in line 6. After all sampled data processed, global parameters are updated using sufficient statistics of all sampled data. Here, we use $\mathbf{s}(x)$ to denote the sufficient statistics for observed random variable x, which is $(||x - \mu||_2^2, 1, x, 1)^{\mathrm{T}}$. The natural parameter of Gaussian component's variational distribution is

$$\psi_t = (\frac{\gamma_t v_t}{\mathbb{E}_q[\lambda_t]}, \frac{\gamma_t}{\mathbb{E}_q[\lambda_t]}, 2b_t, \frac{2(a_t - 1)}{D})^{\mathrm{T}}, \tag{7}$$

in which D is the dimensionality of data. When updating parameters, the statistics are scaled from batch size N_j' and N_D' to full dataset size N_j and N_D.

Algorithm 1. Online Inference for NPGME

1: Initialize global and group parameters randomly, set τ, ρ to 1
2: **while** forever **do**
3:     sample a subset D' for full dataset D randomly
4:     for all ungrouped data $x_{0n} \in D'$, compute $r_{0n}^{(t)} \propto \exp\{\mathbb{E}[\ln \beta_t] + \mathbb{E}[\ln p(x_{jn}|\theta_t)]\}$
5:     **for** for $j = 1, 2, \ldots, J$ **do**
6:         for $n \in N_j'$, compute
$$\phi_{jn}^{(k)} \propto \exp\{\sum_t \mathbb{E}_q[c_{jk}^{(t)}](\mathbb{E}_q[\ln \beta_t] + \mathbb{E}_q[\ln \mathcal{N}(x_{jn}|\theta_t)]) + \mathbb{E}_q[\ln \pi_t]\},$$
$$r_{jn}^{(t)} = \sum_{k=1}^{K} \phi_{jn}^{(k)} \zeta_{jk}^{(t)}$$
7:         for $k = 1, 2, \ldots, K$, update $\hat{\zeta}_{jk}^t \propto \exp\{\sum_{n=1}^{N} \phi_{jn}^k (\mathbb{E}[\ln \beta_t] + \mathbb{E}[\ln p(x_{jn}|\theta_t)])\}$,
$$(\hat{\nu}_{jk}^{(1)}, \hat{\nu}_{jk}^{(2)}) \leftarrow (1 + \frac{N_j}{N_j'} \sum_{n \in N_j'} \phi_{jn}^k, \alpha + \frac{N_j}{N_j'} \sum_{n=1}^{N} \sum_{k=1}^{K} \phi_{jn}^k)$$
8:         for $k = 1, 2, \ldots, K$, update group parameter $\nu_{jk} \leftarrow (1 - \rho)\nu_{jk} + \rho\hat{\nu}_{jk}$ and
$$\zeta_{jk} \leftarrow \exp\{(1 - \rho)\ln \zeta_{jk} + \rho \ln \hat{\zeta}_{jk}\}$$
9:     **end for**
10:     compute temporal global parameters $\hat{\psi}_t \leftarrow \eta + \frac{N_D}{N_{D'}} \sum_{jn \in D'} r_{jn}^{(t)} s(x_{jn})$ and
$$(\hat{\xi}_t^{(1)}, \hat{\xi}_t^{(2)}) \leftarrow (1 + \frac{N_D}{N_{D'}} \sum_{jn \in D'} r_{jn}^{(t)}, \omega + \frac{N_D}{N_{D'}} \sum_{l=t+1}^{T} \sum_{jn \in D'} r_{jn}^{(l)})$$
11:     update global parameters: $\xi \leftarrow (1 - \rho)\xi + \rho\hat{\xi}$, $\psi \leftarrow (1 - \rho)\psi + \rho\hat{\psi}$
12:     update global learning rate $\rho \leftarrow \tau^{-\kappa}, \tau \leftarrow \tau + 1$
13: **end while**

The scaled statistics are used to compute new temporary value of parameters afterwards. At last, parameters are updated to the temporary value by learning rate ρ. It has been showed in [11], updating natural parameters of variational distributions accelerates convergence.

The running time of the algorithm is not proportional to the size of dataset. It does not need full data to update the parameters. Therefore, the algorithm has great scalability and can process incremental dataset.

5 Experiments

Our experiments consist of two parts to serve different purposes. Cluster quality experiment is performed on synthetic and relatively small datasets to compare the clustering quality of different algorithms, while parameter influence experiment is performed on a large-scale original dataset to explore how parameters influence cluster number and also as a demonstration of web page clustering application.

To evaluate the clustering quality, we need manually labeled dataset as ground truth. However, most labeled open datasets are not grouped. To show how effectively our model can exploit the group information, we use synthetic datasets in model comparison experiments. In every labeled dataset, we split each category into 5 partitions equally, then randomly combine 2 partitions to form a group. As a comparison, we also test our model on dataset partitioned by categories. There are five datasets used in this experiment, which are Reuters-21578 titles[1],

---

[1] https://archive.ics.uci.edu/ml/datasets/.

Google news[2], Reuters-21758, TDT2 and 20 Newsgroups[3]. For each of these datasets, we discard some texts to make sure every article belongs to one category and each category contains no less than 50 different articles. The first two datasets are short text datasets and use normalized sum of word vector representation with dimensionality of 100. Others are long text datasets and use 200 dimensional paragraph vectors representation.

We use three metrics to evaluation the cluster quality. *Normalized mutual information* (NMI) is an variant of mutual information, which ranges from 0 to 1 and is widely used in literatures to measure cluster quality. The higher NMI, the more similarity they have. Besides NMI, we also use Homogeneity (Hom) and Completeness (Com) to show more detailed influence of the cluster number. Homogeneity measures how likely members of a cluster are come from the same true cluster while completeness measures how likely a true cluster's members will be assign to one cluster.

We compare the performance of NPGME both with classical clustering algorithms, K-means and Gaussian mixture models (GMM), and recent text clustering algorithms, R2NMF [13] for long text and GSDMM [12] for short text. NPGME and GSDMM are no-parametric models and do not need setting the number of clusters while K-means, GMM and R2NMF need. As we can only control the number of clusters learned by NPGME and GSDMM floating near the real cluster number, we run K-means and GMM with cluster number setting to a range near the real one. For each setting of cluster number, we run K-means and GMM 10 times each. For Google news dataset, we just set cluster number equal to the real one and run K-means and GMM for 3 times due to time reasons. For both Reuters and Google news dataset, we run NPGME and GSDMM for 10 times. The evaluation results are presented in Fig. 1 and Table 2. Considering the space issue, we only display results on Reuters-title and TDT2 in Fig. 1 and only NMI on other datasets is showed in Table 2. In these charts and table, NPGME and NPGME2 denote the results of NPGME model on original and synthetic dataset, respectively.

In Fig. 1, the first row is the results on Reuters-title and the second is on TDT2. From left to right, each column is the chart of NMI, homogeneity and completeness. Every x-axis is the cluster number for parametric models, which are k-means, Gaussian mixture model and NMF. For NPGME and GSDMM, x-axis has no meaning. As a result, the curves for these two models are horizontal lines. The charts show that, for parametric models, homogeneity increases and completeness decreases when cluster number gets greater. The curves for NMI has no significant trends imply that NMI is not very sensitive to cluster numbers. In every chart, the best result comes from NPGME with true clusters as groups. None of these results is close to 1.0. Therefore, we know that NPGME does not simply cluster according to groups. In addition, NPGME2 is still much better than other results.

---

[2] http://news.google.com/, We obtain it from the author of [12].
[3] http://qwone.com/~jason/20Newsgroups/.

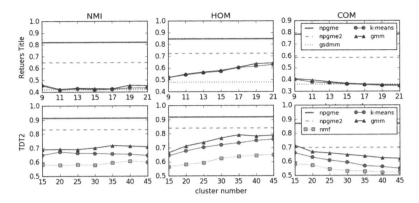

Fig. 1. Quality results (Color figure online)

Table 2. Normalized mutual information

Dataset	npgme	npgme2	kmeans	gmm	gsdmm	nmf
GoogleNews	0.869	0.752	0.738	0.725	0.87	-
Retuers	0.819	0.578	0.450	0.485	-	0.448
20Newsgroup	0.584	0.477	0.339	0.348	-	0.329

Table 2 shows more results of NMI on different datasets. NPGME and NPGME2 achieve much better NMI than others except on Google-News. Since NPGME, k-means and Gaussian mixture model use the same representation, we can conclude the improvement must come from the group information. As for GSDMM, it has completely different performance on two datasets. On Google news, its performance is almost identical to NPGME while on Reuters-21578 its performance drops far behind the other three algorithms. The explanation for this phenomenon lies in the representation of text. Google news dataset is labeled by events, thus texts with the same label shares a lot of same words. GSDMM uses bag-of-words representation, it is easier to capture this kind of similarity. However, bag-of-words representation does not suit for Reuters-titles since country names, such as UK, USA, China, frequently appear in almost all category, and similar words like ship, boat and ferry are treated as distinct words by bag-of-words representation. Normalized sum of word vector representation used in other three algorithms can easily capture the similarity which comes from different but similar words. As a result, NPGME, K-means and GMM perform much better than GSDMM on Reuters-titles dataset. R2NMF is only performed on long text dataset since short text cluster is not its strength. As we can see, even k-means and GMM is slightly better than R2NMF on every long text datasets.

NPGME model is quite complex and has many parameters. Some parameters have been studied in previous literatures. For example, the cluster number in Dirichlet process mixtures is linear with concentration parameter and the logarithm of the size of dataset. However, some other parameters's influence to the model is still unknown. Among them, we are most interested in the variance prior parameters a and b, since they not only influence the posterior variance but also may influence the cluster number. Now we perform experiments on a naturally grouped dataset to unveil how variance prior parameters work in NPGME.

We use AG's corpus[4], which is a collection of more than one million online articles from more than two thousand sources. Each record in this corpus includes source, URL, title, contents and other information. We treat the most common 300 sites as groups. The remaining articles are considered as ungrouped data. We apply word2vec to preprocess the corpus and use normalized sum of word vector representation whose dimensionality is also 100. Because of the abundance of data, we run our stochastic algorithm one pass through the corpus. The global trunking size T is set to 500 and group trunking size K to 200, which should be enough to meet the needs.

From the property of Gamma distribution, the expectation of Gaussian variance is b/a and a represents how strongly we believe in that. Thus we call the b/a the prior variance and a prior belief. It is more straightforward to explore the influence of prior variance and prior belief. Thus, we vary the variance and belief and observe how they influence the number of significant clusters, which are defined as clusters whose global weight is more than 0.001. More specifically, we first vary prior variance when a strong prior belief is fixed and then fix prior variance a set of values and vary prior belief. Each pair of prior parameters is test for 3 times. Fig. 2 shows the results.

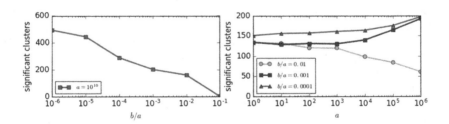

Fig. 2. Influence of variance prior (Color figure online)

The left chart shows how variance influences the cluster number. Prior belief here is set to 10^{10}, which is far greater than the size of corpus. As a result, the inferred variance should be almost the same as the prior. From the chart we can see clearly that small variance leads to large numbers of clusters. In fact, when variance is too large, say 0.1, only one cluster's weight is greater than 0.001. This result should not be surprising. Our data are normalized and all distributed on a unit spherical surface. If the variance is large enough, clusters will overlap heavily and thus be combined into one cluster by the model.

---

[4] http://www.di.unipi.it/~gulli/AG_corpus_of_news_articles.html.

The right chart illustrates more. There are three series of experiments with each prior variance fixed to 0.01, 0.001 and 0.0001, respectively. Prior belief varies in each series from 10^0 to 10^6. In general, small prior variance leads to larger numbers of clusters, especially in the right part, which is consistent with previous chart. Things get more interesting when inspecting each curve individually. When prior belief increases, cluster number increases if prior variance is 0.001 or 0.0001 but decreases if prior variance is 0.01. This can be explained from another perspective. When prior belief is weak, the prior variance takes a small part in the inferred variance. The variance is mainly determined by the dataset itself, which will produce close inferred variances and cluster number. When prior belief gets stronger, the prior variance's influence in inferred variance is also stronger. Therefore, large prior variance gets less clusters while small prior variance gets more clusters.

In summary, the prior distribution of Gaussian's variance also influence the cluster number. Smaller prior variance leads to larger cluster number. The stronger the prior belief is, the more significant the influence is.

6   Discussion and Conclusion

In this paper, we proposed Non-Parametric Gaussian Mixture Experts model to cluster grouped or partially grouped text data. This model takes advantage of group information to achieve better cluster quality. Some other models also use group information, such as HDP, sLDA [14] and HSLDA [15]. NPGME is closely related to HDP, however HDP can only handle datasets in which all data are grouped. SLDA and HSLDA are often used to learn representation of text with labels or groups. It relies on that labels or groups are "correct" while NPGME only require each group concentrates on a few clusters, which is a milder condition. We developed an online variational inference algorithm for NPGME and performed various experiments to test our algorithm's performance and to explore the influence of some parameters. Experiment results demonstrate that NPGME can process large scale datasets efficiently and performs significantly better than many other algorithms in cluster quality.

References

1. Blei, D.M., Ng, A.Y., Jordan, M.I.: Latent dirichlet allocation. J. Mach. Learn. Res. **3**, 993–1022 (2003)
2. Mikolov, T., Sutskever, I., Chen, K., Corrado, G.S., Dean, J.: Distributed representations of words and phrases and their compositionality. In: Advances in Neural Information Processing Systems, pp. 3111–3119 (2013)
3. Mikolov, T., Chen, K., Corrado, G., Dean, J.: Efficient estimation of word representations in vector space. arXiv preprint arXiv: 1301.3781 (2013)
4. Le, Q.V., Mikolov, T.: Distributed representations of sentences, documents. arXiv preprint arXiv: 1405.4053 (2014)

5. Pennington, J., Socher, R., Manning, C.D., Glove: global vectors for word representation. In: Proceedings of the Empiricial Methods in Natural Language Processing (EMNLP 2014), vol. 12 (2014)

6. Rasmussen, C.E.: The infinite gaussian mixture model. In: NIPS, vol. 12, pp. 554–560 (1999)

7. Teh, Y.W., Jordan, M.I., Beal, M.J., Blei, D.M.: Hierarchical dirichlet processes. J. Am. Stat. Assoc. **101**(476), 1566–1581 (2006)

8. Sethuraman, J.: A constructive definition of dirichlet priors. Technical report, DTIC Document (1991)

9. Hoffman, M.D., Blei, D.M., Wang, C., Paisley, J.: Stochastic variational inference. J. Mach. Learn. Res. **14**(1), 1303–1347 (2013)

10. Blei, D.M., Jordan, M.I., et al.: Variational inference for dirichlet process mixtures. Bayesian Anal. **1**(1), 121–143 (2006)

11. Amari, S.-I.: Natural gradient works efficiently in learning. Neural Comput. **10**(2), 251–276 (1998)

12. Yin, J., Wang, J.: A dirichlet multinomial mixture model-based approach for short text clustering. In: Proceedings of the 20th ACM SIGKDD International Conference on Knowledge Discovery and Data Mining, pp. 233–242. ACM (2014)

13. Kuang, D., Park, H.: Fast rank-2 nonnegative matrix factorization for hierarchical document clustering. In: Proceedings of the 19th ACM SIGKDD International Conference on Knowledge Discovery and Data Mining, pp. 739–747. ACM (2013)

14. Blei, D.M.: Mcauliffe, J.D.: Supervised topic models. In: Neural Information Processing Systems (2007)

15. Perotte, A.J., Wood, F., Elhadad, N., Bartlett, N.: Hierarchically supervised latent Dirichlet allocation. In: Advances in Neural Information Processing Systems, pp. 2609–2617 (2011)

Multi-level Occupancy Grids for Efficient Representation of 3D Indoor Environments

Yu Tian, Wanrong Huang, Yanzhen Wang$^{(\boxtimes)}$, Xiaodong Yi, Zhiyuan Wang, and Xuejun Yang

State Key Laboratory of High Performance Computing (HPCL),
College of Computer, National University of Defense Technology,
137 Yanwachi Street, Changsha 410073, Hunan, People's Republic of China
yzwang@nudt.edu.cn

Abstract. Mapping 3D environments is a fundamental yet challenging problem for mobile robot applications. Although 3D sensory data can be efficiently obtained using low-cost commercial RGB-D cameras, direct extension of the widely-adopted occupancy grids to 3D environments would cause problems, such as large storage consumption and intensive computation cost. In this paper, we propose to use a stack of 2D occupancy grids, each of which corresponds to a horizontal slice of the 3D environment at a specific height, as an efficient representation of 3D environments for indoor applications. Moreover, an existing algorithm based on Rao-Blackwellized Particle Filters (RBPF) is modified accordingly to perform simultaneous localization and mapping (SLAM) using the proposed multi-level occupancy grids (M-LOG), the entire codes of which have been made open source at https://github. com/AngelTianYu/micros_mlog. Experimental results from both simulation and real-world tests validate the effectiveness of the proposed approach in indoor environments. Computational cost of the approach scales linearly with the number of 2D map slices, making it the user's choice the trade-off between vertical map resolution and efficiency.

Keywords: Simultaneous localization and mapping · Particle filters · Mobile robots · Multi-level occupancy grids

1 Introduction

Simultaneous localization and mapping (SLAM), which is a fundamental component for almost any mobile robot application, has been recognized as an active research topic in robotics for decades [1–8].

Although SLAM within 2D environments has been well-studied in the literature, extending the existing approaches to 3D is non-trivial [9,10], partially due to the prohibitively large computational cost caused by the extra dimension. On the other hand, 3D sensory data become more and more easy to get due to cheap 3D sensors available in the market, such as Asus Xtion Pro, Microsoft Kinect and so on. Currently, many solutions for mobile robots simply extract

© Springer International Publishing Switzerland 2016
R. Booth and M.-L. Zhang (Eds.): PRICAI 2016, LNAI 9810, pp. 517–528, 2016.
DOI: 10.1007/978-3-319-42911-3_43

a single slice of the 3D sensory data, and consequently use it as a simulated 2D planar laser scan in a typical 2D SLAM process. This situation makes it a waste if we cannot make use of the abundant 3D sensory data during the SLAM process. Moreover, the robot would not be able to perceive any obstacles below the extracted slice and unexpected collisions may occur.

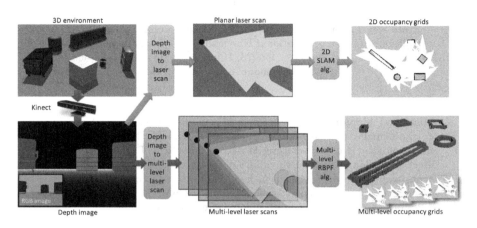

Fig. 1. An overview of the entire pipelines of indoor SLAM using typical 2D algorithms (depicted by gray arrows) and the proposed approach (depicted by blue arrows). The single slice extracted by typical 2D SLAM algorithms is depicted in red line in the *depth image*, while the multi-level slices used by our approach are shown in purple. (Color figure online)

In this paper, we attempt to solve the problem of 3D SLAM for indoor applications, under a reasonable assumption that the robot moves on a planar surface, as shown in Fig. 1. First, we propose a simple and efficient representation method for 3D environments based on multi-level 2D occupancy grids. Compared with direct extension of occupancy grids, the proposed representation scales linearly with the number of levels, and is still compatible with existing path planners and robot navigation algorithms. Second, a widely-adopted 2D SLAM algorithm based on Rao-Blackwellized Particle Filters is modified accordingly to iteratively localize the robot and produces a 3D map in the form of the above-mentioned representation. The modified SLAM algorithm takes multi-level planar scan data, which can be easily extracted from depth images or 3D point clouds, as input and also scales linearly with the number of levels. Finally, the entire pipeline has been implemented in the form of Robot Operating System (ROS) packages. Experimental results carried out on both simulator and real environment demonstrate the capability of new presentation model to capture 3D features with enough accuracy and efficiency. Potential applications include indoor path planning considering overhanging structures, robot trafficability, and path planning for legged robots [11, 22].

2   Related Work

Mapping algorithms for solving 2D SLAM problems can be classified according to their basic principles behind the scene. Typical algorithms include Extended Kalman filters (EKFs), maximum likelihood techniques and Rao-Blackwellized particle filters (RBPF). The advantages of EKF-based approaches rely on the fact that they gauge a completely correlated posterior based on landmark maps and robot poses [12]. The weakness of EKFs lies on the fact that assumptions should be made upon the robot motion model, identifiable landmarks and the sensor noise. The filters tends to be diverge if these assumptions are violated. One of the popular maximum likelihood techniques [14] estimates the most likely map over the history of sensor readings which is given by constructing a network of relations to represent the spatial constraints among the robot poses.

The main idea of Rao-Blackwellized particle filter [13] for SLAM, which we base our approach on, is to estimate the joint posterior probability $p(x_{1:t}, m|z_{1:t}, u_{1:t-1})$ of the map and the trajectory of the robot when the observations and the odometry measurement obtained by the mobile robot are given [20,21]. According to the Bayes' theorem, the recursive computation equation to compute the map m is shown as follows:

$$p(x_{1:t}, m|z_{1:t}, u_{1:t-1}) = p(m|x_{1:t}, z_{1:t}) \cdot p(x_{1:t}|z_{1:t}, u_{1:t-1}), \tag{1}$$

where $x_{1:t} = x_1, \ldots, x_t$ represents the trajectory of the robot, $z_{1:t} = z_1, \ldots, z_t$ illustrates the observations and the odometry information is shown as $u_{1:t-1} = u_1, \ldots, u_{t-1}$. In resource-constrained environment like mobile robots, this method has gained widely adoption as it significantly lowers the computational complexity by reducing the number of used particles with suitable proposal distributions.

However, these algorithms are only applicable to 2D SLAM problems and can only be used to build 2D occupancy grid maps. Since the emergence of novel robotic applications, such as flying robots and rescue robots, has necessitated the need for new 3D model to build 3D representations of environment, several approaches have been proposed among which stand out point clouds, elevation maps, multi-level surface maps and OctoMaps.

Point clouds [15,16] use points to represent the message received from the 3D sensor while each point contains the 3D coordinates. Some points even contain the RGB-D information or the reflective surface intensity information of the object been scanned. An elevation map [17] consists of a 2D grid in which each cell stores the height of the territory, while multi-level surface map [18] is an extension of elevation maps which provides the opportunity to model environments with more than one traversable level. The work done by K.M. Wurm [19] presents OctoMap as an open source framework for 3D mapping. OctoMap needs less space to save 3D maps by using octotree which omits many unnecessary details like shadows and folds on the carpet, it also enables dynamic adjustment on the resolution of maps. Although they provide representations for 3D environments, point clouds, elevation maps, and multi-level surface maps actually cannot be directly used in most popular search-based path and/or motion planning algorithms.

Our work is different from the approaches mentioned above in that we represent the 3D indoor environment using multi-level occupancy grids. We find an efficient way to combine multiple 2D occupancy grids from different heights and greatly save the space cost when storing the final 3D map. Beside reducing resource consumption, our method is also capable of detecting irregular obstacles which may lead to unnecessary detours in robot navigation.

3   Multi-level Occupancy Grids

Multi-level occupancy girds (M-LOG) are used to represent the given 3D environment as a stack of 2D bitmaps. Each pixel in M-LOG can be denoted by $G(x, y, l)$ where (x, y) represents the pixel index within a level and l represents the level number. Each pixel holds a probability value, which illustrates whether the corresponding cell in the grids is occupied by an obstacle or not. The pixel value also reflects the possibility of traversing the 3D environment at the height of corresponding laser scan messages. The height value h of each level of 2D occupancy grids is equivalent to the layer spacing between each 2D bitmap multiplied by the level number, which can be defined by users' demands.

When representing maps in the way of M-LOG, there is a linear relationship with the number of levels and storage cost, which is much smaller than that of direct extension of 2D occupancy grids to 3D grids. Besides, it also facilitates the subsequent mapping process by using M-LOG. Though this method is a kind of approximation, the user can always choose an adequate level numbers to achieve a balance between map accuracy and computational power available on the robot platform.

4   SLAM with M-LOG

4.1   Multi-level Laser Scans from Depth Image

As shown in Fig. 1, there is a popular ROS package used in the community, "*depthimage_to_laserscan*", which transfers a depth image obtained by a 3D sensor to a planar laser scan. Then, any 2D SLAM algorithms that takes 2D laser scans as the input can be used to perform the localization and mapping process. Most commonly, only the center pixels along the vertical direction of the depth image will be used to generate the 2D laser-scan-like range readings. Therefore, most information captured by the 3D sensor will be discarded.

In our approach, the "*depthimage_to_laserscan*" package is modified in our approach so that a given number of planar laser scans can be extracted from the input depth image, by "slicing" the image using several horizontal planes. The output scans can be considered as generated from several planar laser scanners, which are mounted along a vertical line with a known spacing between each other.

4.2   Multi-level RBPF-Based SLAM Algorithm

To build the M-LOG map, we have also made modifications to the original RBPF-based SLAM method. Instead of receiving the planar laser scan message, the new algorithm proactively subscribes to the new multi-level laser scan messages described in Sect. 4.1. In this way the algorithm can use one generation of particles $\{x_t^{(i)}\}$ to process laser scan messages and build maps. The whole workflow can be defined as the following steps:

Step 1. *Initialization*: When it is the first time for RBPF to receive the new form of information, the algorithm will initialize itself to prepare for the following computation.

Step 2. *Updating particles' poses*: After initialization, the scan messages will be added to each particle and the new pose of the particles $\{x_t^{(i)}\}$ will be estimated and updated using the messages received from the odometry. We will call this model the motion model. However, these messages usually are not that accurate due to several reasons such as interferences or the poor quality of the odometry contained in the robot.

Step 3. *Scan-matching*: Afterwards, the RBPF matches the latest multiple scan messages obtained from laser scan with corresponding surrounding maps contained in each particle which have been built according to the latest pose of the particles. During this process, the scan-matching algorithm will compute the likelihood using the multi-level laser scan messages and figure out how likely the particles would be the pose of robot. Then, the pose and the weights are updated and estimated based on the likelihood. But if the scan-matching process fails, the pose and the weights would be computed through the motion model (see Step 2.).

Step 4. *Updating maps*: After generating the new particles, the maps to be published are updated by adding the laser scan messages into the latest maps.

The modified RBPF-based SLAM algorithm is able to generate a M-LOG map, which is an approximate but efficient representation of the 3D environment. With this representation and the corresponding SLAM algorithm, not only the 3D information obtained from the sensor will be utilized, but also the localization and mapping results become more accurate than its 2D counterpart. This is because that multi-level laser scans, which are naturally registered, are used in each iteration of the algorithm, which increases the robustness of the algorithm.

5   Implementation Details

The proposed map representation and modified SLAM algorithm are implemented as ROS packages compatible with the ROS Indigo version [23]. Specifically, the *depthimage_to_laserscan* package receives standard depth images from a depth camera and generates multi-level planar scan data according to the given parameters including number of levels and level heights. Consequently,

```
 1 # (Simulated) multiple laser scans from a single RGB-D camera
 2
 3 Header header        # timestamp in the header is the acquisition time of
 4                      # the scan.
 5                      #
 6                      # in frame frame_id, angles are measured around
 7                      # the positive Z axis (counterclockwise, if Z is up)
 8                      # with zero angle being forward along the X axis
 9
10 sensor_msgs/LaserScan[] laser_scans # range data of multiple levels
11
12 float32[] heights         # height of each scan (according to frame frame_id)
13
```

Fig. 2. The new ROS message format with a message header that contains the time stamp and frame_id, a matrix of sensor messages, and a matrix of height information.

the *slam_gmapping* package subscribes to the multi-level planar laser scans and publishes multi-level occupancy grids. In the implementation, we follow the standard ROS messaging specification whenever possible, which makes it straightforward to incorporate our implementations into existing applications. The only exception is the multi-level plannar scan data, for which we designed a new ROS message format called *"multiscan"*, shown in Fig. 2.

6 Experimental Results

To validate the effectiveness and efficiency of the proposed representation and corresponding mapping approach, both simulation and real-world experiments have been conducted based on the Turtlebot 2 mobile robots. Turtlebot 2 is considered to be an official demonstration platform of ROS, equipped with a Microsoft Kinect RGB-D camera [24] and a differential base.

6.1 Experiment Setup

The experiments have been carried on in two environments. Figure 3 presents the simulation environment in the Gazebo simulator and the real-world environment (a pingpong room). As shown in Fig. 3(a), simulation environment consists of five obstacle: a bookshelf, a cube, a dumpster, a cylinder, and a road barrier. The real-world test environment is set up with several arranged obstacles specially, such as a sofa, a pingpong table, several cabinets, several specially arranged boxes and so on.

6.2 Mapping Results

In order to test the influence of different numbers of RBPF particles on mapping, we performed experiments on both simulator and pingpong room with three numbers of particles, 5, 10 and 20 particles respectively. From the experiments we carried out, the map using 10 particles represents the simulation environment

(a)   (b)

Fig. 3. The two experimental environments. (a) shows a simulation environment with five obstacles: the bookshelf, the cube, the dumpster, the cylinder, and the road barrier. (b) represents a pingpong room with a pingpong table and several manually arranged boxes in it.

best among the three maps built, while the map built with 5 particles gives the best representation of our pingpong room testbed. Besides, it is also obvious that when the particles used in mapping is too small, the maps built would be less accurate since less particles would lead to high uncertainty of robot localization which then results in the inaccuracy of mapping results. On the other hand, too many particles would also cause inaccuracies of mapping due to the reason that the efficiency of the approach reduces when we use large number of particles.

In all the following results, we only present mapping results obtained by using the corresponding optimal particle numbers for both simulation and real-world tests. As shown in Fig. 4, since we separate the depth images into multi-level laser scan messages based on different heights, what is obvious is that the four wheels at the bottom of the dumpster are captured in level one. Besides, the mapping results of barrier are also different significantly due to the reason that the width of barrier narrows when the height of laser scan becomes taller.

A. *Effectiveness of the approach:*

In order to test the effectiveness of our approach, we performed experiments in both environments using a level count of 4.

For the pingpong room environment, we obtained similar results. From the experiment results(shown in Fig. 5), the underneath of sofa is built on the first level of multi-level mapping results while the complex structure of the combination boxes is also captured successfully in the final 3D maps.

B. *Experiments with different levels:*

These experiments are designed to evaluate the influence of number of levels of laser scans on precision of building multi-level maps to represent the 3D gazebo simulation environment as well as pingpong room. Therefore, we use different levels to build maps, which are 4, 6 and 8 levels respectively.

As Fig. 6 illustrates, the more levels we use to build multi-level maps, the more details are added into maps and the more accurately the 3D maps our approach can represent. Furthermore, since we use more levels of laser scans to build maps, the efficiency of our approach reduces, since the more information we need to process, the worse the real time would be.

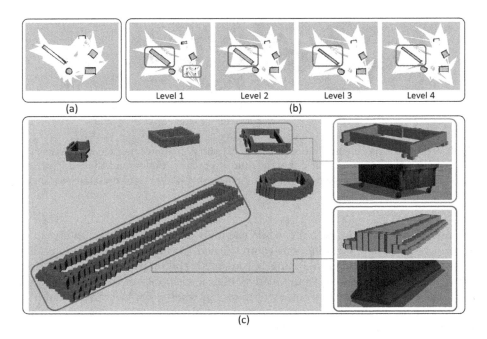

Fig. 4. Results of the Gazebo environment.(a) 2D occupancy grid map obtained by traditional RBPF algorithm; (b) The M-LOG map obtained by our approach shown in a level-wise manner; (c) 3D visualization of the M-LOG, where the zoom-in inlets depict the effectiveness of our approach in capturing complex 3D features, such as the four wheels of the dumpster, and the cross-section of the road barrier.

6.3   Timing Statistics

In order to measure the computational efficiency of our approaches, we have compared the performance of our new approach and the previous RBPF algorithm used in [20,21]. Figure 7 summarizes the time spent on both localization and mapping which are used to providing topologically correct maps by different approaches.

It is significant that time spent on mapping becomes longer when the levels of scans we need to process increase. As shown in Fig. 7(a), take building four levels of 2D occupancy grid maps by using our approach as an example, the time spent on each main function is almost three times of that spent on the main functions of Rao-Blackwellized particle filter approach. Besides, Fig. 7(b) illustrates that when the number of maps we need to build is fixed, the more particles we use, the longer the time is spent on our approach.

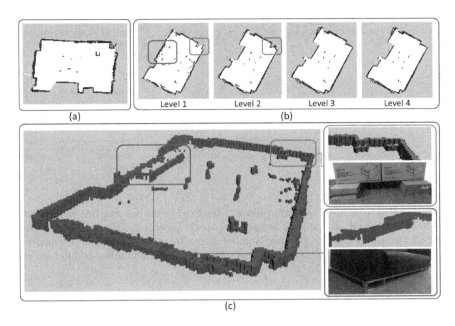

Fig. 5. Results of the pingpong room environment. (a) 2D occupancy grid map obtained by traditional RBPF algorithm; (b) The M-LOG map obtained by our approach shown in a level-wise manner; (c) 3D visualization of the M-LOG, where the zoom-in inlets depict the effectiveness of our approach in capturing complex 3D features, such as the overhanging structures of the stacked boxes and the sofa bottom.

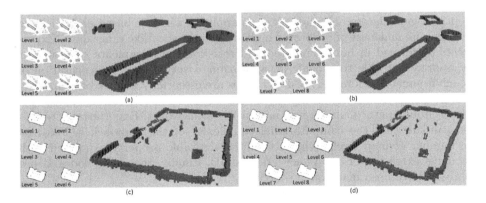

Fig. 6. 3D maps built by using M-LOG algorithm with different numbers of levels. (a) and (b) represent the mapping results with 6 and 8 levels in gazebo simulation while (c) and (d) show the results in pingpong room.

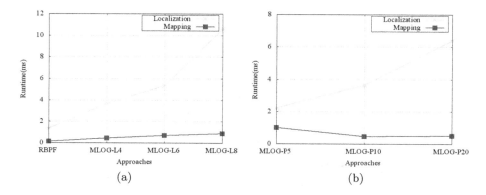

Fig. 7. Line charts of runtime performance. Chart (a) describes runtime performances of localization as well as mapping in each approach based on different levels of maps needed to be built when the particle number is set to to be 10. Chart (b) illustrates runtime performances on the two functions when the particle number is different while the levels of maps is set to be 4.

7   Conclusion

In this paper, we attempt to solve the extremely challenging problem of 3D environment reconstruction within a relatively simple scenario – indoor mobile robot applications. The solution is using a stack of 2D occupancy grids, namely M-LOG, as an efficient representation of the 3D indoor environment. To generate M-LOG based on the 3D sensory data obtained by commercially available depth cameras, we made modifications to popular ROS packages, including "depthimage_to_laserscan" and "slam_gmapping", which is based on RBPF, to generate and make use of calibrated multi-level laser-scan-like range data. Experiments performed in both simulated and real-world environments suggest that our approach is able to provide an approximate yet efficient representation of the environments. Moreover, M-LOG maps are essentially a stack of commonly used probabilistic 2D occupancy grids, hence it can be easily incorporated into existing robot path planning and navigation frameworks. Experimental results also validate that the computational complexity of the approach is linear to the number of map levels, which is quite a desirable property compared with the naive extension of occupancy grids to 3D. Finally, the entire codes of our approach have been made open source.

There are also some limitations of our approach. Currently, the approach is only applicable to indoor applications, or any other applications where the robot is moving on a planar surface, due to the 2D essence of the RBPF-based SLAM algorithm. We would like to explore the possibility of using the representation and the framework in other environments in the future. Another future direction is to parallelize the multi-level RBPF algorithm to utilize the multi-core CPUs or GPUs currently available. Moreover, potential applications of the M-LOG representation, such as path planning considering robot trafficability, as well as step planning for humanoid robots, will also be explored in the future.

Acknowledgment. This work is supported by Research on Foundations of Major Applications, Research Programs of NUDT under Grant No. ZDYYJCYJ20140601 and the National Science Foundation of China under Grant No. 61221491, 61303185 and 61303068.

References

1. Dissanayake, G., Durrant-Whyte, H., Bailey, T.: A computationally efficient solution to the simultaneous localization and map building (SLAM) problem. In: ICRA 2000 Workshop on Mobile Robot Navigation and Mapping (2000)
2. Doucet, A., de Freitas, J.F.G., Murphy, K., Russel, S.: Rao-blackwellized particle filtering for dynamic bayesian networks. In: Proceedings of the Conference on Uncertainty in Artificial Intelligence (UAI) (2000)
3. Eliazar, A., Parr, R.: DP-SLAM: fast, robust simultaneous localization and mapping without predetermined landmarks. In: Proceedings of the International Conference on Artificial Intelligence (IJCAI) (2003)
4. Gutmann, J.-S., Konolige, K.: Incremental mapping of large cyclic environments. In: Proceedings of the International Symposium on Computational Intelligence in Robotics and Automation (CIRA) (2000)
5. Hähnel, D., Burgard, W., Fox, D., Thrun, S.: An efficient FastSLAM algorithm for generating maps of large-scale cyclic environments from raw laser range measurements. In: Proceedings of the IEEE/RSJ International Conference on Intelligent Robots and Systems (IROS) (2003)
6. Montemerlo, M., Thrun, S., Koller, D., Wegbreit, B.: FastSLAM 2.0: an improved particle filtering algorithm for simultaneous localization and mapping that provably converges. In: Proceedings of the International Conference on Artificial Intelligence (IJCAI) (2003)
7. Montemerlo, M., Thrun, S., Koller, D., Wegbreit, B.: FastSLAM: a factored solution to simultaneous localization and mapping. In: Proceedings of the National Conference on Artificial Intelligence (AAAI) (2002)
8. Thrun, S.: An online mapping algorithm for teams of mobile robots. Int. J. Robot. Res. (2001)
9. Henry, P., Krainin, M., Herbst, E., Ren, X., Fox, D.: RGB-D mapping: using depth cameras for dense 3D modeling of indoor environments. In: The 12th International Symposium on Experimental Robotics (ISER) (2010)
10. Handa, A., Whelan, T., McDonald, J., Davison, A.J.: A benchmark for RGB-D visual odometry, 3D reconstruction and SLAM. In: Proceedings of the IEEE International Conference on Robotics and Automation (ICRA), pp. 1524–1531 (2014)
11. Dryanovski, I., Morris, W., Xiao, J.: Multi-volume occupancy grids: an efficient probabilistic 3D mapping model for micro aerial vehicles. In: Proceedings of the IEEE/RSJ International Conference on Intelligent Robots and Systems (IROS) (2010)
12. Frese, U., Hirzinger, G.: Simultaneous localization and mapping - a discussion. In: Proceedings of the International Conference on Artificial Intelligence (IJCAI) (2001)

13. Murphy, K.: Bayesian map learning in dynamic environments. In: Proceedings of the Conference on Neural Information Processing System (NIPS), Denver, CO, USA, pp. 1015–1021 (1999)
14. Hähnel, D., Burgard, W., Wegbreit, B., Thrun, S.: Towards lazy data association in SLAM. In: Proceedings of the International Symposium of Robotics Research (ISRR) (2003)
15. Tam, G.K.L., Cheng, Z., Lai, Y., Langbein, F.C., Liu, Y., Marshall, D., Martin, R.R., Sun, X., Rosin, P.L.: Registration of 3D point clouds and meshes: a survey from rigid to nonrigid. IEEE Trans. Vis. Comput. Graph. **19**(7), 1199–1217 (2013)
16. Wang, J., Xu, K., Liu, L., Cao, J., Liu, S., Yu, Z., Gu, X.D.: Consolidation of low-quality point clouds from outdoor scenes. Comput. Graph. Forum **32**(5), 207–216 (2013). Blackwell Publishing Ltd
17. Hebert, M., Caillas, C., Krotkov, E., Kweon, I.S., Kanade, T.: Terrain mapping for a roving planetary explorer. In: Proceedings of the IEEE International Conference on Robotics and Automation (ICRA), vol. 2, pp. 997–1002 (1989)
18. Triebel, R., Pfaff, P., Eurgard, W.: Multi-level surface maps for outdoor terrain mapping and loop closing. In: Proceedings of the IEEE/RSJ International Conference on Intelligent Robots and Systems (IROS) (2006)
19. Wurm, K.M., Hornung, A., Bennewitz, M., Stachniss, C., Burgard, W.: OctoMap: a probabilistic, flexible, and compact 3D map representation for robotic systems. In: Proceedings of the IEEE International Symposium on Robotics and Automation (ICRA), vol. 2 (2010)
20. Grisetti, G., Stachniss, C., Burgard, W.: Improveing grid-based SLAM with Rao-Blackwellized particle filters by adaptive proposals and selective resampling. In: Proceedings of the IEEE International Symposium on Robotics and Automation (ICRA), Barcelona, Spain, pp. 2443–2448 (2005)
21. Grisetti, G., Stachniss, C., Burgard, W.: Improved techniques for grid mapping with Rao-Blackwellized particle filters. IEEE Trans. Robot. **23**(1), 34–46 (2007)
22. Visual 3D Environment Reconstruction for Autonomous Vehicles. http://ercim-news.ercim.eu/en95/special/visual-3d-environment-reconstruction-for-autonomous-vehicles
23. ROS.org. Ubuntu install of ROS Indigo. http://wiki.ros.org/indigo/Installation/Ubuntu. Accessed Aug 2015
24. Computer Vision Group. http://vision.in.tum.de/research/rgb-d_sensors_kinect

On Partial Features in the $\mathcal{DLF}$ Family of Description Logics

David Toman$^{(\boxtimes)}$ and Grant Weddell

Cheriton School of Computer Science, University of Waterloo, Waterloo, Canada
{david,gweddell}@cs.uwaterloo.ca

Abstract. The $\mathcal{DLF}$ family of description logics are fragments of first order logic with underlying signatures based on unary predicate symbols, called atomic concepts, and unary function symbols interpreted as total functions, called features. We show how computational properties relating to a key reasoning service for dialects of this family are preserved when (a) unary function symbols are now interpreted as partial functions, and when (b) a concept constructor is admitted that can characterize circumstances in which partial functions become total.

1 Introduction

The $\mathcal{DLF}$ family of *description logics* (DLs) have been designed primarily to support reasoning tasks about object relational data sources. This includes the $\mathcal{CFD}$ sub-family that admits PTIME algorithms for many of these tasks. Unlike the usual case of *role*-based DLs [1], dialects in this family are *feature*-based, that is, are fragments of first order logic with underlying signatures that replace binary predicate symbols, called roles, with unary function symbols, interpreted as total functions, called features. Since features are intended to capture the notion of an attribute in a data source, a column for a relational table for example, this practice has led to some cognitive incongruity: one must reconcile that *every attribute is fundamentally defined for every object* and introduce protocols for indirectly saying when an attribute is or is not meaningful for various kinds of objects.

In this paper, we show how computational properties of DLs in the $\mathcal{DLF}$ family are preserved when they are modified to address such incongruity. This modification is as follows. First, features, now called *partial features*, are instead interpreted as partial functions, and second, a concept constructor is added that makes it possible to refer to all objects that *have* a value for a given partial feature. The added constructor yields to an ability for any of the DLs to define cases in which partial functions become total functions, such as to say that *every employee has a salary*, or, for two of the DLs considered, to define cases in which partial functions are not meaningful, such as to say that *departments do not have a salary*.

We consider the particular problem of reasoning about *logical implication* for three representative members of the $\mathcal{DLF}$ family: $\mathcal{DLFD}$ [5–7], $\mathcal{CFD}$ [3,8] and $\mathcal{CFD}_{nc}^{\forall}$ [9,10]. The first is a very expressive dialect for which logical implication is

© Springer International Publishing Switzerland 2016
R. Booth and M.-L. Zhang (Eds.): PRICAI 2016, LNAI 9810, pp. 529–542, 2016.
DOI: 10.1007/978-3-319-42911-3_44

EXPTIME-complete, while the remaining two are also members of the $\mathcal{CFD}$ family and therefore have PTIME decision procedures for this problem. Note that all three include a concept constructor for capturing keys and functional dependencies in object relational data sources and can express, for example, that *no two departments have the same manager*, or that *an employee's pay grade determines her salary*. The constructor is called a *path functional dependency* (PFD).

Our contributions, in the order presented, are as follows. Note that, in presenting them and for the remainder of the paper, we write $partial-\mathcal{DLFD}$, $partial-\mathcal{CFD}$ and $partial-\mathcal{CFD}_{nc}^\forall$ to refer, respectively, to $\mathcal{DLFD}$, $\mathcal{CFD}$ and $\mathcal{CFD}_{nc}^\forall$ when they are presumed to be modified to support partial functions in the above fashion:

1. We introduce a semantics for PFDs when features are partial functions that is entirely neutral on issues of *feature existence*, that is, on whether certain kinds of objects *must have* or even *can have* values for particular features;[1]
2. We show that logical implication for $partial-\mathcal{DLFD}$ reduces to logical implication for $\mathcal{DLFD}$, in the process showing that partial functions can be simulated in a straightforward fashion in this dialect; and
3. We show that logical implication for both $partial-\mathcal{CFD}$ and $partial-\mathcal{CFD}_{nc}^\forall$ remains in PTIME by exhibiting refinements of existing respective PTIME decision procedures for deciding logical implication with $\mathcal{CFD}$ and $\mathcal{CFD}_{nc}^\forall$.

We conclude with summary comments and an outline of possible directions for future work, with a particular focus on $\mathcal{CFD}_{nc}^\forall$ in the latter case: on partial functions for a recent extension, and on relaxing syntactic restrictions to enable straightforward partial function simulation, as we show is possible for $\mathcal{DLFD}$.

2   Background and Definitions

In this section, we begin by reviewing the basic definitions for member dialects of the $\mathcal{DLF}$ family in which features are interpreted as total functions, and then proceed to introduce modifications that yield support for partial features, that is, features that are instead interpreted as partial functions.

Note that DLs in this family do not forgo the ability to capture roles or indeed n-ary relations in general. This can be accomplished with the simple expedient of reification via features, and by using the above-mentioned PFD concept constructor common to these dialects to ensure a set semantics for reified relations. Indeed, the first dialect we consider, $\mathcal{DLFD}$, can capture very expressive role-based dialects, including dialects with so-called qualified number restrictions, inverse roles, role hirarchies, and so on [5].

---

[1] Such issues can be (and we believe should be) explicitly addressed elsewhere in an ontology.

Definition 1 (Feature-Based DLs). Let F and PC be sets of feature names and primitive concept names, respectively. A *path expression* is defined by the grammar "Pf :: $= f.$Pf $\mid id$" for $f \in$ F. We define derived *concept descriptions* by the grammar on the left-hand-side of Fig. 1.

An *inclusion dependency* $\mathcal{C}$ is an expression of the form $C_1 \sqsubseteq C_2$. A *terminology* (TBox) $\mathcal{T}$ consists of a finite set of inclusion dependencies. A *posed question* $\mathcal{Q}$ is a single inclusion dependency.

SYNTAX	SEMANTICS: DEFN OF "$\cdot^{\mathcal{I}}$"	
$C ::=$ A	$A^{\mathcal{I}} \subseteq \triangle$	(primitive concept; $A \in$ PC)
$\mid\ C_1 \sqcap C_2$	$C_1^{\mathcal{I}} \cap C_2^{\mathcal{I}}$	(conjunction)
$\mid\ C_1 \sqcup C_2$	$C_1^{\mathcal{I}} \cup C_2^{\mathcal{I}}$	(disjunction)
$\mid\ \neg C$	$\triangle \setminus C^{\mathcal{I}}$	(negation)
$\mid\ \forall \text{Pf}.C$	$\{x : \text{Pf}^{\mathcal{I}}(x) \in C^{\mathcal{I}}\}$	(value restriction)
$\mid\ C : \text{Pf}_1, ..., \text{Pf}_k \to \text{Pf}_0$	$\{x : \forall y \in C^{\mathcal{I}} . \bigwedge_{i=1}^{k} \text{Pf}_i^{\mathcal{I}}(x) = \text{Pf}_i^{\mathcal{I}}(y)$	(PFD)
	$\qquad\qquad \to \text{Pf}_0^{\mathcal{I}}(x) = \text{Pf}_0^{\mathcal{I}}(y)\}$	
$\mid\ \top$	$\triangle$	(top)
$\mid\ \bot$	$\emptyset$	(bottom)
$\mid\ (\text{Pf}_1 = \text{Pf}_2)$	$\{x : \text{Pf}_1^{\mathcal{I}}(x) = \text{Pf}_2^{\mathcal{I}}(x)\}$	(same-as)

Fig. 1. Syntax and semantics of $\mathcal{DLFD}/\mathcal{CFD}$ concepts.

The *semantics* of expressions is defined with respect to a structure $\mathcal{I} = (\triangle, \cdot^{\mathcal{I}})$, where $\triangle$ is a domain of "objects" and $\cdot^{\mathcal{I}}$ an interpretation function that fixes the interpretations of primitive concepts A to be subsets of $\triangle$ and primitive features f to be total functions $f^{\mathcal{I}} : \triangle \to \triangle$. The interpretation is extended to path expressions, $id^{\mathcal{I}} = \lambda x.x$, $(f.\text{Pf})^{\mathcal{I}} = \text{Pf}^{\mathcal{I}} \circ f^{\mathcal{I}}$ and derived concept descriptions C as defined in the centre column of Fig. 1.

An interpretation $\mathcal{I}$ *satisfies an inclusion dependency* $C_1 \sqsubseteq C_2$ if $C_1^{\mathcal{I}} \subseteq C_2^{\mathcal{I}}$ and is a *model of* $\mathcal{T}$ ($\mathcal{I} \models \mathcal{T}$) if it satisfies all inclusion dependencies in $\mathcal{T}$. The *logical implication problem* asks if $\mathcal{T} \models \mathcal{Q}$ holds, that is, if $\mathcal{Q}$ is satisfied in all models of $\mathcal{T}$. $\square$

In the following, we simplify the notation for path expressions by allowing a syntactic composition $\text{Pf}_1 . \text{Pf}_2$ that stands for their concatenation.

In this presentation, we do *not* consider so-called ABoxes (sets of assertions about membership of individuals in descriptions) and the associated problem of knowledge base consistency. However, these can be reduced to logical implication problems involving posed questions that utilize value restrictions and equational *same-as* descriptions [8].

Also note that the logical implication problem for TBoxes and posed questions characterized so far, that allow arbitrary concepts in inclusion dependencies, is not decidable for a variety of reasons (e.g., see [7] for one case involving arbitrary PFDs and ABoxes encoded in the above manner). However, restrictions on occurrences of concept constructors has led to a number of decidable fragments that range from light-weight to expressive dialects of feature-based DLs. The restrictions that obtain $\mathcal{DLFD}$, $\mathcal{CFD}$ and $\mathcal{CFD}_{nc}^{\forall}$, the focus of our attention, are given in Sect. 3 for the first case and in Sect. 4 for the remaining two cases.

The two definitions that follow now introduce the necessary modifications to our characterization of feature-based DLs to accommodate *partial features*, that is, features that are interpreted as partial functions. Note that their presentation relies on our notational convention given in our introductory comments of qualifying particular dialects with the word *"partial"* whenever we intend such modifications to apply, as in $partial-\mathcal{DLFD}$ for example.

Definition 2 (Partial Features and Existential Restrictions). The syntax of feature-based DLs is extended with an additional concept constructor of the form $\exists f$, called an *existential restriction*. Semantics is updated as follows:

1. features $f \in F$ are now interpreted as *partial* functions on $\triangle$ (i.e., the result can be *undefined* for parts of $\triangle$); and
2. the $\exists f$ concept constructor is interpreted as $\{x : \exists y \in \triangle . f^{\mathcal{I}}(x) = y\}$.

Also, in this setting, path functions (Pf) naturally denote composition of partial functions yielding a partial function, equality ($=$) is true only when both of its arguments are defined (in addition to being equal), set membership ($\in$) requires only defined values to be members of its right hand side argument, etc.[2]   □

Observe that features are still *functional*, and that there is therefore no need for a qualified existential restriction of the form $\exists f.C$, with the standard meaning $(\exists f.C)^{\mathcal{I}}$ given by

$$\{x : \exists y \in \triangle . f^{\mathcal{I}}(x) = y \land y \in C^{\mathcal{I}}\}.$$

Indeed, they can be simulated using the following identity:

$$\exists f.C = \exists f \sqcap \forall f.C.$$

Using this identity, we write $\exists \mathsf{Pf}$ in the following as shorthand for

$$\exists f_1 \sqcap \forall f_1.(\exists f_2 \sqcap \forall f_2.(\ldots (\exists f_k)\ldots)).$$

On interpreting the PFD constructor in the presence of partial features: the minimum necessary (and we believe most natural) circumstance in which one obtains a violation of a *PFD inclusion dependency* of the form

$$C_1 \sqsubseteq C_2 : \mathsf{Pf}_1, \ldots, \mathsf{Pf}_k \to \mathsf{Pf}_0$$

happens when all path functions $\mathsf{Pf}_0, \ldots, \mathsf{Pf}_k$ are defined for a C_1 object e_1 and a C_2 object e_2, and in which $\mathsf{Pf}_i^{\mathcal{I}}(e_1) = \mathsf{Pf}_i^{\mathcal{I}}(e_2)$ holds only for $i > 0$. This yields the

---

[2] This arrangement is common and is referred to as the *strict* interpretation of undefined values.

following modification to the interpretation of PFDs in the presence of partial features that we now adopt:

$$(C : \mathsf{Pf}_1, \ldots, \mathsf{Pf}_k \to \mathsf{Pf}_0)^{\mathcal{I}} = \{x : \forall y.y \in C^{\mathcal{I}} \land x \in (\exists \mathsf{Pf}_0)^{\mathcal{I}} \land y \in (\exists \mathsf{Pf}_0)^{\mathcal{I}} \land$$
$$\bigwedge_{i=1}^{k}(x \in (\exists \mathsf{Pf}_i)^{\mathcal{I}} \land y \in (\exists \mathsf{Pf}_i)^{\mathcal{I}} \land \mathsf{Pf}_i^{\mathcal{I}}(x) = \mathsf{Pf}_i^{\mathcal{I}}(y)) \to \mathsf{Pf}_0^{\mathcal{I}}(x) = \mathsf{Pf}_0^{\mathcal{I}}(y)\}.$$

Observe that this definition coincides with the original semantics of the PFD constructor given in Fig. 1 when features are interpreted as total functions. Also note that, *without* this modification to semantics, the strict interpretation of undefined values would mean that satisfying the left-hand-side of a PFD would *imply the existence of* "Pf_0 *paths*" (and equality of the "endpoint" of these paths), a circumstance that would violate feature existence neutrality of PFDs mentioned in our introductory comments that seems desirable.

We now return to examples of constraints mentioned in our introductory comments to illustrate the use of existential restrictions and PFDs in $\mathcal{DLF}$ dialects with partial features. Each can be expressed in $partial-\mathcal{DLFD}$ and $partial-\mathcal{CFD}_{nc}^{\forall}$, and each but the second mentioning negation in $partial-\mathcal{CFD}$:

1. EMP $\sqsubseteq \exists salary$ (*every employee has a salary*);
2. DEPT $\sqsubseteq \neg\exists salary$ (*departments do not have a salary*);
3. DEPT $\sqsubseteq$ DEPT : *manager* $\to$ *id* (*no two departments have the same manager*); and
4. EMP $\sqsubseteq$ EMP : *paygrade* $\to$ *salary* (*employee pay grades determine salaries*).

3   Expressive Feature Logics: The $\mathcal{DLF}$ Family

In this section, we consider the impact of partial features in expressive feature-based description logics, namely in $\mathcal{DLF}$ and $\mathcal{DLFD}$ [5–7]. $\mathcal{DLF}$ allows both TBox and posed question dependencies to contain concepts formed from primitive concepts and bottom using negation, conjunction, disjunction, and restriction concept constructors. $\mathcal{DLFD}$ in addition allows the PFD concept constructor to appear on the right hand sides of inclusion dependencies. These restrictions on syntax yield an expressive Boolean complete description logic with a logical implication problem that is complete for EXPTIME. Additional extensions, e.g., allowing PFDs on the left-hand sides of inclusion dependencies or equational constraints in the posed questions (or equivalently ABoxes) leads to undecidability [7].

We now proceed to demonstrate that partial features can be effectively *simulated* in the original logics by introducing an auxiliary primitive concept G that stands for *existing or generated* objects, and by using *value restrictions* to assign membership of objects generated by the $\exists f$ constructor to this concept. All remaining inclusion dependencies are then simply preconditioned by this auxiliary concept.

Formally, let $\mathcal{T}$ be a $partial-\mathcal{DLF}$ TBox in which all inclusion dependencies are of the form $\top \sqsubseteq C$. We define a $\mathcal{DLF}$ TBox $\mathcal{T}_{\mathcal{DLF}}$ as

$$\mathcal{T}_{\mathcal{DLF}} = \{G \sqsubseteq C[\exists f \mapsto \forall f.G, \text{ for all } f \in F] \mid \top \sqsubseteq C \in \mathcal{T}\}$$
$$\cup \{\forall f.G \sqsubseteq G \mid f \in F\},$$

where G is a primitive concept not occurring in $\mathcal{T}$. Note that the substitution $[\exists f \mapsto \forall f.G$, for all $f \in F]$ is applied simultaneously to *all* occurrences of the $\exists f$ constructor in the concept C.

Theorem 3. Let $\mathcal{T}$ be a *partial*$-\mathcal{DLF}$ TBox in which all inclusion dependencies are of the form $\top \sqsubseteq C$. Then

$$\mathcal{T} \models \top \sqsubseteq C \text{ if and only if } \mathcal{T}_{\mathcal{DLF}} \models G \sqsubseteq C[\exists f \mapsto \forall f.G, \text{ for all } f \in F],$$

for G a fresh primitive concept.

Proof (sketch): For any $\mathcal{I}$ where $\mathcal{I} \models \mathcal{T}_{\mathcal{DLF}}$, we can define an interpretation $\mathcal{J} = (G^{\mathcal{I}}, \cdot|_{G^{\mathcal{I}}}^{\mathcal{I}})$. It is easy to verify that $\mathcal{J} \models \mathcal{T}$ and also that $\mathcal{J} \models \top \sqsubseteq C$ since $\mathcal{I} \models G \sqsubseteq C[\exists f \mapsto \forall f.G, \text{ for all } f \in F]$.

For the other direction, we need to extend a model $\mathcal{J}$ of $\mathcal{T}$ to a model $\mathcal{I}$ of $\mathcal{T}_{\mathcal{DLF}}$ by setting $G^{\mathcal{I}} = \triangle^{\mathcal{J}}$ and by adding *missing* features connecting $\mathcal{I}$ to complete F^* trees with all nodes in $(\neg G)^{\mathcal{I}}$. This way, either $\mathcal{I}$ coincides with $\mathcal{J}$ or satisfies dependencies in $\mathcal{T}_{\mathcal{DLF}}$ and $G \sqsubseteq C[\exists f \mapsto \forall f.G, \text{ for all } f \in F]$ vacuously.   □

To extend this construction to the full *partial*$-\mathcal{DLFD}$ logic, it is sufficient to *encode* the path function existence preconditions in terms of the auxiliary concept G as follows: if $A \sqsubseteq B : \mathsf{Pf}_1, \ldots, \mathsf{Pf}_k \to \mathsf{Pf}_0 \in \mathcal{T}$ then

$$A \sqcap (\prod_{i=0}^{k} \forall \mathsf{Pf}_i . G) \sqsubseteq B \sqcap (\prod_{i=0}^{k} \forall \mathsf{Pf}_i . G) : \mathsf{Pf}_1, \ldots, \mathsf{Pf}_k \to \mathsf{Pf}_0 \tag{1}$$

is added $\mathcal{T}_{\mathcal{DLFD}}$. Here, we are assuming w.l.o.g. that A and B are primitive concept names ($\mathcal{DLFD}$ allows one to give such names to complex concepts).

Theorem 4. Let $\mathcal{T}$ be a *partial*$-\mathcal{DLFD}$ TBox in which all inclusion dependencies are of the form $\top \sqsubseteq C$ or $A \sqsubseteq B : \mathsf{Pf}_1, \ldots, \mathsf{Pf}_k \to \mathsf{Pf}_0$. Then

$$\mathcal{T} \models \top \sqsubseteq C \text{ if and only if } \mathcal{T}_{\mathcal{DLFD}} \models C \sqsubseteq D[\exists f \mapsto \forall f.G, \text{ for all } f \in F], \text{ and}$$
$$\mathcal{T} \models A \sqsubseteq B : \mathsf{Pf}_1, \ldots, \mathsf{Pf}_k \to \mathsf{Pf} \text{ if and only if } \mathcal{T}_{\mathcal{DLFD}} \models (1),$$

for G a fresh primitive concept.

Proof (sketch): Logical implication in $\mathcal{DLFD}$ can be reduced to logical implication in $\mathcal{DLF}$ [5–7]. Hence the claim holds by observing that (*) captures properly the semantics of PFDs and then by appealing to Theorem 3.   □

Corollary 5. Logical implication is EXPTIME-complete for *partial*$-\mathcal{DLF}$ and for *partial*$-\mathcal{DLFD}$.   □

Similar results can be obtained for other members of the $\mathcal{DLF}$ family.

4   Tractable Logics: The $\mathcal{CFD}$ Family

We now consider how partial features impact logical consequence for light-weight (PTIME) feature-based description logics, namely $\mathcal{CFD}$ [3,8] and $\mathcal{CFD}_{nc}^{\forall}$ [9,10]. Both of these logics allow the use of an ABox. Hence, PFDs must adhere to one of the following two forms to avoid undecidability [7]:

$$\begin{aligned}&1.\ C : \mathsf{Pf}_1, \ldots, \mathsf{Pf} . \mathsf{Pf}_i, \ldots, \mathsf{Pf}_k \rightarrow \mathsf{Pf}\ \ \text{or}\\ &2.\ C : \mathsf{Pf}_1, \ldots, \mathsf{Pf} . \mathsf{Pf}_i, \ldots, \mathsf{Pf}_k \rightarrow \mathsf{Pf} . f\end{aligned} \qquad (2)$$

With this restriction, originally introduced in [3], posed questions can contain inclusion dependencies formed from concepts in Fig. 1 (with a few mild restrictions when tractability in the size of the posed question is required). For simplicity, however, we assume that the concepts in the posed question $\mathcal{Q} = E_1 \sqsubseteq E_2$ adhere to the following grammar:

$$E ::= A \mid \bot \mid E \sqcap E \mid \forall \mathsf{Pf} . E \mid (\mathsf{Pf}_1 = \mathsf{Pf}_2).$$

More complex posed questions, e.g., ones that contain the PFD constructor [8], can be equivalently expressed in the above grammar (perhaps as a sequence of posed questions).

Note that, due to syntactic restrictions on TBox inclusion dependencies in $\mathcal{CFD}$ and $\mathcal{CFD}_{nc}^{\forall}$ (see below), we will *not* be able to directly simulate partial features as was done with $\mathcal{DLF}$ and $\mathcal{DLFD}$ above. However, the approach to extending/modifying existing decision procedures for logical implication in the respective logics is analogous: in both cases we introduce an additional unary predicate $\mathsf{D}(x)$ to *mark* the *necessarily existing* objects and use this predicate to restrict the applications of inclusion dependencies. This in turn simulates partial features.

4.1   *partial*$-\mathcal{CFD}$

To obtain a PTIME decision procedure for $\mathcal{CFD}$, we need to further restrict the inclusion dependencies allowed in the TBox $\mathcal{T}$ as follows:

1. left hand sides must be conjunctions of primitive concepts, and
2. right hand sides must be primitive concepts, conjunctions, value restrictions, existential restrictions, and PFDs (obeying restrictions in (2)).

With these restrictions we can show that the logical implication problem for *partial*$-\mathcal{CFD}$ is in PTIME. Our proof is based on encoding a given problem as a collection of Horn clauses. The reduction introduces terms that correspond to path expressions, and relies on the fact that the number of required terms is polynomial in the size of the problem itself.

$\mathsf{D}(\mathsf{Pf}_1 \,.f) \to \mathsf{D}(\mathsf{Pf}_1)$
$\mathsf{D}(\mathsf{Pf}_1) \to \mathsf{E}(\mathsf{Pf}_1, \mathsf{Pf}_1)$

$\mathsf{E}(\mathsf{Pf}_1, \mathsf{Pf}_2) \to \mathsf{E}(\mathsf{Pf}_2, \mathsf{Pf}_1)$
$\mathsf{E}(\mathsf{Pf}_1, \mathsf{Pf}_2) \land \mathsf{E}(\mathsf{Pf}_2, \mathsf{Pf}_3) \to \mathsf{E}(\mathsf{Pf}_1, \mathsf{Pf}_3)$
$\mathsf{E}(\mathsf{Pf}_1, \mathsf{Pf}_2) \land \mathsf{D}(\mathsf{Pf}_1 \,.f) \land \mathsf{D}(\mathsf{Pf}_2 \,.f) \to \mathsf{E}(\mathsf{Pf}_1 \,.f, \mathsf{Pf}_2 \,.f),$ for $\{\mathsf{Pf}_1 \,.f, \mathsf{Pf}_2 \,.f\} \subseteq \mathsf{PF}(\mathcal{T}, \mathcal{Q})$
$\mathsf{E}(\mathsf{Pf}_1, \mathsf{Pf}_2) \land \mathsf{C}_C(\mathsf{Pf}_1) \to \mathsf{C}_C(\mathsf{Pf}_2)$

$\mathsf{C}_{C_1 \sqcap C_2}(\mathsf{Pf}) \to \mathsf{C}_{C_1}(\mathsf{Pf})$ and $\mathsf{C}_{C_1 \sqcap C_2}(\mathsf{Pf}) \to \mathsf{C}_{C_2}(\mathsf{Pf})$
$\mathsf{C}_{\exists f}(\mathsf{Pf}) \land \mathsf{D}(\mathsf{Pf}) \to \mathsf{D}(\mathsf{Pf} \,.f)$ for all $\mathsf{Pf} \,.f \in \mathsf{PF}(\mathcal{T}, \mathcal{Q})$
$\mathsf{C}_{\forall \mathsf{Pf}' \,.C}(\mathsf{Pf}) \to \mathsf{C}_C(\mathsf{Pf} \,.\mathsf{Pf}')$ for $\mathsf{Pf} \,.\mathsf{Pf}' \in \mathsf{PF}(\mathcal{T}, \mathcal{Q})$
$\mathsf{C}_{(\mathsf{Pf}_1 = \mathsf{Pf}_2)}(\mathsf{Pf}) \land \mathsf{D}(\mathsf{Pf} \,.\mathsf{Pf}_1) \land \mathsf{D}(\mathsf{Pf} \,.\mathsf{Pf}_2) \to \mathsf{E}(\mathsf{Pf} \,.\mathsf{Pf}_1, \mathsf{Pf} \,.\mathsf{Pf}_2)$
$\mathsf{C}_{C:\mathsf{Pf}_1, \ldots, \mathsf{Pf}_k \to \mathsf{Pf}_0}(\mathsf{Pf}) \land \mathsf{C}_C(\mathsf{Pf}') \land (\bigwedge_{0 < i \leq k} \mathsf{E}(\mathsf{Pf} \,.\mathsf{Pf}_i, \mathsf{Pf}' \,.\mathsf{Pf}_i))$
$\qquad\qquad\qquad\qquad \land\, \mathsf{D}(\mathsf{Pf} \,.\mathsf{Pf}_0) \land \mathsf{D}(\mathsf{Pf}' \,.\mathsf{Pf}_0) \to \mathsf{E}(\mathsf{Pf} \,.\mathsf{Pf}_0, \mathsf{Pf}' \,.\mathsf{Pf}_0))$

$\mathsf{C}_{A_1}(\mathsf{Pf}) \land \ldots \land \mathsf{C}_{A_k}(\mathsf{Pf}) \to \mathsf{C}_D(\mathsf{Pf})$ for all $(A_1 \sqcap \ldots \sqcap A_k \sqsubseteq D) \in \mathcal{T}$
$\mathsf{C}_A(\mathsf{Pf} \,.f) \to \mathsf{C}_D(\mathsf{Pf})$ for all $(\forall f.A \sqsubseteq D) \in \mathcal{T}$

Fig. 2. Expansion rules.

Definition 6 (Expansion Rules). Let $\mathcal{T}$ and $\mathcal{Q}$ be a *partial–CFD* terminology and a posed question, respectively. We write $\mathsf{CON}(\mathcal{T}, \mathcal{Q})$ to denote the set of all subconcepts appearing in $\mathcal{T}$ and $\mathcal{Q}$, define $\mathsf{PF}(\mathcal{T}, \mathcal{Q})$ to be the set

$$\{\mathsf{Pf} \,.\mathsf{Pf}' \mid \mathsf{Pf} \text{ is a prefix of a path expression in } \mathcal{Q} \text{ and }$$
$$\mathsf{Pf}' \text{ is a feature occurring in } \mathcal{T} \text{ or } id\},$$

write C_C to denote unary predicates for $C \in \mathsf{CON}(\mathcal{T}, \mathcal{Q})$, and introduce a unary predicate D and a binary predicate E, with all predicates ranging over the universe $\mathsf{PF}(\mathcal{T}, \mathcal{Q})$. The *expansion rules* for a given terminology $\mathcal{T}$, denoted $\mathsf{R}(\mathcal{T})$, are defined in Fig. 2.[3]

A *goal* for each concept E is a set of ground assertions defined as follows:

$$\mathsf{G}_E = \begin{cases} \{\mathsf{C}_A(id), \mathsf{D}(id)\} & \text{for } E = \mathrm{A}; \\ \{\mathsf{C}_\perp(id), \mathsf{D}(id)\} & \text{for } E = \perp; \\ \{\mathsf{E}(\mathsf{Pf}_1, \mathsf{Pf}_2), \mathsf{D}(\mathsf{Pf}_1), \mathsf{D}(\mathsf{Pf}_2)\} & \text{for } E = (\mathsf{Pf}_1 = \mathsf{Pf}_2); \\ \mathsf{G}_{E_1} \cup \mathsf{G}_{E_2} & \text{for } E = E_1 \sqcap E_2; \text{ and} \\ \{\mathsf{C}_C(\mathsf{Pf}' \,.\mathsf{Pf}) \mid \mathsf{C}_C(\mathsf{Pf}) \in \mathsf{G}_{E'}\} & \\ \quad \cup \{\mathsf{D}(\mathsf{Pf}' \,.\mathsf{Pf}) \mid \mathsf{D}(\mathsf{Pf}) \in \mathsf{G}_{E'}\} & \\ \quad \cup \{\mathsf{E}(\mathsf{Pf}' \,\mathsf{Pf}_1, \mathsf{Pf}' \,.\mathsf{Pf}_2) \mid \mathsf{E}(\mathsf{Pf}_1, \mathsf{Pf}_2) \in \mathsf{G}_{E'}\} & \text{for } E = \forall \mathsf{Pf}' \,.E'. \end{cases}$$

Given two concept descriptions E_1 and E_2, we say that

$$\mathsf{R}(\mathcal{T}) \cup \{\mathsf{C}_{E_1}(id)\} \models \mathsf{G}_{E_2}$$

---

[3] The last rule in the figure does not apply to *partial–CFD* and is added w.l.o.g. in preparation for treating *partial–CFD*$_{nc}^\forall$. This rule is neither necessary nor applicable in the *partial–CFD* case.

if $\mathsf{G}_{E_2} \subseteq M$ for every minimal ground model M of $\mathsf{R}(\mathcal{T})$ over $\mathsf{PF}(\mathcal{T}, \mathcal{Q})$ that contains $\mathsf{C}_{E_1}(id)$ and $\mathsf{D}(id)$. $\qquad\square$

Intuitively, $\mathsf{PF}(\mathcal{T}, \mathcal{Q})$ represents a finite graph of objects, predicates $\mathsf{E}(\mathsf{Pf}_1, \mathsf{Pf}_2)$ express equality of the objects at the end of paths Pf_1 and Pf_2, and predicates $\mathsf{C}_{C'}(\mathsf{Pf})$ express that the object at the end of path Pf is in the interpretation of concept C'.

Our PTIME result for the *partial*$-\mathcal{CFD}$ implication problem follows by a simple check for goals occurring in a ground model for expansion rules generated by a polynomial sized collection of path expressions.

Theorem 7. Let $\mathcal{T}$ be a *partial*$-\mathcal{CFD}$ terminology and $\mathcal{Q}$ a posed question of the form $E_1 \sqsubseteq E_2$. Then

$$\mathcal{T} \models \mathcal{Q} \text{ iff } \mathsf{R}(\mathcal{T}) \cup \{\mathsf{C}_{E_1}(id), \mathsf{D}(id)\} \models \mathsf{G}_{E_1} \text{ or}$$
$$\mathsf{R}(\mathcal{T}) \cup \{\mathsf{C}_{E_1}(id), \mathsf{D}(id)\} \models \mathsf{G}_{\forall \mathsf{Pf}.\perp}(id) \text{ for some } \mathsf{Pf} \in \mathsf{PF}(\mathcal{T}, \mathcal{Q}).$$

Proof (sketch): If $\mathsf{C}_\perp(\mathsf{Pf})$ and $\mathsf{D}(\mathsf{Pf})$ for $\mathsf{Pf} \in \mathsf{PF}(\mathcal{T}, \mathcal{Q})$ appear in M, where M is the least model of $\mathsf{R}(\mathcal{T}) \cup \{\mathsf{C}_{E_1}(id)\}$ $\mathsf{R}(\mathcal{T})$, then the concept E_1 is unsatisfiable w.r.t. $\mathcal{T}$ since only implied facts appear in M, and therefore the subsumption holds for any E_1 and $\mathcal{T}$.

Otherwise, if $\mathsf{R}(\mathcal{T}) \cup \{\mathsf{C}_{E_1}(id)\} \not\models \mathsf{G}_{E_1}$, then there must be a model M of $\mathsf{R}(\mathcal{T}) \cup \{\mathsf{C}_{E_1}(id)\}$ such that $G \notin M$ for some $G \in \mathsf{G}_{E_1}$. We construct an interpretation $\mathcal{I}_M$ such that $\mathcal{I}_M \models \mathcal{T}$ but $\mathcal{I}_M \not\models \mathcal{Q}$. The interpretation $\mathcal{I}_M$ contains an object o for each equivalence class defined on the set $\mathsf{PF}(\mathcal{T}, \mathcal{Q})$ by the interpretation of E. The class membership of these objects is determined by the membership of the corresponding path in the interpretations of the C_C predicates in M. Note that, due to the syntactic restriction imposed on PFDs, this is sufficient to satisfy all PFDs in $\mathcal{T}$ since any precondition or a non-trivial consequence of a PFD can only manifest on some path belonging to $\mathsf{PF}(\mathcal{T}, \mathcal{Q})$ and beginning at the distinguished object o. To complete the construction of $\mathcal{I}_M$, we simply attach a unique complete tree F^* to each leaf node (i.e., a node that is missing successors). Nodes of these complete trees belong to all primitive descriptions in $\mathcal{I}_M$ and thus satisfy $\mathcal{T}$.

Conversely, assume $\mathsf{R}(\mathcal{T}) \cup \{\mathsf{C}_{E_1}(id)\} \models \mathsf{G}_{E_1}$ but $\mathcal{T} \not\models \mathcal{Q}$. Then there must be an interpretation $\mathcal{I}$ and an object $o \in \triangle$ such that $\mathcal{I} \models \mathcal{T}$ and $o \in E_1^{\mathcal{I}} - E_2^{\mathcal{I}}$. Thus, there is a model $M_{\mathcal{I}}$ of $\mathsf{R}(\mathcal{T})$ such that $\mathsf{C}_{E_1}(id) \in M_{\mathcal{I}}$. In this model, the element $id \in \mathsf{PF}(\mathcal{T}, \mathcal{Q})$ serves as the counterpart of the object o and the interpretations of the predicates C_C and E is *extracted* from $\mathcal{I}$ by navigating all (pairs of) path functions in $\mathsf{PF}(\mathcal{T}, \mathcal{Q})$. However, since $o \notin E_2^{\mathcal{I}}$, it must be the case that $M_{\mathcal{I}}$ is a strict subset of the least model of $\mathsf{R}(\mathcal{T}) \cup \{\mathsf{C}_{E_1}(id)\}$; a contradiction. $\qquad\square$

Since the expansion rules are Horn clauses over a finite universe $\mathsf{PF}(\mathcal{T}, \mathcal{Q})$ of polynomial size, we have the following:

Corollary 8. Let $\mathcal{T}$ be a terminology and $\mathcal{Q}$ a posed question in *partial*$-\mathcal{CFD}$. Then the implication problem $\mathcal{T} \models \mathcal{Q}$ is complete for PTIME.

Proof (sketch): The least model of $\mathsf{R}(\mathcal{T}) \cup \{\mathsf{C}_{E_1}(id)\}$ can be obtained by using a bottom-up construction of the least fix-point of the rules in time polynomial in $|\mathcal{T}| + |\mathcal{Q}|$ (since all predicates in $\mathsf{R}(\mathcal{T})$ have a fixed arity). Hardness follows from embedding Horn-SAT into reasoning with PFDs.   □

In practice, elements of this set can be constructed on demand by using additional Horn rules in such a way that only path expressions needed to confirm subsumption or non-subsumption are generated [4].

    Note that neither $partial-\mathcal{CFD}$ nor $\mathcal{CFD}$ can be extended to allow disjointness (bottom ($\bot$)), negation (hence $\neg\exists f$ cannot be used), or disjunction on the right hand sides of inclusion dependencies while maintaining PTIME decidability of logical implication [8]. However, a similar technique as in the above development can be used to handle partial features without impacting the complexity of the logical implication problems.

4.2 $partial-\mathcal{CFD}^\forall_{nc}$

$partial-\mathcal{CFD}^\forall_{nc}$ shares the PFD restrictions with $partial-\mathcal{CFD}$. However, it trades the ability to use conjunctions on the left hand sides of TBox inclusion dependencies for the ability to express disjointness and conditional typing:

1. left hand sides must be primitive concepts or value restrictions, and
2. right hand sides must be a primitive concepts, negations of primitive concepts, conjunctions, value restrictions, existential restrictions, and PFDs (again, restricted as in (2)).

It is easy to see that every $partial-\mathcal{CFD}^\forall_{nc}$ TBox $\mathcal{T}$ is consistent (by setting all primitive concepts to be interpreted as the empty set). It is, however, no longer true that all primitive concepts (and their conjunctions) are trivially satisfiable. For example, $\mathsf{A} \sqsubseteq \neg\mathsf{A} \in \mathcal{T}$ forces A to be empty in every model of $\mathcal{T}$.

Concept Satisfiability. The problem of *concept satisfiability* asks, for a given concept C and TBox $\mathcal{T}$, if there exists an interpretation $\mathcal{I}$ for $\mathcal{T}$ in which $C^\mathcal{I}$ is non-empty. Such problems can be reduced to the case where C is a primitive concept A by simply augmenting $\mathcal{T}$ with $\{\mathsf{A} \sqsubseteq C\}$, where A is a fresh primitive concept. Note that concept C can be a *conjunction* of other concepts since it only appears on the right-hand side of an inclusion dependency. We proceed as follows:

Definition 9 (Transition Relation for $\mathcal{T}$). Let $\mathcal{T}$ be a $partial-\mathcal{CFD}^\forall_{nc}$ TBox in normal form. We define a transition relation $\delta(\mathcal{T})$ over the set of states

$$S = \mathsf{PC} \cup \{\neg\mathsf{A} \mid \mathsf{A} \in \mathsf{PC}\} \cup \{\forall f.\mathsf{A} \mid \mathsf{A} \in \mathsf{PC}, f \in \mathsf{F}\} \cup \{\exists f \mid f \in \mathsf{F}\}$$

and the alphabet F as follows:

$$C_1 \xrightarrow{id} C_2 \in \delta(\mathcal{T}), \text{ if } C_1 \sqsubseteq C_2 \in \mathcal{T}, \text{ and}$$
$$\forall f.\mathsf{A} \xrightarrow{f} \mathsf{A} \in \delta(\mathcal{T}), \text{ if } \forall f.\mathsf{A} \xrightarrow{id*} \exists f \in \delta(\mathcal{T}),$$

where id is the empty letter transition, $id*$ is a sequence of id edges, $f \in \mathsf{F}$, $A \in \mathsf{PC}$, and $C_1, C_2 \in S$. □

The transition relation allows us to construct *non-deterministic finite automata* (NFA) that can be used for various reasoning problems on a *partial*–$\mathcal{CFD}_{nc}^{\forall}$ TBox $\mathcal{T}$. Note that, unlike common practice in automata theory, we use id for the empty letter in transition relations. Given a primitive concept A and TBox $\mathcal{T}$, one can test for primitive concept satisfiability by using the following NFA, denoted $\mathsf{nfa}_{\mathrm{B}}^{\mathrm{A}}(\mathcal{T})$:

$$(S, \{\mathrm{A}\}, \{\mathrm{B}\}, \delta(\mathcal{T})),$$

with states induced by primitive concepts, their negations, and value restrictions, with start state A, with the set of final states $\{\mathrm{B}\} \subseteq S$, and with transition relation $\delta(\mathcal{T})$. Intuitively, if $\mathsf{Pf} \in \mathsf{nfa}_{\mathrm{B}}^{\mathrm{A}}(\mathcal{T})$ and $o \in \mathrm{A}^{\mathcal{I}}$ then $\mathsf{Pf}^{\mathcal{I}}(o)$ is defined and $\mathsf{Pf}^{\mathcal{I}}(o) \in \mathrm{B}^{\mathcal{I}}$ in every model $\mathcal{I}$ of $\mathcal{T}$.

Theorem 10 (Concept Satisfiability). A is satisfiable with respect to the TBox $\mathcal{T}$ if and only if

$$\mathcal{L}(\mathsf{nfa}_{\mathrm{B}}^{\mathrm{A}}(\mathcal{T}) \cap \mathcal{L}(\mathsf{nfa}_{\neg\mathrm{B}}^{\mathrm{A}}(\mathcal{T})) = \emptyset$$

for every $\mathrm{B} \in \mathsf{PC}$.

Proof (sketch): Assume A is non-empty and hence there is $a \in \mathrm{A}^{\mathcal{I}}$. For a primitive concept $\mathrm{B} \in \mathsf{PC}$, a word Pf in the intersection language of the two automata above is a witness of the fact that $\mathsf{Pf}^{\mathcal{I}}(a^{\mathcal{I}}) \in \mathrm{B}^{\mathcal{I}}$ and $\mathsf{Pf}^{\mathcal{I}}(a^{\mathcal{I}}) \in \neg\mathrm{B}^{\mathcal{I}}$ must hold in every model of $\mathcal{T}$.

Conversely, if no such word exists, then one can construct a *deterministic* finite automaton from $\mathsf{nfa}_{\mathrm{B}}^{\mathrm{A}}(\mathcal{T})$, using the standard subset construction, in which there is not a state containing both B and ¬B reachable from the start state A. Unfolding the transition relation of this automaton, starting from the state A and labelling nodes by the concepts associated with the automaton's states, yields a tree interpretation that satisfies $\mathcal{T}$ (in particular in which all PFD constraints are satisfied vacuously) and whose root provides a witness for satisfiability of A. □

To test for emptiness of $\mathsf{nfa}_{\mathrm{B}}^{\mathrm{A}}(\mathcal{T})$, we use a graph connectivity algorithm that non-deterministically searches for a $(\mathrm{A}, \mathrm{A}) - (\mathrm{B}, \neg\mathrm{B})$ path in the (virtual) poly-sized product automaton [2]; the following result is then immediate.

Corollary 11. Concept satisfiability with respect to *partial*–$\mathcal{CFD}_{nc}^{\forall}$ TBoxes is complete for NLOGSPACE. □

Note that, as we remarked above, this procedure can be used to test for satisfiability of *conjunctions of concepts* in $\mathcal{CFD}_{nc}^{\forall}$ as follows:

Lemma 12. $\mathrm{A}_1 \sqcap \ldots \sqcap \mathrm{A}_k$ is consistent in $\mathcal{T}$ if and only if A is satisfiable in $\mathcal{T} \cup \{\mathrm{A} \sqsubseteq \mathrm{A}_i \mid 1 \leq i \leq k\}$. □

It is, however, impossible to *precompute* all such inconsistent concepts since this would require consideration of all possible *types* over PC (or finite subsets of primitive concepts), a process essentially equivalent to constructing an equivalent deterministic automaton which can require exponential time [2].

Logical Implication. Logical implication for $partial{-}\mathcal{CFD}^{\forall}_{nc}$ TBoxes $\mathcal{T}$ and posed questions $\mathcal{Q}$ can now be solved similarly to the $\mathcal{CFD}$ case. The main difference lies in detecting inconsistencies caused by object membership in conjunctions of (primitive) concepts that are necessarily *empty* in models of $\mathcal{T}$. This observation yields the following extension to $\mathsf{R}(\mathcal{T})$:

> If $\mathsf{D}(\mathsf{Pf})$ and $\mathsf{C}_{A_1}(\mathsf{Pf}), \ldots, \mathsf{C}_{A_k}(\mathsf{Pf})$ are in $\mathsf{R}(\mathcal{T})$ for some $\mathsf{Pf} \in \mathsf{PF}(\mathcal{T}, \mathcal{Q})$ and $A_1 \sqcap \ldots \sqcap A_k$ is not consistent in $\mathcal{T}$ then add $\mathsf{C}_{\perp}(\mathsf{Pf})$ to $\mathsf{R}(\mathcal{T})$.

Note that $partial{-}\mathcal{CFD}^{\forall}_{nc}$ can be extended to allow $\neg \exists f$ on the right hand sides of TBox dependencies (which would be handled analogously to negated primitive concepts by the NFA in Theorem 10). Alltogether, we obtain following results analogous to those in Sect. 4.1:

Theorem 13. Let $\mathcal{T}$ be a $partial{-}\mathcal{CFD}^{\forall}_{nc}$ terminology and $\mathcal{Q}$ a posed question of the form $E_1 \sqsubseteq E_2$. Then

$$\mathcal{T} \models \mathcal{Q} \text{ iff } \mathsf{R}(\mathcal{T}) \cup \{\mathsf{C}_{E_1}(id), \mathsf{D}(id)\} \models \mathsf{G}_{E_1} \text{ or}$$
$$\mathsf{R}(\mathcal{T}) \cup \{\mathsf{C}_{E_1}(id), \mathsf{D}(id)\} \models \mathsf{G}_{\forall \mathsf{Pf}.\perp}(id) \text{ for some } \mathsf{Pf} \in \mathsf{PF}(\mathcal{T}, \mathcal{Q}).$$

Proof (sketch): The proof is similar to the proof of Theorem 7: the necessary F^* trees are generated by unfolding $\delta(\mathcal{T})$ as in the Proof of Theorem 10. □

Corollary 14. Let $\mathcal{T}$ be a terminology and $\mathcal{Q}$ a posed question in $partial{-}\mathcal{CFD}^{\forall}_{nc}$. Then the implication problem $\mathcal{T} \models \mathcal{Q}$ is complete for PTIME. □

5   Summary and Future Work

In summary, we have shown how partial features coupled with a strict interpretation of undefined values can be incorporated in the feature-based DLs $\mathcal{DLFD}$, $\mathcal{CFD}$ and $\mathcal{CFD}^{\forall}_{nc}$, thus obtaining $partial{-}\mathcal{DLFD}$, $partial{-}\mathcal{CFD}$ and $partial{-}\mathcal{CFD}^{\forall}_{nc}$, respectively. Our primary contributions have been to also show that this can be done without impact on the complexity of their associated logical inference problems. Indeed, with $\mathcal{DLFD}$, this was achieved by showing how $partial{-}\mathcal{DLFD}$ can be fully simulated in $\mathcal{DLFD}$ in an entirely transparent fashion.

One avenue for future work would be to consider the impact on logical inference of alternative semantics for the interpretation of undefined values, in particular, on choosing the so-called Kleene semantics for equality. In this case, "$e_1 = e_2$" is also true when both e_1 and e_2 have undefined values. Note that doing so with $\mathcal{DLFD}$ would in fact *necessitate* changes to the semantics of the PFD concept constructor to avoid undecidability of logical inference.[4]

In our introductory comments, we also hinted at possible directions for future work relating to $\mathcal{CFD}^{\forall}_{nc}$. The first concerns recent work that begins to explore

---

[4] The details for this are beyond the scope of the paper.

how *inverse features* can be added to feature-based DLs, in particular, on adding the $\exists f^{-1}$ concept constructor. For example, logical inference has been shown to be decidable in PTIME for $\mathcal{CFDI}_{nc}^{\forall -}$ [11], a dialect obtained by adding this constructor to $\mathcal{CFD}_{nc}^{\forall}$ and by imposing additional syntactic restrictions, e.g., on the syntax of PFDs, to avoid intractability for this problem. We conjecture that our results for $partial-\mathcal{CFD}_{nc}^{\forall}$ can be extended to $partial-\mathcal{CFDI}_{nc}^{\forall -}$, although the development would be much less straightforward, and that the same applies to the other two dialects that we have considered: that logical consequence for $partial-\mathcal{CFDI}$ and $partial-\mathcal{DLFDI}$ is decidable in PTIME and EXPTIME, respectively.

The second possible direction for future work relating to $\mathcal{CFD}_{nc}^{\forall}$ is an indirect consequence of the ability to easily simulate $partial-\mathcal{DLFD}$ in $\mathcal{DLFD}$. In particular, for this case, there remains little incentive to adopt $partial-\mathcal{DLFD}$: to non-trivially complicate the semantics of feature-based DLs, to add the existential restriction concept constructor, and so on. We conjecture that it is possible to extend the syntax of $\mathcal{CFD}_{nc}^{\forall}$ to allow limited use of conjunction on left-hand-sides of inclusion dependencies to enable simulating $partial-\mathcal{CFD}_{nc}^{\forall}$ in $\mathcal{CFD}_{nc}^{\forall}$ in an analogous fashion, while preserving PTIME decidability for logical consequence.

References

1. Baader, F., Calvanese, D., McGuinness, D.L., Nardi, D., Patel-Schneider, P.F.: The Description Logic Handbook: Theory, Implementation, and Applications. Cambridge University Press, Cambridge (2003)
2. Hopcroft, J.E., Ullman, J.D.: Introduction to Automata Theory, Languages and Computation. Addison-Wesley, Boston (1979)
3. Khizder, V.L., Toman, D., Weddell, G.: Reasoning about duplicate elimination with description logic. In: Rules and Objects in Databases (DOOD, part of CL 2000), pp. 1017–1032 (2000)
4. Ramakrishnan, R.: Magic templates: a spellbinding approach to logic programs. J. Logic Program. **11**(3 & 4), 189–216 (1991)
5. Toman, D., Weddell, G.: On attributes, roles, and dependencies in description logics and the ackermann case of the decision problem. In: Description Logics 2001, CEUR-WS, vol. 49, pp. 76–85 (2001)
6. Toman, D., Weddell, G.: On reasoning about structural equality in XML: a description logic approach. Theor. Comput. Sci. **336**(1), 181–203 (2005)
7. Toman, D., Weddell, G.E.: On keys and functional dependencies as first-class citizens in description logics. J. Aut. Reason. **40**(2–3), 117–132 (2008)
8. Toman, D., Weddell, G.E.: Applications and extensions of PTIME description logics with functional constraints. In: Proceedings International Joint Conference on Artificial Intelligence (IJCAI), pp. 948–954 (2009)
9. Toman, D., Weddell, G.E.: Conjunctive query answering in $\mathcal{CFD}_{nc}$: a PTIME description logic with functional constraints and disjointness. In: AI 2013: Advances in Artificial Intelligence - 26th Australasian Joint Conference, Dunedin, New Zealand, pp. 350–361 (2013)

10. Toman, D., Weddell, G.E.: Answering queries over $\mathcal{CFD}_{nc}^{\forall}$ knowledge bases. Technical report CS-2014-14, Cheriton School of Computer Science, University of Waterloo (2014)
11. Toman, D., Weddell, G.: On adding inverse features to the description logic $\mathcal{CFD}_{nc}^{\forall}$. In: Pham, D.-N., Park, S.-B. (eds.) PRICAI 2014. LNCS, vol. 8862, pp. 587–599. Springer, Heidelberg (2014)

3-D Volume of Interest Based Image Classification

Akadej Udomchaiporn[1]([✉]), Frans Coenen[1], Marta García-Fiñana[2],
and Vanessa Sluming[3]

[1] Department of Computer Science, University of Liverpool, Liverpool, UK
{akadej,coenen}@liv.ac.uk
[2] Department of Biostatistics, University of Liverpool, Liverpool, UK
m.garciafinana@liv.ac.uk
[3] School of Health Science, University of Liverpool, Liverpool, UK
vanessa.sluming@liv.ac.uk

Abstract. This paper proposes a number of techniques for 3-D image classification according to the nature of a particular Volume of Interest (VOI) that appears across a given image set. Three VOI Based Image Classification (VOIBIC) approaches are considered: (i) Statistical metric based, (ii) Point series based and (iii) Tree based. For evaluation purpose, two 3-D MRI brain scan datasets, Epilepsy and Musicians, were used; the aim being to distinguish between: (i) epilepsy patients versus healthy people and (ii) musicians versus non-musicians. The paper also considers augmenting the VOI data with meta data. According to the reported experimental results the Point series based approach, augmented with meta data, is the most effective.

Keywords: Image mining · Image classification · 3-D Magnetic Resonance Imaging (MRI)

1 Introduction

Image mining is concerned with the extraction of useful information and knowledge from image data. The representation of the raw image data in a format that allows for the effective and efficient application of data mining is key to the success of any form of applied image mining. The domain of image mining can be divided into Region Of Interest (ROI) mining and whole image mining. Traditionally work on image mining has been directed at 2-D scenarios, however, increasingly more and more 3-D image data is available. Consequently there is also an increased interest in 3-D image mining; Volume Of Interest (VOI) mining.

This paper proposes and compares a number of techniques for VOI Based Image Classification (VOIBIC). VOIBIC entails a number of challenges. The first is the identification and isolation of the VOI. The second is how best to represent the VOI so that classification can be effectively and efficiently conducted. With regard to the work described in this paper three VOIBIC approaches are considered. The first is founded on the idea of using statistical metrics, the Statistical

© Springer International Publishing Switzerland 2016
R. Booth and M.-L. Zhang (Eds.): PRICAI 2016, LNAI 9810, pp. 543–555, 2016.
DOI: 10.1007/978-3-319-42911-3_45

metrics based representation. This representation offers the advantage that it is straightforward and, although not especially novel, provides a benchmark. The second proposed representation is founded on the concept of point series (curves) describing the perimeter of a VOI, the Point Series representation. Two variations of this representation are considered: (i) Disc based and (ii) Spoke based. The third proposed representation is founded on a Frequent Subgraph Mining (FSM) technique whereby the VOI is represented using an Oct-tree structure to which FSM can be applied. The identified frequent subtrees can then be used to define a feature vector representation compatible with many classifier generation methods. This paper also considers augmenting the VOI data with meta data and determining the effect this has on classification performance. The presented evaluation used two 3-D MRI brain scan datasets: (i) epilepsy patients versus healthy people and (ii) musicians versus non-musicians. The VOI in this case were the lateral ventricles, a distinctive object in MRI brain scan data.

The rest of the paper is organised as follows. Section 2 provides a review of some previous work relating to VOI classification. This is followed in Sect. 3 with a description of the datasets to be adopted in this paper. The three VOI classifications approaches are described in Sect. 4. Section 5 provides a comparative evaluation of the operation of the proposed VOI classification approaches including comparison with some previous related work. Finally, the paper is summarised and concluded in Sect. 6.

2   Related Work

This section provides a review of some previous related work concerning 3-D MRI brain scan classification [22,24,29]. In [24] a method was proposed to classify brain tissue from 3-D MRI brain scans using a partial volume model. A histogram based representation was used to measure brain tissue intensity and noise variance in the image and then classify this into six tissue types using "a posteriori" classifier. Although the result was promising the performance of the classification process was relatively expensive. Both [22] and [29] proposed similar work based on Discrete Wavelet Transform (DWT) representations. Both proposed a classification method to classify normal and abnormal brains. DWT was used to extract features from 3-D MRI images and then Principle Component Analysis (PCA) was applied to reduce the number of dimensions. Their results were excellent in terms of both effectiveness and efficiency but it was argued that the difference between normal and abnormal brains might be too obvious. For work based on Tree based representations, [3] and [4] proposed an approach to classify brain tissue using a minimum spanning tree graph-theoretic approach. This was evaluated using four classes: (i) elderly, (ii) young normal individual brain, (iii) ischemia patients' brain and (iv) Alzheimer' brain. Some good results, in terms of effectiveness, were reported. In [21] an approach to classifying MRI brain scans was proposed according to different levels of education. The lateral and third ventricles of the brain were used and represented using an Oct-tree structure and FSM used to generate feature vectors. The classification results were promising in terms of classification effectiveness, but run time complexity was high.

3 Application Domain

The research presented in this paper is concerned with Volumes Of Interest (VOI) and more specifically at identifying and classifying VOI in 3-D Magnetic Resonance Imaging (MRI) scans of the human brain. The particular focus is the left and right (lateral) ventricles; the cerebrospinal fluid filled spaces at the centre of the brain [20]. Their function is: (i) to act as shock absorbers, (ii) to distribute nutrients to the brain and (iii) remove waste. There are in fact four ventricles in a human brain: two lateral ventricles (referred to as the left and right ventricles), a third smaller ventricle connected to both lateral ventricles and a fourth smaller ventricle that connects the third ventricle to the spinal cord. Only the left and right lateral ventricles are considered with respect to the focus of the presented research. This was because: (i) the lateral ventricles are relatively easy to identify within 3-D MRI brain scans, so facilitating automatic extraction; and (ii) they are much larger than the other two ventricles and consequently can be argued to be more significant. An example of a 3-D MRI brain scan is given in Fig. 1 where the "lateral ventricles" are the dark areas at the centre of the brain. Note that 3-D MRIs comprise a sequence of two dimensional (2-D) "slices" through the brain in each of the three cardinal planes: (a) Sagittal - SAG (left to right), (b) Coronal - COR (front to back) and (c) Transverse - TRA (top to bottom).

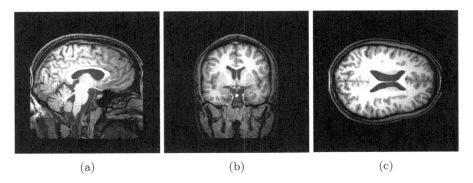

(a)   (b)   (c)

Fig. 1. Example of a 3-D brain MRI scan: (a) Sagittal (SAG) plane, (b) Coronal (COR) plane and (c) Transverse (TRA) plane

Two MRI brain scan datasets were used for evaluation purposes with respect to the research presented in this paper: (i) Epilepsy and (ii) Musician. For the Epilepsy dataset, the MRI brain scan volumes were obtained by the Magnetic Resonance and Image Analysis Research Centre (MRIARC), at the University of Liverpool, between the years 1999 and 2004. The Epilepsy dataset comprised

210 MRI brain scans. Of these 105 were from healthy people and the remaining 105 from epilepsy patients. For the Musician dataset, some of the MRI brain scan volumes were obtained by MRIARC between the years 1999 and 2004, and the remainder by the University of Heidelberg between the years 2000 and 2004. The Musician dataset comprised a total of 160 MRI brain scans. Of these 80 were from musicians and the remaining 80 from non-musicians.

4   Volume of Interest Classification

This section describes the three proposed VOI classification approaches. Note that the thresholding segmentation techniques presented in [26] was used to extract the VOI (lateral ventricles). An example of an extracted lateral ventricle is shown in Fig. 2. The three VOI classification approaches are presented in the following subsections.

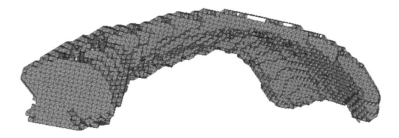

Fig. 2. Example of an extracted right ventricle (figure created using Meshlab [1]).

4.1   Statistical Metrics Based Classification

In this section the Statistical metric based approach to VOIBIC is presented, whereby a number of statistical measures are used to represent VOIs. The measures are used to define a N-dimensional feature space, one dimension per metic, from which a feature vector representation can be extracted, one feature vector per volume (ventricle). The feature vectors generated are then used as input to a classifier generator. It was anticipated that this representation would be unlikely to provide effective results, however the approach would provide a benchmark with which the two alternative representations could be compared.

The usage of statistical techniques is widely reported in the literature. In the context of 2-D, example applications include [12] and [13]; and in 3-D, example applications include [11] and [25]. The metrics considered with respect to the work presented in this paper were:

1. **Axis length** (l): The axis length of the ventricles.
2. **Axis width** (w): The axis width of the ventricles.
3. **Axis depth** (d): The axis depth of the ventricles.
4. **Maximum perimeter on xy plane** (p_{xy}): The maximum perimeter of the ventricles in the xy plane.
5. **Maximum perimeter on yz plane** (p_{yz}): The maximum perimeter of the ventricles in the yz plane.
6. **Maximum perimeter in the xz plane** (p_{xz}): The maximum perimeter of the ventricles on xz plane.
7. **Volume** (vol): The volume of the ventricles (directly relating to the number of voxels, note that in the case of the ventricles 1 voxel = 1 mm^3).
8. **Volume extent** (v_{ext}): The value derived by dividing the volume (vol) by the size of the minimum bounding cube surrounding the volume ($\frac{Vol}{l \times w \times d}$).

Note that two feature vectors were generated, one for each lateral ventricle, per image.

4.2   Point Series Based Classification

The proposed Point series based VOIBIC approach is presented in this section. The point series represents the boundary of the VOI. Two techniques for generating the desired point series are considered: (i) Disc based and (ii) Spoke based. Regardless of which technique is used, the resulting point series can be translated into feature vectors compatible with a number of classification model generators. Alternatively they can be used directly using a K-Nearest Neighbour (KNN) approach. In the context of the first, a signature based approach is proposed for generating the model, founded on Hough signature extraction [14]. The key idea is to transform a shape into a parameter space where the shape can be represented in a spatially compact way, a Straight Line Hough Transform (SLHT) was used for the transformation (see [27] for further detail). For the KNN mechanism a similarity measure is required, the simplest is the Euclidean distance between a labelled comparator series and a previously unseen series. However, this requires that both series are of the same length. Instead it is proposed that the "warping path" distance, generated using Dynamic Time Warping (DTW) [2], can be used.

As noted above, two techniques for generating the desired point series are proposed: (i) Disc based and (ii) Spoke based. The first uses all the boundary voxels while the second uses a representative subset of the complete set of boundary voxels. The input in both cases is a binary-valued 3-D image comprising voxels labelled as being either black (belonging to the VOI) or white (not part of the VOI). The output in both cases is a point series describing the VOI boundary. Both the Disc based and Spoke based techniques operate with reference to a *primary axis* (see Fig. 3). Consequently for each image three point series can be generated for each ventricle. Thus, in total six, curves are generated for each MRI image, three describing the left ventricle and three describing the right ventricle. Each technique is described in further detail in the following two subsections.

Disc Based Technique. The Disc based representation is founded on the idea of generating a point series by considering a sequence of slices, slice by slice (along a primary axis), and collecting point information from the boundary where each slice and the volume of interest intersect. The intersection is usually described by a circular shape, as illustrated in Fig. 3a hence the technique is referred to as the "Disc" based technique. In this manner a sequence of disc boundaries is generated and concatenated together to form a single point series.

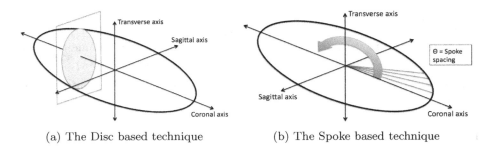

(a) The Disc based technique   (b) The Spoke based technique

Fig. 3. Point series generation techniques

Spoke Based Technique. The Spoke based representation technique is illustrated in Fig. 3b. The technique involves measuring the distance from the geometric centroid of the VOI to points on the boundary, in a given plane. As such it is essentially a 2-D technique. The effect is that of a sequence of spokes of different length radiating from the centroid.

4.3   Tree Based Classification

The tree based VOIBIC approach is presented in this section. The approach uses the concept of Oct-trees decomposition akin to the Quad-trees decomposition used with respect to 2-D [5]. The Oct-tree is constructed by repeatedly subdividing the space into "octants" [17,23]. The decomposition continues until homogeneous sub-regions are arrived at or a user defined "maximum depth" is reached. [8,9]. The decomposition can be conducted according to a variety of image features such as colour or intensity. With respect to the lateral ventricles a binary encoding was used. The VOI was encapsulated in a Minimum Bounding Box (MBB). The voxels representing the VOI were allocated a "1" (black), while the remainder were allocated a "0" (white). The advantage of this representation is that it maintains information about the relative location and size of the VOI.

Trees are not readily compatible with classifier generation, they needs to be processed in some way so that a feature vector representation, compatible with classifier model generation, can be formulated. The idea presented here is to apply Frequent Subgraph Mining (FSM) to the Oct-tree data structure and then used the identified frequent sub-trees to define a feature space. A number of

FSM algorithms have been proposed, such as: (i) gSpan [28], (ii) AGM [16] and (iii) FFSM [15]. The gSpan algorithm, coupled with the Average Total Weighting (ATW) scheme desribed in [18], was adopted. The output from the FSM was a set of frequently occurring sub-graphs together with their occurrence counts. Typically a large number of sub-graphs are generated many of which are redundant (do not serve to discriminate between classes). Feature selection techniques were applied to reduce the overall number of identified frequent sub-graphs.

5   Evaluation

The evaluation of the proposed VOIBIC approaches is presented here. The section is divided into three sub-sections (Sub-sects. 5.1–5.3): (i) comparison of the proposed VOIBIC approaches, (ii) statistical significance testing of the results and (iii) comparison with previous work.

5.1   Comparison of the VOIBIC Approaches

The comparison of the proposed VOIBIC approaches was undertaken in terms of: (i) classification effectiveness and (ii) efficiency. In terms of effectiveness many experiments were conducted, not shown here, using a variety of parameter settings. The results presented here, however, are those generated using best performing parameter settings (so as to consider each technique to its best advantage). The comparison is also divided into two parts: (i) without augmentation and (ii) with augmentation.

Tables 1 and 2 summarise the "best" classification results obtained for the Epilepsy and Musicians datasets respectively. In the tables the acronyms "SMB", "DB", "SB" and "TB" refer to the Statistical Metric Based, Disc Based, Spoke Based, and Tree Based approaches respectively. The abbreviations "Accu.", "Sens." and "Spec." indicate classification accuracy, sensitivity and specificity. Best results are indicated in bold font. From the Tables it can be seen that for all the metrics considered the Tree based approach tended to be most effective. The exception is in the case of the sensitivity value; where the Spoke based approach, when applied to the Musician dataset, produced the best result. Tables 3 and 4 summarise the "best" classification effectiveness results obtained using the Epilepsy and Musician datasets augmented with age and gender data. From the tables it can be seen that the Tree and Spoke based techniques produced the best results.

Figures 4 and 5 show the runtime results obtained, with respect to the best performing VOIBIC approaches considered previously, for the Epilepsy and Musician datasets respectively. The presented run times are the total TCV run time in each case, including: feature extraction, curve generation (for the Point series based approach), Oct-tree generation (for the Tree based approach), training and testing. All the experiments were conducted using a 2.9 GHz Intel Core i7 with 8 GB RAM on OS X (10.9) operating system.

Table 1. Classification effectiveness results for Epilepsy dataset (without augmentation)

Technique	Accu.	Sens.	Spec.
SMB	60.63	63.56	57.16
DB	62.20	67.50	57.14
SB	69.81	71.70	67.92
TB	**72.34**	**75.67**	**70.45**

Table 2. Classification effectiveness results for Musician dataset (without augmentation)

Technique	Accu.	Sens.	Spec.
SMB	69.89	71.77	68.00
DB	77.36	81.13	75.47
SB	82.39	**88.68**	79.25
TB	**86.32**	87.74	**80.19**

Table 3. Classification effectiveness results for Epilepsy dataset with augmentation (Epilepsy+)

Technique	Accu.	Sens.	Spec.
SMB	68.59	70.50	68.17
DB	71.04	72.34	66.04
SB	78.30	76.67	71.70
TB	**78.52**	**81.13**	**75.47**

Table 4. Classification effectiveness results for Musician dataset with augmentation (Musician+)

Technique	Accu.	Sens.	Spec.
SMB	70.36	77.00	64.05
DB	83.96	86.79	83.02
SB	84.91	**90.57**	**85.85**
TB	**86.02**	88.85	83.19

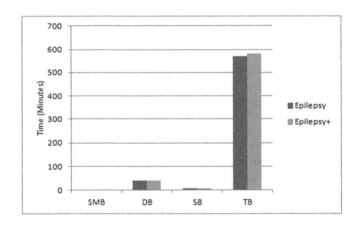

Fig. 4. Run time complexity for the classification process using the Epilepsy and Epilepsy+ datasets (Color figure online)

The run time complexities of the four VOIBIC approaches are presented in Figs. 4 and 5. Figure 4 shows the run time complexity for the classification process when applied to the Epilepsy and Epilepsy+ (augmented) datasets, and Fig. 5 the run time complexity for the classification process when applied to the Musician and Musician+ (augmented) datasets.

From the figures it can be seen that the Statistical metrics based approach, as was to be expected, was the most efficient for all datasets (the run time was less

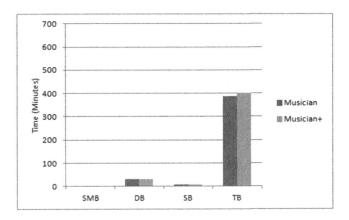

Fig. 5. Run time complexity for the classification process using the Musician and Musician+ datasets (Color figure online)

than one minute; hence the bar is not visible in the figures); while the Tree based approach was the most expensive. It is obvious that there was a trade-off between effectiveness and efficiency. However it is possible to claim that the Spoke based approach produced the best overall results, in terms of both effectiveness and efficiency, compared to the other three approaches. In the following section the statistical significance of the results presented in this subsection is considered.

5.2   Statistical Significance Testing of the Results

This section presents the results obtained from conducting statistical significance testing. Given any competition, at some level of granularity, one of the competitors will always come first, the question is whether this outcome is statistically significant or not? There are a variety of mechanisms that can be used to establish statistical significance. The Friedman rank statistic [10] is well suited for comparing the significance of the results obtained with respect to the proposed VOIBIC approaches and was thus adopted with respect to the work presented in this paper.

The Friedman rank test, F_R, is based on the total ranked performances of a given classification techniques applied to each dataset, and is calculated as follows:

$$F_R = \frac{12}{rc(c+1)} \left[\sum_{j=1}^{c} R_j^2 - 3r(c+1) \right]$$ (1)

where r is the number of datasets used in the study, c is the number of classifiers used in the study and R_j^2 is the square of the Total of the Ranks (TR) for classifier $j, (j = 1, 2, 3, \ldots, c)$.

For the purpose of the evaluation the classification accuracies were used. Table 5 reports the classification accuracies for all four techniques when applied to the two datasets augmented with age and gender. In the table, the techniques achieving the highest classification accuracies with respect to each dataset. The overall highest ranked technique is indicated in bold font. The numbers in the parentheses indicate the rank of each technique. The Friedman test statistic and corresponding p-value are also shown.

If the value of F_R is larger than 6.00, the critical value for the Friedman rank test [10], the null hypothesis that there is no difference among proposed techniques can be rejected. With respect to the work presented in this paper, the calculated F_R from Eq. 1 is "13.5" (which is larger than 6.00). Thus the null hypothesis, at $\alpha = 0.05$, is rejected. It can thus be concluded that there are significant statistical differences in classification accuracies among the proposed VOIBIC approaches.

Table 5. The best classification accuracy results, total rank and average rank for the proposed techniques

Friedman rank test = 13.5 ($\alpha < 0.05$)				
Approach	Epilepsy (Accu.)	Musician (Accu.)	Total Rank (TR)	Average Rank (AR)
SMB	68.59 (4)	70.36 (4)	8	4
DB	71.04 (3)	83.96 (3)	6	3
SB	78.30 (2)	84.91 (2)	4	2
TB	**78.52** (1)	**86.02** (1)	**2**	1

Given that the null hypothesis can be rejected a post hoc Nemenyi test [6] can be applied to highlight significant differences among the individual VOIBIC approaches. The Nemenyi post hoc test states that the performances of two or more classifiers are significantly different if their Average Ranks (AR) differ by at least a Critical Difference (CD), given by:

$$CD = q_{\alpha,\infty,c}\sqrt{\frac{c(c+1)}{12N}} \quad (2)$$

Note that in Eq. 2, the value $q_{\alpha,\infty,c}$ is based on the Studentised range statistic [6].

Figure 6 shows the critical difference diagram for the data presented in Table 5. Note that this is a modified version of the Demsar 2006 significant diagram [19]. This figure shows the classification approaches listed in ascending order of ranked performance on the y-axis, and the image classification approaches' AR across two datasets displayed on the x-axis. The diagram displays the ranked performances of the classification techniques, along with the CD tail, to highlight any techniques which are significantly different to the best performing techniques. The CD value for the figure was calculated as per Eq. 2 and equals to "2.14". The critical difference diagram clearly shows that the TB approach is the best performing classification approach with an AR value of 1.0; however, this result is not significant with respect to SB and DB, while it is significant with respect to SMB.

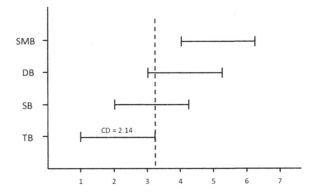

Fig. 6. Critical difference diagram for the proposed image classification approaches

5.3   Comparison with Previous Work

There have been a number of previous studies where data mining has been applied to MRI brain scan data. There are no similar studies using the nature of the lateral ventricles and the same data sets as used with respect to the work presented in this paper, however the work of Elsayed et al. [7,9] and that of Long [21] is of note. The significance of the work presented in [7] and [9] is that this also used the Musician and Epilepsy MRI scan datasets but in the context of 2-D and with respect to corpus callosum, another easily identifiable object found in MRI brain scan data. The significance of the work presented in [21] is that it was directed at the 3-D analysis of the ventricles using a tree structure similar to that used with respect to the third technique presented in this paper, but in the context of Alzheimer's disease detection and level of education. Thus some limited comparisons can be made with respect to the work of Elsayed et al. and Long. In [7,9] the reported results were better than those reported in this paper; however, as already noted, the work of Elsayed et al. was directed at a 2-D representation of the corpus callosum. It might very well be the case that the corpus callosum is a better indicator of epilepsy and musical ability, hence the better results. Long and Holder [21] also reported (slightly) better results than those reported with respect to the work presented in this paper, but as noted above, the work was directed at Alzheimer's disease detection and level of education, not epilepsy and musical ability. Long also considered not only the two lateral ventricles, but also the "third" ventricle, which may also make a difference.

6   Conclusion

In this paper three alternative approaches to conducting VOIBIC have been proposed: (i) Statistical metrics based, (ii) Point series (Disc based and Spoke based) and (iii) Tree based. Two datasets, Epilepsy and Musician, were used for evaluation purpose. The reported results indicated that the Spoke and Tree based approaches were the most effective in the context of classification performance. The results also indicated that there was a trade-off between classification effectiveness and efficiency, the more sophisticated approaches, although providing for a greater accuracy, were also the least efficient. The most efficient was the Statistical metrics based approach which produced the worst classification performance, while the Tree based approach was the least efficient but tended to produced the best classification performance. For most classification applications we are interested in the quality of the classification, not its efficiency, however the Spoke based approach would provide a good compromise if efficiency was of concern. Note that although the Tree based approach tended to produce (slightly) better accuracy values than the Spoke based approach, the statistical significance analysis presented in the paper indicated that there was no statistically significant difference between the Spoke based Point series approach and the Tree based approach in terms of classification accuracy. The evaluation reported in this paper also indicated that by augmenting the proposed representations with meta data improved classification effectiveness could be obtained, regardless of the representation classification paradigm or dataset used.

The research described in this paper has indicated a number of potential research directions for the future work. One potentially fruitful avenue would be to investigate alternative 3-D representation techniques such as meshes, instead of the standard feature vector format used in this paper. It would also be of interest to investigate further alternative classification models to which the proposed representation techniques can be applied.

References

1. http://meshlab.sourceforge.net/
2. Bellman, R., Kalaba, R.: On adaptive control processes. IRE Trans. Autom. Control 4(2) (1959)
3. Cocosco, C.A., Zijdenbos, A.P., Evans, A.C.: Automatic generation of training data for brain tissue classification from MRI. In: Dohi, T., Kikinis, R. (eds.) MICCAI 2002, Part I. LNCS, vol. 2488, pp. 516–523. Springer, Heidelberg (2002)
4. Cocosco, C., Zijdenbos, A., Evans, A.: A fully automatic and robust brain MRI tissue classification method. Med. Image Anal. 7(4), 513–527 (2003)
5. Costa, L.D.F., Cesar Jr., R.M.: Shape Analysis and Classification: Theory and Practice. CRC Press, Boca Raton (2001)
6. Demšar, J.: Statistical comparisons of classifiers over multiple data sets. J. Mach. Learn. Res. 7, 1–30 (2006)
7. Elsayed, A.: Region of interest-based image classification. Ph.D. dissertation, Department of Computer Science, University of Liverpool, November 2011
8. Elsayed, A., Coenen, F., García-Fiñana, M., Sluming, V.: Region of interest based image categorization. In: Bach Pedersen, T., Mohania, M.K., Tjoa, A.M. (eds.) DAWAK 2010. LNCS, vol. 6263, pp. 239–250. Springer, Heidelberg (2010)
9. Elsayed, A., Coenen, F., Jiang, C., García-Fiñana, M., Sluming, V.: Corpus callosum MR image classification. Knowl.-Based Syst. 23(4), 330–336 (2010)

10. Friedman, M.: A comparison of alternative tests of significance for the problem of M rankings. Ann. Math. Stat. **11**(1), 86–92 (1940)
11. Galloway, M.: Texture analysis using gray level run lengths. Comput. Graph. Image Process. **4**(2), 172–179 (1975)
12. Gao, D.: Volume texture extraction for 3D seismic visualization and interpretation. Geophysics **68**(4), 1294–1302 (2003)
13. Gonzalez, R., Woods, R.: Digital Image Processing, 3rd edn. Prentice-Hall, Inc., Upper Saddle River (2006)
14. Hough, P.V.C.: Method and means for recognizing complex patterns. US Patent (1962)
15. Huan, J., Wang, W., Prins, J.: Efficient mining of frequent subgraphs in the presence of isomorphism. In: Proceedings of the Third IEEE International Conference on Data Mining, pp. 549–552. IEEE Computer Society (2003)
16. Inokuchi, A., Washio, T., Motoda, H.: An apriori-based algorithm for mining frequent substructures from graph data. In: Zighed, D.A., Komorowski, J., Żytkow, J.M. (eds.) PKDD 2000. LNCS (LNAI), vol. 1910, pp. 13–23. Springer, Heidelberg (2000)
17. Jackins, C.L., Tanimoto, S.L.: Oct-trees and their use in representing three-dimensional objects. Comput. Graph. Image Process. **14**(3), 249–270 (1980)
18. Jiang, C., Coenen, F.: Graph-based image classification by weighting scheme. In: Allen, T., Ellis, R., Petridis, M. (eds.) Applications and Innovations in Intelligent System XVI, pp. 63–76. Springer, London (2009)
19. Lessmann, S., Baesens, B., Mues, C., Pietsch, S.: Benchmarking classification models for software defect prediction: a proposed framework and novel findings. IEEE Trans. Software Eng. **34**(4), 485–496 (2008)
20. Lin, J., Mula, M., Hermann, B.: Uncovering the neurobehavioural comorbidities of epilepsy over the lifespan. Lancet **380**(9848), 1180–1192 (2012)
21. Long, S., Holder, L.B.: Graph-based shape analysis for MRI classification. Int. J. Knowl. Discovery Bioinform. **2**(2), 19–33 (2011)
22. Rajini, N., Bhavani, R.: Classification of MRI brain images using K-nearest neighbor and artificial neural network. In: 2011 International Conference on Recent Trends in Information Technology (ICRTIT), pp. 563–568. IEEE (2011)
23. Rambally, G., Rambally, R.S.: Octrees and their applications in image processing. In: Southeastcon 1990, Proceedings, pp. 1116–1120. IEEE (1990)
24. Shattuck, D., Sandor-Leahy, S., Schaper, K., Rottenberg, D., Leahy, R.: Magnetic resonance image tissue classification using a partial volume model. NeuroImage **13**(5), 856–876 (2001)
25. Tang, X.: Texture information in run-length matrices. IEEE Trans. Image Process. **7**(11), 1602–1609 (1998)
26. Udomchaiporn, A., Coenen, F., García-Fiñana, M., Sluming, V.: 3-D MRI brain scan feature classification using an oct-tree representation. Adv. Data Min. Appl. **8346**, 229–240 (2013)
27. Udomchaiporn, A., Coenen, F., García-Fiñana, M., Sluming, V.: 3-D MRI brain scan classification using a point series based representation. In: Bellatreche, L., Mohania, M.K. (eds.) DaWaK 2014. LNCS, vol. 8646, pp. 300–307. Springer, Heidelberg (2014)
28. Yan, X.: gSpan: graph-based substructure pattern mining. In: Proceeding of the IEEE International Conference on Data Mining, pp. 721–724. IEEE Computer Society (2002)
29. Zhang, Y., Dong, Z., Wu, L., Wang, S.: A hybrid method for MRI brain image classification. Expert Syst. Appl. **38**(8), 10049–10053 (2011)

An Empirical Local Search for the Stable Marriage Problem

Hoang Huu Viet[1(✉)], Le Hong Trang[1], SeungGwan Lee[2],
and TaeChoong Chung[3]

[1] Department of Information Technology, Vinh University, 182-Le Duan,
Vinh City, Nghe An, Vietnam
{viethh,lhtrang}@vinhuni.edu.vn
[2] Humanitas College, Kyung Hee University, Seoul, South Korea
leesg@khu.ac.kr
[3] Department of Computer Engineering, Kyung Hee University, 1732,
Deogyeong-daero, Giheung-gu, Yongin-si, Gyeonggi-do 446-701, Korea
tcchung@khu.ac.kr

Abstract. This paper proposes a local search algorithm to find the *egalitarian* and the *sex-equal* stable matchings in the stable marriage problem. Based on the dominance relation of stable matchings from the men's point of view, our approach discovers the *egalitarian* and the *sex-equal* stable matchings from the *man-optimal* stable matching. By employing a breakmarriage strategy to find stable neighbor matchings of the current stable matching and moving to the best neighbor matching, our local search finds the solutions while moving towards the *woman-optimal* stable matching. Simulations show that our proposed algorithm is efficient for the stable marriage problem.

1 Introduction

The stable marriage problem (SM) is a well-known problem of matching an equal number men and women to satisfy a certain criterion of stability. This problem was first introduced by Gale and Shapley [3], and has recently received a great deal of attention from the research community due to its important role in a wide range of applications such as the Evolution of the Labor Market for Medical Interns and Residents [12], the Student-Project Allocation problem (SPA)[1] and the Stable Roommates problem (SR) [2,7].

An instance of SM of size n comprises a set of n men and a set of n women and each person has a preference list (PL) in which they rank all members of the opposite sex in strict order. A matching M is a set of n disjoint pairs of men and women. If a man m and a woman w is a pair in M, then m and w are partners in M, denoted by $m = M(w)$ and $w = M(m)$. Matching M is stable if there is no man m and woman w such that m prefers w to $M(m)$ and w prefers m to $M(w)$, otherwise M is unstable. For a stable matching M, we define the *man cost* $sm(M)$ and the *woman cost* $sw(M)$ as follows:

© Springer International Publishing Switzerland 2016
R. Booth and M.-L. Zhang (Eds.): PRICAI 2016, LNAI 9810, pp. 556–564, 2016.
DOI: 10.1007/978-3-319-42911-3_46

$$sm(M) = \sum_{(m,w)\in M} mr(m,w), \quad sw(M) = \sum_{(m,w)\in M} wr(w,m), \tag{1}$$

where $mr(m,w)$ is the rank of woman w in man m's PL and $wr(w,m)$ is the rank of man m in woman w's PL.

Definition 1 (Man-Optimal and Woman-Optimal [11]). *A stable matching M is called man-optimal (respectively woman-optimal) if it has the minimum value of $sm(M)$ (respectively $sw(M)$) over all stable matchings.*

Gale and Shapley proposed an algorithm known as the Gale-Shapley algorithm to find an optimal solution of SM instances of size n in time $O(n^2)$[3]. The Gale-Shapley algorithm is basically a sequence of proposals from men to women to find the *man-optimal* stable matching. If the roles of men and women are interchanged, the matching found by the algorithm is the *woman-optimal* stable matching. It is proved that in the *man-optimal* stable matching, each woman has the worst partner that she can have in any stable matching and that in the *woman-optimal* stable matching, each man has the worst partner that he can have in any stable matching [6]. Therefore, it is appropriate to seek other optimal stable matchings to give more balanced preference for both men and women. For a stable matching M, we define the *egalitarian cost* $c(M)$ and the *sex-equality cost* $d(M)$ as follows:

$$c(M) = sm(M) + sw(M), \quad d(M) = |sm(M) - sw(M)|. \tag{2}$$

Definition 2 (Egalitarian and Sex-Equal [11]). *A stable matching M is called egalitarian (respectively sex-equal) if it has the minimum value of $c(M)$ (respectively $d(M)$) over all stable matchings.*

There are several methods to find the *egalitarian* or *sex-equal* stable matching based on local search approach [4,9,11,14]. Because the number of stable matchings of SM instances grows exponentially in general [8], the above methods either are inefficient for finding solutions or finds only one stable matching of SM instances of large sizes. In this paper, we propose a new local search algorithm to seek an *egalitarian* or *sex-equal* stable matching of a given SM. Start from the *man-optimal* stable matching, the proposed algorithm finds a better solution in the neighbors of the current solution. If a better solution is found, the current solution is moved to the better solution and the local search is repeated for the current solution until no neighbor matchings of the current solution are found. The simulation results show that our approach guarantees to find a stable matching, which is an optimal solution or a near optimal solution, and discovers the solution faster than the ACS approach [14] does.

The rest of this paper is organized as follows: Sect. 2 describes the background, Sect. 3 presents the proposed approach, Sect. 4 discusses the simulations and evaluations, and Sect. 5 concludes our work.

Table 1. Preference lists of eight men and women.

Man	Preference list	Woman	Preference list
m_1	4 3 1 5 2 6 8 7	w_1	4 7 3 8 1 5 2 6
m_2	2 8 4 5 3 7 1 6	w_2	5 3 4 2 1 8 6 7
m_3	5 8 1 4 2 3 6 7	w_3	2 8 6 4 3 7 5 1
m_4	6 4 3 2 5 8 1 7	w_4	5 6 8 3 4 7 1 2
m_5	6 5 4 8 1 7 2 3	w_5	1 8 5 2 3 6 4 7
m_6	7 4 2 5 6 8 1 3	w_6	8 6 2 5 1 7 4 3
m_7	8 5 6 3 7 2 1 4	w_7	5 2 8 3 6 4 7 1
m_8	4 7 1 3 5 8 2 6	w_8	4 5 7 1 6 2 8 3

2   Background

Consider an example of SM consisting of eight men and eight women with their preference lists shown in Table 1.

Definition 3 (Dominance [5]). *Let $M \in \mathcal{M}$ and $M' \in \mathcal{M}$ be two stable matchings. M is said to dominate M' under the men's point of view if and only if every man prefers his partner in M at least as well as to his partner in M'.*

Brute-Force Algorithm. In order to generate all stable matchings and show which one is the *man-optimal, woman-optimal, egalitarian* or *sex-equal* stable matching of the example in Table 1, we give a brute-force search algorithm consisting of three steps: (i) generate all of the permutations of the women set; (ii) pair each man to each woman in the permutations to form 40320 (i.e. 8!) matchings; and (iii) search on the whole matchings and check stable matchings with the given criterion. All stable matchings sorted by the dominance relation from the men's point of view and their cost of the example are shown in Table 2. Obviously, the proposed algorithm is simple to implement and it finds exactly solutions of SM of size n, but its cost is $n!$. Therefore, if the number of men or women is large then this algorithm is inefficient in terms of time complexity.

Gale-Shapley Algorithm [3]. The Gale-Shapley algorithm is basically a sequence of proposals from men to women to find the *man-optimal* stable matching. At the beginning, the algorithm assigns each person to be free. At each iteration step, the algorithm chooses a free man m and finds the most preferred woman w in m's list to whom m has not proposed. If w is free then w and m become engaged. If w is engaged to m' then she rejects the man that she least prefers to engage to the other man. The rejected man becomes free. The algorithm terminates when all men are engaged.

Breakmarriage Operation [10]. Let M be a stable matching and (m, w) be an engaged pair in M. The breakmarriage operation of (M, m) is to find a stable matching. The idea of the breakmarriage is similar to the Gale-Shapley algorithm.

Table 2. Evaluations of stable matchings.

Stable matchings	sm	sw	c	d
$M_0 = \{(1,1),(2,2),(3,5),(4,3),(5,6),(6,7),(7,8),(8,4)\}$	12	33	45	21
$M_1 = \{(1,1),(2,2),(3,5),(4,3),(5,6),(6,4),(7,8),(8,7)\}$	14	30	44	16
$M_2 = \{(1,5),(2,2),(3,1),(4,3),(5,6),(6,7),(7,8),(8,4)\}$	15	27	42	12
$M_3 = \{(1,5),(2,2),(3,1),(4,3),(5,6),(6,4),(7,8),(8,7)\}$	17	24	41	7
$M_4 = \{(1,5),(2,3),(3,1),(4,2),(5,6),(6,7),(7,8),(8,4)\}$	20	23	43	3
$M_5 = \{(1,5),(2,3),(3,1),(4,2),(5,6),(6,4),(7,8),(8,7)\}$	22	20	42	2
$M_6 = \{(1,5),(2,2),(3,1),(4,3),(5,4),(6,6),(7,8),(8,7)\}$	22	21	43	1
$M_7 = \{(1,5),(2,3),(3,1),(4,2),(5,4),(6,6),(7,8),(8,7)\}$	27	17	44	10
$M_8 = \{(1,5),(2,3),(3,2),(4,8),(5,6),(6,7),(7,1),(8,4)\}$	30	19	49	11
$M_9 = \{(1,5),(2,3),(3,2),(4,8),(5,6),(6,4),(7,1),(8,7)\}$	32	16	48	16
$M_{10} = \{(1,5),(2,3),(3,2),(4,8),(5,4),(6,6),(7,1),(8,7\}$	37	13	50	24

●M_0: *man-optimal*, M_{10}: *woman-optimal*.

●M_3: *egalitarian*, M_6: *sex-equal*.

At the beginning, the algorithm assigns the woman w to the partner of the man m and sets m to be free. At each iteration step, the algorithm performs a sequence of proposals, rejects and acceptances as those of the Gale-Shapley algorithm. The algorithm terminates either when some man has been rejected by all women or when the woman w accepts a man m' to whom she prefers to her partner and in this case the algorithm returns a stable matching of n engaged pairs [10]. The following theorem and corollary are the basis of our approach to SM.

Theorem 1 [10]. *Every stable matching $M_i(i = 1, 2, \cdots, t)$ can be obtained by a series of breakmarriage operations starting from the man-optimal stable matching M_0, where M_t is the woman-optimal stable matching.*

Corollary 1 [5]. *If breakmarriage(M, m) results in a stable matching M', then M' dominates all stable matchings which are dominated by M and in which m is not married to his mate in M.*

3   Proposed Algorithm

In this section, we propose a local search algorithm to find an *egalitarian* or *sex-equal* stable matching of SM of size n. Local search algorithms are among the popular methods for solving optimization problems because of two key advantages: (i) they take very little memory; and (ii) they find quickly reasonable solutions in large or infinite state spaces. However, the local search algorithms often fail to find a global optimal solution when one exists because they can get stuck on a local optimum solution.

Our algorithm is shown in Algorithm 1. The basic idea is derived from the Theorem 1, that is, the algorithm finds the best solution while moving the current solution from the *man-optimal* towards the *woman-optimal* one. Procedure GALE-SHAPLEY (line 1) finds the *man-optimal* stable matching using Gale-Shapley algorithm. For each search step, procedure BREAK-MARRIAGE (line 6) discovers a neighbor set of the current solution for every man. The next solution is selected to be the best one among of the neighbor set by means of a cost function $f(M)$, which is *egalitarian cost* (respectively *sex-equal cost*) for finding the *egalitarian* (respectively *sex-equal*) stable matching. The next solution can also be selected randomly in the neighbor set to overcome the stuck on a local optimum. It differs from other local search algorithms such as hill climbing and simulated annealing search [13], our algorithm ends if no neighbors exits, meaning that the algorithm reaches to the *woman-optimal* stable matching. This makes the algorithm increase the run time but it can potentially achieve a global optimum solution.

Algorithm 1. Local Search Algorithm

    Input   : an instance of SM
    Output: a stable matching

1: $M_{current} \leftarrow$ GALE-SHAPLEY(an instance of SM);
2: $M_{best} \leftarrow M_{current}$;
3: **while** *(true)* **do**
4:     $neighbors \leftarrow \emptyset$;
5:     **for** *(each man m in men)* **do**
6:         $matching \leftarrow$ BREAK-MARRIAGE($M_{current}$,m);
7:         add *matching* to *neighbors*;
8:     **end**
9:     **if** *(no neighbors are found)* **then**
10:        break;
11:     **end**
12:     **if** *(small random probability p)* **then**
13:        $M_{next} \leftarrow$ a randomly selected matching in *neighbors*;
14:     **else**
15:        $M_{next} \leftarrow \underset{M \in neighbors}{\arg\min} \ (f(M))$;
16:     **end**
17:     **if** $(f(M_{best}) > f(M_{next}))$ **then**
18:        $M_{best} \leftarrow M_{next}$;
19:     **end**
20:     $M_{current} \leftarrow M_{next}$;
21: **end**
22: **return** M_{best};

An illustration of Algorithm 1 to find a *sex-equal* stable matching for the SM in Table 1 is depicted in Fig. 1. The probability to move the solution to a random neighbor is set to be zero. Initially, the algorithm assigns the current solution and the best solution to the *man-optimal* stable matching M_0. The algorithm then finds the stable neighbor matchings of M_0, which are M_1 and M_2. M_2 is selected to be

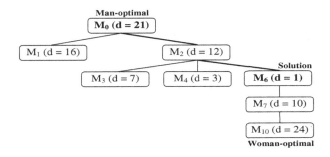

Fig. 1. The tree search of finding the *sex-equal* stable matching of Example 1.

the next solution, since $d(M_1) > d(M_2)$. Because $d(M_0) > d(M_2)$, the best solution is M_2. The algorithm repeats for M_2 to obtain M_6 as the current solution. M_7 and then M_{10} are obtained by the same operation. Since no stable neighbor matchings of M_{10} are found, the algorithm ends and outputs the best solution to be M_6 because M_6 is the best solution found so far. As shown in Table 2, the algorithm finds exactly the *sex-equal* stable matching M_6. Apparently, if we assume that $d(M_5) < d(M_6)$, then the algorithm gives a local optimum M_6 because it does not reach to M_5, M_8 and M_9. However, according to Corollary 1, M_5 must be generated from M_1, M_2, M_3 or M_4 and therefore, if the algorithm selects the next solution to be a random neighbor with a certain probability, then the matching M_1, M_2, M_3 or M_4 can be selected to generate M_5.

4 Simulations

This section presents simulations implemented with the Matlab software on a Core i5-2430M CPU 2.4 GHz with 4 GB RAM computer. The simulations are designed to evaluate the performance of our algorithm for SM instances, in which the preference lists of men and women are generated randomly. The probability of choosing a random stable matching in stable neighbor matchings is set $p = 0.1$.

First, we make simulations in order to compare the solutions found by the proposed algorithm with those found by the brute-force algorithm. The brute-force algorithm is typically used when the SM size is small and therefore, the size of SM instances is set $n = 10$. Table 3 shows the results of 30 simulations on 30 SM instances. The simulation results show that the *egalitarian cost* and the *sex-equal cost* by our algorithm (LS algorithm) is the same as that found by the brute-force algorithm (BF algorithm).

Second, we make simulations on SM instances of large sizes to evaluate the correctness and the run time of our algorithm. We generate randomly 12 SM instances of size n. For each SM instance, we run 10 times and calculate the mean and the standard deviation of the *egalitarian cost* (respectively *sex-equal cost*) for finding an *egalitarian* (respectively *sex-equal*) stable matching. Table 4 shows the simulation results. For each SM instance, the standard deviation value is zero

Table 3. The simulation results of the brute-force algorithm and our algorithm.

Data set	BF algorithm		LS algorithm		Data Set	BF algorithm		LS algorithm	
	$min(c)$	$min(d)$	$min(c)$	$min(d)$		$min(c)$	$min(d)$	$min(c)$	$min(d)$
1	56	6	56	6	16	51	3	51	3
2	61	7	61	7	17	53	5	53	5
3	63	4	63	4	18	51	13	51	13
4	62	0	62	0	19	58	0	58	0
5	80	12	80	12	20	61	1	61	1
6	71	21	71	21	21	58	0	58	0
7	59	9	59	9	22	59	7	59	7
8	60	6	60	6	23	50	4	50	4
9	53	5	53	5	24	60	8	60	8
10	58	6	58	6	25	47	3	47	3
11	56	6	56	6	26	52	2	52	2
12	55	2	55	2	27	60	2	60	2
13	58	20	58	20	28	64	1	64	1
14	53	9	53	9	29	62	0	62	0
15	60	10	60	10	30	59	14	59	14

Table 4. The simulation results of our algorithm for SM instances of large sizes.

Data set	Size	egalitarian cost			sex-equal cost		
		Mean	Deviation	Time (sec.)	Mean	Deviation	Time (sec.)
1	n = 50	678	0	0.661	45	0	0.593
2	n = 75	1214	0	3.128	31	0	3.564
3	n = 100	2050	0	6.764	107	0	6.964
4	n = 125	2748	0	10.095	3	0	15.351
5	n = 150	3630	0	16.589	63	0	18.154
6	n = 175	4514	0	54.497	19	0	48.670
7	n = 200	5348	0	67.438	17	0	77.152
8	n = 225	6413	0	77.319	31	0	88.202
9	n = 250	7847	0	188.323	173	0	238.256
10	n = 300	10094	0	232.674	106	0	220.490
11	n = 350	12565	0	572.411	363	0	578.730
12	n = 400	15607	0	676.545	23	0	680.780

and the mean value is the *egalitarian cost* (respectively *sex-equal cost*) of the *egalitarian* (respectively *sex-equal*) stable matching. Although the probability of choosing a random stable matching in stable neighbor matchings is set $p = 0.1$, the standard deviation of *egalitarian cost* (respectively *sex-equal cost*) in simulations on each SM instance is zero, meaning that our algorithm finds exactly the *egalitarian* (respectively the *sex-equal*) stable matching. Furthermore, the average time shows that our algorithm is efficient in terms of computational time even the size of SM instances is large.

Finally, we make simulations on SM instances to compare the run time of our algorithm with that of ACS algorithm [14]. The simulations show that the ACS algorithm finds an *egalitarian* or *sex-equal* stable matching only for SM instances of small sizes ($n \leq 30$) since the ACS algorithm has to find a large amount of pairs (man,woman) to form a stable matching. For example, given an instance of SM of size $n = 100$, the ACS algorithm has to find $n^2 = 10000$ pairs (man,woman) to form a stable matching of 100 engaged pairs. Moreover, simulations show that our algorithm finds the solution of SM instances more faster than the ACS algorithm does.

5 Conclusions

In this paper, we proposed a local search algorithm to find an *egalitarian* or *sex-equal* stable matching of SM instances by using the Gale-Shapley algorithm, the breakmarriage operation, and a random walk. The simulation results show that the proposed algorithm is efficient for solving SM instances. Moreover, the proposed algorithm is a general approach since with minor modifications of the cost function it can find other optimal stable matchings.

Acknowledgements. The authors are grateful to the Basic Science Research Program through the National Research Foundation of Korea (NRF) funded by the Ministry of Education, Science, and Technology (2014R1A1A2057735), IITP (2015-(R0134-15-1033)) and the Kyung Hee University in 2013 [KHU-20130439].

References

1. Abraham, D.J., Irving, R.W., Manlove, D.F.: The student-project allocation problem. In: Proceedings of the 14th International Symposium, pp. 474–484. Kyoto, Japan, December 2003
2. Fleiner, T., Irving, R.W., Manlove, D.F.: Efficient algorithms for generalized stable marriage and roommates problems. Theor. Comput. Sci. **381**(1–3), 162–176 (2007)
3. Gale, D., Shapley, L.S.: College admissions and the stability of marriage. Am. Math. Mon. **9**(1), 9–15 (1962)
4. Gelain, M., Pini, M.S., Rossi, F., Venable, K.B., Walsh, T.: Local search approaches in stable matching problems. Algorithms **6**(1), 591–617 (2013)
5. Gusfield, D.: Three fast algorithms for four problems in stable marriage. SIAM J. Comput. **16**(1), 111–128 (1987)
6. Gusfield, D., Irving, R.W.: The Stable Marriage Problem: Structure and Algorithms. MIT Press, Cambridge (1989)
7. Irving, R.W.: An efficient algorithm for the "stable roommates" problem. J. Algorithms **6**(1), 577–595 (1985)
8. Irving, R.W., Leather, P.: The complexity of counting stable marriages. SIAM J. Comput. **15**(3), 655–667 (1986)

9. Iwama, K., Miyazaki, S., Yanagisawa, H.: Approximation algorithms for the sex-equal stable marriage problem. ACM Trans. Algorithms **7**(1), 2:1–2:17 (2010)
10. McVitie, D.G., Wilson, L.B.: The stable marriage problem. Commun. ACM **14**(7), 486–490 (1971)
11. Nakamura, M., Onaga, K., Kyan, S., Silva, M.: Genetic algorithm for sex-fairstable marriage problem. In: 1995 IEEE International Symposium on Circuits and Systems, ISCAS 1995, vol. 1, pp. 509–512. Seattle, WA, April 1995
12. Roth, A.E.: The evolution of the labor market for medical interns and residents: a case study in game theory. J. Polit. Econ. **92**(6), 991–1016 (1984)
13. Russel, S., Norvig, P.: Artificial Intelligence: A Modern Approach, 3rd edn. Pearson Education, Upper Saddle River (2010)
14. Vien, N.A., Viet, N.H., Kim, H., Lee, S., Chung, T.: Ant colony based algorithm for stable marriage problem. Adv. Innov. Syst. Comput. Sci. Softw. Eng. **1**, 457–461 (2007)

Instance Selection Method for Improving Graph-Based Semi-supervised Learning

Hai Wang, Shao-Bo Wang, and Yu-Feng Li[(⊠)]

National Key Laboratory for Novel Software Technology,
Nanjing University, Nanjing 210023, China
{wanghai,wangsb,liyf}@lamda.nju.edu.cn

Abstract. Graph-based semi-supervised learning (GSSL) is one of the most important semi-supervised learning (SSL) paradigms. Though GSSL methods are helpful in many situations, they may hurt performance when using unlabeled data. In this paper, we propose a new GSSL method GSSLIS based on instance selection in order to reduce the chances of performance degeneration. Our basic idea is that given a set of unlabeled instances, it is not the best to exploit all the unlabeled instances; instead, we should exploit the unlabeled instances which are highly possible to help improve the performance, while do not take the ones with high risk into account. Experiments on a board range of data sets show that the chance of performance degeneration of our proposal is much smaller than that of many state-of-the-art GSSL methods.

Keywords: Graph-based semi-supervised learning · Performance degeneration · Instance selection

1 Introduction

In many applications, there are plentiful unlabeled training data while the acquisition of class labels is costly and difficult. For example, in webpage categorization (Zhou et al. 2004), manually labeled webpages are always a very small part of the entire web, and unlabeled webpages are in a large part. SSL (Zhu 2007; Chapelle et al. 2006) is now well known as a popular technique that exploits unlabeled data to help improve learning performance, particularly when there are limited labeled examples. During the past decade, SSL has attracted significant attentions in machine learning community. One evidence is that three representative works in SSL (Blum and Mitchell 1998; Joachims 1999; Zhu et al. 2003) have won the 10-Year Best Paper Award by ICML in 2008, 2009 and 2013, respectively.

Among many SSL approaches, GSSL is one of the most important SSL paradigms. This line of methods is generally based upon an assumption that similar instances should be shared by similar labels. It encodes both the labeled and

This research was supported by NSFC (61403186), JiangsuSF (BK20140613), 863 Program (2015AA015406).

© Springer International Publishing Switzerland 2016
R. Booth and M.-L. Zhang (Eds.): PRICAI 2016, LNAI 9810, pp. 565–573, 2016.
DOI: 10.1007/978-3-319-42911-3_47

unlabeled instances as vertices in a weighted graph, with edge weights encoding the similarity between instances. GSSL method aims to assign the labels to unlabeled instances such that the inconsistency with respect to the graph is minimized.

Previous studies generally expected that when the amount of labeled data is limited, GSSL (Zhou et al. 2004; Zhu et al. 2003, 2005; Joachims 2003; Blum and Chawla 2001; Camps-Valls et al. 2007) could be an effective approach to improve the performance by exploiting auxiliary unlabeled data. However, in many cases (Zhou et al. 2004; Belkin and Niyogi 2004; Karlen et al. 2008; Wang and Zhang 2008; Li et al. 2016), GSSL algorithms using auxiliary unlabeled data might even decrease the learning performance. To enable GSSL to be accepted by more users in more application areas, it is desirable to reduce the chances of performance degeneration when using unlabeled data.

In this paper, we propose an instance selection method in order to reduce the chances of performance degeneration when using unlabeled data. Our basic idea is that given a set of unlabeled instances, it is not the best to exploit all the unlabeled instances; instead, we should exploit the unlabeled instances which are highly possible to help improve the performance, while do not take the ones with high risk into account. We propose our GssLIs (Graph Semi-Supervised Learning with Instance Selection) method which exploits both the predictive label and confidence simultaneously. Experiments on a board range of data sets show that the chance of performance degeneration of our proposal is much smaller than that of many state-of-the-art GSSL methods.

We organize the paper as follows. Section 2 briefly introduces the background. Section 3 presents our method. Experimental results are reported in Sect. 4. Finally, Sect. 5 concludes this paper.

2   Background

For the simplicity of notations, let $D = \{\{\boldsymbol{x}_i, y_i\}_{i=1}^l, \{\boldsymbol{x}_j\}_{j=l+1}^{l+u}\}$ denote the training data set where $L = \{\boldsymbol{x}_i, y_i\}_{i=1}^l$ corresponding to the labeled instances and $U = \{\boldsymbol{x}_j\}_{j=l+1}^{l+u}$ corresponding to the unlabeled instances. $y_i \in \{+1, -1\}$ corresponding the label of instance $\boldsymbol{x}_i$, $i = 1, \ldots, l$. In GSSL, a graph $G(V, \boldsymbol{W})$ is constructed with nodes V corresponding to the $l + u$ training instances, with edges $\boldsymbol{W} = [w_{ij}] \in \mathcal{R}^{(l+u) \times (l+u)}$ corresponding to the weighted similarity matrix between training instances. In this following, we briefly introduce two classical GSSL methods. One is the Class Mass Normalization (CMN) method (Zhu et al. 2003) and the other is the Learning with Local and Global Consistency (LLGC) method (Zhou et al. 2004).

CMN defines a function $f : L \cup U \to \mathcal{R}$ over the nodes. According to the intuition of GSSL, similar instances have similar labels and this motivates the choice of the quadratic energy function

$$E(f) = \frac{1}{2} \sum_{i,j=1}^{l+u} w_{ij}(f(\boldsymbol{x}_i) - f(\boldsymbol{x}_j))^2 = \frac{1}{2}\boldsymbol{f}^{\mathrm{T}}\boldsymbol{\Delta}\boldsymbol{f} \tag{1}$$

where Δ is the Laplacian matrix of graph $G(V, W)$ (Belkin and Niyogi 2002). To minimize Eq. 1, since it is a convex quadratic form, its optimal solution can be formulated as a closed-form $f_U = (-\Delta_{UU})^{-1}\Delta_{UL}y_L$, the two matrices ($\Delta_{UU}$ and Δ_{UL}) are partitioned from Δ.

The LLGC method considers a similar idea as CMN but it considers the use of a matrix form rather than a vector form for the predictive results. Besides, rather than the CMN method which enforces that the prediction of GSSL on the labeled data must be the same as the ground-truth label, the LLGC method introduces a loss function for the labeled data, which allows some small losses on the labeled data.

3   Our Proposed Method

Classical GSSL studies (such as, the CMN method and the LLGC method) generally expected that when the amount of labeled data is limited, GSSL could improve the performance by exploiting auxiliary unlabeled data. However, in many empirical cases (Zhou et al. 2004; Belkin and Niyogi 2004; Karlen et al. 2008; Wang and Zhang 2008; Li et al. 2016), GSSL algorithms using auxiliary unlabeled data might even decrease the learning performance. To enable GSSL to be accepted by more users, it is desired to reduce the chances of performance degeneration when using unlabeled data in GSSL.

To address this problem, our basic idea is that given a set of unlabeled instances, it is not the best to exploit all the unlabeled instances without any sanity check; instead, we should exploit the unlabeled instances which are highly possible to help improve the performance, while do not take the ones with high risk into account. Based on this recognition, in the following, we first present two direct approaches based on predictive label aggregation and predictive confidence aggregation respectively, to reduce the chances of performance degeneration. Then, by examining the limitations of these two direct approaches, we propose GSSLIS method with the use of both the predictive label and confidence simultaneously.

3.1   Two Direct Approaches

MV. The first direct approach is the use of MV (Majority Voting) strategy (Kuncheva et al. 2003) which is known as an effective approach to improve the robustness of a learning method. It aggregates multiple predictive labels from multiple GSSL methods (for example, by using multiple graphs). The label of unlabeled instance is assigned to the majority one among multiple predictive labels.

DirA. DirA (Direct Aggregation) is motivated by predictive confidence aggregation, where the confidence obtained by GSSL method can be regarded as a measurement of the reliability of unlabeled data. Formally, let f_m denote the predictive value on a set of weight matrices $\{W_m\}_{m=1}^M$ where M is the number of graphs. The DirA method aggregates the predictive values. The unlabeled instances with a high confidence value (or a high rank) are selected to use and the ones with a low confidence value (or a low rank) are risky and not exploited.

3.2   The GsslIs Method

For the MV method, it only considers the hard label aggregation and may be risky when some hard labels are with low confidences. For the DirA method, it only considers the mean of the predictive values whereas ignores their variance, which might be misled and risky. To alleviate the above deficiencies, we propose the GsslIs method. Our basic observation is that, the MV and the DirA methods are complementary to each other. Specifically, the predictive value aggregation used in the DirA method is able to avoid low confident unlabeled instances and thus could be applied to improve the MV method. On the other hand, the MV method proposes to use the unlabeled data with general consistent labels on multiple graphs, and this could consequently help exclude unlabeled data whose predictive values are with high variance. Based on this observation, the proposed GsslIs method is quite simple and easy to implement. As Algorithm 1 shows, GsslIs first obtains the positive set $\mathcal{P}$ and the negative set $\mathcal{N}$ using the MV method, and then aggregates the predictive confidences on set $\mathcal{P}$ and $\mathcal{N}$, respectively.

Algorithm 1. The Proposed GsslIs Method

Input: $L = \{(\boldsymbol{x}_i, y_i)\}_{i=1}^{l}$, $U = \{\boldsymbol{x}_j\}_{j=l+1}^{l+u}$, multiple weight matrices $\{\boldsymbol{W}_m\}_{m=1}^{M}$, the predictive results of the 1NN algorithm $\hat{\boldsymbol{y}} = [\hat{y}_1, \cdots, \hat{y}_{l+u}]$ and parameter λ;

Output: A label assignment on training data $\tilde{\boldsymbol{y}} = [\tilde{y}_1, \cdots, \tilde{y}_{l+u}]$.

1: Perform classical GSSL methods on a set of weight matrix $\{\boldsymbol{W}_m\}_{m=1}^{M}$, and collect the predictive value $\boldsymbol{F} = [\boldsymbol{f}_1, \cdots, \boldsymbol{f}_M]$ where $\boldsymbol{f}_m = [f^m(\boldsymbol{x}_1), \cdots, f^m(\boldsymbol{x}_{l+u})]$, $\forall m = 1, \cdots, M$.

2: Let $\mathcal{P} = \{i | sign(f^1(\boldsymbol{x}_i)) + \cdots + sign(f^M(\boldsymbol{x}_i)) \geq 0,\ i = 1, \cdots, l+u\}$ and $\mathcal{N} = \{i | sign(f^1(\boldsymbol{x}_i)) + \cdots + sign(f^M(\boldsymbol{x}_i)) < 0,\ i = 1, \cdots, l+u\}$

3: For $\boldsymbol{x}_i \in L \cup U$, calculate the aggregated confidence A_i according to the predictive values $[\boldsymbol{f}_1, \cdots, \boldsymbol{f}_M]$

$$A_i = \frac{1}{M} \sum_{m=1}^{M} f^m(\boldsymbol{x}_i)$$

4: For $\boldsymbol{x}_i \in L \cup U$, assign predictive label $\tilde{y}_i$ according to A_i

$$\tilde{y}_i = \begin{cases} +1 & i \in \mathcal{P}\ \&\ rank(A_i)\ (\text{in a descending order}) \leq \lambda|\mathcal{P}| \\ -1 & i \in \mathcal{N}\ \&\ rank(A_i)\ (\text{in a ascending order}) \leq \lambda|\mathcal{N}| \\ \hat{y}_i & otherwise \end{cases}$$

5: **return** $\tilde{\boldsymbol{y}}$ where $\tilde{\boldsymbol{y}} = [\tilde{y}_1, \cdots, \tilde{y}_{l+u}]$.

---

4   Experiments

4.1   Data Sets

To verify the effectiveness of our proposed method, we evaluate the GSSLIS method on a broad range of data sets[1] (Table 1). For each data set, 10 examples are randomly chosen as the labeled examples, and use the remaining data as unlabeled data. The experiments are repeated for 30 times and the average accuracies with their standard deviations are recorded.

Table 1. Experimental data sets

Data	#Dim	#Pos	#Neg	#Total	Data	#Dim	#Pos	#Neg	#Total
Text	11960	750	750	1500	Liver disorders	6	200	145	345
Credit-approval	15	383	307	690	Spambase	57	1813	2788	4601
Hill-valley	100	606	606	1212	Vehicle	16	218	217	435
Breastw	9	239	444	683	Statlog-heart	13	120	150	270
House-votes	16	267	168	435	House	16	108	124	232
Digit1	241	734	766	1500	German	24	300	700	1000
WDBC	14	357	212	569	Diabetes	8	500	268	768
Isolet	51	300	300	600	Horse-colic	25	136	232	368

4.2   Compared Method

The proposed method is compared with the following methods.

- 1NN: The supervised 1 Nearest Neighbor method, which is used as a baseline supervised approach in classical GSSL (Zhou et al. 2004).
- LLGC: The Learning with Local and Global Consistency method (Zhou et al. 2004).
- CMN: The Class Mass Normalization method (Zhu et al. 2003).
- MV: The majority voting method mentioned in Sect. 3.1.
- DirA: The direct aggregation method mentioned in Sect. 3.1.

For 1NN method, Euclidean distance metric is used to locate the nearest neighbors. For CMN and LLGC method, 5 nearest neighbor graphs under 3 kinds of distance metrics (namely Euclidean distance, Cosine distance, Manhattan distance) are conducted for comparison. The parameter of the CMN method is set to the recommended one in the package[2]. The LLGC method is implemented by ourself and the parameter α is set to 0.99 as recommended in the paper. For MV, DirA and our proposed GSSLIS method, the 5 nearest neighbor graphs used in CMN and LLGC method are employed as the set of graphs. In our proposed GSSLIS method, the parameter λ is set to 0.7 for all the experimental cases.

---

[1] Downloaded from http://archive.ics.uci.edu/ml/datasets.html.
[2] http://pages.cs.wisc.edu/~jerryzhu/pub/harmonic_function.m.

4.3   Comparison Results

Table 2 shows the comparison results based on the implementation of CMN. As can be seen, GssLIs achieves highly competitive performance with compared methods. For example, in terms of average accuracy, GssLIs obtains the best average accuracy. While more importantly, the compared GSSL methods all will significantly decrease the performance in many cases, while our proposed approach never degenerate the performance. Both the MV and the DirA method are capable of reducing the chances of performance degeneration, however, they still degenerate the performance in multiple cases, while our proposed method does not have such kind of phenomena. As for LLGC method, Table 3 shows the comparison results. As can be seen, similar to the cases in Table 2, GssLIs also obtains highly competitive performance with compared GSSL methods.

Overall, these results show that our proposed method is able to reduce the chances of performance degeneration, while still obtains highly competitive performance improvement as state-of-the-art GSSL methods.

Table 2. Accuracy (mean ± std) on 10 labeled examples based on CMN method. For the GSSL methods if the performance is significantly better/worse than 1NN, the corresponding entries are bolded/boxed (paired t-tests at 95 % significance level). The average accuracy is listed for comparison. The win/tie/loss counts are summarized and the method with the smallest number of losses against 1NN is bolded.

Data	1NN	CMN			MV	DirA	GssLIs
	Euclidean	Euclidean	Cosine	Manhattan			
Text	59.5 ± 3.4	**64.0 ± 4.8**	**63.1 ± 4.6**	51.2 ± 3.2	**64.2 ± 4.9**	**62.9 ± 4.2**	**64.5 ± 4.5**
Credit	72.9 ± 6.9	69.9 ± 8.1	68.2 ± 7.1	69.1 ± 8.5	69.6 ± 7.7	73.4 ± 7.3	73.3 ± 6.9
Hill	50.1 ± 1.6	50.0 ± 1.7	**66.9 ± 5.7**	50.0 ± 1.9	**51.8 ± 2.5**	50.6 ± 1.7	50.9 ± 1.6
Breastw	93.2 ± 3.6	**95.6 ± 1.0**	73.1 ± 4.4	**95.7 ± 0.9**	**95.5 ± 1.1**	93.2 ± 3.4	93.5 ± 3.1
House-v	86.7 ± 3.0	**88.7 ± 1.8**	87.7 ± 3.2	**88.7 ± 2.4**	**87.9 ± 2.3**	**87.1 ± 2.4**	**87.5 ± 2.0**
Digit1	78.1 ± 5.3	**86.2 ± 3.5**	**85.3 ± 4.0**	**83.3 ± 3.7**	**86.8 ± 3.3**	80.8 ± 4.3	84.0 ± 3.8
WDBC	80.5 ± 5.5	79.8 ± 4.6	77.3 ± 5.2	73.8 ± 4.3	78.6 ± 5.3	**82.9 ± 4.8**	**85.1 ± 3.8**
Isolet	91.6 ± 3.6	**98.0 ± 0.9**	**98.4 ± 0.8**	**97.7 ± 1.1**	**98.5 ± 0.7**	92.2 ± 3.4	93.2 ± 2.8
Liver	52.6 ± 3.2	52.0 ± 3.3	53.1 ± 4.8	52.4 ± 3.0	52.8 ± 4.1	52.7 ± 3.0	53.2 ± 3.2
Spambase	69.4 ± 8.0	61.5 ± 1.6	61.7 ± 1.2	61.6 ± 1.1	61.4 ± 1.2	**73.9 ± 6.4**	**71.3 ± 7.0**
Vehicle	72.8 ± 6.0	**74.4 ± 7.4**	78.1 ± 9.0	79.5 ± 8.5	78.9 ± 8.7	74.2 ± 6.0	**76.3 ± 7.1**
Statlog	73.3 ± 5.9	74.1 ± 5.8	74.0 ± 4.8	**77.0 ± 5.2**	**77.0 ± 4.6**	75.2 ± 4.5	**76.7 ± 4.0**
House	89.4 ± 2.1	89.8 ± 2.2	88.5 ± 2.8	88.0 ± 2.1	89.6 ± 2.1	89.4 ± 2.1	89.6 ± 1.9
German	63.8 ± 5.2	**69.0 ± 1.3**	**69.3 ± 1.3**	**69.7 ± 0.8**	**69.4 ± 1.0**	62.5 ± 4.5	**65.8 ± 3.0**
Diabetes	64.5 ± 5.3	65.6 ± 2.1	66.5 ± 2.0	65.5 ± 2.5	65.7 ± 2.0	64.3 ± 5.2	66.0 ± 3.3
Horse	65.3 ± 4.6	65.1 ± 4.6	66.6 ± 5.6	64.8 ± 4.0	65.8 ± 5.1	65.6 ± 4.7	**68.1 ± 5.2**
Ave. acc.	72.7	74.0	73.6	73.0	74.6	73.8	**74.9**
W/T/L against 1NN		7/7/2	6/6/4	7/4/5	9/4/3	9/6/1	**13/3/0**

Table 3. Accuracy (mean ± std) on 10 labeled examples based on LLGC method.

Data	1NN	LLGC			MV	DirA	GssLIs
	Euclidean	Euclidean	Cosine	Manhattan			
Text	59.5 ± 3.4	55.8 ± 4.8	57.6 ± 5.9	50.7 ± 1.2	56.2 ± 5.4	**62.9 ± 4.4**	**61.0 ± 4.9**
Credit	72.9 ± 6.9	69.9 ± 7.6	68.8 ± ± 7.7	69.6 ± 8.7	68.8 ± 7.8	73.3 ± 7.4	73.0 ± 6.9
Hill	50.1 ± 1.6	50.0 ± 1.8	**68.5 ± 6.3**	50.0 ± 1.7	**50.6 ± 1.7**	**51.7 ± 2.2**	**51.3 ± 1.9**
Breastw	93.2 ± 3.6	**95.8 ± 0.6**	78.2 ± 7.0	95.9 ± 0.6	**95.6 ± 0.5**	93.3 ± 3.4	**93.6 ± 2.9**
House-v	86.7 ± 3.0	82.9 ± 8.1	84.7 ± 6.5	84.6 ± 7.2	84.9 ± 7.0	**87.2 ± 2.4**	**87.2 ± 3.0**
Digit1	78.1 ± 5.3	**90.0 ± 3.4**	90.1 ± 2.8	88.0 ± 3.1	90.9 ± 3.0	80.7 ± 4.4	84.0 ± 3.7
WDBC	80.5 ± 5.5	71.0 ± 6.8	70.9 ± 6.1	67.1 ± 3.7	69.2 ± 4.9	**83.0 ± 4.7**	**85.4 ± 4.2**
Isolet	91.6 ± 3.6	**97.0 ± 1.9**	97.9 ± 0.8	97.2 ± 1.8	98.3 ± 1.1	92.1 ± 3.5	93.3 ± 2.7
Liver	52.6 ± 3.2	52.5 ± 3.8	53.9 ± 4.7	51.9 ± 4.8	53.2 ± 4.1	53.0 ± 2.8	53.0 ± 3.2
Spambase	69.4 ± 8.0	65.6 ± 4.7	66.3 ± 4.6	64.5 ± 3.3	65.0 ± 4.0	**74.0 ± 6.1**	**72.0 ± 6.8**
Vehicle	72.8 ± 6.0	**74.8 ± 7.9**	77.1 ± 8.5	80.2 ± 8.5	77.2 ± 8.4	73.9 ± 5.9	75.4 ± 6.8
Statlog	73.3 ± 5.9	60.6 ± 5.7	59.2 ± 4.7	59.2 ± 5.3	59.7 ± 5.7	**75.3 ± 4.4**	**75.1 ± 5.2**
House	89.4 ± 2.1	80.8 ± 9.7	83.3 ± 8.4	79.9 ± 9.1	82.4 ± 8.3	89.4 ± 2.1	88.8 ± 2.7
German	63.8 ± 5.2	**69.2 ± 1.4**	69.0 ± 1.5	**69.6 ± 1.0**	**69.4 ± 1.2**	62.6 ± 4.4	**65.7 ± 2.9**
Diabetes	64.5 ± 5.3	65.4 ± 2.1	66.0 ± 1.9	65.4 ± 1.8	65.5 ± 1.7	64.3 ± 5.2	**65.8 ± 3.3**
Horse	65.3 ± 4.6	**63.0 ± 3.4**	63.5 ± 4.0	62.6 ± 3.1	62.4 ± 2.4	65.4 ± 4.8	**68.6 ± 5.4**
Ave. acc.	72.7	71.5	72.2	71.0	71.8	73.9	**74.6**
W/T/L against 1NN		5/3/8	5/5/6	5/4/7	6/3/7	9/6/1	**13/3/0**

4.4 Influence on the Number of Graphs

We further study the influence on the number of candidate graphs. We generate the candidate graphs as followings. For each instance, the number of nearest neighbors is randomly picked up from 3 to 7 with a uniform distribution. The number of candidate graphs $M \in \{3, 5, 7\}$ and the graph is constructed by using different distance metrics. Figure 1 shows the results with different number of candidate graphs. As can be seen, our GssLIs method rarely hurts the performance as the number of candidate graphs varies.

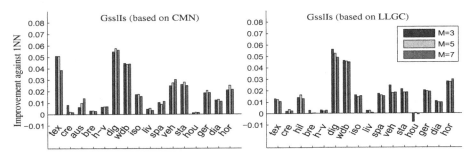

Fig. 1. Influence on the number of graphs M on the performance improvement of GssLIs against 1NN method. (Color figure online)

5   Conclusion

In this paper, we propose a new GSSL method GssLIs based on instance selection in order to reduce the chances of performance degeneration for GSSL. Our basic idea is that given a set of unlabeled instances, it is not the best to exploit all the unlabeled instances; instead, we should exploit the unlabeled instances which are highly possible to help improve the performance, while do not take the ones with high risk into account. Experiments on a board range of data sets show that the chance of performance degeneration of our proposal is much smaller than that of some classical GSSL method, while maintains similar performance improvement as many state-of-the-art GSSL approaches. In future we will study the mathematical foundation of GssLIs and extend it to multi-class scenario.

References

Belkin, M., Niyogi, P.: Semi-supervised learning on Riemannian manifolds. Mach. Learn. **56**, 209–239 (2004)

Belkin, M., Niyogi, P.: Laplacian eigenmaps and spectral techniques for embedding and clustering. In: Advances in Neural Information Processing Systems, vol. 14, pp. 585–591. MIT Press (2002)

Blum, A., Mitchell, T.: Combining labeled and unlabeled data with co-training. In: Proceedings of the 7th Annual Conference on Computational Learning Theory, Madison, WI (1998)

Blum, A., Chawla, S.: Learning from labeled and unlabeled data using graph min-cuts. In: Proceedings of the 18th International Conference on Machine Learning, Williamstown, MA, pp. 19–26 (2001)

Camps-Valls, G., Marsheva, T.V.B., Zhou, D.: Semi-supervised graph-based hyper-spectral image classification. IEEE Trans. Geosci. Remote Sens. **45**(10), 3044–3054 (2007)

Chapelle, O., Schölkopf, B., Zien, A., et al.: Semi-supervised learning (2006)

Joachims, T.: Transductive inference for text classification using support vector machines. In: Proceedings of the 16th International Conference on Machine Learning, Bled, Slovenia, pp. 200–209 (1999)

Joachims, T.: Transductive learning via spectral graph partitioning. In: Proceedings of the 20th International Conference on Machine Learning, Washington, DC, pp. 290–297 (2003)

Karlen, M., Weston, J., Erkan, A., Collobert, R.: Large scale manifold transduction. In: Proceedings of the 25th International Conference on Machine Learning, Helsinki, Finland, pp. 775–782 (2008)

Kuncheva, L.I., Whitaker, C.J., Shipp, C.A., Duin, R.P.: Limits on the majority vote accuracy in classifier fusion. Pattern Anal. Appl. **6**(1), 22–31 (2003)

Li, Y.-F., Wang, S.-B., Zhou, Z.-H.: Graph quality judgement: a large margin expedition. In: Proceedings of the 25th International Joint Confernece on Artificial Intelligence, New York, NY (2016)

Wang, F., Zhang, C.: Label propagation through linear neighborhoods. IEEE Trans. Knowl. Data Eng. **20**(1), 55–67 (2008)

Zhou, D., Bousquet, O., Lal, T.N., Weston, J., Schölkopf, B.: Learning with local and global consistency. In: Advances in Neural Information Processing Systems, vol. 16, pp. 595–602. MIT Press, Cambridge (2004)

Zhu, X.: Semi-supervised learning literature survey. Technical report. University of Wisconsin-Madison (2007)

Zhu, X., Ghahramani, Z., Lafferty, J.: Semi-supervised learning using Gaussian fields and harmonic functions. In: Proceedings of the 20th International Conference on Machine Learning, Washington, DC, pp. 912–919 (2003)

Zhu, X., Lafferty, J., Rosenfeld, R.: Semi-supervised learning with graphs. Ph.D. thesis. Carnegie Mellon University (2005)

Exploring Multi-action Relationship in Reinforcement Learning

Han Wang and Yang Yu[✉]

National Key Laboratory for Novel Software Technology,
Nanjing University, Nanjing 210023, China
{wangh,yuy}@lamda.nju.edu.cn

Abstract. In many real-world *reinforcement learning* problems, an agent needs to control multiple actions simultaneously. To learn under this circumstance, previously, each action was commonly treated independently with other. However, these multiple actions are rarely independent in applications, and it could be helpful to accelerate the learning if the underlying relationship among the actions is utilized. This paper explores multi-action relationship in reinforcement learning. We propose to learn the multi-action relationship by enforcing a regularization term capturing the relationship. We incorporate the regularization term into the *least-square policy-iteration* and the *temporal-difference* methods, which result efficiently solvable convex learning objectives. The proposed methods are validated empirically in several domains. Experiment results show that incorporating multi-action relationship can effectively improve the learning performance.

1 Introduction

Reinforcement learning techniques intend to enable an agent to learn how to behave through trial-and-error interactions with its environment. Reinforcement learning has been an attractive field experiencing progress from theory to practice, with wide applications, including robotics [2], computer Go [10], and combinatorial optimization problems [3].

In many cases, an agent needs to control multiple actions simultaneously. For example, an automated driving agent needs to control multiple components in parallel: when approaching a turning, the agent is supposed to turn the steering wheel, meanwhile release the accelerator and hit the brake. However, most of previous reinforcement learning methods treated each action independently with each other, and thus learned each action separately. The only previous study that explicitly considered multi-action setting is [9], where a concurrent action model was proposed to solve multi-action problems based on SMDP Q-learning.

There are also several studies that are apparently related to this work, but are actually different, including multi-task, multi-agent, and multidimensional

Y. Yu—This research was supported by the NSFC (61375061, 61223003), Foundation for the Author of National Excellent Doctoral Dissertation of China (201451).

R. Booth and M.-L. Zhang (Eds.): PRICAI 2016, LNAI 9810, pp. 574–587, 2016.
DOI: 10.1007/978-3-319-42911-3_48

reinforcement learning. Multi-task reinforcement learning assumes that the agent faces multiple learning tasks within its lifetime [4]. These approaches often solve different tasks sequentially, rather than the same take concurrently as in the multi-action setting. The need for adaptive multi-agent systems has led to the development of a multi-agent reinforcement learning field [13]. While multi-agent methods can be used to learn multi-actions, however, they were developed for solving more general problems [14], but not particularly for the multi-action problem. Reinforcement learning in multidimensional continuous action spaces [8] considered an MDP with an N-dimensional action space, thus provided an effective approach for learning in domains with multidimensional control variables. To the best of our knowledge, few studies tried to utilize the relationship among multiple actions.

In this paper, noticing that actions in a task usually have a significant relationship, we explore utilizing multi-action relationship to improve the learning performance. Inspired by the supervised multi-task learning [15, 16] in *supervised learning*, where the relationship among labels can be learned through a regularization term, we propose to capture the relationship among multiple actions by enforcing a regularization term. Specifically, we model the relationships between actions in a nonparametric manner as a covariance matrix. By utilizing a matrix-variate normal distribution [5] as a prior on learning parameters, we incorporate the regularization term into the *least-square policy-iteration* and the *temporal-difference methods*, which result efficiently solvable convex learning objectives. We then conduct experiments on a 2-action Grid world task, 3-action unmanned aerial vehicle task, and a 3-action helicopter hovering task. Experiment results show that incorporating multi-action relationship can effectively improve the learning performance.

The rest of this paper is organized into 5 sections: Sect. 2 introduces the background; Sect. 3 presents the proposed methods; The experiment results are reported in Sect. 4, and Sect. 5 concludes this paper.

2 Background

This section introduces the notation in the paper and provides further background on the standard framework for solving reinforcement learning problems and then formalizes the problem setting considered in the paper.

2.1 Reinforcement Learning

Based on *Markov Decision Process* (MDP) (S, A, T, R, γ), where S and A are finite sets of states and actions, T a transition function, R a reward function and γ the discount factor, the reinforcement learning (RL) agent's goal is to construct an optimal policy, a mapping from states to actions, that maximizes the expected cumulative reward. Most learning algorithms compute optimal policies by learning value functions, which represent an estimate of how good it is for the agent to be in a certain state (or how good it is to perform a certain action in that state) [13]. In this paper, we consider the state value function and state-action value function defined over all states and all possible combinations of states and actions. The state value function indicates the expected, discounted, total reward thereafter policy π^* while the state-action value function indicates the same reward when taking action a in state s. The value functions can be expressed by recursion according to *Bellman equations*:

$$Q^*(s,a) = \sum_{s'} T(s,a,s')(R(s,a,s') + \gamma \max_{a'} Q^*(s',a')), \tag{1}$$

$$V^*(s) = \sum_{s'} T(s,\pi^*(s),s')(R(s,a,s') + \gamma V^*(s')). \tag{2}$$

Since RL is primarily concerned with learning an optimal policy for an MDP assuming that a (perfect) model is not available, RL can be regarded as model free solution techniques. The transition and reward models are iteratively learned from interaction with the environment when model-based RL methods are used. In contrast, model-free RL methods, which we employ in this paper, step right into estimating values for actions, without even estimating the model of the MDP.

However, in the problem of control, more specifically, MDPs with continuous states and continuous or discrete actions, our aim is to learn an approximation of the optimal policy. Consequently, we can estimate π^* directly, or estimate Q^* to approximate π^* indirectly, or even estimate R and T to construct Q^* and π^* when needed. In this paper, we solve the continuous MDP by applying value approximation which uses samples to approximate Q^* directly. When updating the approximations of value functions, we apply the temporal-difference learning (TD-learning) [12] algorithms to bring some value in accordance to the immediate reward and the estimated value of the next state or the state-action pair.

Modeling concurrent decision making, [9] cast insight in a very general setting, inherited from the ability of *Semi-Markov Decision Process* (SMDP) to model a large class of decision making problems. This is a more realistic view of the agent environment interaction where the agent may not have access to the complete model of the environment. In order to realize this general setting in MDPs, we consider the thought of recursive decomposition of the action space proposed by [8], and decompose the high-dimensional action spaces into a set of sub-action spaces according to the agent's inner structure or physical laws (e.g., degree of freedom) in advance. Thus, the agent can learn the relationships between sub-actions that help the agent improve the future performance.

In the following subsections, we introduce two typical reinforcement learning algorithms. Our work will adapt these algorithms for the multi-action setting.

2.2   LSPI with Value Function Approximation

Given the Bellman equations for state-action values, the reward can be expressed as:

$$r(\boldsymbol{s},a) = Q^\pi(\boldsymbol{s},a) - \gamma \mathbb{E}_{\pi(a'|\boldsymbol{s}')p(\boldsymbol{s}'|\boldsymbol{s},a))} \left[Q^\pi(\boldsymbol{s}',a') \right].$$

As illustrated in [11], if we approximate the state-action value Q by a linear function $\boldsymbol{\theta}^{\mathrm{T}}\boldsymbol{\phi}$, the immediate reward can be approximate as:

$$r(\boldsymbol{s}, a) \approx \boldsymbol{\theta}^{\mathrm{T}}\boldsymbol{\phi}(\boldsymbol{s}, a) - \gamma \mathbb{E}_{\pi(a'|\boldsymbol{s}')p(\boldsymbol{s}'|\boldsymbol{s},a)} \left[\boldsymbol{\theta}^{\mathrm{T}}\boldsymbol{\phi}(\boldsymbol{s}', a')\right],$$

where $\boldsymbol{\theta}$ denotes the adaptable parameter vector and $\boldsymbol{\phi}(\boldsymbol{s}, a)$ is the feature vector in state $\boldsymbol{s}$ when taking action a. Thus, a new basis function vector is defined as:

$$\boldsymbol{\psi}(\boldsymbol{s}, a) \approx \boldsymbol{\phi}(\boldsymbol{s}, a) - \gamma \mathbb{E}_{\pi(a'|\boldsymbol{s}')p(\boldsymbol{s}'|\boldsymbol{s},a)} \left[\boldsymbol{\phi}(\boldsymbol{s}', a')\right],$$

and the expected immediate reward $r(\boldsymbol{s}, a)$ is be approximated as:

$$r(\boldsymbol{s}, a) \approx \boldsymbol{\theta}^{\mathrm{T}}\boldsymbol{\psi}(\boldsymbol{s}, a). \tag{3}$$

The linear approximation problem of state-action value function $Q^{\pi}(\boldsymbol{s}, a_i)$ can be reformed into immediate-reward regression problem according to Eq. (3), and further can be solved by learning $\boldsymbol{\theta}$ in the least-squares framework:

$$\min_{\boldsymbol{\theta}} \frac{1}{|\boldsymbol{D}|} \sum_{t=1}^{|\boldsymbol{D}|} \left(r_t - \boldsymbol{\theta}^{\mathrm{T}}\boldsymbol{\psi}^t\right)^2 + C\|\boldsymbol{\theta}\|_2, \tag{4}$$

where $\boldsymbol{D}$ is the source of samples. Overall, the *LSPI* algorithm is summarized in Algorithm 1 [7].

Algorithm 1. LSPI

Initialize Initialize a policy π, given as $\boldsymbol{\Theta} = \boldsymbol{0}$
            Feature vectors are denoted as $\boldsymbol{\Phi}$
repeat
     Set training set $D = \emptyset$
     Initialize state $\boldsymbol{s}_0$
     for $t = 1$ to T **do**
         Choose action $\boldsymbol{a}$ using policy π, observe $\boldsymbol{s}'_t$ and $\boldsymbol{r}_t = (r_t^1, ..., r_t^N)$
         Add the training tuple $(\boldsymbol{s}_t, \boldsymbol{a}_t, \boldsymbol{r}_t, \boldsymbol{s}'_t)$ to training set D
     end for
     Given training set D, update $\boldsymbol{\Theta}$ by solving objective function Eq.(4)
     Update policy π by $\boldsymbol{\Theta}$
until $\boldsymbol{\Theta}$ converges.

2.3 Gradient Temporal-Difference Learning

Concluded from standard *temporal-difference learning* (*TD-learning*), the tabular TD-learning update is [6],

$$V(s_t) = V(s_t) + \alpha_t \cdot \delta_t,$$

where $\delta_t = r_{t+1} - (V(s_t) - \gamma V(s_{t+1}))$ is the one-step temporal-difference error, and $\alpha_t \in [0,1]$ is a step-size parameter.

When V_t is expressed as $V(s_t) = \boldsymbol{\theta}^{\top} \boldsymbol{\phi}(s_t)$, the parameter of V_t can be solved by minimizing the following equation,

$$E(s_t) = \frac{1}{2}(\delta_t)^2 = \frac{1}{2}(r_{t+1} - (V(s_t) - \gamma V(s_{t+1})))^2$$

and the gradient with respect to the parameters is

$$-\nabla_{\boldsymbol{\theta}} E(s_t, \boldsymbol{\theta}) = -(r_{t+1} + \gamma V(s_{t+1}) - V(s_t)) \cdot \nabla_{\boldsymbol{\theta}}(r_{t+1} + \gamma V(s_{t+1}) - V(s_t)).$$

Treating $r_{t+1} + \gamma V(s_{t+1})$ to be irrelevant with $\boldsymbol{\theta}$ [6], so that the final gradient is:

$$\begin{aligned}
-\nabla_{\boldsymbol{\theta}} E(s_t, \boldsymbol{\theta}) &= -(r_{t+1} + \gamma V(s_{t+1}) - V(s_t)) \cdot \nabla_{\boldsymbol{\theta}} V(s_t) \\
&= -(r_{t+1} + \gamma \boldsymbol{\theta}^{\top} \boldsymbol{\phi}(s_{t+1}) - \boldsymbol{\theta}^{\top} \boldsymbol{\phi}(s_t)) \cdot \boldsymbol{\phi}(s_t)
\end{aligned} \tag{5}$$

Then the parameters $\boldsymbol{\theta}$ is updated by adding this negative gradient with a step size coefficient.

Algorithm 2. Gradient TD Learning

Initialize: Feature function: $\boldsymbol{\phi}$
　　　　　　Step size: α
　　　　　　Parameters: $\boldsymbol{\theta} = \mathbf{0}, t = 0$
for each episode **do**
　　set $\boldsymbol{\delta} = \mathbf{0}$
　　Initialize s_0
　　repeat
　　　　Take action from $\boldsymbol{\theta}$ with exploration, observe s' and r_t
　　　　set $\boldsymbol{\delta} = \boldsymbol{\phi}(r_t + (\boldsymbol{\phi}' - \boldsymbol{\phi})^{\mathrm{T}} \boldsymbol{\theta})$ by Eq. (5)
　　　　$\boldsymbol{\theta} = \boldsymbol{\theta} + \alpha \boldsymbol{\delta}$
　　　　$s \leftarrow s'$
　　until s is terminal
end for

3　Multi-action Relationship Learning

3.1　Problem Setting

A multi-action reinforcement learning problem domain $\mathcal{D}$ is defined as an MDP $M(S, A, T, R, \gamma)$. The MDP is as usual, only that the action set $\mathcal{A} = \mathcal{A}_1 \times \mathcal{A}_2 \times ... \times \mathcal{A}_N$ is multi-dimensional, where each $\mathcal{A}_i$ is a sub-action set. To solve this MDP, an agent can treat each dimensional action as a separated MDP, and learn multiple MDPs simultaneously. Suppose the agent's action space $\mathcal{A}$

is decomposed into N sub-action spaces, $\mathcal{A}_1, \mathcal{A}_2, \ldots, \mathcal{A}_N$. Each sub-action space has a value function to be approximated.

We consider the value function approximation by approximating the state-action value function $Q^\pi(s, a)$ in linear model:

$$Q^\pi(s, a) = \boldsymbol{\theta}^{\mathrm{T}} \boldsymbol{\phi}(s, a), \tag{6}$$

where $\boldsymbol{\theta}$ denotes the adaptable parameter vector and $\boldsymbol{\phi}$ is the feature vector in state s when taking action a. Thus, since there are N sub-action spaces, we formalize N state-action value functions according to Eq. (6). For the i-th sub-action, the state-action value function $Q_i^\pi(s, a_i)$ is approximated by the following vector representation:

$$\boldsymbol{\theta}_i^{\mathrm{T}} \boldsymbol{\phi}_i(s, a_i).$$

However in many applications, these actions usually have a significant relationship, learning separated MDPs ignores the relationship. Inspired by the success in multi-task supervised learning [16], where the relationship among tasks is essential and helpful, we hypothesize that explore the relationship among actions in multi-action reinforcement learning might also be helpful.

However, unlike in multi-task supervised learning, where the relationship among tasks is often assumed, the relationship among actions can be very different in problems, and thus must be learned. In this work, we adopt the regularization idea from [16], and propose to learn and utilize the action relationship as follows:

1. assume that all sub-actions are unrelated initially;
2. receive state-action pairs according to current policy π;
3. for each sub-action space, learn the value function using the multi-action relationship, and update the relationship;
4. loop from step 2.

In the following two subsections, we implement the idea into two reinforcement learning algorithms.

3.2 LSPI with Multi-action Relationship Regularization

Based on Sect. 2.2, the immediate reward $r_i(s, a_i)$ can be approximated as:

$$r_i(s, a_i) \approx \boldsymbol{\theta}_i^{\mathrm{T}} \boldsymbol{\psi}_i(s, a_i).$$

The linear approximation problem of state-action value function $Q_i^\pi(s, a_i)$ can be reformed into immediate-reward regression problem and it can be solved by learning $\boldsymbol{\theta}_i$ in the least-squares framework with multi-action regularization.

We assume that the likelihood for r_t^i at time t, given $\boldsymbol{\psi}_i(s_t, a_t^i)$, $\boldsymbol{\theta}_i$ and ϵ_i can be modeled as:

$$r_t^i | \boldsymbol{\psi}_i(s_t, a_t^i), \boldsymbol{\theta}_i, \epsilon_i \sim \mathcal{N}(\boldsymbol{\theta}_i^{\mathrm{T}} \boldsymbol{\psi}_i(s_t, a_t^i), \epsilon_i^2),$$

where $\mathcal{N}(\boldsymbol{m}, \Sigma)$ denotes the multivariate normal distribution with mean $\boldsymbol{m}$ and covariance matrix Σ.

Inspried by [16], the prior on $\boldsymbol{\Theta} = (\boldsymbol{\theta}_1, \ldots, \boldsymbol{\theta}_N)$ can be defined as:

$$\boldsymbol{\Theta}|\epsilon_i \sim \left(\prod_{i=1}^{N} \mathcal{N}(\boldsymbol{\theta}_i|\mathbf{0}_d, \epsilon_i^2 \boldsymbol{I}_d) \right) q(\boldsymbol{\Theta}),$$

where $\boldsymbol{I}_d$ is the $d \times d$ identity matrix, $\mathbf{0}_d$ is the $d \times 1$ zero vector. While the first term is to penalize the complexity of each column of $\boldsymbol{\Theta}$ separately, the second term is to model the structure of $\boldsymbol{\Theta}$ and is characterized by a matrix-variate distribution:

$$q(\boldsymbol{\Theta}) = \mathcal{MN}_{d \times N} (\boldsymbol{\Theta}|\mathbf{0}_{d \times N}, \boldsymbol{I}_d \otimes \boldsymbol{\Omega}).$$

$\mathcal{MN}(\boldsymbol{M}, \boldsymbol{A} \otimes \boldsymbol{B})$ represents the matrix-variate normal distribution, where row covariance matrix $\boldsymbol{A}$ models the relationships between features and column covariance matrix $\boldsymbol{B}$ models the relationships between each $\boldsymbol{\theta}_i$. Thus, $\boldsymbol{\Omega}$ models the relationships between multi-action.

As a result, the posterior distribution for $\boldsymbol{\Theta}$ is proportional to the product of the prior and the likelihood function:

$$p\left(\boldsymbol{\Theta}|\boldsymbol{\Psi}(\boldsymbol{s}, \boldsymbol{a}), \boldsymbol{r}, \epsilon, \boldsymbol{\Omega}\right) \propto p\left(\boldsymbol{r}|\boldsymbol{\Psi}(\boldsymbol{s}, \boldsymbol{a}), \boldsymbol{\Omega}, \epsilon\right) \cdot p\left(\boldsymbol{\Theta}|\epsilon, \boldsymbol{\Omega}\right) \tag{7}$$

where $\boldsymbol{r} = \left(r_1^1, \ldots, r_T^1, \ldots, r_1^N, \ldots, r_T^N\right)^{\mathrm{T}}$, $\boldsymbol{\Psi}(\boldsymbol{s}, \boldsymbol{a})$ denotes the basis function of all sub-actions. Taking the negative logarithm of Eq. 7, the agent obtain the maximum a posterior (MAP) estimation of $\boldsymbol{\Theta}$, and the maximum likelihood estimation (MLE) of $\boldsymbol{\Omega}$ by solving the following problem:

$$\min_{\boldsymbol{\Theta}, \boldsymbol{\Omega} \succeq 0} \sum_{i=1}^{N} \frac{1}{\epsilon_i^2} \sum_{t=1}^{T} \left(r_t^i - \boldsymbol{\theta}_i^{\mathrm{T}} \boldsymbol{\psi}_i^t\right)^2 + \sum_{i=1}^{N} \frac{1}{\epsilon_i^2} \boldsymbol{\theta}_i^{\mathrm{T}} \boldsymbol{\theta}_i + \mathrm{tr}(\boldsymbol{\Theta} \boldsymbol{\Omega}^{-1} \boldsymbol{\Theta}^{\mathrm{T}}) + d \ln(|\boldsymbol{\Omega}|)$$

where $\mathrm{tr}(\cdot)$ denotes the trace of a square matrix, $|\cdot|$ denotes the determinant of a square matrix, and $\boldsymbol{\Omega} \succeq \mathbf{0}$ means that the matrix $\boldsymbol{\Omega}$ is positive semidefinite due to the fact that $\boldsymbol{\Omega}$ is defined as a covariance matrix.

The squared loss and the revised regularization are expressed as:

$$\min_{\boldsymbol{\Theta}, \boldsymbol{\Omega}} \sum_{i=1}^{N} \frac{1}{T^2} \sum_{t=1}^{T} \left(r_t^i - \boldsymbol{\theta}_i^{\mathrm{T}} \boldsymbol{\psi}_i^t\right)^2 + \frac{\lambda_1}{2} \mathrm{tr}(\boldsymbol{\Theta} \boldsymbol{\Theta}^{\mathrm{T}}) + \frac{\lambda_2}{2} \mathrm{tr}(\boldsymbol{\Theta} \boldsymbol{\Omega}^{-1} \boldsymbol{\Theta}^{\mathrm{T}})$$
$$\text{s.t.} \quad \boldsymbol{\Omega} \succeq \mathbf{0}$$
$$\mathrm{tr}(\boldsymbol{\Omega}) \leq 1, \tag{8}$$

where $\lambda 1$ and $\lambda 2$ are regularization parameters.

Since Eq. 8 is jointly convex with respect to $\boldsymbol{\Theta}$ and $\boldsymbol{\Omega}$, we can optimize the objective function with respect to $\boldsymbol{\Theta}$ when $\boldsymbol{\Omega}$ is fixed, and then optimize the objective function with respect to $\boldsymbol{\Omega}$ when $\boldsymbol{\Theta}$ is fixed alternatively. Then, the parameter $\boldsymbol{\Theta}$ and multi-action relationship matrix $\boldsymbol{\Omega}$ are updated simultaneously to improve the agent's performance eventually. And the LSPI with multi-action relationship regularization term is described in Algorithm 3.

Algorithm 3. LSPI with Multi-action Relationship Regularization

---

Initialize Feature functions $\boldsymbol{\Psi}$
Initialize a policy π, given as $\boldsymbol{\Theta} = \mathbf{0}$
Initialize the relationship matrix $\boldsymbol{\Omega} = \frac{1}{N}I_N$
repeat
   Set training set $D = \emptyset$
   Initialize state $\boldsymbol{s}_0$
   for $t = 1$ to T **do**
      Choose action $\boldsymbol{a}$ using policy π, observe $\boldsymbol{s}_t'$ and $\boldsymbol{r}_t = (r_t^1, ..., r_t^N)$
      Add the training tuple $(\boldsymbol{s}_t, \boldsymbol{a}_t, \boldsymbol{r}_t, \boldsymbol{s}_t')$ to training set D
   end for
   Given training set D, update $\boldsymbol{\Theta}$ by solving objective function Eq. (8)
   Update $\boldsymbol{\Omega}$ by Eq. (11)
   Update policy π by $\boldsymbol{\Theta}$
until $\boldsymbol{\theta}$ converges.

3.3 Gradient TD Learning with Multi-action Relationship Regularization

Generalized from Sect. 2.3, in standard *temporal-difference learning (TD-learning)*, negative gradient can be expressed as:

$$-\nabla_{\boldsymbol{\theta}}\mathrm{E}(\boldsymbol{s}_t, \boldsymbol{\theta}) = -(r_{t+1} + \gamma V_t(\boldsymbol{s}_{t+1}) - V_t(\boldsymbol{s}_t)) \cdot \nabla_{\boldsymbol{\theta}}V_t(\boldsymbol{s}_t).$$

When incorporating the multi-action relationship regularization into gradient temporal-difference learning, we assume that the ith sub-action space maintains the state-action value function $\mathrm{Q}_t^i(\boldsymbol{s}_t, a_t^i) = \boldsymbol{\theta}_i^{\mathrm{T}}\boldsymbol{\phi}_i(\boldsymbol{s}_t, a_t^i)$ at time t. The TD error can be rewritten into

$$\mathrm{E}(\boldsymbol{s}_t) = \sum_{i=1}^{N} \left(r_t^i + \gamma\boldsymbol{\theta}_i^{\mathrm{T}}\boldsymbol{\phi}_t^i(\boldsymbol{s}_{t+1}, a_{t+1}^i) - \boldsymbol{\theta}_i^{\mathrm{T}}\boldsymbol{\phi}_t^i(\boldsymbol{s}_t, a_t^i)\right)^2 +$$
$$\frac{\lambda_1}{2}\mathrm{tr}(\boldsymbol{\Theta\Theta}^{\mathrm{T}}) + \frac{\lambda_2}{2}\mathrm{tr}(\boldsymbol{\Theta\Omega}^{-1}\boldsymbol{\Theta}^{\mathrm{T}}).$$

Following the method in Sect. 2.3, the agent can also obtain a *maximum a posterior* (MAP) estimation of $\boldsymbol{\Theta}$, and a *maximum likelihood estimation* (MLE) of $\boldsymbol{\Omega}$ by solving the following problem:

$$\min_{\boldsymbol{\Theta},\boldsymbol{\Omega}} \sum_{i=1}^{N} \left(r_t^i + \gamma\boldsymbol{\theta}_i^{\mathrm{T}}\boldsymbol{\phi}_t^i(\boldsymbol{s}_{t+1}, a_{t+1}^i) - \boldsymbol{\theta}_i^{\mathrm{T}}\boldsymbol{\phi}_t^i(\boldsymbol{s}_t, a_t^i)\right)^2 +$$
$$\frac{\lambda_1}{2}\mathrm{tr}(\boldsymbol{\Theta\Theta}^{\mathrm{T}}) + \frac{\lambda_2}{2}\mathrm{tr}(\boldsymbol{\Theta\Omega}^{-1}\boldsymbol{\Theta}^{\mathrm{T}}) \tag{9}$$
$$\mathrm{s.t.} \quad \boldsymbol{\Omega} \succeq \mathbf{0}$$
$$\mathrm{tr}(\boldsymbol{\Omega}) \leq 1.$$

This updating process is implemented in every learning step to estimate $\boldsymbol{\Theta}$ and $\boldsymbol{\Omega}$ alternatively.

When optimizing $\boldsymbol{\Theta}$ when $\boldsymbol{\Omega}$ is fixed, we can formulate the optimization problem as

$$G = \sum_{i=1}^{N} \left(r_t^i + \gamma \boldsymbol{\theta}_i^{\mathrm{T}} \boldsymbol{\phi}_t^i(\boldsymbol{s}_{t+1}, a_{t+1}^i) - \boldsymbol{\theta}_i^{\mathrm{T}} \boldsymbol{\phi}_t^i(\boldsymbol{s}_t, a_t^i) \right)^2$$
$$+ \lambda_1 \mathrm{tr}(\boldsymbol{\Theta}\boldsymbol{\Theta}^{\mathrm{T}}) + \lambda_2 \mathrm{tr}(\boldsymbol{\Theta}\boldsymbol{\Omega}^{-1}\boldsymbol{\Theta}^{\mathrm{T}}).$$

Thus the gradient of G with respect to $\boldsymbol{\Theta}$ is

$$\frac{\partial G}{\partial \boldsymbol{\Theta}} = -2 \sum_{i=1}^{N} \left(r_t^i + \gamma \boldsymbol{\theta}_i^{\mathrm{T}} \boldsymbol{\phi}_t^i(\boldsymbol{s}_{t+1}, a_{t+1}^i) - \boldsymbol{\theta}_i^{\mathrm{T}} \boldsymbol{\phi}_t^i(\boldsymbol{s}_t, a_t^i) \right) \cdot \tag{10}$$
$$\boldsymbol{\phi}_t^i \boldsymbol{e}_i^{\mathrm{T}} + 2\boldsymbol{\Theta}(\lambda_1 \boldsymbol{I}_N + \lambda_2 \boldsymbol{\Omega}^{-1}),$$

where each $\boldsymbol{e}_i$ is the ith column vector of $\boldsymbol{I}_N$.

Then we can optimize the $\boldsymbol{\Omega}$ when $\boldsymbol{\Theta}$ is fixed according to [16]:

$$\boldsymbol{\Omega} = \frac{(\boldsymbol{\Theta}^{\mathrm{T}}\boldsymbol{\Theta})^{\frac{1}{2}}}{\mathrm{tr}((\boldsymbol{\Theta}^{\mathrm{T}}\boldsymbol{\Theta})^{\frac{1}{2}})}. \tag{11}$$

Turning this into a control method by always updating the policy to be greedy with respect to the current estimate can be concluded as Algorithm 4.

Algorithm 4. Gradient TD Learning with Multi-action Relationship Regularization

---

Initialize   Feature function $\boldsymbol{\Phi}$
                  Step sizes $\boldsymbol{\alpha}$
                  Parameters $\boldsymbol{\Theta}, t = 0$
for each episode **do**
    $\delta = 0$
    Initialize $\boldsymbol{s}$
    repeat
        Take action $\boldsymbol{a}$ from $\boldsymbol{\theta}$ with exploration, observe $\boldsymbol{s}'$ and r_t
        set $\delta = (r_t + (\boldsymbol{\Phi}' - \boldsymbol{\Phi})^{\mathrm{T}}\boldsymbol{\Theta})\frac{\partial G}{\partial \boldsymbol{\Theta}}$ (Eq. 10)
        $\boldsymbol{\Theta} = \boldsymbol{\Theta} + \boldsymbol{\alpha}\delta$
        Update $\boldsymbol{\Omega}$ using Eq. 11
        $\boldsymbol{s} \leftarrow \boldsymbol{s}'$
    until $\boldsymbol{s}$ is terminal
end for

3.4   Discussions

In general, we set the initial value of $\boldsymbol{\Omega}$ to $\frac{1}{N}\boldsymbol{I}_N$, which is corresponding to the assumption that all sub-actions are unrelated initially. However, in some specific domains, there is some prior knowledge about the relationships between some sub-actions. When this happens, the corresponding $\boldsymbol{\Omega}$ can be represented as equality relations between elements in $\boldsymbol{\Omega}$.

4   Experiments

In this section, we study the multi-action relationship learning in several typical reinforcement learning domains and compare it with other multi-action algorithms which can also be used in multi-action reinforcement learning.

4.1   2-Action GridWorld

The GridWorld domain simulates a path-planning problem for a mobile robot in an environment with obstacles. The goal of the agent is to navigate from the starting point to the goal state. We decompose the action orthogonally into its horizontal and vertical components. In this way, each component represents a sub-action space consisting of three sub-actions: forward, backward, statically.

Thus, there are 2 sub-action spaces, each of them maintains 3 legal sub-actions. We compare the multi-action relationship learning algorithm with normal *LSPI*.

To evaluate the agent's performance, we report the total reward per episode averaged over the test set and the multi-action correlation matrix. The multi-action correlation matrix shows that the two sub-action spaces are correlated closely. On maps of 6-by-9, based on *LSPI*, we observe that learning with the multi-action relationship can converges faster than ignoring the relationship, as showed in Fig. 1(a).

4.2   PST

This domain concerns Persistent Search and Track mission with multiple unmanned aerial vehicle (UAV) agents. The goal is to perform surveillance and communicate it back to base in the presence of stochastic communication and health (overall system functionality) constraints, without losing any UAVs because of running out of fuel. Each UAV has 4 state dimensions: position of a UAV; integer fuel qty remaining; actuator status and sensor status. Each UAV can take one of 3 actions: retreat, loiter, advance.

Namely, this domain can be regarded as a centralized multi-agent system with 3 sub-action spaces (3 UAV agents). Based on *Gradient TD learning*, we evaluate the performance of the PST system by the total rewards that the agents can obtain.

To demonstrate the importance of our idea, the average reward in the first 200 thousand steps are showed in Fig. 1(b). The result shows that multi-action relationship learning is significantly better than ignoring the relationship, as the steps increases.

4.3   Helicopter

An implementation of a simulator that models one of the Stanford autonomous helicopters (an XCell Tempest helicopter) in the flight regime close to hover is

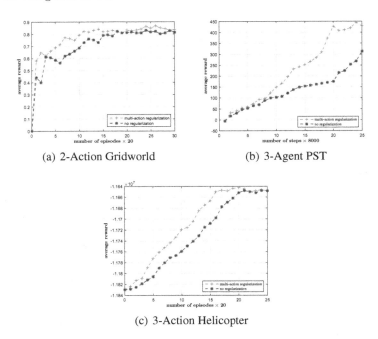

(a) 2-Action Gridworld   (b) 3-Agent PST

(c) 3-Action Helicopter

Fig. 1. Average rewards on 2-Action Gridworld, 3-Agent PST and 3-Action Helicopter.

represented in this domain. Some pilots consider hovering the most challenging aspect of helicopter flight. Generally, the pilot's use of control inputs in a hover is as follows: the cyclic is used to eliminate drift in the horizontal plane; the collective is used to maintain desired altitude; and the tail rotor (or anti-torque system) pedals are used to control nose direction or heading [1]. It is the interaction of these controls that can make learning to hover difficult, since often an adjustment in any one control requires the adjustment of the other two. We decompose the action orthogonally into its forward, sideways and downward components.

Analogically, this domain can be regarded as a centralized control system with 3 sub-action spaces when conducting *Gradient TD learning*. It can be observed that tracking the relationships between helicopter's 3 DOFs outperforms ignoring the relationship, as it converges fasters shown in Fig. 1(c).

To demonstrate the learning effect intuitively, the helicopter's attitude is recorded throughout the two learning processes. As Fig. 2 shows, the helicopter reached to a relatively stable position $(0, 0, 0)$ more quickly. Each point in the Fig. 2 denotes the helicopter's position during the experiment. Apparently, learning the multi-action relationship can improve the performance in the long run and with less fluctuation.

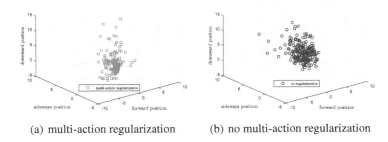

(a) multi-action regularization   (b) no multi-action regularization

Fig. 2. Flight positions on 3-Action Helicopter Hover problem.

4.4   On the Multi-action Relationship

During the experiments described above, we track the relationships matrix's changing by using the sum of absolute values of Ω elements. It can be observed from Fig. 3 that the multi-action correlations are learned along with the reinforcement learning. We can also note that, as long as the correlation is growing, the performance of multi-action methods are superior, by cross-comparison with Fig. 3(a), (b) and (c).

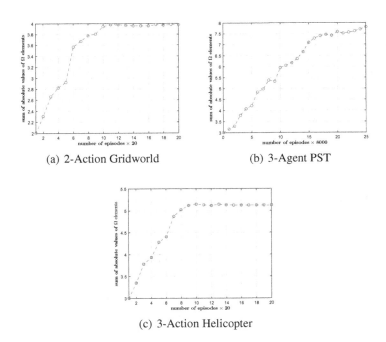

(a) 2-Action Gridworld   (b) 3-Agent PST

(c) 3-Action Helicopter

Fig. 3. Sum of absolute values of Ω elements.

5 Conclusion

In this paper, we explore multi-action relationship in learning multi-action reinforcement learning, where multiple actions are required to be controlled simultaneously. By employing the matrix-variate distribution, we enforce a regularization term capturing the multi-action relationship and obtain efficiently solvable convex learning objectives for value function approximation reinforcement learning algorithms. We incorporate the regularization term into the *LSPI* and the *Gradient TD methods*, which are empirically demonstrated to improve the learning performance. In the future, we will explore the multi-action relationship in more reinforcement learning approaches.

References

1. Abbeel, P., Ganapathi, V., Ng, A.Y.: Learning vehicular dynamics, with application to modeling helicopters. In: Weiss, Y., Schölkopf, B., Platt, J. (eds.) Advances in Neural Information Processing Systems 18, pp. 1–8. MIT Press, Cambridge (2005)
2. Cheng, G., Hyon, S.H., Morimoto, J., Ude, A., Hale, J.G., Colvin, G., Scroggin, W., Jacobsen, S.C.: Cb: a humanoid research platform for exploring neuroscience. In: Proceedings of the 6th IEEE-RAS International Conference on Humanoid Robots, pp. 182–187, Genova, Italy (2006)
3. Dorigo, M., Gambardella, L.M.: Ant colony system: a cooperative learning approach to the traveling salesman problem. IEEE Trans. Evol. Comput. **1**(1), 53–66 (1997)
4. Fumihide, T., Masayuki, Y.: Multitask reinforcement learning on the distribution of MDPs. In: Proceedings of the 2003 IEEE International Symposium on Computational Intelligence in Robotics and Automation, pp. 1108–1113. Kobe, Japan (2003)
5. Gupta, A.K., Nagar, D.K.: Matrix Variate Distributions. Chapman and Hall/CRC, Florida (1999)
6. van Hasselt, H.: Reinforcement learning in continuous state and action spaces (2012)
7. Lagoudakis, M.G., Parr, R.: Least-squares policy iteration. J. Mach. Learn. Res. **4**, 1107–1149 (2003)
8. Pazis, J., Lagoudakis, M.G.: Reinforcement learning in multidimensional continuous action spaces. In: Proceedings of the 2011 IEEE Symposium on Adaptive Dynamic Programming and Reinforcement Learning, Paris, France (2011)
9. Rohanimanesh, K.: Concurrent Decision Making in Markov Decision Processes. Ph.D. thesis, University of Massachusetts Amherst (2006)
10. Silver, D., Sutton, R.S., Müller, M.: Temporal-difference search in computer go. Mach. Learn. **87**(2), 183–219 (2012)
11. Sugiyama, M.: Statistical Reinforcement Learning: Modern Machine Learning Approaches. Chapman and Hall/CRC, Florida (2015)
12. Sutton, R.S., Barto, A.G.: Reinforcement Learning: An Introduction. The MIT Press, Cambridge (1998)
13. Wiering, M., van Otterlo, M. (eds.): Reinforcement Learning: State-of-the-Art. Springer, Berlin (2012)

14. Wunder, M., Littman, M.L., Babes, M.: Classes of multiagent q-learning dynamics with ε-greedy exploration. In: Proceedings of the 27th International Conference on Machine Learning, pp. 1167–1174, Haifa, Israel (2010)
15. Zhang, M.L., Zhou, Z.H.: A review on multi-label learning algorithms. IEEE Trans. Knowl. Data Eng. **26**(8), 1819–1837 (2014)
16. Zhang, Y., Yeung, D.Y.: A regularization approach to learning task relationships in multitask learning. ACM Trans. Knowl. Discovery Data **8**(3), 1–31 (2014)

A Differential Evolution Approach to Feature Selection and Instance Selection

Jiaheng Wang, Bing Xue[✉], Xiaoying Gao, and Mengjie Zhang

School of Engineering and Computer Science, Victoria University of Wellington,
PO Box 600, Wellington 6140, New Zealand
Bing.Xue@ecs.vuw.ac.nz

Abstract. More and more data is being collected due to constant improvements in storage hardware and data collection techniques. The incoming flow of data is so much that data mining techniques cannot keep up with. The data collected often has redundant or irrelevant features/instances that limit classification performance. Feature selection and instance selection are processes that help reduce this problem by eliminating useless data. This paper develops a set of algorithms using Differential Evolution to achieve feature selection, instance selection, and combined feature and instance selection. The reduction of the data, the classification accuracy and the training time are compared with the original data and existing algorithms. Experiments on ten datasets of varying difficulty show that the newly developed algorithms can successfully reduce the size of the data, and maintain or increase the classification performance in most cases. In addition, the computational time is also substantially reduced. This work is the first time for systematically investigating a series of algorithms on feature and/or instance selection in classification and the findings show that instance selection is a much harder task to solve than feature selection, but with effective methods, it can significantly reduce the size of the data and provide great benefit.

Keywords: Differential evolution · Feature selection · Instance selection · Classification

1 Introduction

As hardware technology improves, more and more data is collected at a rate machine learning and data mining techniques cannot deal with. Often the data collected contains redundant or irrelevant features and instances [7,9,14,22,25], which may slow down and hindering the learning process in many tasks such as classification, reduce the learning performance, and/or learn complex models. A pre-processing step is often needed to remove some of the irrelevant or even noisy data, which can be achieved by *feature selection* (FS) for selecting only a small subset of informative features, *instance selection* (IS) for selecting only a small subset of representative examples/instances, or *FS and IS* for removing useless or redundant features and instances [13,17]. However, FS and/or IS is a

© Springer International Publishing Switzerland 2016
R. Booth and M.-L. Zhang (Eds.): PRICAI 2016, LNAI 9810, pp. 588–602, 2016.
DOI: 10.1007/978-3-319-42911-3_49

challenging problem due to two main reasons. The first is the large search space, which grows exponentially with the total number of features and instances. The second is that there are almost always interactions between features, which leads to a complex search space with many local optima and a good fitness function is often needed to guide the search in order to find a good solution. There have been a large number of works on FS, but not much work on IS, or FS and IS [13].

Different search techniques have been used for FS, but existing algorithms still suffer from the problem of stagnation in local optima. Evolutionary computation techniques are capable of searching large dimensions for solutions. Previous work has shown that various evolutionary computation techniques, such as differential evolution (DE) [15], particle swarm optimization [1,11,19], genetic algorithms [18,26] and others [2,8], achieve better performances than traditional FS and IS approaches [20]. This research will be utilizing the DE approach. DE is a simple but effective approach, which has been used for solving a wide range of complex problems, especially the ones with a large search space [24]. Recent works [3,21] also show its capabilities in solving FS problems, but its potential on IS has not been fully investigated.

Based on the evaluation criteria or fitness functions, feature and/or IS approaches can be grouped into wrapper approaches and filter approaches [23], where wrappers involves training a learning/classification algorithm in each evaluation to use the accuracy to show how good the candidate solution is, and filters are independent from any learning/classification algorithm. Due to the direct link between the learning algorithm and the candidate solution, wrappers can often achieve better accuracy than filters, but are computationally expensive. Filters are often very fast, but may not achieve as high accuracy as wrappers [7].

DE has only been used for wrapper FS recently [4,5,12,21]. Compared with the popularity and promising performance achieved by DE in other areas [6], the potential of DE has not been fully investigated. Although most machine learning tasks require FS and/or IS, classification is the area with the most applications, which could be a good starting point for the investigation.

Goals. The aim of this research is to investigate the use of DE for data preprocessing, which includes FS only, IS only, and FS and IS together. The proposed methods are expected to reduce the size of the data and increase or at least without significantly reducing the classification accuracy. More specifically, the overall goal is broken down into the following objectives:

1. develop a new DE based FS algorithm for selecting a subset of features to reduce the dimensionality and maintain or even increase the classification performance,
2. develop a new DE based IS algorithm for selecting a small subset of representative instances to reduce the size of the data without significantly reducing the classification accuracy,
3. develop a new DE based FS and IS algorithm to achieve FS and IS simultaneously, and

4. investigate the performance increase of the new algorithms compared with existing techniques.

2   Proposed Algorithms

In this section, we will investigate the use of DE for FS, IS, as well as FS and IS. Since feature and/or IS are binary tasks, i.e. either select or not, but DE was originally proposed as a continuous search technique, a binary DE algorithm will be needed. Different from most existing approaches using classification accuracy to evaluate the fitness (i.e. wrapper approaches), we will develop a series of filter algorithms based on the interclass and intraclass (IIC) measures to evaluate each candidate solutions.

2.1   Binary Differential Evolution

In DE candidate solutions are represented by vectors, with various operators being performed on them at each generation. The operators can range from mathematical functions such as addition, subtraction, or multiplication, to genetic operators such as crossover and mutation. There are various versions of DE in the literature [16]. One of the most promising one is DE/best/1, which is used in this work. A DE/best/1 iteration is defined as such

$$v_{i,G+1} = x_{best,G} + F.(x_{r_1,G} - x_{r_2,G}) \tag{1}$$

where i indicates i-th solution in the population, G and $G+1$ indicates the current and next generations. F is a scale factor controlling the size of the particle's movement, and $x_{r_1,G}$ and $x_{r_2,G}$ are other random candidate solutions chosen from the population such that $x_{best,G} \neq x_{r_1,G} \neq x_{r_2,G}$. $x_{best,G}$ is the current global best solution, which is a main feature of DE/best/1 that separates it from other implementations such as DE/rand/1. As seen in the equations, the current global best solution is a main factor, or the bases, of all new solutions. In DE/rand/1, three solutions are chosen at random to generate the new solution.

An initial population of candidate solutions are randomly generated. In each generation, a tentative new candidate solution, $v_{i,G+1}$, is generated with the equation above for each solution i of the population. $x_{i,G+1}$ is updated to $v_{i,G}$ at the $G+1$ generation if the fitness $v_{i,G+1}$ is better than $x_{i,G}$, i.e. an improvement on the old solution. Otherwise $x_{i,G+1}$ is the same as $x_{i,G}$.

Due to similarities between DE and PSO, previous work on binary PSO [10] can be used here. A conversion must be made from the continuous vector representation of the candidate solutions to the binary solutions required for the selection problems. The conversion is given by:

$$output_{i,d} = \begin{cases} 1, \text{ if } rand() < \frac{1}{1+e^{-x_{i,d}}} \\ 0 \text{ otherwise} \end{cases} \tag{2}$$

where $output$ is the d-th bit of the i-th solution, $rand()$ is a random number between 0 and 1, and $x_{i,d}$ is the d-th value of the i-th vector of the candidate

---

Algorithm 1. Interclass Distance

1: **for all** classes **do**
2:     find all instances belong to this class
3:     construct mean instance from all instances belong to this class
4:     find and store a representative set of instances (*ReS*s) of this class
5: **end for**
6: distance := 0
7: **for all** *ReS*s **do**
8:     classDistance := 0
9:     links := 0
10:    **for all** other *ReS*s **do**
11:       **for all** Instance $i1$ in *ReS* **do**
12:          **for all** Instance $i2$ in other *ReS* **do**
13:             classDistance += distanceBetween($i1,i2$)
14:             links += 1
15:          **end for**
16:       **end for**
17:    **end for**
18:    distance += classDistance / links
19: **end for**
20: interclass distance := distance / number of classes

---

solution, normalized by the sigmoid function. The values in *output* determine the selection of features or instances.

2.2   Fitness Function

The fitness function is one of the key components in the proposed algorithms, which is based on IIC measures. The IIC measures can be broken down to two parts, which are the interclass distance and the intraclass spread.

Interclass Distance. The interclass distance is a measure of the separability of classes in a dataset. The larger the distance, the further apart and separated the classes are. Therefore, a big distance means that the classes are more distinguished and there is less overlap between classes, which is expected to have a better classification performance. To achieve the goal of performing classification, we propose to build a prototype, which is a mean instance, or centroid, for each class based on a set of representative instances. The reasons for not using all instances here are to avoid outliers and long computational cost.

Algorithm 1 shows how the interclass distance is calculated. For each class, a mean instance, or centroid, is constructed. The mean instance is a feature vector that each value is the mean of all instances belonging to that class. A representative set of instances are the ones nearest to the constructed mean instance, which is found to represent the class. Each representative set has a size of 10 % of the total number of instances belonging to that class plus one, i.e.

Algorithm 2. Intraclass Spread

1: {Part 1: Calculating spread of each feature}
2: **for all** features **do**
3:     featureSpread := 0
4:     **for all** class **do**
5:         featureSpread += $\sigma_{f,c}$
6:     **end for**
7: **end for**
8: {Part 2: Calculating overall spread}
9: intraclass spread := 0
10: **for all** featureSpread **do**
11:     intraclass spread += featureSpread
12: **end for**

---

the mean instance. The representative sets are used to calculate the Euclidean distance between classes, which can be seen from Lines 1–5 in Algorithm 1. Then, the Euclidean distance between the representative sets are found. The average distance between two classes is defined as the average distance of each instance in one representative set to each instance in the other representative set. The average is taken here due to the different numbers of instances belonging to each class. In Lines 10–17, the average distance between two classes is calculated for between every class. Then the average of the averages between every two classes is used as the distance between all classes as shown in Line 20.

The following equations provides a mathematical form of this calculation.

$$Distance = \frac{\sum_{C_a,C_b \in C}^{a \neq b} \frac{\sum_{i \in C_a} \sum_{j \in C_b} |i-j|}{|Ca| \times |Cb|}}{|C|}, \tag{3}$$

where C is the set of all representative sets, C_a, C_b are any two different representative sets, and i, j are individual instances.

Intraclass Spread. The intraclass spread is a measure of how spread out a particular class is. The further spread a class is, the more likely it is to overlap with other classes, providing a more cohesive representation of the class. Therefore a smaller spread is preferred. The spread of a class is given by the spread of its features, particularly by all the feature values of the instances in each class. This allows easier calculation, but does not change the total spread of a given dataset due to the associative properties of addition.

The spread of a particular feature is given by the sum of the standard deviation of each class's set of values for that feature. The spread of the set of features is given by the sum of each feature's spread.

$$Spread = \sum_{f \in F'} \sigma_f, \text{ where } \sigma_f = \sum_{c \in C} \sigma_{f,c} \tag{4}$$

where F' is a set of features, c is an instance belonging to the class C, and $\sigma_{f,c}$ is the standard deviation of the values representing feature f in class c.

Fitness Function. To achieve good classification performance, ideally, the intraclass spread should be minimized and the interclass distance should be maximized. Therefore, a (minimization) fitness function is formed and shown by Eq. 5.

$$Fitness = \frac{Spread + \alpha.|F|}{Distance} \tag{5}$$

where $|F|$ is the number of features, and α is a coefficient. The constant $\alpha.|F|$ is added to the spread in the numerator to control the weight ratio between the spread and the distance. A smaller constant would give more weight to the spread, and a larger constant gives more weight to the distance. This also means that the number of features selected in FS can be controlled, as the number of features directly affect the spread and distance, i.e. intraclass spread wants fewer selected features, whereas interclass distance wants more features. Therefore, by adjusting the weights of spread and distance, the number of features can be adjusted.

2.3   New Algorithms

We will investigate the use of DE for FS, IS, as well as FS and IS together. Since the fitness function, Eq. 5, eventually shows how well different classes can be separated, it is used in all the three algorithms, to form IIC-FS, IIC-IS, and IIC-FIS, for FS, IS, and FS and IS, respectively.

The goal of the three algorithms are the same, i.e. minimizing the fitness value. They all follow the basic DE process. The key difference between them is the representation since the candidate solutions are different, i.e. a subset of features, a subset of instances, and a subset of instances with selected features only for IIC-FS, IIC-IS, and IIC-FIS, respectively. In IIC-FS, the representation of each individual in DE is a m-dimensional boolean vector for a dataset with m features, where each dimension determines whether the corresponding feature is selected. 1 means the feature is selected and 0 otherwise. In IIC-IS, the representation is a n-dimensional boolean vector for a dataset with n instances, where each dimension determines whether the corresponding instance is selected. In IIC-FIS, the representation is a $(n + m)$-dimensional boolean vector, where each dimension determines whether the corresponding feature or instance is selected.

In IIC-FS, as the instances do not change, each feature has a particular intraclass spread value associated with it that also does not change. These values only need to be calculated once. Training times are improved since each feature's spread is stored in memory and is simply read for each fitness evaluation, as opposed to recalculating each value every time it is needed. Therefore the first part of Algorithm 2 is only performed once at the beginning. Further evaluations only need to perform the second part. The same cannot be achieved for interclass distances, as changing the dimensions (features) of instances also changes their relative distances. Therefore Algorithm 1 is performed in full for every fitness evaluation for FS. In IIC-IS and IIC-FIS, due to the changing instances, and thus both the spread and distances of the data, both algorithms' calculations are performed in full for every fitness evaluation.

In addition, since DE has never been used for IS, and FS and IS, we investigate two wrapper based methods using KNN as the classification algorithm to evaluate the classification performance as the fitness function for IS only (KNN-FS), and for FS and IS (KNN-FIS). Both KNN-FS and KNN-FIS are also new to some extent.

3   Experiment Design

The proposed algorithms are run against 10 datasets taken from the UCI machine learning repository shown in Table 1. These datasets are selected to represent a range of feature and instance counts, as well as being widely used datasets such that the new algorithms can be compared against existing ones. Data is normalized as they are loaded, ensuring that distance and standard deviation measures are on the same scale for all features.

Table 1. Experiment datasets

Dataset	NO. of features	NO. of instances	NO. of classes	α
Wine	13	178	3	0.4
Australian	14	690	2	2
Zoo	17	101	7	0.65
Vehicle	18	846	4	0.38
German	24	1000	2	0.16
Wbcd	30	569	2	0.27
Ionosphere	34	351	2	0.2
Lung	56	32	2	0.41
Sonar	60	208	2	0.2
Movementlibras	90	360	15	1.2

For each selection process, 30 runs are conducted for each dataset. The DE has a population of 80 candidate solutions, and is run for 100 generations. Since an optimal solution cannot be easily determined and classification rate is not part of the training process, there is no early stopping criteria. The data is resplit every 10 runs for a total of 3 different splits per dataset. The split is done randomly, with each instance having a 70 % chance of being used for training, and 30 % chance of being used for testing.

In ICC-FS, IIC-IS and IIC-FIS, a search was conducted before the experiments for α. The coefficient values for α in Table 1 were found to give a similar number of features to KNNFS and were used for the experiments. IIC-FIS has two specific implementations. The first one, marked with "200", is run with 200 candidate solutions of DE instead of 80. This is to accommodate for the larger search space due to the dimension size being the sum of number of features and instances. The second, marked with "ICC-Half", uses a modified KNN for classification after using IIC-FS to reduce the features. This modified KNN only uses half the instances. For each class, the centroid, or mean instance, is calculated from every instance of that class in the training set. Then half the instances of that class, the half closest to the centroid, are used in the KNN for classification. Although only the features are selected in the training process, this modified KNN selects instances, putting it under FS and IS. In KNN-FS, KNN-IS, and KNN-FIS, the average classification accuracy of a 10-fold validation on the training set is used as the fitness value, where 10-fold validation is used to make sure that no FS bias is involved and the test set is completely unseen for the FS methods.

After the DE generations, the solutions with the best fitness are evaluated for its classification accuracy on the test set, where KNN (K = 5) is used as the classifier. A non-parametric test, the Mann?-Whitney U test, is then used to compare the testing accuracy and number of features/instances selected by the IIC measure against using all features, as well as the standard KNN technique.

4   Results and Discussions

Tables 2, 3, and 4 show the results of the three sets of experiments. Table 2 shows the results of the FS using KNN-FS, and IIC-FS. The first two columns show the dataset name and the methods. The third column shows the average and standard deviation of the number of selected features. The fourth column shows the average, standard deviation, and best accuracy on the test sets. The column "Test 1" shows the statistical significance tests between the method in the corresponding row against All, where "○", "⋆", and "=" means the corresponding method is significantly better than, worse than, and similar to that of All, respectively. The column "Test 2" shows the same information against KNN-FS. Note that "better" means larger for accuracy, but means smaller for the number of features. The last column shows the average training time for a single run, where the number is shown in seconds. Table 3 shows the results of IS, and Table 4 shows the results of FS and IS together, where the meanings of symbols are the same as in Table 2.

4.1   Results of Feature Selection

According to Table 2, it can be seen that comparing IIC-FS with All, the number of features is reduced to around one third of the total number of features. With the reduced feature subsets, IIC-FS achieved better or at least similar classification accuracy than using all the original features on nine out of the ten datasets. The results show that proposed IIC-FS can be successfully used for FS to evolve a small number of features, which can maintain or even increase the classification performance.

Table 2. Experimental results for feature selection

Dataset	Method	NO. of Features	Accuracy		Test 1		Test 2		Average
			Mean (Std)	Best	Acc	Size	Acc	Size	Time
Wine	All	13	0.948 (0.03)	0.979					
	KNN-FS	6.4 (1.13)	0.936 (0.05)	0.98	=	○			566.57
	IIC-FS	5.3 (0.47)	0.959 (0.02)	0.981	=	○	=	○	5.47
Aus.	All	14	0.859 (0.01)	0.867					
	KNN-FS	5.47 (0.94)	0.867 (0.02)	0.903	=	○			8652.8
	IIC-FS	2.9 (0.66)	0.858 (0.01)	0.862	=	○	=	○	95.67
Zoo	All	17	0.909 (0.05)	0.946					
	KNN-FS	9 (1.58)	0.938 (0.04)	1	=	○			264.7
	IIC-FS	8.33 (1.42)	0.904 (0.03)	0.968	=	○	★	=	2.27
Vehicle	All	18	0.667 -0	0.667					
	KNN-FS	8.6 (1.54)	0.693 (0.03)	0.751	○	○			20937.43
	IIC-FS	8.63 (1.22)	0.645 (0.04)	0.719	★	○	★	=	125.27
German	All	24	0.697 (0.01)	0.709					
	KNN-FS	10.4 (1.92)	0.715 (0.03)	0.766	○	○			43018.27
	IIC-FS	8.67 (1.99)	0.718 (0.02)	0.759	○	○	=	○	413.1
WBCD	All	30	0.959 (0.01)	0.969					
	KNN-FS	13.2 (2.16)	0.956 (0.02)	0.982	=	○			18815.8
	IIC-FS	14.13 (1.93)	0.952 (0.01)	0.982	=	○	=	=	189
Ionos.	All	34	0.839 (0.01)	0.843					
	KNN-FS	9.8 (2.50)	0.876 (0.03)	0.933	○	○			8652.6
	IIC-FS	10.57 (2.03)	0.852 (0.02)	0.899	○	○	★	=	87.17
Lung	All	56	0.747 (0.04)	0.8					
	KNN-FS	23 (4.34)	0.707 (0.09)	0.923	=	○			129.53
	IIC-FS	24.27 (2.88)	0.719 (0.08)	0.846	=	○	=	=	2.7
Sonar	All	60	0.809 (0.06)	0.895					
	KNN-FS	26.07 (2.80)	0.792 (0.05)	0.895	=	○			9331.53
	IIC-FS	26.33 (3.46)	0.798 (0.06)	0.912	=	○	=	=	148
Movement libras	All	90	0.707 (0.03)	0.745					
	KNN-FS	38.9 (6.15)	0.699 (0.04)	0.764	=	○			51255.67
	IIC-FS	39.37 (4.23)	0.682 (0.05)	0.764	=	○	=	=	154.37

Comparing IIC-FS with KNN-FS, the number of features and the classification performance are similar in most of the cases, with three cases of IIC-FS selecting a smaller feature subsets and KNN-FS achieving better classification accuracy. KNN-FS is expected to achieve better accuracy since it is a wrapper approach while IIC-FS is a filter approach.

In terms of the training time, there is a huge difference between IIC-FS with KNN-FS, where IIC-FS always used a substantial shorter time (48 to 167 times faster) than KNN-FS, with the Vehicle dataset having the biggest difference.

In summary, the proposed ICC-FS methods can be successfully used for FS. As a filter approach, IIC-FS is able to achieve similar FS performance to the wrapper method, KNN-FS, but the computational time is much shorter.

Table 3. Experimental results for instance selection

Dataset	Method	NO. of instances	Accuracy		Test 1		Test 2		Average
			Mean(Std)	Best	Acc	Size	Acc	Size	Time
Wine	All	128 (3.61)	0.948 (0.031)	0.979					
	KNN-IS	50.9 (5.82)	0.943 (0.031)	1	=	○			255.33
	IIC-IS	47.57 (6.02)	0.943 (0.02)	0.98	=	○	=	=	3.2
Australian	All	487.67 (10.97)	0.859 (0.009)	0.867					
	KNN-IS	209.47 (20)	0.862 (0.015)	0.888	=	○			4327.37
	IIC-IS	209.9 (20.52)	0.862 (0.021)	0.898	=	○	=	=	40.17
Zoo	All	69.67 (5.51)	0.909 (0.046)	0.946					
	KNN-IS	31.83 (6.79)	0.856 (0.043)	0.968	⋆	○			120.2
	IIC-IS	26.13 (4.03)	0.817 (0.079)	0.968	⋆	○	=	○	0.03
Vehicle	All	596 (10.54)	0.667 (0)	0.667					
	KNN-IS	277 (25.63)	0.62 (0.032)	0.699	⋆	○			11258.87
	IIC-IS	255.13 (22.4)	0.629 (0.037)	0.707	⋆	○	=	○	35.07
German	All	704 (14.53)	0.697 (0.009)	0.709					
	KNN-IS	311.77 (25.42)	0.711 (0.019)	0.756	○	○			26422.23
	IIC-IS	300.47 (25.19)	0.7 (0.022)	0.745	=	○	=	=	275.43
WBCD	All	400 (10.39)	0.959 (0.011)	0.969					
	KNN-IS	185.2 (20.76)	0.957 (0.013)	0.975	=	○			11477.57
	IIC-IS	168.67 (14.59)	0.946 (0.013)	0.969	⋆	○	⋆	○	149.47
Ionosphere	All	241.33 (8.62)	0.839 (0.005)	0.843					
	KNN-IS	104.5 (13.01)	0.86 (0.017)	0.892	○	○			5608.13
	IIC-IS	99.77 (8.6)	0.707 (0.057)	0.866	⋆	○	⋆	=	59.07
Lung	All	21.67 (2.52)	0.747 (0.045)	0.8					
	KNN-IS	6.77 (2.39)	0.647 (0.038)	0.7	⋆	○			32.77
	IIC-IS	2.5 (0.9)	0.693 (0.124)	0.9	=	○	○	○	1.63
Sonar	All	147.67 (3.51)	0.809 (0.063)	0.895					
	KNN-IS	63.4 (9.05)	0.678 (0.054)	0.817	⋆	○			4899.07
	IIC-IS	57.23 (6.88)	0.668 (0.058)	0.767	⋆	○	=	○	68.33
Movement libras	All	247 (9.85)	0.707 (0.033)	0.745					
	KNN-IS	122.7 (9.62)	0.511 (0.057)	0.6	⋆	○			36993.93
	IIC-IS	87.37 (8.68)	0.42 (0.045)	0.482	⋆	○	⋆	○	143.6

4.2 Results of Instance Selection

Table 3 shows the results of IS, where both KNN-IS and IIC-FS are newly investigated in this paper. The results show that both KNN-IS and IIC-IS selected only a much smaller number of instances compared with the total number of instances on all the datasets. Although the number of instances to be selected by IIC-IS was not controlled, the number of instances selected by IIC-FS is significantly smaller than that of KNN-IS on six out of the ten datasets, and similar on the other four datasets. Compared to using all instances, both KNN-IS and IIC-IS performed significantly better or similar in around half of the cases, but in general the difference is not too big, and the best accuracy of KNN-IS and IIC-IS is often better than using all instances. This is different from the good performance of their corresponding FS methods, as shown in Table 2. This is not too surprised given that IS could change the original pattern and distribution of the data, which is probably why there has been much more work on FS than IS, although IS can benefit classification in many ways as FS. We will further investigate effective IS methods in the future.

Regarding the training times, both KNN-IS and IIC-IS have a faster training time than their respective FS counterparts as shown in Table 2. This is due to the highly reduced number of instances in each fitness evaluation, resulting in fewer calculations of the distances between instances. Once again the IIC technique is much faster than the KNN technique. The speed increase ranges from 20–321 times faster. The Zoo dataset is a special case, most of the training times where recorded as 0 (seconds) since the entire training process took less than one second. This results in an extremely low average training time, which was 4000 times faster than KNN-IS.

In summary, the two IS methods cannot in most cases maintain or increase the classification performance, although it can substantially reduce the size of the data. The speed of the algorithms is very fast, much faster than the FS methods. How to maintain the speed and simultaneously increase the classification performance is an interesting direction for future work.

4.3   Results of Feature and Instance Selection

Table 4 shows the results for the FS and IS experiments, where "200" is used to represent the implementation of IIC-FIS with 200 generations, and "IIC-Half" is used to represent the version of IIC-FIS with the KNN implementation. All the three methods on this set of experiments are new in this work.

The results from Table 4 show that both the number of features and the number of instances have been significantly reduced, but the price is the lower classification performance, especially on the large datasets. IIC-200 selected significantly more features than KNN-FIS on every dataset, and they are similar in the number of instances on most datasets. IIC-Half has a similar number of features selected on seven of the ten datasets as KNN-FIS, and a similar classification performance. Compared to using all feature and instances, KNN-FIS achieves a better classification accuracy on three datasets, and worse on six. Both IIC-200 and IIC-Half achieves similar results on three (Wine and German for both, then Australian for 200 and Zoo for IIC-Half) datasets. Neither achieves a significantly better result than using all features and instances.

For the training time, the IIC methods have a much faster time than the KNN based method. Although the improvement here is not as high as in FS and IS, with the range of reduction at 4–80 times faster.

4.4   Analysis on the Computational Time

The ICC methods are orders of magnitude faster than KNN-FIS in terms of the training time while still achieving similar results. According to Tables 2, 3 and 4, the average training speed is roughly 4–400 times faster (on average 120 times faster) using IIC than the KNN technique. This is due to the number of distance

Table 4. Experimental results for feature and instance selection

Dataset	Method	Features used	Instances used	Accuracy mean(Std)	Test 1 Acc	Ins	Feas	Test 2 Acc	Ins	Feas	Average Time
Wine	All	13	128(3.61)	0.948(0.03)							
	KNN-FIS	7.17(2.13)	52(5.62)	0.935(0.04)	=	○	○				84.9
	IIC-200	10.93(1.17)	54.93(7.98)	0.94(0.04)	=	○	○	=	⋆	=	5
	IIC-Half	6.97(1.13)	66.67(1.27)	0.941(0.03)	=	○	○	=	=	⋆	4.07
Aus.	All	14	487.67(10.97)	0.859(0.01)							
	KNN-FIS	7.03(1.77)	207.57(21.50)	0.866(0.02)	○	○	○				1563.4
	IIC-200	11.1(1.18)	223.17(23.03)	0.858(0.02)	=	○	○	=	⋆	⋆	63.57
	IIC-Half	4.7(1.29)	245(4.32)	0.813(0.03)	⋆	○	○	⋆	○	⋆	86.23
Zoo	All	17	69.67(5.51)	0.909(0.05)							
	KNN-FIS	8.43(1.85)	31.17(6.18)	0.835(0.04)	⋆	○	○				38.1
	IIC-200	12.57(1.74)	30.77(4.16)	0.836(0.08)	⋆	○	○	=	⋆	=	1.13
	IIC-Half	8(1.26)	40.67(2.54)	0.916(0.04)	=	○	○	○	=	⋆	1.33
Veh.	All	18	596(10.54)	0.667 -0							
	KNN-FIS	9.5(1.85)	283.03(19.67)	0.652(0.04)	⋆	○	○				4082
	IIC-200	15(1.84)	287.9(24.69)	0.627(0.03)	⋆	○	○	⋆	⋆	=	78.2
	IIC-Half	9.57(1.14)	301.33(4.18)	0.572(0.03)	⋆	○	○	⋆	=	⋆	104.17
Germ.	All	24	704(14.53)	0.697(0.01)							
	KNN-FIS	9.97(2.24)	316.5(28.43)	0.712(0.03)	○	○	○				7683.53
	IIC-200	15.93(2.08)	326.63(21.46)	0.696(0.02)	=	○	○	⋆	⋆	=	273.7
	IIC-Half	10.47(2.21)	353.67(6.23)	0.687(0.04)	=	○	○	⋆	=	⋆	397.47
WBCD	All	30	400(10.39)	0.959(0.01)							
	KNN-FIS	14.23(2.99)	186.13(17.47)	0.947(0.02)	⋆	○	○				4215.1
	IIC-200	21.13(2.87)	185.33(16.04)	0.95(0.01)	⋆	○	○	=	⋆	=	178.67
	IIC-Half	14.83(2.52)	201.33(4.57)	0.941(0.01)	⋆	○	○	=	=	⋆	215.87
Ionos.	All	34	241.33(8.62)	0.839(0.01)							
	KNN-FIS	14.53(2.96)	106.97(12.08)	0.864(0.03)	○	○	○				1633.83
	IIC-200	19.63(2.24)	105.33(11.30)	0.763(0.04)	⋆	○	○	⋆	⋆	=	61.27
	IIC-Half	11.93(2.42)	122.33(3.36)	0.708(0.04)	⋆	○	○	⋆	○	⋆	116.47
Lung	All	56	21.67(2.52)	0.747(0.04)							
	KNN-FIS	23.87(3.66)	6.23(2.06)	0.647(0.04)	⋆	○	○				10.37
	IIC-200	27.73(4.70)	2.43(1.07)	0.704(0.12)	⋆	○	○	○	⋆	○	1.87
	IIC-Half	24.6(3.63)	12.67(1.27)	0.715(0.09)	⋆	○	○	○	=	⋆	2.27
Sonar	All	60	147.67(3.51)	0.809(0.06)							
	KNN-FIS	26.57(4.32)	64.07(9.66)	0.722(0.05)	⋆	○	○				1648
	IIC-200	30.53(4.07)	61.43(6.37)	0.679(0.05)	⋆	○	○	⋆	⋆	=	62.03
	IIC-Half	26.3(4.46)	75.67(1.27)	0.707(0.04)	⋆	○	○	=	=	⋆	86
Move. libras	All	90	247(9.85)	0.707(0.03)							
	KNN-FIS	43.77(5.77)	125.67(9.83)	0.497(0.06)	⋆	○	○				11808.07
	IIC-200	54.17(5.11)	108.67(9.77)	0.46(0.05)	⋆	○	○	⋆	⋆	○	190.83
	IIC-Half	40.3(4.02)	135.33(4.18)	0.553(0.04)	⋆	○	○	○	○	⋆	142.63

calculations between instances, a costly operation, is much lower in IIC than KNN.

Assuming instances are equally distributed between classes, for each fitness evaluation the number of calculations between values in IIC can be roughly calculated by the following equation:

$$n + 0.01 \left(\frac{n}{c}\right)^2 \cdot \frac{c(c-1)}{2}, \tag{6}$$

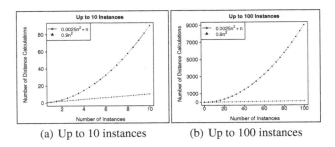

(a) Up to 10 instances   (b) Up to 100 instances

Fig. 1. Number of distance calculations in KNN and IIC

where $c \in [1, n]$ is the number of classes and n is the number of instances. The initial n at the beginning is to find the distance from each instance to its mean instance to identify the representative set. The second part is the distance calculations between representative set. A fully connected graph (where each vertex is a class representative set) has $\frac{c(c-1)}{2}$ edges, with each edge consisting of $0.1\frac{n}{c} \times 0.1\frac{n}{c}$ calculations. This is largest at $c = 2$ for all $n > 2$, giving $n + 0.0025n^2$ calculations.

For the KNN based techniques, each evaluation requires $10 \times 0.9n \times 0.1n$ ($0.9n^2$) distance calculations. This is from the 10-fold cross validation, where in each fold 10 % of instances are compared with the other 90 %. So the rough number of calculations is

$$n + 0.0025n^2 > 0.9n^2 \text{ for all } n > 1, \tag{7}$$

Figure 1 shows the number of distance calculations in KNN and IIC when the number of instances is 10 and 100. One can see that even at only 10 instances, KNN has greatly separated from IIC. When there is 100 instances, IIC is negligible compared to KNN. Note that are more than 100 instances in the training set in all but two datasets.

5   Conclusions and Future Work

The goal of this paper was to investigate the use of DE for feature and/or IS in classification, which a new binary DE algorithm and a new fitness function. The experiments and comparisons on ten datasets show that the proposed DE based FS algorithm is successful in terms of the number of features, the classification accuracy and the training time. However, when the IS task involved, the algorithms are good at reducing the size of the data, but the classification accuracy may suffer, which is a critical problem. The reason for this is due to the large search space, which is also probably why there has been much more work on FS than IS.

This paper investigates a series of different feature and/or IS methods, which have not been done before. Although it is only a preliminary work, the findings are very useful, especially when both feature selection and instance selection are becoming increasingly important for big data tasks. There is still a lot of work should be done in this filed. For example, a novel representation of solutions is needed, which can effectively reduce the search space and also form a more smooth landscape to be more easily searched. A computationally cheap fitness measure is also of key component, especially on datasets with a large number of features and instances. We will focus on these directions in the future.

References

1. Ahmad, S.S.S.: Feature and instances selection for nearest neighbor classification via cooperative PSO. In: 2014 Fourth World Congress on Information and Communication Technologies (WICT), pp. 45–50. IEEE (2014)
2. Ahmed, S., Zhang, M., Peng, L., Xue, B.: Multiple feature construction for effective biomarker identification and classification using genetic programming. In: Proceedings of the 2014 Annual Conference on Genetic and Evolutionary Computation (GECCO), pp. 249–256. ACM (2014)
3. Al-Ani, A., Alsukker, A., Khushaba, R.N.: Feature subset selection using differential evolution and a wheel based search strategy. Swarm Evol. Comput. **9**, 15–26 (2013)
4. Bharathi, P.T., Subashini, P.: Differential evolution and genetic algorithm based feature subset selection for recognition of river ice type. J. Theor. Appl. Inf. Technology **7**(1), 254–262 (2014)
5. Bharathi, P.T., Subashini, P.: Optimal feature subset selection using differential evolution and extreme learning machine. Int. J. Sci. Res. (IJSR) **3**, 1898–1905 (2014)
6. Das, S., Suganthan, P.N.: Differential evolution: a survey of the state-of-the-art. IEEE Trans. Evol. Comput. **15**(1), 4–31 (2011)
7. Guyon, I., Elisseeff, A.: An introduction to variable and feature selection. J. Mach. Learn. Res. **3**, 1157–1182 (2003)
8. Hancer, E., Xue, B., Karaboga, D., Zhang, M.: A binary ABC algorithm based on advanced similarity scheme for feature selection. Appl. Soft Comput. **36**, 334–348 (2015)
9. John, G.H., Kohavi, R., Pfleger, K., et al.: Irrelevant features and the subset selection problem. In: Machine Learning: Proceedings of the Eleventh International Conference, pp. 121–129 (1994)
10. Kennedy, J., Eberhart, R.C.: A discrete binary version of the particle swarm algorithm. In: IEEE International Conference on Systems, Man, and Cybernetics. Computational Cybernetics and Simulation, vol. 5, pp. 4104–4108 (1997)
11. Lane, M.C., Xue, B., Liu, I., Zhang, M.: Gaussian based particle swarm optimisation and statistical clustering for feature selection. In: Blum, C., Ochoa, G. (eds.) EvoCOP 2014. LNCS, vol. 8600, pp. 133–144. Springer, Heidelberg (2014)
12. Li, Z., Shang, Z., Qu, B., Liang, J.: Feature selection based on manifold-learning with dynamic constraint handling differential evolution. In: IEEE Congress on Evolutionary Computation (CEC), pp. 332–337 (2014)
13. Liu, H., Motoda, H.: Instance Selection and Construction for Data Mining, vol. 608. Springer Science & Business Media, US (2013)
14. Liu, H., Yu, L.: Toward integrating feature selection algorithms for classification and clustering. IEEE Trans. Knowl. Data Eng. **17**(4), 491–502 (2005)

15. Qin, A., Huang, V., Suganthan, P.: Differential evolution algorithm with strategy adaptation for global numerical optimization. IEEE Trans. Knowl. Data Eng. **13**(2), 398–417 (2009)
16. Storn, R.: On the usage of differential evolution for function optimization. In: 1996 Biennial Conference of the North American Fuzzy Information Processing Society, pp. 519–523. IEEE (1996)
17. Tsai, C.F., Chen, Z.Y.: Towards high dimensional instance selection: an evolutionary approach. Decision Support Syst. **61**, 79–92 (2014)
18. Tsai, C.F., Eberle, W., Chu, C.Y.: Genetic algorithms in feature and instance selection. Knowl. Based Syst. **39**, 240–247 (2013)
19. Unler, A., Murat, A.: A discrete particle swarm optimization method for feature selection in binary classification problems. Eur. J. Oper. Res. **206**(3), 528–539 (2010)
20. Xue, B., Zhang, M., Browne, W., Yao, X.: A survey on evolutionary computation approaches to feature selection. IEEE Trans. Evol. Comput. PP(99) (2015). doi:10. 1109/TEVC.2015.2504420
21. Xue, B., Fu, W., Zhang, M.: Multi-objective feature selection in classification: a differential evolution approach. In: Dick, G., Browne, W.N., Whigham, P., Zhang, M., Bui, L.T., Ishibuchi, H., Jin, Y., Li, X., Shi, Y., Singh, P., Tan, K.C., Tang, K. (eds.) SEAL 2014. LNCS, vol. 8886, pp. 516–528. Springer, Heidelberg (2014)
22. Xue, B., Zhang, M., Browne, W.N.: Particle swarm optimization for feature selection in classification: a multi-objective approach. IEEE Trans. Cybern. **43**(6), 1656–1671 (2013)
23. Xue, B., Zhang, M., Browne, W.N.: A comprehensive comparison on evolutionary feature selection approaches to classification. Int. J. Comput. Intell. Appl. **14**(02), 1550008 (2015)
24. Yang, Z., Tang, K., Yao, X.: Scalability of generalized adaptive differential evolution for large-scale continuous optimization. Soft Comput. **15**, 2141–2155 (2011)
25. Zhu, P., Zuo, W., Zhang, L., Hu, Q., Shiu, S.C.: Unsupervised feature selection by regularized self-representation. Pattern Recogn. **48**(2), 438–446 (2015)
26. Zhu, Z., Ong, Y.S., Dash, M.: Wrapper-filter feature selection algorithm using a memetic framework. IEEE Trans. Syst. Man Cybern. Part B: Cybern. **37**(1), 70–76 (2007)

Facial Age Estimation by Total Ordering Preserving Projection

Xiao-Dong Wang$^{(\boxtimes)}$ and Zhi-Hua Zhou

National Key Laboratory for Novel Software Technology,
Nanjing University, Nanjing 210023, China
{wangxd,zhouzh}@lamda.nju.edu.cn

Abstract. Facial age estimation is one of the unsolved challenging issues in automatic face perception. Previous studies usually formulated it as a classification problem, where each age is regarded as a class, or a regression problem where the age is regarded as a variable spanning in a real-valued interval. In this paper, we propose to formulate this task as an ordinal regression problem. On one hand, the new formulation emphasizes the fact that the age estimation problem is inherently a classification problem (ordinal regression is a special kind of classification task); on the other hand, the new formulation allows to take into account the order information between different ages, which has been ignored by previous classification formulation. We develop the TOPP (Total Ordering Preserving Projection) approach, by identifying the low-dimensional subspace which preserves the ordinal relations to the best, and experiments show that TOPP significantly outperforms state-of-the-art age estimation methods.

Keywords: Ordering preserving projection · Ordinal regression · Facial age estimation · Face perception · Machine learning

1 Introduction

Face perception has been studied for many decades, and significant progress has been made in recent years due to the success of deep learning methods and the availability of very large-scale training datasets. For example, the accuracy of face verification has been improved to higher than 99 % (Schroff et al. 2015), which is comparable or even superior to the ability of human beings. Other than the identity, age is also an important attribute of the human face and facial age estimation is essential to many applications such as age-specific access control, age-specific service etc. Fig. 1 shows an example of human facial aging. However, facial age estimation is one of the unsolved challenging issues in automatic human face perception. One important reason lies in the fact that there does not exist "big data of facial age images", because collecting facial images under different ages is not easy, and getting the accurate facial age information is quite difficult even when there are lots of facial images. Thus, smart methods are needed for automatic facial age estimation.

© Springer International Publishing Switzerland 2016
R. Booth and M.-L. Zhang (Eds.): PRICAI 2016, LNAI 9810, pp. 603–615, 2016.
DOI: 10.1007/978-3-319-42911-3_50

| 3 | 14 | 25 | 42 | 56 | 68 |

Fig. 1. An example of human facial aging process (images collected from the Internet).

Previous studies usually formulated this task as a classification problem, or a regression problem. Under a typical classification setting, each age or age group is treated as a single class, while the order information among different classes is dropped. Exchanging class names will never lead to different models for classification methods. However, for the task of facial age estimation, ignoring the order information is dangerous, e.g. classifying 5 year old face to age 6 is tolerable while classifying it to age 50 is not. For regression-based methods, the age is regarded as a variable spanning in a real-valued interval, but it is usually hard to determine the mapping from the age to the variable in this interval.

When identifying a low-dimensional subspace of face images, an example is given to illustrate the dilemma of traditional classification methods in Fig. 2. Figure 2(a) shows some face images of age 10, 30 and 50 in the original feature vector space. As can be seen there is no obvious clustering structure in this original space. Figure 2(b) and (c) show 2D plots of transformed feature vectors obtained by two different linear transformations. As shown in both Figures, faces of same age are close to each other while faces of different ages are far away from each other. Thus, both linear transformations are good under the view of classification methods. However, the situation may change when the order information is considered, and it is clear that Fig. 2(c) is preferred to Fig. 2(b) for age estimation.

In this paper, we propose to formulate the facial age estimation task as an ordinal regression problem, where ages or age groups are viewed as ordered classes. On one hand, it is a special kind of classification problem. On the other hand, it takes into account the order information between different classes which has been ignored by classification formulation. Then, we develop the TOPP (Total Ordering Preserving Projection) approach, by identifying the low-dimensional subspace which not only presents the clear clustering structure but also preserves the ordinal relations to the best (as the case in Fig. 2(c)). Experiments are conducted on three benchmark databases and it is shown that TOPP significantly outperforms state-of-the-art facial age estimation methods.

The rest of this paper is organized as follows. In the next section, previous work on facial age estimation is reviewed. In Sect. 3, the TOPP approach is described in detail. Then, the experiment results are shown in Sect. 4. Finally, the conclusions are drawn in Sect. 5.

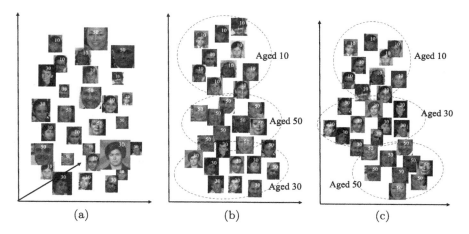

Fig. 2. (a) shows some faces of three different ages (i.e. 10, 30, 50) in original high-dimensional space. (b) and (c) display the faces projected onto two different 2D spaces. Under the view of classification methods, both (b) and (c) show good clustering structures, but (c) is preferred for the task of facial age estimation.

2   Related Work

A variety of facial age estimation methods have been proposed over the last decade. Usually, the design of these methods consists of two parts: (1) the representation of facial images and (2) the design of age estimators. First, facial features which can represent human age or appearance change are extracted from images. Then, an age determination function is fitted to estimate the ages based on these features.

For facial age estimation, the goal of image representation is to capture the aging features on human faces. It is known that facial aging is a process related to both shape and texture of face. Motivated by this observation, Lanitis et al. (2002) suggested to use the active appearance model (AAM) (Cootes et al. 2001) to extract features from face images with 68 landmarks, because the main advantage of AAM is that the extracted features combine both the shape and texture of the face images. However, landmarks are not always available. To deal with such case, Guo et al. (2009) introduced the biologically inspired features (BIF) (Riesenhuber and Poggio, 1999) to age estimation and a couple of improvements over the original biologically inspired models were proposed to ensure a good performance for age estimation. Later, Luu et al. (2011) proposed the contourlet appearance model (CAM) to represent face images, which is claimed to be more accurate and faster at localizing facial landmarks than AAM.

Due to the natural characters of the age attribute, the age determination problem has been formulated in many ways. (1) *regression* Lanitis et al. (2002) viewed the age as a quadratic function of the extracted features and built the weighted appearance specific aging function (WAS) (Lanitis et al. 2002). This idea was further extended to the appearance and age specific aging function

(AAS) (Lanitis et al. 2004) in which aging functions are built for different age groups. Yan et al. (2007) designed a regressor from training samples with uncertain nonnegative labels and learned the regressor through semidefinite programming. Guo et al. (2008) proposed the locally adjusted robust regression (LARR) for age estimation, which takes the support vector regression (SVR) as a global robust regressor and the support vector machine (SVM) as a local adjuster. (2) *multi-class classification.* Geng et al. (2007) proposed the aging pattern subspace (AGES) algorithm based on the subspace trained on constructed aging pattern vectors for each person. After the proposal of AGES, subspace learning techniques have been widely applied to facial age estimation (Fu and Huang 2008; Fu et al. 2007; Geng et al. 2007). Besides, many classical multi-class classification methods have been tested for age estimation problem, such as the kNN (Chang et al. 2010; Geng et al. 2013, 2007), the SVM (Chang et al. 2010, 2011; Geng et al. 2007; Guo et al. 2008) and the backpropagation neural network (Chang et al. 2010, 2011; Geng et al. 2007; Lanitis et al. 2004). (3) *label distribution learning* To overcome problems brought by insufficient and incomplete training data, Geng et al. (2010) associated each face image with a label distribution, and put forward the label distribution learning paradigm. The improved iterative scaling-learning from label distribution (IIS-LLD) (Geng et al. 2010) and the conditional probability neural network (CPNN) (Geng et al. 2013) were proposed successively to solve this label distribution learning problem. (4) *cost-sensitive learning* Recently, Chang et al. (2011) treated it as a cost-sensitive classification problem and proposed the ordinal hyperplanes ranker (OHRank) to improve the estimation accuracy.

3   Total Ordering Preserving Projection

Suppose we have a set of m samples $\{x_1, x_2, \ldots, x_m\}$ taking values in a d-dimensional space, and each sample being associated to one of the c ordered classes in $\{L_1, L_2, \ldots, L_c\}$. Consider a linear transformation mapping the original d-dimensional space into a d'-dimensional space, where $d' < d$, then the transformed vector $\{y_k\}$ are defined as:

$$y_k = \mathbf{W}^{\mathrm{T}} x_k, k = 1, 2, \ldots, m \tag{1}$$

where $\mathbf{W} \in \mathbb{R}^{d \times d'}$ is the corresponding transformation matrix.

The goal is to identify such a projection that in the transformed space, (1) faces of same age should be very close to each other, (2) two faces with small age gap should not be far away from each other while (3) two faces with huge age gap should be far away from each other. From another point of view, if faces in the transformed space present a clear order changing trend, then overlap between younger faces and elder ones should be small. Note that 'younger' and 'elder' are comparative terms, e.g., faces aged from 20 to 30 are younger compared to those from 40 to 50, but elder compared to those from 1 to 10. Different choices of separation age will lead to different partitions of younger faces and elder ones, and overlap between younger faces and elder faces in the transformed space should be small in every possible partition for a good projection. Inspired by this, we develop the new objective function to identify the projection that preserves the ordinal relations.

More specifically, suppose "$<$" indicates the ordinal relationship between different classes (or ages), and we can assume $L_1 < L_2 < \cdots < L_c$ without loss of generality. Let $l_i \in \{L_1, L_2, \ldots, L_c\}$ be the label for instance x_i. For any class L_k, we separate the data set into two subsets, X_k^+ and X_k^- as follows:

$$X_k^+ = \{x_i | l_i > L_k\} , \quad X_k^- = \{x_i | l_i \le L_k\} \tag{2}$$

By assigning the positive labels to the samples in X_k^+ and assigning the negative labels to the samples in X_k^-, the problem is converted to a standard binary classification problem. To avoid the appearance of empty set, we take k among $\{1, 2, \ldots, c-1\}$. In this way, we can convert the original ordinal regression problem into $c - 1$ binary classification problems.

For a given subproblem which is induced by L_k, samples of low-order are included in X_k^- while samples of high-order are included in X_k^+. In order to present a distinct order changing trend, we expect a small overlap between samples in X_k^- and samples in X_k^+. Let m_i denote the number of samples in class L_i, then $m_k^- = \sum_{i=1}^k m_i$ and $m_k^+ = \sum_{i=k+1}^c m_i$ are the number of samples in X_k^- and X_k^+ respectively. Let μ_i and μ denote the center of samples of order i and the center of all samples respectively and let μ_k^- and μ_k^+ denote the centers of samples in X_k^- and X_k^+, then:

$$\mu_k^- = \frac{1}{m_k^-} \sum_{x_i \in X_k^-} x_i = \frac{1}{m_k^-} \sum_{i=1}^k m_i \mu_i \tag{3}$$

$$\mu_k^+ = \frac{1}{m_k^+} \sum_{x_i \in X_k^+} x_i = \frac{1}{m_k^+} \sum_{i=k+1}^c m_i \mu_i \tag{4}$$

Thus, the corresponding centers in the transformed space are $\mathbf{W}^T \mu_k^-$, $\mathbf{W}^T \mu_k^+$ and $\mathbf{W}^T \mu$, and distance between X_k^- and X_k^+ in the transformed space can be characterized by

$$m_k^- (\mathbf{W}^T \mu_k^- - \mathbf{W}^T \mu)^T (\mathbf{W}^T \mu_k^- - \mathbf{W}^T \mu) + m_k^+ (\mathbf{W}^T \mu_k^+ - \mathbf{W}^T \mu)^T (\mathbf{W}^T \mu_k^+ - \mathbf{W}^T \mu). \tag{5}$$

Let $dist_k$ denote this distance, by using the trace trick, it can be simplified as

$$dist_k = \mathrm{tr}\Big(\mathbf{W}^T \big(m_k^-(\mu_k^- - \mu)(\mu_k^- - \mu)^T + m_k^+(\mu_k^+ - \mu)(\mu_k^+ - \mu)^T\big)\mathbf{W}\Big). \tag{6}$$

Finally, such distances for all $c - 1$ subproblems are summed up to measure the overall degree of preservation of ordinal information, i.e. $\sum_{k=1}^{c-1} dist_k$. Define $\mathbf{S}_o$ as the *ordinal-class scatter matrix*

$$\mathbf{S}_o = \sum_{k=1}^{c-1} m_k^- (\mu_k^- - \mu)(\mu_k^- - \mu)^T + m_k^+ (\mu_k^+ - \mu)(\mu_k^+ - \mu)^T, \tag{7}$$

then $\sum_{k=1}^{c-1} dist_k$ can be simplified as

$$\sum_{k=1}^{c-1} dist_k = \sum_{k=1}^{c-1} \mathrm{tr}\Big(\mathbf{W}^{\mathrm{T}}\big(m_k^-(\boldsymbol{\mu}_k^- - \boldsymbol{\mu})(\boldsymbol{\mu}_k^- - \boldsymbol{\mu})^{\mathrm{T}} + m_k^+(\boldsymbol{\mu}_k^+ - \boldsymbol{\mu})(\boldsymbol{\mu}_k^+ - \boldsymbol{\mu})^{\mathrm{T}}\big)\mathbf{W}\Big)$$

$$= \mathrm{tr}\Big(\mathbf{W}^{\mathrm{T}}\sum_{k=1}^{c-1}(m_k^-(\boldsymbol{\mu}_k^- - \boldsymbol{\mu})(\boldsymbol{\mu}_k^- - \boldsymbol{\mu})^{\mathrm{T}} + m_k^+(\boldsymbol{\mu}_k^+ - \boldsymbol{\mu})(\boldsymbol{\mu}_k^+ - \boldsymbol{\mu})^{\mathrm{T}})\mathbf{W}\Big)$$

$$= \mathrm{tr}(\mathbf{W}^{\mathrm{T}}\mathbf{S}_o\mathbf{W}) \tag{8}$$

In order to obtain a normalized distance, $\sum_{k=1}^{c-1} dist_k$ is divided by $dist_{\text{total}}$ which is defined as

$$dist_{\text{total}} = \sum_{i=1}^{m}(\mathbf{W}^{\mathrm{T}}\boldsymbol{x}_i - \mathbf{W}^{\mathrm{T}}\boldsymbol{\mu})^{\mathrm{T}}(\mathbf{W}^{\mathrm{T}}\boldsymbol{x}_i - \mathbf{W}^{\mathrm{T}}\boldsymbol{\mu})$$

$$= \mathrm{tr}\Big(\mathbf{W}^{\mathrm{T}}\big(\sum_{i=1}^{m}(\boldsymbol{x}_i - \boldsymbol{\mu})(\boldsymbol{x}_i - \boldsymbol{\mu})^{\mathrm{T}}\big)\mathbf{W}\Big)$$

$$= \mathrm{tr}(\mathbf{W}^{\mathrm{T}}\mathbf{S}_t\mathbf{W}), \tag{9}$$

where $\mathbf{S}_t$ is known as the total scatter matrix. In fact, $dist_{\text{total}}$ is the total distance from all samples to the sample center in the transformed space. As a result, the final objective is

$$\mathbf{W}^* = \arg\max_{\mathbf{W}} \frac{\sum_{k=1}^{c-1} dist_k}{dist_{\text{total}}}$$

$$= \arg\max_{\mathbf{W}} \frac{\mathrm{tr}(\mathbf{W}^{\mathrm{T}}\mathbf{S}_o\mathbf{W})}{\mathrm{tr}(\mathbf{W}^{\mathrm{T}}\mathbf{S}_t\mathbf{W})} \tag{10}$$

It's hard to solve this trace ratio optimization problem directly, so we take the approximate solution by solving the generalized eigenvalue problem. More specifically, if $\mathbf{W} = [\boldsymbol{w}_1, \boldsymbol{w}_2, \ldots, \boldsymbol{w}_{d'}]$, then $\boldsymbol{w}_i$ is the generalized eigenvector corresponding to the i-th largest generalized eigenvalue λ_i for the following problem:

$$\mathbf{S}_o\boldsymbol{w}_i = \lambda_i\mathbf{S}_t\boldsymbol{w}_i \tag{11}$$

As can be seen, faces of same age is classified into the same group in all $c-1$ subproblems. Furthermore, if two faces are of large age gap, then few subproblems will classify them into the same group. In an extreme case, an L_1-year-old face and an L_c-year-old face will never be classified into the same group in all subproblems. Note that the form of Eq. 10 is very similar to the objective function of the classical linear discriminant analysis (LDA) (Belhumeur et al. 1997), but LDA is a typical classification method of which the drawback is explained in Sect. 1. In LDA, distances of different class centers to the center of all samples are treated equally which makes it not appropriate for the task of facial age estimation.

Our approach[1] for facial age estimation, i.e., TOPP, is summarized as follows. At the training stage, features are extracted from images and the optimal

---

[1] The MATLAB implementation of our approach can be obtained via http://lamda.nju.edu.cn/code_TOPP.ashx.

projection $\mathbf{W}^*$ is chosen by solving Eq. 10. At the test stage, the same type of features are extracted from unseen images and mapped onto the low-dimensional space by $\mathbf{W}^*$. Finally, the mapped training feature vectors and the mapped test feature vectors are passed to the median kNN which is widely used for ordinal regression (Hechenbichler and Schliep 2004), to output age predictions.

4 Experiments

4.1 Data Sets

Experiments are performed on three benchmark databases: the FG-NET Aging Database (Lanitis et al. 2002), the MORPH Album 2 Database (Ricanek and Tesafaye 2006) and the ChaLearn Apparent Age Database (Escalera et al. 2015). The FG-NET contains 1002 color or gray facial images from 82 subjects. Besides age variation, other variations in pose, expression and lighting are also presented in this database. Since the landmarks for each image are available in this database, the AAM is often chosen as the feature extractor. In the experiments, we use about 250 AAM features to preserve 99 % of the variability in the training data.

The MORPH Album 2 Database (hereinafter referred to as 'MORPH') is a large-scale database which contains 55134 facial images from more than 13000 subjects. Due to the absence of landmark information, the biologically inspired features (BIF) (Guo et al. 2009) instead of the AAM features are extracted from these images. We use about 1000 BIF features which preserve 95 % of the variability in the training data, which are sufficient for age estimation suggested by (Guo et al. 2009).

The ChaLearn Apparent Age Database (hereinafter referred to as 'ChaLearn') is a mid-scale database, and it has 3615 face images with their apparent ages and standard deviations provided. A pre-processing step suggested by (Yang et al. 2015) is taken to obtain aligned faces. Then the first 300 BIF features are extracted and served as the input of all compared methods.

4.2 Experiment Setup

We use MAE (mean absolute error) (Lanitis et al. 2002) to measure the performance, where $\hat{l}_i$ and l_i are the estimated age and the ground truth for the test image i, respectively:

$$\text{MAE} = \frac{1}{m} \sum_{i=1}^{m} |\hat{l}_i - l_i|. \tag{12}$$

First, we compare our method to the state-of-the-art methods, including WAS (Lanitis et al. 2002), AAS (Lanitis et al. 2004), AGES (Geng et al. 2007), LARR (Guo et al. 2008), IIS-LLD (Geng et al. 2010), CPNN (Geng et al. 2013) and OHRank (Chang et al. 2011). Second, methods such as LDA (Belhumeur et al. 1997), OLPP (Cai et al. 2006) and OPMFA (Lu and Tan 2013) are compared to show the superiority of TOPP over other subspace learning methods.

For those experiments on the FG-NET, the leave-one-person-out (LOPO) (Geng et al. 2007) test scheme is used. In each fold, the images of one individual are treated as the test set and the remainders are treated as the training set. By this means, each individual is treated as test set once. For AAS, the error threshold in the appearance cluster training step is set to 4, and the age groups are set as 0–9, 10–19, 20–39, and 40–69 to train age specific classifiers. For AGES, the dimensionality of the aging pattern subspace is set to 20. In LARR, the RBF kernel function is used to train the global SVR model and the associated parameters, ϵ, C and γ, are set as 0.02, 40 and 12 separately recommended by (Guo et al. 2008). The local search range in LARR is selected by 5-fold cross validation on training set. Both IIS-LLD and CPNN assign Gaussian label distributions to training images, and the standard deviations σ of Gaussian distributions for IIS-LLD and CPNN are 1 and 2 respectively on this database, which is suggested by (Geng et al. 2013). In CPNN, the number of hidden layer units is set to 400. For each subproblem in OHRank, the absolute cost function is applied to weight each instance and the RBF kernel function is used to train binary SVM classifiers. For all the subspace learning methods followed by median kNN, the dimensionality d' of the reduced subspace and the number of nearest neighbors k are determined by 5-fold cross validation on training set, and euclidean distance is used to find nearest neighbors.

Since the number of individuals in MORPH is more than 13000, the LOPO test scheme may be too time consuming to get final results. Instead, in each round we randomly select 25000 images for training, and another 25000 images for testing. After 10 rounds, final results are calculated from all predictions. WAS and AGES are not tested on this database because the average number of images per individual is 4, which may be too few to obtain a reliable aging function or aging pattern for each individual. Most setups for other methods trained on this database are same with those on FG-NET. For the sake of briefness, only the different setups are listed here. For AAS, the error threshold in the appearance cluster training step is reset to 8, and the age groups are adjusted to 16–24, 25–34, 35–44, and 45–77 to match the age range (16–77) on this database. For IIS-LLD, the standard deviation σ of Gaussian distributions is reset to 3 according to (Geng et al. 2013). For OHRank, the linear kernel function instead of the RBF kernel function is used when training SVM classifiers, because the test for OHRank with RBF kernel function did not come to an end in 48 hours on author's computer (Windows 8 with i5 CPU @3.2 GHz and 8 G RAM).

On the ChaLearn database, 10 times 5-fold cross validation is used to assess age estimation methods. Most setups are same with those on MORPH. Note that, the ChaLearn is an apparent age database and each image is labeled by multiple individuals. As a result, each face image has a votes variance. Hence, for label distribution learning methods (IIS-LLD and CPNN), the variance of Gaussian distribution for each image is set to the votes variance in order to well utilize information in training set.

4.3   Results

The MAEs of our proposed method and existing facial age estimation methods mentioned above are tabulated in Table 1. Note that standard deviations of MAEs on the FG-NET are not displayed, because the size of test set in each fold varies dramatically, and all of the methods get unstable performances (Geng et al. 2013).

Table 1. Mean absolute errors (MAEs) of age estimation in comparison with state-of-the-art methods. In each data set, the lowest MAEs are bolded, and in MORPH and ChaLearn, '•/∘' denote respectively that our method is significantly better/worse than the corresponding method by the t-test with significance level 0.05.

Method	Data Set		
	FG-NET	MORPH	ChaLearn
WAS	9.38	——	——
AAS	8.43	5.11 ± 0.03•	8.68 ± 0.19•
AGES	6.68	——	——
LARR	5.11	5.52 ± 0.03•	7.17 ± 0.02•
IIS-LLD	5.87	6.63 ± 0.09•	8.99 ± 0.09•
CPNN	5.34	4.94 ± 0.22•	8.80 ± 0.25•
OHRank	4.50	4.34 ± 0.02•	6.62 ± 0.02•
Ours	**4.25**	**4.23 ± 0.02**	**6.28 ± 0.03**

As can be seen, our proposed method outperforms all these compared methods on all three aging databases. The good performance of our method is supposed to come from the full utilization of ordinal information among ages in subspace learning phase. If there is only a small difference between the ages of two training images, then most of the binary decompositions will classify them into the same subgroup, which means a relatively large punishment will be received if such two images are projected far away. Thus, the subspace found by TOPP is expected to be 'pure' in local area. Here 'pure' local area means that age differences in this area are relatively small. This is a good preparation for kNN style classifier, since it narrows the age differences in the neighborhood of unseen image.

Furthermore, to show the superiority of TOPP over other subspace learning methods, 3 subspace learning methods are testes on the benchmark databases as well. The corresponding MAEs are listed in Table 2. Comparisons on the FG-NET and ChaLearn show the good performance of TOPP. Nevertheless, it seems that TOPP gets almost the same results with LDA on the MORPH. In fact, projections chosen by TOPP are almost the same with that chosen by LDA in each test round. Thus, when data is sufficient (MORPH is a large scale data set), projections with optimal discriminative power also present the ordinal

Table 2. MAEs in comparison with subspace learning methods ('——' means that no subspace learning method is used). In each data set, the lowest MAEs are bolded, and in MORPH and ChaLearn, '•/○' denote respectively that TOPP is significantly better/worse than the corresponding method by the t-test with significance level 0.05.

Subspace	Data Set		
	FG-NET	MORPH	ChaLearn
——	6.73	6.43 ± 0.03•	8.05 ± 0.02•
OLPP	5.17	4.36 ± 0.02•	6.45 ± 0.03•
LDA	4.46	**4.23 ± 0.02**	6.33 ± 0.03•
OPMFA	6.27	4.43 ± 0.03•	7.18 ± 0.04•
TOPP	**4.25**	4.23 ± 0.02	**6.28 ± 0.03**

relationship among different ordered classes (i.e. the aging trend in facial age estimation problem).

However, sufficient training data are usually not available in practice. To simulate such case, we gradually reduce the training data from MORPH used by subspace learning methods, while the same test data are used to visualize the dimensionality reduction results. More specifically, half of the training data are randomly removed at each round. Images with ages older than 50 (only 5 % of all images) are removed from the very beginning, because these images are too few to ensure a stable partition. After 7 rounds, the last training set contains only 0.8 % $((1/2)^7)$ of the original training data. The results are shown in Fig. 3. In these plots, the first two dimensions are shown for each subspace. To make the aging trend more obvious, points with same age label are characterized by colored area within corresponding Gaussian contour, and different colors indicate different ages. As can be seen, when keeping on reducing the training data,

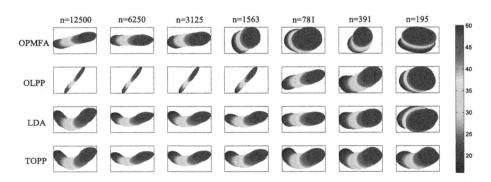

Fig. 3. Visualization for first two projections chosen by OPMFA, OLPP, LDA and TOPP as the training data being reduced. In case of extremely insufficient training data (n = 195), only the subspace found by TOPP still can present a clear aging trend.

projections chosen by OPMFA, OLPP and LDA appear to differ from the original projections. In case of extremely insufficient training data (n = 195), only the subspace found by TOPP still can present a clear aging trend. To quantize the similarity between newly learned projections and the original projections, the cosines of angles between them are calculated and shown in Fig. 4. The cosines for OPMFA, OLPP and LDA drop rapidly while the cosines for TOPP do not. The main reason for this phenomenon is that the prior knowledge of ordinal relationship among age labels makes up for the insufficiency of training data, and becomes more and more important as the training data being reduced. For OLPP and LDA which do not make use of such prior knowledge, performances get worse when fewer and fewer data are available. For OPMFA, the ordinal information is incorporated by introducing label-sensitive weights to each pairwise distance, but the choice of label-sensitive function may affect the exploration of the underlying aging trend. The influence appears to be obvious given insufficient training data.

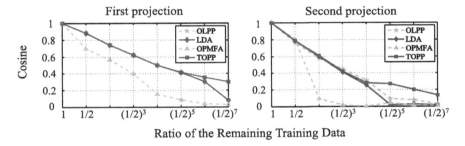

Fig. 4. The cosines of angles between newly learned projections and the projections learned given whole training set. Results are shown for first two projections separately.

5   Conclusion

In this paper, facial age estimation is formulated as an ordinal regression problem. The new formulation not only emphasizes the fact that the facial age estimation problem is inherently a classification problem but also allows to take into account the order information between different ages. The total ordering preserving projection (TOPP) is proposed to identify a subspace that preserves the ordinal relations among different ages to the best. Experiment results demonstrate that our approach for age estimation outperforms state-of-the-art methods. Furthermore, when facing extremely insufficient training data, TOPP still can find projections presenting the order changing trend. Thus, TOPP is very helpful in the situations where only limited training images are available, which is the usual case in practice. Additionally, TOPP can be applied to other applications with ordered labels, such as rating prediction, head pose estimation, and so on, which will be investigated in future work.

Acknowledgments. This research was supported by the NSFC (61333014) and the Nanjing Collaborative Innovation Center of Novel Software Technology and Industrialization.

References

Belhumeur, P.N., Hespanha, J.P., Kriegman, D.J.: Eigenfaces vs. fisherfaces: recognition using class specific linear projection. IEEE Trans. Pattern Anal. Mach. Intell. **19**(7), 711–720 (1997)

Cai, D., He, X., Han, J., Zhang, H.-J.: Orthogonal laplacianfaces for face recognition. IEEE Trans. Image Process. **15**(11), 3608–3614 (2006)

Chang, K.-Y., Chen, C.-S., Hung, Y.-P.: A ranking approach for human ages estimation based on face images. In: Proceedings of the 20th International Conference on Pattern Recognition, pp. 3396–3399, Istanbul, Turkey (2010)

Chang, K.-Y., Chen, C.-S., Hung, Y.-P.: Ordinal hyperplanes ranker with cost sensitivities for age estimation. In: Proceedings of the IEEE Computer Society Conference on Computer Vision and Pattern Recognition, pp. 585–592. Colorado Springs, CO (2011)

Cootes, T.F., Edwards, G.J., Taylor, C.J.: Active appearance models. IEEE Trans. Pattern Anal. Mach. Intell. **23**(6), 681–685 (2001)

Escalera, S., Fabian, J., Pardo, P., Baro, X., Gonzalez, J., Escalante, H., Guyon, I.: Chalearn 2015 apparent age and cultural event recognition: datasets and results. In: ICCV ChaLearn Looking at People Workshop, p. 4, Santiago, Chile (2015)

Fu, Y., Huang, T.S.: Human age estimation with regression on discriminative aging manifold. IEEE Trans. Multimed. **10**(4), 578–584 (2008)

Fu, Y., Xu, Y., Huang, T.S.: Estimating human age by manifold analysis of face pictures and regression on aging features. In: Proceedings of the IEEE International Conference on Multimedia and Expo, pp. 1383–1386, Beijing, China (2007)

Geng, X., Yin, C., Zhou, Z.-H.: Facial age estimation by learning from label distributions. IEEE Trans. Pattern Anal. Mach. Intell. **35**(10), 2401–2412 (2013)

Geng, X., Zhou, Z.-H., Smith-Miles, K.: Automatic age estimation based on facial aging patterns. IEEE Trans. Pattern Anal. Mach. Intell. **29**(12), 2234–2240 (2007)

Geng, X., Smith-Miles, K., Zhou, Z.-H.: Facial age estimation by learning from label distributions. In: Proceedings of the 24th AAAI Conference on Artificial Intelligence, pp. 451–456, Atlanta, GA (2010)

Guo, G., Fu, Y., Dyer, C.R., Huang, T.S.: Image-based human age estimation by manifold learning and locally adjusted robust regression. IEEE Trans. Image Process. **17**(7), 1178–1188 (2008)

Guo, G., Mu, G., Fu, Y., Huang, T.S.: Human age estimation using bio-inspired features. In: Proceedings of the IEEE Computer Society Conference on Computer Vision and Pattern Recognition, pp. 112–119, Miami, Florida (2009)

Hechenbichler, K., Schliep, K.: Weighted k-nearest-neighbor techniques and ordinal classification. Discussion Paper 399 SFB 386 (2004)

Lanitis, A., Draganova, C., Christodoulou, C.: Comparing different classifiers for automatic age estimation. IEEE Trans. Syst. Man Cybern. Part B: Cybern. **34**(1), 621–628 (2004)

Lanitis, A., Taylor, C.J., Cootes, T.F.: Toward automatic simulation of aging effects on face images. IEEE Trans. Pattern Anal. Mach. Intell. **24**(4), 442–455 (2002)

Lu, J., Tan, Y.-P.: Ordinary preserving manifold analysis for human age and head pose estimation. IEEE Trans. Hum. Mach. Syst. **43**(2), 249–258 (2013)

Luu, K., Seshadri, K., Savvides, M., Bui, T.D., Suen, C.Y.: Contourlet appearance model for facial age estimation. In: Proceedings of the IEEE International Joint Conference on Biometrics, pp. 1–8, Washington, D.C. (2011)

Ricanek, K., Tesafaye, T.: Morph: a longitudinal image database of normal adult age-progression. In: Proceedings of the 7th International Conference on Automatic Face and Gesture Recognition, pp. 341–345, Southampton, UK (2006)

Riesenhuber, M., Poggio, T.: Hierarchical models of object recognition in cortex. Nat. Neurosci. 2(11), 1019–1025 (1999)

Schroff, F., Kalenichenko, D., Philbin, J.: Facenet: a unified embedding for face recognition and clustering. In: Proceedings of the IEEE Conference on Computer Vision and Pattern Recognition, pp. 815–823, Boston, USA (2015)

Yan, S., Wang, H., Tang, X., Huang, T.S.: Learning auto-structured regressor from uncertain nonnegative labels. In: Proceedings of the 11th International Conference on Computer Vision, pp. 1–8, Rio de Janeiro, Brazil (2007)

Yang, X., Gao, B.-B., Xing, C., Huo, Z.-W., Wei, X.-S., Zhou, Y., Wu, J., Geng, X.: Deep label distribution learning for apparent age estimation. In: Proceedings of the IEEE International Conference on Computer Vision Workshops, pp. 102–108, Santiago, Chile (2015)

Hybrid Temporal-Difference Algorithm Using Sliding Mode Control and Sigmoid Function

Ke Xu$^{(\boxtimes)}$ and Fengge Wu

Science and Technology on Integrated Information System Laboratory,
Institute of Software Chinese Academy of Sciences, Beijing, China
{xuke13,fengge}@iscas.ac.cn

Abstract. Gradient temporal-different algorithms such as GTD2 and TDC have improved the accuracy of the algorithm to a new level. Unfortunately, these algorithms converge much slower than conventional temporal-different algorithms. In this paper, we present a approach based on sliding mode control to speed up the GTD2 algorithm, and then use sigmoid function to reduce algorithm's jitter. Our experiments on random walk show that our algorithm converges as fast as conventional temporal-different algorithms and as accurate as GTD2 algorithm at the same time. This is an important property for online-learning tasks.

Keywords: Reinforcement learning · Temporal-difference algorithm · Sliding mode control · Sigmoid function

1 Introduction

The ability to learn faster and more accurate at the same time is a challenge task in reinforcement learning(RL) algorithms. In temporal-difference(TD) approach, for example, conventional TD algorithms converge faster with lower accuracy with mean square error(MSE) while gradient temporal-difference(GTD) algorithms such as GTD2 and TDC have higher accuracy with mean square projected bellman error(MSPBE) but converge slower [1]. For large problems, parallelism is a promising approach to speed up the algorithms, parallel RL algorithms of policy evaluation, policy iteration, and off-policy updates based on MapReduce have been developed [2]. Introduce other machine learning technics into RL such as deep learning has been a successful approach to solve complex problems [3]. For more general problems, due to GTD2 and TDC are stochastic gradient algorithms, natural gradient temporal-difference algorithms can achieve that goal of speed up the algorithm without losing the accuracy [4]. There are also improvements for conventional TD algorithms to make conventional TD algorithms' accuracy slightly better but no GTD algorithms are considered [5].

Another well-known approach is least squares temporal-difference(LSTD) algorithm. It was considered as $O(n^2)$ computational complexity algorithm at first, where n is the number of features used in the linear approximator. Incremental methods such as iLSTD reduce per-time-step computation to $O(n)$ when

© Springer International Publishing Switzerland 2016
R. Booth and M.-L. Zhang (Eds.): PRICAI 2016, LNAI 9810, pp. 616–625, 2016.
DOI: 10.1007/978-3-319-42911-3_51

the feature matrix is sparse [6]. Stochastic approximation variant of the LSTD such as fLSTD-SA don't need any sparsity assumption. However, these methods still need $O(n^2)$ memory [7]. From the above discussion, it appears that at present no algorithms could reach convergence speed of conventional TD and convergence accuracy of GTD and the same time with $O(n)$ complexity.

In this paper, a method that using the sliding mode control idea in variable-structure control theory is proposed to reach GTD's convergence accuracy without losing the convergence speed of TD and make the TD algorithm more practical for online learning tasks. It is typically used to achieve accurate, robust, decoupled tracking for a class of nonlinear time-varying multiinput-multioutput systems in the presence of disturbances and parameter variations [8]. This method is used in a different area of RL algorithm with a different way. First, Sliding mode control uses sliding surface equations to keep control system stable around the sliding surface, while our implementation uses sliding surface equations to switch algorithm from a faster iteration(conventional TD algorithm) to a more accurate iteration(GTD algorithm). Second, the design of sliding surface still remains a difficult issue in sliding mode control area, while our sliding surface is much easier to implement. However, this method jitters when switch happens. Then a soft switch method using sigmoid function is proposed to come over that jitter.

The remainder of this paper is organized as follows: Sect. 2 analyses typical TD algorithms. Section 3 explains the proposed switch algorithms. Section 4 describes experimental results. Section 5 concludes the paper.

2   Analysis of Temporal-Difference Algorithms

In this section, conventional TD algorithms and gradient-TD algorithms are both described to show their properties. Then the analysis of these algorithms is also discussed to find the improvement possibility of the algorithms' performance.

2.1   Conventional Temporal-Difference

Conventional temporal-difference algorithms are based on MSE, which have been developed more than twenty years, resulted in lots of improvements. TD(λ) introduced eligibility traces parameter λ to obtain a more general method that may learn more efficiently, TD$_\gamma$(C) proposed a more accurate maximum-likelihood estimator of return model that may learn more accurately but required C times more time and memory than TD(λ) [9]. One direct implementation of these algorithms is so called TD(0) algorithm. It is given by the following update equations [10]:

$$\delta = r + \gamma V_{\theta_t}(S') - V_\theta(S) \tag{1}$$
$$\theta' = \theta + \alpha \delta \Phi \tag{2}$$

where $V_\theta : S \to \mathbb{R}$ is a function approximator with parameter vector θ, $\Phi = \frac{\partial V_\theta(S)}{\partial \theta}$ are basis functions of the linear function approximation, γ is a discount factor, α is a learning rate, S' is the next situation of S.

The main drawbacks of conventional temporal-difference are its low accuracy due to the algorithm is based on MSE and it is not stable with linear function approximation due to the algorithm is not a real gradient-descent algorithm, which implies that this method could be not stable and may have high bias. It still converges very fast that could be used at the preliminary stage of the learning process. Therefore we need a real gradient temporal-difference algorithm to have a more accurate convergence in learning process.

2.2   Gradient Temporal-Difference

Gradient temporal-difference algorithms such as GTD2 and TDC are based on MSPBE so they have higher accuracy than conventional temporal-difference algorithms. They are also linear time algorithms for TD-learning with linear function approximation, support off-policy learning [1]. The GTD2 algorithm is given by the following update equations:

$$\theta' = \theta + \alpha(\Phi - \gamma\Phi')(\Phi^T w) \tag{3}$$

$$w' = w + \beta(\delta - \Phi^T w)\Phi \tag{4}$$

where δ is same as in TD algorithm, β is a learning rate for w_t.

TDC is almost the same algorithm and is given by the following update equations:

$$\theta' = \theta + \alpha\delta\Phi - \alpha\gamma\Phi'(\Phi^T w)$$

$$w' = w + \beta(\delta\Phi - w)$$

Both of these algorithms are proved to be real gradient-descent algorithms. Compare the parameter update equations in TD(0) and GTD2 with Eqs. (1),(2) and Eqs. (3),(4) we can see that GTD2 is a slight change of TD(0) but make a significant result in the accuracy of the algorithm. On one hand, TD(0) can be considered as a junior learner that knows little about the gradient-descent principle, so it is brave enough to learn very fast in the direction given by δ, without worrying about the instability in the algorithm. On the other hand, GTD2 and TDC can be considered as a "senior learner" that knows much about the gradient-descent principle, so it can learn in a more accurate direction, but it always worries about whether the learning direction is correct or not, it uses parameter w to correct its learning direction all the time, which makes its learning speed goes slower than the "junior learner" although it has been speed up a great much than the original GTD algorithm. If an algorithm could be more brave at the first step of learning as a "junior learner" and then be more careful as a "senior learner" at the second step of learning, the algorithm's performance will be improved.

3   Hybrid

There are two ways to have both advantages of following two algorithms, the speed of conventional TD and the accuracy of GTD. Both of these ways are to change the computation from TD to GTD when TD converges slower than GTD. The first way is to combine both of them with sliding mode, this is a directly and hard way to control the learning process. It may be consequently referred to as GTD-HS(hard switch of gradient temporal difference) algorithm. The second way is to use weight to combine two algorithms, this is a soft way to switch TD to GTD gradually. It may be consequently referred to as GTD-SS(soft switch of gradient temporal difference) algorithm.

3.1 Hard Switch

Sliding mode control is a major kind of control method to solve variable structure systems(VSS) control problems. For variable structure systems, they consist of a set of continuous subsystems with a proper switching logic and, as a result, control actions are discontinuous functions of system state, disturbances, and reference inputs [8]. The following functions describe the variable structure systems:

$$\dot{x} = f(x, t, u), x \in \mathbb{R}^n, u \in \mathbb{R}^m \tag{5}$$

$$u = \begin{cases} u^+(x, t) \text{ if } s(x) > 0 \\ u^-(x, t) \text{ if } s(x) < 0 \end{cases} \tag{6}$$

Equation (5) shows the state equation of the VSS, where state $x \in \mathbb{R}^n$, input $u \in \mathbb{R}^m$, consists of 2^m subsystems and its structure varies on m surfaces at the state space. Equation (6) shows that in the neighborhood of segment mn on the switching line s = 0, the input of the system varies such as bang-bang control and the trajectories run in the opposite direction, which leads to the appearance of a sliding mode moves along this line.

The implementation of sliding mode makes control system's trajectory depending only on the sliding surface and invariant to plant parameters and disturbance. Furthermore, since sliding mode trajectory belong to some manifold of a dimension lower than that of the system, the order of a motion is reduced as well.

GTD-HS has two steps. In the first step of the algorithm, TD(0) is used to get a fast convergence. When TD(0) reach its accuracy limit, the algorithm turns to step two: GTD2 is used to find a more accurate convergence result. The implementation of sliding mode in this paper differs greatly from the original sliding mode control in VSS. If the learning trajectory goes along the sliding surface which is shown in figure, the algorithm will constantly switch between TD(0) and GTD2. Which makes the algorithm performs slower and less accurate at the same time. To avoid this situation, *isunchanged* flag is used in algorithm 1 to jump over the sliding surface from TD(0) to GTD2 without more switches.

The fundamental problem of the algorithm is to choose an effective sliding surface so that the algorithm could perform as designed before.

A naive idea is to switch learning algorithms when GTD2 performs better than TD(0). There are several possible ways to decide which algorithm performs better in the iterations of learning, such as learning speed: $\delta\theta$, and cost function: J. Since TD algorithms belong to stochastic gradient descent learning algorithms, the learning speed of the algorithm is unstable during the iterations of learning. Meanwhile, the cost function in RL is a much more stable value than learning speed in iterations of learning and thus the sliding surface chooses cost function as the switch basis of the learning algorithm.

$$s = J_g - J_0 - B; \qquad (7)$$

In function (7), J_o is the cost function of TD(0) and J_g is the cost function of GTD2. B is the bias of the sliding mode that makes the algorithm more flexible. Although cost function is more stable than learning speed, it also has jitter during the iterations. So a bias parameter is used as slight adjustment to reduce the jitter.

3.2   Soft Switch

In GTD-SS, weight parameter is the percentage of update of each algorithm. For example, if weight of TD(0) is 10%, then weight of GTD2 will be 90%, and update of algorithm will contains 10% of TD(0) and 90% of GTD2. At the beginning of the algorithm, TD(0) will contain most of the update to accelerate the algorithm. When TD(0) slows down, weight of GTD2 will become major of update. In this paper, sigmoid function is used to determine the weight:

$$w_{GTD} = \frac{1}{1 + e^{-E(i-B)}} \qquad (8)$$
$$w_{TD} = 1 - w_{GTD}$$

where i is the iteration number of learning process. E is a coefficient of expansion, determine the switch speed of algorithm. B is a coefficient of bias, has same effect in GTD-HS algorithm.

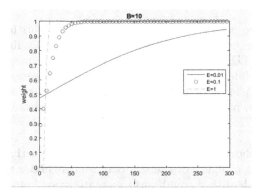

Fig. 1. Soft switch sigmoid function.

When $B = 10, E \in \{0.01, 0.1, 1\}$, switch function is displayed in Fig. 1. It shows bigger E will cause faster switch speed. If E is big enough, switch process will look like GTD-HS algorithm. And if E is given a small value, it will look like linear switch with a gradient of 1. This is a convenient property for adjusting E will get different kinds of switch functions. B is the bias of switch, determines when the switch happens. The bigger B will cause later switch time. Both B and E are always bigger than 0.

Algorithm 1. GTD-SS

1: **Initialize:**$\alpha_0, \alpha_\omega, \alpha_g, B, E, \theta, \omega$
2: GTD2 update:
   $\delta_g = r + \phi_t \theta'_g - \phi_{t-1} \theta'_g$
   $\theta_g = \theta_g + \alpha_g (\phi_{t-1} - \phi_t)(\phi_{t-1} w_t)$
   $w_{t+1} = w_t + \alpha_\omega (\delta_g - \phi_{t-1} w'_t) \phi_{t-1}$
3: TD(0) update:
   $\delta_0 = r + \phi_t \theta'_0 - \phi_{t-1} \theta'_0$
   $\theta_0 = \theta_0 + \alpha_0 \delta_0 \phi_{t-1}$
4: Soft Switch:
   $W = \frac{1}{(1 + e^{-E(i-B)})}$
   $\theta_g = \theta_0 = \theta = W\theta_g + (1 - W)\theta_0$

---

4 Experimental Results

In the experiment, An 100-state Boyan chain problem was implemented to simulate episodic, undiscounted and fixed policy scenario. It is a standard episodic task for comparing TD-style algorithms with linear function approximation. All experiments were initialised with 500 episodes, repeated 50 times for reducing random error, and split with 3 features. And α step size used in these experiments takes the same form as that used in Boyans original experiments [11]:

$$\alpha_t = \alpha_0 \frac{N_0 + 1}{N_0 + Episode\#}$$

The selection of N_0 and α_0 for all tested algorithms was based on experimentally finding the best parameters α_0 from 0.1 to 10 and N_0 from 10 to 10^6 (Fig. 2).

For example, Fig. 4 shows the choice of learning rate. Each curve is the average of 10 times of experiments. The figure also shows that GTD2 still not get to optimum until learning rate reach 2. According to our experiment, GTD2 will get to optimum around the learning rate of 5. And GTD-SS is least affected by learning rate. This probably because the algorithm could always find a better solution from both TD and GTD2. We only report the results for the best set of parameters mentioned above for each algorithm.

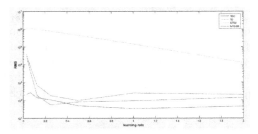

Fig. 2. RMS for different algorithms with different learning rates. (Color figure online)

4.1 Hard and Soft Switch

For GTD-SS algorithm, there are two parameters E and B showed in function
(8) need to be optimized. First, E was set to constant 1 when B was tuned to
find a point where switch is best. Then that B was kept and E was tuned to
decide how fast should the switch goes.

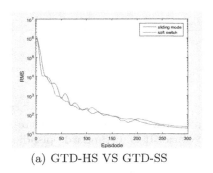

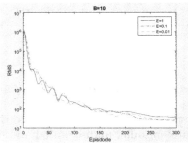

(a) GTD-HS VS GTD-SS

(b) Soft switch in different expansions

Fig. 3. Hard and soft switch results

The performance of two algorithms on 100 states problems, averaged over
20 runs, shown in Fig. 3 respectively. Figure 3(a) shows that GTD-HS algorithm
converged faster than GTD-SS algorithm when episodes are between 0 and 20,
yet when episodes are between 20 and 100, learning process has jitter. Figure 3(b)
shows algorithm's performance in different expansions. The figure shows that
smaller E will get a slower learning speed when episodes are between 0 and 20,
yet when episodes are between 20 and 100, learning process is more stable.

Section 3.2 described that when switch is fast enough, GTD-SS algorithm
will look like GTD-HS algorithm. A common intuition is that GTD-SS algorithm
always gives a more stable fusion process and reduces jitter. If the learning agent
need a more stable learning process, then GTD-SS algorithm would be a better
choice.

Table 1. Run time experiment

Learning rate	TD	TDC	GTD2	GTD-SS
0.03	6.75	9.08	8.45	9.17
0.06	6.39	8.54	9.11	10.56
0.12	7.65	9.87	8.69	10.41
0.25	6.95	9.06	8.59	10.02
0.50	6.67	8.42	7.95	8.91
1.0	6.32	8.40	8.01	8.88
2	6.25	8.34	7.98	8.93
4	6.57	8.38	7.93	9.35
8	6.87	9.03	8.14	9.28
16	6.49	8.56	8.13	9.19
average time	6.69	8.77	8.30	9.47

4.2 GTD-SS Algorithm and Original Algorithms

Now that GTD-SS algorithm is more stable algorithm than GTD-HS algorithm. It should be compared with TD(0) for converging speed and compared with GTD2 for converging accuracy. It was also compared with TDC for overall performance.

GTD-SS algorithm in Fig. 4 chose E = 0.01 to compare it with TDC and GTD2. It shows that TDC performs more like TD(0) with accuracy improvement. GTD-SS algorithm performs more like GTD2 with its higher accuracy. Here GTD-SS converges slightly slower than TDC because it has some GTD2 weight in the calculation. If E is chose more carefully, it will get more close to TDC.

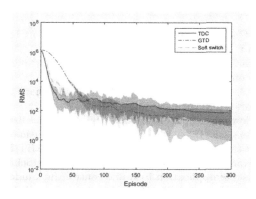

Fig. 4. Contrast experiment.

Table 1 shows actual run time of 300 episodes with different learning rate in different algorithms. The unit of time is second. It shows that learning rate has little impact on run time. And GTD-SS algorithm is only slightly slower than GTD2 algorithm.

5   Conclusion

The experiments show that the switch algorithms could take advantage of TD(0) and GTD to get a faster and more accuracy at the same time. This is an important property for an online learning algorithm because if an agent wishes to get an MSPBE level accuracy it will not need to wait long when using GTD2. This switch algorithm could also uses in other algorithms when mesh up comes to a choice. For example, TD(λ) and TDC can replace TD(0) and GTD2 respectively, it will also works fine.

The results also shows GTD-SS converges more stable than GTD-HS. It may because of TD(0) and GTD converge to different directions, the combination with a continuous changing weight will lead to a mix gradient-descent process and reduce jitter, otherwise a sudden gradient-descent direction change will cause jitter happens.

Acknowledgment. This research was supported and partially sponsored by the National Natural Science Foundation of China Grant No. 61202218.

References

1. Sutton, R.S., Maei, H.R., Precup, D., Bhatnagar, S., Silver, D., Szepesvri, C., Eric Wiewiora: Fast gradient-descent methods for temporal-difference learning with linear function approximation. In: Danyluk Et, pp. 993–1000 (2009)
2. Li, Y., Schuurmans, D.: MapReduce for parallel reinforcement learning. In: Sanner, S., Hutter, M. (eds.) EWRL 2011. LNCS, vol. 7188, pp. 309–320. Springer, Heidelberg (2012)
3. Van Hasselt, H., Guez, A., Silver, D.: Deep reinforcement learning with double q-learning. Comput. Sci. (2015)
4. Dabney, W., Thomas, P.: Natural temporal difference learning. In: Twenty-Eighth AAAI Conference on Artificial Intelligence (2014)
5. Harm Van Seijen, A., Mahmood, R., Pilarski, P.M., Machado, M.C., Sutton, R.S.: True online temporal-difference learning (2015)
6. Geramifard, A., Bowling, M., Zinkevich, M., Sutton, R.S.: ilstd: Eligibility traces and convergence analysis. In: Advances in Neural Information Processing Systems 19: Proceedings of the 2006 Conference, vol. 19, p. 441. MIT Press (2007)
7. Prashanth, L.A., Korda, N., Munos, R.: Fast LSTD using stochastic approximation: finite time analysis and application to traffic control. In: Calders, T., Esposito, F., Hüllermeier, E., Meo, R. (eds.) ECML PKDD 2014, Part II. LNCS, vol. 8275, pp. 66–81. Springer, Heidelberg (2014)
8. Sabanovic, A.: Variable structure systems with sliding modes in motion controla survey. IEEE Trans. Ind. Inform. **2**(7), 212–223 (2011)

9. Konidaris, G., Niekum, S., Thomas, P.S.: Td: Re-evaluating complex backups in temporal difference learning. In: Advances in Neural Information Processing Systems, pp. 2402–2410 (2011)
10. Sutton, S.R., Barto, G.A.: Reinforcement learning : an introduction. IEEE Trans. Neural Netw. **9**(5), 1054 (1998)
11. Boyan, J.A.: Least-squares temporal difference learning. In: Proceedings of the Sixteenth International Conference on Machine Learning, pp. 49–56 (1999)

Modeling of Travel Behavior Processes from Social Media

Yuki Yamagishi[(✉)], Kazumi Saito, and Tetsuo Ikeda

School of Management and Information, University of Shizuoka, Shizuoka, Japan
yamagissy@gmail.com, {k-saito,t-ikeda}@u-shizuoka-ken.ac.jp

Abstract. We attempt stochastic modeling of travel behavior processes from the observed data. To this end, based on the Lévy flight behavior process combined with the popularity of each point of interest, we first propose a probability model and efficient method that estimates the model parameters from the observed user behavior data. Then, we propose two methods for POI ranking by using the probability obtained from our proposed model. In our experiments using user behavior data constructed from a review site dataset, we report our experimental results on parameter estimation and examine the properties of POI ranking methods in comparison to a naive popularity ranking method. As our experimental results, we show that our parameter estimation results are intuitively interpretable, and as a favorable property, our ranking methods naturally give high ranks to POIs located in attractive regions.

Keywords: Travel behavior processes · Social media · Probability model

1 Introduction

The emergence of Social Media, like "TripAdvisor"[1], has provided us with the opportunity to collect a large amount of user behavior data, e.g., through their review articles to point of interest. Such review articles are constantly generated by various types of users based on their individual decisions. However, we can naturally assume some statistical regularities on their travel processes as intrinsic properties of human behavior. Thus, it would be possible to find empirical regularities and develop explanatory accounts of these processes in terms of macroscopic statistical properties. Furthermore, by constructing computational models based on these statistical properties, we can expect to precisely estimate how humans behave in future travel processes. Especially, such predictive capability would be valuable for anticipating social trends, market opportunities and so on. Thus, we propose to conduct research on computational models and methods for uncovering fundamental mechanisms of human behavior over travel processes.

---

[1] http://www.tripadvisor.com/.

© Springer International Publishing Switzerland 2016
R. Booth and M.-L. Zhang (Eds.): PRICAI 2016, LNAI 9810, pp. 626–637, 2016.
DOI: 10.1007/978-3-319-42911-3_52

As pioneering work on user behavior modeling, we focus on the Lévy flight behavior process (Brockmann et al. 2006). Then, based on this process combined with the popularity of each POI, we first propose a probability model and efficient method that estimates the model parameters from the observed user behavior data. We estimate the model parameters by best matching the prediction of the user's next visiting POI with respect to the observed behavior distributions, i.e. by maximizing the logarithmic likelihood function for the observed data (Bishop 2010). The parameter estimation algorithm uses an iterative scheme and it is very efficient taking full advantage of the convexity of this likelihood function (Seber and Wild 1989). As an application of our probability model for user behavior, we construct a POI network by creating each directed link having a conditional probability between two POIs. In order to identify the important nodes in this network as studied in (Wasserman and Faust 1994), we propose two ranking methods based on the conditional probability obtained from our proposed model.

After that, we evaluate the proposed model and ranking methods by applying them to our behavior data constructed from a TripAdvisor dataset. More specifically, after reporting basic statistics of our behavior data, we examine the scale-free properties of POI popularity (Song et al. 2010) and movement distances (Brockmann et al. 2006; Song et al. 2010) in our user behavior data. Then, we report our experimental results on parameter estimation and examine the properties of POI ranking methods in comparison to a naive popularity ranking method. Here, we expect that our proposed model and ranking methods can be core techniques to improve the other methods such as those for the orienteering problem (Vansteenwegen et al. 2011).

The rest of the paper is organized as follows: We describe related work in Sect. 2. We then describe the proposed model, the parameter estimation algorithm and the ranking method in Sect. 3. After explaining our dataset obtained from TripAdvisor, we report the results of our experiments on parameter estimation and evaluate the property of our ranking methods in Sect. 4. We conclude the paper by summarizing the main results and needed future work in Sect. 5.

2   Related Work

Recent technological innovation and popularization of high-performance mobile/smartphones have drastically changed our communication style and the use of various social media has been substantially affected our daily lives. Therefore, studies of the social recommender systems (Ricci et al. 2011) have attracted a great deal of attention by many researchers. Among them, our research directly relates to location-based recommendation methods (Bao et al. 2015). For instance, Zheng et al. (2009) have proposed a method to find interesting POIs from GPS trajectories. However, we consider that most existing methods without assuming the human behavior model must have an intrinsic limitation to improve the predictive performance of user behavior processes.

Studies on human mobility have attracted attention from several research communities recently. Some of them suggested that human mobility pattern shows a scaling property and, in particular, a Lévy flight characteristic. Brockmann et al. (2006) reported the distribution of traveling distances decays as a power law, by analyzing the circulation of bank notes in the US. Jiang et al. (2009) reported that human mobility exhibits Lévy flight behavior using taxi GPS data and that the Lévy flight behavior is mainly attributed to the underlying street network topology. Rhee et al. (2011) reported that human walk patterns contain statistically similar features observed in Lévy walks using 226 daily GPS traces collected from 101 volunteers. In this paper, based on Lévy flight behavior process, we propose our probability model for user behavior combined with the popularity of each POI.

In order to estimate model parameter such as β appearing in the Lévy flight behavior process, we employ a statistical machine learning approach formulated as a maximization problem of logarithmic likelihood function (Bishop 2010). In order to maximize this function with respect to parameters, we utilize an iterative algorithm based on non-linear optimization techniques (Luenberger 2003). Here, we should note that due to the convexity of our objective function, we can guarantee that our model has a globally optimal solution (Seber and Wild 1989).

Studies of the structure and functions of large complex networks have attracted a great deal of attention in many different fields such as sociology, biology, physics and computer science (Newman 2003). Especially, the scale-free properties of these networks have been extensively studied (Song et al. 2010; Easley and Kleinberg 2010), and more intricate properties such as degree correlation (Vázquez 2003) have been proposed. In this paper, by focusing on these properties, we analyze our dataset and experimental results.

For a given network, it is a fundamental task to find important nodes from some aspects. In the field of social network analysis research, a number of centrality measures have been widely studied, which include degree, closeness and betweenness centralities (Wasserman and Faust 1994). On the other hand, in the field of Web information retrieval research, node ranking by PageRank (Brin and Page 1998) and HITS (Kleinberg 1999) have been widely recognized as useful ones. Among these ranking methods, we propose two ranking methods based on the in-degree and PageRank ranking methods because they can be straightforwardly applied to the probability network. In our experiments, we evaluate the some basic properties of our proposed ranking methods in comparison to a naive popularity ranking method.

3   Proposed Model, Parameter Estimation Algorithm and Ranking Methods

First, we propose our probability model for user behavior. Let $\mathcal{U} = \{u, v, w, \cdots\}$ and $\mathcal{S} = \{q, r, s, \cdots\}$ be the sets of users and POIs whose number of elements are denoted by $M = |\mathcal{U}|$ and $N = |\mathcal{S}|$, respectively. Here, we express the distance between two POIs, r and s, as $d(r, s)$. Then, according to the Lévy flight behavior process with an exponent parameter θ_1, the conditional probability $p_1(s \mid r; \theta_1)$ that user u visits POI s after staying at POI r is assumed to be proportional to $d(r, s)^{-\theta_1}$, i.e.,

$$p_1(s \mid r; \theta_1) = \frac{d(r, s)^{-\theta_1}}{\sum_{q \in S} d(r, q)^{-\theta_1}}. \tag{1}$$

Let $f(s)$ be the popularity of POI $s \in S$; then due to a scale-free property of the popularity (Song et al. 2010) as shown in our later experiments, by using an exponent parameter θ_2, the probability $p_2(s; \theta_2)$ that user u visits POI s is assumed to be proportional to $f(s)^{\theta_2}$, i.e.,

$$p_2(s; \theta_2) = \frac{f(s)^{\theta_2}}{\sum_{q \in S} f(q)^{\theta_2}}. \tag{2}$$

Therefore, by combining these probabilities, $p_1(s \mid r; \theta_1)$ and $p_2(s; \theta_2)$, we can obtain the following conditional probability as our basic user behavior model:

$$p(s \mid r; \boldsymbol{\theta}) = \frac{p_1(s \mid r; \theta_1) p_2(s; \theta_2)}{\sum_{q \in S} p_1(q \mid r; \theta_1) p_2(q; \theta_2)} = \frac{d(r, s)^{-\theta_1} f(s)^{\theta_2}}{\sum_{q \in S} d(r, q)^{-\theta_1} f(q)^{\theta_2}}. \tag{3}$$

where $\boldsymbol{\theta} = (\theta_1, \theta_2)^T$ and $\boldsymbol{a}^T$ stands for a transposed vector of $\boldsymbol{a}$. Here, we should emphasize that our model can be easily extended by introducing the other factors as some visiting probabilities $p(s \mid \theta)$.

Next, we describe our learning algorithm to estimate the parameter vector $\boldsymbol{\theta}$. Let $\mathcal{D} = \{\cdots, (u, s, t), \cdots\}$ be the observed user behavior data, where each element (u, s, t) means that user $u \in \mathcal{U}$ visited POI $s \in S$ at time t. Then, from the observed data $\mathcal{D}$, we can know user u's m-th visited POI, denoted by $s(u, m) \in S$. Hereafter, let $M(u)$ be the total number of POIs visited by the user u, and $N(s)$ be the total number of users who visited the POI s. In order to estimate $\boldsymbol{\theta}$ with respect to $\mathcal{D}$, based on a standard machine learning approach (Bishop 2010), we consider the following logarithmic likelihood function as our objective function to be maximized:

$$L(\boldsymbol{\theta}; \mathcal{D}) = \sum_{u \in \mathcal{U}} \sum_{1 \le m < M(u)} \log p(s(u, m + 1) \mid s(u, m); \boldsymbol{\theta}). \tag{4}$$

Then, by introducing a new vector defined by $\boldsymbol{x}(r, s) = (-\log d(r, s), \log f(s))^T$, from Eq. (3), we can transform Eq. (4) as follows:

$$L(\boldsymbol{\theta}; \mathcal{D}) = \sum_{u \in \mathcal{U}} \sum_{1 \le m < M(u)} \left(\boldsymbol{\theta}^T \boldsymbol{x}(s(u, m), s(u, m + 1)) - \log \sum_{q \in S} \exp(\boldsymbol{\theta}^T \boldsymbol{x}(s(u, m), q)) \right). \tag{5}$$

Thus, we can calculate the following gradient vector and Hessian matrix with respect to our objective function defined in Eq. (4).

$$\frac{\partial L(\boldsymbol{\theta}; \mathcal{D})}{\partial \boldsymbol{\theta}} = \sum_{u \in \mathcal{U}} \sum_{1 \leq m < M(u)} \left(\boldsymbol{x}(s(u, m), s(u, m + 1)) - \sum_{q \in \mathcal{S}} p(q \mid s(u, m); \boldsymbol{\theta}) \boldsymbol{x}(s(u, m), q)) \right),$$

$$\frac{\partial^2 L(\boldsymbol{\theta}; \mathcal{D})}{\partial \boldsymbol{\theta} \partial \boldsymbol{\theta}^T} = \sum_{u \in \mathcal{U}} \sum_{1 \leq m < M(u)} - \left(\sum_{q \in \mathcal{S}} p(q \mid s(u, m); \boldsymbol{\theta}) \boldsymbol{x}(s(u, m), q) \boldsymbol{x}(s(u, m), q)^T \right.$$
$$- \left. \left(\sum_{q \in \mathcal{S}} p(q \mid s(u, m); \boldsymbol{\theta}) \boldsymbol{x}(s(u, m), q)) \right) \left(\sum_{q \in \mathcal{S}} p(q \mid s(u, m); \boldsymbol{\theta}) \boldsymbol{x}(s(u, m), q)) \right)^T \right).$$

Here, due to the convexity of our objective function, because the Hessian matrix can be negative-definite under some mild condition, we can guarantee that our model has a globally optimal solution (Seber and Wild 1989). Thus, we can employ any iterative procedure starting from arbitrary initial parameter values. In our experiments, we employed safeguarded Newton's method (Luenberger 2003), whose modification vector is calculated as follows:

$$\boldsymbol{\delta} = -\frac{\partial L(\boldsymbol{\theta}; \mathcal{D})}{\partial \boldsymbol{\theta}} \left(\frac{\partial^2 L(\boldsymbol{\theta}; \mathcal{D})}{\partial \boldsymbol{\theta} \partial \boldsymbol{\theta}^T} \right)^{-1}. \tag{6}$$

Using a small constant $\epsilon = 10^{-8}$, we can summarize our learning algorithm below:

1. Initialize the parameter vector by $\boldsymbol{\theta}_v \leftarrow \boldsymbol{0}$;
2. Calculate the modification vector $\boldsymbol{\delta}$ by Eq. (6), and terminate the iteration if $\|\boldsymbol{\delta}\| < \epsilon$;
3. Update the parameter vector by $\boldsymbol{\theta} \leftarrow \boldsymbol{\theta} + \boldsymbol{\delta}$, and return to step 2.

Finally, we propose two POI ranking methods based on our probability model. Let $\hat{\boldsymbol{\theta}}$ be the estimated parameter value by our algorithm, i.e., $\hat{\boldsymbol{\theta}} = \text{argmax}_{\boldsymbol{\theta}} L(\boldsymbol{\theta}; \mathcal{D})$, then we can obtain the conditional probability of visiting POI s from POI r as $p(s \mid r; \hat{\boldsymbol{\theta}})$. Thus, we can consider a POI network $G = (\mathcal{S}, \mathcal{S} \times \mathcal{S})$ where the conditional probability $p(s \mid r; \hat{\boldsymbol{\theta}})$ is assigned to each link $(r, s) \in \mathcal{S} \times \mathcal{S}$. As noted earlier, one of the fundamental tasks on network analyses is to explore adequate usefulness measures to assign a value to each node (POI) over a given network. For this purpose, we consider two measures referred to as the in-degree and PageRank methods. In the in-degree methods, the usefulness measure of each POI $s \in \mathcal{S}$ is defined as follows:

$$id(s) = \sum_{q \in \mathcal{S}} p(s \mid q; \hat{\boldsymbol{\theta}}). \tag{7}$$

Namely, this method ranks a POI high if its total probability to be visited from other POIs is high. Here note that the total probability of visiting to other POIs is one, i.e., $\sum_{q \in \mathcal{S}} p(q \mid s; \hat{\boldsymbol{\theta}}) = 1$. On the other hand, the PageRank method ranks a POI high if its visiting probability under a random walk process is high. More specifically, let $pr(s)$ be a visiting probability for POI $s \in \mathcal{S}$; then, according to the original PageRank algorithm (Brin and Page 1998), we can consider the following random walk process.

$$pr(s) \leftarrow (1 - \alpha) \sum_{q \in S} p(s \mid q; \hat{\boldsymbol{\theta}})pr(q) + \frac{\alpha}{M}, \tag{8}$$

where α means a uniform jump probability, typically set to $\alpha = 0.15$, as suggested in (Brin and Page 1998). In summary, by performing the above random walk simulation, we can obtain a stationary state value $pr(s)$, which is used as the ranking measure of the PageRank method.

4 Experiments

4.1 Dataset

From "TripAdvisor"[2], we collected review articles about Japanese POIs, and constructed our user behavior data $\mathcal{D}$ from them. More specifically, we collected $441,087$ reviews by $M = 52,355$ users for $N = 19,827$ POIs from 2007/02/07 to 2015/10/21. Thus, the average numbers of reviews per user and item were 8.4 and 22.2 respectively. The review score is an integer value ranging from 1 to 5 and its average of overall ratings was 4.0. Figure 1(a) shows the numbers of reviews per a month on this site since 2008. We can see that these numbers are steadily increasing although there exist some fluctuations. Figure 1(b) shows the frequency of each review score. We can see that most review results are positive to the visited POIs. By assuming that the timestamp ordering of reviews coincides

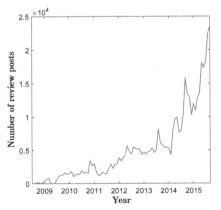

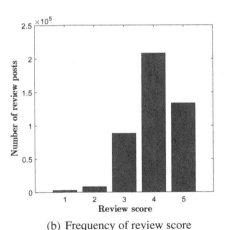

(a) Transition of review posts about Japanese POIs

(b) Frequency of review score

Fig. 1. TripAdvisor dataset

---

[2] http://www.tripadvisor.com/.

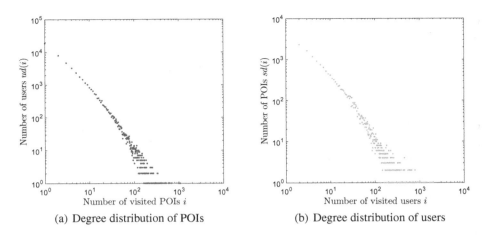

(a) Degree distribution of POIs   (b) Degree distribution of users

Fig. 2. Scale-free property of observed data

with the actual orders of visits, We constructed our behavior data $\mathcal{D}$ by using all of the review data.

Next, we examined the scale-free property of popularity of POIs and user activities (Easley and Kleinberg 2010). To this end, for a given integer i, we define the degree $sd(i)$ of POIs and the degree $ud(i)$ of users as follows:

$$sd(i) = |\{s \in \mathcal{S} : N(s) = i\}|, \quad ud(i) = |\{u \in \mathcal{U} : M(u) = i\}|. \quad (9)$$

Namely, $sd(i)$ the number of POIs whose number of visited users is i, and $ud(i)$ means the number of users whose number of visited POIs is i. Figure 2(a) and (b) show our analysis results about the degree distributions of POIs and users. From these figures, we can confirm that each of the degree distribution is reasonably approximated to a power law, and these results also support our assumption to a scale-free property of popularity of POIs. Thus, in what follows, we define the popularity of POI $f(s)$ by number of visited users at POI s, $f(s) = N(s)$.

Finally, we examined the scale-free property of movement distances (Brockmann et al. 2006; Song et al. 2010) in our user behavior data $\mathcal{D}$. To this end, for a given distance δ, we define the degree $dd(\delta)$ of movement distances as follows:

$$dd(\delta) = |\{\cup_{u \in \mathcal{U}} \cup_{1 \leq m < M(u)} (u, m) : \delta \leq d(s(u, m), s(u, m + 1)) < \delta + \epsilon\}|. \quad (10)$$

Here, based on GRS80 (Moritz 2000), we calculated each distance between POIs each of which is described by a pair of latitude and longitude, and our interval parameter ϵ is set to 1 km. Figure 3(a) shows our analysis result about the degree distribution of movement distances. Again, we can confirm that this degree distribution is also reasonably approximated by a power law, and this result also supports our assumption to the Lévy flight behavior process. Figure 3(b) shows the location of POIs in our data $\mathcal{D}$. We can see that these POI locations used in our behavior data $\mathcal{D}$ are spread in whole Japan.

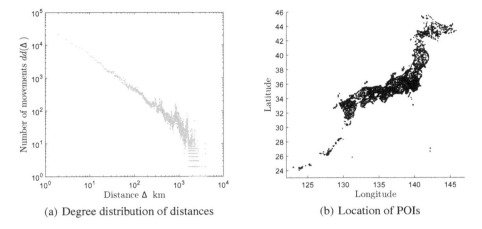

(a) Degree distribution of distances   (b) Location of POIs

Fig. 3. Property of user movements and POI locations

4.2 Experimental Results

First, we estimated parameters θ_1 and θ_2 by selecting the users according to the number of visited POIs $M(u)$, i.e.,

$$\mathcal{D}_\tau = \{(u, v, t) \in \mathcal{D} \ : \ M(u) \geq \tau\}, \tag{11}$$

where τ is a threshold. Figure 3(a) shows our experimental results where vertical and horizontal axes indicate the values of threshold τ and parameters, respectively. Here, we can see that as τ becomes large, θ_1 slightly increased, but θ_2 substantially decreased. This experimental result suggests that for users who visit many POIs, the popularity might not be a so important factor. In order to confirm our claim, we consider the degree correlation (Vázquez 2003) defined by

$$dc(i) = \frac{1}{ud(i)} \sum_{\{u \in \mathcal{U} \ : \ M(u)=i\}} \frac{1}{M(u)} \sum_{1 \leq m \leq M(u)} N(s(u, m)), \tag{12}$$

where recall that $N(s)$ indicates the number of visited users at POI s. Figure 4(b) shows our experimental results. Clearly, this experimental result coincides with that of Fig. 3(a). Namely, we can confirm that user whose numbers of visited POIs are relatively small likely to visit relatively popular POIs. In summary, we can say that our parameter estimation results are intuitively interpretable.

Next, we examined the properties of POI ranking by the PageRank method, in comparison to those by the popularity and in-degree methods. Now, we denote the set of the top-k POIs ranked by the PageRank method by $R(k)$, and those by the popularity and in-degree methods by $R_{pop}(k)$ and $R_{ind}(k)$. Then, we evaluated the ranking similarity among these methods by using the following precision measure:

$$rs(k; x) = \frac{|R(k) \cap R_x(k)|}{k}, \tag{13}$$

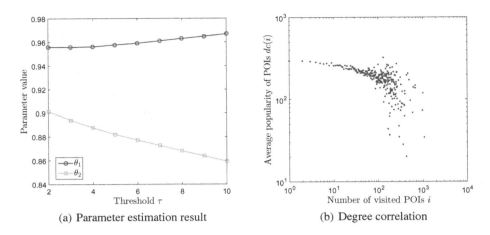

(a) Parameter estimation result   (b) Degree correlation

Fig. 4. Evaluation of parameter evaluation

where $x \in \{pop, ind\}$. Figure 5(a) shows our evaluation results by precision until $k = 1,000$. We can see that the average rates of precision were around 0.70 for the popularity method and 0.80 for the in-degree method. Namely, we can observe substantial numbers of differences among these three methods. In order to visually examine those differences, we plotted those top-1,000 POIs by changing colors according to their ranks. Figures 5(b), (c) and (d) shows our visualization results by the popularity, in-degree and PageRank methods. Then, we can observe that highly ranked colors in Fig. 5(b) are widely spread in whole Japan while those colors in Fig. 5(d) are restricted to a small number of regions; and the distribution of those colors in Fig. 5(c) has the intermediate nature. These experimental results suggest that the PageRank method is likely to give a high rank when the POI locates at some attractive region including a number of highly ranked POIs. In order to more closely examine these ranking properties... the number of elements in a region set $J = |\mathcal{J}|$ (POIs also $M = |\mathcal{S}|$). $\mathcal{S} = \{1, \cdots, s, \cdots, S\}$. $\mathcal{J} = \{1, \cdots, j, \cdots, J\}$. the region where a POI s belongs, $j = f(s)$. the number of POIs belong to a region j, $M_j = |\mathcal{S}_j| = |\{s ; j = f(s)\}|$. the rank of each POI, $1 \leq r_s \leq M$. Mann-Whitney's order statistic (Mann and Whitney 1947). when tied ranks exist, rank r_s is corrected by average ranks. multi-category order statistic, $\mathcal{S}_j$ versus $\mathcal{S} \backslash \mathcal{S}_j$ (where $\cdot \backslash \cdot$ means set difference).

$$z_j = \frac{u_j - \mu_j}{\sigma_j}, \tag{14}$$

$$u_j = M_j(M - M_j) + \frac{M_j(M_j + 1)}{2} - \sum_{s \in \mathcal{S}_j} r_s, \tag{15}$$

$$\mu_j = \frac{M_j(M - M_j)}{2}, \tag{16}$$

$$\sigma_j^2 = \frac{M_j(M - M_j)(M + 1)}{12}. \tag{17}$$

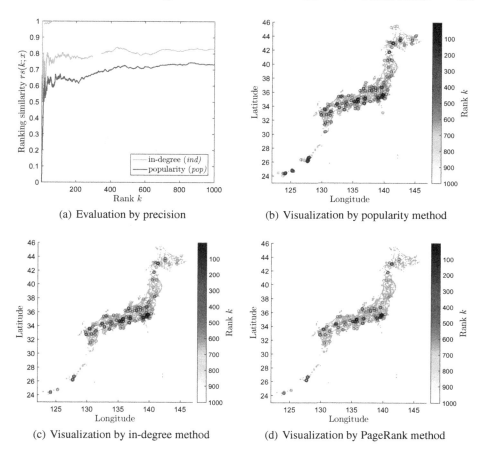

(a) Evaluation by precision

(b) Visualization by popularity method

(c) Visualization by in-degree method

(d) Visualization by PageRank method

Fig. 5. Ranking similarity among popularity, in-degree and PageRank methods (Color figure online)

when tied ranks exist, standard deviation σ_j is corrected by the standard way. Table 1 compares the top-10 prefectures by z-score of order statistic of popularity, in-degree and PageRank methods.

Finally, by removing each link $(r, s) \in \mathcal{S} \times \mathcal{S}$ with a small probability, i.e., $p(s \mid r; \hat{\boldsymbol{\theta}}) < \mu$, we examined this removal effect in terms of the average number of links and ranking similarity. In our experiments, μ is set to $\mu \in \{0.0001 \times 2^\eta : \eta = 0, \cdots 6\}$, and referred to as a link removal rate. Figure 6(a) shows the average number of links per node. We can see that the probabilities assigned to each of the links are relatively quite small. Let $R_\mu(k)$ be the top-k POIs ranked by the PageRank method over the network constructed by using a link removal rate μ; then, we can also use Eq. (13) as our ranking similarity measure. Figure 6(b) shows our evaluation results with precision until $k = 200$. From this figure, we can confirm that since the precision is greater than 0.9 if $\mu < 0.0016$,

Table 1. Top-10 prefectures by multi-category order statistic

	Popurarity		In-degree		PageRank	
Rank	z_j	Prefucture	z_j	Prefucture	z_j	Prefucture
1	14.477	Okinawa	12.605	Tokyo	22.075	Tokyo
2	7.5794	Hokkaido	6.5979	Okinawa	10.186	Kanagawa
3	5.3568	Tokyo	6.2236	Kanagawa	9.4279	Kyoto
4	2.6809	Kanagawa	4.4648	Kyoto	8.7675	Okinawa
5	2.6258	Shizuoka	3.0965	Nagano	3.1010	Chiba
6	2.2201	Kagoshima	2.9030	Shizuoka	2.8854	Shizuoka
7	2.1930	Nagano	2.2823	Aichi	2.6034	Osaka
8	1.9683	Chiba	2.1821	Yamanashi	2.0494	Nagano
9	1.6848	Ishikawa	1.7685	Chiba	1.9495	Yamanashi
10	1.5340	Oita	1.0773	Ishikawa	1.8663	Ishikawa

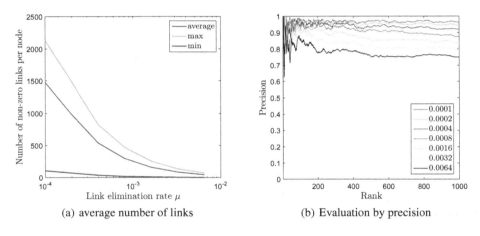

(a) average number of links   (b) Evaluation by precision

Fig. 6. Evaluation of network robustness (Color figure online)

this suggests that the constructed networks are reasonably robust in comparison to the results shown in Fig. 5(a).

5 Conclusions

In this paper, we attempted stochastic modeling of travel behavior processes from the observed data. To this end, based on the Lévy flight behavior process combined with the popularity of each POI, we first proposed our probability model and efficient method that estimates the model parameters from the observed user behavior data. Then, we proposed two methods for POI ranking by using the probability obtained from our proposed model. In our experiments using user behavior data constructed from a review site dataset, we reported our experimental results on parameter estimation and examined the properties of POI ranking methods in comparison to a naive popularity ranking method. As our experimental results, we showed that our parameter estimation results are intuitively interpretable, and as a favorable property, our ranking methods naturally give high ranks to POIs located in attractive regions. In future, we plan to evaluate the proposed method by further experiments using a wider variety of databases.

Acknowledgments. This work was supported by JSPS Grant-in-Aid for Scientific Research (C) (No. 16J11909).

References

Bao, J., Zheng, Y., Wilkie, D., Mokbel, M.: Recommendations in location-based social networks: a survey. GeoInformatica **19**(3), 525–565 (2015)

Bishop, C.M.: Pattern Recognition and Machine Learning. Information Science and Statistics. Springer, New York (2010)

Brin, S., Page, L.: The anatomy of a large-scale hypertextual web search engine. Comput. Netw. ISDN Syst. **30**, 107–117 (1998)

Brockmann, D., Hufnagel, L., Geisel, T.: The scaling laws of human travel. Nature **439**, 462–465 (2006)

Easley, D., Kleinberg, J.: Networks, Crowds, and Markets: Reasoning About a Highly Connected World. Cambridge University Press, New York (2010)

Jiang, B., Yin, J., Zhao, S.: Characterizing the human mobility pattern in a large street network. Phys. Rev. E **80**, 021136 (2009)

Kleinberg, J.: Authoritative sources in a hyperlinked environment. J. ACM **46**(5), 604–632 (1999)

Luenberger, D.G.: Linear and Nonlinear Programming, 2nd edn. Kluwer Academic Publishers, Boston (2003)

Mann, H.B., Whitney, D.R.: On a test of whether one of two random variables is stochastically larger than the other. Ann. Math. Stat. **18**(1), 50–60 (1947)

Moritz, H.: Geodetic reference system 1980. J. Geodesy **74**(1), 128–133 (2000)

Newman, M.: The structure and function of complex networks. SIAM Rev. **45**, 167–256 (2003)

Rhee, I., Shin, M., Hong, S., Lee, K., Kim, S.J., Chong, S.: On the levy-walk nature of human mobility. IEEE/ACM Trans. Netw. **19**(3), 630–643 (2011)

Ricci, F., Rokach, L., Shapira, B., Kantor, P.: Recommender Systems Handbook. Springer, New York (2011)

Seber, G.A.F., Wild, C.J.: Nonlinear Regression. Wiley, Hoboken (1989)

Song, C., Koren, T., Wang, P., Barabási, A.-L.: Modelling the scaling properties of human mobility. Nat. Phys. **6**, 818–823 (2010)

Vansteenwegen, P., Souffriau, W., Oudheusden, D.: The orienteering problem: a survey. Eur. J. Oper. Res. **209**, 1–10 (2011)

Vázquez, A.: Growing network with local rules: preferential attachment, clustering hierarchy, and degree correlations. Phys. Rev. **67**(5), 056104 (2003)

Wasserman, S., Faust, K.: Social Network Analysis. Cambridge University Press, Cambridge (1994)

Zheng, Y., Zhang, L., Xie, X., Ma, W.-Y.: Mining interesting locations and travel sequences from GPS trajectories. In: Proceedings of the 18th International Conference on World Wide Web (WWW 2009), pp. 791–800. ACM, New York (2009)

Fast Training of a Graph Boosting for Large-Scale Text Classification

Hiyori Yoshikawa[✉] and Tomoya Iwakura

Fujitsu Laboratories Ltd., Kawasaki, Japan
{y.hiyori,iwakura.tomoya}@jp.fujitsu.com

Abstract. This paper proposes a fast training method for graph classification based on a boosting algorithm and its application to sentimental analysis with input texts represented by graphs. Graph format is
very suitable for representing texts structured with Natural Language
Processing techniques such as morphological analysis, Named Entity
Recognition, and parsing. A number of classification methods which represent texts as graphs have been proposed so far. However, many of
them limit candidate features in advance because of quite large size of
feature space. Instead of limiting search space in advance, we propose two
approximation methods for learning of graph-based rules in a boosting.
Experimental results on a sentimental analysis dataset show that our
method contributes to improved training speed. In addition, the graph
representation-based classification method exploits rich structural information of texts, which is impossible to be detected when using other
simpler input formats, and shows higher accuracy.

Keywords: Text classification · Feature engineering · Graph boosting

1 Introduction

Text classification is a fundamental task in Natural Language Processing (NLP)
and has applications to a wide variety of tasks including spam filtering, sentimental analysis, topic classification, and profile estimation. While bag-of-words are
widely used as features for classification, a number of researches show that using
richer structure of texts results in better performance (Kudo and Matsumoto
2004; Gee and Cook 2005; Matsumoto et al. 2005; Arora et al. 2010; Jiang
et al. 2010; Iwakura 2013). In other words, features incorporating additional
information about texts such as word dependencies, part of speech (POS) tags,
and named entity types have potential to be key features for classification.

A remarkable approach to classification using rich information of texts is to
represent the texts as graphs. Compared to the other formats such as bag-of-
words, n-grams and trees, graph format has strong power of expression enough
to incorporate almost any kinds of characteristics related to words or texts at
the same time. More precisely, the other formats listed above can be interpreted
as special cases of graphs: bag-of-words correspond to vertexes and n-grams

© Springer International Publishing Switzerland 2016
R. Booth and M.-L. Zhang (Eds.): PRICAI 2016, LNAI 9810, pp. 638–650, 2016.
DOI: 10.1007/978-3-319-42911-3_53

to paths, and trees are graphs in themselves. For a sentimental analysis task, for example, a key feature might be a combination of word order, dependency, and sentiment polarity of each word. Classification methods with graph based features have potential to achieve higher performance especially in such cases that key features might be combinations of different kinds of characteristics. As we refer to in Sect. 5, there are a number of works which use graph representation of texts for text classification. Most of these existing methods convert inputs into subgraph-based feature vectors and then apply a classification algorithm for the vectors such as perceptron (Frank 1958) or Support Vector Machines (SVMs) (Boser et al. 1992). Since the number of potential subgraphs tend to be quite large and it is practically impossible to consider all of them, such methods usually select a part of features in advance using a frequent pattern mining algorithm such as gSpan (Yan and Han 2002). However, infrequent features are sometimes important. Another approach is to deal with the problem as a graph classification problem. One of the most popular graph classification algorithms is perceptron or SVMs with graph kernels (Kashima et al. 2003), which works without previous selection of subgraph features. Although such methods achieve considerably high performance, they have some disadvantages. First, in learning and classification it sometimes requires the calculation of graph kernels for a large number of pairs of graphs. Second, it is difficult to see which subgraphs have strong effect because features do not appear explicitly.

In this paper, we use a graph boosting algorithm originally proposed by (Kudo et al. 2004) for text classification. This boosting method learns subgraph based decision stumps as weak classifiers, and finally constructs a classifier as a linear combination of the stumps. The calculation time for classification does not depend on the size of training dataset but the size of rules, and rules are represented explicitly by subgraphs that constitutes the classifier. In addition, as Kudo et al. (2004) point out, the boosting based method can reflect slight difference of structures of features, while kernel based methods are not good at distinguishing features which have similar structures. It would be an important property for text classification, since the difference of a single word may result in opposite meaning of the whole sentence. A problem is that the graph boosting method requires much learning time, despite using pruning methods suggested in the original paper. We propose two approximation methods to improve training speed of the graph boosting: one is to divide subgraph features into some buckets in order to limit search space of rules, and the other is to expand the search space dynamically according to weak classifiers chosen in previous steps. Experimental results show that our approximation makes it possible to improve the classification accuracy much faster than the original algorithm.

The rest of the paper is organized as follows. In Sect. 2 we define the problem setting, and refer to the graph boosting method. In Sect. 3 we show two approximation methods to calculate weak classifiers efficiently. Section 4 shows experimental results, Sect. 5 discusses the relation to related works, and Sect. 6 concludes this paper.

2 Preliminary

2.1 Problem Setting

In this paper, we focus on binary text classification problems. We are given a set of texts $T = (t_1, t_2, \ldots, t_N)$, each of which is associated with a class label $y_i \in \{\pm 1\}$ $(i = 1, \ldots, N)$. Generally, the class labels are defined based on particular characteristics of the texts such as topics, sentiment, or profiles of writers. The task is to induce a classifier which assigns labels to new texts.

We solve this problem as a graph classification problem by representing the input texts as graphs with NLP techniques to extract syntactic and semantic structure of the original texts. Then the problem reduces to the task to induce a classifier which assigns labels to graphs made from new texts.

2.2 Boosting Based Graph Classification

Our algorithm is based on the graph based classification method by (Kudo et al. 2004). As a boosting method we adopt an improved AdaBoost proposed by (Schapire and Singer 1999), since it showed higher accuracy than the boosting algorithm used in (Kudo et al. 2004). We call the algorithm *Boost-K*. Here we summarize the idea of the general boosting method and Boost-K.

Boosting is one of the well-known meta-algorithms for ensemble learning. Boosting sequentially learns $K(> 0)$ *weak classifiers* and finally constructs a classifier as the linear combination of the weak classifiers. Let h_j be the weak classifier obtained at the jth iteration $(j = 1, 2, \ldots, K)$. Then we eventually obtain the final classifier consisting of the weak classifiers as:

$$f(x) = \text{sgn}\left(\sum_{j=1}^{K} h_j(x)\right), \tag{1}$$

At each iteration in a typical boosting algorithm, a weak classifier is trained to minimize the current weighted error rate. When the classifier is updated by a weak classifier, the weight is recalculated so that misclassified examples have larger weight and correctly classified ones have smaller weight. In this way, the classifier is efficiently trained focusing on the misclassified examples at previous steps.

Boost-K classifies graphs based on their subgraphs. The weak classifiers are decision stumps each of which reflects existence of a particular subgraph in a graph. For a subgraph g and a real number α (confidence value), the subgraph-based decision stump is defined as:

$$h_{\langle g, \alpha \rangle}(x) := \begin{cases} \alpha \text{ if } g \subseteq x \\ 0 \text{ otherwise} \end{cases}, \tag{2}$$

where $g \subseteq x$ means the graph g is the subgraph of the graph x.[1] At each iteration j, the boosting algorithm chooses a weak classifier $h_j = h_{\langle g_j, \alpha_j \rangle}$ with:

$$\langle g_j, \alpha_j \rangle = \arg\min_{\langle g, \alpha \rangle} \sum_{i=1}^{N} d_i^j \exp(-y_i h_{\langle g, \alpha \rangle}(x_i)), \tag{3}$$

where d_i^j is the weight for the input graph x_i at the current iteration j.[2] The right hand side is minimized for a particular g by choosing:

$$\alpha = \frac{1}{2} \log \left(\frac{D_{j,+1}(g)}{D_{j,-1}(g)} \right), \tag{4}$$

where $D_{j,*}(g) := \sum_{i=1}^{N} d_i^j I(g \subseteq x_i \wedge y_i = *)$ ($* \in \{\pm 1\}$) with the indicator function $I(\cdot)$. As shown in (Iwakura and Okamoto 2008), we can minimize (3) by maximizing the following *gain function*, or *gain* simply:

$$gain_j(g) := \left| \sqrt{D_{j,+1}(g)} - \sqrt{D_{j,-1}(g)} \right|. \tag{5}$$

At every step, the algorithm choose a weak classifier which maximizes (5) and then the weight $\boldsymbol{d} = (d_1, d_2, \ldots, d_N)$ is updated by:

$$d_i^{j+1} = d_i^j \exp(-y_i h_{\langle g_j, \alpha_j \rangle}(x_i)) \tag{6}$$

and then normalized to satisfy $\sum_{i=1}^{N} d_i = 1$.

2.3   Efficient Calculation of Weak Classifiers

At each step in the above boosting algorithm, we need to find a weak classifier that maximizes the gain function (5). Generally, the number of possible subgraphs is so large that it is practically impossible to calculate gains for all subgraph features. Thus we need some efficient ways to find the most appropriate subgraph feature. Boost-K addresses this problem based on the following two ideas, both of which do not affect the result of learning.

The first idea, which is by (Kudo et al. 2004), is to search subgraphs on a canonical search space based on *gSpan* algorithm (Yan and Han 2002). gSpan is an efficient method to enumerate subgraphs which appear in a given graph set frequently. The key idea is to retain subgraphs by *DFS codes*. A DFS code is constructed by running depth-first search (DFS) in a search space called *DFS Code Tree*. A node of the DFS Code Tree corresponds to a 5-tuple $(i, j, l_i, l_{i,j}, l_j)$ which represents an edge of a subgraph. Here i and j are the vertex indexes of endpoints of e, and l_i, l_j, and $l_{i,j}$ are labels of the vertices i and j, and the edge $\{i, j\}$, respectively. By running depth-first search in the DFS Code Tree,

---

[1] In (Kudo et al. 2004), a weak classifier is defined to return $-\alpha$ if $g \not\subseteq x$. Considering the results of preliminary experiments, we decided to use the above definition instead.

[2] We may omit the iteration index j when no confusion can arise.

we obtain a DFS code as a sequence of the tuples. Since DFS codes have a lexicographic order, we can use the *minimum DFS code* as the 'canonical' code of a graph. When we find that the current DFS code is not minimum, we can 'prune' the search space to avoid the redundant search. In this way, we can efficiently enumerate all subgraphs.

The second idea, which is also rooted in (Kudo et al. 2004), is to use an upper bound of gain functions. The following is a key observation:

$$\{i : g' \subseteq x_i, y_i = y\} \subseteq \{i : g \subseteq x_i, y_i = y\} \quad (\forall g' \supseteq g). \tag{7}$$

That is, a graph g' appears in a graph x_i only if its subgraph g appears in x_i. Then the following is directly derived from the definition (5): for every graph g' which contains g, the gain $gain(g')$ is bounded by:

$$u(g) := \max\left(\sqrt{D_{j,+1}(g)}, \sqrt{D_{j,-1}(g)}\right). \tag{8}$$

In the depth-first traversal of a DFS Code Tree, the graphs are referred to starting from a single edge graph[3], and then larger graphs are referred to as the search reaches deeper levels. Using the above observation, we can avoid redundant searches for larger graphs: we can prune the search space when we find that the upper bound does not exceed the current maximal gain.

3   Approximation Methods for More Efficient Learning

Despite the processes described above, it still takes much time to search for subgraph features in a large graph set. In addition to the above methods, we adopt two other approximation methods for further efficient learning.

3.1   Dividing Features into Buckets

The first method is to divide features into some buckets, whose idea comes from (Iwakura and Okamoto 2008). They show that distribution of feature into hundreds of buckets results in almost the same or sometimes higher accuracy. We adopt *F-dist*-like distribution, that is, distribution of features in ascending order based on their frequencies. This distribution keeps average frequencies in each bucket roughly the same. Since it is practically impossible to enumerate all possible subgraph features, we modify the method to adapt to our situation as follows.

1. Count frequency of each vertex label in the graph set.
2. Sort vertex labels according to their frequency.
3. Put the vertex labels into $b(> 0)$ buckets in order of their frequency.

---

[3] With a slight modification, we can start searches from single node graphs so that the result may contain single node feature graphs.

4. In the jth iteration we search only for the subgraphs whose start point (id 0 in the DFS Code) has the label in the $(j\%b)$th bucket, where $(j\%b)$ means the remainder of j divided by b.

In this way, one can expect that the total frequency of the feature subgraphs searched in each iteration become roughly the same.

When applying this method, using only the minimum DFS codes results in excessive limitation of the search space, since the search space depends on the first vertex in DFS codes. Thus we omit minimum DFS code tests when applying this approximation. Note that a subgraph feature can appear in more than two buckets. The experimental results in Sect. 4 show that the method reduce the calculation time even though it omits minimum DFS code tests.

3.2 Smaller Rule Priority

The search space of weak classifiers expand explosively with size of subgraphs to search. Usually, however, only a small fraction of subgraph features are significant. In order to avoid unnecessary search, we limit the search space based on the following hypothesis: when a large subgraph feature is important, some of its subgraphs are also important. To realize this idea, we propose to apply the idea of (Freund 1999) to the graph boosting algorithm. (Freund 1999) learns alternating decision tree by boosting. Starting from a set of the simplest rules, the algorithm extends the search space according to the result of each step of boosting. We apply the learning method to our situation of learning subgraph based decision stumps as follows:

1. Initialize $\mathcal{H}$ as the set of all single node graphs.
2. For each jth iteration, do the following:
   (a) Search for the best weak classifier $h_{\langle g_j, \alpha_j \rangle}$ with $g_j \in \mathcal{H}$.
   (b) Update $\mathcal{H}$ by $\mathcal{H} \leftarrow \mathcal{H} \cup \{g' \mid g_j \subseteq g', |g'|_E = |g_j|_E + 1\}$, where $|x|_E$ indicates the number of edges in x.

That is, the algorithm searches for only subgraph features which contains g_j and larger than g_j by one edge for subgraphs g_j chosen in previous iterations.

4 Experiments

To evaluate the proposed method, we used the Amazon review data created by (Blitzer et al. 2007). This dataset contains customer review texts for products available at Amazon. Table 1 shows the product categories we used and the size of each dataset. For each category, we picked positive (4.0 or more score) reviews and negative (2.0 or less score) reviews and learned binary classifiers that distinguish between positive and negative reviews. The construction of graphs from input texts is described in Sect. 4.1. In addition to the above approximation, we also limit the search space to the subgraph features whose size (number of edges) are no more than $ms \in \{0, 1, 2, ...\}$, where $ms = 0$ means the search space is limited to single node subgraph features. We implemented the algorithm by C++.

Table 1. Used categories in Amazon review data. '#Train' and '#Test' mean the number of training and test data.

Category	#Train		#Test	
	positive	negative	positive	negative
video	27489	5074	3054	563
electronics	16165	4544	1796	504
kitchen-housewares	14164	3708	1573	411
toys-games	9522	2312	1057	256
apparel	7111	1216	790	135
camera-photo	5679	990	630	109

4.1 Construction of Input Graphs

Here we describe how we construct graph features from input texts. We construct a graph from each input text, and finally obtain a set of graphs whose number equals that of input texts. From now we call a graph corresponding to a text a *feature graph*.

For a word w, let p_w be the POS tag and n_w be the named entity type of w. Let us write an input text as $t = w_1 w_2 \ldots w_l$, where w_i $(1 \leq i \leq l)$ is the i-th word of the text. The set of vertices of a feature graph consists of some of the following vertices:

- v_w^i: the vertex corresponding to the surface form of the word w_i $(i = 1, \ldots, l)$,
- v_p^i: the vertex corresponding to the POS of w_i $(i = 1, \ldots, l)$,
- v_n^i: the vertex corresponding to the named entity type of w_i $(i \in \{1, \ldots, l\})$, which appears only if w_i is a part of a named entity,
- v_S: the vertex representing the start of the input text, and
- v_T: the vertex representing the end of the input text,

whose labels correspond to w_i, p_{w_i}, n_{w_i}, [S] and [T], respectively. Every edge corresponds to order or dependency of words. Edges with the label ORDER are between v_*^i and v_{**}^{i+1} $(*, ** \in \{w, p, n\})$ if such nodes exist $(i = 1, \ldots, l-1)$, v_S and v_*^1, and v_*^l and v_T $(* \in \{w, p, n\})$. In addition, each pair of vertices v_w^i and v_w^j which has a dependency relation has an edge with a label corresponding to the kind of relation. For graph construction, we used SENNA (Collobert et al. 2011; Collobert 2011)[4]. In this paper, we call a graph which has all the above vertexes and edges 'type A', a graph which has only the information of words and order 'type B', a graph which has only the information of words and dependencies 'type D', and a graph which has the information of words, order and dependencies 'type BD'. Figure 1 shows an example of a feature graph

---

[4] To convert the output of SENNA into tree format, we used Penn2Malt 0.2 (http:// stp.lingfil.uu.se/~nivre/research/Penn2Malt.html) with the following options: head rules in (http://stp.lingfil.uu.se/~nivre/research/headrules.txt), deprel 1, and punctuation 1.

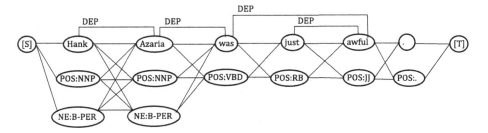

Fig. 1. An example of a feature graph corresponding to a sentence "Hank Azaria was just awful".

(type A) made from a sentence in a negative review in 'video' category, "Hank Azaria was just awful". Words in the vertices and on the edges indicate the labels of the vertices and edges. The edges on which no word appears have the label ORDER.

We also constructed other kinds of graphs: 'type Bs'. These graphs contain information about sentimental polarity of words. We append polarity to the words according to SentiWordNet 3.0 (Baccianella et al. 2010). We simply append 'Sent:p', 'Sent:n', or 'Sent:pn' labels to a word if it has non-zero positive, negative, or both score in SentiWordNet 3.0. The polarity of words is expressed as corresponding vertices: if a word w_i has positive, negative, or both polarity, we append a corresponding node v_P^i, v_N^i, or v_{PN}^i with the label 'Sent:p', 'Sent:n', or 'Sent:pn' respectively and connect these nodes and other nodes v_w^j ($j \in \{1, \ldots, l\}$) according to word order. 'Type Bs' graphs are constructed by adding polarity nodes to 'type B' graphs. Note that 'type Bs' graphs do not contain dependency edges.

Table 2 shows the average number of vertices and edges of training graphs in each category.

4.2   Results

Calculation Time and Accuracy. We conducted a preliminary experiment to evaluate the effect of the proposed approximation. We used 'camera-photo' category from Amazon review dataset. The input is a graph set of type A and we set $ms = 3$. We compared the calculation time and accuracy (F-measure for the 'negative' label) for 4 types of algorithms. The algorithm 'orig' is the original graph boosting method with no approximation, 'B1000' uses only the former approximation with bucket size 1,000, 'S' uses the latter approximation, and 'B1000S' uses both. The result in Fig. 2 shows that the two approximation methods contribute to improvement of accuracy in much shorter time than the original algorithm.

Table 2. The average number of vertices and edges of training graphs in each category. Ave(V) and Ave(E) mean the average number of vertices and edges of the graphs, respectively.

Category	type B		type D		type BD	
	Ave(V)	Ave(E)	Ave(V)	Ave(E)	Ave(V)	Ave(E)
video	184.78	183.78	184.78	137.49	184.78	321.27
electronics	122.49	121.49	122.49	96.18	122.49	217.68
kitchen-housewares	104.02	103.02	104.02	81.17	104.02	184.19
toys-games	106.27	105.27	106.27	79.32	106.27	184.60
apparel	70.54	69.54	70.54	53.69	70.54	123.24
camera-photo	148.34	147.34	148.34	116.60	148.34	263.94

	type A		type Bs	
	Ave(V)	Ave(E)	Ave(V)	Ave(E)
video	379.51	920.62	245.70	305.63
electronics	246.51	593.17	166.11	208.73
kitchen-housewares	207.88	497.07	142.86	180.70
toys-games	212.78	506.06	142.64	178.00
apparel	140.44	333.57	97.36	123.18
camera-photo	298.66	718.98	202.20	255.06

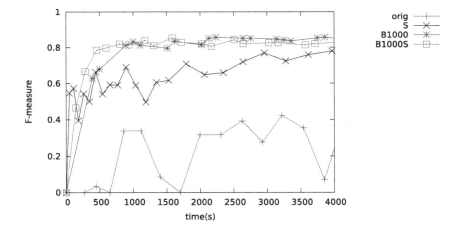

Fig. 2. Effect of the approximation methods

Accuracy Obtained from Different Graph Types. Table 3 shows the results of boosting algorithms measured by F-measure for the 'negative' label. All classifiers are trained through 10,000 iterations with the two approximation methods. We again devided the features into 1,000 buckets. The results show that the graph inputs which include structural information (type BD, type A, and type Bs) performs better than inputs which do not in the most categories. This indicates that some subgraph features with structural information contribute considerably to improved accuracy. Especially, in the most cases the graphs of

Table 3. F-measures for the 'negative' label. The underlined score indicates the best score of the category. 'Ranking average' means the averaged ranking about F-measures for each graph type. '#Best' or '#Worst' indicates the number of categories each type of graph achieved the best or the worst scores, respectively.

(single vertices, strings or trees)	type B, $ms = 0$ (bag-of-words)	type B, $ms = 3$ ($n(\leq 4)$-grams)	type A, $ms = 0$	type D, $ms = 3$
Category				
video	0.882	0.886	0.864	0.878
electronics	0.795	<u>0.824</u>	0.793	0.809
kitchen-housewares	0.755	0.789	0.803	0.806
toys-games	0.746	0.731	0.737	0.741
apparel	0.694	0.776	0.725	0.732
camera-photo	<u>0.847</u>	0.822	0.804	0.814
Average	0.787	0.804	0.787	0.798
Ranking average	4.67	3.17	5.83	4.5
#Best	1	1	0	0
#Worst	2	<u>0</u>	3	<u>0</u>
(general graphs)	type BD, $ms = 3$	type A, $ms = 3$	type Bs, $ms = 3$	SVM with gSpan, type A, $ms = 3$
video	0.881	0.885	<u>0.892</u>	0.753
electronics	0.823	0.822	0.820	0.765
kitchen-housewares	0.768	<u>0.837</u>	0.816	0.739
toys-games	0.724	0.757	<u>0.770</u>	0.683
apparel	0.724	0.771	<u>0.778</u>	0.680
camera-photo	0.810	0.825	0.816	0.746
Average	0.788	<u>0.816</u>	0.815	0.728
Ranking average	5.33	2.33	<u>2.17</u>	-
#Best	0	1	<u>3</u>	-
#Worst	1	<u>0</u>	<u>0</u>	-

type Bs performs better than others. We emphasize that such a kind of representation of texts is not possible by strings or trees but by graphs. Figure 3 shows the examples of extracted features.

Comparing to SVM with Graph Mining. The last column of the Table 3 shows the results of L_1-regularized L_1-loss SVM implemented in Classias (Okazaki 2009). The feature vectors for SVM are frequent subgraphs in the training datasets. To make the vectors, we conducted gSpan with minimum support 0.01. The input graphs are type A with $ms = 3$. The displayed results are the best ones among different coefficients $c \in \{0.01, 0.05, 0.1, 0.5, 1, 5, 10\}$ for L_1-regularization. The total training time (including gSpan and SVM) for the category 'camera-photo' is 1,036.3 s, while the boosting with the two approximation almost converges within 1,000 s (See Fig. 2). The fact that all the results

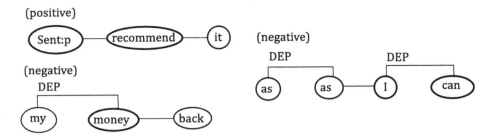

Fig. 3. Examples of extracted features.

of SVM are below those of boosting suggests that feature selection with frequent subgraph mining scrapes off not only unneccesary features but also significant ones.

5   Related Works

There is a number of researches which use graph based features for text classification. Matsumoto et al. (2005) combine word sub-sequences and dependency sub-trees for sentiment classification. Jiang et al. (2010) represent texts as graphs by combining word sequences and syntactic trees. Arora et al. (2010) introduce graph based features representing several linguistic annotations for sentiment classification. This method uses not only unigrams and subgraphs but also newly defined features constructed by combining original features. Because of the large feature space, all of these methods choose features by some mining methods in advance and apply vector based learning algorithms such as perceptron and SVMs. Our method need not mine features previously, since the boosting algorithm find significant subgraph features from the whole search space automatically. The proposed method also has an advantage of easy parameter tuning. The parameters are only number of iteration, maximal size of subgraphs and bucket size, while the mining based algorithms need to decide both parameters for mining such as minimum frequency and for learning algorithms.

A boosting based classification of semi-structured text has already been proposed by (Kudo and Matsumoto 2004). However, the method is applicable only to tree formats, while our algorithm runs on any graph sets. It enables us to make use of rich structure of text more flexibly.

Recent improvement of graph boosting algorithms includes attacks on extended problem settings (Pan et al. 2015a; Wu et al. 2015) and on imbalanced data in the real world (Pan et al. 2015b) and use of additional information (Fei and Huan 2014; Pan et al. 2016). Pan et al. (2015a, 2015b, 2016) and Wu et al. (2015) take similar approaches in that they explore subgraph-based weak classifiers using gSpan and pruning by gain upper bound as we referred to in Sect. 2.3. In addition, Pan et al. (2015a) solves linear programming to minimize the risk function like gBoost (Saigo et al. 2009) and accelerates this step for large scale graphs. These methods can easily be combined with our approximation methods, since our approximation method modifies only the selection step of the discriminative subgraph features.

6 Conclusion

In this paper, we proposed a graph boosting based text classification and efficient approximation methods for the calculation of weak classifiers. The experimental results show that our algorithm extracts significant subgraph features efficiently. Our algorithm helps guess what kinds of information are significant for classifiers. In the case of Amazon review data, the information of sentimental tags seems to be important. It is possible to add any other kinds of nodes or edges to the input graph. It is also a future work to combine our methods with some other text classification methods including the recently-proposed ones we refer to in Sect. 5.

References

Arora, S., Mayfield, E., Rosé, C.P., Nyberg, E.: Sentiment classification using automatically extracted subgraph features. In: Proceedings of the NAACL HLT 2010 Workshop on Computational Approaches to Analysis and Generation of Emotion in Text, pp. 131–139 (2010)

Baccianella, S., Esuli, A., Sebastiani, F.: SentiWordNet 3.0: an enhanced lexical resource for sentiment analysis and opinion mining. In: Proceedings of Seventh International Conference on Language Resources and Evaluation, pp. 2200–2204 (2010)

Blitzer, J., Dredze, M., Pereira, F.: Biographies, bollywood, boom-boxes and blenders: domain adaptation for sentiment classification. In: Proceedings of the 45th Annual Meeting of the Association of Computational Linguistics, pp. 440–447 (2007)

Boser, B.E., Guyon, I.M., Vapnik, V.N.: A training algorithm for optimal margin classifiers. In: Proceedings of the Fifth Annual ACM Conference on Computational Learning Theory, pp. 144–152 (1992)

Collobert, R.: Deep learning for efficient discriminative parsing. In: International Conference on Artificial Intelligence and Statistics (2011)

Collobert, R., Weston, J., Bottou, L., Karlen, M., Kavukcuoglu, K., Kuksa, P.: Natural language processing (almost) from scratch. J. Mach. Learn. Res. **12**, 2493–2537 (2011)

Fei, H., Huan, J.: Structured sparse boosting for graph classification. ACM Trans. Knowl. Discov. Data **9**, 1–22 (2014)

Frank, R.: The perceptron: A probabilistic model for information storage and organization in the brain. Psycholog. Rev. **65**, 386–408 (1958)

Freund, Y.: The alternating decision tree algorithm. In: Proceedings of the Sixteenth International Conference on Machine Learning, pp. 124–133 (1999)

Gee, K.R., Cook, D.J.: Text classification using graph-encoded linguistic elements. In: Proceedings of the Eighteenth International Florida Artificial Intelligence Research Society Conference, pp. 487–492 (2005)

Iwakura, T.: A boosting-based algorithm for classification of semi-structured text using frequency of substructures. In: Proceedings of 9th International Conference on Recent Advances in Natural Language Processing, pp. 319–326 (2013)

Iwakura, T., Okamoto, S.: A fast boosting-based learner for feature-rich tagging and chunking. In: Proceedings of Twelfth Conference on Computational Natural Language Learning, pp. 17–24 (2008)

Jiang, C., Coenen, F., Sanderson, R., Zito, M.: Text classification using graph mining-based feature extraction. Knowl-Bas. Syst. **23**, 302–308 (2010)

Kashima, H., Tsuda, K., Inokuchi, A.: Marginalized kernels between labeled graphs. In: Proceedings of the Twentieth International Conference on Machine Learning, pp. 321–328 (2003)

Kudo, T., Maeda, E., Matsumoto, Y.: An application of boosting to graph classification. Adv. Neural Inf. Process. Syst. **17**, 729–736 (2004)

Kudo, T., Matsumoto, Y.: A boosting algorithm for classification of semi-structured text. In: Proceedings of 9th Conference on Empirical Methods in Natural Language Processing, pp. 301–308 (2004)

Matsumoto, S., Takamura, H., Okumura, M.: Sentiment classification using word sub-sequences and dependency sub-trees. In: Ho, T.B., Cheung, D., Liu, H. (eds.) PAKDD 2005. LNCS, vol. 3518, pp. 301–311. Springer, Heidelberg (2005)

Okazaki, N.: Classias: a collection of machine-learning algorithms for classification (2009). http://www.chokkan.org/software/classias/

Pan, S., Wu, J., Zhu, X.: CogBoost: boosting for fast cost-sensitive graph classification. IEEE Trans. Knowl. Data Eng. **27**, 2933–2946 (2015)

Pan, S., Wu, J., Zhu, X., Long, G., Zhang, C.: Boosting for graph classification with universum. Knowl. Inf. Syst. **47**, 1–25 (2016)

Pan, S., Wu, J., Zhu, X., Zhang, C.: Graph ensemble boosting for imbalanced noisy graph stream classification. IEEE Trans. Cybern. **45**, 940–954 (2015)

Saigo, H., Nowozin, S., Kadowaki, T., Kudo, T., Tsuda, K.: gBoost: a mathematical programming approach to graph classification and regression. Mach. Learn. **75**, 69–89 (2009)

Schapire, R.E., Singer, Y.: Improved boosting algorithms using confidence-rated predictions. Mach. Learn. **37**, 297–336 (1999)

Wu, J., Pan, S., Zhu, X., Cai, Z.: Boosting for multi-graph classification. IEEE Trans. Cybern. **45**, 430–443 (2015)

Yan, X., Han, J.: gSpan: graph-based substructure pattern mining. In: Proceedings of 2002 IEEE International Conference on Data Mining, pp. 721–724 (2002)

On the Gradient-Based Sequential Tuning of the Echo State Network Reservoir Parameters

Sumeth Yuenyong[(✉)]

School of Information Technology, Shinawatra University, 99 Moo 10 Bang Toey,
Sam Khok 12160, Pathum Thani, Thailand
sumeth.y@siu.ac.th

Abstract. In this paper, the derivative of the input scaling and spectral radius parameters of Echo State Network reservoir are derived. This was achieved by re-writing the reservoir state update equation in terms of template matrices whose eigenvalues can be pre-calculated, so the two parameters appear in the state update equation in the form of simple multiplication which is differentiable. After that the paper derives the derivatives and then discusses why direct application of these two derivatives in gradient descent to optimize reservoirs in a sequential manner would be ineffective due to the nature of the error surface and the problem of large eigenvalue spread on the reservoir state matrix. Finally it is suggested how to apply the derivatives obtained here for joint-optimizing the reservoir and readout at the same time.

1 Introduction

Echo State Network (ESN) is a special type of recurrent neural network proposed by [2]. It has found use in problems dealing with nonlinear signal/systems [4,6,8,14]. The basic idea is to use a large, recursive and sparse neural network with random weights as the "reservoir". The state of the reservoir is fed into the output or "readout" layer, generally a linear combiner, that produces the final output. The entire structure is trained by adapting only the weights of the output layer, while the weights of the reservoir are randomly pre-generated and then hold fixed. The basic ESN structure is shown in Fig. 1. The reservoir is characterized by the input weight matrix $\mathbf{W}_{in}$ and the feedback (connections between the neurons themselves) by the weight matrix $\mathbf{W}$. The output layer is characterized by the weight vector $\mathbf{w}_{out}$ (assuming single output for simplicity).

In order to obtain good results, the reservoir of the ESN structure must be generated with appropriate parameters [3]. Originally the process of choosing good parameters[1] for the reservoir was by trial and error which proceeds as follows: choose a set of reservoir parameters, train the readout (batch training), evaluate the performance (usually some sort of mean squared error) by running

---

[1] To avoid confusion, we shall refer to the process of choosing good reservoir parameter as "tuning" of ESN, while "training" means to adapt the weights of the readout layer, given some reservoir.

© Springer International Publishing Switzerland 2016
R. Booth and M.-L. Zhang (Eds.): PRICAI 2016, LNAI 9810, pp. 651–660, 2016.
DOI: 10.1007/978-3-319-42911-3_54

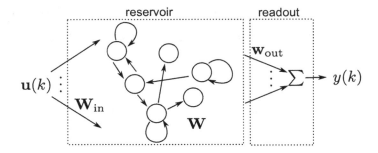

Fig. 1. Basic ESN structure. The input weight $\mathbf{W}_{\text{in}}$ is a full matrix - every input is connected to every neuron. The feedback matrix $\mathbf{W}$ may be dense or sparse. Both of these matrices are randomly generated. The output weight $\mathbf{w}_{\text{out}}$ is also full - every neuron in the reservoir feeds the output layer.

the ESN on another block of signal, adjust the parameter(s) to try to improve the performance, then repeat the process until the performance is acceptable.

It can be seen that tuning can be a very tedious and time-consuming process if done by trial and error. For this reason, many researchers had proposed automatic tuning for ESN [7,15,16]. The approaches vary, but are all based on using a derivative-free optimization algorithm to optimize the reservoir parameters - the difference among these works are the algorithm choice and the details of how to define the cost function to be optimized. The main problem with this type of approach is that it is by nature batch mode - two blocks of signals are needed to evaluate any cost function in the reservoir parameter space, one for training the readout layer and the other for evaluating performance. For a detailed explanation of how reservoir tuning works, please refer to the author's previous work in [15].

For application in situations where the data is streaming, batch mode is not very appropriate. Although they can be made to work using sliding windows, it is much better to have a sequential algorithm with can adapt the reservoir parameters in a sample-by-sample manner. The main contribution of the current paper is to present the derivation of the derivatives of the two most important reservoir parameters that have the most impact on ESN's performance. The paper also discusses why these two derivatives are only a single piece of the puzzle - in order to have an effective sequential tuning algorithm for ESN, more research efforts are still needed.

2   Echo State Network

ESN are basically recurrent neural networks with linear combiners attached. The computation for the network's output is performed in two steps. First the state vector (the output of each neuron at time k) of the reservoir is updated using

$$\mathbf{x}(k) = \tanh\left[\mathbf{W}\mathbf{x}(k-1) + \mathbf{W}_{\text{in}}\mathbf{u}(k)\right], \tag{1}$$

where $\mathbf{x}(k-1)$ is the previous state vector, $\mathbf{x}(k)$ is the updated state vector, $\mathbf{u}(k)$ is the input vector at time k. The reservoir state vector is initialized as $\mathbf{x}(0) = \mathbf{0}$.

In the second step, the output is computed from the updated state vector. The output equation is given by

$$y(k) = \mathbf{w}_{out}^T \mathbf{x}(k), \tag{2}$$

where $\mathbf{w}_{out}^T$ is the vector of readout weights (in this paper we assume single output for simplicity). By inspection of (2), it can be seen that the output $y(k)$ is a linear function of $\mathbf{w}_{out}^T$, which is the only quantity that is adapted in the entire ESN structure (if we are sure that we have a good reservoir). Therefore, the readout layer is structurally and computationally equivalent to a linear finite impulse response (FIR) filter. The only difference being that the input comes from the state vector of the reservoir instead of a tap-delay line. For this reason $\mathbf{w}_{out}^T$ can be adapted sequentially (in least in theory) using any algorithm that works on FIR filters. This is the key advantage of ESN compared to traditional recurrent neural network, in which training is non-convex and the main training algorithm called Real Time Recurrent Learning (RTRL) is $O(N^4)$ [13].

The disadvantage of ESN is that, even though the training of the readout is a simple convex problem, the achievable performance is sensitive to how the reservoir is generated [12]. This is why the above mentioned tuning process for the reservoir is necessary, and this in turn limits the application of ESN to mostly batch off-line setting since all reservoir tuning methods proposed in the literature up till now are all batch mode. See the "reservoir adaptation" section in [10] for a review in this area. When ESN is applied to an on-line setting, the reservoir parameters are usually set the "safe" values based on the user's experience and just left alone [9]. This strategy works, but it is far from optimal.

2.1   Reservoir Parameters

In this section, we define exactly what are the reservoir parameters. In total, there are 5 of them which are directly involved in generating the individual elements of $\mathbf{W}_{in}$ and $\mathbf{W}$:

1. The distribution from which to draw the elements of $\mathbf{W}_{in}$ and $\mathbf{W}$.
2. The spectral radius[2] of $\mathbf{W}$, denoted as $\tilde{\rho}$.
3. The scaling of the input weight matrix $\mathbf{W}_{in}$, denoted as s.
4. The reservoir size N.
5. The sparseness of $\mathbf{W}$.

Once the five parameters have been selected, reservoir generation proceed as follows:

1. Sample the nonzero elements of $\mathbf{W}_{in}$ and $\mathbf{W}$ from the chosen distribution.
2. Scale the input matrix: $\mathbf{W}_{in} = s\mathbf{W}_{in}$

---

[2] The spectral radius of a matrix is the maximum of the magnitudes of its eigenvalues.

3. Let $\rho(\mathbf{W})$ denote the spectral radius of the matrix $\mathbf{W}$. Scale $\mathbf{W}$ to have a specific spectral radius by: $\mathbf{W} = \tilde{\rho}\mathbf{W}/\rho(\mathbf{W})$

The scaled $\mathbf{W}_{\text{in}}$ and $\mathbf{W}$ are respectively the input and the internal reservoir matrices that will be used in the ESN model. Note that compared to regular recurrent neural networks, the reservoir size N is quite large, with values that can often be in the hundreds or even thousands. This makes batch-mode tuning quite problematic from a computational standpoint.

The importance of the five parameters above, according to [9] can be summarized as follows. The reservoir size N is not really a parameter but a design choice balancing between computational cost and performance. The sparseness level, as along as it's below 90%[3] have little impact on performance, so it can just be set to this maximum value to minimize computation if one is using a sparse-aware numerical package. The distribution from which to sample from also have virtually no impact on performance and usually just set to a convenient $[-1, 1]$. Only two parameters: the spectral radius $\tilde{\rho}$ and the input scaling s have strong impact on the performance. These are the two parameter whose derivative with respect to the network output error we will develop in the next section.

3   Derivatives of Spectral Radius and Input Scaling

Before the derivatives of $s, \tilde{\rho}$ can be calculated, we must first re-write the state update Eq. (1) using "templates" like in [16] as

$$\mathbf{x}(k) = \tanh\left[\frac{\widehat{\mathbf{W}}}{\rho(\widehat{\mathbf{W}})}\tilde{\rho}\mathbf{x}(k-1) + s\widehat{\mathbf{W}_{\text{in}}}\mathbf{u}(k)\right], \tag{3}$$

where $\widehat{\mathbf{W}}$ and $\widehat{\mathbf{W}_{\text{in}}}$ are respectively the template for $\mathbf{W}$ and $\mathbf{W}_{\text{in}}$, $\rho(\widehat{\mathbf{W}})$ is the spectral radius of $\widehat{\mathbf{W}}$, pre-calculated when the template was generated. The advantage of (1) over (3) is firstly, the reservoir parameters $\tilde{\rho}$ and s appear directly in the state update equation. The second advantage is the use of templates eliminate the need to perform repeated eigenvalue calculations every time $\tilde{\rho}$ changes and turn it into simple multiplication which is differentiable. From another angle, (3) is simply incorporating reservoir generation outlined in the last section directly into the state update equation.

Now we are ready to start developing the derivatives. First define the cost function to be differentiated as

$$J = \frac{1}{2}e^2(k) = \frac{1}{2}\left[d(k) - y(k)\right],$$

where $y(k)$ is the output of the readout layer and $d(k)$ is the desired response.[4]

---

[3] If the sparseness is above 90% eigenvalue calculation for $\mathbf{W}$ may sometime fail due to numerical problems (using Python's Scipy package).

[4] The desired response is needed for supervised sequential learning. It is defined by what filter configuration the ESN is to be operated in. For details, see [1].

Next, the update rule for s is defined as

$$\Delta s = -\alpha \frac{\partial J(k)}{\partial s}.$$

using the chain rule

$$\frac{\partial J(k)}{\partial s} = \frac{\partial J(k)}{\partial e(k)} \frac{\partial e(k)}{\partial y(k)} \frac{\partial y(k)}{\partial s} = -e(k) \frac{\partial y(k)}{\partial s},$$

so

$$\Delta s = \alpha e(k) \frac{\partial y(k)}{\partial s}, \tag{4}$$

where α is the learning rate. Similarly for the spectral radius $\tilde{\rho}$

$$\Delta \tilde{\rho} = -\alpha \frac{\partial J(k)}{\partial \tilde{\rho}}$$

again by using the chain rule

$$\frac{\partial J(k)}{\partial \tilde{\rho}} = \frac{\partial J(k)}{\partial e(k)} \frac{\partial e(k)}{\partial y(k)} \frac{\partial y(k)}{\partial \tilde{\rho}} = -e(k) \frac{\partial y(k)}{\partial \tilde{\rho}},$$

and thus

$$\Delta \tilde{\rho} = \alpha e(k) \frac{\partial y(k)}{\partial \tilde{\rho}}. \tag{5}$$

Next we have to derive expressions for $\frac{\partial y(k)}{\partial s}$ and $\frac{\partial y(k)}{\partial \tilde{\rho}}$.

3.1 Learning Rule for Input Scaling

We being with $\frac{\partial y(k)}{\partial s}$ as it is easier. From (2) we may write using the chain rule again

$$\frac{\partial y(k)}{\partial s} = \frac{\partial y(k)}{\partial \mathbf{x}(k)} \frac{\partial \mathbf{x}(k)}{\partial s}. \tag{6}$$

To proceed from this point we need to adopt a convention. We define the derivative of a scalar with respect to a vector as a row vector, and the derivative of a vector with respect to a scalar to be a column vector. This is the convention used by many authors including [11]. Under this convention, the right hand side of (6) evaluates to a scalar which matches the left hand side. The first derivative on the right hand side of (6) can be obtained directly from (2) as

$$\frac{\partial y(k)}{\partial \mathbf{x}(k)} = [w_1, w_2, \ldots, w_N], \tag{7}$$

where $w_1, w_2, \ldots, w_N$ are the individual weights of the readout weight vector $\mathbf{w}_{\text{out}}$. The next quantity is $\frac{\partial \mathbf{x}(k)}{\partial s}$ which can be obtained from the modified state update Eq. (3)

$$\frac{\partial \mathbf{x}}{\partial s} = \frac{\partial}{\partial s} \tanh \left[\frac{\widehat{\mathbf{W}}}{\rho(\widehat{\mathbf{W}})} \tilde{\rho} \mathbf{x}(k-1) + s \widehat{\mathbf{W}}_{\text{in}} \mathbf{u}(k) \right].$$

Define $\mathbf{h}(k)$ as

$$\mathbf{h}(k) = \frac{\widehat{\mathbf{W}}}{\rho(\widehat{\mathbf{W}})}\tilde{\rho}\mathbf{x}(k-1) + s\widehat{\mathbf{W}}_{\text{in}}\mathbf{u}(k).$$

Since the tanh function in (3) is applied element-wise, we may write $\mathbf{x}(k)$ as

$$\mathbf{x}(k) = \begin{bmatrix} \tanh(h_1(k)) \\ \tanh(h_2(k)) \\ \vdots \\ \tanh(h_N(k)) \end{bmatrix},$$

where the subscripts indicate the element index of a vector. Differentiating each element of $\mathbf{x}(k)$ with respect to s gives

$$\frac{\partial \mathbf{x}(k)}{\partial s} = \begin{bmatrix} \tanh'(h_1(k))v_1(k) \\ \tanh'(h_2(k))v_2(k) \\ \vdots \\ \tanh'(h_N(k))v_N(k) \end{bmatrix}, \tag{8}$$

where we define $\mathbf{v}(k) = \widehat{\mathbf{W}}_{\text{in}}\mathbf{u}(k)$ for notational compactness and $\tanh'$ denote the derivative of the tanh function. Substituting (8) and (7) into (6) and then substituting that result back into (4) yield the update rule for s.

3.2   Learning Rule for Spectral Radius

For the spectral radius update rule, we have

$$\frac{\partial y(k)}{\partial \tilde{\rho}} = \frac{\partial y(k)}{\partial \mathbf{x}(k)}\frac{\partial \mathbf{x}(k)}{\partial \tilde{\rho}}. \tag{9}$$

$\frac{\partial y(k)}{\partial \mathbf{x}(k)}$ is the same as for the input scaling case. For $\frac{\partial \mathbf{x}(k)}{\partial \tilde{\rho}}$, we have to use recursive differentiation similar to the derivation of RTRL in [13]. The derivative $\frac{\partial \mathbf{x}(k)}{\partial \tilde{\rho}}$ will be updated in a recursive manner starting from $\frac{\partial \mathbf{x}(0)}{\partial \tilde{\rho}} = \mathbf{0}$.

Similar to the input scaling case, we may write

$$\begin{aligned}
\frac{\partial \mathbf{x}}{\partial \tilde{\rho}} &= \frac{\partial}{\partial \tilde{\rho}}\tanh\left[\frac{\widehat{\mathbf{W}}}{\rho(\widehat{\mathbf{W}})}\tilde{\rho}\mathbf{x}(k-1) + s\widehat{\mathbf{W}}_{\text{in}}\mathbf{u}(k)\right] \\
&= \frac{\partial}{\partial \tilde{\rho}}\begin{bmatrix} \tanh(h_1(k)) \\ \tanh(h_2(k)) \\ \vdots \\ \tanh(h_N(k)) \end{bmatrix} \\
&= \begin{bmatrix} \tanh'(h_1(k))\left[\frac{\partial \mathbf{h}}{\partial \tilde{\rho}}\right]_1 \\ \tanh'(h_2(k))\left[\frac{\partial \mathbf{h}}{\partial \tilde{\rho}}\right]_2 \\ \vdots \\ \tanh'(h_N(k))\left[\frac{\partial \mathbf{h}}{\partial \tilde{\rho}}\right]_N \end{bmatrix},
\end{aligned} \tag{10}$$

where the subscript again denote element index of a vector. It can be seen that the key difference between in previous input scaling case and the spectral radius case is the partial derivative of $\mathbf{h}(k)$. In the input scaling case, the partial derivative $\frac{\partial \mathbf{h}(k)}{\partial s}$ is non-recursive and is given by $\mathbf{v}(k)$ defined in the previous subsection. On the other hand, the partial derivative $\frac{\partial \mathbf{h}(k)}{\partial \tilde{\rho}}$ will be recursive in nature because in (3), $\tilde{\rho}$ is multiplied to $\mathbf{x}(k-1)$. In order to differentiate recursively, we must differentiate both $\mathbf{x}(k)$ and $\mathbf{x}(k-1)$. Moreover, the term $\frac{\widehat{\mathbf{W}}}{\rho(\widehat{\mathbf{W}})}\tilde{\rho}\mathbf{x}(k-1)$ must be differentiated as a product because $\mathbf{x}(k-1)$ is also a function of $\tilde{\rho}$ due to the recursion. Denoting $\tilde{\rho}$ as f and $\mathbf{x}(k-1)$ as g, we can write the following, keeping in mind the product rule:

$$\frac{\partial \mathbf{h}(k)}{\partial \tilde{\rho}} = \underbrace{\frac{\widehat{\mathbf{W}}}{\rho(\widehat{\mathbf{W}})}\mathbf{x}(k-1)}_{f'g} + \underbrace{\frac{\widehat{\mathbf{W}}\tilde{\rho}}{\rho(\widehat{\mathbf{W}})}\frac{\partial \mathbf{x}(k-1)}{\partial \tilde{\rho}}}_{fg'}.$$

Applying the last expression to (10) gives

$$\frac{\partial \mathbf{x}(k)}{\partial \tilde{\rho}} = \begin{bmatrix} \tanh'(h_1(k))\left[\frac{\widehat{\mathbf{W}}}{\rho(\widehat{\mathbf{W}})}\mathbf{x}(k-1) + \frac{\widehat{\mathbf{W}}\tilde{\rho}}{\rho(\widehat{\mathbf{W}})}\frac{\partial \mathbf{x}(k-1)}{\partial \tilde{\rho}}\right]_1 \\ \tanh'(h_2(k))\left[\frac{\widehat{\mathbf{W}}}{\rho(\widehat{\mathbf{W}})}\mathbf{x}(k-1) + \frac{\widehat{\mathbf{W}}\tilde{\rho}}{\rho(\widehat{\mathbf{W}})}\frac{\partial \mathbf{x}(k-1)}{\partial \tilde{\rho}}\right]_2 \\ \vdots \\ \tanh'(h_N(k))\left[\frac{\widehat{\mathbf{W}}}{\rho(\widehat{\mathbf{W}})}\mathbf{x}(k-1) + \frac{\widehat{\mathbf{W}}\tilde{\rho}}{\rho(\widehat{\mathbf{W}})}\frac{\partial \mathbf{x}(k-1)}{\partial \tilde{\rho}}\right]_N \end{bmatrix}.$$

This expression gives the recursive update formula we need for $\frac{\partial \mathbf{x}(k)}{\partial \tilde{\rho}}$. Substituting it into (9) and substituting that result into (5) gives the update rule for the spectral radius $\tilde{\rho}$.

4   Discussion of the Result

Having obtained derivatives for s and $\tilde{\rho}$, we may be tempted to conclude that we now have a sequential tuning algorithm by simply applying the gradient descent principle. However, it is not that simple because of two main issues:

1. The adaptation of both s and $\tilde{\rho}$ is dependent on the readout weight $\mathbf{w}_{\text{out}}$.
2. The eigenvalue spread of $\mathbf{x}(k)$ makes gradient descent ineffective.

The reason for the first issue can readily be seen by inspection that $\frac{\partial y(k)}{\partial \mathbf{x}(k)} = [w_1, w_2, \ldots, w_N]$ appears in the update formula for both s and $\tilde{\rho}$. How should $\mathbf{w}_{\text{out}}$ then be set? Obviously, it cannot be set to all zeros because then Δs and $\Delta \tilde{\rho}$ will always be zero. Can it be set to some random values, says sampled from $[-1, 1]$? If we do this then the updates of s and $\tilde{\rho}$ will proceed differently even for exactly the same templates, input signal and desired response each time the algorithm is run - depending on a particular draw for $\mathbf{w}_{\text{out}}$.

This line of reasoning that the update proceeds differently suggests that the error surface seen by any algorithm adapting $s, \tilde{\rho}$ using gradient descent changes with $\mathbf{w}_{\text{out}}$. Now consider the joint optimization of $s, \tilde{\rho}, \mathbf{w}_{\text{out}}$ which has a $N + 2$ dimensional error surface. Setting a particular value for $\mathbf{w}_{\text{out}}$ means that we are taking a 2D slice across the $N + 2$ D hyper-surface. Gradient descent on s and $\tilde{\rho}$ while holding $\mathbf{w}_{\text{out}}$ fixed then can only locate the optimal values for the parameters only if the particular 2D slice it is working on happens to coincide with the error surface seen by a batch mode tuning procedure such as the one in [15] which had been shown to be able to find the optimal reservoir parameters. However, the way such error surface are calculated - by *training* $\mathbf{w}_{out}$ in batch-mode and then running the resulting ESN on a block of test signals for each point $(s, \tilde{\rho})$ means that the error surface seen by batch-mode tuning is not a simple 2D slice through the $N + 2$ D joint error surface by a very complicated and most likely disjoint[5] subset of it. Since it is disjoint, no continuous 2D plane can be cut through the joint error surface to produce it. This means that gradient descent for any fixed value of $\mathbf{w}_{\text{out}}$ will never see the "correct" error surface and therefore unlikely to be able to locate the optimal values for $s, \tilde{\rho}$.

For the second issue, it can be see that the term $\mathbf{x}(k)$ appears in the update rules for both s and $\tilde{\rho}$. While detailed mathematical analysis of this is still needed, we can predict its effect qualitatively by noting that sequential training of $\mathbf{w}_{\text{out}}$ by differentiating (2) with respect to $\mathbf{w}_{\text{out}}$, yielding

$$\Delta \mathbf{w}_{\text{out}} = \alpha \mathbf{x}(k) e(k).$$

While mathematically correct, is not effective at training $\mathbf{w}_{\text{out}}$ because of the enormous (easily $> 10^8$) eigenvalue spread of $\mathbf{x}$. This is a well-known fact in the ESN literature that gradient descent is terribly ineffective a sequential training of the output layer [5]. The error surface is infinitesimally thin in some direction, requiring α to be set to extremely small number for convergence [1], which is useless in practice. Since $\mathbf{x}(k)$ also appears in the update formulas for s and $\tilde{\rho}$, gradient descent using the derivative developed here probably suffers from the same fate.

So where does this leave the derivatives of s and $\tilde{\rho}$ so laboriously obtained? Since the tuning of the reservoir and training of the readout is coupled together and cannot be separated. The most likely useful application of the result in this paper is as part of a joint optimization algorithm adapting both the reservoir and the readout at the same time. Such algorithm cannot be a simple gradient descent, for the reasons we have discussed. It would probably have to be some sort of second order method such as Newton's or Levenberg-Marquardt method which are not sensitive to the curvature of the error surface. Such methods requires in addition to the gradient higher-order derivatives such as the Jacobian or the

---

[5] It is disjoint because, for each point $(s, \tilde{\rho})$, $\mathbf{w}_{\text{out}}$ is solved for by the method of least squares. Since least squares involves matrix inverse, for a slightly different point $s + \Delta s, \tilde{\rho} + \Delta \tilde{\rho}$, a very different $\mathbf{w}_{\text{out}}$ may be produced.

Hessian, the derivation of which can benefit from the derivatives presented in this paper.

5 Conclusion

In this paper, we have derived the derivatives of the two most important parameter for ESN reservoirs: the input scaling s and the spectral radius $\tilde{\rho}$ and discussed why their direct application in gradient descent to optimize $s, \tilde{\rho}$ will not work. Further work include experimentally verifying the correctness of the derivatives by running gradient descent on 2D slices of the joint error surface, as well as deriving higher-order derivatives to be used for joint optimization of reservoir and readout using a second-order method.

References

1. Haykin, S.S.: Adaptive Filter Theory, 4th edn. Pearson Education India, New Delhi (2005)
2. Jaeger, H.: The "echo state" approach to analysing and training recurrent neural networks. Technical report GMD Report 148, German National Research Center for Information Technology (2001)
3. Jaeger, H.: Tutorial on training recurrent neural networks, covering BPPT, RTRL, EKF and the "echo state network" approach. GMD-Forschungszentrum Informationstechnik (2002)
4. Jaeger, H.: Adaptive nonlinear system identification with echo state networks. Networks 8, 9 (2003)
5. Jaeger, H.: Reservoir riddle: suggestions for echo state network research. In: Proceedings of International Joint Conference on Neural Networks, pp. 1460–1462 (2005)
6. Jaeger, H., Haas, H.: Harnessing nonlinearity: predicting chaotic systems and saving energy in wireless communication. Science 304(5667), 78–80 (2004)
7. Jiang, F., Berry, H., Schoenauer, M.: Supervised and evolutionary learning of echo state networks. In: Rudolph, G., Jansen, T., Lucas, S., Poloni, C., Beume, N. (eds.) PPSN 2008. LNCS, vol. 5199, pp. 215–224. Springer, Heidelberg (2008)
8. Küçükemre, A.U.: Echo state networks for adaptive filtering. Ph.D. thesis, University of Applied Sciences (2006)
9. Lukoševičius, M.: A practical guide to applying echo state networks. In: Montavon, G., Orr, G.B., Müller, K.-R. (eds.) Neural Networks: Tricks of the Trade, 2nd edn. LNCS, vol. 7700, pp. 659–686. Springer, Heidelberg (2012)
10. Lukoševičius, M., Jaeger, H.: Reservoir computing approaches to recurrent neural network training. Comput. Sci. Rev. 3(3), 127–149 (2009)
11. Petersen, K.B., Pedersen, M.S., et al.: The matrix cookbook. Technical University of Denmark, vol. 7, p. 15 (2008)
12. Schrauwen, B., Verstraeten, D., Van Campenhout, J.: An overview of reservoir computing: theory, applications and implementations. In: Proceedings of the 15th European Symposium on Artificial Neural Networks, pp. 471–482 (2007)
13. Williams, R.J., Zipser, D.: A learning algorithm for continually running fully recurrent neural networks. Neural Comput. 1(2), 270–280 (1989)

14. Xia, Y., Jelfs, B., Van Hulle, M.M., Príncipe, J.C., Mandic, D.P.: An augmented echo state network for nonlinear adaptive filtering of complex noncircular signals. IEEE Trans. Neural Netw. **22**(1), 74–83 (2011)
15. Yuenyong, S.: Fast and effective tuning of echo state network reservoir parameters using evolutionary algorithms and template matrices. In: 19th International Computer Science and Engineering Conference (ICSEC), November 2015
16. Yuenyong, S., Nishihara, A.: Evolutionary pre-training for CRJ-type reservoir of echo state networks. Neurocomputing **149**, 1324–1329 (2015)

Large Margin Coupled Mapping for Low Resolution Face Recognition

Jiaqi Zhang[1(✉)], Zhenhua Guo[1,3,4], Xiu Li[1], and Youbin Chen[2]

[1] Graduate School at Shenzhen, Tsinghua University, Shenzhen, China
zhang-jq13@mails.tsinghua.edu.cn
[2] Huazhong University of Science and Technology, Wuhan, China
[3] Key Laboratory of Measurement and Control of Complex Systems
of Engineering, Ministry of Education, Southeast University, Nanjing, China
[4] Key Laboratory of Intelligent Perception and Systems for High-Dimensional
Information, Ministry of Education, Nanjing University of Science
and Technology, Nanjing, China

Abstract. Traditional face recognition algorithms can achieve significant performance under well-controlled environments. However, these algorithms perform poorly when the resolution of the face images varies. A two-step framework is proposed to solve the resolution problem through adopting super-resolution (SR) and performing face recognition on the super-resolved face images. However, such method usually has poor performance on recognition tasks as SR focuses more on visual enhancement, rather than classification accuracy. Recently, Coupled Mapping (CM) has been introduced into face recognition framework across different resolutions, which learns a common feature subspace for both high-resolution (HR) and low-resolution (LR) face images. In this paper, inspired by maximum margin projection, we propose Large Margin Coupled Mapping (LMCM) algorithm, which learns projections to maximize the margin between distance of between-class subjects and distance of within-class ones in the common space. Experiments on public FERET and SCface databases demonstrate that LMCM is effective for low-resolution face recognition.

Keywords: Coupled Mapping · Low-resolution face recognition · Large Margin Coupled Mapping · FERET · SCface

1 Introduction

A great number of achievements have been made in the area of automatic face recognition during last decades, especially under well-controlled circumstances. However, the performance of face recognition system in real world always degrades dramatically when the quality of input face images becomes poor, such as low-resolution. This is a specific concern in surveillance environment where the target is far from the sensor, resulting in low-resolution face images.

To solve the low-resolution (LR) problem, a two-step framework is proposed following the intuition of first recovering lost detail information of LR face images and then applying traditional face recognition algorithms on recovered face images. In fact, most

© Springer International Publishing Switzerland 2016
R. Booth and M.-L. Zhang (Eds.): PRICAI 2016, LNAI 9810, pp. 661–672, 2016.
DOI: 10.1007/978-3-319-42911-3_55

proposed two-step algorithms of LR face recognition apply super-resolution (SR) technique as the first step [1–5]. The super-resolved face images are then passed to the second general face recognition pipe. Through the development of last decade, there exists many SR algorithms to reconstruct high-resolution (HR) images from a single LR image [1] or multiple LR images [2]. In many real-world face recognition systems, the intuitive solution is interpolation which are simple and fast, such as bilinear, cubic and so on. The learning-based super-resolution (LSR) algorithms [1, 3–5] recently draw a lot of attention owing to its promising performance. Freeman et al. [1] proposed a patch-wise Markov Random Field as the SR prediction model and recovered HR images by MAP estimation. Baker and Kanade [3] proposed to recover the HR face image from an input LR one by "face hallucination" model based on face priors. Liu et al. [5] proposed to combine a holistic and a local model for SR reconstruction. Inspired by locally linear embedding (LLE) [7], Chang et al. recovered the HR face image from the spatial neighbors of its LR counterpart. Yang et al. [8] proposed to incorporate sparse representation into SR framework which achieves outstanding performance. However, these algorithms aim more at the effect of visual enhancement rather than the performance of the specific face recognition task.

Recently, some algorithms avoiding an explicit SR stage have been introduced into face recognition flow. Gunturk et al. [9] investigated to transfer from pixel domain to eigenface domain for SR reconstruction. Hennings-Yeomans et al. [10, 11] integrated the aims of SR and face recognition simultaneously through a joint objective function. Although these methods improve the recognition rate, their speed even for the speed-up version is slow due to an optimization procedure for each test image. To avoid the super-resolution step, Coupled Mapping (CM) based methods are proposed for LR face recognition. Li et al. [12] proposed Coupled Locality Preserving Mapping (CLPM) based on CM for LR face recognition. Inspired by locality preserving methods [13, 14] for dimensionality reduction, the CLPM brought in a penalty weighting matrix into the objective function to preserve the local relationship of the original space. The CLPM emphasized more on the objective of recognition rather than just reconstruction and thus yielded a better performance. However, it ignored the label information of the training set, which is vital for face recognition. To take advantage of label information, some LDA-like algorithms were introduced into coupled mapping, such as Simultaneous Discriminant Analysis (SDA) [19], Coupled Marginal Fisher Analysis (CMFA) [18]. In [17], Shi et al. first constructed local optimization for each training sample according to the relationship of neighboring data points and then incorporated the local optimizations together for building the global structure. However, these algorithms fail to consider recognition and geometric information of training set simultaneously, thus some valuable information is missing and performance is limited for challenging problems [17].

In this paper, we propose a novel algorithm called Large Margin Coupled Mapping (LMCM) for LR face recognition, which takes both recognition information of the training data and the local geometric relationship of face image pairs into account to maximize the distance of between-class pairs and minimize the distance of within-class pairs in the common subspace. With appropriate constraints, the new-defined optimization problem could be solved in an analytical close-form. So it can be fast enough for real time applications.

The remaining of this paper is organized as follows. Section 2 demonstrates the LR face recognition problem and the formulation of CM. Section 3 describes the details of our proposed algorithm LMCM. Section 4 shows experimental results on FERET and SCface databases. Section 5 draws conclusions of this paper.

2 Low Resolution Face Recognition

In the scenario of LR face recognition, the task could be simplified to find an appropriate distance measure between a LR face image l_i and a HR one h_j, i.e., $d_{ij} = dist(l_i, h_j)$. Here, $l_i \in \mathbb{R}^m$, $i = 1, 2, \ldots, N_p$ and $h_j \in \mathbb{R}^M$, $j = 1, 2, \ldots, N_g$, (m < M) represent the m-dimension feature vectors of the LR query images and the M-dimension HR ones registered in the gallery set, respectively. Due to the dimension mismatch of the feature vectors of LR and HR face images, some common distances (e.g. Euclidean distance) obviously cannot be applied directly. To deal with this problem, traditional two-step algorithms based on explicit SR attempt to find a mapping, $f_{SR} \colon \mathbb{R}^m \mapsto \mathbb{R}^M$, to project the LR image into the target HR space, and then directly calculate the distance in the HR space:

$$d_{ij} = dist(f_{SR}(l_i), h_j) \tag{1}$$

Different from the two-step algorithms, CM based methods intend to establish two coupled mappings: $f_L \colon \mathbb{R}^m \mapsto \mathbb{R}^n$ for LR face images and $f_H \colon \mathbb{R}^M \mapsto \mathbb{R}^n$, to project both the LR and HR feature vectors into a common feature space. Here, n represents the dimensionality of the new common feature space. Then the distance can be measured by:

$$d_{ij} = dist(f_L(l_i), f_H(h_j)) \tag{2}$$

Now the critical problem is to pursue an ideal common feature space. For low-resolution face recognition, the objective of CM algorithm is that the projections of LR and HR face image of the same subject should be as close as possible in the new common feature space. Let $f_L(l) = P_L^T l$ and $f_H(h) = P_H^T h$ be linear mappings, respectively, where P_L and P_H are two projection matrices with size of $m \times n$ and $M \times n$. This principle is formulated as the following objective function:

$$J_{CM}(P_L, P_H) = \sum_{i=1}^{N_t} \left\| P_L^T l_i - P_H^T h_i \right\|^2 \tag{3}$$

N_t represents the number of the training images.

We use $L = [l_1, l_2, \ldots, l_{N_t}]$ and $H = [h_1, h_2, \ldots, h_{N_t}]$ to denote the original LR and HR feature vectors in the training set, respectively. Equation (3) can be reformulated as

$$J_{CM}(P_L, P_H) = tr\left(\left\|P_L^T L - P_H^T H\right\|^2\right) \tag{4}$$

where $tr(\cdot)$ is the matrix trace operator. Furthermore, using some deductions of linear algebra, Eq. (4) can be rewritten as

$$J_{CM}(P_L, P_H) = tr\left(\begin{bmatrix} P_L \\ P_H \end{bmatrix}^T \begin{bmatrix} L & 0 \\ 0 & H \end{bmatrix} \begin{bmatrix} I & -I \\ -I & I \end{bmatrix} \begin{bmatrix} L & 0 \\ 0 & H \end{bmatrix}^T \begin{bmatrix} P_L \\ P_H \end{bmatrix}\right) \tag{5}$$

We can further let $P = \begin{bmatrix} P_L \\ P_H \end{bmatrix}$, $Z = \begin{bmatrix} L & 0 \\ 0 & H \end{bmatrix}$ and $A = \begin{bmatrix} I & -I \\ -I & I \end{bmatrix}$, where I is the identity matrix. Finally, we can get a compact form as

$$J_{CM}(P_L, P_H) = tr\left(P^T Z A Z^T P\right) \tag{6}$$

P_L and P_H can be obtained by minimizing Eq. (6). The details of the optimization procedure can be referred to [12].

3   Proposed LMCM

The CM algorithm described above obtains the projection matrices following the criteria that the distance between each LR face image and the corresponding HR one should be as close as possible. However, it only takes advantage of part of verification information of the training data, e.g. the face image pairs belonging to the same subject. In this paper, we draw an inspiration from Maximum Margin Projection (MMP) [16] and propose LMCM algorithm for LR face recognition, which seeks linear coupled mappings to force a margin between the distance of between-class subjects and the distance of within-class ones in the common feature space, as shown in Fig. 1. To achieve this, we utilize the verification information along with local geometry and identification information of the training data.

Verification Information with Local Geometry: Under this scenario, verification information lies in the distance between face image pairs: ones of identical subjects tend to have small distance and ones of different subjects tend to have large distance.

In order to discover both discriminant and geometrical structures of the face images, we construct two graphs, within-class graph G_w and between-class graph G_b. In graph G_w, face images share the same identities are connected, while in graph G_b, face images belong to different subjects are connected. Let W_w and W_b represent the weight matrices of G_w and G_b, respectively. As HR feature is considered to have more discriminant information, we build these weight matrices in the original HR image space. We define them as the following form

$$W_{w,ij} = \begin{cases} e^{-\frac{\|h_j - h_{i2}\|_2}{\sigma}}, & \text{if } h_i, h_j \text{ connected in } G_w \\ 0 \end{cases} \tag{7}$$

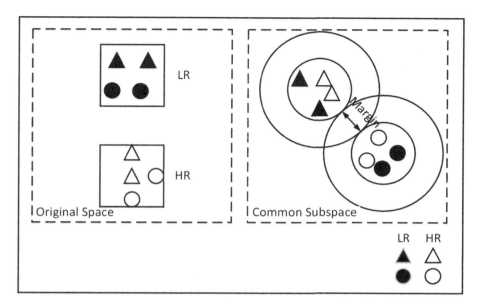

Fig. 1. Overview of the proposed LMCM algorithm. Different shapes represent different subjects.

$$W_{b,ij} = \begin{cases} e^{-\frac{\|h_j - h_{i2}\|_2}{\sigma}}, & \text{if } h_i, h_j \text{ connected in } G_b \\ 0 \end{cases} \qquad (8)$$

where σ is the mean distance between each pair of face images in the training data.

Now, consider the problem of mapping LR and HR face images into a common subspace so that the connected face images of G_w stay as close as possible, while the connected face images of G_b stay as far as possible. Let P_L and P_H represent projection matrices. A reasonable criterion for learning the projection matrices is to optimize the following objective functions:

$$\min_{P_L, P_H} \sum_{i,j} \left\| P_L^T l_i - P_H^T h_j \right\|_2^2 W_{w,ij} + \left\| P_L^T l_i - P_L^T l_j \right\|_2^2 W_{w,ij} + \left\| P_H^T h_i - P_H^T h_j \right\|_2^2 W_{w,ij} \qquad (9)$$

$$\max_{P_L, P_H} \sum_{i,j} \left\| P_L^T l_i - P_H^T h_j \right\|_2^2 W_{b,ij} + \left\| P_L^T l_i - P_L^T l_j \right\|_2^2 W_{b,ij} + \left\| P_H^T h_i - P_H^T h_j \right\|_2^2 W_{b,ij} \qquad (10)$$

where W_w and W_b represents the weight matrices of G_w and G_b respectively. The objective function (9) constructed on the within-class graph G_w imposes a large penalty if neighboring face images of the identical subject in original space are mapped far apart. Similarly, the objective function (10) constructed on the between-class graph G_b imposes a large penalty if neighboring face images belonging to different subjects are mapped close together. The ultimate goal of these objectives is to force a margin between face feature vectors of different subjects.

Following some simple algebraic steps, the objective function (9) can be reduced to the following matrix form

$$\min_{P_L,P_H} Tr\left(P_L^T L\left(2D_w^L + D_w^H - W_w - W_w^T\right)L^T P_L + P_H^T H\left(D_w^L + 2D_w^H - W_w - W_w^T\right)H^T P_H\right)$$
$$- Tr\left(P_L^T L W_w H^T P_H + P_H^T H W_w^T L^T P_L\right) \tag{11}$$

where $D_w^L = \sum_j W_{w,ij}$ and $D_w^H = \sum_i W_{w,ij}$.

Similarly, the objective function (10) can be reduced to a similar matrix form

$$\max_{P_L,P_H} Tr\left(P_L^T L\left(2D_b^L + D_b^H - W_b - W_b^T\right)L^T P_L + P_H^T H\left(D_b^L + 2D_b^H - W_b - W_b^T\right)H^T P_H\right)$$
$$- Tr\left(P_L^T L W_b H^T P_H + P_H^T H W_b^T L^T P_L\right) \tag{12}$$

where $D_b^L = \sum_j W_{b,ij}$ and $D_b^H = \sum_i W_{b,ij}$.

Similar deduction with (5) to (6), we can rewrite Eqs. (11) and (12) as follows

$$\min_{P_L,P_H} Tr(P^T Z A_w Z^T P) \tag{13}$$

$$\max_{P_L,P_H} Tr(P^T Z A_b Z^T P) \tag{14}$$

where $P = \begin{bmatrix} P_L \\ P_H \end{bmatrix}$, $Z = \begin{bmatrix} L & 0 \\ 0 & H \end{bmatrix}$, $A_w = \begin{bmatrix} 2D_w^L + D_w^H - W_w - W_w^T & -W_w \\ -W_w^T & D_w^L + 2D_w^H - W_w - W_w^T \end{bmatrix}$,

$A_b = \begin{bmatrix} 2D_b^L + D_b^H - W_b - W_b^T & -W_b \\ -W_b^T & D_b^L + 2D_b^H - W_b - W_b^T \end{bmatrix}$.

Identification Information as Regularization Term: The identification information classifies the face image into one of the subjects, which encourages the algorithm to learn projection matrix that can map each face image into its own cluster. In this paper, we take advantage of identification information by minimizing the within-class scatter. In learning projection matrices P_L and P_H, we aim to solve the following optimization problem:

$$\min_{P_L,P_H} S_W \tag{15}$$

where S_W represents the within-class scatter. As the overall mean of the training data is zero, the definitions of the scatter matrix are formulated as:

$$S_W = \sum_i (x_i - \mu_{i,c})(x_i - \mu_{i,c})^T \tag{16}$$

where x_i is the n-dimension feature projected by high or low resolution face images into the new common space, $\mu_{i,c}$ is the mean of the projected feature with class label of c which x_i belongs to. With some linear algebra, Eq. (16) can be rewritten in the following matrix form:

$$S_W = (X - U)(X - U)^T \tag{17}$$

where U is the $n \times 2N_t$ mean matrix with column $\mu_{i,c}$, and X is the $n \times 2N_t$ data matrix with column x_i. Let Λ be a $C \times C$ diagonal matrix with element Λ_i. These matrices can be represented by P_L and P_H as:

$$U = P^T Z D \Lambda^{-1} D^T \tag{18}$$

$$X = P^T Z \tag{19}$$

where $P = \begin{bmatrix} P_L \\ P_H \end{bmatrix}$, $Z = \begin{bmatrix} L & 0 \\ 0 & H \end{bmatrix}$ and $D = \{d_{ij}\}_{2N_t \times C}$ with

$$d_{ij} = \begin{cases} 1, & if \ x_i \in class \ j \\ 0, & if \ x_i \notin class \ j \end{cases} \tag{20}$$

With (18) and (19), Eq. (17) can be rewritten as:

$$S_W = P^T Z(I - D\Lambda^{-1}D^T)(I - D^T \Lambda^{-1}D)Z^T P \tag{21}$$

In this paper, the identification information is taken as a regularization term. This is the main difference between our proposed algorithm and CMFA in [18], where identity matrix is taken as the regularization term in the denominator. And the identification term is a key factor for performance improvement. Finally, the optimization problem with objective functions (13) and (14) reduces to

$$\max_{P_L, P_H} \frac{Tr(P^T Z A_b Z^T P)}{Tr(P^T Z A_w Z^T P + \xi S_W)} \tag{22}$$

where ξ is the balance factor between the verification and identification information. In the experiments below, this factor is set to 0.05;

The coupled projection matrices P_L and P_H that maximize the objective function (22) can be obtained by solving the generalized eigenvalue problem

$$(Z A_b Z^T)P = \lambda(Z A_w Z^T + \xi Z(-D\Lambda^{-1}D^T)(I - D^T \Lambda^{-1}D)Z^T)P \tag{23}$$

After obtaining the projection matrices P_L and P_H, we mapped both LR and HR images into the common space and utilize Euclidean norm to measure the distance of each image pair, as described in (24).

$$Dis = \left\| P_L^T l_i - P_H^T h_i \right\|^2 \tag{24}$$

For each probe image, we take as its identity the subject with the smallest distance in the gallery. We use True Positive Identification Rate (TPIR), also refer to as Rank-1 Identification Rate in this circumstances, to measure the performance of our proposed method, as defined in the following

$$TPIR = \frac{\#(correct\ idetified\ images)}{\#(probe)} \tag{25}$$

4   Experimental Results

To evaluate effectiveness of the proposed method, we applied our methods on two public databases: FERET [6] and SCface [15]. Performance is measured by rank-1 identification rate. Before projection, the gray pixel distribution of one image is normalized to have average intensity 0, standard deviation 1 and unit norm.

4.1   Experimental Result on FERET Database

We follow the same test protocol as [17] when we conduct experiments on a subset of FERET database. The subset (ba, bd, be, bf, bg, bj, bk) contains 200 subjects with variations of illumination (bk), expression (bk) and pose (bd, be, bf, bg). We choose 50 subjects for training and the rest 150 subjects are used for test. In the test phase, 4 images of each subject are selected as gallery and the remaining as the probe. In the experiment, the HR face images and corresponding LR ones are scaled with resolution of 32×32 and 8×8. Figure 2 shows some of the HR (top row) and LR (bottom row) face images in FERET database. To evaluate our proposed LMCM algorithm, we compare it with CLPM [12], SDA [19], CMFA [18] and the algorithm proposed in [17].

Table 1 presents the experiment results of LMCM algorithm on FERET database. Our method with 53-D features achieves the recognition rate of 90.00 %, which is higher than 55.22 % for CLPM, 72.09 % for SDA, 75.98 % for CMFA and 80.90 % for coupled mapping method used in [17]. The main reason lies in that our method takes more advantage of the supervised information of the training set than other methods. There are two main differences between CMFA and our proposed algorithm. First, we construct the weight matrices W_w and W_b in a different way, which can capture more discriminant information compared to the method applied in CMFA. Second, we use within-class scatter as the regularization term instead of identical matrix, which can take advantage of the identification in the training data. Our proposed LMCM algorithm also shows its high capability to handle different variations, such as pose and expression, except for low resolution. Table 2 is the test time for each image pair.

Fig. 2. HR (Top row) and LR (Bottom row) face images from FERET database

Table 1. Rank 1 performance on FERET database. The values are rank-1 identification rate (%)

Algorithm	Rank 1 performance (%)
CLPM [12]	55.22 [17]
SDA [19]	72.09 [17]
CMFA [18]	75.98 [17]
Coupled mapping method [17]	80.90 [17]
Proposed LMCM	90.00

Table 2. Test time for each LR and HR image pair

CPU	Memory	Environment	Time (microsecond)
Intel(R) Core(TM) i5-4200U @1.60 GHz	4.00 GB	Windows 10 Matlab 2015B	7.3

4.2 Experimental Result on SCface Database

To show the real recognition performance of our LMCM algorithm under the surveillance circumstances, the SCface database is chosen as a new set to illustrate the recognition performance of LMCM. SCface is a database of static images of human faces [15] captured by surveillance cameras. Images were taken in uncontrolled indoor environment using five video surveillance cameras at three different distances. The database contains 4,160 face images (in visible and infrared spectrum) of 130 subjects, as shown in Fig. 3. Face images from different cameras and distances mimic the real-world conditions. The subset used contains images from surveillance cameras cam1–cam5: (I) distance of 2.6 m (i.e., LR), and (II) distance of 1.0 m (i.e., HR). The resolution of the processed images is 48×48 and 16×16 for the HR and LR, respectively.

For this experiment, the protocol of [17] is implemented. All subjects are used for training and test. In the experiment, LMCM is compared with CLPM, SDA, CMFA and Coupled Mapping Method in [17]. For SCface database, 80 subjects are selected to define the training set. The rest of 50 subjects are used as the test set. This procedure is repeated 10 times. The average results are presented in Table 3. Overall, the rank 1 recognition rates are much lower compared to the FERET database due to the real world challenges posed in SCface database. We can see from the results that our proposed LMCM algorithm improves the LR face recognition significantly on SCface database. The main reason lies in that LMCM learns the discriminant information between HR and LR face images to force a margin between the projection of identical and different subjects according to recognition information. Compared to other algorithms in Table 3, our proposed algorithm apparently can capture more such discriminant feature for LR face recognition (Table 4).

Fig. 3. Examples of face images of one subject with one camera and 3 different distances

Table 3. Experiment on SCface. The values are rank-1 identification rate (%)

Algorithm	Rank 1 performance (%)
CLPM [12]	29.12 [17]
SDA [19]	40.08 [17]
CMFA [18]	39.56 [17]
Coupled mapping method [17]	43.24 [17]
Proposed LMCM	60.40

Table 4. Test time for each LR and HR image pair

CPU	Memory	Environment	Time (microsecond)
Intel(R) Core(TM) i5-4200U @1.60 GHz	4.00 GB	Windows 10 Matlab 2015B	8.5

5   Conclusion

In this paper, we propose a novel algorithm to solve low-resolution face recognition problem without SR procedure. Our method projects both the HR and LR face images into a new common feature subspace by maximizing the distance of features with different labels and minimizing the distance of features with identical label. The objective function attempts to force a margin between different subjects using both the identification and verification information. Experimental results on FERET and SCface databases show that our proposed method can achieve promising performance. In the future, applying nonlinear mappings by kernel methods and using more discriminative features instead of raw intensity will be studied.

Acknowledgement. This work is partially supported by the Key Laboratory of Intelligent Perception and Systems for High-Dimensional Information (Nanjing University of Science and Technology), Ministry of Education (Grant No. 30920140122006).

References

1. Freeman, W., Pasztor, E., Carmichael, O.: Learning low-level vision. Int. J. Comput. Vis. **40**, 25–47 (2000)
2. Elad, M., Feuer, A.: Super-resolution reconstruction of image sequences. IEEE Trans. Pattern Anal. Mach. Intell. **21**(9), 817–834 (1999)
3. Baker, S., Kanade, T.: Hallucinating faces. In: Proceedings of the International Conference on Automatic Face and Gesture Recognition, pp. 83–88 (2000)
4. Chang, H., Yeung, D., Xiong, Y.: Super-resolution through neighbor embedding. In: Proceedings of the IEEE Conference on Computer Vision and Pattern Recognition, pp. 275–282 (2004)
5. Liu, C., Shum, H., Zhang, C.: A two-step approach to hallucinating faces: global parametric model and local nonparametric model. In: Proceedings of the IEEE Conference on Computer Vision and Pattern Recognition, pp. 192–198 (2001)
6. Philips, P., Moon, H., Pauss, P., Rivzvi, S.: The feret evaluation methodology for face-recognition algorithms. In: Proceedings of the IEEE Conference on Computer Vision and Pattern Recognition, pp. 1090–1104 (2000)
7. Roweis, S., Saul, L.: Nonlinear dimensionality reduction by locally linear embedding. Science **290**, 2323–2326 (2000)
8. Yang, J., Wright, J., Huang, T., Ma, Y.: Image super-resolution via sparse representation. IEEE Trans. Image Process. **19**(11), 2861–2873 (2010)
9. Gunturk, B., Batur, A., Altunbasak, Y., Hayes, M., Mersereau, R.: Eigenface-domain super-resolution for face recognition. IEEE Trans. Image Process. **12**(5), 597–606 (2003)
10. Hennings-Yeomans, P., Baker, S., Kumar, B.: Simultaneous super-resolution and feature extraction for recognition of low-resolution faces. In: Proceedings of the IEEE Conference on Computer Vision and Pattern Recognition, pp. 1–8 (2008)
11. Hennings-Yeomans, P., Baker, S., Kumar, B.: Robust low-resolution face identification and verification using high-resolution features. In: 2009 16th IEEE International Conference on Image Processing (ICIP), pp. 33–36. IEEE (2009)
12. Li, B., Chang, H., Shan, S., Chen, X.: Low-resolution face recognition via coupled locality preserving mappings. IEEE Sig. Process. Lett. **17**(1), 20–23 (2010)
13. Belkin, M., Niyogi, P.: Laplacian eigenmaps for dimensionality reduction and data representation. Neural Comput. **15**, 1373–1396 (2003)
14. He, X., Niyogi, P.: Locality preserving projections. Neural Inf. Process. Syst. **16**, 153–160 (2004)
15. Grgic, M., Delac, K., Grgic, S.: SCface–surveillance cameras face database. Multimedia Tools Appl. **51**(3), 863–879 (2011)
16. He, X., Deng, C., Han, J.: Learning a maximum margin subspace for image retrieval. IEEE Trans. Knowl. Data Eng. **20**(2), 189–201 (2008)

17. Shi, J., Qi, C.: From local geometry to global structure: learning latent subspace for low-resolution face image recognition. IEEE Sig. Process. Lett. **22**(5), 554–558 (2015)
18. Siena, S., Boddeti, V.N., Kumar, B.V.K.V.: Coupled marginal fisher analysis for low-resolution face recognition. In: Computer Vision–ECCV Workshops and Demonstrations, pp. 240–249, 2012
19. Zhou, C., Zhang, Z., Dong, Y., Zhen, L., Li, S.Z.: Low-resolution face recognition via simultaneous discriminant analysis. In: International Joint Conference on Biometrics (IJCB), pp. 1–6 (2011)

Topic Detection in Group Chat Based on Implicit Reply

Xinyu Zhang, Ning Zheng, Jian Xu, and Ming Xu$^{(\boxtimes)}$

School of Computer Science and Technology,
Hangzhou Dianzi University, Hangzhou, China
{132050125,nzheng,jian.xu,mxu}@hdu.edu.cn

Abstract. Topic detection in group chat has become a promising research due to the widely usage of Instant Messaging (IM) systems. Previous works mainly focus on improving the text similarity between two related messages by utilizing different weighting factors. However, the text similarity of related texts is likely to be zero (or near zero) due to the characteristics of short text messages in group chat. To solve this problem, an innovative topic detection method based on implicit reply which indicates chat messages interact with each other is proposed in this paper. The comparative experiments results on the datasets gathered from QQ groups demonstrate the superiority of the proposed method as compared to the baseline approaches.

Keywords: Topic detection · Group chat · Multi-topic window

1 Introduction

IM systems are becoming more and more popular among netizens from all walks of life. Online group chat as an import service supported by them brings great convenience for the communication among multiple people. However, due to the high speed of message released and meaningless chatting, group chat logs are filled with large amount but not necessarily useful messages. Topic detection in group chat becomes a significant but challenging research task.

Although topic detection techniques have been well studied in traditional texts [1], they are not applied to group chat texts because of the brief of the chat texts. Meanwhile, unlike the asynchronous nature of microblog texts, chat text streams are synchronous, which result in topic detection methods for microblog mainly based on word frequency [2] are also not applied for group chat texts. The biggest challenges to topic detection in group chat is the sparse eigen-vector of short text messages. Existing algorithms mainly focus on improving the text similarity between two related messages by utilizing different weighting factors to alleviate the sparsity. However, the contexts in group chat considered as weighting factors are not always reliable due to the uncertain changes of group chat features in different groups.

In this paper, an innovative topic detection method based on implicit reply features is proposed to solve the above problems. Messages with reply relations

© Springer International Publishing Switzerland 2016
R. Booth and M.-L. Zhang (Eds.): PRICAI 2016, LNAI 9810, pp. 673–680, 2016.
DOI: 10.1007/978-3-319-42911-3_56

judged by the features are grouped together as a long text to overcome the challenge that the text similarity of two related messages is too low.

2   Related Work

In recent years, many studies have been done for group chat texts analysis. Uthus et al. [3] did a survey on this. According them, topic detection in group chat is one of the high-level researches.

Existing methods for topic detection in group chat can be roughly divided into two categories: supervised [4] and unsupervised. In this paper, unsupervised methods are focused on. Shen et al. [5] represented the messages by a vector space model. They used the similarity of the vectors along with sentence types and personal pronouns to determine the probability of a message belonging to a topic. But their method performed not very well. Adams et al. [6] used the WordNet hypernym augmentation to preprocess the short text and then used the temporal relationship between messages as penalty terms. This method performed better than the above one. Wang et al. [7] expanded the content of messages using user context and explicit reply. This method achieved a better performance. However, when the users' interests varied, the results of this method may be not so good. Moreover, all the above text similarity based methods cannot precisely deal with the case of that text similarity of related messages in group chat is zero (or near zero). Huang et al. [8] proposed an approach to topic detection in Chinese group chat (SPFC) which is similar to this paper. The difference is, they segmented conservation to topics based on contextually correlative characteristics, while our method is based on implicit reply features. Thus, SPFC is sensitive to history information. Our method can better deal with the case of new topics occurring in a group chat.

In this paper, it proposes an unsupervised method and alleviates the sparsity of short text by grouping messages with reply relations together as a long text. The experiment shows the effectiveness of the proposed approach.

3   Basic Concepts and Framework Overview

For ease of description, some basic concepts are given as follows.

Chat Text Message. When chatting with others, a group member sends a text message to the IM system server. The message which initiates a new topic in group is referred to as a start-message. Otherwise, it is a reply-message. For each reply-message, it has one reply-object at least.

Group Chat Log. It is a set of message records which consists of the author, the text, and the timestamp. A chat log usually contains a lot of conversations.

Group Chat Conversation. A set of continuous message records that the message released frequency in it is in line with the trend of "increase, smooth, decrease". A conversation may contain only one topic but maybe more than one.

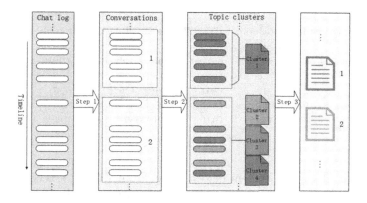

Fig. 1. The overview of our framework.

Figure 1 shows the overall framework of the proposed method. It includes three steps. Firstly, conversations are extracted from group chat text stream. Then, each conversation is segmented to topics based on implicit reply. Finally, topics are summarised by merging similar topics and filtering meaningless topics.

4 Motivating Observations

In group chat, users interact with others via chat text messages, which resulting in a text stream. Intuitively, a reply-message and its reply-objects should belong to the same topic. In this section, it aims to find some motivating observations about reply-messages. Specifically, two questions are proposed,

- If a text message is judged as a reply-message, whether this is useful for topic detection in group chat?
- What characteristics do reply-messages have?

To answer the above questions, four datasets are collected, dormitory dataset (DoD), cycling dataset (CyD), colleagues dataset (CoD), and graduates dataset (GrD). Both of the four datasets are come from real QQ groups, and each is a one-week group chat log. Only topics in CyD are known. Topics in the other three datasets are various, and may change with the trend in physical world. The four datasets are labeled manually by 8 postgraduate volunteers.

To begin with, volunteers summarize topics in dataset and then show the basic statistic data for each dataset as Table 1. To answer the first question, volunteers are asked to count the number of reply-messages in dataset $\#total$, number of reply-messages whose reply-objects cannot be judged by text similarity $\#reply$, and number of reply-messages which is a response to the $i-th$ newest topic for each dataset. The statistical material is show in Table 2. $dist1$ represents the current message is a response to the newest topic, and $dist>3$ represents the current message is a response to the topic which has at least two

Table 1. Statistics for datasets

dataset	#member	#message	#topic
DoD	6	84	5
CyD	41	236	13
CoD	142	920	61
GrD	1036	158992	189

Table 2. Statistics for reply-messages

dataset	#total	#reply	#dist1	#dist2	#dist3	#dist> 3
DoD	77	37	33	2	1	1
CyD	219	110	99	7	4	0
CoD	837	411	289	107	11	4
GrD	149921	88804	82646	4709	1390	59

topics between it and the current newest topic. From Table 2, it is concluded that more than 90 % of chat messages are reply-messages in each chat text stream. Note that almost half of these reply-messages' reply-objects cannot be judged by text similarity. Among these specific reply-messages, most of them are responses to the recent several topics. In detail, 85.6 % of them are responses to the current topic and more than 97 % are related to the newest three topics. All these statistic data give an affirmative to the first question.

For the second question, a further exploration is done on the differences between reply-messages and start-messages. Some interesting things are found. Firstly, several continuous messages sent by the same group user usually belong to the same topic. In this case, the subsequent few messages are the complement of the first one, and the later message is a response to the former one. Furthermore, because of rich context, users tend to chat with others in a very brief way. Specifically, they post a message that contains only one sentence which neither has a subject-predicate structure nor has a predicate-object structure. The omitted part in the sentence can be easily speculated from the chat context. In addition, a message which starts with a word with specific part of-speech (POS), such as conjunctions and demonstrative pronouns, is more likely to be a reply-message. So, the author, the syntax structure and the POS of the first keyword of a chat text which are referred to as implicit reply features can answer the second question and to help recognizing reply-messages in group chat.

5   Proposed Method

5.1   Extracting Conversations

A chat log usually contains a lot of conversations, and a conversation may contains several topics. To improve the accuracy and efficiency of topic detection, the chat logs are segmented to conversations first.

Conversations are in line with the general rule of lifecycle of common things. They will experience four stages of newborn, development, climax and decline. In group chats, this rule is reflected in the change of message released frequency (MRF). According to the change law, it is concluded that the dividing points for conversations in chat logs are the turning points of MRF from increasing to decreasing. That is, the message m_i is a dividing point if it meets the conditions of $f'(\tau(m_i)) < 0$ and $f'(\tau(m_{i+1})) \geq 0$. Where $\tau(m_i)$ presents the timestamp of message m_i, and $f'(\tau(m_i))$ presents the slope of function $f(t)$ at the point of $\tau(m_i)$. Then, a chat log can be segmented to many conversations by m_i.

5.2   Detecting Topics in Conversations

A further analysis on conversations should be made in order to extract fine-grained topic clusters. The proposed method can be divided into two steps.

Predicting Reply-Objects for Chat Messages. According to Table 2, in group chat, more than 97 % of the reply-messages are responses to the topics within the newest three topics. To predict reply-objects for a reply-message precisely, a multi-topics window (MTW) is introduced. Each topic in MTW is composed of messages with reply relations. In the MTW $T = \{T_1, T_2, ..., T_l\}$, the recent l topics in each conversation are saved, where l represents the size of MTW. Now suppose that chat text message m_i is the one to be processed in chat text stream of conversation $C = \{m_1, m_2, \ldots, m_n\}$, the process of finding reply-objects for m_i in MTW T is described as Algorithm 1. The value of t ranges from 0 to l. $t = 0$ indicates m_i not belongs to any topics in MTW, and $t = n$ indicates m_i belongs to the $n-th$ topic in MTW. $maxT$ is index of the most related topic with message m_i in MTW. The messages which belong to the most related topic are the reply-objects of message m_i. $maxTS$ is the text similarity between m_i and the most related topic. The $maxTS$ and $maxT$ are computed using Algorithm 2. Considering that a reply-message must have at least one reply-object, Algorithm 1 predicts reply-objects for m_i in two case. If m_i is judged as a reply-message, its reply-objects should be the messages in the $maxT-th$ topic in MTW, no matter how small the value of $maxTS$ is (lines 4–5, 10). It should be noticed that the initial value of $maxTS$ and $maxT$ are 0 and l in Algorithm 2, respectively. Thus, when the max text similarity between reply-message m_i and topics in MTW is zero because of the sparsity of short text message, Algorithm 1 group m_i into the $l-th$ topic in MTW. This is because a reply-message has the possibility of 85.6 % to be a response to the current topic in group chat. Otherwise, only when the value of $maxTS$ is not smaller than the text similarity threshold of two related messages in group chat can m_i be judged as a response to the $maxT-th$ topic in MTW (lines 7–8).

It uses the implicit reply features to predict whether a message is a reply-message. Here, LTP-Cloud[1] is utilized to analyse the syntax structure of Chinese texts and the POS of Chinese words. It supports the following functions, word segmentation, POS tagging, and dependency parsing [9]. In Algorithm 1, $isBeginWith(m_i)$ judges whether message m_i is a reply-message based on POS tagging, and $isRelaComplete(m_i)$ judges this based on dependency parsing.

---

[1] http://www.ltp-cloud.com/.

Algorithm 1. Reply-Objects Predicting

Input: Message m_i, MTW T, Text similarity threshold α
Output: Index of topic in MTW t
1: $t \leftarrow 0$
2: **if** $m_i.author \neq m_{i-1}.author$ **then**
3:     compute $maxTS$ and $maxT$ using Algorithm 2
4:     **if** $isBeginWith(m_i)$ **or** $!isRelaComplete(m_i)$ **then**
5:         $t = maxT$
6:     **else**
7:         **if** $maxTS \geq \alpha$ **then**
8:             $t = maxT$
9: **else**
10:     the t for m_i is the same with the t for m_{i-1}
11: **return** t

Algorithm 2. Finding the Most Relevant Topic in MTW

Input: Message m_i, MTW T
Output: Max text similarity $maxTS$, Index of the most related topic $maxT$
1: $maxTS \leftarrow 0$
2: $maxT \leftarrow l$
3: **for all** topic $T_i \in T$ **do**
4:     compute $textSim(m_i, T_i)$ by cosine similarity
5:     **if** $textSim(m_i, T_i) > maxTS$ **then**
6:         $maxTS = textSim(m_i, T_i)$
7:         $maxT = i$
8: **return** $maxTS$, $maxT$

Segmenting Conversations. When the current message m_i is a start-message, the MTW should be updated by adding m_i as the newest topic to it and removing the oldest topic from it. Otherwise, the MTW should be updated just by adding m_i to its reply-objects in MTW. These are regarded as the MTW update mechanism (MTWUM). Based on it, the proposed method is displayed as Algorithm 3. Line 7 shows the step of moving the $i-th$ topic to the $(i-1)-th$ window in MTW. The topic set TS is the result of conversation segment method.

Algorithm 3. Conversation Segmenting

Input: Conversation C
Output: Topic set TS
1: **for all** message $m_i \in C$ **do**
2:     compute t using Algorithm 1
3:     **if** $t \neq 0$ **then**
4:         $T_t \leftarrow T_t \cup m_i$
5:     **else**
6:         $TS \leftarrow TS \cup T_1$
7:         move topics window T forward
8:         $T_t \leftarrow m_i$
9: $TS \leftarrow TS \cup T$
10: **return** TS

Table 3. Overall performance of three methods

	Precision	Recall	F measure
TDEC	0.298	0.510	0.375
SPFC	0.466	0.703	0.559
Our method	0.563	0.750	0.644

5.3   Summarizing Topics

In the context of group chat, one topic may be talked about in several conversations. Meanwhile, not all topics obtained by the above steps are meaningful. Thus, topics obtained by the above steps should be further summarized.

Firstly, similar topics are merged by single pass clustering algorithm [10]. Topics that the text similarity between each other is not smaller than a certain threshold β are regarded as similar topics. Secondly, it is concluded that a meaningful topic is related to three factors, the number of its keywords $N_{keyword}$, the number of its related messages $N_{message}$ and the number of its related users N_{user}. In this paper, it simply uses $N_{keyword} \geq a$ and $N_{message} \geq b$ and $N_{user} \geq c$ as the judgment standard of whether a topic is meaningful. After the step in this section, the final topic set for group chat log is got.

6   Experiments

6.1   Experimental Design

For convenience, the four group chat datasets of DoD, CyD, CoD and GrD which have been mentioned in Sect. 4 are used as the experimental datasets. The parameters in this paper are empirically set as follows: $a = 2$, $b = 5$, $c = 2$, $\alpha = 0.25$, $\beta = 0.5$, and $l = 5$.

The proposed method is compared with Adams et al. [6] (TDEC) and Huang et al. [8] (SPFC). The former is a classical method and the latter is similar to this paper. To evaluate the performance of the three topic detection methods, precision, recall and F-measure are used as the evaluating standards.

6.2   Results and Analysis

Figure 2 shows results of the three methods on the same datasets. It is observed that the proposed method performs best against the other two methods on the whole. Table 3 shows the average performance of the three topic detection methods. Likewise, it indicates that the proposed method performs best, and the F-measure of the proposed method reaches 0.644.

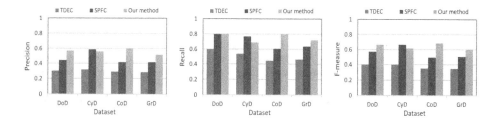

Fig. 2. Performance of topic detection. (Color figure online)

In Fig. 2, it is found that SPFC is better than the proposed method on dataset of CyD. This is because in CyD, there is almost no new topic will appear in this group chat, and SPFC is based on keywords pairs existed in two different messages. The information extracted from history chat logs improves its performance. TDEC has a poorer performance than SPFC and the proposed method because it based on TF-IDF combined with synonyms replacement and temporal relationship information, which cannot deal with the case of the text similarity of related texts is zero (or near zero) correctly. The challenge of text sparsity restricts its accuracy. In addition, it is observed that, on dataset of CoD which has much more entangled topics than the other three group chat datasets, excepting the proposed method, both TDEC and SPFC have a poorer performance against on DoD, CyD and GrD. This suggests that the MTW can help to improve the accuracy of predicting reply-objects for a reply-message, even though there are many entangled topics in the group chat. So the proposed method is more practical than the other two topic detection methods.

7 Conclusion

In this paper, a novel approach is proposed to detect topics in group chat. Firstly, conversations are extracted from chat log. Then, each conversation is segmented to topics based on implicit reply. Finally, the final topic set is optimized by merging similar topics and extracting meaningful topics. The comparative experiments results on the datasets gathered from QQ groups demonstrate the superiority of our proposed framework.

One can envision several directions for future work. While the current work aims to detect topics, the overlapping users' interests and their social ties can be discovered based on the common topics. Another important direction is to build personalized human readable topic summarization for group users.

Acknowledgments. This work was supported by the Natural Science Foundation of China under Grant No.61070212 and 61572165, the State Key Program of Zhejiang Province Natural Science Foundation of China under Grant No. LZ15F020003.

References

1. Allan, J., Papka, R., Lavrenko, V.: On-line new event detection and tracking. In: Proceedings of the 21st Annual International ACM SIGIR Conference on Research and Development in Information Retrieval. ACM (1998)
2. Schubert, E., Weiler, M., Kriegel, H.-P.: Signitrend: scalable detection of emerging topics in textual streams by hashed significance thresholds. In: Proceedings of the 20th ACM SIGKDD International Conference on Knowledge Discovery and Data Mining. ACM (2014)
3. Uthus, D.C., Aha, D.W.: Multiparticipant chat analysis: a survey. Artif. Intell. **199**, 106–121 (2013)
4. Özyurt, Ö., Köse, C.: Chat mining: automatically determination of chat conversations topic in Turkish text based chat mediums. Expert Syst. Appl. **37**, 8705–8710 (2010)
5. Shen, D., et al.: Thread detection in dynamic text message streams. In: Proceedings of the 29th Annual International ACM SIGIR Conference on Research and Development in Information Retrieval. ACM (2006)
6. Adams, P.H., Martell, C.H.: Topic detection and extraction in chat. IEEE International Conference on Semantic Computing. IEEE (2008)
7. Wang, L., Oard, D.W.: Context-based message expansion for disentanglement of interleaved text conversations. In: Proceedings of Human Language Technologies: The 2009 Annual Conference of the North American Chapter of the Association for Computational Linguistics. ACL (2009)
8. Huang, J.-M., et al.: Unsupervised conversation extraction in short text message streams. Ruanjian Xuebao/J. Softw. **23**, 735–747 (2012)
9. Che, W., Li, Z., Liu, T.: Ltp: a chinese language technology platform. In: Proceedings of the 23rd International Conference on Computational Linguistics: Demonstrations. ACL (2010)
10. Papka, R., Allan, J.: On-line new event detection using single pass clustering. UMass Comput. Sci. (1998)

Maximum Margin Tree Error Correcting Output Codes

Fa Zheng[1,2], Hui Xue[1,2(✉)], Xiaohong Chen[3], and Yunyun Wang[4]

[1] School of Computer Science and Engineering, Southeast University,
Nanjing 210096, People's Republic of China
{faaronzheng,hxue}@seu.edu.cn
[2] Key Laboratory of Computer Network and Information Integration,
Southeast University, Ministry of Education, Nanjing, People's Republic of China
[3] College of Science, Nanjing University of Aeronautics and Astronautics,
Nanjing 210016, People's Republic of China
lyandcxh@nuaa.edu.cn
[4] Department of Computer Science and Engineering,
Nanjing University of Posts and Telecommunications,
Nanjing 210046, People's Republic of China
wangyunyun@njupt.edu.cn

Abstract. Encoding is one of the most important steps in Error Correcting Output Codes (ECOCs). Traditional encoding strategies are usually data-independent. Recently, some tree-form encoding algorithms are proposed which firstly utilize mutual information to estimate inter-class separability in order to create a hierarchical partition of the tree from top to down and then obtain a coding matrix. But such criterion is usually computed by a non-parametric method which would generally require vast samples and is more likely to lead to unstable results. In this paper, we present a novel encoding algorithm which uses the maximum margins between classes as the criterion and constructs a bottom-up binary tree based on the maximum margin. As a result, the corresponding coding matrix is more stable and discriminative for the following classification. Experimental results have shown that our algorithm performs much better than some state-of-the-art coding algorithms in ECOC.

Keywords: Multi-class classification · Maximum margin tree · Error Correcting Output Codes

1 Introduction

The multi-class classification problem has attracted a lot attentions in machine learning field. The traditional solutions tend to transform it into multiple binary problems. The corresponding strategies include decision tree, neural networks, and so on.

Error Correcting Output Codes (ECOCs) [1,2] is a widely-used method in these strategies, which was originally proposed by Dietterich and Bakiri [3].

© Springer International Publishing Switzerland 2016
R. Booth and M.-L. Zhang (Eds.): PRICAI 2016, LNAI 9810, pp. 681–691, 2016.
DOI: 10.1007/978-3-319-42911-3_57

It usually involves two parts: encoding and decoding. Encoding part generates a sequence of bits, i.e. a code word for each class. All code words form a coding matrix. Decoding part predicts class labels for unseen data through comparing their output code words with the code words of classes in the coding matrix depending on some specific strategies such as Hamming decoding (HD) [4] and Euclidean decoding [5]. In this paper, we will focus on encoding part.

The goal of encoding is to design a coding matrix M. Each row of M represents one class and each column of M is one binary problem (dichotomizer). For each class, encoding aims to create a corresponding code word where each bit is the prediction of the dichotomizer. Traditionally, the coding matrix is coded by $+1$ and -1. In Table 1(a), $+1$ means that the corresponding dichotomizer takes this class as a positive class and -1 otherwise. However, the length of the code words in this scenario is actually fixed. As a result, a limited number of dichotomizers can be used which would restrict the performance of ECOC to some extent. Allwein et al. [6] further presented a ternary coding matrix which allows some bits of coding matrix to be zero as Table 1(b). The symbol zero denotes that the corresponding class does not participate in the specific classification. Result from the zero symbols, the ternary coding matrix is more flexible and could have much longer code words than the binary one.

Consequently, the core task in encoding has boiled down to how to build such an appropriate coding matrix. The simplest strategy is one-versus-all (OVA) [4] which takes one class as a positive class and all the others as a negative class to build the binary coding matrix. One-versus-one (OVO) [5] forms a ternary coding matrix where each column only considers two classes to be positive and negative classes respectively and the rest are represented by zero symbols. Random codes [6] generate the coding matrix randomly, where the binary coding matrix is called dense random while the ternary one is termed as spare random. Though these traditional strategies are simple, they are all data-independent. As a result, they either perform poorly or get too long code words which would require more dichotomoizers with higher computational costs.

Recent proposed tree-form encoding algorithms [7–10] utilize some criterions to estimate inter-class separability so as to build a tree and obtains a data-dependent coding matrix. Discriminant ECOC (DECOC) [8] applies the sequential forward floating search (SFFS) to generate the tree from top to down

Table 1. Coding matrix for a 4-class problem

(a) Binary

	h_1	h_2	h_3	h_4
C_1	+1	-1	-1	-1
C_2	-1	+1	-1	-1
C_3	-1	-1	+1	-1
C_4	-1	-1	-1	+1

(b) Ternary

	h_1	h_2	h_3	h_4	h_5	h_6
C_1	+1	+1	+1	0	0	0
C_2	-1	0	0	+1	+1	0
C_3	0	-1	0	-1	0	+1
C_4	0	0	-1	0	-1	-1

through maximizing the mutual information (MI) [11] between classes heuristically. Then a ternary coding matrix is constructed according to the hierarchical partition of the tree. Based on DECOC, subclass ECOC (SECOC) [9] further uses a cluster method to create subclasses while the original classification problem is linearly non-separable. However, the MI criterion used in DECOC and SECOC is computed by a non-parametric method which generally requires a large number of samples and further leads to an unstable result. Hierarchical ECOC (HECOC) [12] utilizes support vector domain description (SVDD) [12] as the criterion to estimate inter-class separability, which is more stable than DECOC and SECOC. However, when building the tree, HECOC chooses two classes which have the smallest inter-class separability as a node. As a result, the base dichotomizers will face a relatively difficult binary classification problem which limits the performance of ECOC to some degree.

In this paper, we propose a novel encoding method termed as maximum margin tree ECOC (M^2ECOC). M^2ECOC estimates the maximal inter-class separability by the maximum margins between classes rather than the MI criterion. Consequently, the corresponding coding matrix is more stable and discriminative for the following classification. Concretely, M^2ECOC uses support vector machine (SVM) [13] to compute the maximum margins between classes and then obtains a maximum margin matrix. Depending on this matrix, M^2ECOC further generates a bottom-up binary tree based on choosing the maximal maximum margin. Finally, such maximum margin tree will be converted into a ternary coding matrix according to the hierarchical partition of the tree.

The paper is organized as follows. Section 2 introduces the tree-form encoding algorithms DECOC and SECOC. Section 3 introduces our M^2ECOC algorithm in detail. In Sect. 4, the compared experiments with some state-of-the-art encoding algorithms are shown. Finally, the last section concludes the paper.

2   Tree-Form Encoding Algorithms

2.1   Discriminant ECOC (DECOC)

DECOC [8] firstly applies SFFS to find the hierarchical partition of the tree and builds the tree from top to down. As Fig. 1(a) shown, DECOC separates the original class set $\{C_1, C_2, C_3\}$ into two partitions $\{C_1, C_3\}$ and $\{C_2\}$ until each partition has only one class. SFFS is one kind of suboptimal sequential search methods which dynamically changes the number of forward steps until the resulting subsets are better than the previously ones based on some criterions [8]. MI, which is an often-used metric to compute the relativity between two random variables in information theory, is selected to evaluate the discriminability of class sets in DECOC. MI is defined as follow:

$$I(\boldsymbol{x}, \boldsymbol{y}) = \int \int p(x, y) \log(\frac{p(x, y)}{p(x)p(y)}) \, dx \, dy \qquad (1)$$

where $\boldsymbol{x}$ denotes the sample in the class sets and $\boldsymbol{y}$ denotes the class label. $p(\mathbf{x})$ and $p(\mathbf{y})$ are their probability density functions respectively. DECOC aims to

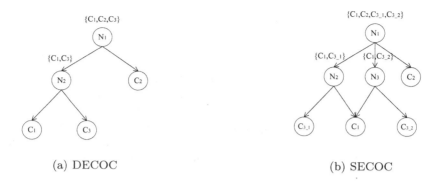

(a) DECOC   (b) SECOC

Fig. 1. Illustration of the trees in DECOC and SECOC

maximize the MI value between the data in the class sets and the class label to maximize the discriminability of class sets. However, when computing the MI value, DECOC uses a non-parametric parzen estimation method which usually requires a large number of samples in order to reach a relatively better performance and is more likely to lead to unstable experimental results.

When the tree has been completely constructed, DECOC further fills a ternary coding matrix based on the tree. Particularly, the classes in the left partition are represented by $+1$ and the classes in the right partition are represented by -1. Meanwhile, classes which are not shown up in the hierarchical partition are represented by 0.

2.2 Subclass ECOC (SECOC)

On the basis of DECOC, SECOC [9] contributes to solving the linearly non-separable problems by dividing the original class into some new subclasses. SECOC also uses SFFS to find the hierarchical partition of the tree by maximizing the MI value. When the original partition is linearly non-separable, SECOC further uses the cluster method K-means to split it into simpler and smaller sub-partitions. Usually, the number of sub-partitions is set to 2 [9]. As Fig. 1(b) shown, SECOC splits the original linearly non-separable problem $\{C_1, C_3\}$ into two linearly separable problems $\{C_1, C_{3\_1}\}$ and $\{C_1, C_{3\_2}\}$ by dividing class C_3 into two new subclasses C_{3_1} and C_{3_2}. Therefore, if the original classification problem is linearly non-separable, SECOC can transform it into a linearly separable one through several times of decompositions.

3   Maximum Margin Tree ECOC (M²ECOC)

3.1   Maximum Margin

Margin, which is defined as the minimum distance between the decision boundary and samples, is one of the most famous concepts in SVM proposed by Vapnik [13].

Specifically, decision boundary which is also called decision hyperplane, is denoted as follows:

$$\boldsymbol{w}^T \boldsymbol{\Phi}(\boldsymbol{x}) + b = 0 \tag{2}$$

where $\boldsymbol{\Phi}(\boldsymbol{x})$ is a fixed feature-space transformation function, $\boldsymbol{w}$ is a weight vector and b is the bias. The functional margin can be formulated as:

$$\hat{r} = \min_i \{y_i(\boldsymbol{w}^T \boldsymbol{\Phi}(\boldsymbol{x}_i) + b)\} \quad i = 1, 2, ..., N \tag{3}$$

where $y_i \in \boldsymbol{y}$ denotes the corresponding class label and N is the number of samples. However, the functional margin does not have the scaling invariance. So, we further get the geometric margin by normalizing (3):

$$\tilde{r} = \min_i \{y_i(\frac{\boldsymbol{w}^T}{||\boldsymbol{w}||}\boldsymbol{\Phi}(\boldsymbol{x}_i) + \frac{b}{||\boldsymbol{w}||})\} \quad i = 1, 2, ..., N \tag{4}$$

According to (3) and (4), we can easily obtain the relationship with the functional margin and the geometric margin as follows:

$$\tilde{r} = \frac{\hat{r}}{||\boldsymbol{w}||} \tag{5}$$

Let $\hat{r}$ equal to 1. The maximum margin can be optimized by solving the following problem:

$$\max_{w,b} \quad \frac{1}{||\boldsymbol{w}||} \tag{6}$$
$$s.t. \quad y_i(\boldsymbol{w}^T \boldsymbol{\Phi}(\boldsymbol{x}_i) + b) - 1 \geqslant 0, \quad i = 1, 2, ..., N$$

It is obvious that the maximization of $||\boldsymbol{w}||^{-1}$ is equivalent to the minimization of $||\boldsymbol{w}||$. So we can transform (6) into the following optimization problem:

$$\min_{w,b} \quad \frac{1}{2}||\boldsymbol{w}||^2 + C\sum_{i=1}^{N}\xi_i \tag{7}$$
$$s.t. \quad y_i(\boldsymbol{w}^T \boldsymbol{\Phi}(\boldsymbol{x}_i) + b) \geqslant 1 - \xi_i, \quad i = 1, 2, ..., N$$
$$\xi_i \geqslant 0, \quad i = 1, 2, ..., N$$

where the parameter $\boldsymbol{\xi}$ is the slack variable and C is used to balance $\boldsymbol{\xi}$ and the margin. (7) can be further changed into a dual problem using Lagrange multipliers with kernel functions:

$$\max_{\boldsymbol{\alpha}} \quad \sum_{i=1}^{N}\alpha_i - \frac{1}{2}\sum_{i,j=1}^{N}\alpha_i\alpha_j y_i y_j K(\boldsymbol{x}_i, \boldsymbol{x}_j) \tag{8}$$
$$s.t. \quad \sum_{i=1}^{N}\alpha_i y_i = 0, \quad i = 1, 2, ..., N$$
$$0 \leqslant \alpha_i \leqslant C, \quad i = 1, 2, ..., N$$

Through (8), we can solve the α. Consequently, the maximum margin can be finally computed as follows [14]:

$$margin = \frac{1}{||\boldsymbol{w}||} \tag{9}$$

where the vector $\boldsymbol{w}$ is determined by

$$\boldsymbol{w} = \sum_{i=1}^{N} \alpha_i y_i \mathbf{x}_i \tag{10}$$

3.2   Maximum Margin Matrix

Given a k-class classification problem, we can compute the maximum margin between each pair of classes according to (9). Then all maximum margins can be combined as a maximum margin matrix:

$$\begin{bmatrix} 0 & m_{12} & \dots & m_{1(k-1)} & m_{1k} \\ m_{21} & 0 & \dots & m_{2(k-1)} & m_{2k} \\ \dots & \dots & \dots & \dots & \dots \\ m_{(k-1)1} & m_{(k-1)2} & \dots & 0 & m_{(k-1)k} \\ m_{k1} & m_{k2} & \dots & m_{k(k-1)} & 0 \end{bmatrix}$$

where m_{ij} is the maximum margin between the ith and jth classes. Obviously, this matrix is symmetric. So we just compute the values of upper triangular matrix elements. As can be seen, the bigger the value of m_{ij} is, the larger the maximum margin between these two classes would be. Furthermore, a larger maximum margin means that the corresponding two classes are more well-separated. Consequently, the maximum margin actually gives us a natural criterion to evaluate the discriminability between classes. In M²ECOC, we will directly use the maximum margin to build the tree.

3.3   Maximum Margin Tree

Traditional tree algorithms such as DECOC and SECOC usually built the tree from top to down as Fig. 1 shown. In fact, such strategy emphasizes more on the discriminability between internal nodes, but ignores the discriminability between leaf nodes. For example, in Fig. 1(a), DECOC firstly separates the original class set $\{C_1, C_2, C_3\}$ into two partitions $\{C_1, C_3\}$ and $\{C_2\}$ and then divides the internal node $\{C_1, C_3\}$ into two leaf nodes $\{C_1\}$ and $\{C_3\}$. As a result, DECOC can guarantee that the internal partition between the internal node $\{C_1, C_3\}$ and $\{C_2\}$ has good discriminability and the corresponding dichotomizer in the internal node can achieve satisfactory performance. However, DECOC can not guarantee that the two leaf nodes $\{C_1\}$ and $\{C_3\}$ also have similarly good discriminability, which leads to the performance of the corresponding dichotomizer uncontrollable.

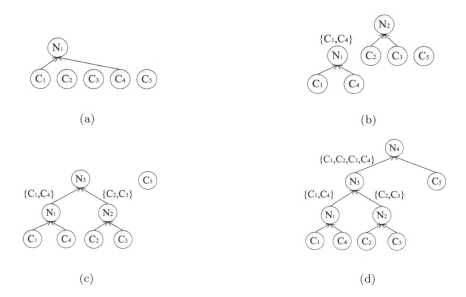

Fig. 2. Construction of a bottom-up maximum margin tree in M^{2}ECOC

In M^{2}ECOC, we adopt a bottom-up strategy [10] to construct the maximum margin tree. In order to illustrate the strategy more clearly, we will take a five-class classification problem as an example in Fig. 2. Concretely, we firstly regard each class as a subclass and use (9) to compute the maximum margin matrix. According to this matrix, $\{C_1\}$ and $\{C_4\}$ have the maximal maximum margin. So we combine them as a new subclass (Fig. 2(a)) but still keep their original classes labels. Then the new subclass and the rest classes generate a new four-class classification problem. Repeating the above process, we take $\{C_2\}$ and $\{C_3\}$ between which have the maximal maximum margin as another new subclass (Fig. 2(b)). Consequently, the subclasses $\{C_1, C_4\}$, $\{C_2, C_3\}$ and $\{C_5\}$ boil down to a new three-class classification problem. Following the same steps, the two subclasses $\{C_1, C_4\}$ and $\{C_2, C_3\}$ are integrated as a new subclass $\{C_1, C_2, C_3, C_4\}$ (Fig. 2(c)). Particularly, when computing the margin between subclasses $\{C_1, C_4\}$ and $\{C_2, C_3\}$, the classes in the same subclass will be considered as one class temporarily. Finally the expected maximum margin tree can be obtained as Fig. 2(d).

After the optimal hierarchical partition of the maximum margin tree has been finished, we obtain the coding matrix M as follows:

$$M(r,l) = \begin{cases} 1 & C_r \in P_l^{left} \\ 0 & C_r \notin P_l \\ -1 & C_r \in P_l^{right} \end{cases} \tag{11}$$

where $M(r,l)$ denotes the element lying in the rth row and the lth column in the coding matrix and the C_r denotes the rth class. P_l^{left} and P_l^{right} are

Table 2. Coding matrix of the example in Fig. 2

	h_1	h_2	h_3	h_4
C_1	+1	0	+1	+1
C_2	0	+1	−1	+1
C_3	0	−1	−1	+1
C_4	−1	0	+1	+1
C_5	0	0	0	−1

the left and right partition of the lth partition respectively (Regardless of the root node). Table 2 lists the corresponding coding matrix of the above example following (11).

4   Experimental Results

In this section, we compare M^2ECOC with some state-of-the-art coding algorithms like OVA [4], OVO [5], dense random [6], sparse random [6], DECOC [8][1], SECOC [9] and HECOC [10] to validate the superiority of our approach.

Ten multi-class datasets from common-used UCI datasets [15] are used in the experiments, that is, Wine (178,13,3), Lenses (24,4,3), Glass (214,9,6), Balance (625,4,3), Cmc (1473,9,3), Ecoli (332,6,6), Iris (150,4,3), Tea (151,5,3), Thyriod (215,5,3), Vehicle (846,18,4), where the numbers of samples, dimension and classes are listed in the bracket. We randomly split each dataset into two non-overlapping training and testing set. The training set contains almost seventy percent of the samples and the rest samples are composed as the testing set. The whole process is repeated ten times. The average accuracies are also reported.

Moreover, in dense and sparse random algorithms, all random matrices are selected from a set of 10000 randomly generated matrices where $P(1) = P(-1) = 0.5$ for the dense random matrix as well as $P(1) = P(-1) = 0.25$ and $P(0) = 0.5$ for the sparse random matrix [6]. In SECOC, the parameter set $\Theta = \{\Theta_{size}, \Theta_{perf}, \Theta_{impr}\}$ is fixed to $\Theta_{size} = \frac{|J|}{50}$, $\Theta_{perf} = 0$ and $\Theta_{impr} = 0.95$ according to [9]. The regularization parameter C and the width σ in radial basis function kernel in HECOC and M^2ECOC are selected from the interval $\{2^{-6}, 2^{-5}, ..., 2^5, 2^6\}$ by cross-validation.

The decoding strategy HD is used to evaluate the performance of different coding algorithms. Two base classifiers Nearest Mean Classifier (NMC) and SVM with radial basis function kernel are applied as the dichotomizers, where the regularization parameter C is set to 1 [9]. Moreover, the width σ in the kernel is also selected from the same interval in HECOC and M^2ECOC.

---

[1] We download the DECOC code from http://jmlr.csail.mit.edu/papers/v11/escalera10a.html which was provided by Sergio Escalera, Oriol Pujol, Petia Radeva in 2010.

The classification results on the ten datasets are reported in Tables 3 and 4. From the tables, we can see that M²ECOC can reach better or comparable performance than compared algorithms on most datasets. Especially, the accuracies of M²ECOC exceed the other algorithms' accuracies beyond 3 % on the Glass and Vehicle sets with NMC in Table 3. In Table 4, Its accuracy even excels the other accuracies nearly 12 % on the Lenses set with SVM. Furthermore, we also list the average accuracies and standard deviations on all the datasets in the bottom of the tables. It obviously can be seen that M²ECOC possesses the best performance compared with the other algorithms, which further indicates the superiority of M²ECOC. On the contrary, DECOC and SECOC perform much poorly with SVM in some datasets. For example, their accuracies are even 10 % lower than the other algorithms' accuracies on the Lenses, Balance, Iris sets

Table 3. Classification results (mean ± std) of NMC and HD on ten datasets (•/○ indicates that our algorithm is significantly better or worse than other algorithms based on the t-test at 95 % significance level)

	OVO	OVA	Dense	Sparse	DECOC	SECOC	HECOC	M²ECOC
Wine	97.55±1.79	94.91±2.82•	93.40±5.85•	94.34±4.71•	97.36±3.11	97.36±3.11	95.09±3.47•	**98.30±1.88**
Lenses	77.50±7.91	78.75±8.44	72.50±9.86	70.00±10.50•	78.75±8.44	78.75±8.44	71.25±11.90	**80.00±8.74**
Glass	46.62±6.07•	23.23±5.69•	40.31±9.22•	47.85±6.62	42.15±11.40•	49.08±6.97	42.31±10.00•	**52.77±6.03**
Balance	73.69±4.63•	**88.07**±1.90○	69.79±20.10	75.72±14.60	80.32±3.72	81.60±5.27	81.60±3.16	81.17±2.77
Cmc	46.74±2.10	45.86±1.58	46.02±2.20	45.81±2.35	46.29±1.19	46.31±1.18	45.48±2.38	**46.97**±1.60
Ecoli	**84.90**±2.88○	70.50±1.90•	77.50±5.46•	78.50±4.14•	74.20±8.27•	77.90±4.07•	71.90±13.40•	81.90±2.60
Iris	86.00±5.55	82.89±2.97	76.67±9.03•	74.44±10.10•	79.78±8.35	85.11±3.93	**86.22**±5.52	86.00±5.55
Tea	**56.52**±5.12	52.83±6.49	54.13±6.60	51.96±4.64	53.26±5.64	55.87±5.80	54.13±6.44	56.09±5.21
Thyriod	92.81±1.98	94.06±2.42	92.50±3.52	91.41±2.97•	92.97±3.40	93.44±3.59	94.06±2.42	**94.53**±2.68
Vehicle	46.10±3.51	43.62±1.52•	40.20±4.22•	38.03±5.40•	40.04±3.37•	43.94±5.22•	40.67±3.40•	**49.25**±4.86
Average	70.84±4.15	67.47±3.57	66.30±7.61	66.81±6.60	68.51±5.69	70.94±4.76	68.27±6.21	**72.70**±4.19
win/tie/loss	2/7/1	4/5/1	5/5/0	6/4/0	3/7/0	2/8/0	4/6/0	/

Table 4. Classification results (mean ± std) of SVM and HD on ten datasets (•/○ indicates that our algorithm is significantly better or worse than other algorithms based on the t-test at 95 % significance level)

	OVO	OVA	Dense	Sparse	DECOC	SECOC	HECOC	M²ECOC
Wine	97.92±1.39•	98.49±1.73	98.11±1.26•	98.11±1.78•	96.23±1.99•	97.36±1.82•	98.30±1.65•	**99.62**±0.80
Lenses	61.25±3.95•	62.50±0.00•	63.75±3.95•	61.25±3.95•	56.25±8.84•	58.75±6.04•	62.50±0.00•	**75.00**±10.20
Glass	68.31±2.43○	53.23±5.04•	**68.46**±3.34○	67.23±2.62○	62.31±9.84	64.15±9.97○	61.85±5.88	62.15±3.78
Balance	90.27±0.97	89.36±0.89•	91.50±2.85	**91.55**±3.07	76.84±1.99•	77.17±2.03•	89.79±1.02•	90.86±0.64
Cmc	54.43±1.46	50.54±1.53•	48.10±1.79•	47.83±1.58•	51.97±1.46•	52.04±1.42•	**54.52**±1.19	54.00±0.98
Ecoli	86.40±3.03	82.40±2.17•	86.50±2.07	86.80±2.04	80.30±15.10	83.10±6.59	70.30±15.50•	**86.90**±2.59
Iris	**95.56**±3.63	93.33±2.34	94.00±3.32	95.11±3.11	72.89±4.42•	74.67±2.81•	95.11±3.60	94.89±1.83
Tea	51.09±4.38	51.09±6.58	49.57±4.44•	51.30±4.25	44.35±5.99•	47.83±5.02•	51.30±3.58•	**55.43**±4.94
Thyriod	**96.09**±1.11	94.84±1.66	95.00±2.31	95.31±1.47	94.37±1.32•	94.37±1.32•	95.78±1.06	**96.09**±1.11
Vehicle	74.92±1.76	66.50±2.48•	72.28±5.20	71.93±2.87•	72.76±2.37•	72.76±2.37•	74.53±1.93	**75.08**±1.76
Average	77.62±2.41	74.23±2.44	76.73±3.05	76.64±2.67	70.83±5.33	72.22±3.94	75.40±3.54	**79.00**±2.86
win/tie/loss	2/7/1	6/4/0	4/5/1	4/5/1	8/2/0	8/1/1	5/5/0	/

in Table 4. The reason lies more on they are more sensitive to different base classifier and their using a non-parametric estimation to compute the MI value, which indeed requires numerous training data to achieve acceptable results.

In order to further statistically measure the significance of performance difference, the pairwise t-tests [16] at 95 % significance level are conducted between the algorithms. Specifically, whenever M^{2}ECOC achieves significantly better/worse performance than the compared algorithms on most datasets, a win/loss is counted and a marker •/∘ are shown. Otherwise, a tie is counted and no marker is given. The resulting win/tie/loss counts for M^{2}ECOC against the compared algorithms are provided in the last line of Tables 3 and 4. As the tables shown, M^{2}ECOC can achieve statistically better or comparable performance on most datasets, which just accords with our conclusion.

5 Conclusion

In this paper, we present a novel encoding algorithm M^{2}ECOC for ECOC. Different from the existing tree-form encoding algorithms, M^{2}ECOC directly utilizes the maximum margin which actually is a natural criterion to evaluate the discriminability between classes to get the optimal hierarchical partition of the tree. Specifically, M^{2}ECOC regards each class as a subclass and computes the maximum margin matrix. According to this matrix, the classes with the maximal maximum margin are selected to combine as a new subclass. Then the new subclass and the rest classes generate a new multi-class classification problem. Repeating the same steps until all classes in one subclass. M^{2}ECOC constructs the maximum margin tree in a bottom-up manner and the corresponding coding matrix can be obtained easily by the tree. The experimental results on several UCI datasets have shown that M^{2}ECOC is superior to some state-of-the-art ECOC encoding algorithms, which further validates that the maximum margin is indeed an effective criterion for building the tree in ECOC.

Acknowledgements. This work was supported by National Natural Science Foundation of China (Grant Nos. 61375057, 61300165 and 61403193) and Natural Science Foundation of Jiangsu Province of China (Grant No. BK20131298). Furthermore, the work was also supported by Collaborative Innovation Center of Wireless Communications Technology.

References

1. Japkowicz, N., Barnabe-Lortie, V., Horvatic, S., et al.: Multi-class learning using data driven ECOC with deep search and re-balancing. In: IEEE International Conference on DSAA, pp. 1–10 (2015)
2. Liu, M., Zhang, D., Chen, S., et al.: Joint binary classifier learning for ECOC-based multi-class classification. IEEE Trans. Pattern Anal. Mach. Intell. **1**, 0162–8828 (2015)
3. Dietterich, T.G., Bakiri, G.: Solving multiclass learning problems via error-correcting output codes. J. Artif. Intell. Res. **2**, 263–286 (1995)

4. Nilsson, N.J.: Learning Machines: Foundations of Trainable Pattern-Classifying Systems. McGraw-Hill, New York (1965)
5. Hastie, T., Tibshirani, R.: Classification by pairwise coupling. Ann. Stat. **26**(2), 451–471 (1998)
6. Allwein, E.L., Schapire, R.E., Singer, Y.: Reducing multiclass to binary: a unifying approach for margin classifiers. J. Mach. Learn. Res. **1**, 113–141 (2001)
7. Escalera, S., Pujol, O., Radeva, P.: Boosted landmarks of contextual descriptors and forest-ECOC: a novel framework to detect and classify objects in cluttered scenes. Pattern Recogn. Lett. **28**(13), 1759–1768 (2007)
8. Pujol, O., Radeva, P., Vitrial, J.: Discriminate ECOC: a heuristic method for application dependent design of error correcting output codes. IEEE Trans. Pattern Anal. Mach. Intell. **28**(6), 1001–1007 (2006)
9. Escalera, S., Tax, D.M.J., Pujol, O., et al.: Subclass problem-dependent design for error-correcting output codes. IEEE. Trans. Pattern Anal. Mach. Intell. **30**(6), 1041–1054 (2008)
10. Lei, L., Wang, X., Luo, X., et al.: Hierarchical error-correcting output codes based on SVDD. J. Syst. Eng. Electron. **37**(8), 1916–1921 (2015). (In Chinese)
11. Principe, J.C., Xu, D., Fisher, J.: Information theoretic learning. In: Unsupervised Adaptive Filtering, vol. 1, pp. 265–319 (2000)
12. Tax, D.M.J., Duin, R.P.W.: Support vector domain description. Pattern Recogn. Lett. **20**(11), 1191–1199 (1999)
13. Cortes, C., Vapnik, V., Guyaon, I.: Support-vector networks. Mach. Learn. **20**(3), 273–297 (1995)
14. Lu, M., Huo, J., Chen, C.L.P., et al.: Multi-stage decision tree based on inter-class and inner-class margin of SVM. In: IEEE International Conference on SYST, pp. 1875–1880 (2009)
15. Asuncion, A., Newman, D.: UCI Machine Learning Repository (2007)
16. Kuncheva, L.I.: Combining Pattern Classifiers: Methods and Algorithms. Wiley, New York (2004)

A Relaxed K-SVD Algorithm for Spontaneous Micro-Expression Recognition

Hao Zheng[1,2,3], Xin Geng[2(✉)], and Zhongxue Yang[1]

[1] Key Laboratory of Trusted Cloud Computing and Big Data Analysis,
School of Information Engineering, Nanjing XiaoZhuang University, Nanjing, China
[2] MOE Key Laboratory of Computer Network and Information Integration,
School of Computer Science and Engineering, Southeast University, Nanjing, China
xgeng@seu.edu.cn
[3] State Key Laboratory for Novel Software Technology,
Nanjing University, Nanjing, China

Abstract. Micro-expression recognition has been a challenging problem in computer vision due to its subtlety, which are often hard to be concealed. In the paper, a relaxed K-SVD algorithm (RK-SVD) to learn sparse dictionary for spontaneous micro-expression recognition is proposed. In RK-SVD, the reconstruction error and the classification error are considered, while the variance of sparse coefficients is minimized to address the similarity of same classes and the distinctiveness of different classes. The optimization is implemented by the K-SVD algorithm and stochastic gradient descent algorithm. Finally a single overcomplete dictionary and an optimal linear classifier are learned simultaneously. Experimental results on two spontaneous micro-expression databases, namely CASME and CASME II, show that the performance of the new proposed algorithm is superior to other state-of-the-art algorithms.

Keywords: Related K-SVD · Dictionary learning · Micro-expression recognition

1 Introduction

Micro-expressions are quick facial expressions that appearing less than 0.5 s. Micro-expression was first discovered by Haggard in 1966. In current literature, there are only three spontaneous micro-expression databases, i.e. SMIC [1], CASME [2], CASME II [3], while there are some works to date on automatic recognition of spontaneous micro-expressions. Among all the available work, Li [1] published the SMIC dataset and released a baseline performance of up to 48.78 % accuracy for 3 classes, adopting LBP-TOP for feature extraction and SVM polynomial kernel for classification with cross validation method. Yan et al. [3] also reported a baseline performance of up to 63.41 % accuracy for a 5-class classification task on CASME II database, adopting LBP-TOP and SVM for feature extraction and classification respectively. Then Wang et al. [4] proposed LBP-SIP method, which incorporated a multi-resolution Gaussian

© Springer International Publishing Switzerland 2016
R. Booth and M.-L. Zhang (Eds.): PRICAI 2016, LNAI 9810, pp. 692–699, 2016.
DOI: 10.1007/978-3-319-42911-3_58

pyramid by concatenating the feature histograms of all four pyramid levels, and obtained the best recognition accuracy of 67.21 %. In addition, Polikovsky et al. [5] used 3D-gradient descriptor for micro-expression recognition. Wang et al. [6] treated a micro-expression gray-scale video clip as a 3rd-order tensor and used Discriminant Tensor Subspace Analysis and Extreme Learning Machine to recognize micro-expressions. Pfister et al. [7] utilized a matrix to normalize the frame numbers of micro-expression video clips. Discriminative KSVD (D-KSVD) [8] has been proposed. D-KSVD is based on extending the K-SVD algorithm by incorporating the classification error into the objective function, so that retains the representational power while making the dictionary discriminative.

Because micro-expression recognition aims to pay attention to a brief and subtle change, which can be well implemented by the different sparse coefficients, we can make full use of sparse coefficients information to distinguish different expression. Motivated by above idea, in this paper, a relaxed K-SVD algorithm (RK-SVD) to learn sparse dictionary for spontaneous micro-expression recognition is proposed. In RK-SVD, the reconstruction error and the classification error are considered, while the variance of sparse coefficients is minimized to address the similarity of same classes and the distinctiveness of different classes in sparse coefficients.

2 Relaxed K-SVD Algorithm (RK-SVD)

We aim to learn a reconstructive dictionary and a linear classifier, while spare coefficients information is important for the dictionary learning. Thus we add a minimization of the variance of sparse coefficients term into objective function for addressing the similarity of same classes and the distinctiveness of different classes in sparse coefficients.

2.1 Relaxed K-SVD Algorithm

Suppose Y is the matrix of all training sample, X is the sparse representation coefficients, $L = [l_1 \ldots l_N]$ is the labels of Y, W is the linear predictive classifier. Consider the reconstruction error and the classification error, an objective function for learning a dictionary D can be defined as follow:

$$< D, X, W >= \underset{D,X,W}{\arg \min} ||Y - DX||_2^2 + \lambda_1 ||L - WX||_2^2 \quad s.t. \forall i, ||x_i||_0 \leq T \quad (1)$$

where the term $||Y - DX||_2^2$ represents the reconstruction error, the term $||L - WX||_2^2$ represents the classification error, T is a parameter to impose the sparsity prior, λ_1 is the scalar.

We aim to minimize the variance of sparse coefficients. Here, we use a term $\left\|\bar{X} - X\right\|_2^2$. So an objective function for learning a dictionary D having reconstructive, classifier power and sparse coefficients information can be defined as follows:

$$< D, X, W >= \underset{D,X,W}{\arg\min} \left\|Y - DX\right\|_2^2 + \lambda_1 \left\|L - WX\right\|_2^2 + \lambda_2 \left\|\bar{X} - X\right\|_2^2 \tag{2}$$
$$s.t. \forall i, \left\|x_i\right\|_0 \leq T$$

The first term represents the reconstruction error, the second term supports learning an optimal classifier, the third term makes full of the information of sparse coefficients between classes, where λ_2 is the scalar controlling the relative contribution of the corresponding term.

In the following section, we describe the optimization procedure for RK-SVD.

2.2   Optimization

The optimization is implemented by the K-SVD algorithm and stochastic gradient descent algorithm. Equation (2) can be rewritten as

$$< D, X, W >= \underset{D,X,W}{\arg\min} \left\| \begin{pmatrix} Y \\ \sqrt{\lambda_1}\,L \\ \sqrt{\lambda_2}\,\bar{X} \end{pmatrix} - \begin{pmatrix} D \\ \sqrt{\lambda_1}\,W \\ \sqrt{\lambda_2}\,I \end{pmatrix} X \right\|_2^2 \tag{3}$$

Let $Y_{new} = (Y^t, \sqrt{\lambda_1}\,L^t, \sqrt{\lambda_2}\,\bar{X}^t)^t$, $D_{new} = (D^t, \sqrt{\lambda_1}\,W^t, \sqrt{\lambda_2}\,I)^t$. The optimization of Eq. (3) is equivalent to solving the following problem:

$$< D_{new}, X >= \underset{D_{new},X}{\arg\min} \left\|Y_{new} - D_{new}X\right\|_2^2 \tag{4}$$

Now, the problem of Eq. (4) can be efficiently solved by updating the dictionary atom by atom with the following method: For each atom d_k and the corresponding coefficient x_k can be computed by

$$< d_k, x_k >= \underset{d_k,x_k}{\arg\min} \left\|E_k - d_k x_k\right\|_F \tag{5}$$

where $E_k = Y_{new} - \sum_{i \neq k} d_k x_k$. This can be solved by original K-SVD, so the solution of Eq. (5) is giving by

$$\begin{aligned} U * \Sigma * V &= SVD(E_k) \\ d_k &= U(:,1) \\ \widetilde{x_k} &= \Sigma(1,1) * V(1,:). \end{aligned} \tag{6}$$

The nonzero values in x_k are replaced by $\widetilde{x_k}$. Thus we obtain D_{new} by solving original K-SVD. Next we learn D and W by minimizing the following objective function with the sparsity constraint using $l1$-norm regularization:

$$< D, W >= \underset{D,W}{\arg\min} \sum_i \ell(D, y_i, W, l_i, \bar{x}_i) + \frac{v}{2} \left\|W\right\|_F^2 \tag{7}$$
$$s.t.\ x_i = \underset{x}{\arg\min} \left\|y_i - Dx\right\|_2^2 + \gamma \left\|x\right\|_1, i \in \{1 \cdots N\}$$

where ℓ is the classification loss function, v is a regularization parameter. Equation (7) is highly nonlinear and highly nonconvex. As shown in [9,10], we use a stochastic gradient descent algorithm to optimize Eq. (7).

$$< D, W >= \arg\min_{D,W} \sum_i (1 - \mu)\ell_s(D, y_i, W, l_i) + \mu\ell_u(D, y_i, \bar{x}_i)$$
$$s.t. \ x_i = \arg\min_x \|y_i - Dx\|_2^2 + \gamma \|x\|_1 , i \in \{1 \cdots N\} \tag{8}$$

$$\frac{\partial\ell}{\partial D} = \mu\frac{\partial\ell_u}{\partial D} + (1 - \mu)\frac{\partial\ell_s}{\partial D} \tag{9}$$

$$\frac{\partial\ell}{\partial W} = (1 - \mu)(Wx_i - l_i)x_i^t + vW \tag{10}$$

$$D^{(t)} = D^{(t)} - \rho_t\frac{\partial\ell}{\partial D} \tag{11}$$

$$W^{(t)} = W^{(t)} - \rho_t\frac{\partial\ell}{\partial W} \tag{12}$$

Finally desired dictionary D and W are computed as follow:

$$D = \left\{ \frac{d_1}{\|d_1\|_2} \cdots \frac{d_k}{\|d_k\|_2} \right\}, W = \left\{ \frac{w_1}{\|d_1\|_2} \cdots \frac{w_k}{\|d_k\|_2} \right\} \tag{13}$$

In addition, we need to initialize the parameters $D^{(0)}$, $\bar{X}^{(0)}$ and $W^{(0)}$. Firstly we obtain $D^{(0)}$ by employing several iterations of K-SVD with each class and combining all the outputs of each K-SVD. Then we apply the original K-SVD to compute the sparse coefficients $X^{(0)}$ of training samples and compute the $\bar{X}^{(0)}$. Finally we use the ridge regression mode and obtain the $W^{(0)}$ by

$$W^{(0)} = LX^t(XX^t + \phi I)^{-1} \tag{14}$$

Thus the proposed RK-SVD algorithm is summarized in Algorithm 1.

Algorithm 1. Relaxed K-SVD algorithm (RK-SVD)

---

1. Input: $\{Y_i, L_i\}$, $i = 1, 2, \cdots, t$, where $Y_i \in \Re^{n \times d}$ is the training samples, L_i is the corresponding label vector.
2. Initialize $D^{(0)}$ and $\bar{X}^{(0)}$ by using original K-SVD, $W^{(0)}$ by Eq. (14).
3. Obtain D_{new} by Eq. (4).
4. Calculate D and W by Eqs. (11) and (12).
5. Go back to step 3 until the condition of convergence is met.
6. update D and W by Eq. (13).

2.3   Classification

With the normalized D, we can find the sparse coefficients for a test face image y by solving

$$< \hat{x} >= \arg\min_{\hat{x}}(||y - Dx||_2^2 + \lambda\, ||x||_1) \qquad (15)$$

Which can be solved by $l1$ optimization method, such as basic pursuit and so on. Finally we obtain the label of the test image:

$$l = \hat{W} * \hat{x} \qquad (16)$$

The label of the test image y is the index corresponding to the largest element of l.

3   Experiments

In this section, we test our proposed algorithm on two publicly available micro-expression datasets, including CASME, and CASME II to verify its effectiveness for micro-expression recognition. These datasets consist of spontaneous micro-expressions which appear in real life. For each database, In the stage of features extraction, we applied LBP-TOP [7] and HOOF [11] respectively, then we compare the proposed algorithm to state-of-the-art algorithms such as SVM, MKL and RF. In all experiments we employed Leave-one-subject-out (LOSO) cross validation. The parameters λ_1 and λ_2 are fixed for each data set and determined by n-fold cross validation on the training data.

3.1   Experiment on CASME Micro-Expression Databases

Micro-Expression (CASME) database includes 195 spontaneous facial micro-expression recorded by two different 60 fps cameras. These samples were selected from more than 1500 facial expressions. The selected micro-expressions either have a total duration less than 500 ms or an onset duration less than 250 ms. These samples are coded with the onset, apex and offset frames, furthermore tagged with AUs. In this database, micro-expressions are classified into 7 categories (happiness, surprise, disgust, fear, sadness, repression and tense). It was argued that the temporal interpolation of frames can extract more statistically interpolation of frames can extract more statistically stable LBP-TOP, the frame numbers of all samples were normalized to 70 by using linear interpolation, and we found that a frame number of more than 70 produced unnecessary redundance, which degraded recognition performance. We applied our proposed algorithm using LBP-TOP feature, and the best recognition rate was 69.04 %, which is better than the performance of other three classifiers. In the setting of LBP-TOP, the radii values in axes X and Y ranged from 1 to 4. To void too many combinations of parameters, we chose $R_x = R_y$. The radius R_t in axis T ranged from 1 to 4. The number of neighboring point in the XY, XT and XT planes were all set to be 4 or 8. The uniform pattern and basic pattern were used in

LBP coding. From Table 1, it can be seen that RK-SVD outperform than other methods, especially RK-SVD is at least 7 % higher than RF classifier. This is because RK-SVD makes full use of sparse coefficients information, which address the similarity of same classes and the distinctiveness of different classes.

We also compare RK-SVD with other classifiers using HOOF features. From Table 2, we can see that recognition rate of RK-SVD achieve maximal accuracy, with 54.20 % for SVM, 56.83 % for MKL, 53.69 % for RF, 60.82 % for RK-SVD. We can concluded that in the CASME database, RK-SVD is better than SVM, MKL, RF classifiers in both LBP-TOP feature and HOOP feature.

We further compared the confusion matrices of RK-SVD using LBP-TOP feature when it obtained the best recognition rates. From Table 3, we can see that the results that RK-SVD had a good recognition rates in all four classes.

Table 1. The best recognition rates for micro-expression recognition using LBP-TOP in CASME database.

Method	SVM(%)	MKL(%)	RF(%)	RK-SVD(%)
Accuracy	63.21	65.83	61.77	**69.04**

Table 2. The best recognition rates for micro-expression recognition using HOOF in CASME database.

Method	SVM(%)	MKL(%)	RF(%)	RK-SVD(%)
Accuracy	54.20	56.83	53.69	**60.82**

Table 3. Confusion matrix of micro-expression recognition obtained by RK-SVD on CASME database

	Positive(%)	Negative(%)	Surprise(%)	Others(%)
Positive	46.23	5.61	3.08	45.08
Negative	4.62	54.93	4.58	35.87
Surprise	3.93	5.09	67.88	23.1
Others	8.12	11.34	2.11	78.43

3.2 Experiment on CASME II Micro-Expression Databases

The CASME II database includes 246 spontaneous facial micro-expressions recorded by a 200 fps camera. In the experiments, the frame numbers of all samples were normalized to 150 by using linear interpolation. The size of each frame is normalized to 163×134 pixels. So, each sample was normalized to a fourth-order tensor with the size of $163 \times 134 \times 150 \times 3$. We applied our proposed algorithm using LBP-TOP feature and HOOF feature, respectively. In the setting of LBP-TOP, its parameter setting is same as in CASME. The uniform pattern and basic pattern were used in LBP coding. λ_1, λ_2 are set to 4.0 and 2.0 respectively.

Tables 4 and 5 show that our proposed algorithm outperformed other three classifiers, achieved that the best recognition rates both LBP-TOP feature and HOOF feature. We can see that the recognition rates in CASME II database is worse than those in CASME database, because compared with CASME, the database is improved in increased sample size, fixed illumination, and higher resolution (both temporal and spatial).

Table 4. The best recognition rates for micro-expression recognition using LBP-TOP in CASME II database.

Method	SVM(%)	MKL(%)	RF(%)	RK-SVD(%)
Accuracy	56.82	58.34	55.21	**63.25**

Table 5. The best recognition rates for micro-expression recognition using HOOF in CASME II database.

Method	SVM(%)	MKL(%)	RF(%)	RK-SVD(%)
Accuracy	51.33	53.01	50.82	**58.64**

4   Conclusion

In this paper, we propose a relax K-SVD algorithm for micro-expression recognition. Not only reconstruction error and classification error are considered, but also the variance of sparse coefficients is minimized, which aims to address the similarity of same classes and the distinctiveness of different classes in sparse coefficients. It is especially meaningful for micro-expressions, which are subtle and brief, and can easily be concealed by appearances irrelevant to the expressions of interest. Furthermore, the optimization is efficiently implemented by the K-SVD algorithm and stochastic gradient descent algorithm. Extensive experimental results on spontaneous micro-expression datasets including CASME, and CASME II demonstrate the effectiveness of the proposed algorithm for micro-expression recognition.

Acknowledgement. This work is partially supported by the Project funded by China Postdoctoral Science Foundation Under grant No. 2014M5615556, supported by the National Science Foundation of China (61273300, 61232007) and Jiangsu Natural Science Funds for Distinguished Young Scholar (BK20140022). And, it is also partially supported by grants 15KJB520024 from Jiangsu University Natural Science Funds, supported by grants KFKT2014B18 from the State Key Laboratory for Novel Software

Technology from Nanjing University, supported by the Collaborative Innovation Center of Wireless Communications Technology, grants 2015NXY05 from Nanjing Xiaozhuang University. Finally, the authors would like to thank the anonymous reviewers for their constructive advice.

References

1. Li, X., Pfister, T., Huang, X., Zhao, G., Pietikainen, M.: A spontaneous microexpression database: inducement, collection and baseline. In: 10th IEEE International Conference and Workshops on Automatic Face and Gesture Recognition (FG), pp. 1–6 (2013)
2. Yan, W.J., Wu, Q., Liu, Y.J., Wang, S.J., Fu, X.: CASME database: a dataset of spontaneous micro-expressions collected from neutralized faces. In: 10th IEEE International Conference and Workshops on Automatic Face and Gesture Recognition (FG), pp. 1–7 (2013)
3. Yan, W.J., Li, X., Wang, S.J., Zhao, G., Liu, Y.J., Chen, Y.H., Fu, X.: CASME II: an improved spontaneous micro-expression database and the baseline evaluation. PloS ONE 9(1), e86041 (2014)
4. Wang, Y., See, J., Phan, R.C.-W., Oh, Y.-H.: Efficient spatio-temporal local binary patterns for spontaneous facial micro-expression recognition. PLoS ONE 10(5), e0124674 (2015)
5. Polikovsky, S., Kameda, Y., Ohta, Y.: Facial micro-expressions recognition using high speed camera and 3D-gradient descriptor. In: 3rd International Conference on Crime Detection and Prevention, pp. 1–6. IET (2009)
6. Wang, S.J., Chen, H.L., Yan, W.J., Chen, Y.H., Fu, X.: Face recognition and micro-expression based on discriminant tensor subspace analysis plus extreme learning machine. Neural Process. Lett. 39, 25–43 (2013)
7. Pfister, T., Li, X., Zhao, G., Pietikainen, M.: Recognising spontaneous facial micro-expressions. In: 12th IEEE International Conference on Computer Vision, pp. 1449–1456. IEEE (2011)
8. Zhang, Q., Li, B.X.: Discriminative k-svd for dictionary learning in face recognition. In: Proceedings of the Computer Vision and Pattern Recognition (CVPR) (2010)
9. Bradley, D., Bagnell, J.: Differential sparse coding. In: Proceedings Conference on Neural Information Processing Systems (2008)
10. Mairal, J., Bach, F., Ponce, J.: Task-driven dictionary learning. PLoS ONE 34(4), 791–804 (2012)
11. Chaudhry, R., Ravichandran, A., Hager, G., Vidal, R.: Histograms of oriented optical flow and binet-cauchy kernels on nonlinear dynamical systems for the recognition of human actions. In: 2009 IEEE Conference on Computer Vision and Pattern Recognition (CVPR), pp. 1932–1939. IEEE (2009)

Set to Set Visual Tracking

Wencheng Zhu, Pengfei Zhu$^{(\boxtimes)}$, Qinghua Hu, and Changqing Zhang

School of Computer Science and Technology, Tianjin University, Tianjin, China
{zhu1992719,zhupengfei,huqinghua,zhangchangqing}@tju.edu.cn

Abstract. Sparse representation has been widely used in visual tracking and achieves superior tracking results. However, most sparse representation models represent the target candidate as a linear combination of target templates and need to solve a sparse optimization problem. In this paper, we propose a novel set to set visual tracking (SSVT) method. Under the particle filter framework, we consider both the target candidates and target templates as image sets, and model them as convex hulls. Then the distance between two image sets is minimized and the tracking result is the target candidate with the maximum coefficient. As the target candidates are modeled as one convex hull, SSVT utilizes the underlying relationship of the target candidates. Moreover, SSVT is very efficient in that it only needs to solve one quadratic optimization problem rather than sparse optimization problems. Both qualitative and quantitative analyses on several challenging image sequences show that the proposed SSVT algorithm outperforms the state-of-the-art trackers.

Keywords: Set to set distance · Visual tracking · Particle filter · Convex hull · Support vector machine

1 Introduction

Visual tracking is one of the fundamental topics in computer vision and has been applied to many applications such as video surveillance, robot navigation, human computer interface, intelligent transportation, activity analysis, human motion analysis, etc. Despite the massive tracking algorithms in the past few years [1–8], visual tracking is still a challenging problem. A tracking algorithm needs to accommodate continuous changes of appearance, e.g. variations in various scales and poses, motion blur, illuminations, occlusions, in/out of plane rotation, background clutters, low resolution etc. [9]. There are two vital components in a tracking system, i.e., a motion model and an observation model. The motion model establishes the temporal relations of states and provides the candidate states in one frame, e.g., particle filter [10]. The observation model represents the states and estimates the probabilities of the candidate states. In this paper, we aim to develop an effective observation model because of its key role in the visual tracking task.

Existing visual tracking methods can be categorized into two classes: generative tracking methods and discriminative tracking methods. Discriminative

© Springer International Publishing Switzerland 2016
R. Booth and M.-L. Zhang (Eds.): PRICAI 2016, LNAI 9810, pp. 700–712, 2016.
DOI: 10.1007/978-3-319-42911-3_59

trackers take visual tracking as a binary classification task and aim to distinguish the area of the target from the background [9]. The typical discriminative tracking algorithms are online multiple instance learning tracking [9], online boosting tracking [11], semi-online boosting tracking [12], ensemble tracking [13], and dense spatio-temporal context learning tracking [14]. For multiple instance learning tracker, the concept of bags is proposed, and a multi-instance classifier is learned [9]. In online multiple instance learning tracking algorithm, it is found that one positive example tracking method may degrade over time, if the target location is not precise and multiple positive examples may cause confusion of boundary. In [14], spatial context information as well as time context information is emphasized and a spatio-temporal context model is proposed. In [11], an online Adaboost algorithm is used to train a discriminant classifier.

In contrast, generative tracking algorithms adopt an appearance model to describe the observation of the target and the tracking result is obtained via searching the most similar observation. An off-line subspace model is employed to represent the region of interests for tracking [15]. The mean-shift tracker uses color histogram as feature and the algorithm iterates mean-shift vector until it converges to the target location [16]. The Frag tracker models the object appearances by using histogram of local patches to deal with the occlusion problem [17]. The IVT tracker introduces an incremental subspace model to adapt to appearance changes [18]. To deal with different variations, a multiple observation model is used to cover a wide range of changes in visual tracking [19]. Recently, due to the successful applications of sparse representation in face recognition and image classification, many sparse representation based generative tracking methods have been developed [1–5, 10, 20–23]. These methods can be categorized into four types: global, local, joint and structural representation based trackers.

Global methods model the target candidates as a single entity and represent it by a sparse combination of the dynamically updated dictionary [4, 10, 24]. The dictionary consists of two parts: the target templates and the trivial templates. The target templates model the appearance of target and the trivial templates model the variations. L_1 tracker has good robustness [10]. However, it needs to solve a large number of L_1 minimization problems and the computation cost is expensive. L_1-APG tracker utilizes minimal error bound to reduce the number of the target candidates and an approximation algorithm is proposed to handle L_1 minimization problem [24]. CEST exploits context information as well as the robustness of L_1 tracker [23]. As the trivial templates cannot well model the occlusions, the performance of the global representation methods may degrade when there are heavy occlusions. Additionally, existing global methods do not consider the structural relations of the target candidates.

Compared with holistic representation, local sparse tracking uses local image patches to model the appearance of the target and the sparse code of each patch is obtained [2, 3]. In [3], a local sparse appearance model is proposed by extracting a sparse coding histogram. In [2], patches of the target candidates are independently represented by the dictionary composed of all the patches and then an alignment pooling method is applied. Unfortunately, the spatial layout structure and correlation among patches are not taken into account.

Because of the dependencies among target candidates, joint sparse appearance models are proposed to employ the intrinsic relationship of different target candidates [1,5,21]. In [5], multi-task tracker takes the representation of each target candidate as one task and learns the joint sparse representation of the target candidates to seek the underlying structure. In [21], the joint representation of all the target candidates is required to be low rank. In [1], the multi-view multi-task representation is proposed for tracking. In [22], considering that the image observations of the target candidates tend to be low rank. Then, low-rank property and sparsity constraint and temporal consistency are exploited to learn robust representation. Joint representation not only exploits relationships of the target candidates but solves a single optimization problem. To combine the advantages of both local and joint representation, structural sparse tracking algorithm [25] seeks the relationship between the target candidates and the joint sparse representation of local patches is obtained by keeping their spatial structure.

The underlying relationships among the target candidates are the key factor to develop a robust and efficient sparse representation based visual tracking model. The target candidates from the current frame have high similarity, because they are randomly sampled according to a zero-mean Gaussian distribution based on the previous state. Most of tracking algorithms learn the representation of the target candidates separately. A large number of L_1-norm minimization problems need to be solved at each frame, as we know the computation cost of L_1-norm minimization problem is expensive. The target candidates should be jointly rather than individually represented to improve the efficiency. Besides, the target candidates should be similarly represented by the target templates due to the high similarity of different target candidates.

Motivated by the underlying relationship among the target candidates, in this paper, we propose a novel robust and efficient set to set visual tracking (SSVT) algorithm. We consider both the target candidates and target templates as image sets, and model either set as one convex hull. The distance between two convex hulls is minimized and the target candidate with the maximum coefficient is considered as the tracking result. Minimal error bound is used to speed up the proposed algorithm by selecting a subset of the target candidates. By minimizing the set to set distance, SSVT requires all the target candidates to achieve the same representation, which utilizes the high similarity of different target candidates. Besides, SSVT solves a single quadratic programming problem, which is quite efficient. Experiments on several benchmark image sequences validate that our tracking algorithm performs well against the state-of-the-art tracking methods.

2   Set to Set Visual Tracking

The objective of visual tracking is to identify the tracking object from the target candidates. In this section, we will present a set to set visual tracking model.

2.1 Motivation

For particle filter based visual tracking methods, by following a zero-mean Gaussian distribution, the target candidates are the particles randomly sampled around the current state of the tracked object [5]. Hence, the target candidates have high similarity and share dependencies. Let $\mathbf{Y} = [\mathbf{y}_1, \mathbf{y}_2, ..., \mathbf{y}_n] \in \mathbb{R}^{d \times n}$ be the target candidates, where each column is one target candidate. The target templates are $\mathbf{D} = [\mathbf{d}_1, \mathbf{d}_2, ..., \mathbf{d}_m] \in \mathbb{R}^{d \times m}$, where each column is one template. The target templates include the ground truth and patches randomly sampled around the ground truth. When the target candidates are noiseless, each target candidate $\mathbf{y}_i \in \mathbb{R}^d$ can be represented by a linear combination of $\mathbf{D}$, i.e., $\mathbf{y}_i = \mathbf{D}\beta_i$. As the target candidates are sampled around the current state, the representation should be very sparse. Hence, L_1 trackers sparsely and independently represent each target candidate on the target templates [10]. Because there are usually many variations in poses, illuminations, occlusions, etc., trivial templates $\mathbf{I} \in \mathbb{R}^{d \times d}$ are introduced to augment the target templates $\mathbf{D}$ and model the noise in visual tracking. As shown in Eq. (1), $\mathbf{y}_i$ is represented on $\mathbf{D}$ and $\mathbf{I}$ as:

$$\mathbf{y}_i = \mathbf{D}\beta_i + \mathbf{I}\gamma_i \tag{1}$$

Then $\mathbf{y}_i$ can be represented by $\mathbf{Z}\mathbf{b}_i$, where $\mathbf{Z} = [\mathbf{D}, \mathbf{I}]$ and $\mathbf{b}_i = [\beta_i, \gamma_i]$. As L_1 tracker ignores the dependencies among the target candidates, multi-task tracker is proposed by jointly representing $\mathbf{Y}$ on $\mathbf{Z}$ as $\mathbf{Y} = \mathbf{Z}\mathbf{B}$, where $\mathbf{B} = \{\mathbf{b}_1, \mathbf{b}_2, ..., \mathbf{b}_n\}$. Group sparsity regularization is imposed on $\mathbf{B}$ to enforce the target candidates to have similar representation on some templates [5].

Most existing sparse representation based visual tracking methods aim to find the target candidate with the minimum representation residual. From the perspective of distance/similarity, the objective is to find the most similar target candidate to the target templates. There are two sets in visual tracking, i.e., the target candidate set and the target template set. The set $\mathbf{Y} \in \mathbb{R}^{d \times n}$ can be represented by a hull $H(\mathbf{Y}) = \mathbf{Y}a$, where $a = [a_1; ...a_i; ...; a_n]$ and $H(\mathbf{Y})$ is defined as follow:

$$H(\mathbf{Y}) = \left\{ \sum \mathbf{y}_i a_i \, \middle| \, \sum a_i = 1, \mu_1 \leq a_i \leq \mu_2 \right\} \tag{2}$$

when $\mu_1 = -\infty$ and $\mu_2 = +\infty$, $H(\mathbf{Y})$ is an affine hull. When $\mu_1 = 0$ and $\mu_2 = 1$, $H(\mathbf{Y})$ is a convex hull. If we model both the target candidate set $\mathbf{Y}$ and the target template set $\mathbf{Z}$ as hulls, the distance between two sets is defined as:

$$d(\mathbf{Y}, \mathbf{Z}) = \left\| \mathbf{Y}\widehat{a} - \mathbf{Z}\widehat{b} \right\|_2^2 \tag{3}$$

where $\widehat{a}$ and $\widehat{b}$ are obtained by Eq. (4).

$$\left(\widehat{a}, \widehat{b} \right) = \arg\min_{\mathbf{a}, \mathbf{b}} \| H(\mathbf{Y}) - H(\mathbf{Z}) \|_2^2 \tag{4}$$

$\mathbf{a}$ and $\mathbf{b}$ are the coefficient vectors of corresponding hulls. By minimizing the distance between two hulls, we can find the most similar target candidate to

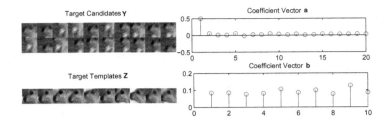

Fig. 1. Set to set visual tracking on deer dataset. The left subfigure shows 20 target candidates and 10 target templates, and the right subfigure shows the corresponding coefficients of target candidates and target templates.

the target templates, i.e., the target candidate with the maximum coefficient. As shown in Fig. 1, there are 20 target candidates and 10 target templates. By solving Eq. (3), we get the coefficient vectors $\hat{a}$ and $\hat{b}$. We can see that the target candidate with the largest coefficient is selected as the tracking result. We call this method set to set visual tracking (SSVT).

2.2   Set to Set Distance

As we use the coefficient values to reflect the importance of the target candidates, the coefficients should be non-negative. Therefore, both the target candidates and templates are modeled as convex hulls. The crucial step of the proposed SSVT method is to solve the distance between two convex hulls. Because the target candidates and templates are sampled from different frames, there is no intersection between two sets. As shown in Fig. 2, the set to set visual tracking becomes finding the nearest points in two convex hulls. It is easy to see that the set to set distance is the same as the geometric interpretation of support vector machines [26].

The distance between two convex hulls is defined as follows:

$$
\begin{aligned}
&\min_{\{a,b\}} \| Ya - Zb \|_2^2 \\
&s.t. \sum a_i = 1, \sum b_j = 1, \\
&0 \le a_i \le 1, i = 1, 2, ... n, \\
&0 \le b_j \le 1, j = 1, 2, ..., m + d.
\end{aligned} \tag{5}
$$

the problem in Eq. (5) can be easily solved by the standard quadratic optimization method [27]. The solution to the problem in Eq. (5) shows global and quadratic convergence [27]. As there are many redundant target candidates in the target candidate set, following the work in [24], minimal error bound is introduced to remove the target candidates with large errors.

2.3   Target Templates Update

Because of variations in various poses, motions, etc., the target templates can only work for a period of time. The template updating is very important since

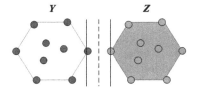

Fig. 2. Set to set distance between target candidates **Y** and target templates **Z**

a fixed template cannot accommodate all the appearance changes over time. Similar to the updating strategy in [10], the template will be replaced with the tracking result if the minimum angle of the tracking result and templates is less than one threshold, and we set the threshold to 40.

2.4 Discussions

Relations with Context Prior Model. In biological vision system, the focus of attention theory shows that the attention is paid to a certain area when human beings observe one image. Generally speaking, the object close to us will get more attention [14]. Hence, context prior model is defined as

$$p(c(\mathbf{y})|\mathbf{o}) = \mathbf{G}(\mathbf{y})w_\sigma(\mathbf{y} - \mathbf{z}^*) \qquad (6)$$

where $\mathbf{G}(\mathbf{y})$ is the image intensity of $\mathbf{y}$, which describes the appearance of $\mathbf{y}$. $w_\sigma(\mathbf{y} - \mathbf{z}^*)$ is a weight function. The weight is inversely proportional to the distance between $\mathbf{y}$ and the template $\mathbf{z}^*$. For set to set visual tracking, the target candidate with the maximum weight is the nearest one to the target templates. Hence, SSVT follows the context prior model as well.

Relations with Sparse Representation Models. From the perspective of representation, SSVT can also be considered as a kind of representation based model. The hull of the target candidates is represented by the templates. Different from L_1 tracker that represents the target candidates independently, SSVT jointly represents all the target candidates. Different from multi-task tracker that requires all the target candidates to have similar representation on some templates, we enforce all the target candidates to have the same representation.

3   Experiments

In this section, a detailed description of our experiments is given and experiments on benchmark image sequences are conducted to prove the effectiveness and efficiency of the proposed SSVT algorithm.

3.1   Datasets

In our experiments, there are 12 challenging video sequences (named as car4, car11, davidgt, deer, faceocc1, faceocc2, girl, jumping, mountainBike, singer1, walking, walking2). These video sequences contain most challenges in visual tracking: illumination variation, fast motion, deformation, occlusion, scale variation, background clutters, motion blur, etc.

3.2   Baselines

Six popular tracking algorithms are selected for comparison, including incremental visual tracking (IVT) [18], multiple instance learning tracking (MIL) [9], L_1 tracking [10], multi-task tracking (MTT) [5], compressive tracking (CT) [28], fast Fourier transform tracking (FFT) [29]. MATLAB codes of these tracking algorithms are available online.

3.3   Implementation Details

We run our tracking algorithm and comparison algorithms on a 3.6 GHz Intel(R) Core(TM) i7-4790 Dual Core PC with 16 GB RAM and the version of MATLAB is MATLAB R2013a. We assume that $p(s_t|s_{t-1}) \sim \mathrm{N}(0, \mathrm{diag}(\sigma))$, where $\sigma = [0.005, 0.005, 0.005, 0.005, 3, 3]^T$. The number of target candidates n is 600 and 10 updated target templates and 1 fixed target template are used (we obtain the templates as the same as L_1 tracker [4]). The number of trivial templates is equal to the template size d. Note that beside the parameters in particle filter model, there is no other parameter for SSVT. In fast Fourier transform tracker, compressive tracker and incremental visual tracker, we set the parameters according to their paper. As to L_1 tracker, the value of λ is 0.01. In multiple instance tracker, the number of negative samples, search window and other parameters are set as default values. For multi-task tracker, there are three different trackers: L_{21}, L_{11}, L_{01}, we use L_{11} tracker and the value of λ is 0.005.

3.4   Computational Cost

Sparse representation based tracking methods have achieved impressive tracking results [24]. However, they are computationally complex and the latent structure between the target candidates is not utilized. MTT tracker mines the relationship of the target candidates and obtains their coefficients together. It has been proven that MTT performs better than L_1 tracker [5], but L_{p1} $(p = 1, 2, \infty)$-norm minimization problems also need to be solved. SSVT models the target candidates and templates as convex hulls and only needs to solve one single QP problem. Experiments show that our tracker is more efficient compared with L_1 tracker and MTT tracker. The average frames per second for L_1 tracker, MTT tracker and SSVT tracker are 0.04, 3.72 and 4.27, respectively.

3.5   Qualitative Evaluation

In this section, the qualitative analyese of tracking algorithms are conducted and the tracking results are shown in Figs. 3 and 4.

In car4 sequence, a car runs on the road and illumination changes. The tracking results of 7 tracking algorithms at frame 20, 190, 236, 300, 470 and 657 are presented in Fig. 3(a). MIL and CT drift from the target around frame 190 and lose the target at frame 300. MTT starts to change tracking scale at frame 190 and L_1 tracker fails to track the target at frame 657. IVT, FFT and SSVT perform quite well in whole sequence.

The background around the car has the similar color in car11 sequence. The tracking results at frame 10, 110, 200, 250, 309 and 392 are shown in Fig. 3(b). L_1 tracker and MIL tracker lose the target at frame 309. MTT, IVT, CT, FFT and SSVT track the target during entire sequence.

In the sequence of davidgt, a man walks around and his face changes. The tracking results at frame 27, 126, 158, 214, 296 and 462 are presented in Fig. 3(c). L_1 tracker starts to drift at frame 126 and loses the target at frame 158. IVT, CT and SSVT have good robustness and track well.

Fig. 3. Tracking results of evaluated algorithms on car4, car11, davidgt, deer, faceocc1, faceocc2 video sequences marked with different colors (Color figure online)

Fig. 4. Tracking results of evaluated algorithms on girl, jumping, mountainBike, singer1, walking, walking2 video sequences marked with different colors (Color figure online)

In deer sequence, a moving deer is tracked in the water. The tracking results at frame 2, 11, 30, 51, 57 and 61 are described in Fig. 3(d). MIL and L_1 tracker drift at frame 11, L_1 and MTT tracker lose the target at frame 30. The other trackers perform well.

In faceocc1 sequence, a changing face is tracked. The tracking results at frame 100, 231, 314, 474, 571 and 665 are shown in Fig. 3(e) to evaluate the robustness of tracking algorithms. All tracking algorithms perform well and have good robustness.

In faceocc2 sequence, the tracking results at frame 168, 276, 423, 582, 658 and 725 are shown in Fig. 3(f). MIL and L_1 tracker drift at frame 725.

In girl sequence, the tracking results at frame 46, 310, 326, 434, 444 and 466 are shown in Fig. 4(a). Most tracking algorithms lose the target at frame 310 except SSVT and MTT, SSVT and MTT track the target during the whole sequence and show good robustness.

In jumping sequence, a man plays a jump rope. The tracking results at frame 2, 83, 147, 206, 265 and 312 are shown in Fig. 4(b). MIL loses the target at frame

Table 1. Success rate of 7 different trackers on 12 different video sequences. The best result is marked in red color.

VID	IVT	MIL	L1	MTT	CT	FFT	SSVT
Car4	1.00	0.27	0.82	0.46	0.28	0.28	1.00
Car11	1.00	0.60	0.67	1.00	1.00	0.99	1.00
Davidgt-indoor	0.92	0.75	0.34	0.45	0.99	0.55	1.00
Deer	1.00	0.45	0.14	0.79	0.93	1.00	1.00
Faceocc1	0.95	0.94	1.00	0.99	0.96	1.00	1.00
Faceocc2	0.99	0.81	0.68	0.96	0.97	1.00	0.97
Girl	0.35	0.14	0.64	0.93	0.47	0.40	0.99
Jumping	0.91	0.40	0.54	0.34	0.67	0.05	0.94
MountainBike	0.98	0.65	0.82	1.00	0.54	1.00	1.00
Singer	1.00	0.36	1.00	0.77	0.26	0.30	0.97
Walking	1.00	0.49	0.93	0.98	0.98	0.52	1.00
Walking2	0.98	0.37	0.99	0.99	0.38	0.39	1.00
average	0.92	0.52	0.71	0.81	0.70	0.62	0.99

Table 2. The average overlap score of 7 different trackers on 12 different video sequences. For each video sequence, the top two results are marked red and blue.

VID	IVT	MIL	L1	MTT	CT	FFT	SSVT
Car4	0.88	0.23	0.64	0.55	0.24	0.47	0.80
Car11	0.83	0.40	0.55	0.63	0.70	0.76	0.85
Davidgt-indoor	0.67	0.59	0.31	0.59	0.73	0.64	0.66
Deer	0.75	0.46	0.12	0.59	0.70	0.75	0.78
Faceocc1	0.73	0.60	0.78	0.70	0.69	0.79	0.73
Faceocc2	0.73	0.61	0.60	0.75	0.73	0.78	0.77
Girl	0.31	0.23	0.61	0.60	0.46	0.37	0.73
Jumping	0.64	0.00	0.41	0.40	0.57	0.05	0.65
MountainBike	0.75	0.51	0.61	0.76	0.49	0.71	0.74
Singer	0.79	0.41	0.81	0.63	0.35	0.36	0.71
Walking	0.79	0.52	0.73	0.71	0.55	0.54	0.70
Walking2	0.71	0.26	0.79	0.80	0.29	0.46	0.77

2 and FFT loses the target at frame 83, L_1 tracker loses the target at frame 147 and MTT loses the target at frame 265. However, IVT and SSVT perform well.

In mountain bike sequence, a man rides a bike and the deformation occurs. The tracking results at frame 38, 68, 86, 102, 198 and 215 are presented in Fig. 4(c). CT and MIL lose the target at frame 198. L_1 tracker drifts the target at frame 198 and loses the target at frame 215. IVT, MTT, FFT and SSVT show good performance.

In singer sequence, a woman sings on the stage and the scale of the target varies. The tracking results at frame 50, 75, 130, 190, 282 and 351 are shown in Fig. 4(d). MIL, CT and FFT change the scale largely at frame 130. IVT, L_1 tracker and SSVT handle the variation easily.

In walking sequence, there is a man walking on the road. The tracking results at frame 4, 52, 100, 170, 296 and 410 are shown in Fig. 4(e). Most tracking algorithms locate the target. However, the tracking bounding box of MIL and FFT is too long.

In walking2 sequence, occlusion occurs when a man walks into the scene. The tracking results at frame 1, 186, 204, 273, 365 and 465 are presented in Fig. 4(f). CT and MIL drift the target at frame 204 and lose the target at frame 273. FFT has a large bounding box and the other tracking algorithms perform well.

3.6   Quantitative Evaluation

Center location error, overlap score and success rate are utilized to evaluate the tracking results. Center location error is the Euclidean distance between the center of the ground truth and the tracking result, overlap score denotes $\frac{Area(groundtruth) \cap Area(trackingresult)}{Area(groundtruth) \cup Area(trackingresult)}$ and success rate is obtained by the number of overlap score greater than 0.5 accounting for the total frame number. Quantitative results of success rate, overlap score, and center location error are listed in Tables 1, 2 and 3, respectively.

In Table 1, we show the success rate of the tracking algorithms. It can be seen that our tracking algorithm achieves excellent results in all the sequences. IVT has shown good performance, but it works bad when long time occlusions occur in girl sequence. FFT can tackle the occlusion and complex environment. However, it drifts from the target in jumping sequence. MTT performs better in deformation and confusing environment.

In Table 2, the average overlap rate is presented. Our tracking algorithm outperforms other algorithms in car11, deer, girl and jumping. SSVT shows good performance in occlusion, deformation, fast movement. IVT achieves competitive results, but it's difficult to deal with a long time of occlusions.

Table 3. Center location error of 7 trackers on 12 video sequences. The best two results are marked in red and blue.

VID	IVT	MIL	L1	MTT	CT	FFT	SSVT
Car4	4.81	65.86	10.51	8.91	75.61	19.13	1.93
Car11	3.72	18.27	18.24	1.10	3.67	3.23	0.97
Davidgt-indoor	1.80	19.03	3.73	16.19	2.22	3.34	4.13
Deer	5.74	21.69	97.94	21.17	7.80	4.97	4.99
Faceocc1	19.01	29.23	13.79	20.25	20.78	11.93	16.25
Faceocc2	7.90	17.78	14.52	9.37	9.69	5.92	7.02
Girl	18.68	21.52	8.07	5.08	11.40	19.34	3.80
Jumping	6.33	10.98	29.89	18.12	8.50	85.97	5.70
MountainBike	6.33	71.92	15.49	6.02	54.38	6.51	5.10
Singer	3.87	20.00	3.86	11.10	15.02	14.01	3.87
Walking	1.78	6.97	2.93	3.36	3.69	7.17	3.33
Walking2	3.23	60.17	2.59	2.85	48.67	17.93	2.51

In Table 3, the center location error is shown, SSVT has small center location error as well as high accuracy. Even though the success rate of SSVT and IVT is 1.00 in car4, car11 and deer sequence, SSVT has smaller center location error. Due to fixed size of bounding box, FFT cannot handle scale variations and deformation, and has large center location error in jumping, singer, girl and walking sequences.

4 Conclusion

In this paper, we proposed a novel robust and efficient set to set visual tracking (SSVT) method. Different from representation based methods, we consider both the target candidates and templates as image sets, and model them as convex hulls. The set to set distance between the target candidate set and the target template set is minimized. The tracking result is the target candidate with the maximum coefficient value. SSVT exploits the underlying relationships among the target candidates, and only needs to solve a single quadratic programming problem. Experiments on several challenging image sequences demonstrate that SSVT achieves superior accuracy and efficiency compared with the state-of-the-art visual tracking methods.

Acknowledgement. This work was supported by the National Program on Key Basic Research Project under Grant 2013CB329304, the National Natural Science Foundation of China under Grants 61502332, 61432011, 61222210.

References

1. Hong, Z., Mei, X., Prokhorov, D., Tao, D.: Tracking via robust multi-task multi-view joint sparse representation. In: ICCV (2013)
2. Jia, X., Lu, H., Yang, M.H.: Visual tracking via adaptive structural local sparse appearance model. In: CVPR (2012)
3. Liu, B., Huang, J., Kulikowski, C., Yang, L.: Robust visual tracking using local sparse appearance model and k-selection. TPAMI **35**(12), 2968–2981 (2013)
4. Mei, X., Ling, H., Wu, Y., Blasch, E., Bai, L.: Minimum error bounded efficient l_1 tracker with occlusion detection. In: CVPR (2011)
5. Zhang, T., Ghanem, B., Liu, S., Ahuja, N.: Robust visual tracking via multi-task sparse learning. In: CVPR (2012)
6. Henriques, J.F., Caseiro, R., Martins, P., Batista, J.: High-speed tracking with kernelized correlation filters. TPAMI **37**(3), 583–596 (2015)
7. Zhang, K., Liu, Q., Wu, Y., Yang, M.H.: Robust visual tracking via convolutional networks without training. TIP **25**(4), 1779–1792 (2016)
8. Zhou, Y., Bai, X., Liu, W., Latecki, L.J.: Similarity fusion for visual tracking. IJCV, 1–27 (2016)
9. Babenko, B., Yang, M.H., Belongie, S.: Robust object tracking with online multiple instance learning. TPAMI **33**(8), 1619–1632 (2011)
10. Mei, X., Ling, H.: Robust visual tracking and vehicle classification via sparse representation. TPAMI **33**(11), 2259–2272 (2011)

11. Grabner, H., Grabner, M., Bischof, H.: Real-time tracking via on-line boosting. In: BMVC (2006)
12. Grabner, H., Leistner, C., Bischof, H.: Semi-supervised on-line boosting for robust tracking. In: Forsyth, D., Torr, P., Zisserman, A. (eds.) ECCV 2008, Part I. LNCS, vol. 5302, pp. 234–247. Springer, Heidelberg (2008)
13. Avidan, S.: Ensemble tracking. TPAMI **29**(2), 261–271 (2007)
14. Zhang, K., Zhang, L., Liu, Q., Zhang, D., Yang, M.-H.: Fast visual tracking via dense spatio-temporal context learning. In: Fleet, D., Pajdla, T., Schiele, B., Tuytelaars, T. (eds.) ECCV 2014, Part V. LNCS, vol. 8693, pp. 127–141. Springer, Heidelberg (2014)
15. Black, M.J., Jepson, A.D.: Eigentracking: robust matching and tracking of articulated objects using a view-based representation. IJCV **26**(1), 63–84 (1998)
16. Comaniciu, D., Ramesh, V., Meer, P.: Real-time tracking of non-rigid objects using mean shift. In: CVPR (2000)
17. Adam, A., Rivlin, E., Shimshoni, I.: Robust fragments-based tracking using the integral histogram. In: CVPR (2006)
18. Ross, D.A., Lim, J., Lin, R.S., Yang, M.H.: Incremental learning for robust visual tracking. IJCV **77**(1–3), 125–141 (2008)
19. Kwon, J., Lee, K.M.: Visual tracking decomposition. In: CVPR (2010)
20. Liu, B., Yang, L., Huang, J., Meer, P., Gong, L., Kulikowski, C.: Robust and fast collaborative tracking with two stage sparse optimization. In: Daniilidis, K., Maragos, P., Paragios, N. (eds.) ECCV 2010, Part IV. LNCS, vol. 6314, pp. 624–637. Springer, Heidelberg (2010)
21. Zhang, T., Ghanem, B., Liu, S., Ahuja, N.: Low-rank sparse learning for robust visual tracking. In: Fitzgibbon, A., Lazebnik, S., Perona, P., Sato, Y., Schmid, C. (eds.) ECCV 2012, Part VI. LNCS, vol. 7577, pp. 470–484. Springer, Heidelberg (2012)
22. Zhang, T., Liu, S., Ahuja, N., Yang, M.H., Ghanem, B.: Robust visual tracking via consistent low-rank sparse learning. IJCV **111**(2), 171–190 (2015)
23. Zhang, T., Ghanem, B., Liu, S., Xu, C., Ahuja, N.: Robust visual tracking via exclusive context modeling. IEEE Trans. Cybern. **46**(1), 51–63 (2016)
24. Bao, C., Wu, Y., Ling, H., Ji, H.: Real time robust L1 tracker using accelerated proximal gradient approach. In: CVPR (2012)
25. Zhang, T., Liu, S., Xu, C., Yan, S., Ghanem, B., Ahuja, N., Yang, M.H.: Structural sparse tracking. In: CVPR (2015)
26. Burges, D., Crisp, C.: A geometric interpretation of v−SVM classifiers. In: NIPS (2000)
27. Coleman, T.F., Li, Y.: A reflective newton method for minimizing a quadratic function subject to bounds on some of the variables. SIAM J. Optim. **6**(4), 1040–1058 (1996)
28. Zhang, K., Zhang, L., Yang, M.-H.: Real-time compressive tracking. In: Fitzgibbon, A., Lazebnik, S., Perona, P., Sato, Y., Schmid, C. (eds.) ECCV 2012, Part III. LNCS, vol. 7574, pp. 864–877. Springer, Heidelberg (2012)
29. Henriques, J.F., Caseiro, R., Martins, P., Batista, J.: Exploiting the circulant structure of tracking-by-detection with kernels. In: Fitzgibbon, A., Lazebnik, S., Perona, P., Sato, Y., Schmid, C. (eds.) ECCV 2012, Part IV. LNCS, vol. 7575, pp. 702–715. Springer, Heidelberg (2012)

BDSCyto: An Automated Approach for Identifying Cytokines Based on Best Dimension Searching

Quan Zou[1(✉)], Shixiang Wan[1], Bing Han[2], and Zhihui Zhan[3]

[1] School of Computer Science and Technology, Tianjin University, Tianjin, China
zouquan@tju.edu.cn
[2] School of Electronic Engineering, Xidian University, Xi'an, China
bhan@xidian.edu.cn
[3] School of Computer Science and Engineering, South China University of Technology,
Guangzhou, China
zhanapollo@163.com

Abstract. We proposed an automated method for distinguishing cytokines from other proteins according to their primary sequences. Two strategies were employed to extract features from protein sequences. The first one is a single method, which includes autocorrelation and pseudo amino acid composition extracted feature methods based on composition and physical–chemical properties of proteins; while the second one is an optimal dimension searching method. Moreover, we developed BDSCyto as a web server to help researchers in classifying protein sequences efficiently and accurately. BDSCyto reduces the processing time and offers high accuracy by a series of efficient methods and multithreading technology based on Spark for large-scale data. Currently, numerous methods exceed 90 % accuracy in cytokine protein prediction, which is better than the existing single methods. BDSCyto is an open-source project and can be freely accessed by the public at http://bdscyto.sinaapp.com/.

1 Introduction

Cytokines are a broad and loose category of small proteins that are important in cell signaling. They act through receptors, and are especially important in the immune system; cytokines modulate the balance between humoral and cell-based immune responses, and they regulate the maturation, growth, and responsiveness of particular cell populations. Cytokines play a key role in health and diseases, specifically in host responses to infection, immune responses, inflammation, trauma, sepsis, cancer, and reproduction. Although extensive work has been conducted to classify and predict cytokines [1], the direct prediction of cytokine protein types still yields unsatisfactory results. Applying experimental methods to define the type of a given protein sequence is time consuming and labor intensive. Thus, a rapid and cost-efficient method for determining cytokine proteins is significant.

Most methods classify or predict cytokine proteins by using machine learning algorithms that consist of two steps. The first step determines whether a given unknown

© Springer International Publishing Switzerland 2016
R. Booth and M.-L. Zhang (Eds.): PRICAI 2016, LNAI 9810, pp. 713–725, 2016.
DOI: 10.1007/978-3-319-42911-3_60

protein sequence is a cytokine protein, and the second step predicts the type of the given cytokine protein sequence. These research methods are as follows:

- Support vector machines (SVM) and Random Forest [2].
- Statistical learning theory, such as Artificial Neutral Network (ANN) and Hidden Markov Model (HMM).
- Similarity theory, including basic local alignment search tool (BLAST) [3] and FASTA [4].
- Emphasizing machine learning techniques, such as CTKPred [1] and a prediction method proposed by Liu *et al.* [5], are both based on SVMs. Some other methods use ensemble classifiers [6] or hierarchical multi-label classifiers to obtain precise prediction [7].
- Probabilistic language model called n-gram [8].

Moreover, several web servers and software have been constructed to provide a powerful and efficient way to operate the prediction [6], but they are ineffective when dealing with large data sets. More importantly, single methods are unable to consider all advantages of various methods. For example, SVM and Random Forest have relatively high efficient speed but produce different results on different data sets, and the accuracy depends on experiment data. ANN encounters over learning or under learning when the algorithm parameters are not optimal, and it cannot help researchers understand learning principles. The function of BLAST and FASTA is limited with regard to visualization tools. Although CTKPred has precise prediction, it has no visualization tools and it is a single method. Meanwhile, n-gram has a 20^n dimensions feature, i.e., there are 8000 dimensions if n equals to 3, which results in the curse of high dimensionality.

This paper proposes a best dimension searching method based on the existing work for improving prediction accuracy. This method is employed on BDSCyto, which not only supports online prediction but also reduces the processing time and offers high accuracy for large-scale data. BDSCyto achieved satisfactory predictive speed and accuracy by a series of efficient methods and multithreading technology based on Spark.

2 Methods   •

2.1 Extracted Features Based on Autocorrelation

(1) Auto Covariance. Suppose a protein sequence P with L amino acid residues:

$$P = A_1A_2A_3A_4A_5A_6A_7A_8 \cdots A_L$$

where A_1 represents the amino acid residue at the sequence position 1, A_2 the amino acid residue at position 2 and so forth.

The AC [9] approach measures the correlation of the same property between two residues separated by a distance of lag along the sequence, which can be calculated as:

$$AC(i, \ lag) = \frac{\sum\limits_{i=1}^{L-lag} \left(P_u(A_i) - \overline{P}_u\right)\left(P_u(A_{i+lag}) - \overline{P}_u\right)}{L - lag}$$

where u is a physicochemical index, L is the length of the protein sequence, $P_u(A_i)$ is the numerical value of the physicochemical index u for the amino acid A_i at position i, and P_u is the average value for physicochemical index u along the whole sequence:

$$\overline{P}_u = \frac{\sum\limits_{j=1}^{L} P_u(A_j)}{L}$$

In such a way, the length of AC feature vector is N × LAG, where N is the number of physicochemical indices extracted from AA index [10], and LAG is the maximum of lag (lag = 1, 2,…, LAG).

(2) Cross Covariance. Given a protein sequence P (same as above), the CC [11] approach measures the correlation of two different properties between two residues separated by a distance of lag along the sequence, which can be calculated by:

$$CC(u_1, \ u_2, \ lag) = \frac{\sum\limits_{i=1}^{L-lag} \left(P_{u_1}(R_i) - \overline{P}_{u_1}\right)\left(P_{u_2}(R_{i+lag}) - \overline{P}_{u_2}\right)}{L - lag}$$

where u_1, u_2 are two different physicochemical indices, L is the length of the protein sequence, $P_{u_1}(A_i) P_{u_2}(A_{i+lag})$ is the numerical value of the physicochemical index $u_1(u_2)$ for the amino acid $A_i(A_{i+lag})$ at position $i(i + lag)$, and $\overline{P}_{u_1}\left(\overline{P}_{u_2}\right)$ is the average value for physicochemical index value $u_1(u_2)$ along the whole sequence:

$$\overline{P}_u = \frac{\sum\limits_{j=1}^{L} P_u(A_j)}{L}$$

In such a way, the length of the CC feature vector is N × (N − 1) × LAG.
ACC [9] is a combination of AC and CC. Therefore, the length of the ACC feature vector is N × N × LAG.

2.2 Extracted Features Based on Pseudo Amino Acid Composition

(1) Parallel Correlation Pseudo Amino Acid Composition. Given a protein sequence P, the PC-PseAAC feature vector of P is defined as:

$$P = \left[x_1,\ x_2,\ \cdots,\ x_{20},\ x_{20+1},\ \cdots x_{20+\lambda} \right]^T$$

where

$$x_u = \begin{cases} \dfrac{f_u}{\sum_{i=1}^{20} f_i + w \sum_{j=1}^{\lambda} \theta_j} & (1 \le u \le 20) \\[4mm] \dfrac{w\theta_{u-20}}{\sum_{i=1}^{20} f_i + w \sum_{j=1}^{\lambda} \theta_j} & (20 + 1 \le u \le 20 + \lambda) \end{cases}$$

where $f_i(i = 1,\ 2,\ \cdots,\ 20)$ is the normalized occurrence frequency of the 20 amino acids in the protein P; the parameter λ is an integer, representing the highest counted rank (or tier) of the correlation along a protein sequence; w is the weight factor ranging from 0 to 1; and $\theta_j(j = 1,\ 2,\ \ldots,\ \lambda)$ is called the j-tier correlation factor reflecting the sequence-order correlation between all the j-th most contiguous residues along a protein chain, which is defined as:

$$\begin{cases} \theta_1 = \dfrac{1}{L-1} \sum_{i=1}^{L-1} \Theta\left(A_i,\ A_{i+1}\right) \\[4mm] \theta_2 = \dfrac{1}{L-2} \sum_{i=1}^{L-2} \Theta\left(A_i,\ A_{i+2}\right) \\[4mm] \theta_3 = \dfrac{1}{L-3} \sum_{i=1}^{L-3} \Theta\left(A_i,\ A_{i+3}\right) \\[2mm] \qquad \cdots\cdots \\[2mm] \theta_\lambda = \dfrac{1}{L-\lambda} \sum_{i=1}^{L-\lambda} \Theta\left(A_i,\ A_{i+\lambda}\right) \end{cases} \quad (\lambda < L)$$

where the correlation function is given by

$$\Theta\left(A_i,\ A_j\right) = \frac{1}{3}\left\{ \left[H_1\left(A_j\right) - H_1\left(A_i\right)\right]^2 + \left[H_2\left(A_j\right) - H_2\left(A_i\right)\right]^2 + \left[M\left(A_j\right) - M\left(A_i\right)\right]^2 \right\}$$

where $H_1\left(A_i\right)$, $H_2\left(A_i\right)$, and $M\left(A_i\right)$ are the hydrophilicity, hydrophilicity, and side-chain mass of the amino acid A_i, respectively. Note that before substituting the values of hydrophilicity, hydrophobicity, and side-chain mass, they are all subjected to a standard conversion, as described by the following equations:

$$
\begin{cases}
H_1(i) = \dfrac{H_1^0(i) - \sum\limits_{i=1}^{20} \dfrac{H_1^0(i)}{20}}{\sqrt{\dfrac{\sum\limits_{i=1}^{20}\left[H_1^0(i) - \sum\limits_{i=1}^{20}\dfrac{H_1^0(i)}{20}\right]^2}{20}}} \\[4em]
H_2(i) = \dfrac{H_2^0(i) - \sum\limits_{i=1}^{20} \dfrac{H_2^0(i)}{20}}{\sqrt{\dfrac{\sum\limits_{i=1}^{20}\left[H_2^0(i) - \sum\limits_{i=1}^{20}\dfrac{H_2^0(i)}{20}\right]^2}{20}}} \\[4em]
M(i) = \dfrac{M^0(i) - \sum\limits_{i=1}^{20} \dfrac{M^0(i)}{20}}{\sqrt{\dfrac{\sum\limits_{i=1}^{20}\left[M^0(i) - \sum\limits_{i=1}^{20}\dfrac{M^0(i)}{20}\right]^2}{20}}}
\end{cases}
$$

where $H_1^0(i)$ is the original hydrophilicity value of the i-th amino acid; $H_2^0(i)$ is the corresponding original hydrophilicity value; and $M^0(i)$ the mass of the i-th amino acid side chain.

(2) Series Correlation Pseudo Amino Acid Composition (SC-PseAAC). SC-PseAAC is a variant of PC-PseAAC. Given a protein sequence P, the SC-PseAAC feature vector of P is defined as:

$$
P = \left[p_1,\ p_2,\ \cdots p_{20},\ p_{20+1},\ \cdots p_{20+\lambda},\ p_{20+\lambda+1},\ \cdots p_{20+2\lambda}\right]^T
$$

Where

$$
p_u = \begin{cases}
\dfrac{f_u}{\sum_{i=1}^{20} f_i + w \sum_{j=1}^{2\lambda} \tau_j} & (1 \leq u \leq 20) \\[2em]
\dfrac{w\tau_{u-20}}{\sum_{i=1}^{20} f_i + w \sum_{j=1}^{2\lambda} \tau_j} & (20 + 1 \leq u \leq 20 + \lambda)
\end{cases}
$$

where $f_i(i = 1, 2, \cdots, 20)$ is the normalized occurrence frequency of the 20 native amino acids in the protein P; the parameter λ is an integer, representing the highest counted rank (or tier) of the correlation along a protein sequence; w is the weight factor ranging from 0 to 1; and τ_j the j-tier sequence-correlation factor that reflects the sequence-order correlation between all the most contiguous residues along a protein sequence, which is defined as:

$$
\begin{cases}
\tau_1 = \dfrac{1}{L-1} \sum_{i=1}^{L-1} H^1_{i,\,i+1} \\[2mm]
\tau_2 = \dfrac{1}{L-1} \sum_{i=1}^{L-1} H^2_{i,\,i+1} \\[2mm]
\tau_3 = \dfrac{1}{L-2} \sum_{i=1}^{L-2} H^1_{i,\,i+2} \\[2mm]
\tau_4 = \dfrac{1}{L-2} \sum_{i=1}^{L-2} H^2_{i,\,i+2} \quad (\lambda < L-1) \\[2mm]
\cdots\cdots \\[2mm]
\tau_{2\lambda-1} = \dfrac{1}{L-\lambda} \sum_{i=1}^{L-\lambda} H^1_{i,\,i+\lambda} \\[2mm]
\tau_{2\lambda} = \dfrac{1}{L-\lambda} \sum_{i=1}^{L-\lambda} H^2_{i,\,i+\lambda}
\end{cases}
$$

Where $H^1_{i,j}$ and $H^2_{i,j}$ are the hydrophobicity and hydrophilicity correlation functions given by

$$
\begin{cases}
H^1_{i,j} = h^1(A_i) \cdot h^1(A_j) \\
H^2_{i,j} = h^2(A_i) \cdot h^2(A_j)
\end{cases}
$$

where $h^1(A_i)$ and $h^2(A_i)$ are respectively the hydrophobicity and hydrophilicity values for the i-th (i = 1, 2,..., L) amino acid; the dot ($\cdot$) means the multiplication sign.

Note that before substituting the values of hydrophobicity and hydrophilicity, they are all subjected to a standard conversion as described by the following equation [12]:

$$
\begin{cases}
h^1(A_i) = \dfrac{h^1_0(A_i) - \sum_{i=1}^{20} \dfrac{h^1_0(B_k)}{20}}{\sqrt{\dfrac{\sum_{i=1}^{20}\left[h^1_0(A_i) - \sum_{i=1}^{20} \dfrac{h^1_0(B_k)}{20} \right]^2}{20}}} \\[8mm]
h^2(A_i) = \dfrac{h^2_0(A_i) - \sum_{i=1}^{20} \dfrac{h^2_0(B_k)}{20}}{\sqrt{\dfrac{\sum_{i=1}^{20}\left[h^2_0(A_i) - \sum_{i=1}^{20} \dfrac{h^2_0(B_k)}{20} \right]^2}{20}}}
\end{cases}
$$

where we use the $B_i(i = 1, 2, \cdots, 20)$ to represent the 20 native amino acids. The symbols h_0^1 and h_0^2 represent the original hydrophobicity and hydrophilicity values of the amino acid in the brackets right after the symbols.

All the aforementioned features can be generated by a recently proposed web-server called Pse-in-One [13], which can be accessed at http://bioinformatics.hitsz.edu.cn/Pse-in-One/.

2.3 Optimal Dimension Search of High Dimension

In addition to the above method based on the physical and chemical properties of proteins, certain pure mathematical method is typically applied, such as N-Gram. An N-Gram model is a type of probabilistic language model for predicting the next item in such a sequence in the form of a (n − 1)-order Markov model. N-Gram models are widely used in computational biology (e.g., biological sequence analysis) and data compression. The two core advantages of N-Gram models are relative simplicity and the ability to scale up by simply increasing n. This model can be used to store more contextual information with a well-understood space–time tradeoff, enabling small experiments to scale up very efficiently. For protein sequence consisting of 20 types amino acid, there are 8000D ($20^3 = 8000$) feature vectors if n is 3. However, this method produces very large dimensions for feature extraction, which causes high dimension disaster, and thus computation will be explosive [14].

To solve high dimension disaster, our experiment reduced 8000D features by MRMD [15] and search the optimal dimension based on ACC metrics. Similar as the feature selection algorithm, MRMD is determined by two main parts: one is the characteristic and the correlation among instance classes, and the other is the redundancy among features. MRMD calculates the correlation between features and class standards by Pearson's correlation coefficient, as well as the redundancy among features by three distance functions (Euclidean distance, cosine distance, and Tanimoto coefficient). Higher Pearson correlation coefficient indicates closer relationship between features and class standards; in addition, greater distance means lower redundancy among features. Finally, MRMD selects feature subsets which exhibit strong correlation and low redundancy between features. Hence, MRMD dimension reduction is simple and rapid.

Subsequently, we developed an optimal dimension search method to achieve our goal, as shown in Fig. 1.

Based on Fig. 1, we first determine the dimensions N given by the current dataset and the initial dimension. Iteration is performed n times starting from the initial dimension, and ACC is calculated through 10-fold cross validation by LibD3C [16]. If ACC is higher than the last experiment, then all current results and accuracy are recorded, and the current dimension is reduced according to reduction iteration; otherwise, the next search is performed. If the dimension of next experiment exceeds raw dimension N, or achieves the best ACC of multiple dimensions search, then a local optimal linear search is initiated again from the current dimension, and a dimension is dropped by integer i. Similarly, all related data are recorded until algorithm the best ACC is found.

1. Global multiple search

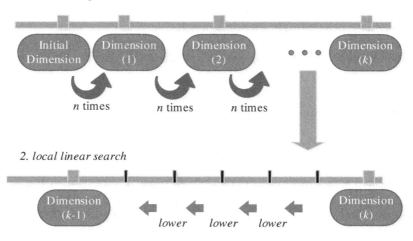

Fig. 1. Searching the best dimension based on MRMD

3 Results

3.1 Dataset Description and Performance Metrics

A series of experiments are performed to confirm the innovativeness of our methods. First, we analyzed the effectiveness of the extracted feature vectors based on autocorrelation and pseudo amino acid composition, as well as comparison with 188D and Pseb feature vectors. Second, we showed the performance of optimal dimension search under high dimension, and compared findings with the performance of the ensemble classifier. Finally, we tested all known proteins and for the determination of cytokines. This section provides details of the experiments.

Two important measures are used to assess the performance of individual class: sensitivity (SN) and specificity (SP):

$$SN = \frac{TP}{TP + FN} \times 100\%$$

$$SP = \frac{TP}{TN + FP} \times 100\%$$

In addition, overall accuracy (ACC) is defined as the ratio of correctly predicted samples to all tested samples:

$$ACC = \frac{TP + TN}{TP + TN + FP + FN}$$

where TP, TN, FN and FP are the number of true positives, true negatives, false negatives and false positives, respectively.

Based on above, we consider the receiver operating characteristic (ROC) curve to express the relation between SN and SP; larger area under ROC indicates better effectiveness of classifier.

The training and testing datasets were downloaded from Uniport database. The datasets contain 50365 cytokine protein sequences and 10094 non-cytokine protein sequences. We cluster the raw data set by CD-HIT [17] before the following experiment because of redundant information in the raw data (even repeat sequences). When the clustering threshold value reaches 70 %, the number of positive instances (denoted as Ψ_{cyto}) is similar to that of the negative instances (denoted as $\Psi_{non-cyto}$), i.e., 9338 instances and 9293 instances, respectively.

3.2 Performance Evaluation Based on Autocorrelation

The raw dataset for the two-class predictor contains Ψ_{cyto} and $\Psi_{non-cyto}$. We extracted 9D, 18D, 27D and 188D feature vectors of positive and negative instances based on the distribution of autocorrelation. The validity was verified by Experiments 1 and 2.

Experiment 1—Autocorrelation Method. The raw datasets with 9D, 18D, 27D, and 188D features were trained, and the results of 10-fold cross-validation were analyzed by Weka (version 3.7.13). We calculated the SN, SP, and ACC values of the five classifiers; the results are illustrated in Fig. 2.

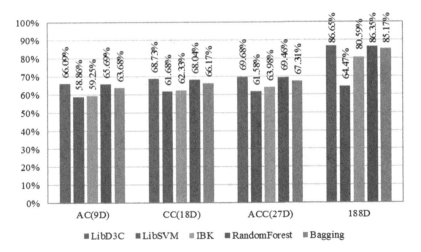

Fig. 2. Accuracy (ACC) of experiments (Color figure online)

Experiment 2—Contrast with 188D. The sampled dataset with 188D feature was trained, and the results of 10-fold cross-validation were analyzed by Weka (version 3.7.13). Based on Zou's experiment [18], the rebuilt sampled data set contained 2019 positive feature samples and 1996 negative feature samples; this finding is the same as that of Zou's dataset. To demonstrate that the performance of the 188D feature in the present study is better than that of Cai's 188D feature [19] for classifying cytokines, we conducted Experiment 2 and compared their effects. A comprehensive comparison of results illustrated the superiority of our method for cytokine identification.

The calculated SN, SP, and ACC of the unsampled dataset are 84.80 %, 88.50 %, and 86.65 %, respectively, as shown in Fig. 3. These findings demonstrate that the classification works well, and that larger dataset provides more comprehensive data, resulting in better effectiveness.

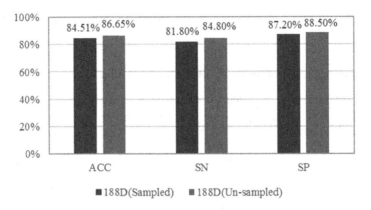

Fig. 3. Comparison of validation on sampled and unsampled datasets. (Color figure online)

3.3 Performance Based on Pseudo Amino Acid Composition

To validate the classification effect of the methods based on pseudo amino acid composition, we conducted Experiment 3 using Weka (version 3.7.13).

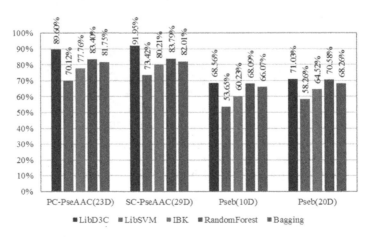

Fig. 4. Accuracy (ACC) of experiments (Color figure online)

In this experiment, we choose the same dataset (Ψ_{cyto} and $\Psi_{non-cyto}$), and compared the four methods based on pseudo amino acid composition, including PC-PseAAC (23D), SC-PseAAC (29D), Pseb (10D), and Pseb (20D). Similarly, our test used five different classifiers, namely, LibD3C, LibSVM, IBK, Random Forest, and Bagging for each method. The calculated SN, SP, and ACC are shown in Fig. 4.

3.4 Optimal Dimension Search Results

We designed next experiment to illustrate the improvement of the ACC and efficiency of optimal dimension search. The input dataset (Ψ_{cyto} and $\Psi_{non-cyto}$) includes 9338 positive and 9293 negative instances, respectively, and then we obtain 8000D feature vectors by N-Gram (Sect. 2.3). To search for dimensions more efficiently, the initial dimension is set as 10; the multiple iteration is 2; and the reduction iteration is 10; the classifier is LibD3C; and the upper dimension is set as 500. The optimal dimension search algorithm (Sect. 2.3) is then performed; the best accuracy is achieved at 400D, which is higher than that of 188D for PC-PseAAC.

We concluded that the ACC increases with the increase in dimension until 500D, at which a downward trend is found. Experiments were run at an interval of 10 dimensions. Result shows that ACC increases again until 390D; hence, 400D is the optimal dimension for the search. In particular, time efficiency is achieved when coding is run in a server of parallel computing, which is necessary in bioinformatics research.

We combined dimension reduction and 188D, as well as PC-PseAAC (23D) and SC-PseAAC (29D) to develop a web server with BDSCyto, by which the optimal dimension can be determined.

BDSCyto allows users to submit multiple sequences for online cytokine prediction and supports offline optimal dimension search on personal computers that run on Windows or Linux/Unix operating systems. BDSCyto provides three kinds of features to the training model: 23D feature based on PC-PseAAC, 29D feature based on SC-PseAAC, and 188D feature based on physical–chemical properties. The prediction results show whether the sequence is a cytokine protein or not. If so, further search will be conducted to predict its type and obtain its prediction accuracy.

Extraction of the 23D and 29D feature is more time efficient than the 188D feature. The 23D and 29D feature is time efficient based on pseudo amino acid composition because the program only need to obtain less feature vectors than the 188D feature. However, the ACC of 188D is better than that of 23D and 29D feature because the 188D feature contains more data.

The LibD3C classifier with optimal dimension has higher prediction accuracy than that with the single 188D feature. Although the 188D feature is theoretically more comprehensive, it is not as effective as the optimal dimension feature in actual classification of cytokine proteins. For the optimal dimension feature, the highest accuracy of LibD3C in discriminating cytokine and non-cytokine proteins is 89.24 %, which is considerably better than that of 188D and Pseb methods.

4   Discussion and Conclusion

Our work aimed to identify positive and negative families from the PFAM database more efficiently and accurately. First, we extracted the longest protein sequence in each family. We then removed redundant sequences based on a sequence consistency standard for data cleaning. We extracted 9D, 18D, 27D, 188D, 23D, and 29D feature vectors of positive and negative sequences based on the autocorrelation and certain physicochemical properties. Five different classifiers, namely, LibD3C, LibSVM, IBK, Random Forest, and Bagging are applied to improve the stability and accuracy of cytokine classification. More importantly, we developed BDSCyto as a web server and software to classify and predict the type of cytokine protein. The proposed method demonstrates high accuracy and efficiency even when processing large-scale data.

Seven features are introduced in this paper, namely, 9D, 18D, 27D, 188D, 23D, 29D, and 8000D, which are based on the autocorrelation, physical–chemical properties, and protein composition. We improved the 8000D feature by the optimal dimension search method, which achieved a higher ACC than other six methods. The performance of our classification strategy and predictor demonstrates that our method is feasible with high prediction efficiency, thereby enabling appropriate large-scale data prediction. An online server with open-source software that supports massive data processing was developed. Our project can be freely accessed by the public at http://bdscyto.sinaapp.com/. Currently methods provide at least 90 % accuracy in cytokine protein prediction. However, with the rapid increase in quantity of research dataset, an automated platform with high accuracy rate and high prediction efficiency is urgently needed. BDSCyto is a pioneer work that instantly classifies and predicts cytokine proteins. With the continuous improvement of the proposed method, more researchers can undertake research on prediction of proteins.

Acknowledgments. The work was supported by the Natural Science Foundation of China (No. 61370010, 61572384, 61402545).

References

1. Huang, N., Chen, H., Sun, Z.: CTKPred: an SVM-based method for the prediction and classification of the cytokine superfamily. Protein Eng. Des. Selection **18**(8), 365–368 (2005)
2. Zou, Q., et al.: BinMemPredict: a web server and software for predicting membrane protein types. Curr. Proteomics **888**(1), 2–9 (2013)
3. Altschul, S.F., et al.: Basic local alignment search tool. J. Mol. Biol. **215**(3), 403–410 (1990)
4. Pearson, W.R.: Searching protein sequence libraries: comparison of the sensitivity and selectivity of the Smith-Waterman and FASTA algorithms. Genomics **11**(3), 635–650 (1991)
5. Liu, B., et al.: Prediction of protein binding sites in protein structures using hidden Markov support vector machine. BMC Bioinformatics **10**(2), 1–14 (2009)
6. Chen, L., et al.: Hierarchical classification of protein folds using a novel ensemble classifier. PLoS ONE **8**(2), e56499 (2013)
7. Zou, Q., et al.: Identifying multi-functional enzyme with hierarchical multi-label classifier. J. Comput. Theor. Nanosci. **10**(4), 1038–1043 (2013)

8. Zeng, X., et al.: Identification of cytokine via an improved genetic algorithm. Front. Comput. Sci. **9**(4), 643–651 (2015)

9. Dong, Q., Zhou, S., Guan, J.: A new taxonomy-based protein fold recognition approach based on autocross-covariance transformation. Bioinformatics **25**(20), 2655–2662 (2009)

10. Kawashima, S., Kanehisa, M.: AAindex: amino acid index database. Nucleic Acids Res. **28**(1), 374 (2000)

11. Cao, D.-S., Xu, Q.-S., Liang, Y.-Z.: propy: a tool to generate various modes of Chou's PseAAC. Bioinformatics **29**(7), 960–962 (2013)

12. Lin, H., et al.: iPro54-PseKNC: a sequence-based predictor for identifying sigma-54 promoters in prokaryote with pseudo k-tuple nucleotide composition. Nucleic Acids Res. **42**(21), 12961–12972 (2014)

13. Liu, B., et al.: Pse-in-One: a web server for generating various modes of pseudo components of DNA, RNA, and protein sequences. Nucleic Acids Res. **W1**, W65–W71 (2015)

14. Cai, R.C., Zhang, Z.J., Hao, Z.F.: Causal gene identification using combinatorial V-structure search. Neural Networks **43**, 63–71 (2013)

15. Zou, Q., et al.: A novel features ranking metric with application to scalable visual and bioinformatics data classification. Neurocomputing **173**, 346–354 (2016)

16. Lin, C., et al.: LibD3C: ensemble classifiers with a clustering and dynamic selection strategy. Neurocomputing **123**, 424–435 (2014)

17. Huang, Y., et al.: CD-HIT suite: a web server for clustering and comparing biological sequences. Bioinformatics **26**(5), 680–682 (2010)

18. Zou, Q., et al.: An approach for identifying cytokines based on a novel ensemble classifier. Biomed. Res. Int. **2013**(8), 616–617 (2013)

19. Cai, C.Z., et al.: SVM-Prot: web-based support vector machine software for functional classification of a protein from its primary sequence. Nucleic Acids Res. **31**(13), 3692–3697 (2003)

Special Track: Smart Modelling and Simulation

Optimization of Road Distribution for Traffic System Based on Vehicle's Priority

Wen Gu$^{(\boxtimes)}$ and Takayuki Ito

Nagoya Institute of Technology, Gokiso-cho, Showa-ku, Nagoya,
Aichi, Japan
koku.bun@itolab.nitech.ac.jp, ito.takayuki@nitech.ac.jp

Abstract. Instead of making the traffic system work fluently by focusing on each car's way to choose their routes, in this paper, we proposed a way to make the vehicles avoid being involved into the traffic congestion by allocating the roads which are regarded as one kind of resources to the vehicles. In order to make the road allocation fair, we introduce the parameter to show each vehicle's priority. We allocate the roads by regarding it as a linear programming problem and use linear programming to solve it. The experiment was done by using simulator SUMO and we testified that our proposal can make the vehicles avoid getting involved into traffic congestion and verified the usefulness of the vehicle's priority.

Keywords: Traffic simulation · Road allocation · Priority · Linear programming · Optimization

1 Introduction

With the economical development, the number of the vehicles in the world is increasing continuously. As a result, traffic congestion has become a significant traffic problem in many cities [1]. In order to reduce the happening of the traffic congestion, many researches have been done from different aspects.

In this paper, we propose a framework to find the optimal route set which includes each vehicle's own route of the traffic system based on vehicle's priority. We regard the roads as one kind of resources and try to use the idea of resource and allocation [2] to find the optimal way to allocate the roads to the vehicles that need to use the roads. At the same time, in order to insure that the road allocation is fair to each vehicle, we introduce the parameter which represents the priority of vehicle.

In the proposed framework, first, we need to find the possible routes that a vehicle can use to travel from its origin to its destination. The routes for each vehicle in the traffic system should be found respectively. Second, we need to find the optimal route combination from each vehicle's possible routes by using linear programming. After all the vehicles reaching their destination, the priority of each vehicle will be updated.

© Springer International Publishing Switzerland 2016
R. Booth and M.-L. Zhang (Eds.): PRICAI 2016, LNAI 9810, pp. 729–737, 2016.
DOI: 10.1007/978-3-319-42911-3_61

The remainder of this paper is organized as follows: Sect. 2 introduces the related work of this paper. Section 3 describes the details of the optimization of road distribution. Section 4 shows the experiment we designed to testify the proposed framework and the results we got from the experiment. Section 5 shows the conclusion of this research and the future work of this research.

2   Related Work

Many approaches have been done to try to alleviate traffic congestion. For example, in-vehicle route guidance systems are designed to help the vehicles choose a more efficient route [3, 4]. New parking system [5] is developed to help drivers save more time during the parking process.

In traffic simulation, new simulation model is designed to in order to apply in more complex urban networks and get more accurate travel information [6]. Recently, multi-agent technologies [7] also have been used in different systems such as distributed resource allocation [8]. They can also be well used in transportation management such as intersection control [9]. Negotiation system also has been used in order to acquire efficient traffic flow [10].

At the same time, it should be noted that sometimes it will make the traffic congestion become worse [11] if we don't choose a effective method.

Some of the papers are focusing on searching the optimal route for each individual vehicle. It is also necessary to consider multiple vehicles in the traffic system simultaneously. Most of the papers are considering one time traffic situation. In order to control the traffic in a fair way, it is necessary to consider the traffic control in a long time way because there will be conflicts if multiple vehicles want to use the same roads at the same time and we need to decide which vehicle should use the road this time.

In this paper, we proposed a framework using the idea of resource and allocation to distribute roads to multiple vehicles and try to improve the fairness of the road distribution by considering vehicle's priority.

3   Optimization of Road Distribution

In this section, first we introduce the algorithm that we use to find each vehicle's possible routes. After that the way to find the optimal route combination is introduced. The method to update vehicle's priority is introduced in the end.

3.1   Find Each Vehicle's Possible Routes

Before calculating the optimal route combination for the traffic system. We need to find each vehicle's possible routes at first. In order to make vehicle avoid being involved in the traffic congestion which may happen in part of particular routes, instead of one possible route, multiple possible routes should be searched for each vehicle. In this paper, we adopt Yen's Algorithm [12] to find each vehicle's multiple possible routes. Yen's Algorithm is one kind of K Shortest Path Algorithm. The number of the possible routes that we want to find can be set in the algorithm. The possible routes that found by Yen's algorithm are without loops and in an ascending order which starts from the shortest route.

3.2   Find Optimal Route Set

In this paper, we consider a route is made up of one road or multiple roads. Every possible route set of the traffic system is made up by choosing one route from each vehicle's possible routes. The optimal route set is chosen by the sum cost of the route set. The way to calculate the cost of the route set is to use the objective function which is defined by Eq. (1). The restrictions are defined by Eqs. (2), (3) and (4).

$$\min \sum_{i \in I} \sum_{r_i \in R_i} C_d(r_i) \times C_{p,i} \times x(r_i) \tag{1}$$

$$s.t. \sum_{r_i \in R_i} x(r_i) = 1 \quad \forall i \in I \tag{2}$$

$$x(r_i) = \{0, 1\} \quad \forall r_i \in R_i \quad \forall i \in I \tag{3}$$

$$\sum_{i \in I} \sum_{r_i \in R_i} x(r_i) \times Num(r_i, e) \leq n(e) \tag{4}$$

In the objective function, there are three main parts which are route cost, priority cost and the restriction of choosing only one route for each vehicle from the vehicle's possible route set.

I represents the set of all the vehicles in the traffic system. i is one individual vehicle. R_i represents the set of the possible routes of vehicle i. r_i is one possible route of vehicle i. $C_d(r_i)$ represents the route cost which is decided by the length of the route. $C_{p,i}$ represents the priority cost which is decided by the priority of the vehicle.

$\sum_{r_i \in R_i} x(r_i)$ is equal to 1 because each vehicle can be allocated with only one route eventually. $x(r_i)$ shows whether route r_i is chosen by vehicle. As a result, r_i can just be 0 or 1. 1 means that the route is chosen by the vehicle and 0 means not chosen. $Num(r_i, e)$ represents the number of road e in route r_i. $n(e)$ represents the restrictive vehicle's number of the vehicles that can pass through road e in one unit time. $n(e)$ can be used to represent the traffic control of the road which is caused by construction or disaster. If $\sum_{i \in I} \sum_{r_i \in R_i} x(r_i) \times Num(r_i, e)$ is bigger than $n(e)$, it means there will be traffic congestion in road e.

In this paper, we use linear programming to solve the problem of finding the optimal route set.

3.3   Update the Priority of Vehicle

In this paper, the priority of the vehicle is updated when the vehicle get allocated to a route which has a longer length than the shortest route. In other words, it shows that how much did the vehicle sacrifice itself when actually there is a shorter route that can be used to get to the destination.

After all the vehicles reaching their destinations, the priority of each vehicle will be updated. If the vehicle is allocated with a route that is longer than its shortest route, it will be updated. And the way to update the priority is shown in Eq. (5). Otherwise, it will be the same as before.

$$C_{p,i} = C_{p,i} + \frac{Length(r_i) - Length(r_0)}{Length(r_0)} \tag{5}$$

where $C_{p,i}$ represents the priority of the vehicle. $Length(r_i)$ represents the length of the route that is been allocated to vehicle i. And $Length(r_0)$ represents the length of the shortest route in the possible route set of vehicle i.

4   Experiment

In order to testify whether the proposed framework can be useful to help the vehicles to avoid the traffic congestion and verify the usefulness of the introduced parameter, vehicle's priority, we did the experiment by using the simulator SUMO (Simulation of Urban MObility) [13]. SUMO is a free and open traffic simulation suite which allows modelling of intermodal traffic systems including road vehicles, public transport and pedestrians. In the experiment, we use Gurobi Optimizer [14] as the tool to find the optimal route set which is regarded as a linear programming problem.

4.1   Setting of the Experiment

The map we use in the experiment is part of the Nagoya City's Open Street Map (OSM) [15]. Simulator SUMO can import OSM format data to use. And it is shown in Figs. 1 and 2.

As is shown in Fig. 3, in the experiment, we let 20 vehicles travel from the origin point A to destination point B. The experiment is divided into three cases which differ from each other in whether there is traffic control in the road and whether the priority is introduced.

- In case1, we do not set traffic control in the road and we do not introduce the priority of the vehicle.
- In case2, we set traffic control in the road which is red in the map but we do not introduce the priority of the vehicle.
- In case3, we set traffic control in the road which is red in the map and introduce the priority of the vehicle.

The details of the three cases are shown in Table 1.

Fig. 1. The map shown in OSM

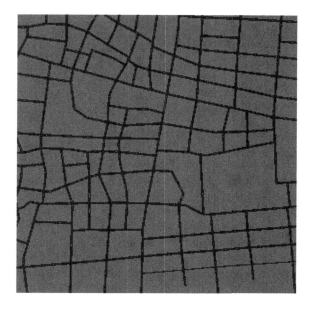

Fig. 2. The map shown in SUMO

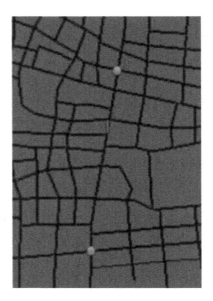

Fig. 3. The map used in experiment (Color figure online)

Table 1. Experiment setting

Case number	Vehicle number	Simulation number	Traffic control (car/min)	Priority introduced
1	20	10	-	No
2	20	10	6	No
3	20	10	6	Yes

4.2 Results of the Experiment

After finish 10 iterations of the experiment, we collect the statistics of the sum of each vehicle's movement distance which is shown in Figs. 4 and 5.

From Fig. 4, we can see the contrast between Case1 and Case2. In Case2, part of the vehicles traveled longer distance than in Case1. From that statistics, we confirmed that if under the condition that there is traffic control in the shortest route, some of the vehicles will be allocated with a route which is longer than their shortest route in order to avoid leading traffic congestion. But the vehicles that are allocated with a longer route than their shortest route are always some particular vehicles. As a result, we confirmed that our framework can split the vehicles from the roads where may have traffic congestion but we can not say it is fair for all vehicles.

From Fig. 5, we can see the contrast between Case2 and Case3. From that statistics, we confirmed that if under the condition that there is traffic control in the shortest route, just like Case2, some of the vehicles will be allocated with

Sum of the Movement Distance

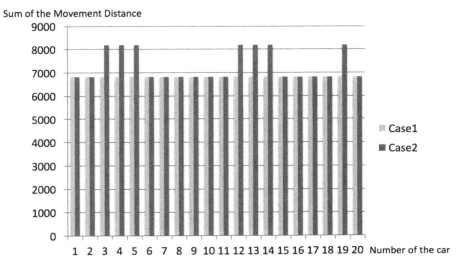

Fig. 4. The statistics of the sum of the movement distance (Color figure online)

Sum of the Movement Distance

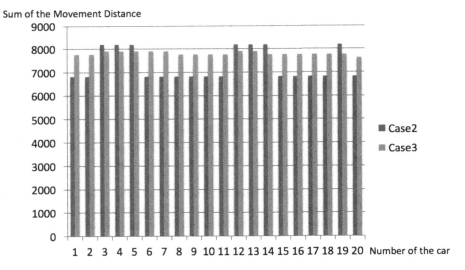

Fig. 5. The statistics of the sum of the movement distance (Color figure online)

a route which is longer than their shortest route in order to avoid leading traffic congestion in Case3. But in Case3, by calculating the standard deviation of each vehicle's sum of movement distance, as shown in Fig. 6, we can know that the variation of each vehicle's sum of movement distance is much smaller than that in Case2. It means that instead of particular vehicles, all the vehicles have been

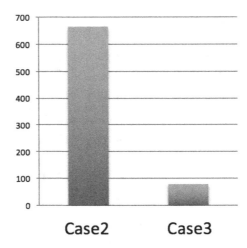

Fig. 6. Deviation of each vehicle's sum of movement distance

allocated with the route which is longer than their shortest route. As a result, we confirmed the usefulness of the priority of the vehicles that we introduced.

5   Conclusion and Future Work

In this paper, we proposed a centralized control framework and testified that it can be used to split the traffic flow to help the vehicles from being involved into traffic congestion and we verified the usefulness of the priority of vehicle which can maintain the fairness of the road allocation for the vehicles in the traffic system.

In the future, we are going to testify the framework in larger scale and improve the way to define the cost function. What's more, it will be more meaningful if we can introduce the negotiation between the vehicles into our framework. As our goal, we want our proposed framework will be useful as a part of the policy that can be applied in the automatic driving.

Acknowledgement. The research results have been achieved by "Congestion Management based on Multiagent Future Traffic Prediction, Researches and Developments for utilizations and platforms of social big data", the Commissioned Research of National Institute of Information and Communications Technology (NICT).

References

1. Arnott, R., Rave, T., Schob, R.: Alleviating Urban Traffic Congestion. MIT Press, Cambridge (2005)

2. Jihang, Z., Minjie, Z., Fenghui, R., Jiakun, L.: A multiagent-based domain transportation approach for optimal resource allocation in emergency management. In: The Proceedings of the 2nd International Workshop on Smart Simulation and Modelling for Complex Systems, Buenos Aires, Argentina, 25 July 2015
3. Wei, D.: An overview of in-vehicle route guidance system. In: Australasian Transport Research Forum Proceedings (2011)
4. Zou, L., Xu, J.M., Zhu, L.X.: Application of genetic algorithm in dynamic route guidance system. J. Transp. Syst. Eng. Inf. Technol. **7**(3), 45–48 (2007)
5. Geng, Y., Cassandras, C.: New "smart parking" system based on resource allocation and reservations. IEEE Trans. Intell. Transp. Syst. **14**(3), 1129–1139 (2013)
6. Tokuda, S., Kanamori, R., Ito, T.: Development of traffic simulator based on stochastic cell transmission model for urban network. In: Dam, H.K., Pitt, J., Xu, Y., Governatori, G., Ito, T. (eds.) PRIMA 2014. LNCS, vol. 8861, pp. 150–165. Springer, Heidelberg (2014)
7. Ito, T., Kanamori, R., Chakraborty, S., Otsuka, T., Hara, K.: A survery of multi-agents research that supports future societal systems(1)-economic paradigm, negotiating agents, and transportation management. J. JSAI **28**(3), 360–367 (2013)
8. Zhang, C., Lesser, V., Shenoy, P.J.: A multi-agent learning approach to online distributed resource allocation. In: Proceedings of the 21st International Joint Conference on Artificial Intelligence, IJCAI 2009, Pasadena, California, USA, 11–17 July 2009
9. Dresner, K., Stone, P.: Multiagent traffic management: an improved intersection control mechanism. In: 4th International Conference on Autonomous Agents and Multiagent Systems, AAMAS 2005, Utrecht, Netherlands, 25–29 July 2005
10. Takahashi, J., Kanamori, R., Ito, T.: Evaluation of automated negotiation system for changing route assignment to acquire efficient traffic flow. In: 2013 IEEE 6th International Conference on Service-Oriented Computing and Applications, Koloa, HI, pp. 351–355, 16–18 December 2013
11. Braess, D., Nagurney, A., Wakolbinger, T.: On a paradox of traffic planning. Transp. Sci. **39**, 446–450 (2005)
12. Yen, J.Y.: Finding the k-shortest loopless paths in a network. Manag. Sci. **17**, 712–716 (1971)
13. Krajzewicz, D., Erdmann, J., Behrisch, M., Bieker, L.: Recent development and applications of SUMO - Simulation of Urban MObility. Int. J. Adv. Syst. Meas. **5**(3&4), 128–138 (2012)
14. Gurobi Optimizer. http://www.octobersky.jp/products/gurobi/
15. Open Street Map. http://www.openstreetmap.org/

Evaluation of Deposit-Based Road Pricing Scheme by Agent-Based Simulator

Ryo Kanamori[(✉)], Toshiyuki Yamamoto, and Takayuki Morikawa

Nagoya University, Nagoya, Japan
{kanamori.ryo,morikava}@nagoya-u.jp,
yamamoto@civil.nagoya-u.ac.jp

Abstract. Parking Deposit System (PDS) is proposed to improve public acceptance of road pricing (RP), which can reduce effectively traffic congestion and air pollution. This study examines the characteristics of PDS and conventional cordon-based RP in terms of efficiency and equity. In order to evaluate income regressive effects as equity, we develop a multi-agent based simulator which can consider user's mode and route choice behaviors for several income classes. The results of empirical analysis at the Nagoya Metropolitan Area suggest the followings in comparison with conventional RP: (1) PDS gives sufficient environmental improvement by reducing car use, although the effect of decreases by refund, and (2) PDS is more equitable because it is a kind of revenue allocation scheme.

Keywords: Traffic problem · Road pricing · Agent based simulation

1 Introduction

Road pricing (RP) is a traffic policy that is regaining attention as a result of advancements in ICT/ITS technologies and proven overseas examples of reduced traffic congestion and improved road environment. However, a major barrier to an introduction of RP is social acceptance. Mainly, the public feels burdened by new charges for using cars, and businesses feel anxiety that such a measure will reduce the number of shoppers and visitors to the area where RP is applied. Therefore, it is crucial to devise a pricing scheme with higher acceptability, evaluate the introduction of RP using more detailed and elaborate methods for representing traffic demand, and closely consult all stakeholders.

A realistic RP is the second-best pricing that targets some links and areas [1]. Pricing schemes in actual use today include the cordon-based pricing (e.g. in Singapore and Norway), where a toll is charged each time for entering a restricted area, and the area-based pricing (e.g. in London), where all traveler within a targeted area is charged on a day-to-day basis. As a result of the spread of probe cars equipped with GPS, a system that charges a toll within a targeted area based on the distance and time traveled is considered to be more efficient. Also, recently proposed pricing schemes that are expected to improve acceptability include Parento-improvement-based [2], credit-based [3, 4], and deposit-based systems [5]. In addition to measuring changes in traffic conditions to evaluate RP, we need to conduct a comparative analysis of the features of various RP schemes from multiple perspectives, including efficiency and acceptability.

© Springer International Publishing Switzerland 2016
R. Booth and M.-L. Zhang (Eds.): PRICAI 2016, LNAI 9810, pp. 738–749, 2016.
DOI: 10.1007/978-3-319-42911-3_62

Compared with research on traffic conditions and efficiency, there are relatively few studies on RP schemes' acceptance when evaluating them. This is because the dominant view is that if compensation is carried out by redistributing toll revenue without reducing social surplus (because charges collected from motorists are transferred as surplus to the road administrator), then problems with the acceptance of RP do not arise. The mechanisms of redistribution include programs to maintain or improve roads, measures to improve the service level of alternative modes of transportation, subsidies for work commute, and tax deductions [6]. However, because it takes time for these benefits to return to users, the possibility is high that when evaluating an RP scheme, these benefits may not be recognized. From results of questionnaire surveys, researchers have reported that by presenting the purpose of the toll revenue, a pricing scheme's acceptance (approval rate) becomes higher [7, 8]. Thus, it is critical to consider the redistribution of the toll revenue when designing an RP scheme. Meanwhile, equity between stake-holders may change as a result of redistribution. Thus, as part of analyzing an RP system's acceptance, it is critical to study the system itself and the optimal method of redistribution from the standpoint of equity.

As mentioned above, many pricing systems have been proposed with the expectation of improved acceptance. "Parking Deposit System" (PDS) [5] has been found to have higher acceptance (approval rate) compared with conventional pricing schemes such as cordon and area-based systems [9]. As shown in Fig. 1, PDS applies to visitors who enter a pricing targeted area, like conventional pricing schemes. By incorporating a deposit system that can return a portion of the charge to motorists who use parking lots and who shop, motorists can immediately feel the benefits of charge redistribution (a sense of saving money). Furthermore, using a combined equilibrium model to calculate the results of traffic conditions by time period with the introduction of PDS, researchers confirmed that when the pricing targeted area was set as the Nagoya Metropolitan Area, PDS mitigated the trend of reduction in the number of visitors to the target area, although the results varied to some degree depending on the amount of charge-refund. Also, researchers showed that PDS led to environmental improvement as a result of reduction in excess traffic to an extent similar to that of conventional pricing schemes [10]. Thus, we have scientific knowledge about PDS in terms of changes in traffic conditions and

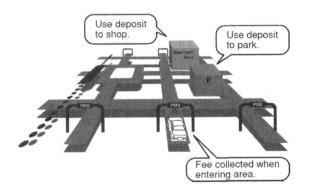

Fig. 1. Parking Deposit System (PDS) scheme [5]

its acceptance. By carrying out a comparative analysis of its efficiency and equity, we can further understand the characteristics of PDS from multiple perspectives.

2   Review of Research on Equity of Road Pricing

It is difficult to define equity uniformly and quantitatively [11]. Measures such as the Gini coefficient, Theil index, and Atkinson index have been proposed as indicators to evaluate equity. One study constructed an optimization problem to minimize the disparity in accessibility and conducted a comparative analysis of transportation policies using these indicators [12]. However, the optimal scheme and reproduced traffic conditions differed depending on the indicator used. Maruyama and Sumalee [13] used the Gini coefficient as an indicator of equity in their comparative analysis of the features of cordon- and area-based RP schemes. Also, by discretizing value of time (VOT) distribution and dividing users into several groups to express disparities in income, Maruyama et al. [14] showed income regressive effects in road pricing and the effectiveness of redistribution of toll revenue as a relief measure.

In this study, we learn from the methods of Maruyama et al. [14] which is relatively simple and easy to interpret, to evaluate income and regional equity. The targeted region of the evaluation is the Nagoya Metropolitan Area, and the RP targeted area is the urban center of Nagoya City. The pricing schemes are general cordon-based schemes. PDS is introduced as cordon-based PDS. We conduct a comparative analysis on traffic conditions, efficiency, and equity to understand the characteristics of PDS.

3   Agent-Based Evaluation Simulator

Changes to individuals' activities and travel patterns as a result of RP introduction include the content and frequency of their activities, the time of their departures and the duration of their activities or staying at same location, their mode of transportation, and routes. By taking into account these changes in travel behavior, we can understand the trend in number of visitors to the pricing targeted area and the change in automobile traffic volume. However, the evaluation model then becomes extremely complex. Because we focus on the analysis of efficiency and equity in our study, the traveler's transportation mode and route choice should be at minimum considered. We also assume that the analysis can be described with the nested logit model. This is because this model is appropriate for travel patterns that assume the place of activity to be fixed (i.e. workplace), the departure time to be habitual, and traffic congestion to take place during rush hour (morning peak time: 7 a.m.–10 a.m.).

The transportation modes which each traveler can choose are assumed to be automobiles, trains, buses and bicycles or walking. Route selection is only considered for automobiles and trains. Next section explains in detail the VOT class that each traveler belongs to and the parameters of the travel behavior model.

From the description above, the framework of this study's evaluation model is basically the same as Maruyama et al. [14]. A multi-class equilibrium model that consistently handles changes in the network service level due to users' travel behaviors (transportation mode and route choice model) by VOT class. The travel time on each link in the road network is expressed as the link performance function (i.e. the BPR function). Link travel time varies according to the link traffic flow and the path flow, which is the result of travelers' behaviors. At the same time, the generalized travel time in the behavioral model varies according to the link travel time. Thus, we need to seek an equilibrium state between demand and supply. This equilibrium state can be obtained by solving the following equivalent convex minimization problem.

$$
\text{min. } Z = \sum_a \int_0^{x_a} t_a(\omega)d\omega + \sum_{i,a} x_a^i p_a^i / \tau^i
$$

$$
+ \sum_{i,rs,m,k} \frac{1}{\theta_1^{i,m}} f_{m,k}^{i,rs} \ln(f_{m,k}^{i,rs}/q_m^{i,rs}) + \sum_{i,rs,m',k} f_{m',k}^{i,rs} C_{m',k}^{i,rs} \tag{1a}
$$

$$
+ \sum_{i,rs,m} \frac{1}{\theta_2^i} q_m^{i,rs} \ln(q_m^{i,rs}/Q_{rs}^i) + \sum_{i,rs,m} q_m^{i,rs} V_m^{i,rs}
$$

subject to

$$
x_a = \sum_{i,rs,k,a} f_{m,k}^{i,rs} \cdot \delta_{a,k}^{i,rs}, \ \forall a \tag{1b}
$$

$$
\sum_i x_a^i = x_a, \ \forall a \tag{1c}
$$

$$
\sum_k f_{m,k}^{i,rs} = q_m^{i,rs}, \ \forall i, rs, m \tag{1d}
$$

$$
\sum_m q_m^{i,rs} = Q_{rs}^i, \ \forall i, rs \tag{1e}
$$

$$
f_{m,k}^{i,rs} \geq 0, \quad q_m^{i,rs} \geq 0, \quad Q_{rs}^i \geq 0 \tag{1f}
$$

where Q_{rs}^i is the O-D trips of each class, $q_m^{i,rs}$ is the O-D trips of each class by each mode, $f_{m,k}^{i,rs}$ is the path flow of each class by each mode, x_a is the link flow in road network, x_a^i is the link flow of each class, $t_a(\cdot)$ is the link performance function, $\delta_{a,k}^{i,rs}$ is 1 if the link is on path k between an O-D pair of each class by each mode and 0 otherwise, p_a^i is the charge on the link for each class, τ^i is the value of time for each class traveler, $C_{m',k}^{i,rs}$ is generalized travel time on the route between an O-D pair by mode m' (not including car), $V_m^{i,rs}$ are systematic components of mode choice, and $\theta_1^{i,m}, \theta_2^i$ are scale parameters.

It can be proved easily that this problem has a unique solution under these conditions (1b–1f). The Kuhn-Tucker conditions for the problem lead to the aforementioned nested logit model with stochastic user equilibrium conditions.

$$f_{m,k}^{i,rs} = \frac{\exp[-\theta_1^{i,m} C_{m,k}^{i,rs}]}{\sum_k \exp[-\theta_1^{i,m} C_{m,k}^{i,rs}]} q_m^{i,rs} \tag{2a}$$

$$q_m^{i,rs} = \frac{\exp[-\theta_2^i(V_m^{i,rs} + S_m^{i,rs})]}{\sum_m \exp[-\theta_2^i(V_m^{i,rs} + S_m^{i,rs})]} Q_{rs}^i \tag{2b}$$

$$S_m^{i,rs} = -\frac{1}{\theta_1^{i,m}} \ln \sum_k \exp[-\theta_1^{i,m} C_{k,m}^{i,rs}] \tag{2c}$$

where $S_m^{i,rs}$ are inclusive values.

Here, efficiency and equity can be calculated using expected maximum utility (inclusive values) by VOT class. The partial linearization algorithm can be used to efficiently solve this problem. Even though the problem includes a path-flow entropy term, the model can be applied to large networks using entropy decomposition [15].

4   Estimation of Parameters of Travel Behavior Model

To evaluate the introduction of RP and PDS in the Nagoya Metropolitan Area, we estimate parameters in the travel behavior model (i.e. nested logit model), which are included in the evaluation model (formula (2a, 2b, 2c)). Person Trip (PT) survey data were used in estimating the parameters. We also present other necessary data and condition settings, and discuss the estimation results. Finally, we investigate the suitability of the evaluation model for accurately reproducing current conditions.

4.1   Estimation of Value of Time Distribution and User Class Division

Value of time (VOT), an important explanatory variable in our travel behavior model, is calculated by preference-based and income-based approaches. However, because the PT survey data does not contain the data on individual/household income, income and related VOT cannot be calculated endogenously from the behavior model as preference-based approach. In this study, VOT is set exogenously as the income-based approach using income-related data from sources besides PT surveys.

We assume that nation wide income distribution in Japan (average: 5.70 million yen annually, std. dev: 3.24 million yen annually; obtained from the Ministry of Health, Labor and Welfare's Survey on the Redistribution of Income [16]) follows a log-normal distribution. Each individual's number of working hours is the statutory working hours set by the Labor Standards Law (2,080 h per year). To obtain VOT (yen/minute/person), we divide income by working hours. VOT also follows a log-normal distribution.

The distribution of calculated VOT is shown in Fig. 2. The average is 45.7 yen/minute. Compared with the VOT used in Japan for the automobile traffic assignment (62.9 yen/minute/car), the calculated distribution is somewhat low, but we consider it to be within permissible range.

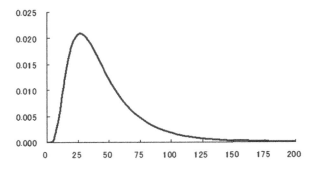

Fig. 2. Probability density distribution of calculated value of time (unit: yen/minute)

To set the VOT of individuals, we discretize the VOT distribution. We set the basic interval as 10 yen/minute. So that the occurrence probability became 1 % or greater, we aggregated the intervals. The final result was 12 classes with the intervals 0–20, 20–30, 30–40, …, 100–110, 110–140, 140– (yen/minute). We assume that each class's VOT contains the average value of each interval.

Moreover, we assume that the income distribution of residents within the Nagoya Metropolitan Area is the same as the national distribution. Random values were generated in accordance with the occurrence probability of the 12 classes' VOT, and added to the PT survey data. Therefore, we could not take into account the relationship between attributes of individuals and their income. The calculation of VOT distribution by household/occupation is future work.

4.2   Estimation of Parameters in Nested Logit Model

(1) Travel Behavior Data and Level of Service (LOS) Data.

For travel behavior data, we used data from the 4th Nagoya Metropolitan Area PT Survey (peak morning time-period only). Because of the entire data set's large sample size during the target time period (about 3 million people), we selected a data set of 10,000 trip makers at random to estimate parameters.

Also the LOS of each mode of transportation (i.e., automobiles, trains, buses, bicycles/walking) was created using the same method as that of Kanamori et al. [17].

(2) Parameter Estimation Method.

We estimated the parameters of travel behavior model besides the VOT established for individuals. However, because each individual's VOT was provided by random values, it differs from their true VOT. Because the data consists of just 10,000 randomly selected people, their VOT may not suitably explain travel behavior. That is, if only people who were established with high VOT were selected even though the true VOT of individuals tended to be low, this may result in the estimation of parameters different from the original travel behavior's preference.

Therefore, as a way to study the reduction of the effects of VOT settings, we apply a latent class model. The parameters in travel behavior model are estimated as given in the following equation:

$$P(m, k) = \sum_n P\left(m, k \mid \tau_n\right) \cdot Q\left(\tau_n\right) \tag{3}$$

where $P(m, k)$ is probability of selecting transportation mode m, route k, $P\left(m, k \mid \tau_n\right)$ is probability of selecting mode m, route k when an individual has VOT τ_n, $Q\left(\tau_n\right)$ is probability of having VOT τ_n. In this study, the number of the latent class n is set 30 and sampled from the 12 VOT's classes.

(3) Result of Parameter Estimation.
The result of estimating the travel behavior model's parameters is shown in Table 1. Parameters related to choosing automobile routes (scale: 0.5, generalized time: -1.0) was provided exogenously in this study. Expenses of driving a car such as fuel and toll are converted to travel time using VOT and considered generalized time.

Table 1. Estimation results of route and mode choice

Mode	Variable	parameter	t-value
<Route Choice>			
Car	Scale parameter	0.5	--
	Travel time(min)	-1	--
Train	Scale parameter	0.299	17.8
	Travel cost(yen)	-0.253	-16.8
	Out of vehicle time(min)	-0.741	-13.0
	Access distance(km)	-3.635	-19.8
	Egress distance(km)	-4.887	-18.8
	Main station Dummy	0.746	3.0
	Number of transfers	-1.623	-5.1
<Mode Choice>			
	Scale parameter	0.092	24.9
Car	Travel cast(yen)	-1	--
	Driver's license Dummy	2.452	28.6
	Car owner Dummy	1.287	10.3
Train	Male Dummy	0.431	7.0
	Constant_inner-inner Nagoya	2.908	19.3
	Constant_inner-outer Nagoya	1.993	9.4
	Constant_outer-inner Nagoya	3.400	18.3
	Constant_outer-outer Nagoya	1.071	6.1
Bus	Constant	2.432	8.9
	Travel cost(yen)	-0.253	-16.8
	Out of vehicle time(min)	-0.397	-3.0
	Access distance(km)	-7.065	-2.8
	Egress distance(km)	-4.852	-2.1
	Elderly Dummy	0.540	2.5
Bicycle & Walking	Constant_inner Nagoya	4.811	28.3
	Constant_outer Nagoya	4.034	22.0
	Student Dummy	1.540	14.4
	Travel time(min)	-1.133	-22.5
	Number of samples	10,000	
	Adjusted Rho-squared	0.43	

In addition to each mode of transportation's generalized time, explanatory variables of the travel behavior model are the distance of access and egress on a train or bus and attributes of individuals. All of the estimated parameters satisfy the 5 % significance level, and they have the expected sign. Also, a rho-squared is 0.43, indicating a good fit.

4.3   Reproducibility of Current Traffic Condition

We estimate the traveler's behavior model to establish VOT for 10,000 randomly selected people, to use as the evaluation model. This is done by calculating the equilibrium state using the PT data of about 3 million people and testing reproducibility.

For creating LOS data of trains, buses, automobiles, and walking, we used the same data set used in estimating the parameters of the travel behavior model. The automobile network, the link cost function, and the traffic volume of freight cars between origins and destinations (OD) were created using the same method as Kanamori et al. [17].

The convergence conditions to achieve an equilibrium state was set as less than 1 % of the variation in traffic volume before and after calculation. Afterwards, uniform random numbers were generated and the travel behavior of each individual was finally calculated discretely.

Our models could reproduce the share of the transportation modes in the entire metropolitan area with a high degree of accuracy. To test the accurate reproduction of automobile and train along the OD distance group, we calculated R^2 value (auto: 0.99, train: 0.85) and RMS (root mean square) value (auto: 4301.1, train: 3087.3). On the other hand, when all traveler has the same average VOT (i.e. 45.7 yen/minute) without using a latent model, R^2 value (auto: 0.99, train: 0.86) and RMS value (auto: 4509.0, train: 3105.1) are slightly worse. Our model has better RMS for both modes of transportation, can be said to have the better reproducibility.

5   Evaluation of PDS in the Nagoya Metropolitan Area

5.1   PDS Settings

The pricing targeted area is the urban center of Nagoya City (surface area: 25.5 km^2). The pricing schemes are cordon-based RP and cordon-based PDS. The names of each case and the amount of charges/refunds are shown in Table 2. The charges are considered on the basis of the average fare when using trains within Nagoya City. Base_0 represents the current traffic conditions of the transportation network, and codn_50 is the conventional pricing schemes. PDS is applied in the rest of the cases.

The methods of reflecting each pricing scheme to our agent-based evaluation simulator are as follows:

Cordon-based RP [codn_50].   Because the charge is levied each time a car enters the pricing targeted area, the charge is set at the automobile inflow links in the targeted area.

Table 2. Charge and refund in each case

	base_0	codn_50	codn_53	codn_55
Charge (yen/enter)	0	500		
Refund (yen/enter)	0	0	300	500
Net Charge	0	500	200	0

Cordon-based PDS [codn_53], [codn_55]. OD types are classified as inside/outside of the targeted area. Net charges are imposed on car users who depart from outside the targeted area and arrive inside the targeted area (outside-inside OD). For all other OD types, charges as set in the same manner as the cordon-based RP scheme.

5.2   Analysis of Traffic Conditions and Efficiency

To evaluate each pricing scheme, we calculate the share of transportation modes, user benefits and social surplus.

From Fig. 3, which shows the modal share under each pricing scheme, the conventional pricing schemes (codn_50) have an extremely great effect on reducing automobile use (about 30 % reduction). For PDS, as the amount of refund becomes greater, the reduction in automobile use becomes smaller. In the case of codn_55, the entire charge is refunded (net charge for visitors is 0 yen/enter), so the number of motorists increases slightly, because they can obtain the benefits of an improved level of automobile service within the targeted area that reduction of excessive traffic would bring. As a result, the use of automobiles actually increases. However, PDS has a greater effect on CO_2 reduction because the traffic congestion in the suburbs of the targeted area as motorists take detours.

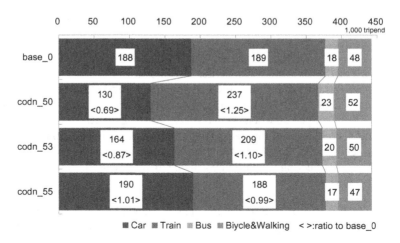

Fig. 3.   Change of mode usage under each pricing scheme (Color figure online)

Next, several benefit indicators of each case, such as user benefit and social surplus, are shown in Table 3. For our research, in addition to user benefit (i.e. consumer surplus), the toll revenue due to an increase in expressway users and fare revenue due to an increase in users using public transit are calculated. Also, for RP-related expenses, we referred to Singapore's case and estimate about 12.5 million yen per year, including construction and operational expenses. The benefit to users (general residents) is the logsum variable in an individual's travel model divided by the parameter -1.0 (minutes), which is for automobile's generalized travel time. The value is then converted to a monetary term by each person's VOT and then summed for all individuals.

Table 3. Benefit Indicators in each case

	User Benefit	Revenue	Toll & Fare	Management Cost	Social Surplus
codn_50	-67.1	56.4	26.8	-2.5	13.6
codn_53	-41.7	36.0	18.9	-2.5	10.7
codn_55	-22.8	11.2	13.0	-2.5	-1.2

From Table 3, user benefits in all cases are greatly reduced due to the introduction of road pricing. However, as the amount of PDS refund increases, user benefits also improve. When the entire charge is refunded (codn_55), benefits improve 30 % compared with the conventional RP scheme (codn_50). Focusing on social surplus, we confirm that it is positive for the conventional RP scheme without refunds. Cordon-based PDS has smaller social surplus compared with codn_50. Codn_55, which refunds the entire charge, has negative benefits, and is considered an inefficient scheme.

From the results above, with the exception of codn_55, all RP schemes analyzed in this study are efficient. Future studies are needed to set the optimal pricing scheme, using a detailed model that expands the target time periods and the travel behavior model. In other words, more analysis is required on topics such as the charge-refund amount and pricing target areas.

5.3 Analysis of Equity

Figure 4 shows the average change of the expected maximum utility by VOT class (unit of time, base_0 as standard). In the conventional RP (codn_50), compared with high-income classes, the level of utility is greatly reduced for low-income classes after the introduction of the pricing schemes. Therefore, road pricing in this study also has the problem of being income regressive, and raises the problem of equity. The reason an income regressive effect arises is that while low-income travelers are forcibly removed from road services due to the charges, high-income travelers receive the benefits of improved road services due to alleviating congestion.

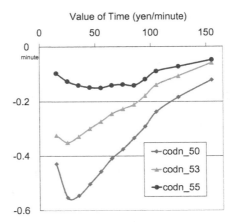

Fig. 4. Change of user benefit of each VOT class

Figure 5 shows the effects of cordon-based PDS refund by OD patterns. From this figure, we confirm that the utility level of outside-inside OD, which levies a substantive charge, changes greatly. For codn_55, which refunds the entire charge, in addition to income equity, regional equity (equity between OD patterns) also improved.

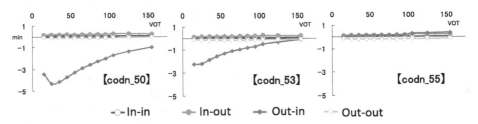

Fig. 5. Change of user benefit of each VOT class in OD patterns

6   Conclusion

In this paper, we examined PDS as a novel road pricing scheme to understand the characteristics to improve acceptance. We conducted a comparative analysis of PDS and conventional pricing schemes from the viewpoint of traffic conditions, efficiency, and equity. Our findings are summarized as follows:

- When applying a multi-class equilibrium model that can take into account differences in individuals' VOT, the application of a latent class model was effective as a method for estimating the parameters of travel behavior that uses PT survey data and income-related data.
- For PDS, as the refund rate increases, it becomes less effective in reducing automobile use. However, its benefit of improving the environment is sufficient.

- Conventional pricing schemes are problematic when it comes to equity between income levels. PDS, which is a type of scheme that redistributes toll revenue, has a mitigating effect.
- A tradeoff exists between efficiency and equity. There is a need to discuss with stakeholders on which to emphasize and to what extent. PDS may be effective as a compromise plan.

As future study, evaluating by equity indicators besides income equity and regional equity, elaborating the travel behavior model, and understanding characteristics of pricing schemes are needed. Combining the redistribution of toll revenue, which was recognized as effective with PDS in this study, with other policy options (e.g. discount of fare of public transportations) besides refund directly is also needed.

References

1. Noordegraaf, D.V., Annema, J.A., Wee, B.V.: Policy implementation lessons from six road pricing cases. Transp. Res. Part A **59**, 172–191 (2014)
2. Daganzo, C.F., Gracia, R.C.: A pareto improving strategy for the time-dependent morning commuter problem. Transp. Sci. **34**, 303–310 (2000)
3. Nakamura, K., Kockelman, K.M.: Congestion pricing and road space rationing: an application to the San Francisco Bay Bridge corridor. Transp. Res. Part A **36**, 403–417 (2002)
4. Kockelman, K.M., Kalmanje, S.: Credit-based congestion pricing: a policy proposal and the public's response. Transp. Res. Part A **39**, 671–690 (2005)
5. Morikawa, T.: Eco-transport cities utilizing ITS. IATSS Res. **32**, 26–31 (2008)
6. Small, K.A.: Using the revenues from congestion pricing. Transportation **19**, 359–381 (1992)
7. Jones, P.: Gaining public support for road pricing through a package approach. Traffic Eng. Control **32**, 194–196 (1991)
8. Schuitema, G., Steg, L.: The role of revenue use in the acceptability of transport pricing policies. Transp. Res. Part F **11**, 221–231 (2008)
9. Ando, A., Morikawa, T., Miwa, T., Yamamoto, T.: Study of the acceptability of a parking deposit system as an alternative road pricing scheme. IET Intell. Transp. Syst. **4**, 61–75 (2010)
10. Kanamori, R., Morikawa, T., Yamamoto, T., Miwa, T.: An evaluation of parking deposit system by semi-dynamic combined stochastic equilibrium model. Infra Struct. Plan. Rev. **24**, 915–925 (2007). (in Japanese)
11. Levinson, D.: Equity effects of road pricing: a review. Transp. Rev. **30**, 33–57 (2010)
12. Feng, T., Zhang, J.: Multicriteria evaluation on accessibility-based transportation equity in road network design problem. J. Adv. Transp. **48**, 526–541 (2014)
13. Maruyama, T., Sumalee, A.: Efficiency and equity comparison of cordon-and area-based road pricing schemes using a trip-chain equilibrium model. Transp. Res. Part A **41**, 655–671 (2007)
14. Maruyama, T., Harata, N., Ohta, K.: A study on income regressive effect of road pricing policy and its alleviation. J. City Plan. Inst. Jpn. **37**, 253–258 (2002). (in Japanese)
15. Akamatsu, T.: Decomposition of path choice entropy in general transport networks. Transp. Sci. **31**, 349–362 (1997)
16. Ministry of Health, Labor and Welfare: Survey on the Redistribution of Income (2008)
17. Kanamori, R., Miwa, T., Morikawa, T.: Evaluation of road pricing policy with semi-dynamic combined stochastic user equilibrium model. Int. J. ITS Res. **6**, 67–77 (2008)

Stigmergy-Based Influence Maximization in Social Networks

Weihua Li[1(✉)], Quan Bai[1], Chang Jiang[1], and Minjie Zhang[2]

[1] Auckland University of Technology, Auckland, New Zealand
{weihua.li,quan.bai}@aut.ac.nz, rky9795@autuni.ac.nz
[2] University of Wollongong, Wollongong, Australia
minjie@uow.edu.au

Abstract. Influence maximization is an important research topic which has been extensively studied in various fields. In this paper, a stigmergy-based approach has been proposed to tackle the influence maximization problem. We modelled the influence propagation process as ant's crawling behaviours, and their communications rely on a kind of biological chemicals, i.e., pheromone. The amount of the pheromone allocation is concerning the factors of influence propagation in the social network. The model is capable of analysing influential relationships in a social network in decentralized manners and identifying the influential users more efficiently than traditional seed selection algorithms.

Keywords: Influence maximization · Ant algorithm · Stigmergy

1 Introduction

With the development of social networks, on-line marketing has developed in an unprecedented scale. One of the typical on-line sales strategies is viral marketing, which propagates influence through 'word-of-mouth' effect [2]. It is capable of increasing brand awareness and achieving marketing objectives effectively. One of the critical tasks is to understand how to select a set of influential users from the network to propagate influence as much as possible with limited resources, namely, influence maximization, and the solution is NP-hard [3,7]. Thus, approximation approaches are considered as a replacement. In general, if a set of influential users, i.e., seed set, can be selected properly and completely, we regard that the influence spread has been achieved. Most researchers seek solutions for influence maximization problem based on the centralized influence diffusion models, such as, the classic Independent Cascade (IC) model and Linear Threshold (LT) model [7]. However, these centralized approaches are normally not efficient, especially when the network is large-scale and dynamic. Specifically, these approaches require a central component to complete all tasks alone. Furthermore, the seed selection algorithms under the traditional influence diffusion models are time-consuming. By contrast, decentralized approaches tend to share the workload by distributing the computational tasks to individuals.

© Springer International Publishing Switzerland 2016
R. Booth and M.-L. Zhang (Eds.): PRICAI 2016, LNAI 9810, pp. 750–762, 2016.
DOI: 10.1007/978-3-319-42911-3_63

There are two kinds of decentralized approaches in terms of communications. One relies on the direct communications among the individuals, such as cellular automata [9], where each cell in the grid adapts its state by looking at the adjacent neighbours based on a set of rules. While, the other focuses on the indirect communications by reading or analysing the messages left by the peers. One of the typical approaches is ant and stigmergy algorithm [4]. The French Entomologist, Pierre-Paul Grasse defines stigmergy as "stimulation of workers by the performance they have achieved", which is associated with two major features of ants [1]. First, the communication among the ants is indirect. To be more specific, stigmergy is a particular indirect communication mechanism that ants exploited to harmonize their daily tasks with each other. Their indirect communication is conducted through leaving 'pheromone' on the trails, which is a kind of chemical substance and evaporates over time. Second, ants' activities are self-organized. They can complete a complicated task independently without any control. With the development of stigmergy, it has been applied for communication network routing, exploratory data analysis, and diagram drawing etc.

In this paper, we exploit a novel decentralized approach, the Stigmergy-based Influence Maximization approach (SIM), to tackle the influence maximization problem. In SIM, influence propagation process is modelled as ants crawling across the network topology. Furthermore, the ant's key behaviours, including path selection and pheromone allocation, have been modelled for selecting suitable nodes to achieve influence maximization. The former aims to identify the next node to walk when an ant faces multiple options. While, the objective of the latter is to allocate pheromone on the specific nodes based on the heuristics when an ant explores a possible influence-diffusion path. Experiments have been conducted to evaluate the performance of SIM by comparing with the traditional seed selection algorithms, such as greedy selection, degree-based selection and random selection. The results demonstrate that the proposed model is more advanced by considering both efficiency and effectiveness, and can dramatically reduce computational overhead compared with centralized approaches.

The rest of this paper is organized as follows. Section 2 reviews the literatures related to this research work. Section 3 systematically elaborates the SIM approach, including problem description, formal definitions, path selection and pheromone operations. In Sect. 4, experiments are conducted to evaluate the performance of SIM. Finally, the paper is concluded in Sect. 5.

2   Related Work

In on-line marking, it is critical to investigate how to propagate influence in a social network with limited budgets. Motivated by this background, influence maximization aims to select a set of influential users from the network to diffuse influence as much as possible with finite resources [7]. Many studies on influence maximization problem are conducted on the basis of two fundamental influence propagation models, i.e., IC and LT [7]. Both models have two key properties, i.e., propagation and attenuation. The influence initiates from the seed set, i.e., activated nodes. They transfer their influence through the correlation graph, whereas the power of this effect decreases when hopping further and further away from the activated nodes.

There are a couple of popular seed selection approaches, such as greedy selection, degree-based selection and random selection. Many research works have been conducted to improve the efficiency and effectiveness of seed selection algorithms on influence maximization. Chen et al. study the efficient influence maximization by improving the original greedy selection and proposing a novel seed selection approach, namely, degree discount heuristics for the uniform IC model, where all edge probabilities are the same [2]. Goyal et al. design and propose a novel CELF algorithm, i.e., CELF++, to reduce running time [6]. Zhang et al. research the least Cost Influence Problem (CIP) in multiplex network, and the CIP is alleviated by mapping a set of networks into a single one via lossless and lossy coupling schemes [12]. However, all these approaches only can be applied in a static network and the network topology must be discovered. Specifically, they cannot handle the dynamics of social networks. Meanwhile, the traditional approaches are not applicable when the global view is unavailable.

Ant and stigmergy-based algorithms do not rely on the network typology, and the computation is decentralized. Stigmergy consists in the main body of ant colony knowledge, as it is a particular mechanism exploited for indirect communication among ants to control and coordinate their tasks. In natural environments, stigmergy-based systems have been demonstrated that they can be utilized for generating complicated and robust behaviours in the systems even if each ant has limited or no intelligence. Nest building is the representative example of stigmergy. Some researchers has applied stigmergy for computer science fields. Dorigo et al. introduce how to solve the Travel Salesman Problem (TSP) [11] by leveraging ant and stigmergy-based algorithms, where the pheromone allocation is concerning the distances among the cities [4]. Ahmed et al. propose a stigmergy-based approach for modelling dynamic interactions among web service agents in decentralized environments [8]. Takahashi et al. proposed anticipatory stigmergy model with allocation strategy for sharing near future traffic information related to traffic congestion management in a decentralized environment [10].

3   Stigmergy-Based Influence Maximization Modelling

SIM tends to select appropriate influential candidates by considering both influence strengths among users and the assembled influential effect. In this model, numerous ants walk simultaneously and update the shared environment by distributing pheromone, and the influence propagation process is simulated as crawling behaviours of ants. The influential users can be identified when the pheromone distribution in the network starts to converge, and the seed selection is based on the pheromone amount of each node. The SIM will be elaborated in the following subsections.

3.1 Problem Description

Suppose an organization plans to promote a particular product in a large-scale on-line social network. Due to limited budgets and insufficient time, the organization needs to select k initial candidates as influential users to experience the product as soon as possible, hoping that these users can recommend it in their social circle. Ideally, the k influential users can produce maximum influence in the social network.

3.2 Formal Definitions

Definition 1: A **Social network** is defined as a weighted graph $G = (V, E)$ with a clear topological structure, where $V = \{v_1, v_2, ..., v_n\}$ stands for the nodes (users) in the network, $E = \{e_{ij}|v_i \in V \wedge v_j \in V, v_i \neq v_j\}$ denotes the edges (relationships) among nodes. A particular edge can be represented as a three-tuple, i.e., $e_{ij} = (v_i, v_j, w_{ij})$, where w_{ij} is the weight of e_{ij} which represents the influence strength. Each node v_i has a set of neighbours $\{v_j|v_j \in \Gamma(v_i), e_{ij} \in E\}$. While, $v_i.q$ indicates the pheromone amount (see Definition 4) accumulated on corresponding node v_i, which can be regarded as an attribute of v_i. Similarly, since w_{ij} represents the weight of edge e_{ij}, it is denoted by using the notation $e_{ij}.w$ in this paper.

Definition 2: An **Ant** a_m is defined as an autonomous agent in the network G, which crawls across G based on the network topology. An ant can be represented as a three-tuple, i.e., $a_m = (m, q_m^n, T_m^n)$, which means ant a_m carries q_m^n pheromone in tour T_m^n (see Definition 3). There exist a number of ants, $A = \{a_1, a_2, ..., a_n\}$, in the social network, and they keep crawling in the network. Moreover, they are capable of discovering and evaluating the amount of pheromone on the current node and the ones nearby. However, the ants cannot communicate directly with each other.

Definition 3: A **tour** $T_m^n =< v_1, v_2, ..., v_n >$ is defined as the path that ant a_m walks through in the n round. Specifically, ant a_m randomly selects a starting point. Next, it crawls from one node to the adjacent neighbours and eventually ceases when reaches the end point v_e, where $\Gamma(v_e) \subset T_m^n \cup |\Gamma(v_e)| = 1$.

Definition 4: Pheromone represents the information and heuristics passed by an ant to the peers based on its experience. q_m^n denotes the total amount of artificial pheromone carried by ant a_m in the n round, which will be distributed to each node of T_m^n after a_m completes the tour.

3.3 Path Selection

In this context, path selection is one of the ant's basic behaviours, which describes how a particular ant a_m selects the next node to walk when standing at v_i and facing multiple choices $V_c = \{v_j|v_j \in \Gamma(v_i) \wedge e_{ij} \in E\}$.

Basically, the path selection decision is based on two aspects, which are the pheromone amount of v_j, i.e., $v_j.q$, and the weight of corresponding edge, i.e., $e_{ij}.w$.

The path selection behaviour has been modelled as a probabilistic event by using Eq. 1, where p_{ij} denotes the probability that an ant walks from node v_i to v_j.

$$p_{ij} = \begin{cases} \dfrac{e_{ij}.w \cdot v_j.q}{\sum_{v_x \in \Gamma(v_i)} e_{ix}.w \cdot v_x.q}, & e_{ij} \in E \\ 0, & e_{ij} \notin E \end{cases} \tag{1}$$

Here, we demonstrate the path selection by giving two concrete examples. In Fig. 1, ant a_i starts from node v_i and confronts three options, i.e., v_k, v_j and v_n. The decision is made by considering both targeting nodes' pheromone amount and the influence strength / weight of the corresponding edges. In this diagram, the probability of choosing node v_j is calculated as: $p_{ij} = e_{ij}.w \cdot v_j.q/(e_{ij}.w \cdot v_j.q + e_{ik}.w \cdot v_k.q + e_{in}.w \cdot v_n.q) = 0.8 \times 0.5/(0.4 \times 0.6 + 0.8 \times 0.5 + 1.0 \times 0.7) = 29.85\%$

Figure 2 demonstrates another example, where two ants, i.e., a_i and a_j, walk in the same network. Based on the path selection principles, they cannot choose the nodes which have been walked through within the same tour, but they can choose the ones that other ants have passed before in the either current or previous iterations.

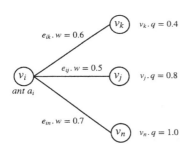

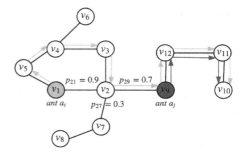

Fig. 1. Path selection of an ant   **Fig. 2.** Path selection of multiple ants

Each ant keeps performing an iterative process: walking and selecting path, whereas, the action stops when the ant reaches the end point. In other words, the iterative process triggered by ant m in round n produces a path vector, i.e., tour T_m^n. The tour formation is described in Algorithm 1.

Algorithm 1 presents the process of how a particular ant completes a tour. The input of this algorithm includes ant a_m and the round index n, while the output is tour T_m^n. Line 3 shows the criteria of walking to the next node. Lines 5–10 demonstrate the targeting candidates selection, where σ is a predefined threshold to filter out those candidates with low probability. Lines 11–17 indicate the path selection process. The iterative walking process ends when all of the current node v_s's neighbours reside in the tour list T_m^n.

Algorithm 1. Tour Formation Algorithm

Input: a_m, n

Output: $T_m^n, T_m^n \subseteq V$

1: Initialize a_m and random select a starting point $v_s, v_s \in V$
2: Initialize a tour list $T_m^n := \emptyset$
3: **while** $\exists \Gamma(v_s) \wedge \Gamma(v_s) \not\subseteq T_m^n$ **do**
4:      Initialize candidate list $V_c := \emptyset$
5:      **for** $\forall v_i \in \Gamma(v_s) \wedge v_i \notin T_m^n$ **do**
6:          Compute the probability p_{si} using Equation 1.
7:          **if** $p_{si} > \sigma$ **then**
8:              $V_c := V_c \cup \{v_i\}$
9:          **end if**
10:     **end for**
11:     **if** $V_c \neq \emptyset$ **then**
12:         Determine the next node $v_n \in V_c$ using Equation 1.
13:         $T_m^n := T_m^n \cup \{v_n\}$
14:         $v_s := v_n$
15:     **else**
16:         $v_s := null$
17:     **end if**
18: **end while**

3.4   Pheromone Operations

Sub-network Generation. Sub-network generation is the preliminary step of pheromone operations. After ant a_m completes a tour T_m^n, a corresponding sub-network $G_m^n = (V_m^n, E_m^n)$ will be generated based on the path that a_m walked through. V_m^n incorporates all nodes in tour T_m^n and their valid first-layer neighbours $\Gamma(T_m^n)$, thus, $V_m^n = T_m^n \cup \Gamma(T_m^n)$. While, the edge set E_m^n includes all the links among V_m^n.

The total amount of pheromone q_m^n carried by ant a_m for tour T_m^n depends on the total number of nodes in the sub-network, i.e., $|V_m^n|$. Each node in the sub-network contributes one unit of pheromone. Figure 3 presents an example of a generated sub-network. An ant walked from node v_a to node v_e sequentially. By walking pass each of them, the ant searches for the valid first-layer neighbours. In this way, a sub-network is generated.

Pheromone Allocation. Pheromone allocation in general refers to how ants leave the biological information on the nodes that they have walked through. The distribution of pheromone plays an important role in the stigmergy algorithms, since it updates the context by considering the relevant impact factors. Therefore, the solution is continuously being optimized.

In the current setting, the pheromone distribution is based on size of the sub-network. The shorter length path and larger sub-network size, the more pheromone will be allocated on each node of the tour. Equation 2 aims to compute the number of connected neighbours of node v_i in the sub-network G_m^n.

Equation 3 describes the pheromone accumulation of node v_m in tour T_m^n, which is calculated by adding up all the pheromone contributions given by the direct neighbours $\Gamma(v_m)$.

$$v_i.N = |\{v_i|v_i \in V_m^n \wedge \Gamma(v_i) \in T_i^n\}| \tag{2}$$

$$v_m.\Delta q = \begin{cases} \sum_{v_i \in \Gamma(v_m)} \frac{1}{v_i.N}, & v_m \in T_m^n, v_i.N \neq 0 \\ 0, & v_m \in T_m^n, v_i.N = 0 \end{cases} \tag{3}$$

Figure 3 shows an example of a specific sub-network $G_m^n = (V_m^n, E_m^n)$, where the tour travelled by ant a_m is represented as $T_m^n = <v_a, v_b, v_c, v_d, v_e>$, $V_m^n = \{v_a, v_b, v_c, v_d, v_e, v_f, v_k, v_h, v_i\}$ and E_m^n includes all the edges among the nodes in V_m^n, $|E_m^n| = 12$ in this diagram. Node v_f is the direct neighbour of two nodes in tour T_m^n, hence both v_a and v_b obtain half of a unit pheromone from v_f. Meanwhile, node v_b contributes 0.5 unit pheromone to v_a and v_c, but v_f and v_k are not considered in this scope. Therefore, we can derive that the pheromone gain for node v_a and v_b are 1.5 and 2.0 respectively.

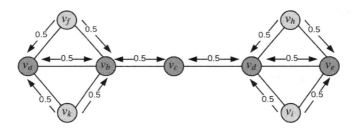

Fig. 3. Pheromone allocation in a tour with five nodes

Algorithm 2 shows the pheromone allocation process initiated by ant a_m in tour T_m^n. The distribution is based on the explored sub-network G_m^n's topology. The input is a specific tour T_m^n. Whereas, the output is pheromone amount update. Specifically, this algorithm aims to change the context by updating the pheromone amount located in each node of the tour path. Lines 1–9 initialize and construct the sub-network G_m^n. The objective of Lines 10–12 is to obtain the denominator for each node which supposes to contribute pheromone to the nodes in tour path. Lines 13–14 show the variations of pheromone.

Pheromone Evaporation. Pheromone evaporation is a common phenomenon, where the amount of allocated pheromone decreases over time. In ant and stigmergy algorithms, it helps to avoid the convergence to a locally optimal solution. Pheromone evaporates from each node within the scope of the whole network at the same time. At a justified time, all of the nodes in the network will evaporate a predefined unit of pheromone. The pheromone evaporation is quantified

---

Algorithm 2. Pheromone Allocation Algorithm

Input: T_m^n
Output: *pheromone changes for all the nodes in T_m^n*
1: Initialize sub-network graph $G_m^n := (V_m^n, E_m^n), V_m^n := \emptyset, E_m^n := \emptyset$
2: **for** $\forall v_i \in T_m^n$ **do**
3:     **for** $\forall v_j \in (\Gamma(v_i) \cup v_i)$ **do**
4:         $V_m^n := V_m^n \cup \{v_j\}$
5:         **if** $p_{ij} > 0 \land i \neq j$ **then**
6:             $E_m^n := E_m^n \cup \{e_{ij}\}$
7:         **end if**
8:     **end for**
9: **end for**
10: **for** $\forall v_n \in V_m^n$ **do**
11:     Compute $v_n.N$ using Equation 2
12: **end for**
13: **for** $\forall v_m \in T_m^n$ **do**
14:     $v_m.q := v_m.q + v_m.\Delta q$, using Equation 3
15: **end for**

---

by using Eq. 4, where the amount of pheromone evaporated from each node is associated with the time difference Δt and the evaporation speed λ.

$$EQ = e^{\frac{\Delta t}{\lambda}}, \lambda \neq 0 \tag{4}$$

3.5   Seed Selection

Seed selection aims to select a set of influential users from a specific network, so that they can propagate influence to others. There are quite a few classic seed selection approaches. More specifically, degree-based seed selection tends to select the nodes with high node degree. Intuitively, the users with large friend circle can influence more users in the social network. However, this does not hold in general, e.g., two connected users with very high degree may have a lot of common friends, in other words, the impact generated by both may be pretty much close to choosing either of them. Another well-known approach is greedy selection, which aims to obtain the maximum influence marginal gain in selecting each seed. However, this approach is not applicable in large-scale networks due to the computational overhead. Random selection is also applied in some cases, but its performance is normally the worst since it is not based on any heuristics.

The seed selection in stigmergy-based algorithm relies on the amount of pheromone allocated on each node. The selection is similar to degree-based approach, but it identifies the influential users by ranking the pheromone degree of each node.

In Algorithm 3, the input includes the number of ants n, seed set size k, evaporation speed λ, time difference Δt and the network $G = (V, E)$. Lines 1–2 initialize the ants and seed set. Line 3 indicates the ants' autonomous behaviours in the network by using Algorithms 1 and 2. Lines 4–10 show the global

Algorithm 3. Seed Selection Algorithm
___

Input: $n, k, \lambda, \Delta t, G = (V, E)$
Output: V_s
1: Initialize ant set $A := \{a_1, a_2, ..., a_n\}$ which contains n ants.
2: Initialize seed set $V_s := \emptyset$
3: All the n ants start to crawl in network G in the distributed servers.
4: **while** !convergence **do**
5:      Compute EQ using Equation 4.
6:      **for** $v_i \in V$ **do**
7:          $v_i.q := v_i.q - EQ$
8:      **end for**
9:      Sleep for Δt
10: **end while**
11: Sort V order by q descend
12: **for** $\forall v_i \in V$ **do**
13:     **if** $|V_s| < k$ **then**
14:         $V_s := V_s \cup \{v_i\}$
15:     **end if**
16: **end for**
___

pheromone evaporation process. Lines 11–16 indicate the seed selection from the updated environment.

4   Experiments and Analysis

4.1   Experiment Setup

MovieLens[1] dataset has been used for the experiments. It is a stable benchmark dataset, which contains around one million ratings for 3,900 movies given by 6,040 users. To filter noise data, users whose number of ratings are less than 50 have been removed from the dataset. There are no explicit links among the users, but the implicit links can be generated according to the ratings to items. Moreover, in order to control the computing time, we select three sub-graphs of the network with different scales, i.e., size of 500, 750 and 1000 respectively, for the experiments.

The node degree distributions of three sub-graphs are represented as Figs. 4, 5 and 6. All of them follows the power-law distribution pattern which is satisfied by most real networks [5].

4.2   Global Pheromone Distribution

As explained in Sect. 3.4, all the artificial ants crawl in the social network and allocate pheromone after completing tours, whereas the allocated pheromone keeps evaporating over time. The total amount outstanding pheromone in the social network is regarded as the global pheromone.

___
[1] http://grouplens.org/datasets/movielens/.

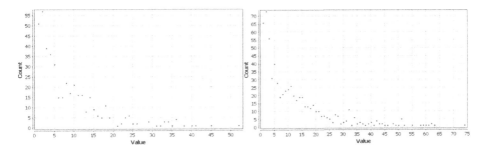

Fig. 4. Degree distribution (size = 500)   **Fig. 5.** Degree distribution (size = 750)

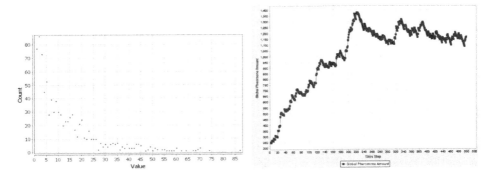

Fig. 6. Degree distribution (size = 1000)   **Fig. 7.** Global pheromone distribution
(size = 500)

The global pheromone distributions of three sub-graphs are demonstrated in Figs. 7, 8, and 9. As we could observe from these three diagrams that the pheromone amount increases steadily and starts to oscillate when reaching a certain level. Thus, the pheromone allocation and evaluation almost achieve a balance. At this phase, it implies the network starts to converge, since the sequential pheromone ranking list does not vary a lot.

4.3   Experimental Results

We conducted two experiments by using the same social network of three different sizes, which are 500, 750 and 1000 respectively. The first experiment aims to evaluate the influence effectiveness of stigmergy-based algorithm, i.e., the total number of users activated by the seed set. While, the second tends to compare the efficiencies, i.e., the running time of selecting seed set. The counter parts of stigmergy-based algorithm include greedy selection, degree-based selection and random selection.

In the first experiment, seeds are selected from the proposed model, and input into the IC model to evaluate the influence effectiveness by comparing with

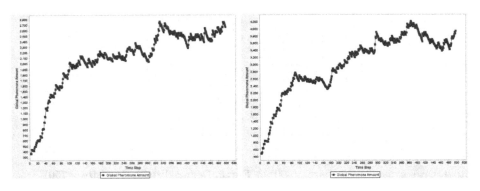

Fig. 8. Global pheromone distribution (size = 750)

Fig. 9. Global pheromone distribution (size = 1000)

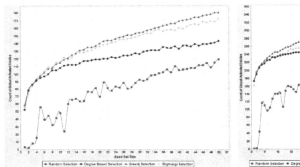

Fig. 10. Influence effectiveness comparison (size = 500) (Color figure online)

Fig. 11. Influence effectiveness comparison (size = 750) (Color figure online)

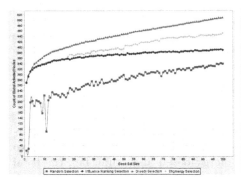

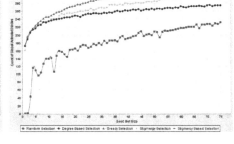

Fig. 12. Influence effectiveness comparison (size = 1000) (Color figure online)

Fig. 13. Efficiency comparison (size = 500) (Color figure online)

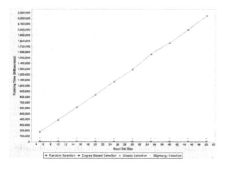 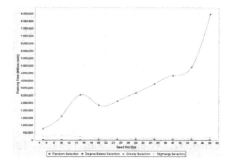

Fig. 14. Efficiency comparison (size = 750) (Color figure online)

Fig. 15. Efficiency comparison (size = 1000) (Color figure online)

the other classic algorithms. Figures 10, 11 and 12 demonstrate the influence effectiveness comparison among the four algorithms in three sub-graphs. The stigmergy-based algorithm performs better than both degree-based selection and random selection, and its performance is even closer to the greedy selection when the network size is 500. With the expansion of the graph, stigmergy-based selection's influence effectiveness drops a little bit but still outperforms the rest.

The second experiment analyses the efficiency of four seed selection algorithms by comparing the running time. Running time required by stigmergy-based algorithm includes ants initiation and pheromone operations. While, the other three algorithms are evaluated in the IC model. Figures 13, 14 and 15 show the efficiency comparison among the four algorithms in different sub-graphs. It is clear that the greedy selection is the most computational expensive of all, the running time increases dramatically when the seed set size enlarges. Both random and degree-based selection are very similar to each other in terms of efficiency. The stigmergy-based appear a little bit higher than degree-based selection, but it is much more efficient than the greedy selection and computational cost does not increase a lot with the expansion of the network.

In summary, observing from the experimental results, we can conclude that the stigmergy-based algorithm performs better than the traditional algorithms by considering both efficiency and effectiveness.

5   Conclusion and Future Work

In this research, we introduced a novel approach, i.e., stigmergy-based algorithm, to tackle the influence maximization problem in a decentralized environment. In the meanwhile, SIM model has been proposed and systematically elaborated. Experiments have been conducted to evaluate the performance of SIM. Experimental results reveal that SIM outperforms the traditional seed selection approaches, including greedy selection, degree-based selection and random selection, by considering both effectiveness and efficiency. Moreover, SIM is applicable for large-scale networks and even functions without a global view.

In the future, learning algorithms will be employed to improve the performance of the stigmergy-based algorithm in influence maximization problem. Meanwhile, we will consider a hybrid approach for developing a more practical model.

References

1. Bonabeau, E.: Editor's introduction: stigmergy. Artif. Life **5**(2), 95–96 (1999)
2. Chen, W., Wang, Y., and Yang. S.: Efficient influence maximization in social networks. In: Proceedings of the 15th ACM SIGKDD International Conference on Knowledge Discovery and Data Mining, pp. 199–208. ACM (2009)
3. Domingos, P., Richardson, M.: Mining the network value of customers. In: Proceedings of the Seventh ACM SIGKDD International Conference on Knowledge Discovery and Data Mining, pp. 57–66. ACM (2001)
4. Dorigo, M., Bonabeau, E., Theraulaz, G.: Ant algorithms and stigmergy. Future Gener. Comput. Syst. **16**(8), 851–871 (2000)
5. Easley, D., Kleinberg, J.: Networks, Crowds, and Markets: Reasoning about a Highly Connected World. Cambridge University Press, Cambridge (2010)
6. Goyal, A., Lu, W., Lakshmanan, L.V.: Celf++: optimizing the greedy algorithm for influence maximization in social networks. In: Proceedings of the 20th International Conference Companion on World Wide Web, pp. 47–48. ACM (2011)
7. Kempe, D., Kleinberg, J., Tardos, É.: Maximizing the spread of influence through a social network. In: Proceedings of the Ninth ACM SIGKDD International Conference on Knowledge Discovery and Data Mining, pp. 137–146. ACM (2003)
8. Mostafa, A., Zhang, M., Bai, Q.: Trustworthy stigmergic service composition and adaptation in decentralized environments. IEEE Trans. Serv. Comput. **9**(2), 317–329 (2016). doi:10.1109/TSC.2014.2298873
9. Shiffman, D.: Cellular automata. In: Shiffman, D., Fry, S., Marsh, Z. (eds.) The Nature of Code, pp. 323–330 (2012)
10. Takahashi, J., Kanamori, R., Ito, T.: A preliminary study on anticipatory stigmergy for traffic management. In: 2012/WIC/ACM International Conferences on Web Intelligence and Intelligent Agent Technology (WI-IAT), vol. 3, pp. 399–405. IEEE (2012)
11. Wang, K., Huang, L., Zhou, C., Pang, W., et al.: Traveling salesman problem. Conf. Mach. Learn. Cybern. **3**, 1583–1585 (2003)
12. Zhang, H., Nguyen, D.T., Zhang, H., Thai, M.T.: Least cost influence maximization across multiple social networks (2015)

Towards Exposing Cyberstalkers in Online Social Networks

Jiamou Liu[1(✉)], Yingying Tao[2], and Quan Bai[2]

[1] The University of Auckland, Auckland, New Zealand
jiamou.liu@auckland.ac.nz
[2] Auckland University of Technology, Auckland, New Zealand
{kdx8538,quan.bai}@aut.ac.nz

Abstract. This paper presents work-in-progress towards a computational approach for capturing a type of deviant behaviors, that are characterised by persistent monitoring and information gathering through online social medias. Such behaviors are commonly associated with stalking on the cyberspace. We present a network-based framework for describing online user interactions. Based on this framework we provide a description of excessive and unreciprocated attention an agent pays to another agent. We conclude with a discussion on limitation of the current work and a guideline for future extensions.

Keywords: Cyberstalking · Online social network · Agent behaviors

1 Introduction

This paper aims to provide a formal behavioral-oriented study on certain situations, where an online user persistently monitors and gathers information from another user through online social medias. Such deviant behaviors embody, to a large extent, the main characteristics of the so-called behaviors of *cyberstalking* and *cyberbullying*. Both types of deviant behaviors have become prevalent problems that are associated with the online social medias, and have serious social and psychological implications, which hamper the safe usage of the Internet [1, 10–12].

This research falls into the broader area of cyber-safety and privacy in computer science: In the last decade, there are major efforts towards the detection of phising, spamming, cloning, and bots on online social medias [5]. However we have found no technical breakthrough specifically focusing on cyberstalking. This may be due to the complex nature of the type of online user interactions and the actions generally associated with cyberstalking, such as online harassment and identity theft. As pointed out in [11], despite decades of criminological research, there has not been a generally agreeable definition of cyberstalking. The goal of our result is to bridge this gap by investigating cyberstalking in a formal, computational perspective and hopefully develop useful technologies

© Springer International Publishing Switzerland 2016
R. Booth and M.-L. Zhang (Eds.): PRICAI 2016, LNAI 9810, pp. 763–770, 2016.
DOI: 10.1007/978-3-319-42911-3_64

that help with the prediction, detection, and forensics of cyberstalking. Indeed, this pilot work presents work-in-progress towards this goal.

The basic premises of this work relies on tools and techniques provided by network science in the modeling, simulation, detection and prevention of cyberstalking behaviors. Network science is an interdisciplinary area that aims to explore properties of complex relational structures such as information networks, social networks and biological networks. The main object of investigation consists of a crowd of autonomous agents that are interconnected through ties. The actions of individual agents are affected by actions of other agents as well as the interactions between them. The field focuses on developing models of such networks and explores structural properties of complex networks through algorithmic analysis [2,6,9]. When viewing the Internet as a large information network, one can extract and analyze traits and trends of Internet users using various statistical and graph-theoretical tools.

The main goal of the work is to introduce a formal framework for agents in a social network, that could be used to describe a range of user behaviors. We then would like to develop tools that are able to detect abnormalities in such simulated environment, that is, the tool should help to distinguish benevolent behaviors and abnormal behaviors. We provide a mathematical model, which has the potential to be developed into an automatic mechanism for analyzing and detecting cyberstalkers through analyzing network logs.

Organization. The rest of the paper is organized as follows: Sect. 2 discusses existing literatures of cyberstalking in the social sciences. In particular, we revisit existing descriptions of cyberstalking behaviors and key challenges. Section 3 presents a general framework for describing user behaviors in an online social network. A crucial component of our model aims to capture the amount of attention a user pays to another. Based on this, Sect. 4 provides a brief discussion on our behavioral model of cyberstalking.Finally, Sect. 5 concludes the paper with discussion of limitation of the model and future directions.

2   Understanding Cyberstalking

The conventional, physical act of *stalking* refers to the stealthy and persistent pursuit of a specific person with unwanted and obsessive attention which results in harassment. The person who carries out stalking is referred to as a *stalker* or the *perpetrator*, and the person being stalked is the *victim*. *Cyberstalking* is the form of stalking that occurs in the cyberspace. It is believed that cyberstalking may result in a similar, if not higher, level of threatening as conventional stalking; indeed, these two types of behaviors share a number of common traits [11]:

(a) Both behaviors are characterized by obsessive attention from the perpetrator to the victim, which include monitoring and information gathering.
(b) Both behaviors are mainly driven by the perpetrator's desire and need to gain power, control and influence over the victim.

(c) Both behaviors are *victim-defined*, that is, the level of seriousness of a particular stalking incident is determined by how much intimidation the victim perceives [12].

However, cyberstalking should not be regarded simply as an extension of conventional stalking, but rather a different deviant behaviors in its own form [4]:

1. Firstly, conventional stalking includes some clearly defined, detectable actions such as physically following the victim home, vandalizing victims properties and sending gifts, leaving a physical trail of evidences. Cyberstalking, on the other hand, are much harder to define: Here, a victim may purposely expose private information on online social networks, making monitoring extremely easy; any online user may derive information regarding another's occupation, age, and address with little effort. Furthermore a stalker may also use identity masks, making information gathering unnoticeable.
2. Secondly, in most conventional stalking incidents, the stalker and the victim are within each other's social or physical periphery: Either they have prior relationship (whether real or perceived), or they live or work within relatively close proximity of each other. For cyberstalking, the victim is more often chosen at random and may occur between two people with arbitrary physical distance.
3. Thirdly, cyberstalkers often employ tools and techniques that are unique to the use of the Internet. For example, a number of incidents involve stalkers carrying their deeds using Trojan softwares, email spamming or phishing techniques [5], all of which are challenging technical online threats themselves.

Due to the above reasons, the detection, prevention, and forensics of cyberstalking become considerably more challenging than for conventional stalking. Despite serious research efforts, there has not been major technical advancements on the detection and prevention of cyberstalking. The most widely used methodology is *profiling*: by gathering common characteristics of cyberstalkers and victims (such as age, gender, occupation, etc.), the method aims to identify the likelihood of an individual to be a stalker or be stalked [1]. While profiling provides certain statistical information, it is far from an effective method for prediction and detection. In view that anyone has the potential to cyberstalk, one needs a *behavioral* approach, which focuses on describing behaviors, rather than summarising personal attributes of cyberstalkers.

One of the most comprehensive definitions of cyberstalking was offered by Bocij and McFarlane in [3]:

"A group of behaviours in which an individual, group of individuals or organisation uses information technology to harass one or more individuals. Such behaviour may include, but are not limited to, the transmission of threats and false accusations, identity theft, data theft, damage to data or equipment, computer monitoring and the solicitation of minors for sexual purposes."

The definition entails a large sum of deviant behaviors, which vary by nature and require different countermeasures. As a pilot study, we can only focus on a particular aspect of cyberstalking. In an earlier study [7], Meloy provides a more "strip-down" definition, which asserts cyberstalking as consisting of two major functions

1. the stalker gathers private information of the target to further a pursuit; and
2. the stalker communicate (in real time or not) with the target to implicitly or explicitly threaten or to induce fear.

In this paper we focus on the first function above. We argue that stalkers' gathering of information is the prerequisite and precursor to all cyberstalking behaviors as laid out by Bocij and McFarlane. It is the stalker's ability to collect personal information of the victim that places the victim in the danger of threats and harassment. By appropriate monitoring and control of the gathering of information, one may develop a feasible countermeasure for cyberstalking.

3   A Model for Online Behaviorals

3.1   A Model of User Interactions

People interact in an online social network in a variety of ways, which range from posting profiles, publishing blogs and photos, forming groups, leaving comments, messaging, sending emails or simply clicking `like` button on others posts. To capture this diverse range of interactions, we need an abstraction of cyberspace that serves as a general platform for our investigation. Indeed, we provide a high level description which implicitly embodies all the mentioned activities.

Naturally, an online social network consists of ties between its members; such ties could mean mutual friendship, or the relationship when a person "follows" another which indicates a directed interest from one person to another. Such ties consist of visible links among users of the network and form the basis of the dissipation of information.

However, establishing interpersonal ties are not the only form of interactions. A published message can be viewed by anyone on the web, regardless of whether the viewer has an established tie with the publisher. Similarly, the cyberspace enables communication between two users, who may or may not have a friendship on the online social network. It is the mechanism that allow information sharing through distant parties fundamentally distinguishes physical interaction from online interaction, and hence any model of the cyberspace should take into account not only the visible connections between users, but also interactions among arbitrary users on the network.

With this view in mind, we introduce the following model of an online social network: The carcass of the model consists of a directed network where nodes represent agents (i.e., users). Directed links allows modeling of the relationship when one agent "follows" another. Note that by restricting to directed links, we also do not rule out mutual relations as they can be represented by a pair of directed links. Each agent in this network has the ability to carry out two actions: *posting* messages and *retrieving* messages. Once a message is posted, it can be retrieved by others in the cyberspace. For simplicitly, we assume that there is a universal set of messages that could be posted and hence retrieved over the network and all posted and retrieved messages are taken from this set. Furthermore, we require that an agent should not post and retrieve the same message at the same time.

Definition 1 (Network and agents). A *network* is a directed graph $G = (V, E)$ where V is a set of *nodes* (or *agents*) and $E \subseteq V^2$ is a set of *directed edges* denoting interpersonal ties.Let M be a set of *messages*. An *instance* of G is a pair of functions $s = (\mathsf{post}, \mathsf{retrieve})$, where $\mathsf{post} : V \to 2^M$ is called the *post function* and $\mathsf{retrieve} : V \to 2^M$ is called the *retrieve function*. We further requires that for all $v \in V$, $\mathsf{post}(v) \cap \mathsf{retrieve}(v) = \varnothing$.

An instance $s = (\mathsf{post}, \mathsf{retrieve})$ captures the posting and retrieving behavior of agents in a single time instance: We say that a message $m \in M$ is *posted* by an agent v if $m \in \mathsf{post}(v)$; a message $m \in M$ is *retrieved* by v if $m \in \mathsf{retrieve}(v)$. Hence $\mathsf{post}(v)$ contains all messages in M that are posted by agent v and $\mathsf{retrieve}(v)$ contains all messages that are retrieved by v.

Remark. It may seem that the actions of "posting" and "retrieving" are too restrictive as they only give users limited capabilities in an social network. Nevertheless, they can be used as a high-level abstractions of a wide range of activities. For example, when one person sends an email to another person, this can be seen as a process where the sender posts a message, while the receiver – and only the retriever – retrieves the message. It is therefore not hard to see that this general set up can be used as a simple abstraction of online social networks.

3.2   A Model of Attention

Attention refers to an invested interest from one person to another. It comes often as the result of a desire to know about and establish connection with the target person. The amount of attention an agent a pays to another agent b can be viewed from two perspectives. The first is a micro perspective: the attention a pays to b depends on the posted and retrieved messages of a and b. The second is a macro perspective: the link structure of the network affects attention from a to b. In our model we need to incorporate both types of attentions to truthfully capture the attention one pays to another in the cyberspace.

Firstly, in a micro-level, we measure attention by comparing the set $\mathsf{post}(b)$ against the set $\mathsf{retrieve}(a)$. If $\mathsf{post}(b)$ and $\mathsf{retrieve}(a)$ are similar, then the agent a retrieves mostly the posted messages of b, which naturally implies that a pays attention to the messages posted by b. To measure similarity between two sets, we adopt the well-known *Jaccard distance*, which is the ratio of the size of the intersection of the two sets over the size of their union. Thus the Jaccard distance has a range between 0 and 1 where a higher value implies a higher level of similarity. We define the *message-based attention index* (M-index) from

a to b as the Jaccard distance between retrieve(a) and post(b). A special case is when $a = b$, that is, we need to define the amount of attention a person gives to her own posted messages. We set the M-index from a to a to 1, i.e., a pays full attention to her own posted messages, which is clearly reasonable.

Then, in a macro-level, the second factor for measuring attention is the link structure of the network $G = (V, E)$. Indeed, the M-index only measures an agent's interest in the *information* sent by another agent, which is not necessarily the same as the interest on the target agent herself. For example, an agent a may retrieve blogs posted by another agent b in the hope to monitor a close friend c of b; while no message posted by c is retrieved by a, a's primary intention is to get to know about c. This method of "indirect" information gathering has been shown to pose a great privacy threat on the Internet [8]. We measure this *linked-based attention* using a Markov chain model.

We briefly outline the necessary steps in defining linked-based attention as follows: Given an instance s of the network G, we define an *attention transfer Markov chain* $T_s(a)$ for every node a by taking into account the local network structure of a and neighbors of a. The linked-based attention of a is then given by the stationary distribution vector of this Markov chain.

4   Capturing Stalking-Like Behaviors

As argued in [11], cyberstalkers tend to be emotionally distant loners who is eager to seek the attention and companionship of another. The problem lies in the fact that these individuals often become obsessed or infatuated with the victim, but such a feeling is not reciprocated. Hence it is reasonable to view stalking as *excessive, unreciprocated attention over a lengthy period of time*. We need to make clear of the following notions.

To capture the above notions mathematically, we use a measure for the extend of attention an agent pays to another in the network. The judgement whether an agent i pays obsessive attention to another agent j depends not only on the attention from i to j alone, but also how much attention j generally receives from all other users in the network. At the same time it should also depend on how much attention j pays to i. Furthermore, the type of watchful behavior that we want to measure here happens over a lengthy period of time. Thus, assuming that the network G does not change with time. Given a sequence of instances $s_0, s_1, \ldots$ of G, we accumulate the attention between i and j over all the instances in this sequence. The result is a numerical indication of how likely the agent i pays to the agent j in the network.

5   Discussions on Limitations

This work-in-progress amounts to our first step towards building a mathematical framework of cyberstalking. It is still a speculation that our definition leads to automated technologies for the identification, detection and prevention of cyberstalking. Nevertheless, we believe that by proposing a precise definition, it becomes possible to tackle these problems from a computational perspective.

Given the simplicity of the current model, some natural limitations exist, which we will try to address in our future works.

Justification of the Attention Model. The model of link-based attention using Markov chains provide a powerful measure on a latent intention from one agent to another in the network. Through our initial experiments, we identify a clear indication that cyberstalking behaviors result in a very high level of attention. It is therefore imperative to carry out more theoretical and empirical studies regarding this model to further verify its validity. To accomplish this task, one would need to design appropriate simulation models and scenarios and identify how much different factors (such as network density, level of interactions, network structural properties, etc.) affects the level of attentions among agents. One should also consider verifying this model using real-world data collected from online networks.

Message Types and Meanings. In real life networks, any message carries a meaning, i.e., a piece of information that relates to certain topics or people. For example, an online user normally maintains a profile page which contains important personal information. A stalker in collecting information about the user, typically first attempts to access the user profiles of the victim and friends of the victim. Another possible scenario is when the stalker pays a high level of attention on the topic of interest to the victim, rather than the specific messages. Such intention cannot be precisely described by our model. To deal with this issue, one would consider not only the interpersonal relations among agents, but also semantic relations among messages and topics, as well as attributes of messages.

Harassments and Threats. Cyberstalking in most legal contexts entails a level of intimidation or harassment to the victim. Indeed, such harassment must come from communication between the stalker and the victim. Numerous real-world cases of cyberstalking involve the stalker making certain early contacts (such as sending friend request) to the victim. The definition in this paper, however, merely capture the process of information collection, but not communication between agents. Hence more novel approaches based on cognitive or behavioral science may be employed to provide a useful solution.

Multiple Online Identities. One important property of real life social networks is that a user can hide behind multiple anonymous identities. When stalking a victim, each online identify of the stalker may engage in a collective effort in collecting information, hence making the problem much harder. This also requires enriching the current model with mechanisms that identify user accounts that are associated with the same identity.

Incorporating Historical Records. When stalking a victim, a perpetrator usually not only collect current post of the victim but also retrieve historical messages (old blogs, diaries, photos, etc.). However, the current definition of message-based attention only take into account the current retrieval and posts. This requires enriching the current model with a knowledge base which stores logs to all historical messages posted by users, and updating the attention model by taking into account all historical data.

As discussed above, an automated mechanism for the prediction, detections and forensics of cyberstalking behaviors relies on resolving all the issues above by enriching and validating our model. Nevertheless, we believe that the final outcome will be an important technical breakthrough towards solving the increasingly-significant problem of online stalking and bullying. The current work-in-progress provides us with a clear guideline of how this could be achieved.

References

1. Al-Khateeb, H., Alhaboby, Z., Barnes, J., Brown, A., Brown, R., Cobley, P., Gilbert, J., McNamara, N., Short, E., Shukla, M.: A Practical Guide To Coping With Cyberstalking, 1st edn. National Centre for Cyberstalking Research (NCCR), University of Bedfordshire, Luton (2011). Observation of strains. Infect Dis Ther. 3(1), 35–43. Andrews UK Limited (2015)
2. Barrat, A., Barthelemy, M., Vespignani, A.: Dynamical Processes on Complex Networks. Cambridge University Press, Cambridge (2008)
3. Bocij, P., McFarlane, L.: Online harassment: towards a definition of cyberstalking. Prison Serv. J. **139**, 31–38 (2002)
4. Bocij, P., McFarlane, L.: Cyber stalking: genuine problem or public hysteria? Prison Services J. **140**(1), 32–35 (2002)
5. Fire, M., Goldschmidt, R., Elovici, Y.: Online social networks: threats and solutions. IEEE Commun. Surv. Tutor. **16**(4), 2019–2036 (2014)
6. Liu, J., Moskvina, A.: Hierarchies, ties and power in organisational networks: model and analysis. In: Proceedings of ASONAM, pp. 202–209 (2015)
7. Meloy, J.: Stalking. An old behavior, a new crime. Psychiatr. Clin. North Am. **22**(1), 85–99 (1999)
8. Mislove, A., Viswanath, B., Gummadi, K., Druschel, P.: You are who you know: inferring user profiles in online social networks. In: Proceedings of the Third ACM International Conference on Web Search and Data Mining, WSDM 2010, pp. 251–260 (2010)
9. Moskvina, A., Liu, J.: How to build your network? a structural analysis. In: IJCAI, 2016 to appear
10. Parsons-Pollard, N., Moriarty, L.: Cyberstalking: utilizing what we do know. Vict. Offenders: Int. J. Evid. Based Res. Policy Pract. **4**(4), 435–441 (2009)
11. Pittaro, M.: Cyber stalking: an analysis of online harassment and intimidation. Int. J. Cyber Criminol. (IJCC) **1**(2), 180–197 (2007)
12. Reno, J.: Cyber stalking: A new challenge for law enforcement and industry. United States Department of Justice (1999)

Capability-Aware Trust Evaluation Model in Multi-agent Systems

Tung Doan Nguyen$^{(\boxtimes)}$, Quan Bai, and Weihua Li

School of Engineering, Computer and Mathematical Sciences,
Auckland University of Technology, Auckland, New Zealand
{tung.nguyen,quan.bai,weihua.li}@aut.ac.nz

Abstract. Modeling trust in a real time of dynamic multi-agent systems is important but challenging, particularly when agents frequently join and leave, and the structure of the society may often change. With the increasing complexity of services, some simplified assumptions, e.g., unlimited processing capability, adopted by several trust models have shown their limitations which restrict the application of trust model in real-world situations. This paper attempts to relax the unlimited processing capability assumption of agents by introducing a capability-aware trust evaluation with temporal factor using hidden Markov model. The experimental results show that the approach not only can improve the accuracy of trust computation but also benefit the trust-aware decision making for both individual and agent group context.

Keywords: Multi-agent system · Trust · Composite services · Capability-aware

1 Introduction

In open, dynamic environments of Multi-Agent Systems, agents have high possibility to be exposed to risk when interacting with strangers. Trust has become an essential tool for selecting interaction partners by reducing the uncertainty of interactions, promoting robustness and vitality of diverse social interactions [8]. Several computational trust models have been proposed to address different situations. Many of them assume agent to be rational with unlimited processing capability (UPC), i.e., the performance of agents is not affected by the number of requests [11]. However, the survey [3] also points out that the trustee capability is closely related to timeliness of task completion. It is part of the quality metric for truster agents. The practice has shown the defect of not including agents' capability into trust model [9,11]. The issue exists in many popular trust models, for example, [2,4,10]. It can also lead to reputation damage problem [13].

Recently, the context-aware trust evaluation has been receiving a great attention [5,6]. In [5], the authors tried to connect the feature set of a trustee agent to its performance for computing the trust value. The accuracy of the approach depends on training data and it is not suitable for distributed environment.

© Springer International Publishing Switzerland 2016
R. Booth and M.-L. Zhang (Eds.): PRICAI 2016, LNAI 9810, pp. 771–778, 2016.
DOI: 10.1007/978-3-319-42911-3_65

Taking limited capability into account, the work of Yu et al. [11,12] benefits the trustee agents by improving the request management. However, the approach does not address the quality of trust evaluation for truster agents.

In many practical applications, e.g. composite services, trustee agents can group up to deliver more sophisticated services to truster agents. Single-tasking is not preferable in a group context since it significantly reduces the productivity the group. For example, when one agent is processing requests, another agent can process other tasks rather than waiting. Since the quality of service of a group is closely related to the agent coordination, it is better to consider the agents' capability in trust model. From the perspective of truster agents, single or multi-tasking model of a service is uncertain to most of them. Truster agents do not have to know internal activities within a group to evaluate its trustworthiness. Nevertheless, considering agent or group of agents as an entity with limited processing capability can have a significant impact on the evaluation process. Because the current state (resource states) of a service provider can be used to improve the accuracy of the trust evaluation.

In this paper, we try to address the relation between the number observable requests and the outputs with trustee's capability. So how can we take the agents capability into the trust evaluation model? We find that the hidden Markov model (HMM) can be used to reveal the relation between the temporal states and the capabilities of agents, therefore, can be used for computing trust values and relax the UPC assumption. This paper firstly tries to determine capabilities of agents and then evaluates their trustworthiness based on HMM.

2 Definitions

This work assumes that trustee agents or groups can accept multiple requests while processing other tasks (see Fig. 1). The number of requests can impact the output, i.e., Quality of Service (QoS). A trustee agent delegates a task not only based on reputation value, but also current situation of each trustee candidate based on our proposed trust evaluation method.

Fig. 1. Tasks handling of a trustee agent

Definition 1. *The **capability-aware trust** $T_{(c_i,p_j)}$ represents the evaluation of truster c_i over the trustworthiness of trustee p_j. It is a real value indicating the probability that trustee p_i will produce an expected output when p_j is processing k concurrent tasks, i.e., $T_{(c_i,p_j)} = P(QoS_{d_{p_j}} \geq QoS_{r_{c_i}} | CT_{p_j} = k)$.*

$QoS_{r_{c_i}}$ and $QoS_{d_{p_j}}$ are the expected QoS of c_i and the QoS of delivered service of p_j, respectively. Trust is considered to be subjective; different truster agents can have different trust values over the same trustee agent. Thus, this paper models trust as a conditional probability of producing the expected output when given the certain number of requests. The trust value is in 0 and 1, which stands for the most untrustworthy and the most trustworthy values, respectively.

Definition 2. *The **historical record** H_{p_i} of agent p_i is a set of transaction (see Definition 4) records, i.e., $H_i = \{tran_{i_1}, tran_{i_2}, tran_{i_3}, \dots\}$, which is kept at the local database of agent.*

Each transaction contains public verifiable information from starting request to rating stage. This assumption can boost the confidence of evidence collecting, especially in distributed environments.

Definition 3. *A **request** or a task is a contract between a truster c_i and a trustee p_j respecting a service. A task can be represented by a double $req_{c_i, p_j} = \{c_i, p_j, st, t_{dl}\}$, where st is service type and t_{dl} is the deadline of the task.*

Definition 4. *The **transaction record** is data of one transaction session, represented by a 3-tuple $tran_{(c_i, p_j)} = \{t_r, t_d, r_{c_i, p_i}\}$, where t_r, t_d, r_{c_i, p_i} represent the timestamp of the request, timestamp of the delivered service, and the rating of c_i (see Definition 5), respectively.*

Definition 5. *The **rating** r_{c_i, p_i} is a binary value of either 0 or 1, indicating the unsatisfied and satisfied evaluation of the output of p_i, respectively.*

Since the task includes both the QoS and deadline, truster agents satisfy when both conditions are met:

$$r_{c_i, p_j} = \begin{cases} 0 & \text{if } QoS_{r_{c_i}} > QoS_{d_{p_j}} \text{ or } t_d > t_{dl} \\ 1 & \text{if } QoS_{r_{c_i}} \le QoS_{d_{p_j}} \text{ and } t_d \le t_{dl} \end{cases} \tag{1}$$

3   The Capability-Aware Trust Evaluation

To take the capability into account, this paper adopts the following assumptions: (1) The requests for services from different trustee agents are observable by other trustee agents; (2) The historical records of agents are visible and verifiable; (3) The internal activities of an agent (or agent group) are unknown to truster agents, all coordination are handled privately by service providers.

3.1   Capability Measure of Trustee Agents

To find the capability of a trustee agents, this paper utilizes the available information obtained from historical transactions including ratings and timestamps. A truster agent can observe the state of a trustee agent at the time of making request and the time of delivering service. A truster c_i can calculate the number of concurrent tasks (CT) of trustee p_i when accepting a request at the time tr_k:

$$CT_{tr_k} = 1 + \left|\{req_m : tr_k \in [tr_m, td_m]\}\right|, \qquad (2)$$

where req_m is a request of the mth transaction of H_{p_i}; td_m is the delivery timestamp of the mth transaction. Because the CT is bounded, i.e., $0 < CT_{tr_n} \leq Max\{CT_{tr_n}\}$, we can use the Bayesian discrete probability [1] to estimate the probability of getting good output given the number of concurrent tasks. We compute the posterior probability according to Bayes' theorem:

$$P(r = 1 \mid CT = k) = \frac{P(CT = k \mid r = 1) \cdot P(r = 1)}{P(CT = k)}, \qquad (3)$$

where prior probability $P(r = 1)$ is the probability of $r = 1$ before $CT = k$ is observed. $P(r = 1 \mid CT = k)$ is the posterior probability, is the probability of $r = 1$ given $CT = k$, i.e., after $CT = k$ is observed. The prior probability $P(CT)$ can be calculated by using Eq. 4:

$$P(CT = k) = \frac{\left|\{tran_j \in p_i : CT = k\}\right|}{N}, \qquad (4)$$

where N is total number of transactions in historical records. From the calculated CT_k with the associated outcome of the kth transaction, we have a set of prior probability. $P(r \mid CT)$ can be interpreted as for a given number of concurrent tasks, the probability of the outcome r is $P(r \mid CT)$.

3.2   Capability-Aware Trust Value Calculation

To calculate the capability-aware trust value, firstly, the paper estimates the capability of a trustee from collective of evidence. We also use the Δ_t denotes the forgetting factor for outdated evidence, $\delta = \lambda^{\Delta_t}$. Having $\lambda = 1$ is equivalent to not having Δ_t, i.e., nothing is forgotten. Whereas, having $\lambda = 0$ results in only the last feedback value to be counted and all others to be completely forgotten. This time discount factor can assure that the outdated will be omitted from evaluation. A valid transaction evidence satisfies the following:

$$tran_i = \begin{cases} 1 & \text{if } \delta \geq 0.5 \\ 0 & \text{otherwise} \end{cases}, \qquad (5)$$

where 0 and 1 represent invalid and valid transaction record to use. As truster agents can observe the state of trustee agent the time of making the request but can not sure about the result at the time of delivering, we can model the trust evaluation using HMM with the memory of size 1 (first-order Markov model) [7].

Figure 2 shows the model of HMM of current problem using Trellis diagram. To a truster agent c_i, state S_k of trustee p_j is the number of $(k - 1)$ concurrent tasks when a new request is made. $x = \{x_1, x_2, \ldots, x_n\}$ is a sequence of

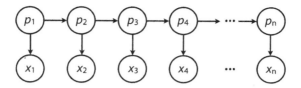

Fig. 2. HMM for capability-aware trust evaluation

observed output. Like Markov chain, edges capture conditional independence. For example, x_2 is conditionally independent of everything else given s_2.

As mentioned in Subsect. 3.1, the probability of the output is based on the processing requests of trustee agent p_i. We can incorporate the probability with a two-step trust evaluation. Firstly, it uses a traditional approach to evaluating the trustworthiness of an agent based on the rating evidence. Secondly, the value will be adjusted based on the situational evaluations by considering the capability and current processing tasks.

Suppose that at the time of evaluation, trustee agent p_j has k ongoing requests. The number of concurrent tasks will increase by one if p_j accepts the request from c_i. In Subsect. 3.1, trustee agent c_i has the table of transition state of p_j at time "$t + 1$" (see Table 1).

Table 1. Transition table for states of trustee agent

	s_i	s_j	s_k
s_i	0.2	0.6	0.2
s_j	0.3	0.1	0.6
s_k	0.5	0.3	0.2

	$r = 0$	$r = 1$
s_i	0.3	0.7
s_j	0.6	0.4
s_k	0.2	0.8

The problem can be described as evaluation problem of HMM, given the observation sequence x and a formal HMM

$$\lambda = (A, B, \pi), \tag{6}$$

where A is a transition array, storing the probability of state j following state i.

$$A = \{p_{ij} | p_{ij} = P(CT_{t+1} = j | CT_t = i)\} \tag{7}$$

and B denotes the observation array, storing the probability of observation k being produced from the state j, independent of t:

$$B = \{bi(k)|, bi(k) = P(xt = vk | qt = si)\} \tag{8}$$

How do we compute $P(x|\lambda)$, i.e., the probability of the observation sequence given the model. We can use the equation below to calculate the probability of the observations x for a specific state sequence state S:

$$P(x|Q, \lambda) = \prod_{t=1}^{P} (x_t|q_t, \lambda) \tag{9}$$

Finally, we can calculate the probability of the observation given the model as:

$$P(O|\lambda) = \sum_{Q} P(O|Q, \lambda)P(Q|\lambda) \tag{10}$$

However, due to the fact that our model is first-order HMM. So we can obtain the capability-trust evaluation with:

$$T(c_i|S_{p_j} = k) = \sum p(S_i|S_k)p(r = 1|S_i) \tag{11}$$

4   Experimental Results

The experiment evaluates the trust values of providers with different timestamps and states. However, we use agent groups in the experiments for better simulating the multi-tasking activities. Ten composite services with different profiles were created. All groups provide the same service type to simplify the experimental settings. We compared the approach with a traditional probability-based trust evaluation in two different scenarios: (1) trustee groups with a handleable number of requests, and (2) trustee groups with an excessive number of requests.

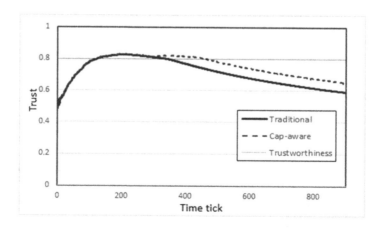

Fig. 3. Capability-aware vs. a traditional trust values

In the first experiment, the initial phase set agents with some transactions; the number of requests at a time was set to be handleable. We then monitor the trust value of each entity in case of no incoming request to see the effect of the temporal factor and trust. Figure 3 shows the result of two approaches in two phases. At first, under the handleable number of requests, both trustee groups produce the expected outputs. Thus, there is no significant difference between ours and the traditional approach. However, when there is no incoming request, the HMM can keep the trust value relatively close to the real trustworthiness.

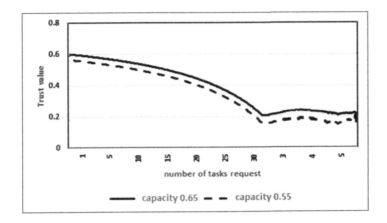

Fig. 4. Capability-aware trust value with varied number of requests

In the second experiment, we varied the number of requests each time step by increasing and then decreasing the number to see the effect of the model. Figure 4 illustrates the trust values of a truster agent observes two different trustee agents with different capability profiles. The results show that when the number of concurrent tasks exceeded the handleable capability, the trust values to both agents decreased, and the one with higher capability are perceived as more trustworthy. Interestingly, when the number of requests decreased to a handleable value, the trust value started to increase. This perfectly demonstrates the effect of the visible requests on capability-aware trust.

The above results also show the intuition that in practical situations, the reputation may be decreased because of performance affected by the excessive requests. However, the trust value for the trustee starts to increase when the completed tasks are released. Namely, lower reputation agents do not always mean lower expectation of the output.

5   Conclusions

The paper proposes a trust model considering agents' limited capability to relax the UPC assumption by using HMM approach. The model can produce promising results of trust evaluation for truster agents at the time of making requests. Our future work is set to improve the experiments and discover different QoS influencing factors for the model rather than using only the number of requests.

References

1. Gelman, A., Carlin, J.B., Stern, H.S., Rubin, D.B.: Bayesian Data Analysis, vol. 2. Taylor & Francis, Boca Raton (2014)
2. Huynh, T.D., Jennings, N.R., Shadbolt, N.R.: An integrated trust and reputation model for open multi-agent systems. Auton. Agent. Multi-Agent Syst. **13**(2), 119–154 (2006)
3. Josang, A., Ismail, R., Boyd, C.: A survey of trust, reputation systems for online service provision. Decis. Support Syst. **43**(2), 618–644 (2007). Emerging Issues in Collaborative Commerce
4. Jsang, A., Haller, J.: Dirichlet reputation systems. In: The Second International Conference on Availability, Reliability and Security, ARES 2007, pp. 112–119, April 2007
5. Liu, X., Datta, A.: Modeling context aware dynamic trust using hidden Markov model. In: Twenty-Sixth AAAI Conference on Artificial Intelligence (2012)
6. Naseri, M., Ludwig, S.A.: Evaluating workflow trust using hidden Markov modeling and provenance data. In: Liu, Q., Bai, Q., Giugni, S., Williamson, D., Taylor, J. (eds.) Data Provenance and Data Management in eScience. SCI, vol. 426, pp. 35–58. Springer, Heidelberg (2013)
7. Rabiner, L.R., Juang, B.-H.: An introduction to hidden Markov models. IEEE ASSP Mag. **3**(1), 4–16 (1986)
8. Sabater, J., Sierra, C.: Social regret, a reputation model based on social relations. SIGecom Exch. **3**(1), 44–56 (2001)
9. Sen, S.: A comprehensive approach to trust management. In: Proceedings of the International Conference on Autonomous Agents and Multi-agent Systems, AAMAS 2013, pp. 797–800. International Foundation for Autonomous Agents and Multiagent Systems, Richland (2013)
10. Wang, Y., Singh, M.P.: Evidence-based trust: a mathematical model geared for multiagent systems. ACM Trans. Auton. Adapt. Syst. **5**(4), 14:1–14:28 (2010)
11. Yu, H., Miao, C., An, B., Leung, C., Lesser, V.R.: A reputation management approach for resource constrained trustee agents. In: Proceedings of the Twenty-Third International Joint Conference on Artificial Intelligence, IJCAI 2013, pp. 418–424. AAAI Press (2013)
12. Yu, H., Miao, C., An, B., Shen, Z., Leung, C.: Reputation-aware task allocation for human trustees. In: Proceedings of the International Conference on Autonomous Agents and Multi-agent Systems, AAMAS 2014, pp. 357–364. International Foundation for Autonomous Agents and Multiagent Systems, Richland (2014)
13. Yu, H., Shen, Z., Leung, C., Miao, C., Lesser, V.: A survey of multi-agent trust management systems. IEEE Access **1**, 35–50 (2013)

A Concurrent Multiple Negotiation Mechanism Under Consideration of a Dynamic Negotiation Environment

Lei Niu[(✉)], Fenghui Ren, and Minjie Zhang

School of Computing and Information Technology, University of Wollongong,
Wollongong, NSW, Australia
ln982@uowmail.edu.au, {fren,minjie}@uow.edu.au

Abstract. Concurrent Multiple Negotiation (CMN) mechanism is necessary for agents to achieve agreements in multiple negotiations, and it has become a very important research topic in multi-agent systems in recent years. However, in the open and dynamic negotiation environment, negotiations may be dynamically and concurrently initialized or terminated during the process of other existing negotiations. Therefore, how to process dynamic CMN becomes a serious challenge in agent negotiation research. The motivation of this paper is to propose an adaptive mechanism for handling dynamic CMN by considering the possible changes of concurrent negotiations. First, a formal mechanism for modeling and representing dynamic CMN is presented. Then, a novel Colored Petri Net-based CMN protocol for processing CMN with unexpected negotiation changes is presented. We also demonstrate the performance and procedure of the proposed approach in handling the dynamism of negotiations in CMN, and the experimental results show that the proposed approach can effectively handle unexpected changes in the CMN dynamically, and successfully lead the CMN to agreements.

1 Introduction

Concurrent Multiple Negotiation (CMN) indicates the situation that a series of related negotiations are conducted in parallel or concurrently by an agent. These concurrent negotiations which may start and end at different time are somehow related and impact each other, and the outcomes of these negotiations will contribute together for an agent to achieve its overall negotiation goal [4]. By comparing with single goal negotiation [7,10,11], CMN is trying to reach a series of agreements on concurrently processed negotiations with different goals, and these concurrent negotiations are somehow related and the success of their individual goals will contribute to the success of a global goal of the series of negotiations.

For instance, a customer intends to loan from a banker and buy a house with the loan, thus, he/she has to negotiate with the banker and the house seller. In this scenario, these two negotiations have to be performed concurrently. Otherwise, the customer will lose his/her utility since whether the customer can

© Springer International Publishing Switzerland 2016
R. Booth and M.-L. Zhang (Eds.): PRICAI 2016, LNAI 9810, pp. 779–792, 2016.
DOI: 10.1007/978-3-319-42911-3_66

afford a house depends on the negotiation result with the banker. If the loan is lower than the house's price, the customer does not have enough money to buy the house, while if the loan is higher than the house's price, the customer has to pay extra interest. Both situations will cause the loss of utility to the customer.

Through our studies, we notice that in the real world, agents probably make changes of negotiations during negotiating (i.e. uncertainties when negotiations will be added into or removed from an existing CMN). Therefore, achieving CMN when a negotiation environment is open and dynamic has a great significance and is well worth of study.

In dynamic CMN, dependency relationships will be altered with the changing of negotiations, and the change of negotiations with dependency relationships will be the chain reaction when a negotiation environment is dynamic. Therefore, how to conduct the CMN with changing of negotiations is a big challenge. On the other hand, achieving concurrency, which is another crucial feature of CMN, would also meet a serious challenge by considering a dynamic negotiation environment. CMN has been an active research topic in recent years, and handling CMN by the consideration of a dynamic environment is one of the most challenging research issues.

According to the number of negotiation goals, negotiation can be classified into two levels, which are single negotiation level and multi-negotiation level. Most of the existing researches related to dynamic negotiations focus on single negotiation level [9,13]. Based on our knowledge, limited work on the multi-negotiation level with consideration of a dynamic negotiation environment has been presented by researchers.

In the previous work, we proposed a CMN model and introduced the formal definitions to handle CMN. In order to effectively address the challenging problems and overcome the limitations of existing work in dynamic CMN, this paper proposes a mechanism for handling dynamic CMN based on the previous work. The contributions of this paper are that: (1) a mechanism for handling dynamic CMN with consideration of three different dynamic negotiation scenarios (i.e. add a single negotiation, remove a single negotiation and add & remove negotiations) is proposed; and (2) an innovative negotiation protocol for handling dynamic CMN with changing multiple negotiations by employing Colored Petri Net (CPN) is presented.

The rest of this paper is organised as follows. Section 2 gives a mechanism of CMN including formal definitions and the basic model of CMN. A mechanism of dynamic CMN is presented in Sect. 3. Section 4 proposes a CPN-based protocol for dynamic CMN, which is able to handle the scenarios of changing multiple negotiations. Section 5 gives the experimental results and analysis. Section 6 presented related work and Sect. 7 concludes the paper and outlines the future work.

2   Mechanism of CMN

2.1   Definitions on CMN

Definition 1 *(Concurrent Multiple Negotiation (CMN)). A CMN is defined as a set* $\mathbb{N} = \{A_0, \cdots, A_i, \cdots, A_n\}$ $(i \geq 0)$, *where* A_i *indicates an individual negotiation in CMN with one particular single negotiation goal.*

A CMN describes a situation that an agent performs a number of negotiations concurrently where each negotiation has own negotiation goal, negotiation opponent, negotiation issues. However, these separated negotiations might be somehow related and might impact each other towards a global goal of the CMN. In order to describe the relationships of negotiations, the dependency of negotiations is defined as follows.

Definition 2 *(Dependency). For Negotiations* $A_i, A_j \in \mathbb{N}$, $A_i \propto A_j$ *indicates that Negotiation* A_j *depends on Negotiation* A_i *and Negotiation* A_i *must start before Negotiation* A_j *in each multi-negotiation round (see* Definition 3*). Dependency has two properties and one lemma as follows.*

Property 1 (Unidirectionality). For $\forall A_i, A_j \in \mathbb{N}$, *if* $A_i \propto A_j$ *is held, then* $A_j \propto A_i$ *is not held.*
This property indicates that Negotiation A_i *and* A_j *cannot depend on each other simultaneously.*

Property 2 (Transitivity). For $\forall A_i, A_j, A_k \in \mathbb{N}$, *it holds that:*

$$A_i \propto A_j, A_j \propto A_k \Rightarrow A_i \propto A_k$$

This property indicates that the dependency between multiple negotiations can be transferred.

Lemma 1. For $\forall A_i, A_j, A_k \in \mathbb{N}$, *it holds that:* $A_i \propto A_j$, $A_j \propto A_k \nRightarrow A_k \propto A_i$.
proof. *According to Property 2, it holds that* $A_i \propto A_j$, $A_j \propto A_k \Rightarrow A_i \propto A_k$ *and based on Property 1, if* $A_i \propto A_k$ *is held, then* $A_k \propto A_i$ *is not held, so Lemma 1 is held.*

Based on the dependency of two negotiations, the connection between a series of related negotiations in a CMN can also be defined, and we name such a connection as multi-negotiation round.

Definition 3 *(Multi-Negotiation Round (MNR)). A MNR is defined as a set* $\mathbb{R}_i = \{r_{i,0}, \cdots, r_{i,j}, \cdots, r_{i,k_i-1}\}$, *where* $\mathbb{R}_i \neq \varnothing$, $\bigcup_{i=0}^{l-1} \mathbb{R}_i = \mathbb{N}$, $|\mathbb{R}_i| = k_i$ *and* l *is the total number of MNRs in a CMN.* $r_{i,j-1}$ *indicates the* j*th negotiation (i.e., negotiation* A_k*) in a MNR* $\mathbb{R}_i$.

In a MNR $\mathbb{R}_i = \{r_{i,0}, \cdots, r_{i,j}, \cdots, r_{i,k_i-1}\}$, it satisfies $r_{i,0} \propto \cdots \propto r_{i,j} \propto \cdots \propto r_{i,k_i-1}$, and there is no such $r_{i,j}$ that for $\forall r_{i,j} \in \mathbb{R}_i$, $r_{i,j} \propto r_{i,0}$ and $r_{i,k_i-1} \propto r_{i,j}$ hold.

In a MNR $\mathbb{R}_i$, if all involved negotiations, i.e. $r_{i,j} \in \mathbb{R}_i$, reach a successful negotiation outcome, the MNR $\mathbb{R}_i$ is success. Obviously, the success of a CMN is based on the success of all involved MNR $\mathbb{R}_i$, then the global goal of a CMN is defined as follows.

Definition 4 *(Multi-Negotiation Goal (MNG)). The MNG of a CMN is classified as Complete Success Goal, Partial Success Goal and No Success Goal based on the expected success on the number of $\mathbb{R}_i$.*

- *Complete Success Goal: The CMN achieves the Complete Success Goal if all $\mathbb{R}_i$ in a CMN are success.*
- *Partial Success Goal: The CMN achieves a Partial Success Goal if not all but at least one $\mathbb{R}_i$ in a CMN is success.*
- *No Success Goal: The CMN achieves the No Success Goal if none of $\mathbb{R}_i$ involved in a CMN is success.*

2.2 Basic Model of CMN

• Graph Representation of CMN

Definition 5 *(Graph Representation). A CMN $\mathbb{N}$ can be represented by a directed graph $G = (V, E)$, where V indicates vertexes in G ($V = \mathbb{N}$), and E indicates dependencies between negotiations in $\mathbb{N}$. For any two Negotiations A_i and A_j, if $A_i \propto A_j$, $e_{ij} = (A_i, A_j) \in E$.*

• Logic Representation of CMN

Definition 6 *(Logic Representation). Let $L(A_i) = \{True, False, Unsure\}(i \geq 0)$ be logic representation of Negotiation A_i, where $L(A_i) = true$ indicates that Negotiation A_i is success, $L(A_i) = false$ indicates that Negotiation A_i is failure and $L(A_i) = unsure$ indicates that Negotiation A_i is still ongoing.*

Definition 7 *(Logic Conjunction). Let $L(\mathbb{R}_i)$ be the logic representation of a MNR $\mathbb{R}_i$, then $L(\mathbb{R}_i) = \sqcap_{\forall A_j \in \mathbb{R}_i} L(A_j)$ ($i \geq 0, j \geq 1$). $L(A_i) \sqcap L(A_j)$ is logic representation of $A_i \propto A_j$ where $L(A_i) \sqcap L(A_j)$ indicates that a conjunction logic relationship of $A_i \propto A_j$. $L(A_i) \sqcap L(A_j)$ has the same truth value table as $L(A_i) \wedge L(A_j)$ (logic "and"), but the symbol "$\sqcap$" does not satisfy associativity and commutativity.*

Definition 8 *(Logic Disjunction). Let $L(\mathbb{N}) = \bigvee_{\forall \mathbb{R}_i \in \mathbb{N}} L(\mathbb{R}_i)$ ($i \geq 0$) indicate the logic representation of a CMN $\mathbb{N}$, where "$\bigvee$" indicates disjunction of MNR $\mathbb{R}_i$ in a CMN $\mathbb{N}$. The symbol "$\bigvee$" and logic symbol "$\vee$" (logic "or") have the same truth value table.*

Based on the logic representations introduced above, the possibility of a successful CMN $\mathbb{N}$ can be estimated as follows.

Let $V(A_i)$ indicate the success possibility of Negotiation A_i, and $V(A_i)$ can be calculated by Eq. (1) based on A_i's logic representation.

$$V(A_i) = \begin{cases} 0 & \text{if } L(A_i) = False, \\ 1 & \text{if } L(A_i) = True, \\ x \in (0,1) & \text{if } L(A_i) = Unsure. \end{cases} \tag{1}$$

Based on Eq. (1), the possibility of a successful MNR $\mathbb{R}_i$ and a successful CMN $\mathbb{N}$ can be calculated by Eqs. (2) and (3) as follows.

$$V(\mathbb{R}_i) = \prod_{\forall A_j \in \mathbb{R}_i} V(A_j) \ (i \geq 0, j \geq 1) \tag{2}$$

$$V(\mathbb{N}) = \frac{\sum_{\forall \mathbb{R}_i \in \mathbb{N}} V(\mathbb{R}_i)}{l} \ (i \geq 0) \tag{3}$$

Let $\Omega_{\mathbb{N}} \in [0,1]$ donate the value of multi-negotiation goal of a CMN defined in Definition 4, where $\Omega_{\mathbb{N}} = 1$ indicates the Complete Success Goal, and $\Omega_{\mathbb{N}} \in (0,1)$ indicates the Partial Success Goal, and $\Omega_{\mathbb{N}} = 0$ indicates the No Success Goal. The outcome of a CMN $\mathbb{N}$ is defined as follows.

$$Out(\mathbb{N}) = \begin{cases} success & \text{if } U(\mathbb{N}) \geq \Omega_{\mathbb{N}}, \\ failure & \text{if } U(\mathbb{N}) < \Omega_{\mathbb{N}}, \end{cases} \tag{4}$$

where $\Omega_{\mathbb{N}}$ should be specified by the agent before the CMN $\mathbb{N}$ starts.

- **Utility Calculation of CMN**

In this subsection, a method for utility calculation of a CMN is presented based on the utility calculations on MNR $\mathbb{R}_i$ as follows.

Let $U(A_i)$ be the utility of Negotiation A_i. Based on the definition of $V(A_i)$, the utility of a MNR $\mathbb{R}_i$ can be calculated as

$$U(\mathbb{R}_i) = \sum_{\forall A_j \in \mathbb{R}_i} \left(\omega_j \times U(A_j) \times V(A_j) \right) \ (i \geq 0, j \geq 1), \tag{5}$$

where $U(\mathbb{R}_i) \in [0,1]$, and $\omega_j \in [0,1]$ represents an agent's preference of Negotiation A_j, where $\omega = (\omega_0, \cdots \omega_j, \cdots \omega_{n-1})$, and $\sum_{j=0}^{n-1} \omega_j = 1$.

Based on the value of $V(\mathbb{R}_i)$ and the value of $U(\mathbb{R}_i)$, the utility of a CMN $\mathbb{N}$ can be calculated as follows.

$$U(\mathbb{N}) = \frac{\sum_{\forall \mathbb{R}_i \in \mathbb{N}} \left(U(\mathbb{R}_i) \times V(\mathbb{R}_i) \right)}{l}, \tag{6}$$

where $U(\mathbb{N}) \in [0,1]$ and $i \geq 0$.

2.3   Connection Between CPN and CMN

Colored Petri Net (CPN) is a language for the modelling and validation of systems in which concurrency, communication, and synchronisation play a major role [5]. A CPN model of a system is an executable model representing the states of the system and the events (transitions) that can cause the system to change states [6]. A CPN is defined by a nine-tuple [6], where $CPN = (P, T, A, \Sigma, V, C, G, E, I)$. P is the set of places, and T indicates the set of transitions. A is the set of arcs and $A \subseteq P \times T \cup T \times P$. Σ is a finite non-empty colour set, and V shows a set of variables. C indicates the set of colour set functions, which assigns a colour set to each place, and $C : P \rightarrow \Sigma$ and $C(p) \in \Sigma$. G is a guard function set, where $Type[G(t)] = Bool$. E and I are arc expression function set and initialisation function set, respectively.

In the basic CMN model, transitions represent negotiations, and places represent states of negotiations. The inputs and outputs of negotiations are shown by arc directions, and Token $1'(A, m)$ indicates one offer from the mth round of Negotiation A (i.e., enable transition t_A).

In dynamic CMN, any single negotiation could be added into or removed from the original CMN. Therefore, how to handle dynamic CMN based on the basic CMN model is the major problem that this paper tries to solve.

3   Mechanism of Dynamic CMN

In this section, we introduce the mechanism of dynamic CMN. The proposed approach is able to capture the dynamic changes of CMN, and allows agents to add or remove negotiations during the process of CMN.

3.1   Graph Updating

- **Adding Negotiation A^***

    When a new negotiation needs to be added into a CMN, the corresponding MNRs will be changed as well. Figure 1 shows the graph updating of adding Negotiation A^* whose dependency relationships will be $A_0 \propto A^* \propto A_1$.

Fig. 1. Adding Negotiation A^*

In Fig. 1, two new Edges (A_0, A^*) and (A^*, A_1) will be added, and Edge (A_0, A_1) needs to be removed in the meantime. The reason for removing Edge (A_0, A_1) is that the dependency relationship of $A_0 \propto A_1$ satisfies the property of *Transitivity* which is *Property 2* in Definition 2 (see Sect. 2.1), so it is not necessary to keep the Edge (A_0, A_1) after the creation of Edges (A_0, A^*) and (A^*, A_1).

- **Removing Negotiation A^***
  Figure 2 shows the graph updating of removing Negotiation A^*.

Fig. 2. Removing Negotiation A^*

In Fig. 2, Edges (A_0, A^*) and (A^*, A_1) will be removed, and Edge (A_0, A_1) needs to be added in the meantime. Based on the property of *Transitivity* in Definition 2 (see Sect. 2.1), the dependency relationship $A_0 \propto A_1$ holds. After removing the dependency $A_0 \propto A^*$ and $A^* \propto A_1$, the other existing dependency relationships should not be impacted and Edge (A_0, A_1) is added to indicate other existing dependency relationships.

3.2 Preference Updating

- **Adding Negotiation A^*** (preference is ω_{A^*})
  In a CMN $\mathbb{N} = \{A_1, \cdots, A_j, \cdots, A_n\}$, preferences of all negotiations are $\omega = (\omega_1, \cdots, \omega_j, \cdots, \omega_n)$, and $\sum_{j=1}^n \omega_j = 1$. We assume that new CMN $\mathbb{N}' = \{A_1, \cdots, A_j, \cdots, A_n, A^*\}$, where the preferences of all negotiation are $\omega' = (\omega'_1, \cdots, \omega'_j, \cdots, \omega'_n, \omega'_{A^*})$, and ω_{A^*} is specified by the agent during adding Negotiation A^*. After adding Negotiation A^*, new weighting for each existing negotiations can be updated as follows.

$$\omega'_j = \frac{\omega_j}{1 + \omega_{A^*}} \qquad (7)$$

- **Removing Negotiation A^*** (preference is ω_{A^*})
  In a CMN $\mathbb{N} = \{A_1, \cdots A_j, \cdots A_n, A^*\}$, the preferences of all negotiations are $\omega = (\omega_1, \cdots \omega_j, \cdots \omega_n, \omega_{A^*})$. We assume that after removing an existing Negotiation A^*, the weighting distribution among the remaining negotiations will not be impacted, and the preference of the new CMN $\mathbb{N}$ is $\omega' = (\omega'_1, \cdots, \omega'_j, \cdots \omega'_n)$. It is held that:

$$\omega'_j = \frac{\omega_j}{1 - \omega_{A^*}} \qquad (8)$$

3.3 CPN Updating

As a part in CMN, CPN is updated as well to achieve dynamic CMN. In detail, nine tuples in a CPN are updated based on the Graph Updating. Due to the limitation of pages, this part is omitted. Readers can refer to the related work of CPN [5,6].

4   CPN-Based Protocol for Dynamic CMN

In this section, we introduce a CPN-based protocol of dynamic CMN which is represented by Algorithm 1.

Algorithm 1. CPN-based Protocol of Dynamic CMN

Input: a CPN $C_{\mathbb{N}} = (P, T, A, \Sigma, V, C, G, E, I)$ based on the CMN $\mathbb{N} = \{\mathbb{R}_1, \cdots, \mathbb{R}_i, \cdots, \mathbb{R}_l\}$, utility function U, preferences of negotiations ω.
Output: success or failure.
1: Updating the graph,preference and CPN;
2: **while** at least one MNR $\mathbb{R}_i$ is finished **do**
3:     calculate $V(\mathbb{N})$ and $U(\mathbb{N})$;
4:     **if** $U(\mathbb{N}) < \Omega_{\mathbb{N}}$ **then**
5:         keep executing other unfinished MNRs;
6:     **else if** $U(\mathbb{N}) \geq \Omega_{\mathbb{N}}$ **then**
7:         terminate whole CPN and quit;
8:         **return** success;
9:     **end if**;
10: **end while**;
11: **while** whole CPN is not completed **do**
12:     calculate $V(\mathbb{N})$ and $U(\mathbb{N})$;
13:     **if** $U(\mathbb{N}) \geq \Omega_{\mathbb{N}}$ **then**
14:         **return** success;
15:     **else if** $U(\mathbb{N}) < \Omega_{\mathbb{N}}$ **then**
16:         **return** failure;
17:     **end if**;
18: **end while**.
19: **return** success or failure.

The protocol illustrates the process of handling dynamic CMN when negotiations are added or removed. The inputs of Algorithm 1 are a CPN $C_{\mathbb{N}} = (P, T, A, \Sigma, V, C, G, E, I)$, utility function U, and preferences of negotiations ω, and the output is the result of updated CMN (i.e. success or failure). When the changes of negotiations happen, the algorithm updates graph representations, preferences of negotiations, the CPN (Line 1). It computes $V(\mathbb{N})$ and $U(\mathbb{N})$ as soon as one MNR $\mathbb{R}_i$ has been completed (Lines 2–3). If the utility of a CMN $\mathbb{N}$ is less than MNG $\Omega_{\mathbb{N}}$, the agent keeps executing CPN of other unfinished MNRs while if the utility of a CMN $\mathbb{N}$ is no less than MNG $\Omega_{\mathbb{N}}$, the agent terminates whole CPN and quits from the CMN (Lines 4–10). The algorithm shows that if a completed MNR $\mathbb{R}_i$ achieves MNG $\Omega_{\mathbb{N}}$, then it outputs "success" for the whole CMN $\mathbb{N}$. Because there is no need to execute other unfinished MNRs, it can improve efficiency in some extent. Then the agent keeps on calculating $V(\mathbb{N})$ and $U(\mathbb{N})$ until the whole CPN is completed. It compares the utility of a CMN $\mathbb{N}$ with MNG $\Omega_{\mathbb{N}}$ to make different decisions (Lines 11–18). In the end, it returns the result of CMN.

5   Experiment

5.1   Experimental Settings

Suppose that Agent a concurrently performs more than one negotiation with different agents. In the experiment, a CMN structure is shown in Fig. 3 and

Agent a's preferences for different negotiations are generated randomly. For simplification, each single negotiation is a bilateral single-issue negotiation. Agent's negotiation deadlines are selected randomly between the 10th and 20th round.

A single-issue negotiation model [8] is employed to process each single negotiation and every agent randomly chooses a concession strategy from Conceder, Linear and Boulware strategies [12]. Utility function U_o^a for each single negotiation is described as follows, where $U^a(o)$ indicates the Agent a's utility based on the Offer o.

$$U^a(o) = \frac{\text{Reserved Offer} - o}{\text{Reserved Offer} - \text{Initial Offer}} \tag{9}$$

In the experiment, we take the CMN in Fig. 3 as the test case. The CMN contains five negotiations, i.e. $\mathbb{N} = \{A_0, A_1, A_2, A_3, A_4\}$ and $\mathbb{R}_0 = \{A_0, A_1\}$, $\mathbb{R}_1 = \{A_0, A_2, A_4\}$, and $\mathbb{R}_2 = \{A_3, A_4\}$.

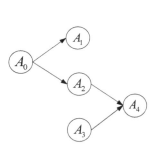

Fig. 3. Basic CMN Structure in the Experiment

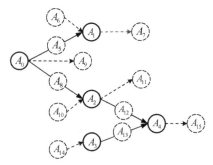

Fig. 4. CMN's Final Utilities and Number of Rounds in the Adding Scenarios

The parameters for each single negotiation are the negotiation *preference*, participant's negotiation *deadlines*, and participant's initial and reserved offers. The experiment was conducted for 100 times, where all parameters were generated randomly in each time. In order to test the performance of the proposed approach in handling dynamic CMNs, the experiment was conducted in following scenarios.

1. Adding/Removing a Single Negotiation

In the experiment settings, all possible adding and removing cases of the existing CMN in Fig. 3 were considered. In the adding scenario, most possible cases for adding a single Negotiation $A_i(i \in [5, 15])$ were considered. In the removing scenario, all cases for removing a single Negotiation $A_i(i \in [0, 4])$ were considered as well. The possible CMN by adding/removing a single negotiation is shown in Fig. 4, where the broken circle indicates a possible negotiation added into the existing CMN.

2. Adding and Removing Negotiations

There are a number of cases in adding & removing scenario for basic CMN. Due to page limitation, we carried only one case study in this paper, where Negotiation A_7 will be added into CMN in the 10th negotiation round, and Negotiation A_3 will be removed from CMN in the 15th negotiation round (see Fig. 4).

5.2   Experimental Results

In Fig. 5, the x-axis indicates three dynamic scenarios, and the y-axis indicates percentage of each results in the experiment. From the figure, we can see that the percentage of no success are all below 15%. Figure 5 indicates that the proposed approach is able to handle dynamic CMN in three scenarios to achieve different MNGs (see Definition 4).

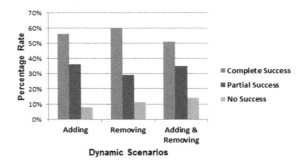

Fig. 5. The Percentage Rate of CMN results in three dynamic scenarios (Color figure online)

Particularly, two types of experimental results which are final utility and number of rounds are recorded to further valuate the performance of the proposed approach. The final utility of CMN $\mathbb{N}$ is the $U(\mathbb{N})$, which is the utility achieved by the agent when CMN $\mathbb{N}$ finishes. Number of rounds of the CMN $\mathbb{N}$ are defined as follows by Eq. (10).

Let $T = \{t_0, \cdots t_i, \cdots t_k\}$ indicate the CMN $\mathbb{N}$'s round sequence, which means that the CMN $\mathbb{N}$ will finish its negotiation round at Rounds $t_0, \cdots t_i, \cdots t_k$. The number of rounds of the CMN $\mathbb{N}$ is defined as $g(T)$, and

$$g(T) = |T|. \tag{10}$$

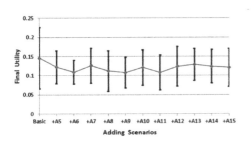

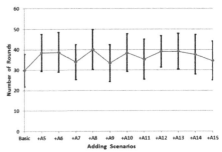

Fig. 6. CMN's Final Utilities in the Adding Scenarios

Fig. 7. CMN's Number of Rounds in the Adding Scenarios

(1) Adding a Single Negotiation

Figure 6 shows the average final utilities of the basic CMN and all cases in adding a single negotiation scenario. The x-axis indicates that each single negotiation (from A_5 to A_{15}) was added to the CMN in each case, and the y-axis indicates the CMN's average final utility after adding the corresponding negotiation. The black vertical lines indicate the standard deviation of average final utilities in all cases. From Fig. 6, we can see that the new CMNs in all the adding cases achieved final utilities, which indicates that the proposed approach is able to handle dynamic changes by considering adding new negotiations.

Figure 7 shows the average numbers of rounds of all cases in the adding scenarios. The x-axis indicates that each single negotiation (from A_5 to A_{15}) was added into the CMN in each case. The y-axis indicates the CMN's average number of rounds after adding the corresponding negotiation. The black vertical lines indicate standard deviation of CMN's average numbers of rounds in all the cases. Figure 7 shows that the average numbers of rounds in all adding cases are greater than those of the basic CMN.

(2) Removing a Single Negotiation

Figure 8 shows the average final utilities of the basic CMN and all the cases in the removing scenarios. The x-axis indicates that each single negotiation (from A_0 to A_4) was removed from the CMN in each case, and the y-axis indicates the CMN's average final utility after removing the corresponding negotiation. The black vertical lines indicate the standard deviation of CMN's average final utilities in all cases. We take the case of removing Negotiation A_0 as an example. In the case of removing Negotiation A_0, we can see that all the negotiations in Fig. 3 will be removed when removing Negotiation A_0. For this reason, the average final utilities of CMN in the case of removing Negotiation A_0 are much less than those of other cases.

Figure 9 shows the average numbers of rounds of all the cases in the removing scenarios. The x-axis indicates that each single negotiation (from A_0 to A_4) was removed from the CMN in each case. The y-axis indicates the CMN's average

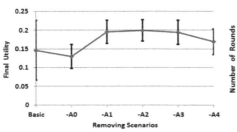 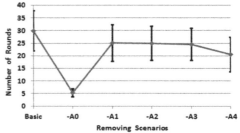

Fig. 8. CMN's Final Utilities in the Removing Scenarios

Fig. 9. CMN's Number of Rounds in the Removing Scenarios

number of rounds after removing the corresponding negotiation. The black vertical lines indicate the standard deviation of CMN's average numbers of rounds in all cases. For instance, for the same reason of average utilities in the case of removing Negotiation A_0, the average number of rounds are much less than those of other cases.

(3) Adding and Removing Negotiations

Figure 10 shows the utilities of each single negotiation and the CMN. The x-axis indicates negotiation round, and the y-axis indicates the utilities of each single negotiation in CMN. In this case, Negotiation A_5 was added into CMN in the 10th negotiation round, and Negotiation A_3 was removed from CMN in the 15th negotiation round. In Fig. 10, all the negotiations are performed concurrently. Moreover, Negotiation A_5 started in the 10th negotiation round, and Negotiation A_3 only negotiated three rounds before it was removed from CMN in the 15th negotiation round. The new CMN achieved the final utility at 0.14 in the end, which indicates that the approach is able to handle dynamic CMN in the adding & removing scenario.

6   Related Work

To date, some work [1–3,13,14] has been done in the research area of single negotiation by the consideration of a dynamic environment. Li et al. [3] presented a model for bilateral negotiations by consideration of the uncertain and dynamic outside options. In [1], Moon et al. introduced a dynamic negotiation mechanism, which is suitable for the dynamic environment following a rudimentary electronic market (e-market) feature allowing agents to freely enter or leave the e-market and negotiate with each other to obtain the benefit. In [13], Mansour et al. proposed a novel dynamic negotiation strategy that works in a more complicated negotiation scenario. An et al. [2] proposed a distributed negotiation mechanism for dynamic resource allocation problem in cloud computing. The buyer and seller agents are able to dynamically enter and leave the market.

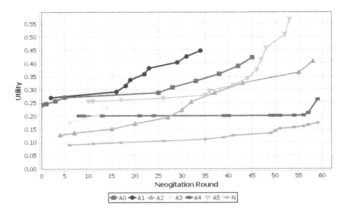

Fig. 10. Utilities of Each Single Negotiation in the Adding & Removing Scenario (Color figure online)

All the work above related to dynamic negotiation focuses on single negotiation level. Based on our knowledge, limited approaches on dynamic negotiation has been proposed in multi-negotiation level. In order to overcome limitations of present work, the approach in this paper proposed a dynamic negotiation mechanism in multi-negotiation level by the consideration of a dynamic negotiation environment, where negotiations could be added into or removed from CMN.

7   Conclusion and Future Work

Concurrent multiple negotiation in a dynamic environment is an exciting and a challenging research topic. The goal of this work is to study negotiation mechanism for handling dynamic CMN in multi-negotiation level. In this paper, a negotiation mechanism for handling dynamic CMN was proposed. We also presented a CPN-based protocol of dynamic CMN which is able to handle the changing of multiple negotiations. Experimental results showed the proposed approach is able to handle the dynamics in CMN by the consideration of adding and removing negotiations.

Our future work will focus on the studies of the optimization of dynamic CMN, in which to optimize the mechanism of dynamic CMN. Therefore, agents are able to make adaptive decisions on multiple negotiations by the consideration of maximizing effectiveness and efficiency of negotiations.

References

1. Moon, S.K., Park, J., Simson, T.W., Kumara, S.R.T.: A dynamic multiagent system based on a negotiation mechanism for product family design. IEEE Trans. Autom. Sci. Eng. **5**(2), 234–244 (2008)

2. An, B., Lesser, V., Irwin, D., Zink, M.: Automated negotiation with decommitment for dynamic resource allocation in cloud computing. In: Proceedings of the 9th International Conference on Autonomous Agents and Multiagent Systems (AAMAS 2010), pp. 981–988 (2010)
3. Li, C., Sycara, K., Giampapa, J.: Dynamic outside options in alternating-offers negotiations. In: Proceedings of the 38th Annual Hawaii International Conference on system Sciences (HICS 2005), pp. 1–10 (2005)
4. Ren, F., Zhang, M., Miao, C., Shen, Z.: Optimization of multiple related negotiation through multi-negotiation network. In: Bi, Y., Williams, M.-A. (eds.) KSEM 2010. LNCS, vol. 6291, pp. 174–185. Springer, Heidelberg (2010)
5. Jensen, K., Kristensen, L.M., Wells, L.: Coloured petri nets and CPN tools for modelling and validation of concurrent systems. Int. J. Softw. Tools Technol. Transf. 9(3–4), 213–254 (2007)
6. Jensen, K.: Coloured Petri Nets: Basic Concepts, Analysis Methods and Practical Use, vol. 1. Springer, Heidelberg (2013)
7. Delecroix, F., Morge, M., Routier, J.-C.: Bilateral negotiation of a meeting point in a maze. In: Demazeau, Y., Zambonelli, F., Corchado, J.M., Bajo, J. (eds.) PAAMS 2014. LNCS, vol. 8473, pp. 86–97. Springer, Heidelberg (2014)
8. Fatima, S., Shaheen, S., Wooldridge, M., Jennings, N.R.: Multi-issue negotiation under time constraints. In: Proceedings of the 1st International Joint Conference on Autonomous Agents and Multiagent Systems, pp. 143–150 (2002)
9. Li, J., Wang, Y., Lin, X., Nazarian, S., Pedram, M.: Negotiation-based task scheduling and storage control algorithm to minimize user's electric bills under dynamic prices. In: The 20th Asia and South Pacific design Automation Conference (ASP-DAC 2015), pp. 261–266 (2015)
10. Baarslag, T., Gerding, E., Aydogan, R., Schraefel, M.: Optimal negotiation decision functions in time-sensitive domains. In: International Joint Conferences on Web Intelligence (WI) and Intelligent Agent Technologies (IAT) (2015)
11. Cao, M., Luo, X., Luo, R., Dai, X.: Automated negotiation for e-commerce decision making: a goal deliberated agent architecture for multi-strategy selection. Decis. Support Syst. 73, 1–14 (2015)
12. Faratin, P., Sierra, C., Jennings, N.R.: Negotiation decision functions for autonomous agents. Robot. Auton. Syst. 24(3), 159–182 (1998)
13. Mansour, K., Kowalcayk, R.: On dynamic negotiation strategy for concurrent negotiation over distinct objects. In Novel Insights in Agent-based Complex Autom. Negot. 535, 109–124 (2014)
14. Ren, F., Zhang, M., Bai, Q.: A dynamic, optimal approach for multi-issue negotiation under time constraints. In Novel Insights in Agent-based Complex Autom. Negot. 535, 85–108 (2014)

L_1-Regularized Continuous Conditional Random Fields

Xishun Wang[1(⊠)], Fenghui Ren[1], Chen Liu[2], and Minjie Zhang[1]

[1] School of Computing and Information Technology,
University of Wollongong, Wollongong, Australia
xw357@uowmail.edu.au, {fren,minjie}@uow.edu.au
[2] School of Science, RMIT University, Melbourne, Australia
s3481556@student.rmit.edu.au

Abstract. Continuous Conditional Random Fields (CCRF) has been widely applied to various research domains as an efficient approach for structural regression. In previous studies, the weights of CCRF are constrained to be positive from a theoretical perspective. This paper extends the definition domains of weights of CCRF and thus introduces L_1 norm to regularize CCRF, which enables CCRF to perform feature selection. We provide a plausible learning method for L_1-Regularized CCRF (L_1-CCRF) and verify its effectiveness. Moreover, we demonstrate that the proposed L_1-CCRF performs well in selecting key features related to the various customers' power usages in Smart Grid.

Keywords: Continuous Conditional Random Fields · Regularization · Feature selection

1 Introduction

Conditional Random Fields (CRF) [5] is an efficient structural learning tool which has been used in image recognition, natural language processing and bioinformatics etc. CRF is an undirected graphical model that supplies flexible structural learning framework. There are two kinds of potentials in CRF, which are state potentials and edge potentials. The state potentials model how the inner states are influenced by a series of outside factors, while the edge potentials model the interactions of inside states (under the influences of outside factors). With the two arrays of potentials, CRF is capable of simultaneously modeling outside influences and inside interactions for the structural outputs to be predicted. Due to the above favorable advantages, CRF promotes various research domains in computational intelligence.

In 2009, Qin et al. [9] proposed Continuous CRF (CCRF), which extends CRF to be able to output continuous values. Since then, CCRF has been widely applied to different research fields. Qin et al. [9] applied CCRF in learning to rank, and gained superior performance to the baseline learning to rank algorithms. Baltrusaitis et al. [2] used CCRF as a structural regression tool for

© Springer International Publishing Switzerland 2016
R. Booth and M.-L. Zhang (Eds.): PRICAI 2016, LNAI 9810, pp. 793–804, 2016.
DOI: 10.1007/978-3-319-42911-3_67

expression recognition. In their work, a facial expression was defined by four measurements. CCRF was employed to regress an unseen expression to the four measurements. Xin et al. [11] developed multi-scale CCRF to build a social recommendation framework. Guo [3] used CCRF for energy load forecasting (the usage of electricity and gas) of a building. His experimental results demonstrated that CCRF gained better performance against state-of-art regression methods.

In previous work [2,9], the weight parameters of CCRF were stipulated to be positive in a theoretical view. These constraints in fact limit the applications of CCRF. In real-world problems, there may be some features (in the whole feature set) that are not related to the final predictions, or even there are sparse feature problems that the predicted results are determined by a small fraction of all the features. For the above problems, feature selection should be done before using CCRF. To surmount the limit of CCRF caused by the positive weights, we do not constrain the weights of CCRF from a practical perspective. In this scenario, L_1 norm can be introduced to regularize CCRF, which enables CCRF to select effective features. However, as L_1 norm is not differentiable at zero, it brings extra difficulty in L_1-CCRF learning. We then provide a plausible learning method for the proposed L_1-CCRF using Orthant-Wise Limited-memory Quasi-Newton (OWL-QN) algorithm [1].

The paper has three major contributions: 1. The definition domains of the weights of CCRF are extended from a practical view. This enable CCRF to perform feature selection with an L_1 norm and consequently extends the applicable areas of CCRF. 2. We provide a practical learning method for L_1-CCRF and demonstrate its effectiveness. Experimental results show that the new learning method is more efficient than the previous learning method. 3. Experimental results demonstrate that the proposed L_1-CCRF is effective in feature selection in the power market domain.

The rest of the paper is organised as follows. Section 2 gives a brief introduction to CCRF. Section 3 proposes L_1-CCRF and describes its learning and inference process. Experiments are reported in Sect. 4 to verify the learning process of L_1-CCRF and the feature selection capacity of L_1-CCRF. Finally, the conclusion is given in Sect. 5.

2   Introduction to CCRF

The Continuous Conditional Random Fields [9] (CCRF) was initially proposed for structural regression. CCRF extended CRF to be able to output real-value predictions. The chain-structured CCRF, as illustrated in Fig. 1, is widely used.

Assume $X = \{x_1, x_2, \cdots, x_m\}$ is the given sequence of observations, and $Y = \{y_1, y_2, \cdots, y_n\}$ is the value sequence to be predicted. CCRF defines the conditional probability $P(Y|X)$ in Eq. 1.

$$P(Y|X) = \frac{1}{Z(X)} exp(\Psi), \tag{1}$$

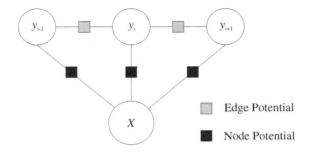

Fig. 1. An illustration of a CCRF with a chain structure.

where Ψ is the energy function, and $Z(X)$ is the partition function that normalizes $P(Y|X)$. The energy function Ψ is further defined as

$$\Psi = \sum_i \sum_{k=1}^{K_1} \alpha_k f_k(y_i, X) + \sum_{i,j} \sum_{k=1}^{K_2} \beta_k g_k(y_i, y_j, X), \tag{2}$$

where function $f_k(y_i, X)$ is called node potential and function $g_k(y_i, y_j, X)$ is called edge potential, and α_k and β_k are corresponding weight parameters. In the energy function, the node potential captures the associations between inputs and outputs, and the edge potential captures the interactions between related outputs. The partition function $Z(X)$ is defined in Eq. 3.

$$Z(X) = \int_Y exp(\Psi) \tag{3}$$

CCRF explicitly defines $P(Y|X)$, which means Y is determined by the whole observation X. Therefore, CCRF gains the capacity of considering the whole observed sequence for the output.

To learn a CCRF model, maximum log-likelihood is used to find the fittest weights α and β. Given training data $D = \{(X, Y)\}_1^Q$, where Q is the total number of training samples, the log-likelihood $L(\alpha, \beta)$ is maximized:

$$(\hat{\alpha}, \hat{\beta}) = \operatorname{argmax}_{(\alpha, \beta)}(L(\alpha, \beta)), \tag{4}$$

where

$$L(\alpha, \beta) = \sum_{l=1}^Q log P(Y_q | X_q) \tag{5}$$

After the weights α and β are obtained, inference for a CCRF is to find the most likely value for Y_k, provided an observed sequence X_k:

$$\hat{Y}_k = \operatorname{argmax}_{Y_k}(P(Y_k | X_k)) \tag{6}$$

In practical use, we usually use a vector $\mathbf{y}$ to represent the array of values to be predicted, and a matrix $\mathbf{X}$ to represent the array of features.

Radosavljevic et al. [10] have shown that $P(\mathbf{y}|\mathbf{X})$ with quadratic potentials can be transformed into a multivariate Gaussian, which facilitates the learning and inference processes. The quadratic form is widely used in CCRF, and consequently, a universal partition function is shown as follows.

$$Z(\mathbf{X}) = \int_{\mathbf{y}} exp(\sum_i \sum_{k=1}^{K_1} -\alpha_k(y_i - \mathbf{X}_{i,k})^2 + \sum_{i,j} \sum_{k=1}^{K_2} -\delta_k\beta_k(y_i - y_j)^2) \tag{7}$$

In Eq. 7, δ_k is an indicator function. When a certain assertion holds, δ_k takes the value "1", otherwise, it is "0". In this scenario, $P(\mathbf{y}|\mathbf{X})$ could be transformed to a multivariate Gaussian form, resulting in a concise inference process [10]. This is further shown in Subsect. 3.3.

3 L_1-Regularized Continuous Conditional Random Fields

We propose L_1-Regularized Continuous Conditional Random Fields (L_1-CCRF) in this section. We first extend definition domains of the weights for CCRF from a perspective of practical use. As a consequence, L_1 norm can be introduced to regularize CCRF. Then we provide a plausible learning method for L_1-CCRF based on OWL-QN algorithm [1].

3.1   Introducing L_1 Norm to Regularize CCRF

We first extend the definition domains of weights for CCRF. In previous CCRF [9,10], the partition function had the form shown in Eq. 7. In this equation, we do not pay much attention to any specific parameters, but focus on the quadratic terms. When the variables $\mathbf{X}$ and $\mathbf{y}$ are defined in infinite domains, both $\boldsymbol{\alpha}$ and $\boldsymbol{\beta}$ are required to be positive to ensure that the partition function is integrable. However, in practical use, data preprocessing is a necessary step before learning. In this step, the observed feature $\mathbf{X}$ is preprocessed to be within a certain domain, and $\mathbf{y}$ is also targeted in a finite range. Thus, the partition function is integrable regardless of the domains of $\boldsymbol{\alpha}$ and $\boldsymbol{\beta}$. Therefore, we do not have to constrain $\boldsymbol{\alpha}$ and $\boldsymbol{\beta}$.

In machine learning, regularization has been commonly used in learning process to achieve a model that generalizes to unseen data. L_2-norm regularization has been used in [2,3,10] in learning CCRF. L_1 norm regularizer is theoretically studied in [8] by Ng, and in practice, the L_1 norm regularizer has gained roughly the same accuracy as the L_2-norm regularizer [6]. Besides, L_1 norm has a favorable property of capable of selecting effective features, which can be utilized to analyze customer behaviours in our research. Therefore, we introduce L_1 norm to regularize the CCRF in the learning procedure. $\boldsymbol{\lambda} =< \boldsymbol{\alpha}, \boldsymbol{\beta} >$ is introduced to compactly represent the weights. The objective function to be minimized for L_1-CCRF is designed in Eq. 8,

$$F(\boldsymbol{\lambda}) = -L(\boldsymbol{\lambda}) + \rho\|\boldsymbol{\lambda}\|_1, \tag{8}$$

where $\|\|_1$ stands for L_1 norm. In the objective function, the first term is the loss function, which is a negative of log-likelihood of the training set (see Eq. 5), while the second term is the L_1 norm of $\boldsymbol{\lambda}$, used as a regularization term. The parameter ρ compromises the loss and the regularization term.

3.2   Learning L_1-CCRF

As L_1 norm is not differentiable at zero, the previous learning method for CCRF [9,10] is no longer applied to L_1-CCRF learning. Some special methods have been proposed to tackle the learning with L_1 norm regularizer [1,12]. Orthant-Wise Limited-memory Quasi-Newton (OWL-QN), proposed by Andrew and Gao [1], has been verified an advantageous algorithm for L_1-regularized log-linear model in [6]. We therefore introduce the OWL-QN algorithm to learn an L_1-CCRF.

For the purpose of comparison and experimental evaluation, we briefly review the learning process for CCRF. In CCRF, each weight parameter λ_i is constrained to be positive. Authors in [9,10] maximize $L(\boldsymbol{\lambda})$ with respect to $\log \lambda_i$. With this transform, the constrained optimization problem becomes unconstrained. As a consequence, Stochastic Gradient Descent (SGD) [13] algorithm can be used to optimize the unconstrained problem. In CCRF learning using SGD, in each iteration, two major steps are taken:

(1) Compute gradient $\nabla_{\log \lambda_i}$ (gradient with respect to $\log \lambda_i$) for the objective function $P(\mathbf{y}|\mathbf{X})$ with a random training sample.
(2) Update weight parameter $\log \lambda_i$: $\log \lambda_i = \log \lambda_i + \eta \times \nabla_{\log \lambda_i}$, where η is the learning rate.

After T iterations, SGD outputs the optimized weights $\boldsymbol{\lambda}$.

For L_1-CCRF learning, we introduce the Orthant-Wise Limited-memory Quasi-Newton (OWL-QN) algorithm, which in fact extends L-BFGS [7] algorithm to be capable of optimizing convex function with L_1 norm. The quasi-Newton algorithms gain the advantage of second-order convergence rate with a small computation cost. They construct an approximation of the second-order Taylor expansion of the objective function, and then try to minimize the approximation. In the approximated Taylor expansion, the Hessian matrix is constructed with the first-order information gathered from previous steps. OWL-QN, which modifies L-BFGS, is motivated by the following basic idea. For the L_1 norm, when the orthant is given, its sign can be determined and become differentiable. Furthermore, L_1 norm is not related to the Hessian, which can be approximated by the loss term alone. Thus, OWL-QN in fact imitates L-BFGS steps in a chosen orthant.

The process of using OWL-QN to learn L_1-CCRF is briefly described as follows. OWL-QN iteratively changes $\boldsymbol{\lambda}$ to obtain an optimized value. In the

kth iteration step, OWL-QN first computes the pseudo-gradient of $F(\boldsymbol{\lambda})$ as a basement to choose the appropriate orthant and search direction, then chooses an orthant $\boldsymbol{\xi}^k$, computes the search direction $\mathbf{p}^k$, and searches the next objective point $\boldsymbol{\lambda}^{k+1}$. In each step, a displacement $\mathbf{s}^k = \boldsymbol{\lambda}^{k+1} - \boldsymbol{\lambda}^k$ and the change in gradient $\mathbf{r}^k = -\nabla L(\boldsymbol{\lambda}^{k+1}) + \nabla L(\boldsymbol{\lambda}^k)$ are updated and recorded. The previous m displacements and changes in gradient are used to construct an approximate $\mathbf{H}^k$, the inverse of Hessian matrix of $F(\boldsymbol{\lambda})$, which is the key to compute the search direction. After T iterations, OWL-QN converges and outputs the optimal weights $\boldsymbol{\lambda}$. The learning procedure for L_1-CCRF using OWL-QN is summarized in Algorithm 1.

Algorithm 1. L_1-CCRF learning using OWL-QN

Input: Training samples $D = \{(X, Y)\}_1^Q$;
Output: Weight parameter $\boldsymbol{\lambda}$;
 1: **Initialize:** Initial point $\boldsymbol{\lambda}^0$; $S \Leftarrow \{\}$, $R \Leftarrow \{\}$.
 2: **for** $k = 0$ to T **do**
 3:     Compute the pseudo-gradient of $F(\boldsymbol{\lambda})$
 4:     Choose an orthant $\boldsymbol{\xi}^k$
 5:     Construct $\mathbf{H}_k$ using S and R
 6:     Compute search direction $\mathbf{p}^k$
 7:     Find $\boldsymbol{\lambda}^{k+1}$ with constrained line search
 8:     **if** termination condition satisfied **then**
 9:         Stop and return $\boldsymbol{\lambda}^{k+1}$
10:     **end if**
11:     Update S with $\mathbf{s}^k = \boldsymbol{\lambda}^{k+1} - \boldsymbol{\lambda}^k$
12:     Update R with $\mathbf{r}^k = -\nabla L(\boldsymbol{\lambda}^{k+1}) + \nabla L(\boldsymbol{\lambda}^k)$
13: **end for**

Before the explanation of Algorithm 1, we introduce two special functions [1] for convenience. The first one is *sign function* σ: $\sigma(x)$ results in a value in $\{-1, 0, 1\}$ according to whether x is negative, zero or positive. The second one is *project function* π: $\pi(\mathbf{x}; \mathbf{y})$, $\mathbb{R}^n \mapsto \mathbb{R}^n$, is parameterized by $y \in \mathbb{R}^n$, where

$$\pi_i(x; y) = \begin{cases} x_i \ if & \sigma(x_i) = \sigma(y_i) \\ 0 & otherwise \end{cases} \tag{9}$$

It can be interpreted as projecting $\mathbf{x}$ onto the orthant defined by $\mathbf{y}$.

In Algorithm 1, Step 1 chooses and initial $\boldsymbol{\lambda}$, and initialize sets S and R. Step 2–13 are the main iteration loop. Step 3 calculates the pseudo-gradient of $F(\boldsymbol{\lambda})$ at $\boldsymbol{\lambda}$, denoted $\diamond F(\boldsymbol{\lambda})$, according to the following equation:

$$\diamond_i F(\lambda) = \begin{cases} \partial_i^- F(\lambda) \ if & \partial_i^- F(\lambda) > 0 \\ \partial_i^+ F(\lambda) \ if & \partial_i^+ F(\lambda) < 0 \ , \\ 0 & otherwise \end{cases} \tag{10}$$

where

$$\partial_i^{\pm} F(\lambda) = -\frac{\partial}{\partial \lambda_i} L(\lambda) + \begin{cases} \rho \sigma(\lambda_i) \ if & \lambda_i \neq 0 \\ \pm \rho & if \quad \lambda_i = 0 \end{cases} . \tag{11}$$

In Eq. 11, the term $\partial L(\lambda)/\partial\lambda_i$ is derived with respect to the specified model. In Step 4, an orthant $\boldsymbol{\xi}^k$ is chosen based on $\diamond F(\boldsymbol{\lambda})$,

$$\xi_i^k = \begin{cases} \sigma(\lambda_i^k) & if \quad \lambda_i^k \neq 0 \\ \sigma(-\diamond_i F(\boldsymbol{\lambda}^k)) & if \quad \lambda_i^k = 0 \end{cases}. \tag{12}$$

Step 5 constructs the inverse of Hessian $\mathbf{H}_k$, which is constructed the same as the traditional L-BFGS [7], not shown here again. Step 6 then determines the search direction $\mathbf{p}^k$, formulated by

$$\mathbf{p}^k = \pi(\mathbf{H}_k v^k; v^k), \tag{13}$$

where $v^k = -\diamond F(\boldsymbol{\lambda}^k)$. Step 7–10 aims to find the next point $\boldsymbol{\lambda}^{k+1}$ using constrained line search, in which each point explored is projected back onto the chosen orthant: $\boldsymbol{\lambda}^{k+1} = \pi(\boldsymbol{\lambda}^k + \alpha\mathbf{p}^k; \boldsymbol{\xi}^k)$, where α controls the search step. Step 11 and 12 update sets S and R, respectively.

The proposed learning method for L_1-CCRF has a semi-second-order convergence rate in theory. We will show it converges faster than the CCRF learning method in Subsect. 4.1.

3.3   Inference

In the prediction for a learned L_1-CCRF model, we find the most likely $\mathbf{y}$ given the observed feature $\mathbf{X}$, as formulated in Eq. 6. We first derive $P(\mathbf{y}|\mathbf{X})$ into multi-variant Gaussian form.

$$P(\mathbf{y}|\mathbf{X}) = \frac{1}{(2\pi)^{n/2}|\boldsymbol{\Sigma}|^{1/2}} \cdot \\ exp(-\frac{1}{2}(\mathbf{y} - \mu(\mathbf{X}))^T\boldsymbol{\Sigma}^{-1}(\mathbf{y} - \mu(\mathbf{X}))) \tag{14}$$

In this Gaussian form, the inverse of the covariance matrix $\boldsymbol{\Sigma}^{-1}$, is the sum of two $n \times n$ matrices, further expressed as follows.

$$\boldsymbol{\Sigma}^{-1} = 2(\mathbf{M}^1 + \mathbf{M}^2), where$$
$$M_{i,j}^1 = \begin{cases} \sum_{k=1}^{K_1} \alpha_k & if \quad i = j \\ 0 & otherwise \end{cases}$$
$$M_{i,j}^2 = \begin{cases} \sum_{j=1}^n \sum_{k=1}^{K_2} \delta_k\beta_k - \sum_{k=1}^{K_2} \delta_k\beta_k & if \quad i = j \\ -\sum_{k=1}^{K_2} \delta_k\beta_k & if \quad i \neq j \end{cases} \tag{15}$$

The diagonal matrix M^1 represents the contribution of $\boldsymbol{\alpha}$ terms (node potentials), and the symmetric matrix M^2 represents the contribution of $\boldsymbol{\beta}$ terms (edge potentials). The mean $\mu(\mathbf{X})$ is computed by

$$\mu(\mathbf{X}) = \boldsymbol{\Sigma}\theta \tag{16}$$

Here, $\boldsymbol{\theta}$ is an n-dimension vector, where each element is calculated by

$$\theta_i = 2 \sum_{k=1}^{K_1} \alpha_k \mathbf{X}_{i,k} \tag{17}$$

Benefiting from the multivariate Gaussian form, the inference becomes quite tractable. To maximize $P(\mathbf{y}|\mathbf{X})$ in the multivariate Gaussian (see Eq. 14), we simply make $\mathbf{y}$ equal to $\mu(\mathbf{X})$,

$$\hat{\mathbf{y}} = \mathrm{argmax}_{\mathbf{y}}(P(\mathbf{y}|\mathbf{X})) = \mu(\mathbf{X}) = \boldsymbol{\Sigma}\boldsymbol{\theta} \tag{18}$$

4 Experiment

We conducted two experiments for the proposed L_1-CCRF. Experiment 1 evaluated the learning process of L_1-CCRF. This experiment aimed to demonstrate the proposed learning method for L_1-CCRF was correct and more efficient than the previous learning method for CCRF. Experiment 2 aimed to verify the feature selection capacity of L_1-CCRF. This experiment used L_1-CCRF to predict the hourly load of some specific customers in a Smart Grid to demonstrate L_1-CCRF was effective in feature selection and load forecasting. In the following, the two experiments are reported, respectively.

4.1 Experiment 1: Learning of L_1-CCRF

In this experiment, we compared the proposed learning method for L_1-CCRF with the conventional CCRF learning. The experimental data we used here are from the work of Baltrusaitis et al. [2], whose source code is publicly available[1]. In their work, they used CCRF as a structural regression tool to regress a test expression to a four-dimension measurement defined to measure facial expressions. For page limits, their model is not introduced here. Interested reader may refer to their work [2]. In their CCRF learning, they used the conventional SGD algorithm (introduced in Subsect. 3.2) and introduced L_2 norm for regularization. We first repeated their results, and then built the same CCRF model, and used the proposed L_1-CCRF learning algorithm (see Algorithm 1) to learn the same data.

We repeat the dimensional expression recognition experiment using CCRF model, not the CA-CCRF [2]. For the experimental settings, the convergence criteria for both methods were: $\Delta\lambda = \lambda_{i+1} - \lambda_i < 10^{-3}$. The other parameters in learning process of CCRF used the default settings in [2]. In L_1-CCRF learning, parameter ρ was set as 100, which was optimized by cross-validations. The convergence curves of the two methods for the "valence" dimension are shown in Fig. 2.

We first make some explanations to Fig. 2. The convergence rate is measured by the total iterations of all the training samples. In CCRF learning, SGD derives

---

[1] http://www.cl.cam.ac.uk/research/rainbow/projects/ccrf/.

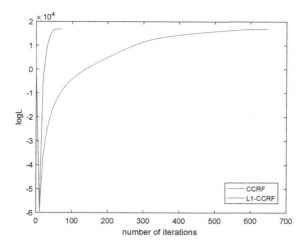

Fig. 2. The convergence of curves of CCRF and L_1-CCRF learning. (Color figure online)

gradient of each sample, while L_1-CCRF derives gradient based on all the training samples. We have divided the total loops by the total number of samples to obtain the number of iterations for CCRF learning, which is comparable with the total iterations of L_1-CCRF. CCRF takes 624 iterations to converge, while L_1-CCRF iterates 68 times to reach convergence. From Fig. 2, we can see that the convergence rate of L_1-CCRF is much faster than that of CCRF. In the theoretical view, the convergence rate of SGD is first-order, while that of OWL-QN is semi-second-order. Thus, the learning process of L_1-CCRF takes much less iterations than that of CCRF. When converged, the final log-likelihood of CCRF is 1.687×10^4, and that of L_1-CCRF is 1.672×10^4.

Table 1. The learning performance comparison of CCRF and L_1-CCRF

Method	Learning time (seconds)	Precision (average correlations)
CCRF	1532	0.304
L_1-CCRF	257	0.301

We compared the learning time and prediction results between CCRF and L_1-CCRF in Table 1. In Table 1, the learning time is measure by seconds. In the prediction results, both methods calculates the correlations of each predicted values to each of the four dimensions of an expression. For the prediction precision, we use the average correlations to measure both methods. In our desktop with Duo CPU 3.33 GHz and 4 GB RAM, CCRF learning takes 1532 s to reach convergence, while L_1-CCRF learning converges in 257 s. For the prediction precision, we can see that the two methods show similar results. The predicted results demonstrate two points, which are: (1) the L_1-CCRF learning is correct, and (2) L_1 norm and L_2 norm regularization result in similar precisions.

From this experiment, we draw two major conclusions: 1. The convergence rate of L_1-CCRF learning is faster than that of CCRF. 2. L_1 norm shows similar performance as L_2 norm in prediction precision.

4.2 Experiment 2: Feature Selection Using L_1-CCRF

This experiment was conducted on the platform of Power Trading Agent Competition (Power TAC) [4]. Power TAC has drawn wide attentions and has become a benchmark in the Smart Grid research community. We here utilized this platform to evaluate the performance of L_1-CCRF in feature selection and short-term load forecasting (the future hourly power usage). Power TAC simulates a variety of customers with various behaviours in a Smart Grid. Moreover, there are rich features, including real-world weather conditions and real-time market status. Besides, the Power TAC server supplies rich logs of customers' hourly power usages, which are regarded as the ground-truth to evaluate the prediction precision of L_1-CCRF.

Table 2. Features used in Experiment 2

Feature	Content	Index
Temporal feature	Hour of a day	t_1
	Day of a week	t_2
Weather feature	Temperature	w_1
	Wind strength	w_2
	Wind direction	w_3
	Cloudiness	w_4
Market feature	Lowest price	m_1
	Average price	m_2

Representative features, which may relate to the customers' power consumption behaviors, are studied in our work. These features include temporal features, weather features and market features. Table 2 lists the contents of the three features, and the indexes provide convenience for further discussions.

We configured the Power TAC server and weather data server, and utilized Power TAC games to generate training and test data. Six games were run to roughly cover the year of 2009. The server logs, which contained features and customers' energy usages, were used as the training data. Another six games were run for year 2010. The logged customers' usages were regarded as the ground-truths of loads to be predicted. The L_1-CCRF model for load forecasting is described in details. L_1-CCRF modeled 24 h power usages under the influences of 24 hourly features. For the features in each hour, 8 node potentials were generated. There were 23 intervals between every two adjacent hours in a day. Edge potentials were generated in every intervals. To ensure the convergence of L_1-CCRF, we set 50 iterations for OWL-QN.

Four typical customers are selected to test the performance of L_1-CCRF. Customer 1 is a village householder, Customer 2 is a photovoltaic energy generator, Customer 3 is an office building and Customer 4 is a cold storage company with power storage capacity. We first analyze the feature selection capacity of L_1-CCRF. Table 3 shows effective features selected by L_1-CCRF for the four customers.

Table 3. Feature selection for the four typical customers

Customer	t_1	t_2	w_1	w_2	w_3	w_4	m_1	m_2
$C1$	1	1	1	0	0	0	1	0
$C2$	1	0	1	0	0	1	0	0
$C3$	1	1	1	0	0	1	0	0
$C4$	0	0	1	1	0	0	1	1

In Table 3, "1" indicates that the feature is related to the power usage of the customer, and "0" otherwise. Customer 1 is a village user. Intuitively, the power usage of Customer 1 may be influenced by time, temperature and market price. The feature selection result of Customer 1 in Table 3 is quite reasonable according to our intuitions. Customer 2, who is a photovoltaic energy produce, is affected by hours in one day, the cloudiness and temperature. The result shown in Table 3 is also correct. Similarly, the feature selection result for Customer 3 is reasonable. Customer 4 is a cold storage company, and the power usage is not likely affected by wind strength. In this case, the wind strength feature might be selected improperly. In contrast, if we used conventional CCRF learning, as parameters were confined to be positive, it could not perform feature selection. Thus, L_1-CCRF gains the advantage to build a concise (using only effective features) and explanatory (harmony with intuitions or commonsense) model.

At the same time, L_1-CCRF predicts the hourly power usages of the four customers. Mean Absolute Percentage Error (MAPE) for each hour is calculated to measure the prediction results. The average MAPE for each of the four customers are 7.32 %, 5.63 %, 6.41 % and 3.36 %. L_1-CCRF shows good performances in short-term load forecasting.

Two points are verified in this experiment. 1. L_1-CCRF is effective in feature selection for customers' power consumption behaviors in the Smart Grid market. 2. L_1-CCRF shows good performance in load forecasting. Thus, the proposed L_1-CCRF can be an efficient structural regression tool with feature selection capacity.

5   Conclusion

This work first used L_1 norm to regularize CCRF and proposed L_1-CCRF. The definition domains of weights of CCRF were extended, so that L_1 norm could be introduced for CCRF regularization in the learning process. We then provided a plausible learning method for L_1-CCRF, and experimental results demonstrated its correctness and efficiency. We used L_1-CCRF for feature selection for the customers in a Smart Grid and demonstrate that L_1-CCRF is effective in selecting key features in the power market domain. In a nutshell, the proposed L_1-CCRF is effective in learning and feature selection, and we suggest it be applied to wide research domains.

References

1. Andrew, G., Gao, J.: Scalable training of L_1-regularized log-linear models. In: Proceedings of the 24th International Conference on Machine Learning, pp. 33–40. ACM (2007)
2. Baltrusaitis, T., Banda, N., Robinson, P.: Dimensional affect recognition using continuous conditional random fields. In: IEEE Automatic Face and Gesture Recognition, pp. 1–8 (2013)
3. Guo, H.: Accelerated continuous conditional random fields for load forecasting. IEEE Trans. Knowl. Data Eng. **27**(8), 2023–2033 (2015)
4. Ketter, W., Collins, J., Reddy, P., Weerdt, M.: The power trading agent competition. ERIM Report Series Reference No. ERS–004-LIS (2014)
5. Lafferty, J.: Conditional random fields: probabilistic models for segmenting and labeling sequence data. In: Proceedings of the 18th International Conference on Machine Learning (ICML-2001) (2001)
6. Lavergne, T., Cappé, O., Yvon, F.: Practical very large scale CRFs. In: Proceedings of the 48th Annual Meeting of the Association for Computational Linguistics, pp. 504–513. Association for Computational Linguistics (2010)
7. Liu, D., Nocedal, J.: On the limited memory BFGS method for large scale optimization. Math. Program. **45**(1–3), 503–528 (1989)
8. Ng, A.: Feature selection, L_1 vs. L_2 regularization, and rotational invariance. In: Proceedings of the Twenty-first International Conference on Machine Learning, pp. 78–85. ACM (2004)
9. Qin, T., Liu, T., Zhang, X., Wang, D., Li, H.: Global ranking using continuous conditional random fields. In: Advances in Neural Information Processing Systems, pp. 1281–1288 (2009)
10. Radosavljevic, V., Vucetic, S., Obradovic, Z.: Continuous conditional random fields for regression in remote sensing. In: ECAI, pp. 809–814 (2010)
11. Xin, X., King, I., Deng, H., Lyu, M.R.: A social recommendation framework based on multi-scale continuous conditional random fields. In: Proceedings of the 18th ACM Conference on Information and Knowledge Management, pp. 1247–1256. ACM (2009)
12. Yu, J., Vishwanathan, S., Günter, S., Schraudolph, N.: A quasi-newton approach to nonsmooth convex optimization problems in machine learning. J. Mach. Learn. Res. **11**, 1145–1200 (2010)
13. Zhang, T.: Solving large scale linear prediction problems using stochastic gradient descent algorithms. In: Proceedings of the Twenty-first International Conference on Machine Learning, pp. 116–123. ACM (2004)

Adaptive Learning for Efficient Emergence of Social Norms in Networked Multiagent Systems

Chao Yu[1], Hongtao Lv[1], Sandip Sen[2], Fenghui Ren[3], and Guozhen Tan[1(✉)]

[1] School of Computer Science and Technology, Dalian University of Technology,
Dalian 116024, China
cy496@uowmail.edu.au, lvhongtao@mail.dlut.edu.cn, gztan@dlut.edu.cn
[2] Department of Mathematical and Computer Sciences,
University of Tulsa, Tulsa, OK 74104, USA
sandip-sen@utulsa.edu
[3] School of Computer Science and Software Engineering,
University of Wollongong, Wollongong 2500, Australia
fren@uow.edu.au

Abstract. This paper investigates how norm emergence can be facilitated by agents' adaptive learning behaviors in networked multiagent systems. A general learning framework is proposed, in which agents can dynamically adapt their learning behaviors through social learning of their individual learning experience. Extensive verification of the proposed framework is conducted in a variety of situations, using comprehensive evaluation criteria of efficiency, effectiveness and efficacy. Experimental results show that the adaptive learning framework is robust and efficient for evolving stable norms among agents.

Keywords: Norm emergence · Learning · Multiagent systems

1 Introduction

Coordination of agent behaviors is central in Multiagent Systems (MASs). Social norm is an effective technique to achieve coordination in MASs by placing social constraints on agent action choices [1]. Understanding how social norms can emerge through local interactions has gained increasingly high attention in the research of MASs. Numerous investigations of norm emergence have been done in recent years under different assumptions about agent interaction protocols, societal topologies and observation capabilities [2–4].

Learning from individual experience has been shown to be a robust mechanism to enable norm emergence in MASs [5]. A great deal of work has studied norm emergence achieved through agent learning behaviors [6–13]. The focus of these existing studies is to examine general mechanisms behind efficient emergence of social norms while agents interact with each other using basic learning (particularly reinforcement learning) methods. These mechanisms include the social learning strategy [6,7], the collective interaction protocol [11–13],

© Springer International Publishing Switzerland 2016
R. Booth and M.-L. Zhang (Eds.): PRICAI 2016, LNAI 9810, pp. 805–818, 2016.
DOI: 10.1007/978-3-319-42911-3_68

the utilization of topological knowledge [8,9], and the observation capability of agents [10], etc. Although these studies provide us with a deep understanding of efficient mechanisms of norm emergence, they share the same limitation inevitably. That is, learning parameters in these studies are often fine-tuned by hand and thus cannot be adapted according to the varying norm emerging situations. A key question then arises that how agents can adapt their learning behaviors dynamically during the process of norm emergence, and how this kind of adaptiveness can influence the final emergence performance?

To this end, this paper provides another perspective in the research of norm emergence by simply focusing on the role of learning itself in affecting the process of norm emergence. A double-layered adaptive learning framework is proposed, in which agents interact with each other using basic Reinforcement Learning (RL) methods [14] in the local layer learning, and generate supervision policies by exploiting historical learning experience in the upper layer learning. The supervision policies are then passed down to the local layer learning in order to adapt agents' learning behaviors based on the consistency between agents' behaviors and the supervision policies. In the framework, two challenging technical issues are needed to be carefully resolved: (1) how to generate supervision policies simply based on agents' historical learning experience? and (2) how to adapt agents' local learning behaviors based on the supervision policies from the upper layer learning? To solve the former problem, the historical learning experience of each agent is synthesized into a strategy that competes with other strategies in the population. The strategies that have better performance are more likely to survive and thus be accepted by other agents. The competing process of the agents' strategies then can be carried out through a social learning process based on the principle of Evolutionary Game Theory (EGT) [15]. For the latter, the concept of "winning" or "losing" in the well-known Multi-Agent Learning (MAL) algorithm WoLF (Win-or-Learn-Fast) [16] is elegantly borrowed to indicate whether an agent's behavior is consistent with the supervision policy. According to the "winning" or "losing" situation, agents then can dynamically adapt their learning behaviors in local layer learning. Experiments show that the proposed framework enables norms to emerge more efficiently and with higher convergence levels than the static learning framework, and some critical factors such as size of norm space and network topologies can have significant influences on norm emergence.

The remainder of the paper is organized as follows. Section 2 briefly introduces social norms and RL. Section 3 describes the proposed framework. Section 4 presents experimental studies. Finally, Sect. 5 concludes the paper.

2   Social Norms and RL

As most previous studies do, this paper also uses learning "rules of the road" [6,17] as a metaphor to study emergence of norms, which can be viewed as a Coordination Game (CG) in Table 1. The abstraction of coordination given by the CG covers a number of practical scenarios, such as distributed robots coordinating on which object to work on together, and wireless nodes coordinating

Table 1. The CG

	L	R
L	1, 1	1, −1
R	1, −1	1,1

Table 2. The extended CG

	a_1	a_2	...	a_{N_a}
a_1	1,1	1, −1	...	−1, −1
a_2	1, −1	1,1	...	1, −1
⋮	⋮	⋮	⋱	⋮
a_{N_a}	1, −1	1, −1	...	1,1

on which channel to transmit message [18]. The problem to deal with the CG is that there is nothing in the structure of the game itself that allows players (even purely rational players) to infer what they ought to do [17]. Therefore, social norms can be used to guide agent behaviors towards specific ones when moral or rational reasoning does not provide a clear guidance of how to behave. To study the impact of norm space on norm emergence, the CG can be extended to a general form by considering N_a actions in the norm space (Table 2).

In this paper, agent interactions are purely local and are constrained by the network structures. Three different kinds of topologies are considered: grid networks, small-world networks and scale-free networks [19].

RL algorithms [14] has been widely used for agent interactions in previous work on norm emergence, and Q-learning [20] is the most adopted one. In Q-learning, an agent makes a decision through the estimation of a set of Q-values, which can be updated by:

$$Q(s,a) = Q(s,a) + \alpha_i[r(s,a) + \gamma \max_{a'} Q(s',a') - Q(s,a)] \tag{1}$$

where $\alpha_i \in (0,1]$ is a learning rate of agent i, and $\gamma \in [0,1)$ is a discount factor, $r(s,a)$ and $Q(s,a)$ are the immediate and expected reward of choosing action a in state s at time step t, respectively, and $Q(s',a')$ is the expected discounted reward of choosing action a' in state s' at time step $t+1$.

Q-values of each state-action pair are stored in a table for a discrete state-action space. At each time step, agent i chooses the best-response action with the highest Q-value with a probability of $1 - \epsilon$ (i.e., exploitation), or chooses other actions randomly with a probability of ϵ (i.e., exploration).

3   The Proposed Learning Framework

To better reflect the nature of adaptive learning behaviors during norm emergence, a general adaptive RL framework is proposed, which extends the traditional RL framework by equipping an agent with a double-layered learning ability. In the framework, the local layer learning represents an individual learning process, in which the agent interacts with the environment using traditional RL algorithms. The upper layer learning represents a social learning process, in which the agent synthesizes its learning experience from the local layer learning and generates a supervision policy through exchanging its learning experience with other agents. The supervision policy is then utilized to guide the local layer learning process by tuning the critical learning parameters heuristically.

3.1 Generating Supervision Policies

In the proposed learning framework, each agent is equipped with a capability to memorize a certain period of learning experience in terms of the chosen action and the corresponding reward. Assuming a memory capability of agent is well justified and quite common in the MAS research [1,8]. Formally, let M denote an agent's memory length. At step t, the agent can memorize the historical information in the period of M steps prior to t. A memory table of agent i at time step t, MT_i^t, then can be denoted as $MT_i^t = \{(a_i^{t-M+1}, r_i^{t-M+1}), ..., (a_i^{t-1}, r_i^{t-1}), (a_i^t, r_i^t)\}$. Based on the memory table, agent i then synthesizes its past learning experience into two tables $TA_i^t(a)$ and $TR_i^t(a)$. $TA_i^t(a)$ denotes the frequency of choosing action a in the last M steps and $TR_i^t(a)$ denotes the overall reward of choosing action a in the last M steps. $TA_i^t(a)$ can be calculated by Eq. 2,

$$TA_i^t(a) = \sum_{j=1}^{j=M} s_i(a, a_i^{t-j+1})$$ (2)

where $s_i(a, a_i^{t-j+1})$ is 1 if $a = a_i^{t-j+1}$ and 0 otherwise.

$TA_i^t(a)$ stores the historical information of how often action a has been chosen in the past. To exclude those actions that have never been chosen, set $X(i, t, M)$ is defined to contain all the actions that have been taken at least once in the last M steps by agent i, i.e., $X(i, t, M) = \{a | TA_i^t(a) > 0\}$. The average reward of choosing action a, $TR_i^t(a)$, then can be given by:

$$TR_i^t(a) = \frac{1}{TA_i^t(a)} \sum_{j=1}^{j=M} f_i(a, a_i^{t-j+1}), \quad \forall a \in X(i, t, M)$$ (3)

where $f_i(a, a_i^{t-j+1})$ is r_i^{t-j+1} if $a = a_i^{t-j+1}$ and 0 otherwise.

The past learning experience in terms of table $TA_i^t(a)$ and $TR_i^t(a)$ indicates how successful the strategy of choosing action a is in the past. The upper layer learning makes use of this information in order to generate a supervision policy for local layer learning. To realize the supervision policy generation, each agent learns from other agents by comparing their learning experience. The motivation of this comparison comes from the EGT, which provides a powerful methodology to model how strategies evolve overtime based on their performance. In the context of EGT, an individual's payoff represents its fitness or social success. The dynamics of strategy change in a population is governed by social learning, that is, the most successful agents will tend to be imitated by the others. Two different approaches are proposed in this paper to realize the EGT concept in the upper layer learning process, depending on how to define the competing strategy and the corresponding performance evaluation criteria (i.e., fitness) in EGT.

Reward-based approach: This approach is a performance-driven approach in that agents are aiming at maximizing their own rewards. If an action has brought about the highest reward among all the actions in the past, this action is the most profitable one and thus should be more likely to be imitated by the others in the population. Therefore, the strategy in EGT is represented by the most profitable action, and the fitness is represented by the corresponding reward of that action. More formally, let a'_i denote the most profitable action. It can be given by Eq. 4:

$$a'_i = arg\ max_{a \in X(i,t,M)}\ TR_i(a) \tag{4}$$

Action-based approach: Unlike the reward-based approach, the action-based approach is a behavior-driven approach. If an agent has chosen the same action all the time, it considers this action to be the most successful one (being the norm accepted by the population). Therefore, action-based approach considers the action which has been most adopted in the past to be the strategy in EGT, and the corresponding reward of that action to be the fitness in EGT. Let a'_i denote the most adopted action. It can be given by:

$$a'_i = arg\ max_{a \in X(i,t,M)}\ TA_i(a) \tag{5}$$

After synthesizing the historical learning experience, agent i then gets a strategy a'_i and its corresponding fitness of $TR(a'_i)$. It then interacts with other agents through social learning based on the pairwise comparison rule, i.e., learning from a randomly selected neighbor in the neighborhood. Imitation rules in EGT then model how the strategies of these two agents evolve overtime based on their performance, which can be realized by the famous Fermi function:

$$p_{i \to j} = \frac{1}{1 + exp[-\beta(TR_i^t(a'_i) - TR_j^t(a'_j))]} \tag{6}$$

where $p_{i \to j}$ denotes the probability that agent i switches to the strategy of agent j and β is the selection bias.

3.2 Adapting Local Learning Behaviors

Based on the principle of EGT, the upper layer learning process generates a supervision policy represented as the new strategy a'_i. The new strategy a'_i indicates the most successful strategy in the neighborhood and therefore should be integrated into the local layer learning in order to entrench its influence. By comparing its action at time step t, a_i^t, with the supervision strategy a'_i, agent i can evaluate whether it is performing well or not so that its learning behavior can be dynamically adapted to fit the supervision strategy. Depending on the consistency between the agent's behavior and the supervision policy, the local layer learning can be adapted based on the following mechanisms:

Supervision-α: In RL, the learning performance heavily depends on the learning rate parameter, which is difficult to tune. This mechanism adapts the learning rate α in the local layer learning. When agent i has chosen the same action with the supervision policy, it decreases its learning rate to maintain its current state, otherwise, it increases its learning rate to learn faster from its interaction experience. Formally, learning rate α_i can be adjusted as follows:

$$\alpha_i = \begin{cases} \alpha_i - \Delta\alpha_i & if \ a_i^t = a_i', \\ \alpha_i + \Delta\alpha_i & otherwise. \end{cases} \tag{7}$$

where $\Delta\alpha_i$ is the changing rate of α_i and can be given by:

$$\Delta\alpha_i = \begin{cases} \lambda\alpha_i & if \ a_i^t = a_i', \\ \lambda(1 - \alpha_i) & otherwise. \end{cases} \tag{8}$$

where $\lambda \in [0, 1]$ is a variable to control the adaption rate.

Supervision-ϵ: Exploration-exploitation trade-off has a crucial impact on the learning process. Therefore, this mechanism adapts the exploration rate ϵ in the local layer learning. The motivation of this mechanism is that an agent needs to explore more of the environment when it is performing poorly and explore less otherwise. Similarly, the exploration rate ϵ can be adjusted according to Eq. 9:

$$\epsilon_i = \begin{cases} \epsilon_i - \Delta\epsilon_i & if \ a_i^t = a_i', \\ \epsilon_i + \Delta\epsilon_i & otherwise. \end{cases} \tag{9}$$

in which $\Delta\epsilon_i$ is $\lambda\epsilon_i$ if $a_i^t = a_i'$, and $\lambda(1 - \epsilon_i)$ otherwise. In RL, exploration rate ϵ_i is always set to a small value in order to indicate a small probability of exploration. Thus, a variable $\overline{\epsilon_i}$ is introduced to confine the exploration rate to a maximal value of $\overline{\epsilon_i}$.

Supervision-both: This mechanism adapts the learning rate and the exploration rate at the same time based on Supervision-α and Supervision-ϵ.

The above mechanisms are based on the concept of "winning" and "losing" in the well-known MAL algorithm WoLF. Although the original meaning of "winning" or "losing" in WoLF and its variants is to indicate whether an agent is doing better or worse than its Nash-Equilibrium policy, this heuristic is gracefully introduced into the proposed framework to evaluate an agent's performance against the supervision policy. Specifically, an agent is considered to be winning (i.e., performing well) if its action is the same with the supervision strategy and losing (i.e., performing poorly) otherwise. The different situations of "winning" or "losing" thus indicate whether the agent's behavior is complying with the norm in the society. If an agent is in a losing state (i.e., its action is against the norm in the society), it needs to learn faster or explores more of the environment in order to escape from this adverse situation. On the contrary, it should decrease its learning and/or exploration rate to stay in the winning state.

4   Experiments and Results

Experiments are carried out to demonstrate the performance of the proposed framework, based on metrics of **effectiveness** (i.e., possibility of convergence),

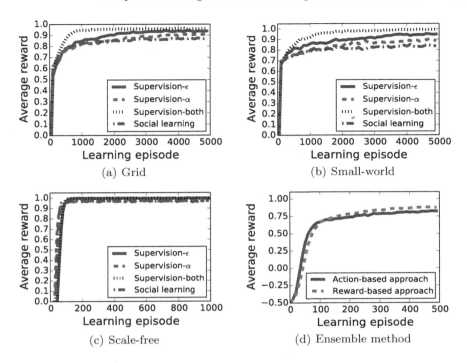

Fig. 1. Norm emergence using the different approaches.

efficiency (i.e., speed of convergence) and **efficacy** (i.e., level of convergence). Effectiveness indicates the percentage of runs in which a norm can successfully emerge, efficiency represents how many steps are needed for a norm to emerge, and efficacy denotes the ratio of agents in the system that can achieve the norm emergence. The Watts-Strogatz model [22] is used to generate a small-world network and the Barabasi-Albert model [19] is used to generate a scale-free network. Stateless version of Q-learning algorithm is used, thus $\lambda = 0$ in Eq. 1. Unless specified otherwise, the rewiring probability in Watts-Strogatz model is set to 0.1 and average number of neighbors in the small-world network is set to 12; learning rate α is set to 0.1 initially, exploration rate ϵ is set to 0.01 initially and $\overline{\epsilon_i}$ is 0.3 by default; each agent can choose from 4 actions and have a moderate memory length of 4 steps; moreover, the parameter β and λ are both set to 0.1.

Emergence of Social Norms. Figure 1(a)–(c) plots the learning curves of the three adaptive learning approaches under the proposed framework, and the static learning approach under the social learning framework [6,21]. In the social learning framework, each agent interacts with one of its neighbors and learns directly from this interaction. Comparing with this approach thus enables to demonstrate the merits of agents' adaptive learning behaviors in influencing the norm emerging process. The results show that the three adaptive learning approaches outperform the static social learning approach in all three networks in terms of a higher level of convergence and a faster convergence speed. Through dynamically adapting their learning behaviors during the norm emerging process, agents are able to reach an agreement more easily, and thus norms can emerge more quickly to a higher level of convergence in the adaptive learning framework. The distinction of performance between the adaptive learning and the social learning approaches is less apparent in the scale-free network because this kind of network is easier to emerge a norm compared with other networks. This is due to the small graph diameter of scale-free networks, as reported in various previous studies [3,23]. Figure 1(d) shows the performance of the two different kinds of approaches adopted in the upper layer learning. As can be seen, norms emerge faster with the action-based approach at the beginning. As the process proceeds, the reward-based approach catches up with the action-based approach, and then brings about a higher level of norm emergence afterwards. This result implies that it is more reasonable to use the most profitable action rather than the most adopted action in the past as the competing strategy in EGT.

Table 3. Overall performance in three different networks

Grid	Efficacy	90 % convergence		100 % convergence	
		Effectiveness	Efficiency	Effectiveness	Efficiency
Supervision-ϵ	0.922	74.7 %	1087	74.7 %	1180
Supervision-α	0.919	74.8 %	1509	66.1 %	4113
Supervision-both	0.957	86.7 %	970	86.7 %	1029
Social learning	0.861	55.0 %	1617	46.6 %	4288
Small-world	Efficacy	90 % convergence		100 % convergence	
		Effectiveness	Efficiency	Effectiveness	Efficiency
Supervision-ϵ	0.972	91.7 %	1692	91.6 %	1735
Supervision-α	0.937	84.2 %	1969	71.6 %	4077
Supervision-both	0.992	98.4 %	818	98.4 %	862
Social learning	0.871	54.9 %	2212	46.5 %	4450
Scale-free	Efficacy	90 % convergence		100 % convergence	
		Effectiveness	Efficiency	Effectiveness	Efficiency
Supervision-ϵ	0.997	100 %	181	100 %	246
Supervision-α	0.969	99.9 %	183	93.1 %	3075
Supervision-both	0.996	100 %	114	100 %	162
Social learning	0.967	99.1 %	331	90.4 %	3204

Table 3 summarizes the final performance of the different approaches in 1000 independent runs. In order to better demonstrate the different performance of these approaches, we also include the results when 100 % agents have chosen the same action as the norm. Achieving 100 % norm emergence is an extremely challenging issue due to the widely recognized existence of subnorms [18]. Clearly, the proposed learning approaches outperform the social learning approach in all aspect of comparison. For example, in the grid network, social learning can only enable averagely 86.1 % agents in the population to achieve a consensus over the social norm. This performance is upgraded to as high as 92.2 %, 91.9 % and 95.7 % using the three proposed approaches, respectively. Regarding effectiveness, the possibility that a norm can successfully emerge using social learning is quite low (i.e., 55.0 % for 90 % convergence, and 46.6 % for 100 % convergence). The adaptive learning approaches, however, can greatly increase the possibility of norm emergence (e.g., 86.7 % for 90 % and 100 % convergence using supervision-both). As for efficiency, it takes averagely 4288 steps for 100 % convergence using social learning approach, against 4113, 1180 and 1029 steps using the three adaptive learning approaches, respectively. To sum up, the adaptive learning approaches can promote the emergence of social norms to higher levels of convergence with fewer steps.

Learning dynamics of ϵ and α. To have a better understanding of the learning dynamics under the proposed framework, it is necessary to see how the critical learning parameters of learning rate and exploration rate evolve during the norm emerging process. The dynamics of ϵ and α with different sizes of norm space are shown in Fig. 2.

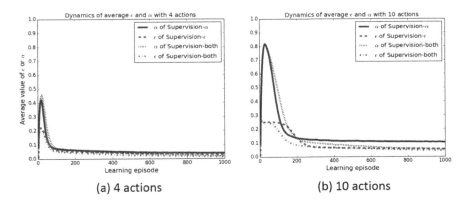

(a) 4 actions   (b) 10 actions

Fig. 2. Learning dynamics of ϵ and α.

In both action sizes, the values of α and ϵ increase sharply at the beginning, and then drop to nearly zero gradually. This is because the whole agent system is still in chaos at the beginning of the norm emerging process and it is more likely that the stochastic learning process cause the agents to be in a "losing" state. Therefore, agents increase their learning rate and/or exploration rate to learn faster and/or explore more, so as to get over the "losing" state. As the norm is emerging, the action choice is more and more consistent with supervision policy. Thus, ϵ and α decrease accordingly to indicate a "winning" state of the agents. Comparing Fig. 2(a) with (b), we can also see that, in 10-action scenario, the values change more dramatically at first and then it takes a longer time for these values to decrease to zero. This is because agents are more likely to choose the same action as the norm with a smaller action size. When the norm space gets larger, the probability to find the right action as the norm is greatly reduced. The large number of conflicts among the agents thus causes the agents to be in a "losing" state more often in a large norm space, and thus the norm emerging process is greatly prolonged.

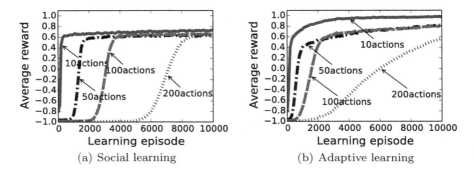

(a) Social learning   (b) Adaptive learning

Fig. 3. Influence of number of actions.

Influence of Norm Space and Population Size. The influence of norm space on norm emergence is given by Fig. 3. We can see that a larger number of available actions results in a delayed convergence of norms. This is because a larger number of actions are more likely to produce local sub-norms, leading to diversity across the society. It thus takes a longer time for the agents to eliminate this diversity and achieve a final consensus. In all cases, the adaptive learning approach performs better than the social learning approach in terms of a faster convergence speed and a higher convergence level. This result shows that the proposed adaptive learning framework is indeed effective for norm emergence in a large norm space.

The influence of population size on norm emergence is shown in Fig. 4. In both learning approaches, the norm emergence process is hindered as the population is growing larger. This result occurs because the larger the society, the more difficult to diffuse the effect of local learning to the whole society. This phenomenon can be observed in human societies where small groups can more easily establish social norms than larger groups [6]. The proposed adaptive learning approach, however, can greatly facilitate norm emergence in different population sizes. In cases of 100, 500 and 1000 population size, the adaptive learning approach can achieve almost 100 % convergence, which is a great promotion from the low convergence levels using the static social learning approach. In a population of 5000 agents, the norm emerging process is steadily facilitated to a level of 90 % during 10000 steps using the adaptive learning approach, against a convergence level close to 70 % using the social learning approach.

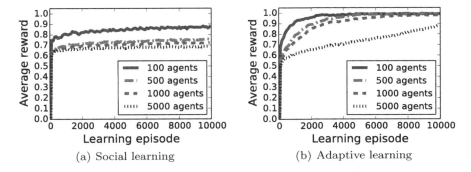

(a) Social learning   (b) Adaptive learning

Fig. 4. Influence of population size.

Influence of Network Topology. Table 4 summarizes the performance of 100 % norm emergence with various network topologies in small-world networks. The rewiring possibility p indicates different levels of network randomness. The results show that it is more efficient for a norm to emerge in a network with higher randomness. This is because the increase in randomness can reduce the network diameter (i.e., the largest number of hops in order to traverse from one vertex to another [8]), and the smaller a network diameter, the more efficient for the network to evolve a social norm [23]. The results also show that a minor increase of the rewiring possibility (from 0 to 0.1, especially from 0.01 to

Table 4. Performance of the learning approaches in various topological settings

Approach	Criterion	Randomness (rewiring possibility p)									Neighborhood Size (k)				
		0	0.001	0.01	0.1	0.2	0.4	0.6	0.8	1.0	4	8	12	16	20
Effectiveness	Social learning	2.8%	4.5%	7.3%	46.5%	81.5%	92.7%	92.8%	92.9%	93.2%	1.8%	22.0%	46.5%	59.8%	77.0%
	Supervision-ϵ	16.5%	19.4%	30.4%	91.6%	98.7%	100%	100%	100%	100%	15.4%	71.1%	91.6%	97.6%	99.0%
	Supervision-α	12.2%	14.1%	17.8%	71.6%	87.6%	93.0%	93.8%	94.1%	94.9%	6.9%	42.9%	71.6%	79.4%	86.0%
	Supervision-both	45.3%	50.2%	67.5%	98.4%	100%	100%	100%	100%	100%	38.9%	90.6%	98.4%	100%	100%
Efficiency	Social learning	5432	4975	4628	4450	3537	3120	3083	3001	2984	6336	5371	4450	3923	3832
	Supervision-ϵ	4400	4041	4023	1374	473	134	121	118	117	5321	2761	1374	979	642
	Supervision-α	4769	4494	3913	4077	3494	3081	3005	2998	2932	5658	4420	4077	3835	3649
	Supervision-both	4389	4153	4082	862	283	122	114	113	112	4071	1796	862	509	325
Efficacy	Social learning	0.790	0.790	0.799	0.871	0.941	0.969	0.970	0.970	0.970	0.794	0.827	0.871	0.897	0.932
	Supervision-ϵ	0.829	0.836	0.854	0.972	0.992	0.997	0.997	0.997	0.997	0.837	0.929	0.972	0.989	0.993
	Supervision-α	0.837	0.839	0.851	0.937	0.965	0.970	0.970	0.970	0.970	0.844	0.893	0.937	0.952	0.962
	Supervision-both	0.894	0.904	0.935	0.992	0.997	0.997	0.997	0.997	0.997	0.907	0.976	0.992	0.997	0.997

0.1) can bring about significant improvement of norm emergence, while further increasing the rewiring possibility (from 0.2 to 1.0) cannot cause a further significant improvement. This is due to the fact that the network randomness is already quite high when the rewiring possibility p is in-between $[0.01, 0.1]$. In all scenarios, the proposed learning approaches outperform the social learning approach in all three comparison criteria. Specially, when the randomness is high, approach supervision-ϵ and supervision-both can achieve norm convergence with 100 % possibility. This robust norm emergence, however, only takes very short time (e.g., 117 and 112 steps for supervision-ϵ and supervision-both, respectively, againt 2984 steps for social learning, when $p = 1.0$.).

As for neighborhood size, norm emergence is steadily promoted when the average number of neighbors is increased. This effect is due to the clustering coefficient of the network [10]. When the average number of neighbors increases, the clustering coefficient also increases. Therefore, agents located in different parts of the network only need a smaller number of interactions to reach a consensus. On the other hand, when agents have a small neighborhood size, they only interact with their neighbors, which account for a small proportion of the whole population. This results in diverse sub-norms formed at different regions of the network. Such sub-norms conflict with each other in the network, and thus more interactions are needed to solve these conflicts and achieve a uniform norm for the whole society. In all cases of neighborhood sizes, the adaptive learning approaches can bring about more robust norm emergence with a faster convergence speed and a higher convergence level than the social learning approach.

5   Conclusion

In this paper, a novel learning framework was proposed to investigate how agents' adaptive learning behaviors can facilitate the process of norm emergence in network MASs. The highlight of the proposed model is the integration of social learning into the local individual learning in order to dynamically adapt agents' learning behaviors for a better performance of norm emergence. Our work thus bridges the gap between the two distinct research paradigms for learning of norm emergence by coupling a social learning process (through imitation in EGT) with a local individual learning process (i.e., RL). Although it can be expected that requiring communication among agents or additional information through social learning can facilitate norm emergence, this is not straightforward in the proposed model as the synthesized information used in social learning is generated from trail-and-error individual learning interactions, and this information is then utilized as a guide to heuristically adapt the local learning further. Tight coupling between these two learning processes can make the whole learning system rather dynamic. However, by synthesizing the individual learning experience into competing strategies in EGT and adapting local learning behaviors based on the principle of "Win-or-Learn-Fast", our work has illustrated that this kind of interplay between individual learning and social learning is indeed helpful in facilitating the emergence of social norms among agents.

Acknowledgments. This work is supported by the National Natural Science Foundation of China under Grant 61502072, Fundamental Research Funds for the Central Universities of China under Grant DUT14RC(3)064, and Post-Doctoral Science Foundation of China under Grants 2014M561229 and 2015T80251.

References

1. Shoham, Y., Tennenholtz, M.: On the emergence of social conventions: modeling, analysis, and simulations. Artif. Intel. **94**(1–2), 139–166 (1997)
2. Hao, J., Sun, J., Huang, D., Cai, Y., Yu, C.: Heuristic collective learning for efficient and robust emergence of social norms. In: Proceeedings of 14th AAMAS, pp. 1647–1648 (2015)
3. Hasan, M., Raja, A., Bazzan, A.: Fast convention formation in dynamic networks using topological knowledge. In: Proceedings of 29th AAAI, pp. 2067–2073. IEEE (2015)
4. Brooks, L., Iba, W., Sen, S.: Modeling the emergence and convergence of norms. In: Proceedings of 22nd IJCAI, pp. 97–102 (2011)
5. Savarimuthu, B.: Norm learning in multi-agent societies (2011)
6. Sen, S., Airiau, S.: Emergence of norms through social learning. In: Proceedings of 20th IJCAI, pp. 1507–1512 (2007)
7. Mukherjee, P., Sen, S., Airiau, S.: Norm emergence under constrained interactions in diverse societies. In: Proceedings of 7th AAMAS, pp. 779–786 (2008)
8. Villatoro, D., Sen, S., Sabater-Mir, J.: Topology and memory effect on convention emergence. In: Proceedings of WI-IAT 2009, pp. 233–240 (2009)
9. Yu, C., Lv, H., Ren, F., Bao, H., Hao, J.: Hierarchical learning for emergence of social norms in networked multiagent systems. In: Proceedings of AI 2015, pp. 630–643 (2015)
10. Villatoro, D., Sabater-Mir, J., Sen, S.: Social instruments for robust convention emergence. In: Proceedings of 22nd IJCAI, pp. 420–425 (2011)
11. Yu, C., Zhang, M., Ren, F., Luo, X.: Emergence of social norms through collective learning in networked agent societies. In: Proceedings of AAMAS, pp. 475–482 (2013)
12. Yu, C., Zhang, M., Ren, F.: Collective learning for the emergence of social norms in networked multiagent systems. IEEE Trans. Cybern. **44**(12), 2342–2355 (2014)
13. Shibusawa, R., Sugawara, T.: Norm emergence via influential weight propagation in complex networks. In: Proceedings of ENIC, pp. 30–37. IEEE (2014)
14. Sutton, R.S., Barto, A.G.: Reinforcement Learning: An Introduction. The MIT Press, Cambridge (1998)
15. Szabo, G., Fáth, G.: Evolutionary games on graphs. Phys. Rep. **446**(4–6), 97–216 (2007)
16. Bowling, M., Veloso, M.: Multiagent learning using a variable learning rate. Artif. Intell. **136**, 215–250 (2002)
17. Young, H.P.: The economics of convention. J. Econ. Pers. **10**(2), 105–122 (1996)
18. Mihaylov, M., Tuyls, K., Now, A.: A decentralized approach for convention emergence in multi-agent systems. Auton. Agent. Multi-Agent Syst. **15**(2), 1–30 (2013)
19. Barabási, A.L., Albert, R.: Statistical mechanics of complex networks. Rev. Modern Phys. **74**, 47–97 (2002)

20. Watkins, C.J.C.H., Dayan, P.: Q-learning. Mach. Learn. **8**(3), 279–292 (1992)
21. Airiau, S., Sen, S., Villatoro, D.: Emergence of conventions through social learning. Auton. Agent. Multi-Agent Syst. **28**(5), 779–804 (2014)
22. Watts, D.J., Strogatz, S.H.: Collective dynamics of small-world networks. Nature **393**, 440–442 (1998)
23. Delgado, J.: Emergence of social conventions in complex networks. Artif. Intell. **141**(1), 171–185 (2002)

Author Index

Printed in the United States
By Bookmasters